# Advances in Intelligent Systems and Computing

Volume 1108

The series "Advances in Intelligent Systems and Computing" contains publications on theory, applications, and design methods of Intelligent Systems and Intelligent Computing. Virtually all disciplines such as engineering, natural sciences, computer and information science, ICT, economics, business, e-commerce, environment, healthcare, life science are covered. The list of topics spans all the areas of modern intelligent systems and computing such as: computational intelligence, soft computing including neural networks, fuzzy systems, evolutionary computing and the fusion of these paradigms, social intelligence, ambient intelligence, computational neuroscience, artificial life, virtual worlds and society, cognitive science and systems, Perception and Vision, DNA and immune based systems, self-organizing and adaptive systems, e-Learning and teaching, human-centered and human-centric computing, recommender systems, intelligent control, robotics and mechatronics including human-machine teaming, knowledge-based paradigms, learning paradigms, machine ethics, intelligent data analysis, knowledge management, intelligent agents, intelligent decision making and support, intelligent network security, trust management, interactive entertainment, Web intelligence and multimedia.

The publications within "Advances in Intelligent Systems and Computing" are primarily proceedings of important conferences, symposia and congresses. They cover significant recent developments in the field, both of a foundational and applicable character. An important characteristic feature of the series is the short publication time and world-wide distribution. This permits a rapid and broad dissemination of research results.

**** Indexing: The books of this series are submitted to ISI Proceedings, EI-Compendex, DBLP, SCOPUS, Google Scholar and Springerlink ****

More information about this series at http://www.springer.com/series/11156

S. Smys · João Manuel R. S. Tavares ·
Valentina Emilia Balas · Abdullah M. Iliyasu
Editors

# Computational Vision and Bio-Inspired Computing

ICCVBIC 2019

Set 1

*Editors*
S. Smys
Department of CSE
RVS Technical Campus
Coimbatore, India

João Manuel R. S. Tavares
Faculty of Engineering
Faculdade de Engenharia
da Universidade do Porto
Porto, Portugal

Valentina Emilia Balas
Faculty of Engineering
Aurel Vlaicu University of Arad
Arad, Romania

Abdullah M. Iliyasu
School of Computing
Tokyo Institute of Technology
Tokyo, Japan

ISSN 2194-5357 ISSN 2194-5365 (electronic)
Advances in Intelligent Systems and Computing
ISBN 978-3-030-37217-0 ISBN 978-3-030-37218-7 (eBook)
https://doi.org/10.1007/978-3-030-37218-7



This Springer imprint is published by the registered company Springer Nature Switzerland AG
The registered company address is: Gewerbestrasse 11, 6330 Cham, Switzerland

*We are honored to dedicate the proceedings of ICCVBIC 2019 to all the participants and editors of ICCVBIC 2019.*

# Foreword

It is with deep satisfaction that I write this foreword to the proceedings of the ICCVBIC 2019 held in RVS Technical Campus, Coimbatore, Tamil Nadu, on September 25–26, 2019.

This conference was bringing together researchers, academics, and professionals from all over the world, experts in computational vision and bio-inspired computing.

This conference particularly encouraged the interaction of research students and developing academics with the more established academic community in an informal setting to present and to discuss new and current work. The papers contributed the most recent scientific knowledge known in the field of computational vision, soft computing, fuzzy, image processing, and bio-inspired computing. Their contributions helped to make the conference as outstanding as it has been. The local organizing committee members and their helpers put much effort into ensuring the success of the day-to-day operation of the meeting.

We hope that this program will further stimulate research in computational vision, soft computing, fuzzy, image processing, and bio-inspired computing and provide practitioners with better techniques, algorithms, and tools for deployment. We feel honored and privileged to serve the best recent developments to you through this exciting program.

We thank all authors and participants for their contributions.

S. Smys
Conference Chair

# Preface

This conference proceedings volume contains the written versions of most of the contributions presented during the conference of ICCVBIC 2019. The conference provided a setting for discussing recent developments in a wide variety of topics including computational vision, fuzzy, image processing, and bio-inspired computing. The conference has been a good opportunity for participants coming from various destinations to present and discuss topics in their respective research areas.

ICCVBIC 2019 conference tends to collect the latest research results and applications on computational vision and bio-inspired computing. It includes a selection of 147 papers from 397 papers submitted to the conference from universities and industries all over the world. All of accepted papers were subjected to strict peer-reviewing by 2–4 expert referees. The papers have been selected for this volume because of quality and the relevance to the conference.

ICCVBIC 2019 would like to express our sincere appreciation to all authors for their contributions to this book. We would like to extend our thanks to all the referees for their constructive comments on all papers; especially, we would like to thank to organizing committee for their hardworking. Finally, we would like to thank the Springer publications for producing this volume.

S. Smys
Conference Chair

# Acknowledgments

ICCVBIC 2019 would like to acknowledge the excellent work of our conference organizing the committee, keynote speakers for their presentation on September 25–26, 2019. The organizers also wish to acknowledge publicly the valuable services provided by the reviewers.

On behalf of the editors, organizers, authors, and readers of this conference, we wish to thank the keynote speakers and the reviewers for their time, hard work, and dedication to this conference. The organizers wish to acknowledge Dr. Smys, Dr. Valentina Emilia Balas, Dr. Abdul M. Elias, and Dr. Joao Manuel R. S. Tavares for the discussion, suggestion, and cooperation to organize the keynote speakers of this conference. The organizers also wish to acknowledge for speakers and participants who attend this conference. Many thanks are given to all persons who help and support this conference. ICCVBIC 2019 would like to acknowledge the contribution made to the organization by its many volunteers. Members contribute their time, energy, and knowledge at local, regional, and international levels.

We also thank all the chair persons and conference committee members for their support.

# Contents

# Psychosocial Analysis of Policy Confidence Through Multifactorial Statistics

Jesús Silva[1(✉)], Darwin Solano[2], Claudia Fernández[2], Ligia Romero[2], and Omar Bonerge Pineda Lezama[3]

[1] Universidad Peruana de Ciencias Aplicadas, Lima, Peru
jesussilvaUPC@gmail.com
[2] Universidad de la Costa, St. 58 #66, Barranquilla, Atlántico, Colombia
{dsolano1, cfernand10, lromero11}@cuc.edu.co
[3] Universidad Tecnológica Centroamericana (UNITEC), San Pedro Sula, Honduras
omarpineda@unitec.edu

**Abstract.** Trust is defined as widespread belief or rooted value orientation in evaluative standards of technical and ethical competence, and in the future actions of a person (interpersonal trust) or an institution (institutional trust). From a psychosocial perspective, trust transcends positive or negative affectivity, alludes to the belief that the behavior of others can be predicted and implies a positive attitude and expectation regarding the behavior of the person or institution. This belief refers to the likelihood that individuals or institutions will take certain actions or refrain from inflicting harm, for the sake of personal or collective well-being. The objective of this study is to examine psychosocial factors related to the interaction between the police and the public that predict the perception of trust in police groups in Colombia.

**Keywords:** Trust · Police · Colombia · Honesty · Performance

## 1 Introduction

In the field of security institutions, the police are one of the closest to the public and one of the most relevant to the perception of insecurity and, consequently, to the maintenance of democracy. Trust in the police strengthens the sense of security in citizens, to the extent that police work is perceived to be part of a set of policy actions planned by the rulers in order to improve citizen security in the community [1]. The link between trust, closeness and security stimulates the climate of public collaboration with public institutions, such as the police [2, 3]. In this sense, as some authors point out, criminal incidence, violence and police performance provide information about the trust in the police and the democratic governance of a country [4, 5].

On the contrary, high criminality has been found to undermine confidence in political institutions in general and towards political institutions in particular [6]. When this mistrust occurs in communities with high crime, the sense of vulnerability is even higher and, consequently, the loss of confidence tends to worsen, which is evident, among other aspects, in a decrease in rather than the procurement of private security or


S. Smys et al. (Eds.): ICCVBIC 2019, AISC 1108, pp. 1–7, 2020.
https://doi.org/10.1007/978-3-030-37218-7_1

the exercise of justice outside public proceedings [7–9]. As [10], the inconsistency between democratic norms and poor police action, characterized by poor performance, high corruption, and involvement in crimes, in contexts where criminal incidence has increased, have led to an increase in distrust of the police and a sense of insecurity widely shared by the public. In this regard, [11] he states that distrust of the police sharpens resistance to reporting, particularly in communities with a high crime incidence.

Considering the previous history, and given that there are still few work in the Latin American context that examines how trust in the police is defined [12], incorporating psychosocial variables, such as perception of performance and honesty of the police, the fear of victimization and the perception of insecurity, this study analyses the role of the perception of the performance and honesty of the police, the perception of insecurity, and the fear of victimization, as determinants of trust towards the police in Bogotá, Colombia. In this sense, the experience of victimization, performance and honesty are expected to be predictors of trust in the police.

## 2 Methodology

### 2.1 Data

Random sampling and stratified sampling proportional to population density was carried out, in order to select participants from all regions and from the eight delegations that make up the Colombian capital (Bogotá) [13]. To do this, the number of participants in each area was proportional to the number of dwellings inhabited, with a minimum of five participants per area being established to ensure the representativeness of all, especially the least populated.

Under these criteria, 12,458 citizens among men and women participated, residents of the city of Bogotá at the time of the study. In addition, by the sample size it is possible to make predictions with the variables selected here, with a coefficient of determination of .05 and a power of .90 [14].

### 2.2 Methods

The battery of questionnaires was administered per individual, in interview format, by 102 surveyors previously trained by experts and members of the research group of a university in the Region. This application strategy was chosen to ensure that all items are understood among all respondents. The pollsters are randomly assigned in a balanced way to the four sectors in which the municipality was conventionally divided (north, south, east and west). A supervisor/supervisor and a trainer/trainer coordinated each sector and ensured that at least one trainer and one supervisor were involved in each delegation to resolve any doubts. The completion of the instruments was carried out by the pollsters. In addition, the participation of pollsters, supervisors and participants was voluntary. Participants were informed of the objective and the confidentiality of the data was ensured. 1.46% of respondents refused to participate in the study. In these cases, other participants were selected according to the same criteria show them. The time of application of the questionnaire was between 40 and 45 min [15, 16].

## 3 Results

The following describes the variables used in the study, the selected instruments and their properties [17].

- *Victimization.* To assess direct and indirect victimization, he wondered whether both respondents and their families had been victims of any crime in Bogota during the past year (2019). To this end, the question was asked “In the last year, have you or the people living in your household been the victims of any crime in Bogota?” The question was coded with two answer options (1 × Yes, 2 × No).
- *Fear of* victimization. To assess the fear of being a victim of crime, the adoption of measures to protect against crime has been used as an indicator.

Measures to protect against crime. This dichotomous scale consists of thirteen reagents that assess the frequency of use of different measures to protect against the possibility of crimes. An exploratory factor analysis with oblimin rotation was calculated that yielded a two-factor structure. The first, physical protection measures, explains 55.45% of the variance and refers to how often measures have been used such as buying and carrying a weapon, installing alarms at home, hiring personal security, taking joint actions with the neighborhood, hire private security on the street or in the region, buy a dog, place fences or fences and increase security on doors or windows.

The second factor explains 12.12% of the variance and alludes to the control of personal information such as avoiding giving telephone information, avoiding giving keys or personal data over the Internet, not providing information to strangers and using caller ID Telephone. The results of the factorial analyses performed are presented in Table 1.

**Table 1.** Exploratory factor analysis and confirmative factorial analysis for the questionnaire measures to protect against insecurity

| Protective measures | Exploratory factor analysis | | Confirmatory factor analysis*: standardized solution*** | |
|---|---|---|---|---|
| | *F1* | *F2* | *F1* | *F2* |
| Hire personal security | .86 | .44 | .31 | |
| Hire private security on the street or colony | .82 | .39 | .37 | |
| Don’t give phone information | .34 | .82 | | 0.60 |
| Place fences, fences | .81 | .48 | .39 | |
| Get a dog | .82 | .45 | .29 | |
| Putting extra locks on doors/windows | .79 | .57 | .39 | |
| Don’t give keys or data online | .33 | .73 | | 0.73 |
| Take joint action with your neighbors | .85 | .44 | .51 | |
| Install alarms at home or work | .84 | .45 | .63 | |
| Not giving information to strangers | .52 | .72 | | 0.59 |
| Buy and carry a gun | .88 | .52 | .24 | |
| Use phone call ID | .51 | .75 | | 0.53 |

*Adjustment rates of confirmatory factor analysis: [S-B s2 s 375.82, gl .47, p < .001, CFI. .93, RMSEA .05 (.04, .05)]. ***Significant coefficients (p < .001)

**Table 2.** Exploratory factor analysis and confirmatory factorial analysis for the Crime frequency scale in the colony

| Protective measures | Exploratory factor analysis | | Confirmatory factor analysis*: standardized solution*** | |
|---|---|---|---|---|
| | *F1* | *F2* | *F1* | *F2* |
| Vandalism | .84 | −.39 | .75 | |
| Drug sales | .83 | −.52 | .81 | |
| Robbery of people | .76 | −.43 | .67 | |
| Murder | .54 | −.82 | | .84 |
| Assault | .76 | −.43 | .65 | |
| fraud | .44 | −.81 | | .76 |
| Sex offences | .44 | −.88 | | .78 |
| Kidnapping | .46 | −.87 | | .75 |
| Drug use | .81 | −.50 | .73 | |

*Adjustment rates of confirmatory factor analysis: [S-B s2 x 173.13, gl .22, p < .001, CFI .98, RMSEA .05 (0.04, 0.06) ***Significant coefficients (p < .001).

– *Type of crime in the region.* The frequency of different crimes in the region was assessed on a Likert scale consisting of eleven reagents, with a range of five response options (not frequent, rare, regular, frequent and very frequent). An exploratory factor analysis with oblimin rotation was calculated, resulting in two factors. The first, called violence and drug use, explains 54.22% of the variance and refers to the frequency with which drug-dealing, drug use, vandalism, robbery and assault occur.

The second factor, serious crimes, explains 12.31% of the variance and groups together criminal acts of seriousness: sexual offences, kidnapping, fraud and homicide. Reagents "threats" and "injuries and quarrels" have been eliminated because they have similar saturations in both factors, therefore they are complex items saturated with uncertainty. The final scale consists of nine reagents. Cronbach's alpha coefficient for factors is .84 and .88, respectively. Table 2 contains the results of exploratory and confirmatory factorial analyses.

– *Factors that promote crime.* This scale is composed of twenty reagents that measure the perception of citizens of possible causes that favor the increase of crime. The Likert scale presents a range of five response options (totally disagree, disagree, disagreement, agreement, agreement, and total agreement). An exploratory factor analysis was performed with oblimin rotation that yielded a three-factor structure. The first, psychosocial factors, explains 42.93% of the variance and alludes to the following causes: [crime] is a way to obtain easy resources, bad companies, economic needs, loss of values and lack of empathy or concern for other. The second factor, corruption and institutional impunity, explains 11.62% of the variance and groups the reagents as corruption, impunity and decomposition of the institutions.

Finally, the third, socialization factors, explains 8.43% of the variance and encompasses the reagents that allude to socialization processes: lack of recreation spaces, learning from childhood, influence of the media, and discrimination. The Cronbach alpha obtained for the factors is .82, .80 and 0.63, respectively.

- *Trust in security institutions.* It assesses the public's trust in public security institutions. The Likert scale consists of seven items with five answer options (unreliable, regular, reliable, very reliable). An exploratory factor analysis was performed with oblimin rotation that showed a single-factor structure that explains 61.77% of the variance. Cronbach's alpha obtained is .90.

Performance and Honesty in security institutions. This Likert scale consists of seven reagents to measure the performance of security institutions, with five response options (very poor performance, poor performance, regular, good performance and very good performance), and seven that evaluate honesty, with five response options (nothing honest, dishonest, regular, honest and very honest). A factor analysis with oblimin rotation showed the presence of two factors that explained 66.44% of the variance. The first, which explains 53.80% of the variance, groups the reagents related to the performance of the Preventive Police, the Transit Police, Judicial Agents, Local and Neighborhood Round, State Police, Federal Preventive Police and Army. The second factor explains 11.64% of the variance and groups the reagents that measure the perception of honesty towards these institutions.

Involvement in security. This Likert-type scale with five response options (completely disagree, disagree, neutral, agree, and completely agree) is composed of five reagents that assess the weight of the involvement of the following agents in the safety of Bogota: One self; neighbors, government, public security institutions, and society. A factor analysis with oblimin rotation was performed showing a one-factor structure that explains 53.67% of the variance. Cronbach's alpha obtained is .80.

Willingness to participate in actions that promote safety. This Likert scale with five response options (nothing, little, from time to time, much and always) is configured by eight reagents that evaluate availability to participate in the following actions or activities associated with security: organization neighborhood, promoting more values in the family, recovering public spaces, generating spaces of coexistence, promoting sport and recreation, instilling culture, promoting arts and crafts workshops, and conducting community work. A factorial analysis with oblimin rotation that has shown a one-factor structure explaining 64.89% of the variance was calculated. Cronbach's alpha obtained is .96.

Provision to do justice outside the law. This dichotomous questionnaire has two answer options (1 × Yes, 2 No) asking what crimes you are willing to do justice outside the law. This questionnaire consists of the twelve reagents that constitute crimes such as theft/assault, express kidnapping, abuse of authority, homicide, abuse of trust, threats, extortion, property damage, injury, corruption, fraud, and sexual offense. A factor analysis with oblimin rotation was performed showing a single-factor structure that explains 62.77% of the variance. Cronbach's alpha obtained is .93.

Subsequently, a logistical linear regression was performed with the variables included in the t-test, in order to know their specific weight in predicting trust towards police groups, such as security institutions. Trust in the various security institutions that coexist in the region was used as a dicomic dependent variable.

The bus test on the coefficients of the model indicates that the variables introduced in the regression improve the fit of the model ($*^2$ (19) × 1365.70; p .000). The value of the −2 logarithm of likelihood (−2LL) is 644.69. In addition, the R-squared coefficient of Cox and Snell (.60) indicates that 60.1% of the dependent variable variation is explained by the variables included in the analysis. Similarly, the value obtained with the R squared of Nagelkerke (.83) indicates that the calculated regression explains a variance rate of 80.3%. Both values indicate that the model calculated in the logistic regression shows an appropriate fit and validity. The proportion of cases correctly classified by the regression model has resulted in 94.1% for low confidence and 93.3% for high confidence, so it can be verified that the model has a high specificity and sensitivity.

The variable that most affects trust in security institutions is the perception of honesty of these institutions (Wald x 214.56; p < .00; B x 2.80; Exp. (B) x 19.23), to the extent that those who consider security institutions to be honest are 18.72 times more likely to present high confidence towards them. Second, it has been found that the appreciation of the performance of security institutions by citizens predicts confidence, in the sense that those who positively value the performance of these institutions are 4.88 times likely higher than have high confidence. (Wald s 68.14; p < .000; B x 1.45; Exp. (B) x 4.88). Similarly, citizens who are less willing to exercise justice outside the action of security institutions are 2.10 times more likely to have high confidence in the institutions (Wald 5.44; p < .05; B…73; Exp. (B) x 2.50). In addition, having higher education increases by 1.77 the likelihood of having confidence in institutions (Wald s 3.74; p < .05; B…53; Exp. (B) −1.45).

## 4 Conclusions

The objective of this study has been to examine the psychosocial factors, related to the interaction between police groups and citizens, involved in the process of building trust towards the police in the municipality of Cuernavaca. A poor assessment of the various security agents can be seen from the previous analyses, a trend that has been found in previous work [5, 8, 10, 16].

With regard to the determinants of the institutions' trust through logistic regression analysis, the results partially confirm the expected relationships. Thus, it shows that the main predictive factor of trust is honesty; in other words, people who perceive security institutions as honest and very honest show them greater confidence. In fact, honesty carries more weight than performance in predicting trust in police groups, something that has not been highlighted in previous studies. This finding is suggestive, since it makes it possible to underline the importance of the values and the ethical-moral dimension of the institutions and their implications, both in corruption and in the relationship with citizens.

**Compliance with Ethical Standards.**

- All authors declare that there is no conflict of interest.
- No humans/animals involved in this research work.
- We have used our own data.

## References

1. Newton, K., Norris, P.: Confidence in public institutions, faith, culture and performance? In: Pharr, S.J., Putnam, R.D. (eds.) Disaffected Democracies Princeton, pp. 52–73. Princeton University Press, New Jersey (2000)
2. León Medina, F.J.: Mecanismos generadores de la confianza en la institución policial. Indret: Revista para el Análisis del Derecho (2), 15–30 (2014)
3. Lunecke, A.: Exclusión social, tráfico de drogas y vulnerabilidad barrial. In: Lunecke, A., Munizaga, A.M., Ruiz, J.C. (eds.) Violencia y delincuencia en barrios: Sistematización de experiencias, pp. 40–52 (2009). http://www.pazciudadana.cl/wp-content/uploads/2013/07/2009-11-04_Violencia-y-delincuencia-en-barriossistematizaci%C3%83%C2%B3n-de-experiencias.pdf
4. Malone, M.F.T.: The verdict is in the impact of crime on public trust in Central American justice systems. J. Polit. Lat. Am. (2), 99–128 (2010)
5. Ávila Guerrero, M.E., Vera Jiménez, J.A., Ferrer, B.M., Bahena Rivera, A.: Un análisis psicosocial de la confianza en los grupos policiales: el caso de Cuernavaca (México). Perfiles Latinoamericanos **24**(47), 151–174 (2016). https://doi.org/10.18504/pl2447-009-2016
6. Bento, N., Gianfrate, G., Thoni, M.H.: Crowdfunding for sustainability ventures. J. Clean. Prod. **237**(10), 117751 (2019)
7. Lis-Gutiérrez, J.P., Reyna-Niño, H.E., Gaitán-Angulo, M., Viloria, A., Abril, J.E.S.: Hierarchical ascending classification: an application to contraband apprehensions in Colombia (2015–2016). In: International Conference on Data Mining and Big Data, pp. 168–178. Springer, Cham (2018)
8. Amelec, V., Carmen, V.: Relationship between variables of performance social and financial of microfinance institutions. Adv. Sci. Lett. **21**(6), 1931–1934 (2015)
9. Wang, W., Mahmood, A., Sismeiro, C., Vulkan, N.: The evolution of equity crowdfunding: insights from co-investments of angels and the crowd. Res. Policy **48**(8), 103727 (2019). https://doi.org/10.1016/j.respol.2019.01.003
10. Bergman, M., Flom, H.: Determinantes de la confianza en la policía. una comparación entre argentina y México. Perfiles Latinoamericanos **20**(40), 97–122 (2012)
11. Buendía, J., Somuano, F.: Participación electoral en nuevas democracias: la elección presidencial de 2000 en México. Política y Gobierno **2**(2), 289–323 (2003)
12. Costa, G.: La inseguridad en América Latina ¿Cómo estamos? Foro Brasileño de Seguridad Pública. Revista Brasileña de Seguridad Pública **5**(8), 6–36 (2011)
13. Segovia, C., Haye, A., González, R., Manzi, J.: Confianza en instituciones políticas en Chile: un modelo de los componentes centrales de juicios de confianza. Revista de Ciencia Política **28**(2), 39–602 (2008)
14. Tankebe, J.: Police effectiveness and police trustworthiness in Ghana: an empirical appraisal. Criminol. Crim. Justice (8), 185–202 (2008)
15. DANE: Proyecciones de Población [databasc]. DANE, Bogotá (2019)
16. Oviedo, C., Campo, A.: Aproximación al uso del coeficiente alfa de Cronbach. Revista Colombianan de Psiquiatría **34**(4), 572–580 (2005)
17. Demsar, J., Curk, T., Erjavec, A., Gorup, C., Hocevar, T., Milutinovic, M., Mozina, M., Polajnar, M., Toplak, M., Staric, A., Stajdohar, M., Umek, L., Zagar, L., Zbontar, J., Zitnik, M., Zupan, B.: Orange: data mining toolbox in python. J. Mach. Learn. Res. **14**(Aug), 2349–2353 (2013)

# Human Pose Detection: A Machine Learning Approach

Munindra Kakati and Parismita Sarma(✉)

Department of Information Technology, Gauhati University, Guwahati, India
kakatimunindra@gmail.com, parismita.sarma@gmail.com

**Abstract.** With the recent advancements in computer science and information technology, computers can now leverage human-like abilities in their operations. Image processing and computer vision technology made many practical applications easy and thus making our life comfortable with machine aid. Medical diagnosis and automatic car driving are two prominent examples in this field. Different types of human poses will be observed in our day to day life. The poses are distinguished from each other at the joints of body parts. In this paper we will discuss the methods to identify four fundamental human poses viz. sitting, standing, handshaking and waving. We are proposing here a framework which can automatically recognize the aforementioned four human poses in any angle or direction from the digital image. For working in this frame work we have created a image database of our own, which is described in subsequent chapter. Our working methodology involves image processing and neural network techniques. Finally, the performance of the proposed system is illustrated.

**Keywords:** Human pose · Computer vision · Neural network · Image processing

## 1 Introduction

### 1.1 Importance of Human Pose

Human beings will quite often express their mental and physical health with different types of poses framed with different body parts. Proper recognition of those poses will help to understand the current mental as well as physical condition of the particular person. Human pose detection can be marked as the initial stage of human behavior analysis [1]. Now a days researchers are interested to recognize human poses with the help of computer. If any human pose can be automatically recognized with the help of computer then many other applications like medical diagnosis, crime detection will be easier for full machine implementation. This project has so many applications in our day to day life. We can consider it as advanced level reasoning activity identification in the field of human computer interaction. This work mainly focuses on localizing the parts or joints of human body from the digital image.

**Pose Estimation**

As already we have mentioned that human pose will be straight or bent of different parts present in the human body. This topic is a study on the most emerging research


S. Smys et al. (Eds.): ICCVBIC 2019, AISC 1108, pp. 8–18, 2020.
https://doi.org/10.1007/978-3-030-37218-7_2

area called Computer vision. This area of research, studies the imitation and duplication of what the human eye see and analyse. That is why to achieve this level of reasoning we should make the machine to be intelligent enough to retrieve the posture information from digital image data. Thus the way of evaluating the posture of human body from image or video is termed as pose estimation. Pose estimation is generally determined by spotting location and direction of an object.

### 1.2 Machine Learning in Computer Vision

Learning about the information hidden inside an image and analyzing them is an efficient way for understanding the image [2]. Computer vision technology embraces image understanding and machine learning technologies to understand a digital image. The root of computer vision lies on pattern recognition and image processing, by yielding model oriented and knowledge base vision with more data extraction features. This system has capability of reasoning and aptitude of learning. People are interested in this field due to commercial product that can be created from different projects of computer vision. Different machine learning technologies can be used in the field of computer vision. In two different ways machine learning training can be incorporated in computer vision. In case of supervised learning future is predicted from past facts which are known. After a rigorous training the model becomes ready to compare the result with the correct one and evaluate errors, which then rectify the model accordingly. In case of unsupervised learning, training is done on the unlabeled data. This learning method tries to find out different patterns present in the different unlabeled data.

This paper is organized into five sections. Section 2 describes Related Study. Section 3 is on Proposed Problem and Methodology. Section 4 is all about Results and Discussion. Section 5 is on Conclusion and Future Work.

## 2 Related Study

To get a clear idea on the topic we have gone through a number of journals and research papers. A few important among them are discussed herewith.

According to Chaitra et al. [3] human pose can be detected using a map based and self categorizing retrieval manner. In the year 2018 Liu et al. [4] published a review paper and put emphasis on deep learning used for estimation of human pose from digital image, they did some experimental work on single and multi stage convolution neural network. They also experimented with multi branch, recurrent neural methods. Belagiannis and Zisserman [5] in the year 2017 proposed another method based on heat map depiction for human figure key point detection. In this method the system was able to learn and symbolize body component appearances and testimonial of the human part arrangement. Another paper by Chu et al. [6] proposed a new model for structured feature learning which can verify correlation among the body joints. Zhuang et al. [7] in the year 2018, a new approach was seen with a direction map. This approach rendered directional information on different human body components. This model was divided into two distinct parts, first part learns about direction of every component, with reference to central body part, the second part works as a collector by combining places as well as direction information to acquire an ideal posture. In 2018, Geetha and Samundeeswari [8] had a review paper where they put light on human actions recognition methods. Recently new ideas on pose regression is applied by Ionescu et al. [9], they are very much interested on 3D pose.

# 3 Proposed Problem and Methodology

Human pose detection by machine is helpful in different inspection application like activity or behaviour analysis. In Human machine interaction control systems like athlete performance analysis take help of pose identification techniques. Figures 1 and 2 show flowchart for training and testing phases for the whole human pose detection method. The input digital image was preprocessed for more enhancement. Next important and necessary features of the interested regions are extracted. In the last phase recognition is done from the features of extracted image.

## 3.1 Flowchart of the Proposed System

As shown in the flowchart for training purpose, training images are preprocessed with different preprocessing techniques like grayscaling, resizing, filtering, gray to binary conversion etc. Then from the preprocessed images, the required features are extracted using Principal Component Analysis and Discrete Cosine transform. After that the extracted features are trained using a neural network classifier.

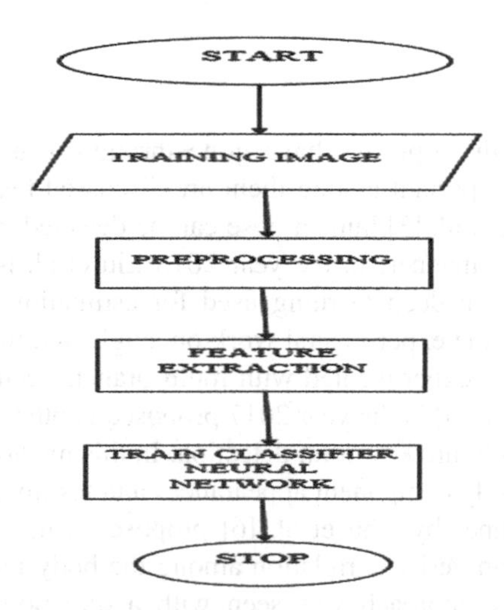


**Fig. 1.** Flowchart of the training process

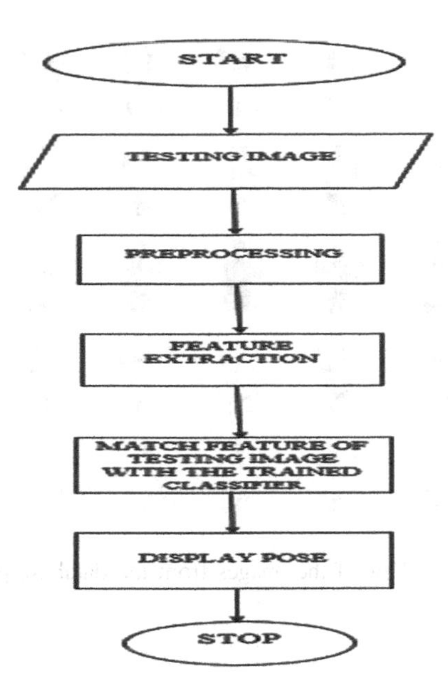


**Fig. 2.** Flowchart of the testing process

According to flowchart for testing purpose, at first testing images are preprocessed with different preprocessing techniques, which are also used for training the image. Then from the preprocessed images features are extracted. After that extracted features are matched with the trained neural network classifier. In the last phase, the pose will be displayed according to the matching percentage.

### 3.2 Image Database

We have created the database of our own for implementation. Thus we were able to collect almost 500 images with various human poses. The format of the photograph is in .jpg. The input images are divided into training image and testing image. The image database is created for the recognition of human pose and is classified into four classes for distinct training and testing purposes. During training purpose almost 500 images are used, Fig. 3 shows a preview of the database for the training image.

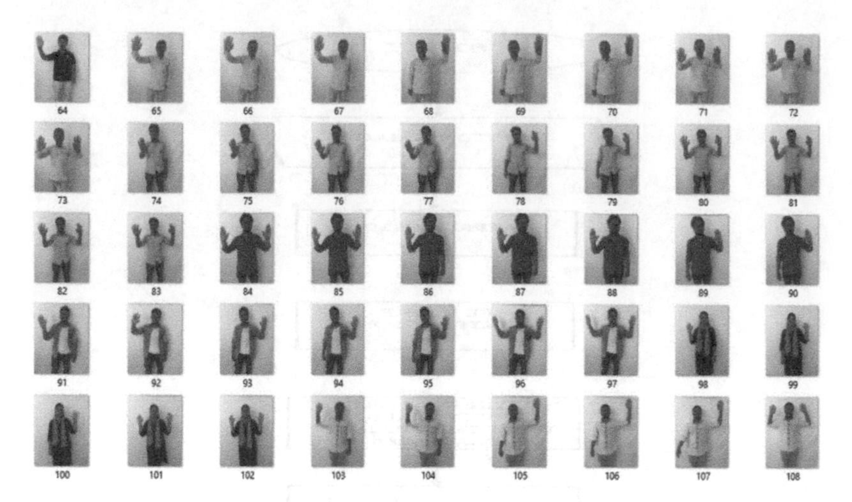


**Fig. 3.** Few of the images from the database [6]

### 3.3 Preprocessing

The aim of pre-processing of digital image is to improve the image data which may be suppressed unwilling or for removal of salt and pepper as well as Gaussian noises. Pre-processing performs two tasks, they are calculation of Region of Interest (ROI) and conversion of colored image to gray scale image [10]. Enhancement of digital image is important as this process make the image clear, bright and distinct but no extra information is added to it. Some preprocessing techniques that are used in this process are mentioned below.

**Grayscaling:** A grayscale image is one which has only shades of gray colors scaling from 0 to 255. In black and white image 0 means total black color and 1 means absolute white color. Using gray scale in digital image processing amount of information to be processed is reduced. Which in turn reduces processing cost. Gray scale poses equal intensity for all the components of Red, Green and Blue in the common RGB scale and hence single intensity value for every pixel is enough for processing.

Figure 4 shows one of our example image which is converted from RGB to grayscale image.

**Filtering:** Filtering is necessary in digital image enhancement phase. We can find three types of filters high pass, band pass and low pass. High pass filter allows to pass high frequency region like line, border etc. low pass filter permits low frequency components like background or region with minute variation in gray values. Image smoothing, blurring or contrasting are done with the help of these filters.

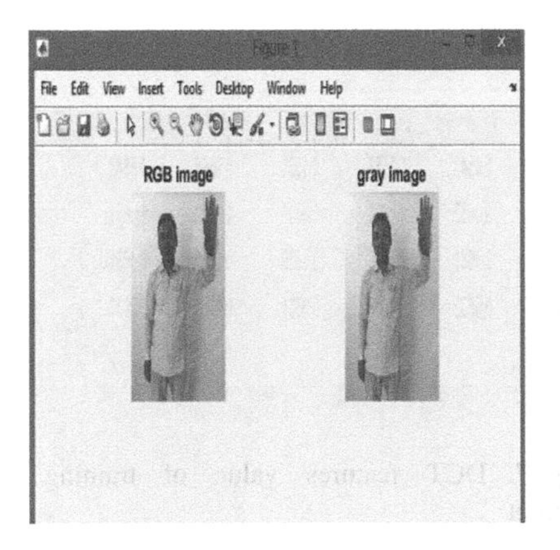


**Fig. 4.** RGB to gray conversion

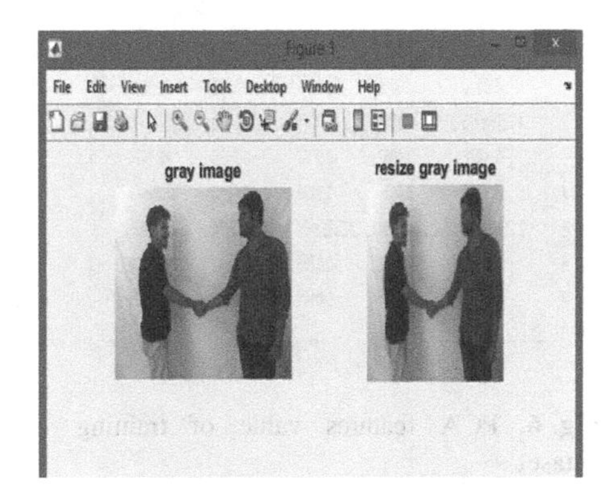


**Fig. 5.** Resized image

**Image Scaling:** Image scaling is changing the existing coordinates from older one to new one. After scaling a new image must be generated which are composed of higher or lower number of pixels in comparison with original image. Figure 5 shows a resized image which is prepared for further processing.

**Gray Image to Binary Image Conversion:** A binary image assigns two values for each pixel, 0 and 1 for black and white respectively. The image contains two types of color namely foreground and background color. The foreground color is used for the object in the image while the rest of the image uses the background color.

### 3.4 Feature Extraction

In the field of computer vision, feature play a very important role. Various image preprocessing methods such as gray scaling, thresholding, resizing, filtering etc. are applied on the image before feature is extracted. The features are extracted using specific feature extraction methods that can later be used for classification and recognition of images.

**Principal Component Analysis (PCA)**
PCA minimizes the broad dimensionality of the data space to a smaller fundamental dimensionality of feature space. PCA is basically used for feature extraction, data compression, redundancy removal, prediction etc. PCA function determines the eigen vectors of a covariance matrix with the largest eigenvalues and then uses this value to project the data into a new subspace of less or equal dimensions. PCA converts matrix of n features into a new dataset of smaller than n. It minimizes the number of features by forming a new, lesser number variables which captures an important portion of the information which are found in the original features.

PCA

| | 1 | 2 | 3 | 4 | 5 |
|---|---|---|---|---|---|
| 1 | -0.0012 | 0.0035 | -0.0351 | -0.0312 | -0.0058 |
| 2 | 4.2886e-15 | -0.0030 | 0.0079 | 0.0219 | -9.6144e-15 |
| 3 | 0.0476 | -0.0067 | -0.0067 | -0.0470 | -0.0470 |
| 4 | -0.0135 | -0.0135 | 0.0038 | 0.0038 | 2.2258e-15 |
| 5 | 2.2365e-04 | 2.2365e-04 | 2.2365e-04 | 0.0074 | -0.0269 |
| 6 | 0.0526 | -0.0030 | -0.0255 | 0.0042 | 0.0526 |
| 7 | -0.0345 | 0.0404 | 0.0168 | 0.0041 | 0.0601 |

**Fig. 6.** PCA features value of training dataset

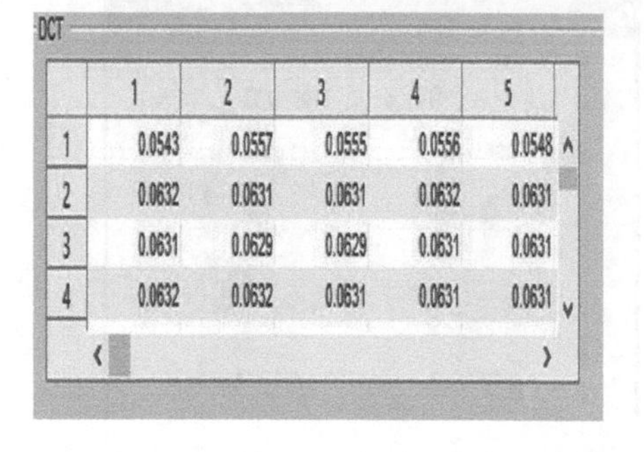

DCT

| | 1 | 2 | 3 | 4 | 5 |
|---|---|---|---|---|---|
| 1 | 0.0543 | 0.0557 | 0.0555 | 0.0556 | 0.0548 |
| 2 | 0.0632 | 0.0631 | 0.0631 | 0.0632 | 0.0631 |
| 3 | 0.0631 | 0.0629 | 0.0629 | 0.0631 | 0.0631 |
| 4 | 0.0632 | 0.0632 | 0.0631 | 0.0631 | 0.0631 |

**Fig. 7.** DCT features value of training dataset

As shown in the Figs. 6 and 7, The column of the tables represent number of extracted features from the image and the row of the tables represent number of images.

**Discrete Cosine Transform (DCT)**

The DCT is a transform that maps a block of pixel color values in the spatial domain to values in the frequency domain [11]. DCT is defined by finite sequence of data points as sum of cosine functions rolling at various frequencies. The DCT applied to the digital image gives a DCT matrix with different coefficients. The coefficients are divided into two blocks. The white blocks is about the illumination with detail information of the image. Whereas the dark blocks represent the edges of the image. Here block $8 \times 8$ DCT is used, the reason behind this is to have enough sized blocks to provide enough compression, though it keeps the transform simple.

### 3.5 Training Neural Network Classifier

For training purpose, we have collected human images in different poses. We are interested to classify the poses like sitting, standing, waving and handshakeing. After extracting the input features, training set is used to train using feedforward neural network and the output is divided into four classes i.e., Sitting, Standing, Waving and Handshake.

**Feed-Forward Neural Networks:** Feed-forward neural networks are the simplest type of neural network, including an input layer, one or more hidden layers and an output layer. It has no back propagation only front propagated wave by using a classifying activation function. FF Neural networks are usually applied in fields where classification of target classes are found to be complicated. This may include field such as speech recognition and computer vision. Not only such neural networks are responsive towards noiseless data but also easier to maintain.

**Levenberg-Marquardt Algorithm (L-M):** The learning phase is one of the more critical phases in the building of a successful neural network application [12]. In order to reduce network error the training algorithm is designed to compute the best network framework. For ANN model training various optimization methods can be used, one of which is the L-M algorithm.

### 3.6 Testing Image

The trained classifier are then tested with a dataset of images which weren't used for training the classifiers. The images to be tested were first pre processed by following the same method as those with the trained images. Then the features were extracted using PCA and DCT which is also used for training. These features were then tested on the neural network classifier that we have previously trained.

### 3.7 Tools/Environment/Experimental Platform

MATLAB is a 4th generation programming language. MATLAB allows different types of features which are implementation of algorithms, matrix manipulations, plotting of functions and data, creation of user interfaces etc. During the whole project windows 8 operating system was used. For capturing images an android phone was used.

## 4 Result and Discussion

We have trained this system to detect the Sitting, Standing, Waving and Handshake position of humans using ANN. We have tested the system with random images.

As shown in the Fig. 8, the interface of the proposed system has been designed. The load button is used to load the testing of RGB image. The preprocessing method is there and it process the loaded RGB image. The feature extraction button is used to extract the features from the preprocessed binary image. The ANN classifier button is used to classify the testing image.

As seen in Fig. 8, it is a screen shot of our proposed system. It is the main interface of our system. Different buttons are there and functions of those buttons are already discussed earlier. Figure 9 shows a screen shot of a hand shaking pose with the PCA and DCT values calculated. Now we have got an idea on how the recognition system works. Figures 10, 11, 12 and 13 are screen shots of human pose detected by our system with tagging handshaking, waving, standing and sitting respectively. Every screen shot has correct tagging with their DCT and PCA values. Table 1 shows the overall performance of our implementation.

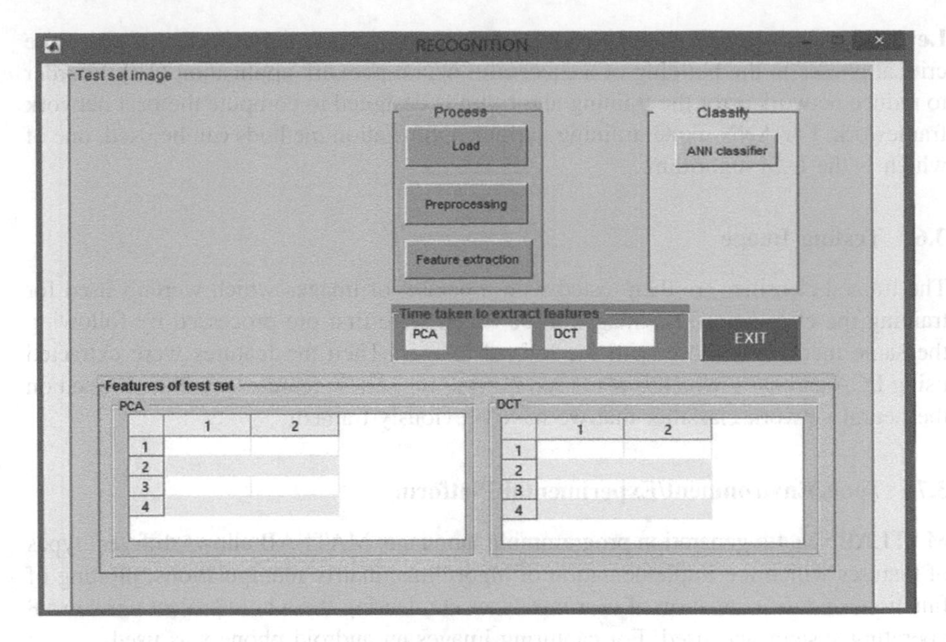


Fig. 8. Interface of the proposed system

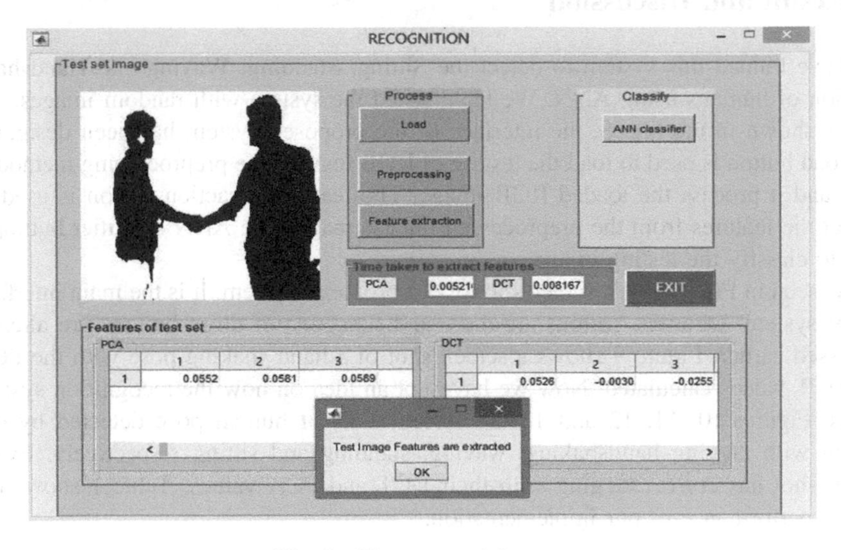


Fig. 9. The extracted features

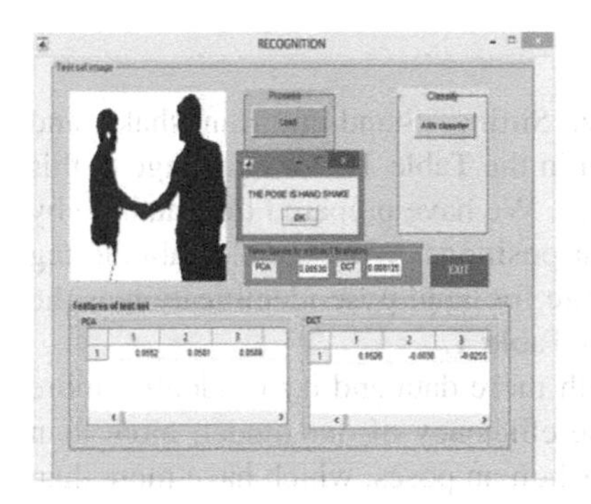


**Fig. 10.** Screenshot of hand shake output

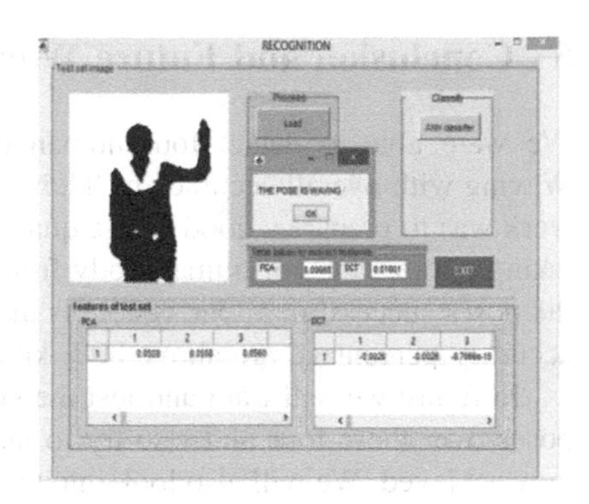


**Fig. 11.** Screen shot of waving

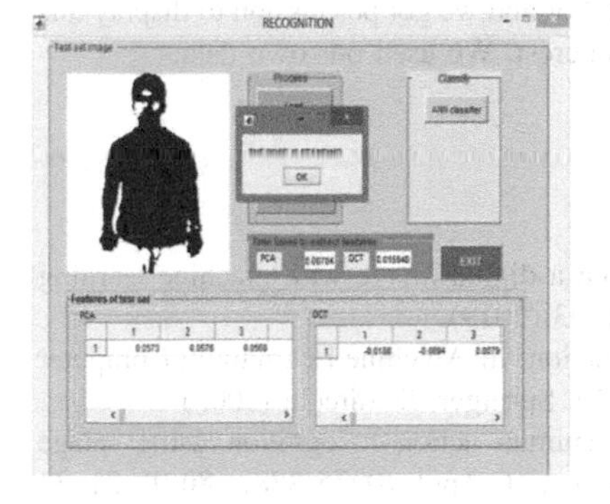


**Fig. 12.** Screenshot of standing pose

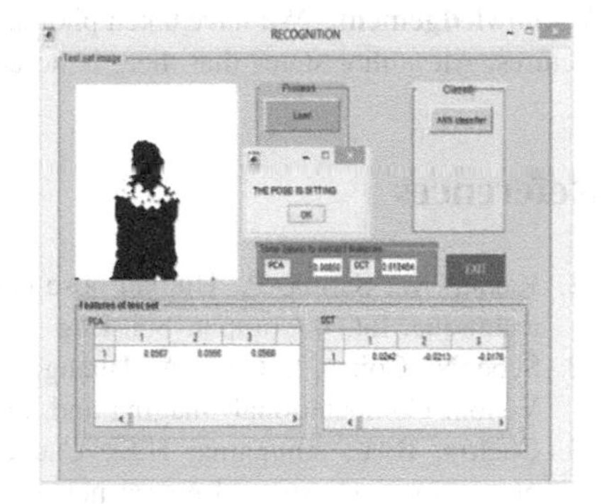


**Fig. 13.** Screen shot of sitting pose

### 4.1 Accuracy of the Proposed System

We have tested the trained Feed-Forward Neural Network classifier with the help of different images. We have tested random images containing frequently used human poses. An acceptable means good accuracy of detection was obtained. The following Table 1 gives an idea about the accuracy of the system.

**Table 1.** Accuracy of the proposed system

| Class | No. of testing image | Correctly identified | Accuracy in percentage |
|---|---|---|---|
| Sitting | 30 | 22 | 73% |
| Standing | 40 | 31 | 77% |
| Waving | 25 | 22 | 88% |
| Handshake | 40 | 37 | 92% |

So, a over all of 82.5% accuracy was found when tested with 135 random images.

## 5 Conclusion and Future Work

We were able to detect four human poses viz. Sitting, Standing, Handshake and Waving with overall accuracy of 82.5% as shown in the Table 1. Our challenge to this work was to prepare a good image quality database. We have prepared the database by taking photographs of human body from different positions. We were also able to tag the poses successfully. Our system is more accurate for hand pose identification as the accuracy percentage are above 80% known from Table 1.

In future we will train and test the system with more data and try to identify more poses over these four and also try to increase the efficiency of our model, more than now achieved. We will also look into tag complex human poses, which have more than one pose. The work will be extended more by using CNN machine learning techniques in the near future.

**Acknowledgement.** We have taken photographs of author and we got permission to display this pictures. All author states that there is no conflict of interest. We used our own data.

## References

1. Kakati, M., Sarma, P.: Human pose detection from a digital image with machine learning technique. Int. J. Comput. Sci. Eng. **7**(5), 1389–1393 (2019)
2. Sebe, N., Cohen, I., Garg, A., Huang, T.S.: Introduction. In: Machine Learning in Computer Vision. Computational Imaging and Vision, vol. 29. Springer, Dordrecht (2005)
3. Chaitra, B.H., Anupama, H.S., Cauvery, N.K.: Human action recognition using image processing and artificial neural networks. Int. J. Comput. Appl. (0975-8887) **80**(9), 31–34 (2013)
4. Liu, Y., Xu, Y., Li, S.B.: 2-D human pose estimation from images based on deep learning: a review. In: 2018 2nd IEEE Advanced Information Management, Communicates, Electronic and Automation Control Conference, pp. 462–465 (2018)
5. Belagiannis, V., Zisserman, A.: Recurrent human pose estimation. In: 2017 12th IEEE International Conference on Automatic Face and Gesture Recognition, pp. 468–475 (2017)
6. Chu, X., Ouyang, W., Li, H., Wang, X.: Structured feature learning for pose estimation. In: IEEE Conference on Computer Vision and Pattern Recognition, pp. 4715–4723 (2016)
7. Zhuang, W., Xia, S., Wang, Y.: Human pose estimation using direction maps. In: 33rd Youth Academic Annual Conference of Chinese Association of Automation, pp. 977–982 (2018)
8. Geetha, N., Samundeeswari, E.S.: A review on human activity recognition system. Int. J. Comput. Sci. Eng. **6**(12), 825–829 (2018)
9. Ionescu, C., Li, F., Sminchisescu, C.: Latent structured models for human pose estimation. In: ICCV, pp. 2220–2227 (2011)
10. Barman, K., Sarma, P.: Glaucoma detection using fuzzy C-means clustering algorithm and thresholding. Int. J. Comput. Sci. Eng. **7**(3), 859–864 (2019)
11. Robinson, J., Kecman, V.: Combining support vector machine learning with the discrete cosine transform in image compression. IEEE Trans. Neural Netw. **14**(4), 950–958 (2003)
12. Lera, G., Pinzolas, M.: Neighborhood based Levenberg-Marquardt algorithm for neural network training. IEEE Trans. Neural Netw. **13**(5), 1200–1203 (2005)

# Machine Learning Algorithm in Smart Farming for Crop Identification

N. D. Vindya(✉) and H. K. Vedamurthy(✉)

Department of CSE, SIT, Tumakuru, India
vindyamadappa@gmail.com, vedahk@rediffmail.com

**Abstract.** Agriculture is the main supporting sector of Indian economy and most of the rural population's livelihood depends on it. Crop yielding has been decreasing day by day because it's growth mainly depends on monsoon which is highly unpredictable in India. Further, the crop growth also depends on soil parameters, which varies due to ground contamination. For increasing the crop productivity, there is a need to incorporate new technologies in the field of agriculture. By utilizing new technologies like Machine Learning and Internet of Things applications in the field of agriculture will be enhance the financial growth of the country. Here, we have presented a framework which mainly focuses on agriculture in Tumakuru district of Karnataka that uses Naive Bayes classification method to suggest the farmers with the best suitable crop for sowing in that particular environmental condition.

**Keywords:** Machine learning · Gaussian Naive Bayes · Data analytics · Prediction

## 1 Introduction

Agriculture is the backbone of India; around 61.5% of Indian population in rural area depends on agriculture. In crop productivity, India is ranked second in world wide. Agriculture contributed 17–18% of GDPA and employed 50% of Indian work force [1]. In India, crop productivity mainly depends on monsoon. When there is good monsoon then crop yield will be more, when there is poor monsoon crop yield will be less. But nowadays monsoon is getting varied, farmers may go wrong in forecasting the monsoon. Along with this, crops productivity also depends on soil conditions, but contamination of ground, reduction in water body, green house emission are all affecting the productivity. Therefore, integrating new technologies like Machine Learning and Internet of Things (IoT) can able to predict the monsoon and to decide which crop is best for the present environmental condition.

To develop new openings for recognising the intensity of data, quantify and unravel its processes in agriculture sector, machine learning has been merged with IoT, cloud computing, and big data. It is a scientific field which gives ability for machines to learn without being explicitly programmed [2]. This technology is now applied in many fields like medical, meteorology, economic science, robotics, agriculture, food, climatic, etc.


S. Smys et al. (Eds.): ICCVBIC 2019, AISC 1108, pp. 19–25, 2020.
https://doi.org/10.1007/978-3-030-37218-7_3

Internet of Things is also a main part in machine learning because that helps in gathering of data from different fields. Real time soil moisture, humidity and temperature data are taken by integrating sensor elements with Raspberry pi for crop prediction [3]. Data analytics (DA) is a method for inspecting datasets in order to obtain the hidden information related to the data, which is increasingly used with the support of specific software and system [4].

Smart farming is an idea that focuses on data involvement and proportionate inventions in the physical, administration and digital cycle. Smart farm helps farmers for taking good decision in order to gain the high quality output by applying data analysis and automation in the agriculture field. In agriculture, crops are mainly lossed due to variations in monsoon. Farmers are able to predict the monsoon based on past years' experience, since it is continuously varying, prediction made by them may also fail. Once crops related data and rainfall data are gathered, then prediction can be made and decision is given to the farmers. In this paper, we focused agriculture in Tumakuru district of Karnataka and collected dataset related to crops and rainfall from Krishi Vignana Kendra Hirehalli and Karnataka State Disaster Monitoring Centre Bangalore. Gaussian Naive Bayes classifier model is developed to provide information about which is the best crop for sowing in particular season in each taluk of Tumakuru district.

## 2 Literature Survey

Using sensors real time weather and farm related data are collected and analysis is made on collected data then result is given to the farmers. Data from irrigation related reports, sensor recorded field data and satellite images, weather data and crops data are collected from various sources. If there are any missing values in collected data, then it is filled by calculating the mean values. By using Naive Bayes classifier for collected data sets a model is developed to recommend about crops to the farmers [5].

Soil parameters like, Crop rotation, Phosphorus, Soil moisture, Surface temperature, Nitrogen, Potassium, etc. are important for growing crops. Using these parameters agriculture field can be monitored, which will increase the productivity in excessive amount. From IMD (Indian Metrological Department) weather data and soil parameters are integrated and developed a model by applying a machine learning algorithm. This provides crops related information to the farmers [6].

A model for discovering and controlling soil parameters and five types of diseases in cotton leaf is developed using Support Vector Machine algorithm. A motor driver is developed having raspberry pi with four different sensors containing LM 35 soil moisture, DHT-22 humidity-temperature and water sensors, which helps to manage soil conditions at different places [7].

To increase the growth of different types of crops in a closed environment usually in a Greenhouse is done by controlling the changing factors like humidity, soil moisture, temperature is done by connecting sensors with Raspberry pi and also a mobile application is developed which helps to alerts the user when there is a variation in these factors. The sensor values are transferred to the ThingSpeak cloud which will be used for analysing [3].

By applying data mining techniques on soil data sets can predict the better yielding crops. Naïve Bayes and K-Nearest Neighbour algorithms are applied to the soil data sets for predicting better yielding crops [8].

Machine learning is currently used in multidisciplinary agricultural area using computing technologies and analysis, which contains (1) management of crops like disease detection, crop yield analysis, species identification; (2) management of soil; (3) management of animal welfare; (4) management of water etc. Farm management systems are developed for identification and taking decision for the farm by the farmers using data analysis, machine learning and artificial intelligence [9].

# 3 Process Description

## 3.1 Modality

Tumakuru is a district in Karnataka state having 2,678,980 populations as per 2011 census [10]. It falls under agriculture zone 4, 5 and 6 with Central dry, Eastern dry and Southern dry zones respectively. Six taluks namely Chikkanayakanahalli, Madhugiri, Tiptur, Sira, Pavagada and Koratagere are come under zone 4. Tumakuru and Gubbi taluks comes under zone 5 and the zone 6 contains Turuvekere and Kunigal. In Tumakuru agriculture is considered as one of the parameters for economic growth. It is having 6.1% GSDP (Gross State Domestic Product) for Karnataka.

District has average annual rainfall around 593.0 mm in the district. Three major types of soil found in Tumakuru, that are Red loamy soil, mixed red and black soil, Red sandy soil. Red loamy soil found in Koratagere, Tumakuru, Madhugiri and Kunigal. Red sandy soil found in Tiptur, Turuvekere, Sira, Pavagada and Gubbi. Mixed red and black soil found in Chikkanayakana Halli.

There are many crops have grown in Tumakuru like Paddy, Maize, Bajra, Ground nuts, etc. Paddy can be grown with temperature ranges from 21–37, humidity 60–80, soil type black soil, red loamy, red soil with PH 4–6. Like that there are many crops have grown in Tumakuru, here we considered eighteen crops that are Paddy, Ragi, Jowar, Maize, Bajra, Groundnut Mustard, Sunflower, Castor, Green gram, Horse gram, Red gram, Tomato, Beans, Cabbage, Brinjal and Turmeric.

By collecting required environmental conditions needed for above mentioned crops, the proposed model is done. Rainfall data of each taluk from the year 2014–2018 are collected from Karnataka State Disaster Monitoring Centre. Some crop details are taken from Krishi Vignana Kendra Hirehalli, Tumakuru. Along with this real time soil moisture, temperature and humidity data are taken from sensors. Attributes containing temperature, humidity, rainfall, soil type and soil PH which helps for sowing is considered.

## 3.2 Methodology

Bayesian models (BM) works based on analysis framework of Bayesian interface, which is a family of conditional probability [11]. For solving regression and classification problem Naïve Bayes algorithm is used and it comes under supervised learning of machine learning algorithm [12]. Bernoulli, Gaussian and Multinomial are different types of Naïve Bayes algorithm. This algorithm is having low computational cost because, it works by taking each class value is independent to other and to find maximum likelihood closed form expression is taken. Initially using training datasets classifier model is trained. The training datasets having tuples X with features $Y_1$, $Y_2$, $Y_3$, …, $Y_n$. These tuples are checked against each class.

Figure 1, shows the crop identification system which is trained using crops data sets contain major crops grown in Tumakuru. Before loading the crops data, import the packages like numpy, seaborn, pandas and sklearn libraries to the algorithm. Numpy is used for the mathematical functions, pandas library is used for analytics purpose, seaborn is a library for plotting of graphs, scikit library is used for creating different models. Then import the crop datasets which contains crops name, humidity, rainfall, soil type, temperature and soil PH features with eighteen classes by using read_csv() function. Header function is used to identify the header in first row, in our dataset first row contains crop names.

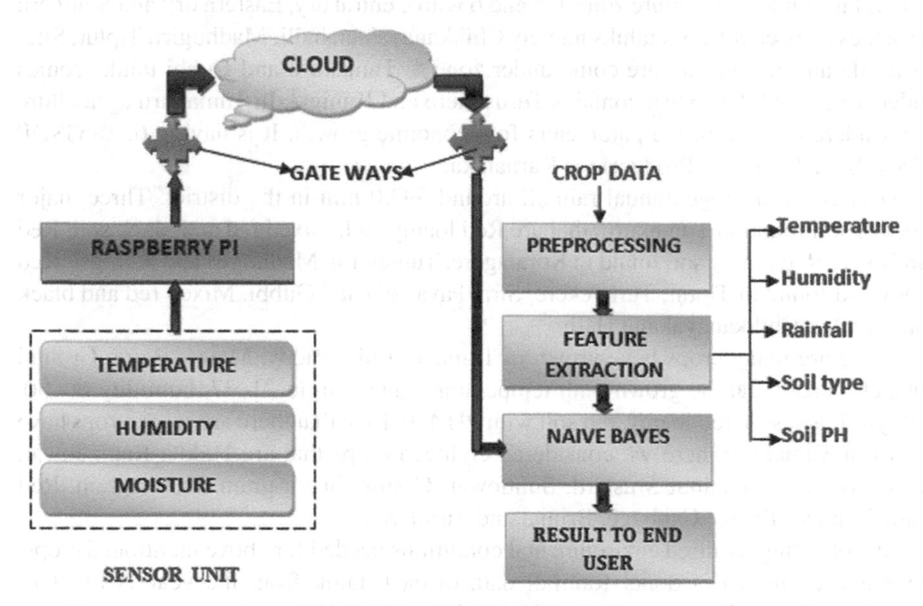


**Fig. 1.** Crop identification system

After this missing values are filled by calculating mean values of data. In Naïve Bayes method all the data should be in a single format. But our dataset contains thirteen types of soils with string format, therefore One-Hot Encoding [13] method is performed by taking each soil type in separate columns with assigning binary values for each. Then perform the reshaping of datasets to convert it into the 3D format. After converting it into a single format, data pre-processing is done on the updated datasets and split it into test and train data, which is done by using method train_test_split().

After splitting the datasets, Gaussian Naïve Bayes [14] model is selected using scikit learn method and fit() method is used to train the model with training dataset.

Using predict() method, prediction is made by passing test data to the trained classifier model. Then accuracy of the model is done by comparing predicted values with target values by accuracy_score() method.

As shown in Fig. 1, the model is used for the prediction by passing new real time data which is obtained by connecting YL 69 soil moisture and DHT22 humidity and temperature sensor components with Raspberry pi 2 Model-B. From this sensor unit, real time data for every 5 s are taken. Then data are directly uploaded into ThingSpeak IoT cloud using Raspberry pi as a gateway between sensor unit and cloud. This stored data along with rainfall data from past five years of each taluk, soil PH and soil types are passed to the model.

The final output consists of predicted crop i.e., which is suitable for sowing in the given environmental conditions.

## 4 Results

Figure 2 shows, that user select the Tumakuru taluk and April month for the prediction of crop in that particular region and time.

**Fig. 2.** Selection of taluk and month

Figure 3 shows that, when taluk and month are selected, it will display the humidity, soil PH, soil type, rainfall and temperature of that particular region.

**Fig. 3.** Displaying of parameter values

Figure 4 shows, if the field having temperature 32, rainfall 38, soil Ph 6, humidity 41 and black soil then Horse Gram can be grown in that area.

**Fig. 4.** Display of predicted crop

## 5 Conclusion

In this research work, we have designed a model of crop identification system using Gaussian Naïve Bayes classifier. Based on the parameters like humidity, soil PH, temperature, rainfall and soil type, we predicted the suitable crop for the given region. When a user request for crop recommendation in the particular region and month, he will get the real time data on the humidity and temperature values which is taken by the deployed sensors in the different regions of each taluk. Along with this he will get the rainfall prediction and the crop name which is suitable for that particular environmental condition. The presented work archives 70% of accuracy. In further, we have planned to extend our work to identify different crops in various regions of Karnataka.

**Compliance with Ethical Standards.** All author states that there is no conflict of interest. We used our own data. Animals/human are not involved in this work.

## References


1. Agriculture in India. https://en.wikipedia.org/wiki/Agriculture_in_India
2. Samuel, A.L.: Some studies in machine learning using the game of checkers. IBM J. Res. Dev. **44**, 206–226 (1959)
3. Dedeepya, P., Srinija, U.S.A., Krishna, M.G., Sindhusha, G., Gnanesh, T.: Smart greenhouse farming based on IOT. In: 2018 Second International Conference on Electronics, Communication and Aerospace Technology (ICECA), pp. 1890–1893 (2018)
4. Data analytics. https://searchdatamanagemsent.techtarget.com/definition/data-analytics
5. Priya, R., Ramesh, D.: Crop prediction on the region belts of india: a Naïve Bayes MapReduce precision agricultural model. In: International Conference on Advances in Computing, Communications and Informatics (ICACCI), pp. 99–104 (2018)
6. Zingade, D.S., Buchade, O., Mehta, N., Ghodekar, S., Mehta, C.: Crop prediction system using machine learning. Int. J. Adv. Eng. Res. Dev. Spec. Issue Recent Trends Data Eng. **4**(5), 1–6 (2017)
7. Sarangdhar, A.A., Pawar, V.R.: Machine learning regression technique for cotton leaf disease detection and controlling using IoT. In: International Conference on Electronics, Communication and Aerospace Technology, ICECA, vol. 2, pp. 449–454 (2017)
8. Paul, M., Vishwakarma, S.K., Verma, A.: Analysis of soil behaviour and prediction of crop yield using data mining approach. In: 2015 International Conference on Computational Intelligence and Communication Networks (CICN), pp. 766–771 (2015)
9. Liakos, K.G., Busato, P., Moshou, D., Pearson, S., Bochtis, D.: Machine learning in agriculture: a review. Sensors **18**, 2674 (2018). https://doi.org/10.3390/s18082674
10. Census in Tumakuru. https://www.census2011.co.in/census/district/267-tumkur.html
11. Alipio, M.I., Cruz, A.E.M.D., Doria, J.D.A., Fruto, R.M.: A smart hydroponics farming system using exact inference in Bayesian network. In: 2017 IEEE 6th Global Conference on Consumer Electronics, pp. 1–5 (2017)
12. Chen, J., Huang, H., Tian, S., Qu, Y.: Feature selection for text classification with Naïve Bayes. Expert Syst. Appl. **36**(3), 5432–5435 (2009)
13. Tahmassebi, A., Gandomi, A.H., Schulte, M.H.J., Goudriaan, A.E., Foo, S.Y., Meyer-Baese, A.: Optimized Naive-Bayes and decision tree approaches for fMRI smoking cessation classification **2018**, 24 p. Article ID 2740817
14. One-Hot Bit Encoding. https://machinelearningmastery.com/why-one-hot-encode-data-in-machine-learning/

# Analysis of Feature Extraction Algorithm Using Two Dimensional Discrete Wavelet Transforms in Mammograms to Detect Microcalcifications

Subash Chandra Bose[1(✉)], Murugesh Veerasamy[2], Azath Mubarakali[3], Ninoslav Marina[1], and Elena Hadzieva[1]

[1] University of Information Science and Technology (UIST) "St. Paul the Apostle", Ohrid, Republic of Macedonia
subash.jaganathan@uist.edu.mk

[2] Department of Computer Science, Cihan University-Erbil, Kurdistan Region, Iraq
murugesh.veerasamy@cihanuniversity.edu.iq

[3] College of Computer Science, Department of CNE, King Khalid University, Abha, Saudi Arabia
mailmeazath@gmail.com

**Abstract.** The paper focusing on the analysis of the feature extraction (FE) algorithm using a two-dimensional discrete wavelet transform (DWT) in mammogram to detect microcalcifications. We are extracting nine features namely Mean(M), Standard deviation(SD), Variance(V), Covariance(Co_V), Entropy(E), Energy(En), Kurtosis(K), Area(A) and Sum(S). FE is a tool for dimensionality reduction. Instead of too much information the data should be reduced representation of the input size. The extraction has been done to construct the dataset in the proper format of the segmented tumor, then only it can be given to the classifier tools and achieving high classification accuracy. The nine statistical features are extracted from the mammogram is determined the effectiveness of the proposed technique. This experiment has been conducted for 322 mammogram images and in this paper, we listed only a few. Actually, the detection of microcalcifications has been done with various algorithms discussed below. The analysis of the FE algorithm using two-dimensional discrete wavelet transform in mammogram to detect macrocalcification technique has been analyzed using MATLAB.

**Keywords:** Breast cancer · Mammogram · Microcalcifications · Feature extraction · Discrete wavelet transform

## 1 Introduction

In Macedonia, breast cancer is leading cancer among all other types of cancer. Breast cancer in Macedonia is alarming. The morality of breast cancer in Macedonia is higher than all other cancer types. According to the latest WHO data published in 2017 Breast Cancer deaths in Macedonia reached 328 or 1.78% of total deaths. The death rate based


S. Smys et al. (Eds.): ICCVBIC 2019, AISC 1108, pp. 26–39, 2020.
https://doi.org/10.1007/978-3-030-37218-7_4

on the age-adjusted is 20.76 per 100,000 of population ranks Macedonia #50 in the world [1].

Here, a new method for Detection of microcalcifications in mammograms using soft segmentation technique is presented [2]. Different image processing techniques have been done like preprocessing enhancement of the image. Segmentation using Fuzzy c-means clustering (FCM), FE using two-dimensional discrete wavelet transform, and finally normal or abnormal (benign or malignant) images have been classified by using artificial neural network (ANN).

## 2 Literature Survey

Last several decades many authors had been conducted research in this field here we listed few detections of breast cancer techniques.

Li [3] used a Markov random field (MRF) in image analysis for detection of tumors. Vega-Corona et al. [4] using a method CAD system to detect microcalcifications in mammograms using a general regression neural network (GRNN). D'Elia et al. [5] used a MRF and support vector machine classification (SVM) to detect early stage of breast cancer. Fu and Lee [6] used the GRNN and SVM to detect microcalcifications. Osta et al. [7] Used wavelet-based FE for dimensionality reduction. Jiji et al. [8] used Region of Interest (ROI) and Backpropagation Neural Network (BPNN) for detection of microcalcification.

## 3 Image Acquisition

The mammograms are available on the website: http://peipa.essex.ac.uk/info/mias.html. This database contains 322 mammograms [9, 11].

Detection of microcalcifications has been done in four stages Preprocessing, Segmentation, FE, and classification are shown in Fig. 1.

**Enhancement and Preprocessing:** It has been done using adaptive median filtering for noise removal.
**Segmentation:** It has been done using Fuzzy c-means clustering [12].
**Feature extraction:** The features have been extracted using two-dimensional discrete wavelet transforms discussed and the experimental results have been discussed below in detail.
**Classification:** The extracted features have been feed to the ANN and the benign or malignant image has been classified.

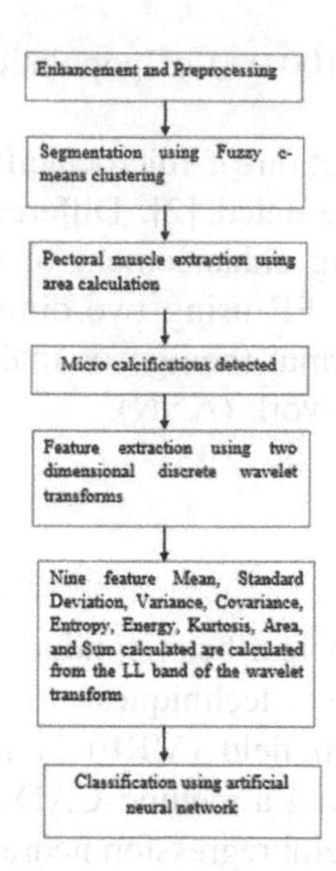


**Fig. 1.** Flow chart

# 4 Experimental Results

The peak to noise ratio (PSNR), Mean Square Error (MSE) values for 322 mammogram image has been evaluated and listed few has been shown in Figs. 2 and 3.

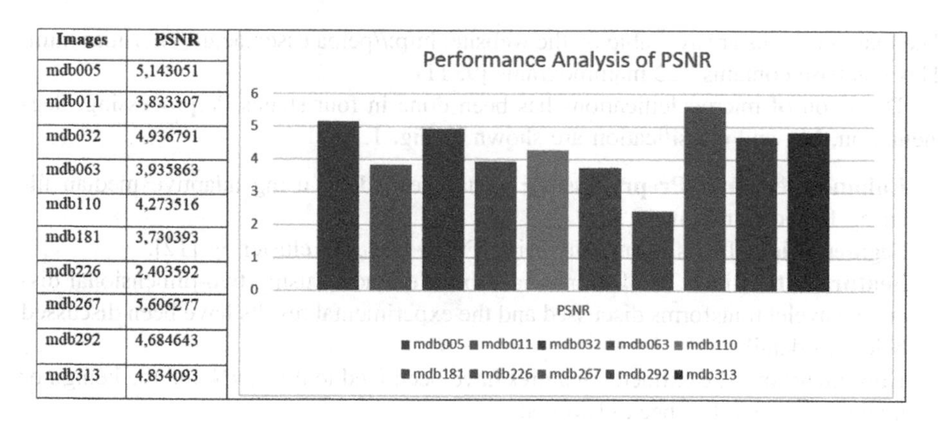

| Images | PSNR |
|---|---|
| mdb005 | 5,143051 |
| mdb011 | 3,833307 |
| mdb032 | 4,936791 |
| mdb063 | 3,935863 |
| mdb110 | 4,273516 |
| mdb181 | 3,730393 |
| mdb226 | 2,403592 |
| mdb267 | 5,606277 |
| mdb292 | 4,684643 |
| mdb313 | 4,834093 |



**Fig. 2.** Performance analysis of PSNR

To get a better performance result the image should be in better quality. If the PSNR is high and MSE is valued is less then, the image quality will be good and the performance will be better.

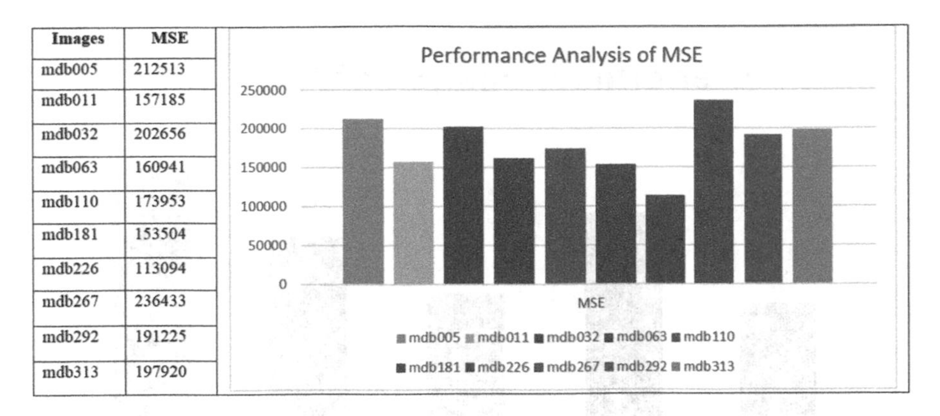

| Images | MSE |
|---|---|
| mdb005 | 212513 |
| mdb011 | 157185 |
| mdb032 | 202656 |
| mdb063 | 160941 |
| mdb110 | 173953 |
| mdb181 | 153504 |
| mdb226 | 113094 |
| mdb267 | 236433 |
| mdb292 | 191225 |
| mdb313 | 197920 |



**Fig. 3.** Performance analysis of MSE

Here, PSNR and ASNR, the values c and d respectively is larger, then the enhancement and preprocessing method is better compared to the existing method.

The PSNR value has been calculated using the formula: $PSNR = (c - a)/\sigma$.
The ASNR value has been calculated using the formula: $ASNR = (c - a)/\sigma$.
The MSE value has been calculated using the formula: $MSE = (c - d)/m * n$.

Where a is the mean gray level value of the original image, c is the maximum gray level value and d is the average gray level value of the enhanced image, m, n is the pixel values of an enhanced image and σ is the SD of the original image.

The enhanced mammogram image is evaluated using PSNR values and SNR values. The result is shown in Figs. 4 and 5.

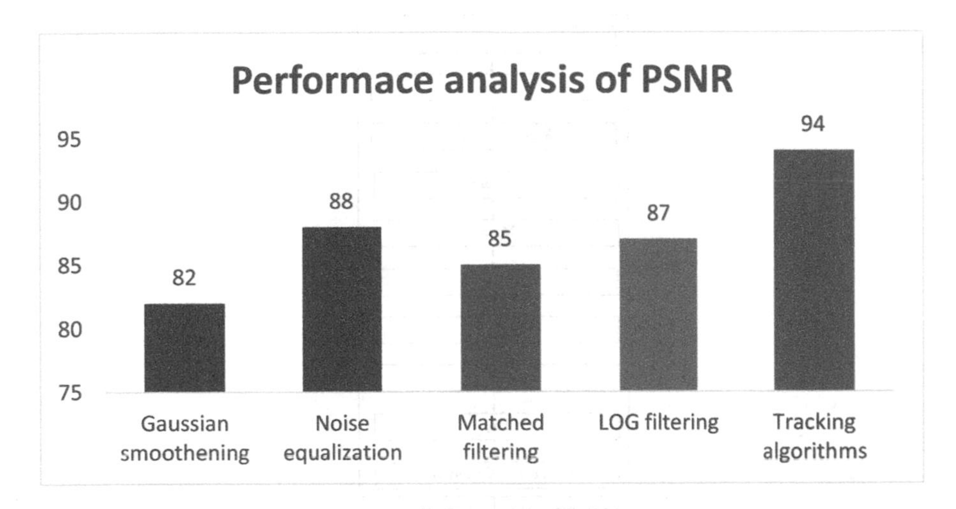


**Fig. 4.** Performance analysis of PSNR.

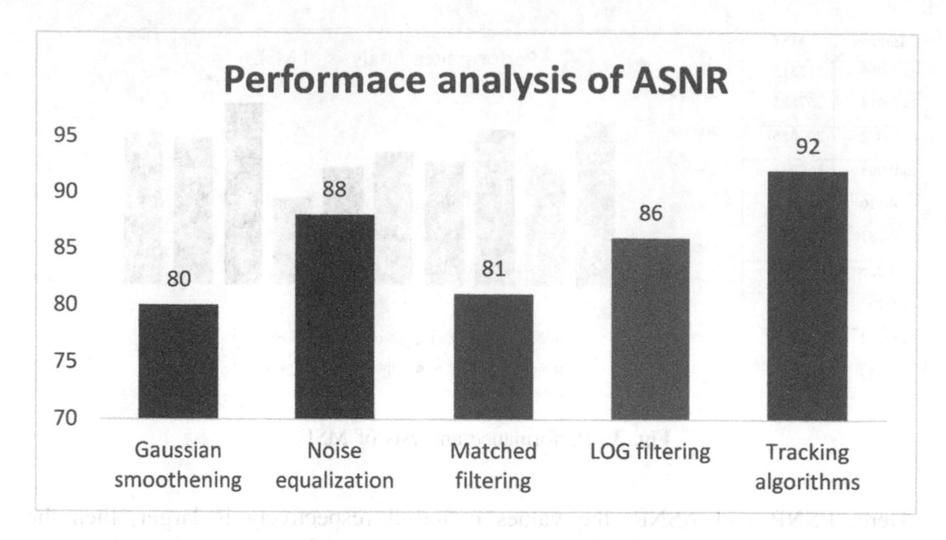


**Fig. 5.** Performance analysis of ASNR.

## 5 Feature Extraction

FE had been conducted by using discrete wavelet transforms. The block diagram for FE has been shown in the Fig. 6.

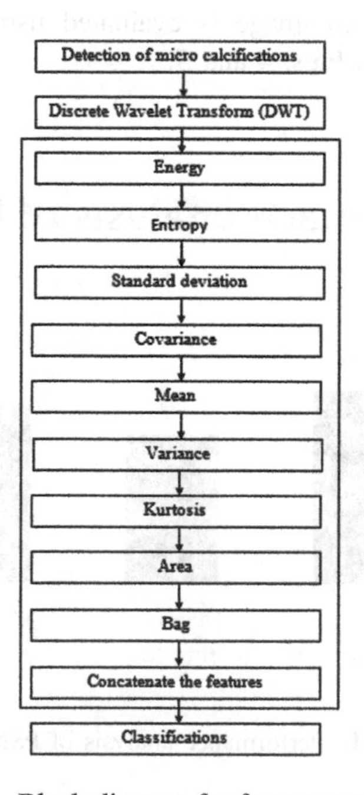


**Fig. 6.** Block diagram for feature extraction

### 5.1 Discrete Wavelet Transforms

Discrete wavelet transforms (DWT) technique has been used to classify the system [10].

**Statistical Features Calculation:**
There are nine features extracted from the Low, Low band of discrete wavelet transform such as, M, SD, V, Co_V, E, En, K, A, and S.

***Equivalent Matlab code***

```
[a1, h1, v1, d1] = dwt2 (STI, 'db1');
[a2, h2, v2, d2] = dwt2 (a1, 'db1');
imshow ([a2, h2, v2, d2]);
title ('Two dimensional discrete wavelet transform');
```

**a. Mean (M)**
The mean value has been calculated using the formula:

$$\mu - \frac{1}{MN}\sum_{i=1}^{M}\sum_{j=1}^{N} p(i,j)$$

MATLAB code for calculating M

```
mean1 = mean (double (a (:)))
```

The performance analysis of mean for various mammogram image has been shown in the Fig. 7.

| Images | Mean |
|---|---|
| mdb005 | 7,649230957 |
| mdb011 | 6,183532715 |
| mdb032 | 3,236877441 |
| mdb063 | 8,348754883 |
| mdb110 | 6,516784668 |
| mdb181 | 9,463256836 |
| mdb226 | 8,667724609 |
| mdb267 | 9,094543457 |
| mdb292 | 4,030273438 |
| mdb313 | 4,325561523 |

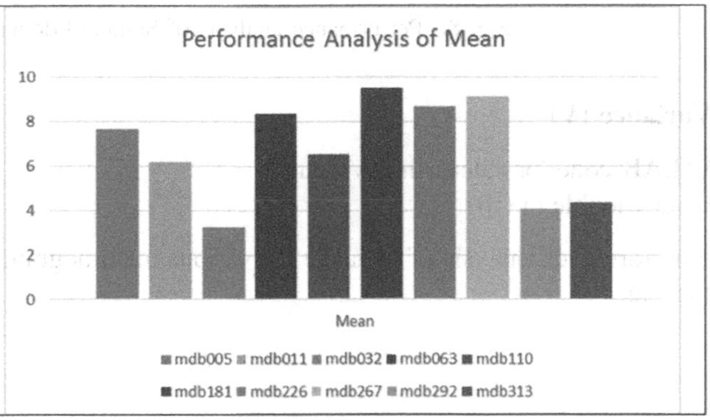

**Fig. 7.** Performance analysis of Standard deviation

**b. Standard Deviation (SD)**

It has been evaluated of the mean square deviation of pixel value p (i, j) from its M value.

MATLAB code for calculating the SD
SD = std (std (double (a)))
Steps to calculate the SD
SD = sqrt (V) = sqrt (8.7) = roughly 2.95

The performance analysis of SD for various mammogram image has been shown in the Fig. 8.

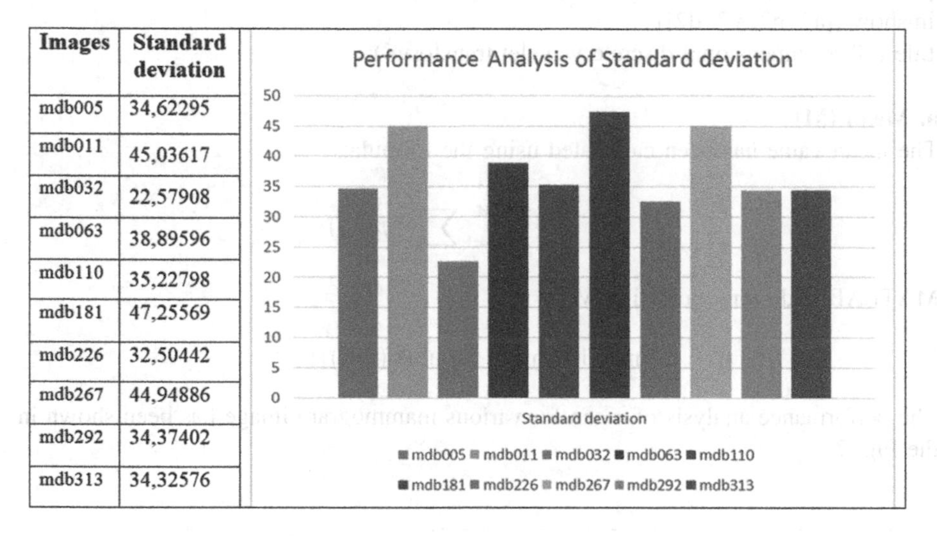

| Images | Standard deviation |
|---|---|
| mdb005 | 34,62295 |
| mdb011 | 45,03617 |
| mdb032 | 22,57908 |
| mdb063 | 38,89596 |
| mdb110 | 35,22798 |
| mdb181 | 47,25569 |
| mdb226 | 32,50442 |
| mdb267 | 44,94886 |
| mdb292 | 34,37402 |
| mdb313 | 34,32576 |



**Fig. 8.** Performance analysis of Standard deviation

**c. Variance (V)**

MATLAB code for calculating Variance
V = var (double (a (:)))

The performance analysis of variance for various mammogram image has been shown in the Fig. 9.

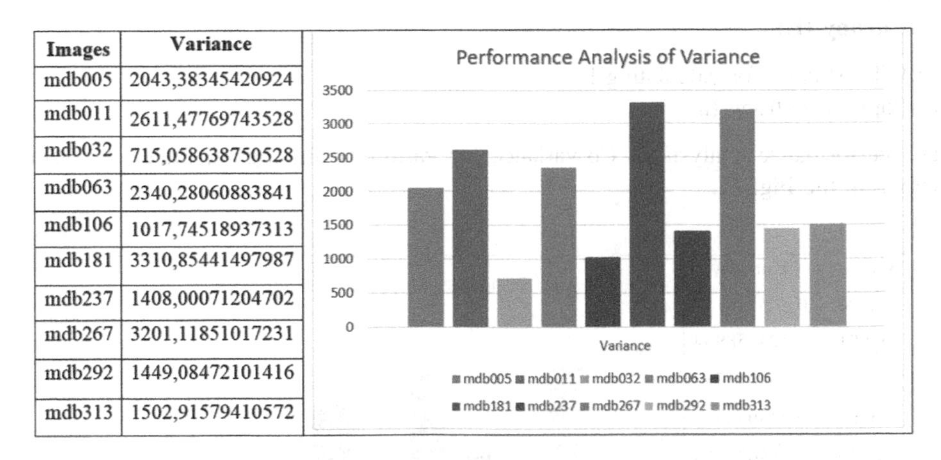

| Images | Variance |
|---|---|
| mdb005 | 2043,38345420924 |
| mdb011 | 2611,47769743528 |
| mdb032 | 715,058638750528 |
| mdb063 | 2340,28060883841 |
| mdb106 | 1017,74518937313 |
| mdb181 | 3310,85441497987 |
| mdb237 | 1408,00071204702 |
| mdb267 | 3201,11851017231 |
| mdb292 | 1449,08472101416 |
| mdb313 | 1502,91579410572 |



**Fig. 9.** Performance analysis of Variance

**d. Covariance (Co_V)**

MATLAB code for calculating Co_V

```
Co_V = Co_V (double (a));
Co_V = sum (sum (Co_V))/(length (Co_V) * 1000)
```

The performance analysis of Co-variance for various mammogram image has been shown in the Fig. 10.

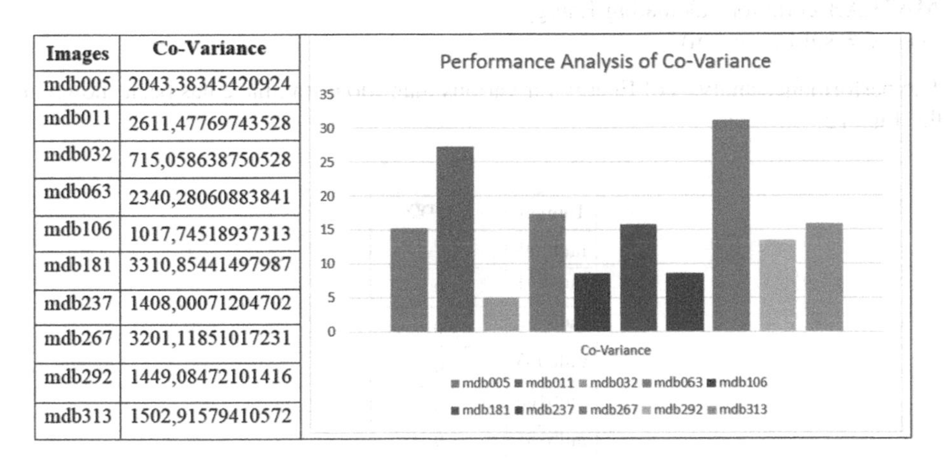

| Images | Co-Variance |
|---|---|
| mdb005 | 2043,38345420924 |
| mdb011 | 2611,47769743528 |
| mdb032 | 715,058638750528 |
| mdb063 | 2340,28060883841 |
| mdb106 | 1017,74518937313 |
| mdb181 | 3310,85441497987 |
| mdb237 | 1408,00071204702 |
| mdb267 | 3201,11851017231 |
| mdb292 | 1449,08472101416 |
| mdb313 | 1502,91579410572 |



**Fig. 10.** Performance analysis of Co-variance

**e. Entropy (E)**

MATLAB code for calculating E
Entropy1 = entropy (a)

The performance analysis of Co-variance for various mammogram image has been shown in the Fig. 11.

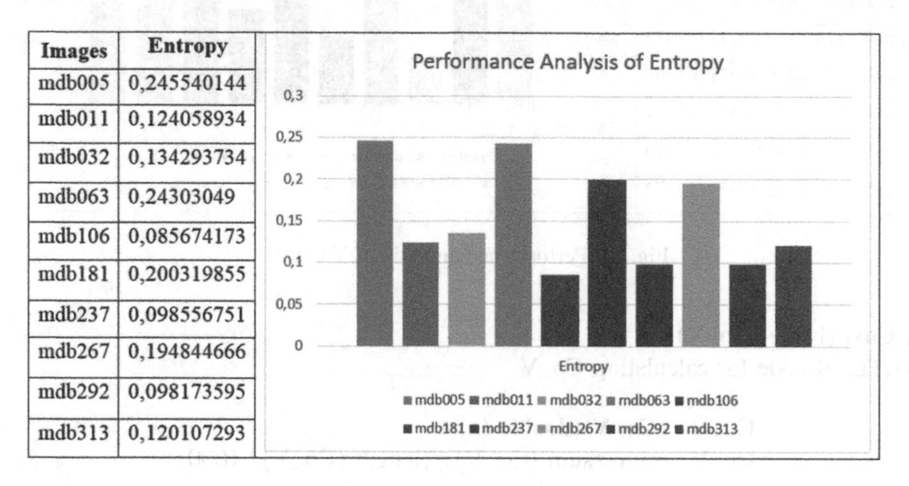

| Images | Entropy |
|---|---|
| mdb005 | 0,245540144 |
| mdb011 | 0,124058934 |
| mdb032 | 0,134293734 |
| mdb063 | 0,24303049 |
| mdb106 | 0,085674173 |
| mdb181 | 0,200319855 |
| mdb237 | 0,098556751 |
| mdb267 | 0,194844666 |
| mdb292 | 0,098173595 |
| mdb313 | 0,120107293 |



**Fig. 11.** Performance analysis of Entropy

**f. Energy**

MATLAB code for calculating Energy
energy = sum (h_norm)

The performance analysis of Energy for various mammogram image has been shown in the Fig. 12.

| Images | Energy |
|---|---|
| mdb005 | 0 |
| mdb011 | 0 |
| mdb032 | 0 |
| mdb063 | 0 |
| mdb106 | 0 |
| mdb181 | 0 |
| mdb237 | 0 |
| mdb267 | 0 |
| mdb292 | 0 |
| mdb313 | 0 |

**Fig. 12.** Performance analysis of Energy

**g. Kurtosis**

MATLAB code for calculating Kurtosis
kurtosis1 = kurtosis (double (a (:)))

The performance analysis of kurtosis for various mammogram image has been shown in the Fig. 13.

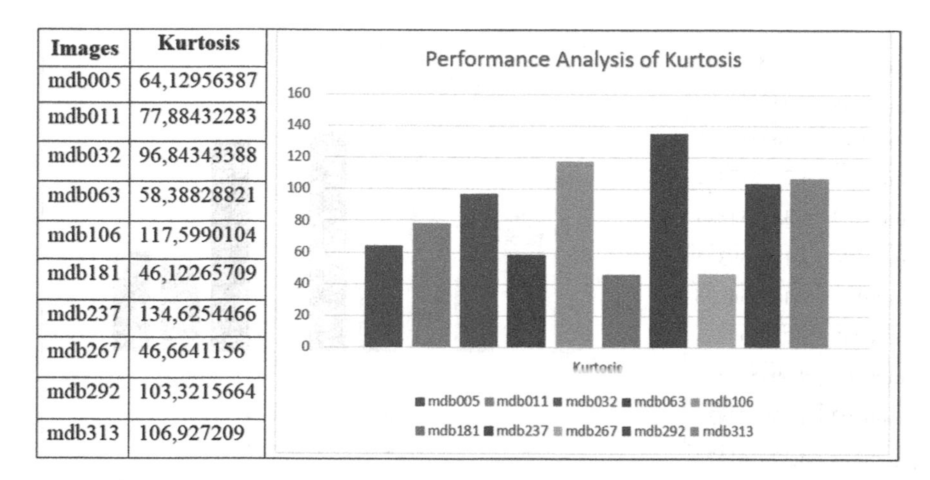

| Images | Kurtosis |
|---|---|
| mdb005 | 64,12956387 |
| mdb011 | 77,88432283 |
| mdb032 | 96,84343388 |
| mdb063 | 58,38828821 |
| mdb106 | 117,5990104 |
| mdb181 | 46,12265709 |
| mdb237 | 134,6254466 |
| mdb267 | 46,6641156 |
| mdb292 | 103,3215664 |
| mdb313 | 106,927209 |



**Fig. 13.** Performance analysis of Kurtosis

**h. Area**

MATLAB code for calculating Area
area1 = bwarea (double (a (:)))

The performance analysis of kurtosis for various mammogram image has been shown in the Fig. 14.

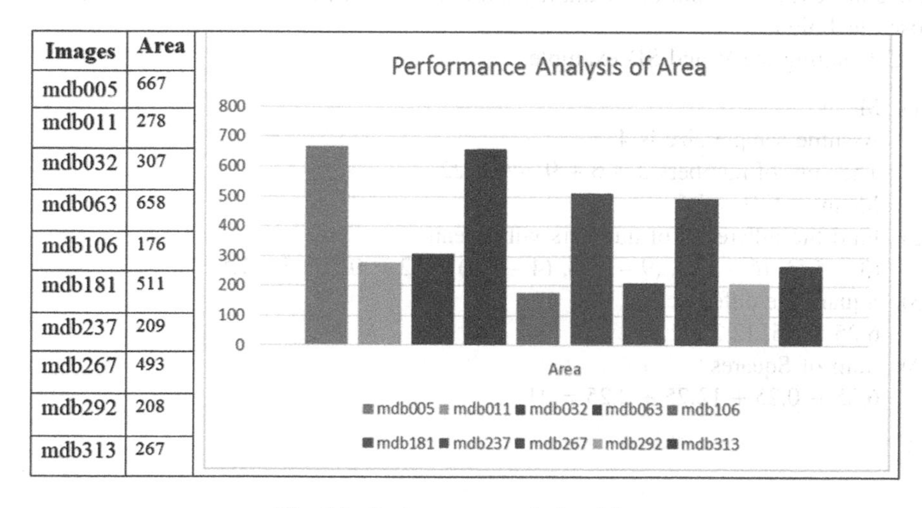

| Images | Area |
|---|---|
| mdb005 | 667 |
| mdb011 | 278 |
| mdb032 | 307 |
| mdb063 | 658 |
| mdb106 | 176 |
| mdb181 | 511 |
| mdb237 | 209 |
| mdb267 | 493 |
| mdb292 | 208 |
| mdb313 | 267 |



**Fig. 14.** Performance analysis of Area

**i. Sum**

MATLAB code for calculating Sum
Bags = sum (sum (a))/(sze (1) * sze (2))

The performance analysis of Bags for various mammogram image has been shown in the Fig. 15.

| Images | Bag |
|---|---|
| mdb005 | 7,649230957 |
| mdb011 | 6,183532715 |
| mdb032 | 3,236877441 |
| mdb063 | 8,348754883 |
| mdb106 | 3,130859375 |
| mdb181 | 9,463256836 |
| mdb237 | 3,719909668 |
| mdb267 | 9,094543457 |
| mdb292 | 4,030273438 |
| mdb313 | 4,325561523 |

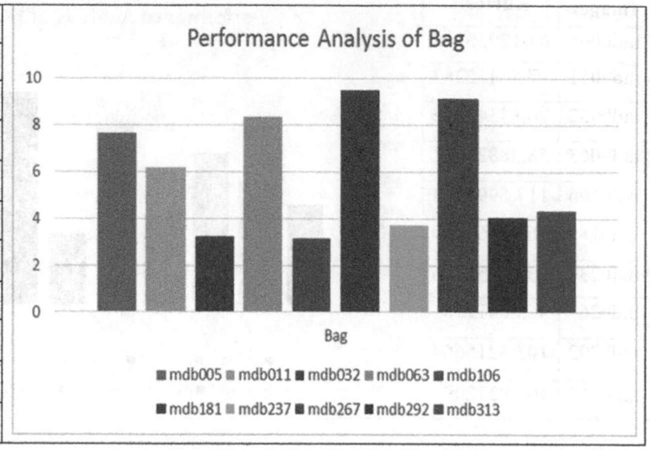

**Fig. 15.** Performance analysis of bag

MATLAB code for combining the features
data = [ag;ahg;energy;entropy1;standarddeviation;covariance;mean;variance;kutosis; area 1;bags]/100.

The above mentioned all these nine statistical features M, SD, V, Co_V, E, En, K, A and S are extracted from the mammogram is determined the effectiveness of the proposed technique.

Calculating the V and SD example.

(1) Mean
Assume sample size is 4
The sum of numbers 3 + 6 + 9 + 4 = 22
Mean = 22/4 = 5.5
(2) Find the difference of numbers with mean
(3 – 5.5), (6 – 5.5), (9 – 5.5), (4 – 5.5) = –2.5, 0.5, 3.5, –1.5
(3) Square the differences
6.25, 0.25, 12.25, 2.25
(4) Sum of Squares
6.25 + 0.25 + 12.25 + 2.25 = 21

(5) V = Divide Sum of squares by sample size
V = 21/4 = 5.25
(6) SD = Sqrt(V)
Sqrt(5.25) = 2.29 (approximately)

The various methods of detection rate and the comparison is shown in the Table 1 and Fig. 16 below.

**Table 1.** Performance analysis comparison

| Authors | Methods | Detection rate |
|---|---|---|
| Lau et al. [20] | Asymmetric measures | 85% |
| Nishikawa et al. [21] | Linear classifier | 70% |
| Sallman et al. [22] | Unwrapping technique | 86.60% |
| Kim et al. [23] | Neural networks | 88% |
| Cordella et al. [24] | Multiple expert system | 78.6% |
| Ferrari et al. [25] | Directional filtering with Gabor wavelets | 74.4% |
| Thangavel et al. [26] | MRF-ACO | 94.8% |
| Thangavel et al. [27] | Ant colony optimization | 96.40% |
| Thangavel et al. [28] | Genetic algorithm | 93.21% |
| Bose et al. [17] | Bilateral subtraction using PSO | 94.6% |
| Bose et al. [18] | MRF-PSO segmentation | 98.3% |
| Bose et al. [2] | Two dimensional discrete wavelet transforms | 97.3% |
| Karnan et al. (2011) | Particle Swarm Optimization | 96.4% |

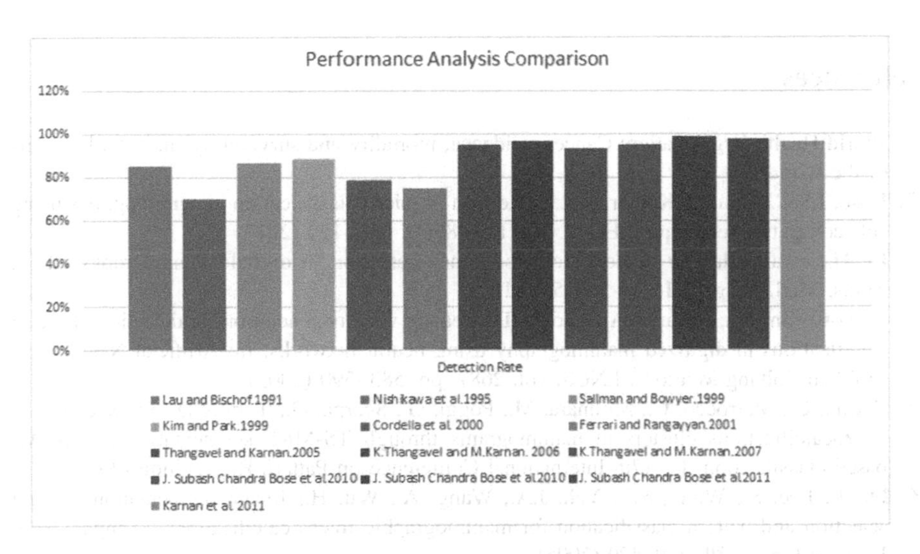


**Fig. 16.** Performance analysis comparison

In Discrete wavelet transformation (DWT) method the true positive (TP) and false positive (FP) detection rate for different threshold values of images are used to measure the performance. These rates have been found using the Receiver Operating Characteristic (ROC) curves ($A_z$ value) ROC curve range should lie between [0, 1]. The $A_Z$ value is 1.0 the detection of diagnosis is perfect, then TP is 100% and FP is 0%. In the FE using the discrete wavelet transformation (FE-DWT) method, the Az value is 94%.

In other methods like Sallam and Bowyer (1999), Lau and Bischof (1991) the overlap is about 40% TP. In the FE-DWT method, the overlap is about 75%. Bilateral Subtraction and MRF – Ant colony optimization method if the threshold value is low then the TP is merged with FP regions.

## 6 Conclusion

The analysis of FE algorithm using two-dimensional discrete wavelet transform (DWT) in mammogram to detect macrocalcification techniques has been discussed in detail and the results have been displayed. Also, this method can be used in mammogram after lossless compression, encryption techniques and for corrupted mammogram images in future [13–18]. This detection of microcalcifications technique can be also used to detect the microparticle in firefighter robotic industries [19].

**Compliance with Ethical Standards.**

✓ All authors declare that there is no conflict of interest.
✓ No humans/animals involved in this research work.
✓ We have used our own data.

## References

1. World Health Organization: Cancer incidence, mortality and survival by site for 14 regions of the world
2. Bose, J.S.C., Shankar Kumar, K.R.: Detection of micro classification in mammograms using soft computing techniques. Eur. J. Sci. Res. 86(1), 103–122 (2012)
3. Li, H., et al.: Markov random field for tumor detection in digital mammography. IEEE Trans. Med. Imaging **14**(3), 565–576 (1995)
4. Vega-Corona, A., Álvarez, A., Andina, D.: Feature vectors generation for detection of micro calcifications in digitized mammography using neural networks. In: Artificial Neural Nets Problem Solving Methods, LNCS, vol. 2687, pp. 583–590 (2003)
5. D'Elia, C., Marrocco, C., Molinara, M., Poggi, G., Scarpa, G., Tortorella, F.: Detection of microcalcifications clusters in mammograms through TS-MRF segmentation and SVM-based classification. In: 17th International Conference on Pattern Recognition (2004)
6. Fu, J., Lee, S., Wong, S.J., Yeh, J.A., Wang, A., Wu, H.: Image segmentation, feature selection and pattern classification for mammographic micro calcifications. Comput. Med. Imaging Graph. **29**, 419–429 (2005)
7. Osta, H., Qahwaji, R., Ipson, S.: Wavelet-based feature extraction and classification for mammogram images using RBF and SVM. In: Visualization, Imaging, and Image Processing (VIIP), Palma de Mallorca, Spain (2008)

8. Dheeba, J., Jiji, W.G.: Detection of microcalcification clusters in mammograms using neural network. Int. J. Adv. Sci. Technol. **19**, 13–22 (2010)
9. MIAS database. www.mias.org
10. Juarez, L.C., Ponomaryov, V., Sanchez, R.J.L.: Detection of micro calcifications in digital mammograms images using wavelet transform. In: Electronics, Robotics and Automotive Mechanics Conference, September 2006, vol. 2, pp. 58–61 (2006)
11. Suckling, J., Parker, J., Dance, D., Astley, S., Hutt, I., Boggis, C., et al.: The mammographic images analysis society digital mammogram database. Exerpta Med. Int. Congr. Ser. **1069**, 375–378 (1994)
12. Mohanad Alata, M., Molhim, M., Ramini, A.: Optimizing of fuzzy C-means clustering algorithm. World Acad. Sci. Eng. Technol. **39**, 224–229 (2008)
13. Bose, J.S.C., Gopinath, G.: A survey based on image encryption then compression techniques for efficient image transmission. J. Ind. Eng. Res. **1**(1), 15–18 (2015). ISSN 2077-4559
14. Bose, J.S.C., Gopinath, G.: An ETC system using advanced encryption standard and arithmetic coding. Middle-East J. Sci. Res. **23**(5), 932–935 (2015). https://doi.org/10.5829/idosi.mejsr.2015.23.05.22233. ISSN 1990-9233, © IDOSI Publications
15. Bose, J.S.C., Saranya, B., Monisha, S.: Optimization and generalization of Lloyd's algorithm for medical image compression. Middle-East J. Sci. Res. **23**(4), 647–651 (2015). https://doi.org/10.5829/idosi.mejsr.2015.23.04.103. ISSN 1990-9233 © IDOSI Publications
16. Sebastian, L., Bose, J.S.: Efficient restoration of corrupted images and data hiding in encrypted images. J. Ind. Eng. Res. **1**(2), 38–44 (2015). ISSN 2077-4559
17. Bose, J.S.C.: Detection of ovarian cancer through protein analysis using ant colony and particle swarm optimization. Proc. Int. J. Multimed. Comput. Vis. Mach. Learn. **1**, 59–65 (2010)
18. Bose, J.S.C.: Detection of masses in digital mammograms. Int. J. Comput. Netw. Secur. **2**, 78–86 (2010)
19. Bose, J.S.C., Mehrez, M., Badawy, A.S., Ghribi, W., Bangali, H., Basha, A.: Development and designing of fire fighter robotics using cyber security. In: Proceedings of IEEE - 2nd International Conference on Anti-Cyber Crimes (ICACC 2017), 26–27 March 2017, pp. 118–122 (2017)
20. Lau, T.-K., et al.: Automated detection of breast tumors using the asymmetry approach. Comput. Biomed. Res. **24**, 273–295 (1991)
21. Nishikawa, R.M., et al.: Computer-aided detection of clustered microcalcifications on digital mammograms. Med. Biol. Eng. Comput. **33**(2), 174–178 (1995)
22. Sallam, M.Y., et al.: Registration and difference analysis of corresponding mammogram images. Med. Image Anal. **3**(2), 103–118 (1999)
23. Kim, J.K., et al.: Statistical textural features for detection of microcalcifications in digitized mammograms. IEEE Trans. Med. Image **18**(3), 231–238 (1999)
24. Cordella, L.P., et al.: Combing experts with different features for classifying clustered microcalcifications in mammograms. In: Proceedings of 15th International Conference on Patten Recognition, pp. 324–327 (2000)
25. Ferrari, R.J., et al.: Analysis of asymmetry in mammograms via directional filtering with Gabor wavelets. IEEE Trans. Med. Imaging **20**(9), 953–964 (2001)
26. Thangavel, K., et al.: Ant colony system for segmentation and classification of microcalcification in mammograms. Int. J. Artif. Intell. Mach. Learn. **5**(3), 29–40 (2005)
27. Karnan, M., et al.: Ant colony optimization for feature selection and classification of microcalcifications in digital mammograms. In: International Conference on Advanced Computing and Communications (2006)
28. Karnan, M., et al.: Hybrid Markov random field with parallel Ant Colony Optimization and fuzzy C means for MRI brain image segmentation. In: 2010 IEEE International Conference on Computational Intelligence and Computing Research (2010)

# Prediction of Fetal Distress Using Linear and Non-linear Features of CTG Signals

E. Ramanujam[1(✉)], T. Chandrakumar[2], K. Nandhana[1], and N. T. Laaxmi[1]

[1] Department of Information Technology, Thiagarajar College of Engineering, Madurai 625015, Tamil Nadu, India
erit@tce.edu, nandhanak3@gmail.com, laaxmi15@gmail.com

[2] Department of Computer Applications, Thiagarajar College of Engineering, Madurai, Tamil Nadu, India
tckcse@tce.edu

**Abstract.** Cardiotocography records the fetal heart rate and Uterine contractions which is used to monitor the fetal distress during delivery. This signal supports the physicians to assess the fetal and maternal risk. During the last decade, various technique have proposed computer-aided assessment of fetal distress. The drawback of these techniques are complex in extraction of features and costlier in classification algorithm utilized. This paper proposes a feature selection technique Multivariate Adaptive Regression Spline and Recursive Feature Elimination to evaluate 48 numbers of linear and non-linear features and to classify using Decision tree and *k*-Nearest Neighbor algorithms. Experimental results shows the performance of the proposed with state-of-the-art techniques.

**Keywords:** Cardiotocography · Feature extraction · Feature selection · Decision tree · Fetal heart rate · Uterine contractions

## 1 Introduction

Electronic Fetal Monitoring (EFM) from Cardiotocography (CTG) signals is a widely used technique to identify the fetal distress. CTG has two signals - Fetal Heart Rate (FHR) and Uterine Contraction (UC). Obstetricians use these recorded signals to assess the condition of the fetus during antepartum and intra-partum periods [1]. In general, lack of sufficient oxygen during the pregnancy results in various fetal distress conditions which can only be diagnosed through CTG signals. Manual Interpretation of CTG signals by a gynecologist for analyzing those abnormalities depend on the expertise level which shows high false positive rate [2]. In this case there, is a more chance of misclassifying the normal recordings as pathological. To overcome this issue, there requires a computer-aided diagnosis to provide better classification of CTG signals. In recent years, various techniques have proposed a computer-aided system which follows the process of data collection, pre-processing, feature extraction and classification. The data collections are often obtained from various benchmark data [3] available online, preprocessing is to clear the noise and signal attenuations. Feature extraction is the core for the classification process. The research work [1, 4–10] has utilized various numbers


S. Smys et al. (Eds.): ICCVBIC 2019, AISC 1108, pp. 40–47, 2020.
https://doi.org/10.1007/978-3-030-37218-7_5

of features to show up their performance in classification. The major challenge with existing techniques is the efficient selection of minimal features and their evaluation. This paper uses feature selection technique Multivariate Adaptive Regression Spline (MARS) and Recursive Feature Elimination (RFE) to select minimal number of linear and non-linear features collected from state-of-the-art techniques for better classification performance. The paper is structured as follows. Section 2 describes the existing feature extraction techniques with their corresponding linear and non-linear features, Sect. 3 discusses the dataset used, Sect. 4 deals with features of various state-of-the-art techniques, Sect. 5 proposes feature selection and Sect. 6 discusses the evaluation of features using various classification algorithms and Sect. 7 provides the conclusion.

## 2 Literature Review

In the fetal distress classification, researchers have categorized features as linear and non-linear features. Most of the researchers used the combination of both linear and non-linear features. Iraj et al. in [4] have used Multi-layer of sub ANFIS topology technique with 21 features of CTG signal. In addition, deep stacked sparse auto-encoders and deep ANFIS were utilized to achieve the accuracy of 99.503%. Ramla et al. in [5] have utilized 20 statistical features to predict the CTG signals using CART decision tree algorithm and achieved accuracy of 90.12% in 5-fold cross validation. Comert et al. in [6] have utilized 21 set of features from CTG signals. The proposed work achieved high classification performance through ensemble and neural network classifiers. Deb et al. in [7] has proposed a strong and robust mathematical analysis model by applying Hilbert Huang Transform (HHT) and extract 6 numbers of linear statistical features to classify CTG signals. The proposed achieved 98.5% accuracy for Multi-Layer Perceptron (MLP) and 98.2% accuracy for Support Vector Machine (SVM). Comert et al. in [1] have used various 6 linear features and 4 Non-linear features and achieved 92.40%, 82.39%, 79.22% accuracy through Artificial Neural Network (ANN) algorithm at three stages of analysis. Karabulut et al. in [8] provided a computer based classification approach for determining a fetus to be normal or pathological. 5 numbers of linear and non-linear features are extracted to evaluate with ensemble of decision trees using Adaptive Boosting, Bayesian network, and RBFN. The proposed achieved to the maximum of 92.6% accuracy for Bayesian Network. State-of-the-art techniques have utilized various Linear and non-linear features to show up their performance in classification. However, the major part of proposed performance depends on costly classifiers as discussed above. Most of classifiers are ensemble based and neural network based techniques. Naturally, the ensemble and neural network based algorithms provides better performance while training more number of time than simpler classification algorithms. To avoid costliness and to provide better accuracy, this paper proposes a simple feature selection algorithm MARS [9] and RFE [10] to provide better performance in classification through *k*-NN and Decision tree (Information Gain). Overall architecture of the proposed paper is shown in Fig. 1.

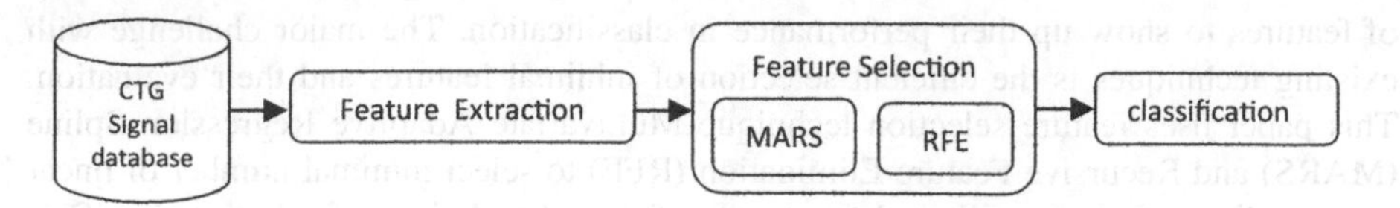


**Fig. 1.** Overall architecture of the proposed work.

# 3 Data Set Collection

The fetal CTG data set used in the proposed evaluation technique is available as a public source downloadable from Phsyiobank [3]. The database contains 552 CTG recordings which have both normal and pathological signals recorded during the period of 2010 and 2012. The signals are of duration 90 min prior to delivery and some has lesser duration due to immediate delivery and are sampled at 4 Hz digitized form.

# 4 Feature Extraction

The primary concern of the proposed technique is to select/evaluate the linear and non-linear features to achieve higher accuracy using simple classification algorithms. These features are collected through a complete research on various papers [1, 4–10] that deals with ECG, EEG and fetal distress classification. The Linear features of FHR are listed in Table 1 and non-linear features are listed in Table 2 with their decriptions.

**Table 1.** Linear features

| S. No. | Feature | Description |
|---|---|---|
| 1. | Mean | Arithmetic average of range of values or quantities |
| 2. | Standard deviation | Measure to quantify the amount of variation of set of values |
| 3. | Variance | Measurement of spread between numbers in a dataset |
| 4. | Long term irregularity | Describes the irregularity conditions over a particular period |
| 5. | Range | Difference between maximum and minimum value |
| 6. | Interval index | Defines the gross change |
| 7. | Short term variability | Variability condition over a short period |
| 8. | Median | Midpoint of the range of values |
| 9. | Skewness | Extent to which a distribution differs from normal distribution |
| 10. | Kurtosis | Sharpness of the peak of frequency distribution curve |
| 11. | Mean absolute deviation | Average of the absolute deviation from the central point |
| 12. | Bandwidth | The range of frequencies within a given band |
| 13. | 1st quartile | Middle number between the smallest number and the median of the dataset |

(*continued*)

**Table 1.** (*continued*)

| S. No. | Feature | Description |
|---|---|---|
| 14. | $3^{rd}$ quartile | Middle value between the median and the highest value of the data set |
| 15. | Frequency | Number of occurrences in a dataset |
| 16. | bw.nrd0 | Rule-of-thumb for choosing the bandwidth of a Gaussian kernel density estimator |
| 17. | bw.nrd | More common variation given by Scott by a factor of 1.06 |
| 18. | Root mean square | Square root of the mean of the squares of values |
| 19. | Harmonic mean | Reciprocal of the mean of the reciprocals |
| 20. | Geometric mean | The central number in a geometric progression |
| 21. | Sterling | Computes the Sterling ratio of the univariate time series |
| 22. | Sharpness | Return per unit risk |
| 23. | Roughness | Total curvature of a curve or a time wave of spectrum |
| 24. | Rugosity | Small scale variations in the height of the surface |

**Table 2.** Non-linear features

| S. No. | Feature | Description |
|---|---|---|
| 1. | Approximate entropy | Quantifies the amount of regularity |
| 2. | Sample entropy | Measures the complexity of time series |
| 3. | Fast entropy | Calculates ApEn in an efficient way |
| 4. | Fast sample entropy | Calculates SampEn in an efficient way |
| 5. | 1-norm | Sum of absolute value of the columns |
| 6. | Spectral norm | Maximum singular value of a matrix |
| 7. | Amplitude | Angular peak amplitude of the flapping motion |
| 8. | Shannon entropy | Information-theoretic formulation of entropy |
| 9 | Temporal entropy | Entropy of a temporal envelope |
| 10 | Hs | Simple R/S estimation |
| 11 | Hrs | Corrected R over S exponent |
| 12 | He | Empirical hurst exponent |
| 13 | Hal | Corrected empirical hurst exponent |
| 14 | Ht | Theoretical hurst exponent |
| 15 | Fractal dimension | A dim-dimensional array of fractal dimensions |
| 16. | Fractal dimension | A dim-dimensional array of scales |
| 17. | Burg$aicc | Autocorrelation or co-variance |
| 18. | Fractal dimension | Size of the actual sliding window |
| 19. | Burg$phi | Vector of AR coefficients |
| 20. | Burg$se.phi | Standard error of AR coefficients |
| 21. | Burg$se.theta | Defaults to 0 |
| 22. | Burg$theta | Defaults to 0 |
| 23. | Burg$sigma2 | White noise variance |
| 24. | SFM | Ratio between the geometric and the arithmetic mean |

# 5 Feature Selection/Evaluation

To evaluate the linear and non-linear features collected from various research works, the proposed uses MARS and RFE to select the best features. Feature selection is one of the core concepts in machine learning that have huge impacts in the performance of the model. The irrelevant features may degrade the performance of the algorithm also it may increase the computation time of the classification algorithm. Selection of relevant features has following advantages such as reduces overfitting, improves accuracy and reduced training time.

## 5.1 Multivariate Adaptive Regression Spline (MARS)

MARS [11] is a variety and extension of linear regression analysis also termed as non-parametric regression technique. The model automatically categorizes the linearity and interaction between the features using recursive partitioning method as like regression trees and decision tree. MARS is more flexible and scalable than linear regression models, easy to understand and interpret, can handle both categorical and continuous data.

$$\hat{f}(x) = \sum\nolimits_{i=1}^{k} c_i B_i(x) \tag{1}$$

The model is a weighted sum of basis function $B_i(x)$ where each $c_i$ is a constant coefficient.

## 5.2 Recursive Feature Elimination (RFE)

RFE [12] is a feature selection method that fits a model using ranking of feature coefficient and feature importance to remove the weakest feature until the number of features are reached. RFE uses the model accuracy to determine the most important feature as well as to identify the weak features. As a result, least important features are eliminated from the dataset. RFE procedure is recursively iterated on the pruned dataset until the desired result is obtained. To identify the optimal solution of features cross-validation is used with RFE to produce best features.

**Table 3.** Classification performance (accuracy) of extracted 48 set of features

| Methods | Decision tree | *k*-NN |
|---|---|---|
| Training-testing (60–40) | 84.14 | 85.41 |
| Training-testing (80–20) | 87.67 | 90.57 |
| 10-fold cross validation | 89.03 | 93.46 |
| 5-fold cross validation | 90.16 | 94.15 |

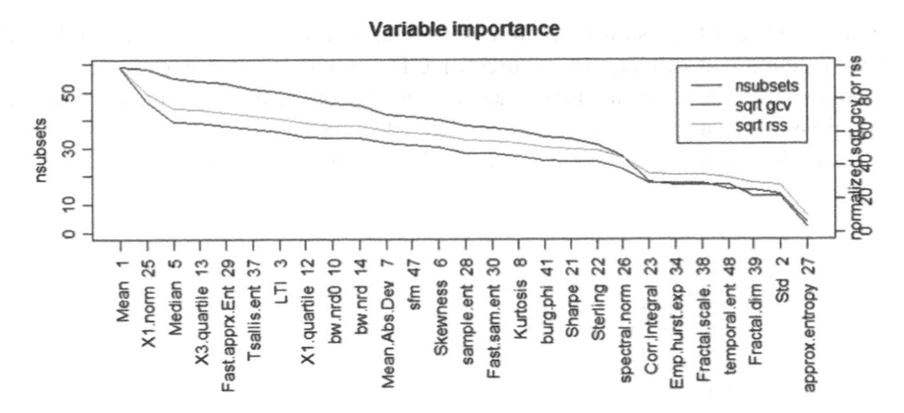


**Fig. 2.** Plotting of 20 best features by MARS feature selection

**Table 4.** Classification performance of best 16 (8-linear and 8 non-linear) features selected through MARS and RFE.

| Methods | MARS | | RFE | |
|---|---|---|---|---|
| | Decision tree | *k*-NN | Decision tree | *k*-NN |
| Training-testing (60–40) | 89.03 | 90.19 | 88.24 | 90.16 |
| Training-testing (80–20) | 93.91 | 93.67 | 93.35 | 93.67 |
| 10-fold cross validation | 96.89 | 95.82 | 97.41 | 94.48 |
| 5-fold cross validation | 94.5 | 95.12 | 91.8 | 90.76 |

## 6 Results and Discussions

The 48 number of linear and non-linear features (24 each) of CTG signals are collected from fetal distress dataset of Physionet. These features are evaluated and selected using the proposed MARS and RFE to produce efficient accuracy through classification algorithms such as Decision tree using Information Gain and *k*-NN. The features are evaluated using both training-testing and cross-validated performance of classification algorithms. Initially the extracted original 48 features are tested for accuracies using classification algorithms and the accuracies are reported in Table 3. The performance of extracted 48 features shows higher accuracy of 93.46% and 94.15% for 10-fold and 5-fold cross validation of *k*-NN algorithm. To provide best accuracy, the 48 features are further reduced by applying MARS and RFE feature selection techniques. The best 20 features based on the importance measure of MARS and RFE selection techniques are shown in Figs. 2 and 3. In Fig. 2, X-axis shows the best features and Y-axis the repetitive subset of iterations performed to extract the features. In Fig. 3, X-axis shows the importance measure and Y-axis shows the list of best features.

For best accuracy, 16 best features of top 8 linear and 8 non-linear features from both MARS and RFE are evaluated individually for their performance using the classification algorithms on various training-testing (80–20, 60–40) and cross-validations of 5 and 10 as shown in Table 4. On comparing the performance of selected 16 best features and original 48 features of CTG signals from Tables 3 and 4, the selected features shows better performance than the original 48 features. From the observations of Table 4, the RFE and MARS selection technique with decision tree achieves higher accuracy of 96.89% and 97.41% for 10-fold cross validation.

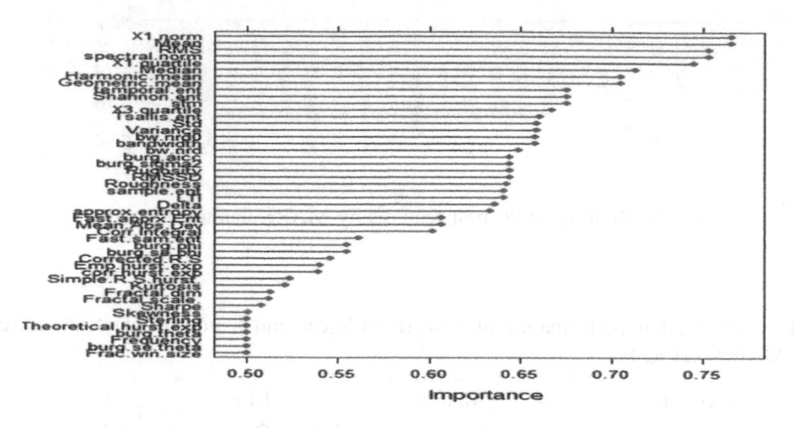


**Fig. 3.** Plotting of 20 best features by RFE feature selection

**Table 5.** Performace comparison of proposed with various state-of-the-art techniques for CTG signal classification.

| Reference | Method | # of features | Accuracy (%) |
|---|---|---|---|
| [4] | ANFIS | 21 | 99.503 |
| [5] | CART | 20 | 90.12 |
| [6] | ANN | 21 | 99.73 (Sensitivity) |
| [7] | HHT + MLP | 6 | 98.5 |
| [1] | ANN | 10 | 92.40 |
| [8] | Adaptive Boosting + Bayesian | 5 | 92.6 |
| Proposed | MARS + Decision Tree | 16 | 96.89 |
| | RFE + Decision Tree | 16 | 97.81 |

To justify the performance of the proposed feature selection with the state-of-the-art techniques, the comparison has been made in Table 5 with the features and accuracies obtained with various research techniques. On comparing the performance of the proposed with the state-of-the-art techniques, the proposed ranked 3 where research work [4] ranked 1 and research work [7] ranked 2. However, these techniques used Neural network based classifier and dimensionality reduction techniques for their better performance. This makes the algorithm costlier in terms of execution than the proposed technique. The performance of the proposed certainly outperforms the state-of-the-art technique using simple feature selection technique with less complex classification algorithms.

## 7 Conclusion

In this proposed work, the linear and non-linear features of various state-of-the-art techniques are efficiently selected through feature selection techniques such as MARS and RFE to produce better classification performance with minimal features. Out of 48 extracted features of CTG signals 16 best features are selected through proposed and evaluated using classifiers such as *k*-Nearest Neighbor and Decision tree. The proposed achieved 96.89% and 97.81% of accuracy for the combination of MARS and RFE with Decision Tree algorithm to the CTG signals. In future, the proposed can be implemented in realtime CTG signal machine for classifications.

**Compliance with Ethical Standards.** All author states that there is no conflict of interest. We used our own data. No animals/human are not involved in this work.

## References


1. Cömert, Z., Fatih, A.: Evaluation of fetal distress diagnosis during delivery stages based on linear and nonlinear features of fetal heart rate for neural network community. Int. J. Comput. Appl. **156**(4), 26–31 (2016)
2. Sameni, R.: A review of fetal ECG signal processing issues and promising directions. Open Pacing Electrophysiol. Ther. J. **3**, 4–20 (2010)
3. Goldberger, A.L., Amaral, L.A.N., Glass, L., et al.: PhysioBank, PhysioToolkit, and PhysioNet. Circulation **101**(23), E215–E220 (2000)
4. Iraji, M.S.: Prediction of fetal state from the cardiotocogram recordings using neural network models. Artif. Intell. Med. **96**, 33–44 (2019)
5. Ramla, M., Sangeetha, S., Nickolas, S.: Fetal health state monitoring using decision tree classifier from cardiotocography measurements. In: 2018 Second International Conference on Intelligent Computing and Control Systems (ICICCS), pp. 1799–1803. IEEE (2018)
6. Cömert, Z., Kocamaz, A.F.: Comparison of machine learning techniques for fetal heart rate classification. Acta Phys. Pol. A **132**(3), 451–454 (2017)
7. Deb, S., Islam, S.M.R., Johura, F.T., Huang, X.: Extraction of linear and non-linear features of electrocardiogram signal and classification. In: 2017 2nd International Conference on Electrical and Electronic Engineering (ICEEE), pp. 1–4. IEEE (2017)
8. Karabulut, E.M., Ibrikci, T.: Analysis of cardiotocogram data for fetal distress determination by decision tree based adaptive boosting approach. J. Comput. Commun. **2**(09), 32 (2014)
9. Padmavathi, S., Ramanujam, E.: Naïve Bayes classifier for ECG abnormalities using multivariate maximal time series Motif. Procedia Comput. Sci. **47**, 222–228 (2015)
10. Balayla, J., Shrem, G.: Use of artificial intelligence (AI) in the interpretation of intrapartum fetal heart rate (FHR) tracings: a systematic review and meta-analysis. Arch. Gynecol. Obstet. **300**, 1–8 (2019)
11. Friedman, J.H.: Multivariate adaptive regression splines. Ann. Stat. **19**(1), 1–67 (1991)
12. Guyon, I., Weston, J., Barnhill, S., Vapnik, V.: Gene selection for cancer classification using support vector machines. Mach. Learn. **46**(1–3), 389–422 (2002)

# Optimizing Street Mobility Through a NetLogo Simulation Environment

Jesús Silva[1(✉)], Noel Varela[2], and Omar Bonerge Pineda Lezama[3]

[1] Universidad Peruana de Ciencias Aplicadas, Lima, Peru
jesussilvaUPC@gmail.com

[2] Universidad de la Costa (CUC), Calle 58 # 55-66, Baranquilla, Atlantico, Colombia
nvarela1@cuc.edu.co

[3] Universidad Tecnológica Centroamericana (UNITEC), San Pedro Sula, Honduras
omarpineda@unitec.edu

**Abstract.** The routes and streets make it possible to drive and travel through the cities, but unfortunately traffic and particularly congestion leads to drivers losing time while traveling from one place to another, because of the time it takes to transit on the roads, in addition to waiting times by traffic lights. This research introduces the extension of an agent-oriented system aimed at reducing driver waiting times at a street intersection. The simulation environment was implemented in NetLogo, which allowed comparison of the impact of smart traffic light use versus a fixed-time traffic light.

**Keywords:** Multi-agent systems · Agent-oriented programming · Traffic · NetLogo

## 1 Introduction

Various disciplines have studied the problem of mobility in cities in order to understand the phenomenon [1, 2, 3], and achieve a viable solution [4, 5, 6]. Some authors have found that drivers stuck in high congestion conditions have high levels of stress [7], which can lead them to modify their behavior while driving their vehicle. As a result, it's important to change the way people drive, and find better solutions to deal with traffic, such as smart traffic lights or even cars communicating with each other. The rapid growth of cities, the location of educational institutions and the diversity of jobs lead to an increase in the daily flow of traffic, leading to an increase in the number of vehicles and means of transport on the streets in and in cities in general.

While several domains have addressed the problem of traffic congestion [8–12] we consider that the implementation of a simple approach, implementing agent-based modeling and simulation techniques, with a low-medium difficulty level of implementation is able to be effective and viable for its application in real life. There are a large number of tools and platforms such as NetLogo [13] and Repast, and a large number of applications, where the most common uses take place in social simulations


S. Smys et al. (Eds.): ICCVBIC 2019, AISC 1108, pp. 48–55, 2020.
https://doi.org/10.1007/978-3-030-37218-7_6

and optimization problems such as human behavior, urban simulation, traffic management, among others [14].

In [15], the authors present a traffic simulation in NetLogo, at an intersection for optimization through an agent-based approach, which decreases the wait time of vehicles, preventing them from waiting for an indeterminate time or excessively long; with the ultimate goal of reducing traffic congestion at a two-lane intersection - double-hand and double-lane, where vehicles circulate. In this context, the objective pursued by this article is to introduce an extension of [15], adding to the simulation, left turn traffic lights, taxis and pedestrians; and adapting the smart traffic light to the new scenario.

## 2 Description of Modeled Agents

The agents implemented are those seen in Tables 1, 2 and 3, which describe the agents by their attributes, by the information of the environment and their behavior.

**Table 1.** Description of the intersection agent.

| Description | Intersection agent |
|---|---|
| Attributes | Traffic light: current state of traffic lights,<br>Intelligence (Yes or No): Defines the algorithm to use for handling traffic lights, Minimum green time,<br>Time since the last light change,<br>Direction, It's turn,<br>It is car: If you apply cars or pedestrians |
| Information censored of the environment | Number of cars behind the limits of the intersection in all directions, Car wait times |
| Behavior | Change traffic light:<br>Without intelligence: it is done for defined times, the traffic light is green for a certain time and then changes to red,<br>With intelligence: it does so from the censed data of the environment. From an algorithm that counts how many cars are waiting on the other side and how long they've been waiting decides whether to change or continue in the current state |

### 2.1 Smart Traffic Light Algorithm

The intelligent traffic light algorithm developed coordinates two traffic lights (X and Y) one for each direction of the intersection of the streets, with their respective left turn traffic lights and two pedestrians. In addition, consider 4 pedestrian traffic lights, which at the same time are all green or red, so the algorithm considers it as a unit. 5 different states are distinguished, described in Table 4.

The algorithm will cause the heavier traffic light to turn green, unless the minimum green time set through the user controls has not yet been completed. It is worth noting

**Table 2.** Description of agent auto

| Description | Auto agent |
|---|---|
| Attributes | Speed: current speed of the car, Patience: the remaining tolerance level that the driver has to wait for, Direction: where the car is going, Current Lane, Current Coordinates, Turn right (Yes or no): Indicates whether the car will rotate at the intersection, Turn left (Yes or no): Indicates whether the car will rotate in the intersection, Standby time at traffic light, Acceleration, Deceleration, Taxi (yes or no): Indicates if it is a taxi, It is free (Applies only for taxis): Indicates if you have passengers, Shocked |
| Information censored of the environment | Nearby cars: using a 15° viewing angle and a radius that varies depending on the speed of the car, Maximum street speed, Light color of traffic light, Cars in adjacent lane, Position within the environment, Pedestrians waiting for taxis |
| Behavior | Accelerate: It does so if you don't have any cars in your view spectrum going at a speed lower than yours, if you don't exceed the speed, and if you're not at the pre-traffic light limit, while it's red, Brake: It does so if you have a car in your vision spectrum going at a speed lower than yours and if you are at the pre-traffic light limit, while it is in red, Turn right: it does if you are in the right lane when you reach the critical area of the intersection with a probability defined by a variable and the traffic light is green, Turn left: It does if it is in the left lane when you reach the critical area of the intersection with a probability defined by a variable and the turn light is green, Change lanes: it does if the driver's patience reaches 0 (it is decreasing as the waiting time increases) and if there is no car in its possible future position, Lift a passenger: It does if it is a taxi and is free and passes by a pedestrian who is waiting for a taxi, Crash: Overlapping with another vehicle is considered shocked and stays still in place until it is removed after a time limit |

that the green light corresponds to the direction with the highest weight and the weights i, are calculated according to (1):

$$Pi = Ai + T_{sri} * Fa \tag{1}$$

Where:

- i, are the weight for the x direction, Y, Left turn axis X, Left turn axis Y and Pedestrians
- Ai, agents (cars or pedestrians) arriving at the intersection at the traffic light (braking and heading to)
- $T_{sri}$, red light time
- F, it's the adjustment factor

**Table 3.** Description of the pedestrian agent

| Description | Agent pedestrian |
|---|---|
| Attributes | Speed: pedestrian's current speed, Direction: where the pedestrian goes, Current coordinates, Turn right (Yes or no), Turn left (Yes or not), Rotate Before: Indicates whether to rotate before crossing, rotate after: Indicates whether to rotate after crossing, Waiting time at the traffic light |
| Information censored of the environment | Pedestrians nearby, Traffic light color, Position within the environment, Taxi waiting for the pedestrian to come up |
| Behavior | Accelerate: It does so if you don't have any pedestrians exactly in front of you, Brake: It does so if you have a pedestrian in front of you going at a speed lower than yours and if you are at the pre-traffic light limit, while it is in red, Turn right: it does if you are in a corner with the probability defined to bend, Wait for a taxi: It does so if it meets the defined probability to wait for the taxi and is in an area where one can be stopped, Get in a taxi: It does if you are waiting for a taxi and one stops to get it up |

## 3 Results

In order to explicitly determine the feasibility of the developed algorithm and demonstrate emerging behavior, the following six (6) scenarios have been raised, in which the frequency of agents and the adjustment factor for the smart semaphore will be varied. An adjustment factor 0.001 and 0.01 were selected. Increasing the factor value for turn traffic lights we manage to increase the weight for these traffic lights, modifying their priority, since the selection of the next traffic light to change to green, it will be the traffic light with the greatest weight. In addition, each scenario will be evaluated with a turn frequency of 20 and 80.

- Scenario 1: Same frequency of agents in all directions, same smart traffic light factor (0.001) at all traffic lights.

**Table 4.** States of traffic lights

| Traffic lights | E1 | E2 | E3 | E4 | E5 |
|---|---|---|---|---|---|
| X-axis | V | r | r | r | r |
| X turn traffic light | r | V | r | r | r |
| Y-axis | r | r | V | r | r |
| Turn traffic light Y | r | r | r | V | r |
| Pedestrians | r | r | r | r | V |

- Scenario 2: Same frequency of agents in all directions. Smart traffic light factor 0.001 at X axis traffic lights, Y axis, and pedestrians. Smart traffic light factor 0.01 at turn traffic lights.
- Scenario 3: Agent generation frequency of 10 on the Y axis, and agent generation frequency of 3 on the X axis and pedestrians. Same smart traffic light factor (0.001) at all traffic lights.
- Scenario 4: Agent generation frequency of 10 on the Y axis, and agent generation frequency of 3 on the X axis and pedestrians. Smart traffic light factor 0.001 at X axis traffic lights, Y axis, and pedestrians. Smart traffic light factor 0.01 at turn traffic lights.
- Scenario 5: Agent generation frequency of 15 on the Y axis, and agent generation frequency of 1 on the X axis and pedestrians. Same smart traffic light factor (0.001) at all traffic lights.
- Scenario 6: Agent generation frequency of 15 on the Y axis, and agent generation frequency of 1 on the X axis and pedestrians. Smart traffic light factor 0.001 at X axis traffic lights, Y axis, and pedestrians. Smart traffic light factor 0.01 at turn traffic lights.

In the case of scenario 1, the average wait time is greatly reduced with the smart traffic light. The problem is that it increases the maximum wait value. In the case of the frequently 22 spin, the maximum timeout was not found in the 10010 ticks. This is because having such a low turn factor and so few cars can bend, the weight of that traffic light never gets higher than others. With the frequency of 82 turn, cars double faster, but wait longer than cars controlled by a fixed-time traffic light.

In scenario 2, with another semaphore factor, the values change because, although the timeout value is not reduced as much as with the previous scenario, the maximum wait value takes a dimensioned value and less than the value in scenario 1.

For scenario 3, the average wait value is significantly reduced with the intelligent system, by more than 69% compared to the smart semaphore. The maximum waiting time increases, because there are few cars that rotate, and considering that the turn factor is equal to the rest this takes a long time to occur; 10010 ticks never happened. A higher turning frequency allows for better turning flow, and although the average wait is not reduced as much as with 22, the maximum wait value is similar to that with non-smart traffic light.

However, in scenario 4, smart traffic lights also improve the average wait relative to the smart traffic light for both turning frequencies. The lowest average wait time of all simulations is obtained for a frequency of 82 rotations; it is reduced by approximately 59% while also reducing the maximum waiting time by 33%, compared to the fixed traffic light. With the 21 turn the average wait time is reduced less and the maximum wait time increases too much.

Then, in Scenario 5, something similar to Scenario 3 happens, the average wait times show an even greater improvement. At the same time, you can see that in situations with very different flows the factor 0.0015 can leave a car waiting a long time. Finally, in scenario 6, for a larger car flow gap (1–15), factor 0.015 also fails to make much difference with the other factor 0.001, but manages to reduce the maximum wait value when the turn flow is low. However, the average wait improvement obtained between the smart semaphore and the fixed time in all scenarios is remarkable (average 43%, varying [14%–69%]).

## 4 Analysis of Scenarios

Smart traffic lights in all scenarios improve the average waiting of vehicles (cars and taxis) at the traffic light. A better result was obtained with a higher turning frequency (80 compared to 20), preventing vehicles from waiting for excessive time. The 0.0015 factor introduced in the authors' previous article [15] is not recommended for this environment as if there are not many cars that bend, these cars will have a long wait. Decreased the factor to 0.01 at turn traffic lights and keeping 0.0015 in the rest, improves the result. The system is optimized when more cars want to turn left and exert more weight to do so. When defining what kind of vehicle flows works best, you should analyze how the average wait time influences.

In the case of a similar flow in all directions, the average wait time is slightly lower but the cost of the extra waiting time is minimal so, at an intersection with this type of flow it would be convenient if there are many cars spinning. In the event that the flow is higher at an intersection, but not excessively higher, it is convenient no matter how many cars turn left because the average improvement is very high (for a turn of 20 low from 1258 to 686 and for a turn of 82 low from 914 to 388). In the scenario where the flows are very disparate is not convenient since, although the average time drops almost three times in both cases, the maximum wait time goes up too high (for a turn of 20 from 4978 to 8281 and for a turn of 80 from 3312 to 7478), so it's not convenient.

We conclude that the intelligent algorithm manages to reduce the average and maximum waiting time for a medium frequency difference between the two directions and a 0.015 factor. If the frequency of agents in the two directions is similar or greater, the average waiting time with respect to the fixed time traffic light is reduced, but the maximum waiting time will be a little longer for similar agent frequencies and a little more double for very different frequencies between the X and Y axes.

## 5 Conclusions

Traffic is a reality that affects many people, and there is evidence of decreased driver wait times if dynamic and adaptive traffic control systems are implemented. The agent-oriented approach has essential features that make it easy to model and simulate. There are various applications, systems and simulations that allow the area to remain active, because the increase in society's demand for improvement is not only in infrastructure but also in management systems. The traffic light control system at a complex intersection described, manages to optimize the average wait time of vehicles at a intersection of streets with double lanes applying an algorithm that evaluates weights for traffic directions (X and Y), turns (right and left), and pedestrians.

Reducing the waiting time at the traffic light is achieved with an adjustment factor of 0.015 for right and left turns and an adjustment factor of 0.0015 for all other directions and a turning frequency of 82. The best-optimized scenario is achieved with vehicle flows in the X and Y direction, which is moderately different. Analysis of the different scenarios makes it possible to understand how vehicle flow variation (cars and taxis) affects pedestrian traffic, and rotation frequency variation. The agent-oriented environment provides an appropriate context for performing this analysis. It allows the extension of agents such as bicycles, public transport, ambulances, motorbikes, among others. We emphasize that the simulation carried out shows that it is feasible to make a sensitive reduction in the wait times of a traffic light, at an intersection of complex streets, applying a simple algorithm, aspect that has a positive impact on the life of a City.

**Compliance with Ethical Standards**

✓ All authors declare that there is no conflict of interest

✓ No humans/animals involved in this research work.

✓ We have used our own data.

## References

1. Banos, A., Lang, C., Marilleau, N.: Agent-Based Spatial Simulation with NetLogo, vol. 1. Elsevier, Amsterdam (2015)
2. Amelec, V.: Validation process container for distribution of articles for home. Adv. Sci. Lett. **21**(5), 1413–1415 (2015)
3. Nagatani, T.: Vehicular traffic through a sequence of green-wave lights. Physica A Stat. Mech. Appl. **380**, 503–511 (2007)
4. Bui, K.H.N., Jung, J.E., Camacho, D.: Game theoretic approach on real-time decision making for IoT-based traffic light control. Concurr. Comput. Pract. Exp. **29**(11), e4077 (2017)
5. Bui, K.H.N., Camacho, D., Jung, J.E.: Real-time traffic flow management based on inter-object communication: a case study at intersection. Mob. Netw. Appl. **22**(4), 613–624 (2017)
6. Rao, A.M., Rao, K.R.: Measuring urban traffic congestion-a review. Int. J. Traffic Transp. Eng. **2**(4), 286–305 (2012)

7. Hennessy, D.A., Wiesenthal, D.L.: Traffic congestion, driver stress, and driver aggression. Aggress. Behav. **25**(6), 409–423 (1999)
8. Putha, R., Quadrifoglio, L., Zechman, E.: Comparing ant colony optimization and genetic algorithm approaches for solving traffic signal coordination under oversaturation conditions. Comput. Aided Civ. Infrastruct. Eng. **27**(1), 14–28 (2012)
9. Teodorović, D., Dell'Orco, M.: Mitigating traffic congestion: solving the ride-matching problem by bee colony optimization. Transp. Plan. Technol. **31**(2), 135–152 (2008)
10. Bazzan, A.L., Klügl, F.: A review on agent-based technology for traffic and transportation. Knowl. Eng. Rev. **29**(3), 375–403 (2014)
11. Oviedo, C., Campo, A.: Aproximación al uso del coeficiente alfa de Cronbach. Revista Colombianan de Psiquiatría **34**(4), 572–580 (2005)
12. Demsar, J., Curk, T., Erjavec, A., Gorup, C., Hocevar, T., Milutinovic, M., Mozina, M., Polajnar, M., Toplak, M., Staric, A., Stajdohar, M., Umek, L., Zagar, L., Zbontar, J., Zitnik, M., Zupan, B.: Orange: data mining toolbox in Python. J. Mach. Learn. Res. **14**(1), 2349–2353 (2013)
13. Amelec, V., Alexander, P.: Improvements in the automatic distribution process of finished product for pet food category in multinational company. Adv. Sci. Lett. **21**(5), 1419–1421 (2015)
14. Tan, F., Wu, J., Xia, Y., Chi, K.T.: Traffic congestion in interconnected complex networks. Phys. Rev. E **89**(6), 062813 (2014)
15. Battolla, T.F., Fuentes, S., Illi, J.I., Nacht, J., Falco, M., Pezzuchi, G., Robiolo, G.: Sistema dinámico y adaptativo para el control del tráfico de una intersección de calles: modelación y simulación de un sistema multi-agente. En: Simposio Argentino de Inteligencia Artificial (ASAI) – Jornadas Argentinas de Informática, Universidad de Palermo, Septiembre de 2018

# Image Processing Based Lane Crossing Detection Alert System to Avoid Vehicle Collision

B. S. Priyangha[1(✉)], B. Thiyaneswaran[2], and N. S. Yoganathan[2]

[1] Communication Systems, Sona College of Technology, Salem, India
bspriyangha@gmail.com

[2] Department of Electronics and Communication, Sona College of Technology, Salem, India

**Abstract.** In recent years, the lackadaisical lane crossing of vehicles causes major accidents on the highway roads. Particularly, in India different lane systems are available, they are 4-way, 6 way, and 8-way lane systems. The proposed work helps to identify the unwanted lane crossing activities and send an alert to the vehicle driver. A smart mobile is fixed in the dashpot and captures the lane video. The video is then converted into frames and the frame is detected using median filter, thresholding, and Hough transforms. In the lane, track marking is detected using pattern match and Kalman filter. The detected lane track is marked in different coloring and shown in the smart phone display. The proposed work is tested and compared with the existing lane crossing algorithms.

**Keywords:** Lane · Hough transform · Median filter · Android

## 1 Introduction

In India, careless and rash driving can cause other drivers and passengers in risky situations. On highways one of the most reasons for the accident is careless lane crossing. A detailed research work is going to improve safety in the road transport system. Lane with separations regulates the traffic from preventing accidents. The lane crossing system helps the driver while crossing the lane. The driver may change the lane by intention or due to their careless attitude. These systems are designed to minimize the accidents caused by disturbances, collision and drowsiness. Lane warning/keeping systems are based on video sensors, laser sensors and infrared sensors. The lane departure warning system detects lane lines from real-time camera images fed from the front-end camera of the automobile by employing the principle of Hough transform. Advanced driver-assistance system helps driver in the driving process with a safe human-machine interface. ADAS is dependent on inputs from multiple data sources that includes automotive imaging, image processing, LiDAR, computer vision, radar. Additional inputs are from the vehicular ad-hoc network.

## 2 Related Work

The lane detection is done by Edge Drawing (ED) algorithm and works faster with a particular Region of Interest. The algorithm was deployed in intel 3.3 GHz processor.


S. Smys et al. (Eds.): ICCVBIC 2019, AISC 1108, pp. 56–61, 2020.
https://doi.org/10.1007/978-3-030-37218-7_7

The final result analysis shows that for processing of each image takes up a time of 13 ms [1].

A robust wiped out point detection in the lanes was proposed by Youjin et al. [2]. Initially the vertical lines are detected in the frames. The priority-based directional approach is used to the remove noises present in the frame.

A computable vision-built lane finding approach is proposed by Yurtsever et al. [3]. In this approach, closely related 2 frames are selected. The ROI of corresponding binary are extracted. A AND operation was imposed on binary images. A bird's eye view was extracted from the resultant image.

A nature and causes of human distractions which leads accident are discussed in this paper. The author proposed statute based approach to determine the causes [4]. A method to forecast a vehicle's path and to sense the lane changes of adjacent vehicles. This technique is adopted as two parts such as intentional driving approximation and vehicle path forecast [5].

The use of smartphones to sense the driving condition of vehicles on highways using lane-level localization and implementation of phone hardware to capture the change of lane behaviors. The built-in sensors are available in the mobile phone provides the required data [6].

The author used adaptive traffic management system, Hough transforms, angles based on maximum likely hood, dynamic based pole identification, and region of interest. Further they are given suggestions to improve the system using Geo-based information and google maps [7]. A histogram-oriented gradient, region of interest, and Haar features based target detection system was proposed for two-way lane system. The classification of system is performed using support vector machine [8].

An accurate lane detection method was proposed using constrained disappearing points. The algorithm was suitable for various road conditions such as misty conditions, constrained lane detection method which performs accurately in a variety of road conditions based on stereo vision, echo images [9]. An effective method for sensing, tracking the lanes in spatiotemporal frames, which are constructed based on scanlines and the consecutive scan lines. It provides the consistent parallel vertical lines [10].

A constantly varying functionality and parting of vehicle based on the distance was proposed to detect the lanes. A warning signal was issued when the vehicle exceeds threshold values. The distance measure is based on the Euclidean distance transform [11].

The top view of the forward lane frame was computed using arbitrary consent approach. Further lanes are detected using region of interest using curvature lane model [12].

A monocular camera is used for detecting lanes and ROI is performed using computer vision technologies and the adjacent lanes are also estimated in this method [13]. Fourier based line detector method is used for detecting the lanes on the roads. It uses the Hough transform to detect lanes and finds the orientation of the potential lines [14].

The author used region of interest approach for detecting bending lanes, and edges are identified using a canny edge detector and Hough transform. A RANSAC and Kalman approach if further applied to detect the lanes effectively [15]. A Snake part-B type was used in the lane identification and tracking. The author used CHEVP algorithm to find the left and right end of the road [16].

# 3 Proposed Work

Our proposed system is based on detecting the lane when the vehicle approach departure unintentionally. The proposed system captures the video using smart phone. The proposed implementation efficiently detects the lane crossing and also provides a voice-based announcement.

## 3.1 Block Diagram

In this section we split the road and lane detection task into different blocks, and enumerate the possible approaches for the implementation of each block. The block diagram of the model consists of input video, lane finding model, lane tracing, parting warning and output. The input to the lane detection is live streaming video which is fragmented into frames and processed to remove noise in the video.

The lane identification is required for lane crossing announcement system. The lane identification system detects lane points in a complicated atmosphere. The lane detection model consists of median filter and Auto-thresholding blocks to detect the lane boundaries in the current video frame. The lane points form the straight lines.

Hough transform with which the lane lines are detected based on the parameterization i.e., finding a parametric equation for the curve are lined. Hough transform uses 2D-array called accumulator array to detect the existence of line by the equation $\rho = x\cos\theta + y\sin\theta$. For each pixel and its neighborhood pixel $(\rho, \theta)$ value is determined and stored in bin. Compares the $(\rho, \theta)$ value with the pixel and determines the straight line. This system uses the limited highest block to find the Polar coordinate position of the lane points (Fig. 1).

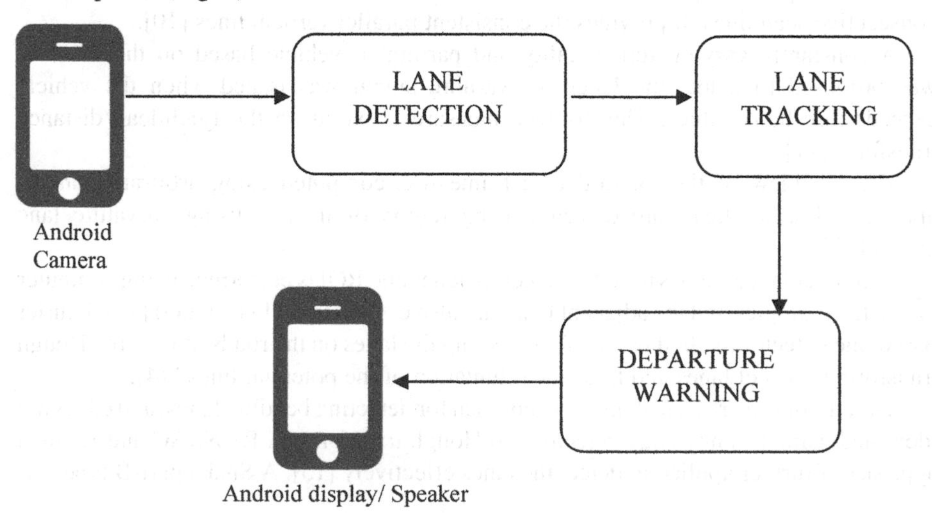


**Fig. 1.** Block of proposed system

Lane tracking model consist of the repository in which the identified lanes are stored based on number of epochs each lane is detected. Then matching the current lane with the data stored in the repository. If a current lane matches with the data base the lane is identified, or otherwise it updates new data in the database. Furthermore, Kalman filter is used to track the lane accurately.

The departure notice system uses the Hough positions which converts the Polar quantity to Cartesian. Departure warning system uses coordinate Cartesian model to compute the distance from the markings of the lane to the camera centric. If the system boundary is smaller than the particular edge value, a cautionary signal is initiated. Then the warning is provided by smartphone display and an external speaker or smartphone speaker.

## 4 Results

The output is obtained from the blocks created in the Simulink model with video as the input to the lane departure warning system. The images symbolise the output with detecting the lane on road.

The original image is shown in the Fig. 2a and the lane detected image is shown in the Fig. 2b with lane markings on both the side of the lane. The Fig. 2c represents the model detecting middle lane in the video stream. The Simulink model finds drawback in case of detecting a vehicle at the middle of the two-way lane indicated in the Fig. 3a and b and doesn't not indicate the departure at specific locations. The Fig. 3c represents the model not detecting the barriers and sign boards that are created temporarily in Indian highways. Further, the implement of lane detection model that includes the design of an android application to detect the lanes on the roads.

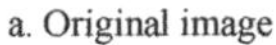

a. Original image b. Lane detected image c. The middle lane detected

**Fig. 2.** Demonstration of the algorithm [7]

a. Departure not identified b. Middle lane not detected c. Barriers not identified

**Fig. 3.** Drawbacks of the existing algorithm [6]

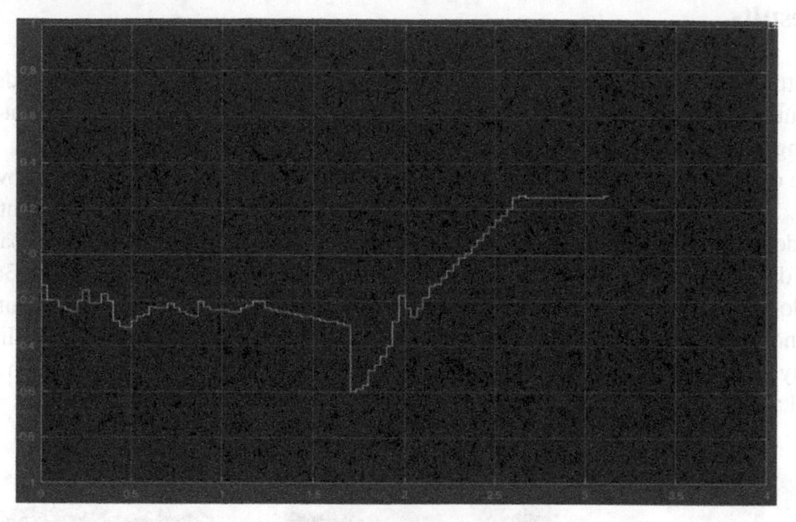

**Fig. 4.** Lane change of vehicle

The above Fig. 4 indicates the fall of signal for lane change of vehicles from middle lane to left lane and back to the middle lane and the stable signal indicates the vehicle travelling in particular lane with no departure. The implementation of the system is by using an android app with the android camera as the input device and the output device as the android display. The warning to the system is based on the smartphone/android display or from the external speakers or by android's speaker.

## 5 Conclusion

Based on the camera centric view point, the lane crossing is identified. The lane tracking is carried out using pattern matching and Kalman filter and the departure warning is indicated by android display with coloured markings on the lane and an audio output is provided externally by the android speaker. The implementation is tested in 4-way, 6-way, and 8-way lanes.

**Compliance with Ethical Standards**

✓ All authors declare that there is no conflict of interest
✓ No humans/animals involved in this research work.
✓ We have used our own data.

## References

1. Nguyen, V., Kim, H.: A study on real-time detection of lane and vehicle for lane change assistance system using vision system on highway. **21**(5) (2018). 978-1-5090
2. Youjin, T., Wei, C.: A robust lane detection method based on vanishing point estimation. Procedia Comput. Sci. **131**, 354–360 (2018)
3. Yurtsever, E., Yamazaki, S.: Integrating driving behaviour and traffic context through signal symbolization for data reduction and risky lane change detection. IEEE Trans. Intell. Veh. **3**(3), 242–253 (2018)
4. Yeniaydin, Y., Schmidt, K.W.: A lane detection algorithm based on reliable lane markings. In: Signal Processing and Communications Applications Conference (SIU). IEEE, May 2018. 978-1-5386
5. Woo, H., Ji, Y., Kono, H.: Lane change detection based on vehicle-trajectory prediction. IEEE Robot. Autom. **2**(2), 1109–1116 (2017)
6. Xu, X., Yu, J., Zhu, Y.: Leveraging smartphones for vehicle lane-level localization on highways. IEEE Trans. Mob. Comput. **17**(8), 1894–1907 (2018)
7. Song, W., Yang, Y., Ful, M.: Lane detection and classification for forward collision warning system based on stereo vision. IEEE Sens. J. **18**(12), 5151–5163 (2018)
8. Wei, Y., Tian, Q.: Multi-vehicle detection algorithm through combined Harr and HOG features. Math. Comput. Simul. **155**, 130–145 (2017)
9. Su, Y., Zhang, Y., Lu, T., Yang, J., Kong, H.: Vanishing point constrained lane detection with a stereo camera. IEEE Trans. Intell. Transp. Syst. **19**(8), 2739–2744 (2018)
10. Jung, S., Youn, J.: Efficient lane detection based on spatiotemporal images. IEEE Trans. Intell. Transp. Syst. **17**(1), 289–295 (2016)
11. Gaikwad, V., Lokhande, S.: Lane departure identification for advanced driver assistance. IEEE Trans. Intell. Transp. Syst. **16**(2), 910–918 (2015)
12. Yi, S.-C., Chen, Y.-C., Chang, C.-H.: A lane detection approach based on intelligent vision. Comput. Electr. Eng. **42**, 23–29 (2015)
13. Shin, J., Lee, E., Kwon, K., Lee, S.: Lane detection algorithm based on top-view image using random sample consensus algorithm and curve road model. In: Sixth International Conference on Ubiquitous and Future Networks. IEEE (2014). 978-1-4799
14. Kim, H., Kwon, O., Song, B., Lee, H., Jang, H.: Lane confidence assessment and lane change decision for lane-level localization. In: 14th International Conference on Control, Automation and Systems (ICCAS), October 2014. 978-89-93215
15. Rahmdel, P.S., Shi, D., Comley, R.: Lane detection using Fourier-based line detector. In: IEEE 56th International Midwest Symposium on Circuits and Systems, August 2013. 978-1-4799
16. Wang, Y., Teoh, E.K., Shen, D.: Lane detection and tracking using B-Snake. Image Vis. Comput. **22**(4), 269–280 (2004)

# Stratified Meta Structure Based Similarity Measure in Heterogeneous Information Networks for Medical Diagnosis

Ganga Gireesan(✉) and Linda Sara Mathew

Computer Science and Engineering, Mar Athanasius College of Engineering, Kothamangalam 686666, Kerala, India
gangagireesan@gmail.com, lindasaramathew@gmail.com

**Abstract.** Electronic Health Records (EHR) offers point by point documentation on various clinical occasions that crop up amid a patient's stay in the emergency clinic. Clinical occasions can be spoken to as Heterogeneous Information Networks (HIN) which comprises of multi-composed and interconnected articles. A focal issue in HINs is that of estimating the likenesses between articles by means of basic and semantic information. The proposed strategy utilizes a Stratified Meta Structure-centred Resemblance measure named SMSS in heterogeneous data systems. The stratified meta-structure can be made consequently and seizure opulent semantics. At that point, the com-quieting framework of the stratified meta-edifice is characterized by the prudence of the transforming matrix of meta-ways and meta-structures. The primary point is to decipher EHR information and its rich connections into a heterogeneous data for robust therapeutic analysis. This demonstrating approach takes into account the direct handling of missing qualities and heterogeneity of information.

**Keywords:** Heterogeneous Information Networks · Electronic Health Records · Meta path · Meta structure · Stratified Meta Structure

## 1 Introduction

Numerous genuine frameworks, for example, biological frameworks and social medium can be formed utilizing networks. Therefore, network analysis turns into a warm research point in the field of information mining. Most genuine frameworks by and largely comprised of countless acting, numerous composed constituents, similar to human social exercises, correspondence, and computer frameworks, and biological systems. In such frameworks, the participating constituents portray unified systems, that be able to be call information networks or data systems lacking forfeiture of review [1]. Nonetheless, the genuine data organizes mostly include interconnected and various composed segments. This sort of data systems is for the most part called Heterogeneous Information Networks (HIN) [2]. Contrasted with generally utilized homogeneous data organize, the heterogeneous data system can effectively join more data and contain rich semantics in hubs and joints, and therefore it figures another advancement of information mining. Deciding the closeness between items assumes essential and fundamental jobs in heterogeneous data arrange digging tasks.


S. Smys et al. (Eds.): ICCVBIC 2019, AISC 1108, pp. 62–70, 2020.
https://doi.org/10.1007/978-3-030-37218-7_8

In situations of restorative field, Electronic Health Records (EHR) give nitty gritty reported data on a few clinical occasions that crop up all through a patient's stopover in the emergency clinic [3]. Research Centre tests, prescriptions, nurture notes, and diagnoses are occasions of assorted kinds of clinical minutes. The primary point of numerous examinations is illuminating clinical basic leadership and malady analysis. Clinical occasions can be described as Heterogeneous Information Networks. Likewise, the rate of numerous illnesses can cause trouble in perceptions and their relations. In this way, fabricating a computer supported determination framework is of incredible significance in dropping mistake and convalescing social insurance.

At that point create Stratified Meta-Structure to bring composite connection semantics into the system and capture those that are useful for critical thinking purposes [4]. This structure need not be indicated ahead of time, and it consolidates numerous meta-ways (meta-paths) and meta-erections (meta-structures). This gathering guarantees that: (1) purchasers don't have to stress over how to pick the meta-way or meta erection; (2) opulent semantics can be secured. Accordingly, the SMS is fundamentally a coordinated non-cyclic graph involving item sorts with modified stratum marks. The SMS be able to naturally created by means of monotonously checking the item sorts in the netting schema or outline. The SMS as an unpredictable structure is in this way a composite connection. This is the reason the SMS can seizure deep semantics. The meaning contained in meta-ways are typically noble via their commuting milieus. Generally, the meta-structures have indistinguishable nature from the meta-paths since they all have hierarchical structures. Along these lines, the driving networks of meta-structures are demarcated by the prudence of the Cartesian item, and the driving lattice of the SMS is defined by normally blending the boundless quantity of the driving frameworks of meta-erections. SMSS is outlined by the driving framework of the SMS. Utilizing HIN for demonstrating EHR and determination expectation beats best in class models in two dimensions of general illness gathering and explicit finding forecast.

## 2 Related Works

Similarity measure is an old activity in database and web indexes. It is essential to contemplate likeness seek in huge scale heterogeneous data systems, for example, the bibliographic systems and online life systems [2]. Intuitively, two articles are comparable on the off chance that they are connected by a few ways in the system. All things considered, most existing likeness strategies are clear for homogeneous systems. Disparate semantic implications behind ways are not taken into dread. In this manner they can't be unswervingly connected to heterogeneous systems. By taking into consideration modified linkage ways in a system, one could get diverse closeness semantics. In this manner, start the idea of meta path (MP) centered comparability, where a MP is a route comprised of a keep running of associations particular amongst different item sorts. A large portion of the current measurements relies upon client indicated meta-paths (MP) or meta-structures (MS). For representation, PathSim and Biased Path Con-stressed Random Walk (BPCRW) [2] take a MP determined by clients as information, and Biased Structure Constrained Subgraph Expansion

(BSCSE) takes a MS indicated by clients as information. These measurements are defenseless to the pre-indicated meta-ways or meta-erections. It is much trying for handlers to imply MPs or MSs. For instance, an organic data system may include a few distinct sorts of items. Moreover, the meta-paths can just restrict one-sided and generally inconvenience free semantics. Along these lines, the meta-structure are acquainted all together with catch progressively complex semantics. Indeed, the meta-structure can just internment one-sided semantics too. The MPs and MSs are fundamentally two sorts of schematic structures.

A heterogeneous information network (HIN) is an outline demonstrate in which entities and relations are recognized by sorts. Gigantic and multifaceted databases, for example, YAGO and DBLP, are able to symbolized as HINs. An essential issue in HINs is the calculation of vicinity, or pertinence, among two HIN articles [5]. Significance procedures are able to reused in a few solicitations like substance goals, suggestion, and data recovery.

## 3 Schematic Structures

An information network is a coordinated diagram G = (V, E, A, R) where V speaks to a lot of items and E speaks to a lot of connections. A and R separately demonstrate the arrangement of article types and connection types. G is entitled as a heterogeneous info net (HIN) if $|A| > 1$ or $|R| > 1$. If not, it's known as homogeneous data net.

Heterogeneous information systems contain multi-composed articles and their interconnected relations. For any item $v \in V$, it fits to an article type $\phi(v) \in A$. For each connection $e \in E$, it has a place with a connection type $\Psi(e) \in R$. Basically, $\Psi(e)$ describes a connection from its source object type to its target item type [1]. On the off chance that two connections have a place with the indistinguishable connection type, they share a similar beginning object type just as the target object type [6].

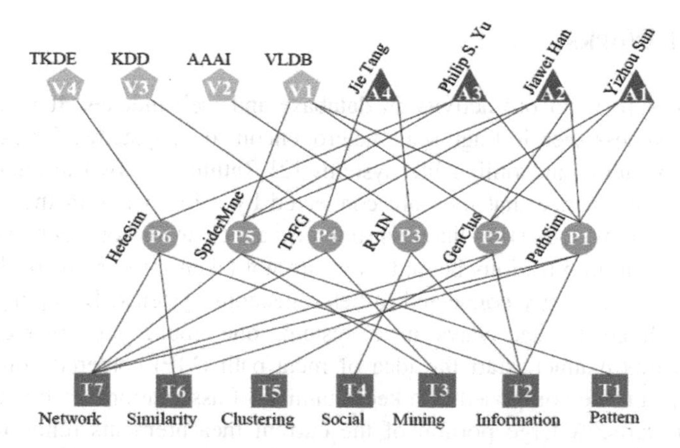


**Fig. 1.** A Realistic Bibliographic information network with clear papers, author, terms and venues. The triangles, circles, squares, and pentagons separately emblematize authors or creators, papers or journals, terms or topics and venues or places.

Figure 1 exhibits a realistic bibliographic data connect with four real article types, that is, Author or creator (A), Paper or journal (P), Venue or place (V) and Term or topic (T). The sort Author encases four events: Yizhou Sun, Jiawei Han, Philip S. Yu, and Jie Tang. The sort Venue involves four cases: VLDB, AAAI, KDD, TKDE. The sort Paper contains six events: PathSim, GenClus, RAIN, TPFG, SpiderMine and HeteSim. The type Term includes six examples: Pattern, Information, Mining, Social, Clustering, Similarity, and Network. Each paper distributed at a setting basically has its creators and its related terms. From now on, they contain three sorts of connections: P< = > A, P< = > V and P< = > T.

### 3.1 Network Schema

The Schema offers a meta-glassy delineation for heterogeneous data systems. Unequivocally, the system outline $T_G$ = (A, R) of G is a coordinated diagram containing the item sorts in A and the connection sorts in R [7].

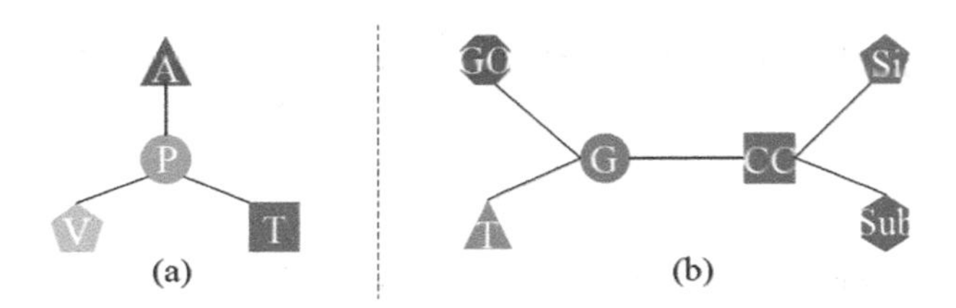


**Fig. 2.** (a) Bibliographic network schema. (b) Biological network schema.

Figure 2 illustrates the network schema (NS) for the heterogeneous data system in Fig. 1. The biological system is an elective case of heterogeneous information network. An organic information organize includes six entity forms, that is, Gene or Genetic material (G), Tissue or matter (T), GeneOntology (GO), ChemicalCompound (CC), Substructure (Sub) and SideEffect (Si), and five connection forms, that is, GO< = > G, T< = > G, G< = > CC, CC< = > Si, CC< = > Sub. Its system schema appears in Fig. 2(b).

### 3.2 Metapaths

Presently are two classes of graphic arrangements (MPs and MSs) for the systematic mapping of HIN. The two sorts pass on some semantics [1]. The semantic basically be demonstrated by clients ahead of time when utilizing these structures. There is rich semantics in Heterogeneous data organize G. This semantics can be taken by metapaths, meta structures or considerably increasingly complex schematic structures in a

Network pattern. A meta-paths is essentially a substitute grouping of article types and connection types, that is:

$$\mathscr{P} = O_1 \xrightarrow{R_1} O_2 \xrightarrow{R_2} \ldots \xrightarrow{R_{l-2}} O_{l-1} \xrightarrow{R_{l-1}} O_l$$

$R_j \in R$, $j = 1, \ldots, l_l$ & $R_j$ is a connection form initiating from Oj to $O_{j+1}$ where, $j = 1, \ldots, l_1$. Fundamentally, the MP involves few amalgamated meaning on the grounds that it represents a multifaceted connection. The MP, P is able to minimally signified as $O_1, O_2, \ldots O_{l-1}, O_l$. There are sure significant ideas identified with the meta-way, that is, length of P, way example succeeding P, switch MP of P, symmetric meta-way and commuting matrix $M_P$ of P.

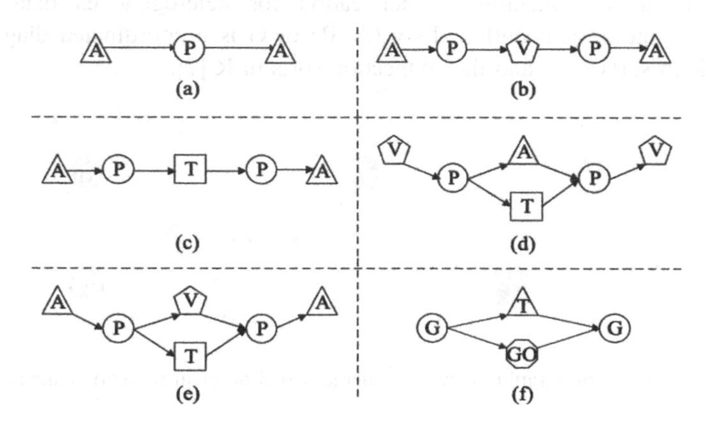


**Fig. 3.** Some meta-paths and meta-structures.

For instance, Fig. 3(a) to (c) demonstrates MPs in the NS appeared in Fig. 3(a). The meta-ways can be thickly assigned as (A, P, A), (A, P, V, P, A) and (A, P, T, P, A). These MPs separately depict unique semantics. (A, P, A) communicates "Two authors collaborate on a paper". (A, P, V, P, An) indicates "Two authors distribute two papers on a similar venue". (A, P, T, P, A) depict "Two authors distribute two papers containing similar terms".

### 3.3 Meta Structures

The meta-structure S = ($V_S$, $E_S$, $T_s$, $T_t$) is fundamentally a coordinated non-cyclic diagram by means of a solitary cradle entity form $T_s$ and a solitary destination entity form $T_t$. $V_s$ represents group of entity sorts, and $E_S$ indicates group of connection sorts. Figure 4(d)–(e) indicates different sorts of MSs for the system mapping uncovered in Fig. 3(a). These MSs are able to productively assigned as (VP (AT) PV) and (AP (VT) PA). Figure 4(f) speaks to a meta-erection aimed at the system mapping appeared in Fig. 3(b). The MS be able to trimly signified as (G (GO T) G). The MS (VP (AT) PV) communicates the most mind-boggling meaning "Two venues acknowledge two papers both containing similar terms and composed by similar authors".

The meta-erection (AP (VT) PA) represents the most mind-boggling meaning "Two authors distribute two papers both containing similar terms and distributed on a similar venue".

### 3.4 Stratified Meta Structure

Meta-ways and meta-erections can just catch generally straightforward semantics and their meanings can be communicated via amalgamated associations [9]. The complex associations are as a progressive structure accomplished by the return of entity forms. An epic schematic erection called Stratified Meta-Structure (SMS) be able to utilized as a comparability measure in HINs. The SMS will catch multifaceted meaning since it is mainly made out of MSs through various ranges or potentially the meta-ways via various ranges. Here, every single entity form is rehashed an unbounded amount of epochs, since catching rich semantics need to consolidate the MPs as well as MSs with various extents [10].

A SMS is fundamentally a coordinated non-cyclic diagram including entity forms utilizing diverse stratum names. The noticeable favourable position of the SMS is that it tends to be mechanically developed by every now and again visiting article sorts during the time spent navigating the system pattern. Stated a heterogeneous info net G, initially concentrate its net scheme $T_G$, and after that pick a starting entity form and an objective item form and ruminate the circumstance, where the starting entity form is the equivalent as the objective sort. In the event that the starting entity form isn't equivalent with objective, despite everything unique practice the development principle of SMS to build a graphic arrangement.

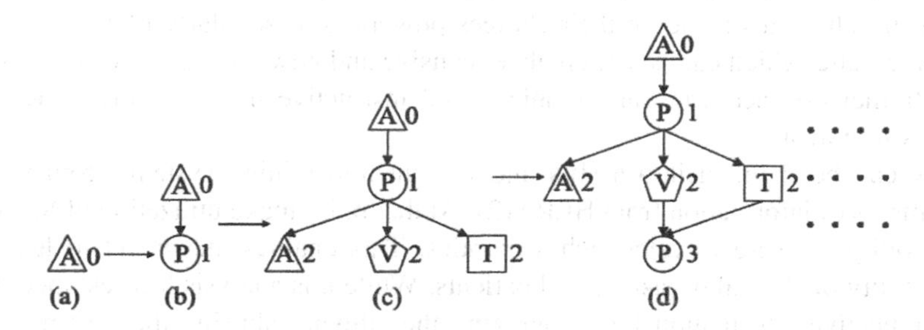


**Fig. 4.** Construction of the SMS of the toy bibliographic information network.

The development principle of SMS $D_G$ of Heterogeneous info net G is spoken to in Fig. 4 and it incorporates the ensuing strategy [1]. The starting entity form is situated in 0th stratum. The article forms on the stratum $l = 1, 2, \ldots \infty$ are com-presented of the neighbours of article forms in stratum $(l - 1)$ in $T_G$. The contiguous entity forms are connected via a bolt directing as of $(l - 1)^{th}$ stratum despondent to the $l$th stratum. So, now the wake of gaining the objective article form in stratum $l \geq 1$, their active connections are removed.

Subsequent to getting the SMS the following stage is to discover the similitude between objects. Stratified meta-structure based similitude can be characterized dependent on the transforming milieus of MPs and MSs [11]. The driving network for a

SMS is for the most part characterized as the precis of the driving grids of MPs and MSs. By utilizing SMS based closeness it is conceivable to anticipate the complex relationship between various sorts of articles. In this way, SMS can be utilized to recognize the closeness between different items in Electronic Health Records. Such a strategy can be connected, all things considered, for making the therapeutic finding process less demanding.

## 4 Methodology

Ailment examination has turned into an imperative procedure in social insurance. Relations among clinical occasions convey different semantics and sponsor oppositely to infection analysis. To talk these issues, this work connotes interpretation of high-dimensional EHR information and its rich connections into HIN for extreme restorative finding [12]. This evil presence starting approach allows direct treatment of missing qualities and heterogeneity of information. For the conclusion reason, it abuses SMS to catch larger amount and semantically essential relations identified with sickness determination.

The symptomatic procedure incorporates watchful thought of clinical perceptions (side effects and demonstrative tests), mining of applicable data, and more prominently paying thought to its associations [11]. A medical reflection is regularly non-explicit towards a solitary sickness. The aforementioned indicates its connection or co-event through different perceptions that are able to emblematic upon sickness. In addition, the event of numerous sicknesses be able to become a reason inconvenience in perceptions too their associations. These impediments alongside an immense measure of data to be analysed by clinicians settle on their choices powerless to scholarly blunder and as a rule problematic, which can be extremely expensive and now and again deadly. To plan such a framework, acquiring an organized and instructive model of the EHR information is important.

EHR can be changed into a HIN and present hub mining systems from various arrangements of information from HER [12]. At that point make utilization of SMS as a comparability measure to catch rich semantics. This enables the model to learn the comparability of clinical occasions and patients. While this analysis forecast model just works indicative information for recognizing the ailment, abusing the treatment data next to the season upon unsubstantiated showing to enhance cultured embedding and discover comparability of scientific occasions as far as result.

## 5 Conclusion

Closeness measure as a fundamental assignment in heterogeneous information networks analysis has been helpful to numerous fields, for instance, item suggestion, bunching, and Web search. Loads of the overarching measurements dependent upon the MP or MS demonstrated by clients ahead of time. These measurements are in this manner fragile to the previously determined MP or MS. The SMS centred comparability SMSS can be utilized among heterogeneous info nets. The SMS as a compound graphic arrangement be able to unexpectedly amassed via redundantly navigating the system pattern. This pattern implies that its not at all compulsory for clients to trouble on in what way to take a proper MP or MS. First, portray the transforming milieu of the Stratified Meta-Structure by conjoining all the transforming milieus of the critical or meta-ways, then afterward utilize this relation milieu to prompt the closeness degree SMSS. Stratified meta-structure put together similitude with respect to the entire overwhelms the baselines as far as grouping and positioning. SMSS can be spread in EHR datasets for therapeutic analysis reason. When utilizing HIN for therapeutic finding, it is proficient of getting educational associations for the analysis objective and utilizes the superlative connection inspecting procedure when learning clinical occasion portrayals. It comparatively takes into consideration the simple treatment of missing qualities and learning embeddings customized to the sickness forecast objective utilizing a joint inserting framework.

**Compliance with Ethical Standards**

✓ All authors declare that there is no conflict of interest
✓ No humans/animals involved in this research work.
✓ We have used our own data.

## References

1. Zhou, Y., Huang, J., Sun, H., Sun, Y.: Recurrent meta-structure for robust similarity measure in heterogeneous information networks. arXiv:1712.09008v2 [cs.DB], 23 May 2018, @ 2010 Association for Computing Machinery
2. Sun, Y., Han, J., Yan, X., Yu, P.S., Wu, T.: PathSim: meta path-based top-k similarity search in heterogeneous information networks. Proc. VLDB Endow. **4**(11), 992–1003 (2011)
3. Zhoua, Y., Huang, J., Lia, H., Sunc, H., Pengd, Y., Xu, Y.: A semantic-rich similarity measure in heterogeneous information networks. Knowl.-Based Syst. **154**, 32–42 (2018)
4. Huang, Z., Zheng, Y., Cheng, R., Zhou, Y., Mamoulis, N., Li, X.: Meta structure: computing relevance in large heterogeneous information networks. In: Proceedings of the ACM SIGKDD International Conference on Knowledge Discovery and Data Mining, pp. 1595–1604. ACM, San Francisco (2016)
5. Sun, Y., Barber, R., Gupta, M., Aggarwal, C.C., Ha, J.: Co-author relationship prediction in heterogeneous bibliographic networks. In: 2011 International Conference on Advances in Social Networks Analysis and Mining (2011)
6. Shi, C., Li, Y., Zhang, J., Sun, Y., Yu, P.S.: A survey of heterogeneous information network analysis. IEEE Trans. Knowl. Data Eng. **29**(1), 17–37 (2017)
7. Sun, Y., Han, J.: Mining heterogeneous information networks: a structural analysis approach. SIGKDD Explor. **14**(2), 20–28 (2012). https://doi.org/10.1145/2481244.2481248
8. Shi, C., Wang, R., Li, Y., Yu, P.S., Wu, B.: Ranking-based clustering on general heterogeneous information networks by network projection. In: Proceedings of the ACM CIKM International Conference on Information and Knowledge Management, pp. 699–708. ACM, Shanghai (2014)
9. Gupta, M., Kumar, P., Bhasker, B.: HeteClass: a meta-path based framework for transductive classification of objects in heterogeneous information networks. Knowl.-Based Syst. **68**(1), 106–122 (2017)

10. Shi, C., Kong, X., Huang, Y., Yu, P.S., Wu, B.: HeteSim: a general framework for relevance measure in heterogeneous networks. IEEE Trans. Knowl. Data Eng. **26**(10), 2479–2492 (2014)
11. Jeh, G., Widom, J.: SimRank: a measure of structural-context similarity. In: Proceedings of the 8th ACM SIGKDD International Conference on Knowledge Discovery and Data Mining, pp. 538–543. ACM, Edmonton (2002)
12. Hosseini, A., Chen, T., Wu, W., Sun, Y., Sarrafzadeh, M.: HeteroMed: heterogeneous information network for medical diagnosis. In: WOODSTOCK 1997, El Paso, Texas, USA, July 1997

# An Effective Imputation Model for Vehicle Traffic Data Using Stacked Denoise Autoencoder

S. Narmadha(✉) and V. Vijayakumar

Sri Ramakrishna College of Arts and Science (Autonomous), Coimbatore, India
narmadhas17@gmail.com, veluvijay20@gmail.com

**Abstract.** Modern transportation systems are highly depend on quality and complete source of data for traffic state identification, prediction and forecasting processes. Due to device (sensor, camera, and detector) failures, communication problems, some sources inevitably miss the data, which leads to the degradation of traffic data quality. Data pre processing is an important one for transport related applications. Imputation is the process of finding missing data and make available as complete data. Both Spatial and temporal information has been a high impact on impute the traffic data. In this paper deep learning based stacked denoise autoencoder (one autoencoder at a time) is proposed to impute the traffic data with less computational complexity and high performance. Experimental results demonstrate that autoencoder performs well in random corruption aspect with less complexity.

**Keywords:** Denoise autoencoder · Traffic flow · Imputation · Train one autoencoder

## 1 Introduction

Smart city has emerged all over the world due to urban growth, better assessment of all facilities, safety and economic growth etc. Transportation systems is a main component in a crowded city which has limited space constraint of road and resources. But every time it is complicated to make a new road or extending the road structure for flexible migration of vehicles and people because of economy and environment [13]. In recent year's usage of surveillance systems, fast growing technology, and limited cost of storage and computing sources vast amount of data has been collecting in all spatial locations for every time serious. However, still missing traffic data problem exists in many traffic information systems and it is not sensible for further utilization [9]. Many transport information systems from different countries gradually suffers from missing traffic data [6]. For example, PEMS (Performance measurement system) is a California based database system to maintain real time traffic data and incidents collection which has been using by all researchers in the transportation field, has higher than 10% of missing data [16]. The missing ratio [6, 9] of the daily traffic flow in Beijing, china has around 10%.


S. Smys et al. (Eds.): ICCVBIC 2019, AISC 1108, pp. 71–78, 2020.
https://doi.org/10.1007/978-3-030-37218-7_9

The consequence of missing data for traffic prediction and estimation can be divided into two ways: (i) data loss of a certain location and particular time periods may be the important to solve the transportation issue. (ii) statistical information loss.

## 2 Related Work

Auto Regressive Integrated Moving Average (ARIMA) model was [12] investigated the application of box Jenkins analysis techniques for freeway traffic prediction and imputation. It was more accurate in terms of representing time serious data. Least square support vector machine technique (LS-SVM) proposed to predict the missing traffic flow data in the arterial road with spatio temporal information [17]. Cokriging methodology [3] is used to impute multisource data (Radar detecter data, Probe vehicle data). Matrix and tensor based methods used [4, 5, 8] for traffic data imputation.

Hybrid models were effectively imputed the data. Fuzzy c-means (FCM) [11] applied to estimate the missing values in a multi attribute data set and parameters were optimized by Genetic algorithm (GA). An iterative approach based on the integration and time serious algorithm was proposed [7] to fill small and large missing values. Most of the imputation methods fail for real time and historical data which may contains unavailable and inappropriate neighbouring data. Deep learning is used to extract high and complex representation of data. An imputation approach deep stacked denoise autoencoder [10, 14, 15] proposed based on statistical learning. It imputes simple structured data (single location, single period) to complex structure (single period multiple location, multiple period multiple location etc.). Stacked autoencoders and their variations have been used in various applications of noise removal and prediction. But it takes higher running time for larger datasets. It increased the attention to stacked autoencoder (run one at a time) to reduce the computational complexity and save the running time.

## 3 Methodology

In this work, stacked denoise autoencoder (run one autoencoder at a time) is used to impute the traffic flow data.

### 3.1 Autoencoder (AE)

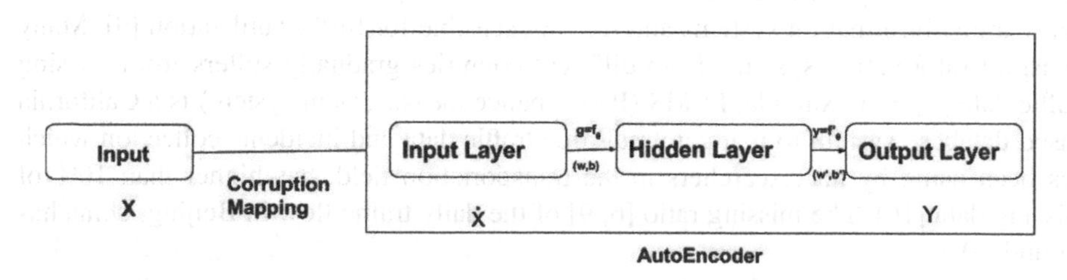


**Fig. 1.** Simple denoise autoencoder

Autoencoder is a neural network with backpropagation. It works by encoding and decoding as shown in Fig. 1. An encoder [10] takes an corrupted input vector ẋ and map into hidden representation g through a mapping function $g = f_\theta(x) = s(wx + b)$ where Θ = {w, b}. Algorithm is summarized in Table 1. w is a weight matrix and b is a bias vector. The resulting latent representation g is mapped back to a reconstructed vector $y = f'_\theta(g) = s(w'g + b')$ The weight matrix w′ and b′ may be the reverse mapping of w, b. The model is optimize the parameters upto the number of iterations to minimize the error between x (input) and y (output). Final Θ and $\Theta'$ are the updated parameters of a model.

**Table 1.** Algorithm 1 - Autoencoder

| |
|---|
| **Input:** data X={$x_1$…$x_n$}, Iterations I={1,2..n} |
| **Initialize** weights and bias randomly (w,b) |
| Pass X into hidden layer ; |
| Reconstruct the output from hidden layer through transpose T'; |

### 3.2 Stacked Denoise Autoencoder (SDAE) (One Autoencoder at a Time)

A SDAE is a stack of multiple AE's trained to reconstruct the clean output y from the noisy or corrupted version of input ẋ. The sequence of process as follows, corrupting the input x into ẋ by means of indiscriminate mapping ẋ ~ M(ẋ|x) [15]. Corrupted input is passed into the first auto encoder, and then it passed through the successive auto encoders to remove the noise value and learn the internal representation of data from the input layer. All auto encoders trained, copy the weights and biases from each autoencoder and use them to build SAE.

One autoencoder at a time in SDAE (Fig. 2) runs faster than training whole SAE [2]. Training algorithm summarized in Table 2. While first phase of training, first AE learns to reconstruct the inputs. Hidden layer 1 is freeze (updation false) when processing phase 2, its output will be same for any training instance. It will avoid the re compute the output of hidden layer 1 at every epoch. Transpose the weights for hidden layer 3 & 4. Stack the weights of all hidden layers from top to bottom at the end of phase. The deep autoencoder greatly reduces the computational complexity and helps to increase the performance.

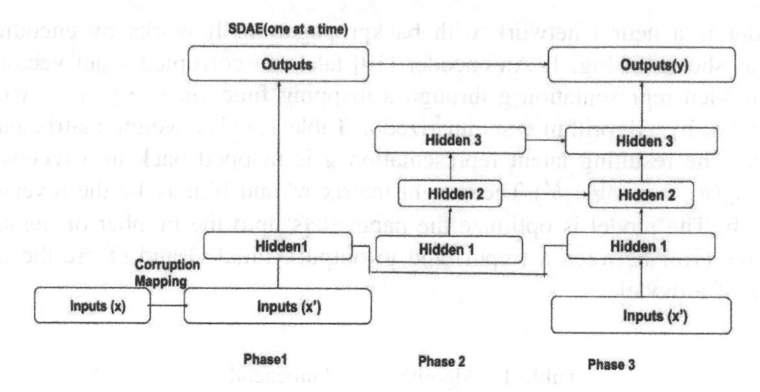


**Fig. 2.** Stacked denoise autoencoder (OAE)

# 4 Result Analysis

## 4.1 Data Description

Traffic flow data is collected from California based PeMS (Performance measurement system) [18] which contains traffic related data of all freeways in the state. PeMS has 39000 individual detectors which spreads all over the metropolitan areas and it collects the data at every 30 s and cumulated into 5 min interval. The data collected from the year 2017 of district 5. Total 243 vehicle detector stations (VDS) in the district. In 365 days feb 1st has some improper data except that have 364 days data and each day is

**Table 2.** Algorithm 2 - Stacked denoise autoencoder (One Auto Encoder (AE) at a time)

**Input:** traffic flow data X= $\{x_1, x_2 \ldots x_n\}$ with missing values Ø and noise ¥;
**Output:** Reconstructed data without Ø and ¥ values;
**Initialization:** No of hidden layers H, No of hidden nodes h= {1,2…H}, no of iterations I, epoch= Ω;
**Step1:**Map the input data into noise $X = \bar{X}$;
Scale the value and pass into hidden layer1 (AE1)
Freeze the weight and bias of first AE1;
Extract the hidden layer1 weights(w ε h1),bias(b ε h1) and passed into the hidden layer 2( AE2)
Transpose weights and bias to hidden layer 3(AE3) $h_i = (h_{i-1,} w^T) + b^T$ and layer4 (AE4) $O_i = (h_{i-2,} w^T) + b^T$
Using optimizer to reduce the error rate e;
**Step 2**:**Fine tune** the model
Initialize weights and bias of all pre-trained auto encoders from top to bottom(in reverse order);
Perform forward propagation to compute y;
Perform backward propagation to compute x-y;
Using optimizer to reduce the error rate;

represented as vector K. Each vector has 288 dimensions L as it is 5 min data. Weekdays, non-weekdays (weekends and holidays), both weekdays and non-weekdays are taken as temporal factors. 115 non-weekdays and 250 weekdays in 2017.

Train and test ratio is 80:20 for all experiments. To avoid overfitting early stop method is used. 20% to 40% of missing rate is considered as smoother [1], so here 30% of missing rate is fixed for all experiments. Random Corruption ratio (RC) = $\frac{\sum_{i=1}^{K}\sum_{j=1}^{L} O_{ij}}{KL} * 100\%$ {where o = observed value [0 or 1]}. Tesla k80 12 GB GPU RAM machine is used to train and test the model. Sigmoid and elu activation function is used in input and output layer respectively. Have 3 hidden layers with size 144,72,144. Performance measures [15] are,

$$\text{Root mean square error (RMSE)} = \sqrt{\frac{\sum_{i=1}^{k}\sum_{j=1}^{l} O_{ij}(x_{ij} - y_{ij})2}{\sum_{i=1}^{k}\sum_{j=1}^{l} O_{ij}}}$$

$$\text{Mean absolute error (MAE)} = \frac{\sum_{i=1}^{k}\sum_{j=1}^{l} O_{ij}|x_{ij} - y_{ij}|}{\sum_{i=1}^{k}\sum_{j=1}^{l} O_{ij}}$$

$$\text{Mean Relative error (MRE)} = \frac{\sum_{i=1}^{k}\sum_{j=1}^{l} o_{ij}\frac{|x_{ij} - y_{ij}|}{x_{ij}}}{\sum_{i=1}^{k}\sum_{j=1}^{l} O_{ij}}$$

### 4.2 Results

Spatial and temporal correlation are the crucial for imputation and prediction of traffic data. In this imputation process current VDS (single), current VDS with upstream and downstream (augmented) and all VDS are considered as spatial factors to impute single station. Week days (WK), Non-weekdays (N-WK) and both weekdays and weekdays temporal data are evaluated with random corruption scenario.

**Single VDS Imputation**

VDS 500010092 is taken as current station for imputation and analysis of result. Find the missing values of single station based on the same station. Single Location and multiple periods of data are imputed and evaluate the result (Table 3). Both weekdays and non-weekdays gives better result for imputation.

**Table 3.** Comparision based on single VDS

| Temporal type | RMSE | MAE | MRE |
|---|---|---|---|
| Week days | 12.51 | 10.68 | 0.45 |
| Non week days | 23.3 | 14.6 | 0.47 |
| Both | **12.3** | **10.2** | **0.41** |

**Impact of Upstream (US) and Downstream (DS) on single VDS Imputation**
Both US and DS locations are highly correlated with current station. Current station data is augmented with upstream and/or downstream. From the analysis single VDS with downstream gives better imputation result than others (Table 4).

**Table 4.** Single VDS with upstream (US) and downstream (DS)

| Temporal type | RMSE | MAE | MRE |
|---|---|---|---|
| Single VDS | 10.41 | 9.03 | 0.47 |
| Single VDS with US | 10.62 | 9.56 | 0.46 |
| Single VDS with DS | **9.9** | **8.91** | **0.46** |
| Single VDS with US & DS | 10.3 | 9.38 | 0.49 |

**Impact of All VDS on Single VDS**
One autoencoder train well in all stations data and fill the missing values very effectively. In this part result compare with linear stacked autoencoder. Train one autoencoder reconstruct the observed value with minimal error rate and high performance (Table 5).

**Table 5.** All VDS on single station

| Temporal type | RMSE | | MAE | | MRE | |
|---|---|---|---|---|---|---|
| | SDAE | **SDAE (OAE)** | SDAE | **SDAE (OAE)** | SDAE | **SDAE (OAE)** |
| Weekdays (WD) | 17.1 | 14.03 | 14.3 | 8.22 | 0.47 | 0.47 |
| Non-weekdays | 23.3 | 12.01 | 14.6 | 10.56 | 0.48 | 0.47 |
| Both | 12.3 | **11.4** | 10.2 | **9.93** | 0.46 | **0.46** |

From all the observations both weekdays and non-weekdays gives more accurate value than considering single VDS and also Downstream and upstream.

**Realization of Processing Time**
Impact of all VDS running time is evaluated and shown in Fig. 3. SDAE one autoencoder runs faster (since takes only 6 s's than SDAE 33 s) Autoencoder not only improves the performance, it reduces the processing time in both training and testing. In all timeline SDAE (OAE) process well in both small and large datasets.

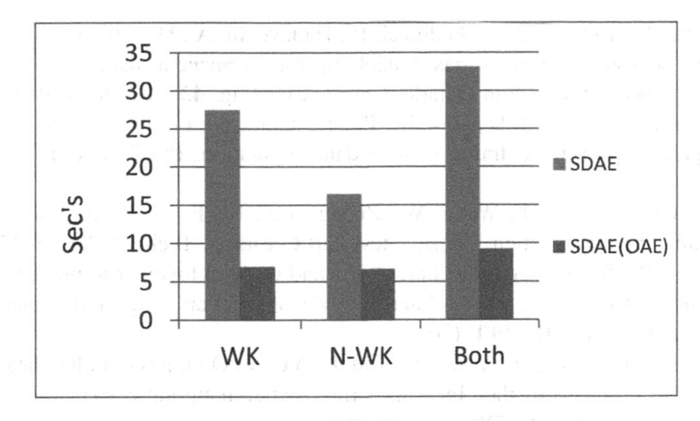


**Fig. 3.** Comparision of running time

## 5 Conclusion

Stacked denoise autocncoder run one at a time outperforms well in large datasets in terms of capturing hidden layer parameters without updating at each iteration. Fine tune the trained model again in top down (reverse order) fashion greatly improve the accuracy of reconstructed value and reduce the processing time. Simple and complex structure of data (both spatial and temporal) are evaluated with random corruption strategy. İn future the model may evaluated based on continuous corruption with various missing rates.

**Compliance with Ethical Standards**

✓ All authors declare that there is no conflict of interest

✓ No humans/animals involved in this research work.

✓ We have used our own data.

## References

1. Costa, A.F., Santos, M.S., Soares, J.P.: Missing data ımputation via denoising autoencoders: the untold story. In: IDA 2018, pp. 87–98. Springer (2018)
2. Geron, A.: Hands on Machine Learning with Scikit-Learn and TensorFlow. O'Reilly Media, Sebastopol (2017)
3. Bae, B., Kim, H., Lim, H., Liu, Y., Han, L.D., Freeze, P.B.: Missing data imputation for traffic flow speed using spatiotemporal cokriging. Transp. Res. Part C **88**, 124–139 (2018)
4. Ran, B., Tan, H., Feng, J., Liu, Y., Wang, W.: Traffic speed data ımputation method based on tensor completion. Comput. Intell. Neurosci. **2015**, 9 pages (2015). Article ID 364089
5. Acar, E., Dunlavy, D.M., Kolda, T.G., Mørup, M.: Scalable tensor factorizations for incomplete data. Chemometr. Intell. Lab. Syst. **106**(1), 41–56 (2011). arXiv:1005.2197v1
6. Chang, G., Ge, T.: Comparison of missing data imputation methods for traffic flow. In: Proceedings 2011 International Conference on Transportation, Mechanical, and Electrical Engineering (TMEE), Changchun, China, 16–18 December. IEEE (2011)

7. Abdelgawad, H., Abdulazim, T., Abdulhai, B., Hadayeghi, A., Harrett, W.: Data imputation and nested seasonality time series modelling for permanent data collection stations: methodology and application to Ontario. Can. J. Civ. Eng. **42**, 287–302 (2015)
8. Yang, H., Yang, J., Han, L.D., Liu, X., Pu, L., Chin, S., Hwang, H.: A Kriging based spatiotemporal approach for traffic volume data imputation. PLoS ONE **13**(4), e0195957 (2018)
9. Tan, H., Feng, G., Feng, J., Wang, W., Zhang, Y.J., Li, F.: A tensor-based method for missing traffic data completion. Transp. Res. Part C Emerg. Technol. **28**, 15–27 (2013)
10. Liang, J., Liu, R.: Stacked denoise autoencoder and dropout together to prevent over fitting in deep neural network. In: 2015 8th International Congress on Image and Signal Processing (CISP 2015), pp. 697–701. IEEE (2015)
11. Tang, J., Wang, Y., Zhang, S., Wang, H., Liu, F., Yu, S.: On missing traffic data ımputation based on fuzzy C-means method by considering spatial–temporal correlation. Transp. Res. Rec. J. Transp. Res. Board **2528**, 86–95 (2015)
12. Ahmed, M.S., Cook, A.R.: Analysis of freeway traffic time-series data by using Box-Jenkins techniques. Transp. Res. Rec. **722**, 116 (1979)
13. Shang, Q., Yang, Z., Gao, S., Tan, D.: An imputation method for missing traffic data based on FCM optimized by PSO-SVR. J. Adv. Transp. **2018**, 21 pages (2018). Article ID 2935248
14. Duan, Y., Lv, Y., Kang, W., Zhao, Y.: A deep learning based approach for traffic data imputation. In: 17th International IEEE Conference on Intelligent Transportation Systems (ITSC), pp. 912–917. IEEE (2014)
15. Duan, Y., Lv, Y., Liu, Y.L., Wang, F.Y.: An efficient realization of deep learning for traffic data imputation. Transp. Res. Part C **72**, 168–181 (2016)
16. Li, Y., Li, Z., Li, L.: Missing traffic data: comparison of imputation methods. IET Intell. Transp. Syst. **8**(1), 51–57 (2014)
17. Yang, Z.: Missing traffic flow data prediction using least squares support vector machines in urban arterial streets. In: 2009 IEEE Symposium on Computational Intelligence and Data Mining. IEEE Xplore (2009)
18. http://pems.dot.ca.gov/

# Fruit Classification Using Traditional Machine Learning and Deep Learning Approach

N. Saranya[1(✉)], K. Srinivasan[2], S. K. Pravin Kumar[3], V. Rukkumani[2], and R. Ramya[2]

[1] Department of Information Technology, Sri Ramakrishna Engineering College, Coimbatore, India
saranya.pravin@srec.ac.in

[2] Department of Electronics and Instrumentation Engineering, Sri Ramakrishna Engineering College, Coimbatore, India
{hod-eie,rukkumani.v,ramya.r}@srec.ac.in

[3] Department of Electronics and Communication Engineering, United Institute of Technology, Coimbatore, India
skpk.87@gmail.com

**Abstract.** Advancement in image processing techniques and automation in industrial sector urge its usage in almost all the fields. Fruit classification and grading with its image still remain a challenging task. Fruit classification can be used to perform the sorting and grading process automatically. A traditional method for fruits classification is manual sorting which is time consuming and involves human presence always. Automated sorting process can be used to implement Smart Fresh Park. In this paper, various methods used for fruit classification have experimented. Different fruits considered for classification are five categories of apple, banana, orange and pomegranate. Results were compared by applying the fruit-360 dataset between typical machine learning and deep learning algorithms. To apply machine learning algorithms, basic features of the fruit like the color (RGB Color space), size, height and width were extracted from its image. Traditional machine learning algorithms KNN and SVM were applied over the extracted features. The result shows that using Convolutional Neural Network (CNN) gives a promising result than traditional machine learning algorithms.

**Keywords:** Fruit classification · Machine learning · CNN

## 1 Introduction

Growing population all over the world increases the need for huge amount of food products. Fruits are sources of many essential nutrients like potassium, dietary fiber, vitamin C, and folic acid. Computer vision based system can be designed to replace the manual sorting process and also the usage of the barcode system. Due to the Big Data era and advancement in Graphics Processing Unit (GPU), deep learning gains more popularity and acceleration. Application of image processing and deep learning in the field of agriculture and food factories helps to implement smart farms using machine vision based system. Deep learning techniques were applied in various agricultural and


S. Smys et al. (Eds.): ICCVBIC 2019, AISC 1108, pp. 79–89, 2020.
https://doi.org/10.1007/978-3-030-37218-7_10

food production based industries [1]. These techniques lead to the creation of more UAV (Unmanned Aerial Vehicle) designs. In this paper, automatic sorting of fruits is done after capturing the images of the fruits. Features of the fruits were extracted from the image and classification process is done using the classifier model. Classifier model is built using images of the fruit-360 dataset. Sorted fruits are then collected into different buckets and the weight is automatically calculated for billing. This kind of smart fresh park based system automates the process and involves only less human efforts. In this paper, we compare the process involved in building a classifier model using traditional classification algorithm with deep learning techniques.

## 2 Literature Survey

In traditional machine learning approach, image classification can be done after extracting the features of the image. It involves various processing steps like image preprocessing, feature extraction, building a classifier model and validation. Several works related to fruit classification have been proposed. Feed Forward Neural Network along with Fitness Scaled Chaotic Artificial Bee Colony (FFNN-FSCABC) based system has been proposed [2] to recognize multi-class fruit. Image preprocessing techniques used are image segmentation, split and merge procedure for background removal [3]. Features considered for classification process are color, texture and shape. The accuracy obtained through FFNN-FSCAB system was 89.1%. Another classifier system has been proposed based on Multiclass SVM along with Gaussian Radial Basis with 88.2% accuracy [4]. A shape-based fruit recognition and classification system were proposed [8] and the author used morphological operations for image preprocessing [9]. Several shape descriptors have been considered from the parameters area, perimeter, major axis length and minor axis length. Classifier model was built after implementing Naïve Bayes, KNN and Neural Network algorithms. Fruit defect detection system was proposed [10] to detect defects from the sample apple images. YCbCr color space and K-Means clustering were used to distinguish calyx region and the other defected regions, Classifiers like SVM, Multi-Layer perceptron and KNN were implemented. The authors discussed various image preprocessing and feature extraction techniques [11] used in machine vision based systems.

Deep learning is an advanced machine learning which aims to achieve high accuracy in image recognition and classification. This can be used to predict the type of fruit by building a convolution neural network. A classifier was build using VGGNET [12] with an accuracy of 92.5% were image saliency is extracted to reduce the interference and noise due to a complex background. In another work [13] for fruit recognition using CNN was constructed with 11 layers such as 3-convolutional, 3-ReLU, 3-pooling and 2-fully connected layer and it gives an accuracy of 94.8%. Control system for fruit classification with CNN [14] has been developed with only 8 different layers such as convolution, max pooling and a fully connected layer with softmax activation function. CNN based system [15] was used in defect identification of mangosteen on its surface area using convolution, sub-sampling, fully connected and softmax layers were used. The accuracy of the system is 97.5 on average. An approach [16] was proposed to detect fungus in fruits with its image were features extracted from CNN is combined with the selective search algorithm. The accuracy of the system was 94.8%.

## 3 Dataset

In this paper fruit classification is done on sample images of fruit-360 dataset. Initially, four fruit categories considered for classifications are apple, orange, banana and pomegranate. Later eight categories of fruits were considered such as five different varieties of Red apple, orange, banana and pomegranate. Sample images were shown in Fig. 1. Images in fruit-360 dataset are of size 100 × 100.

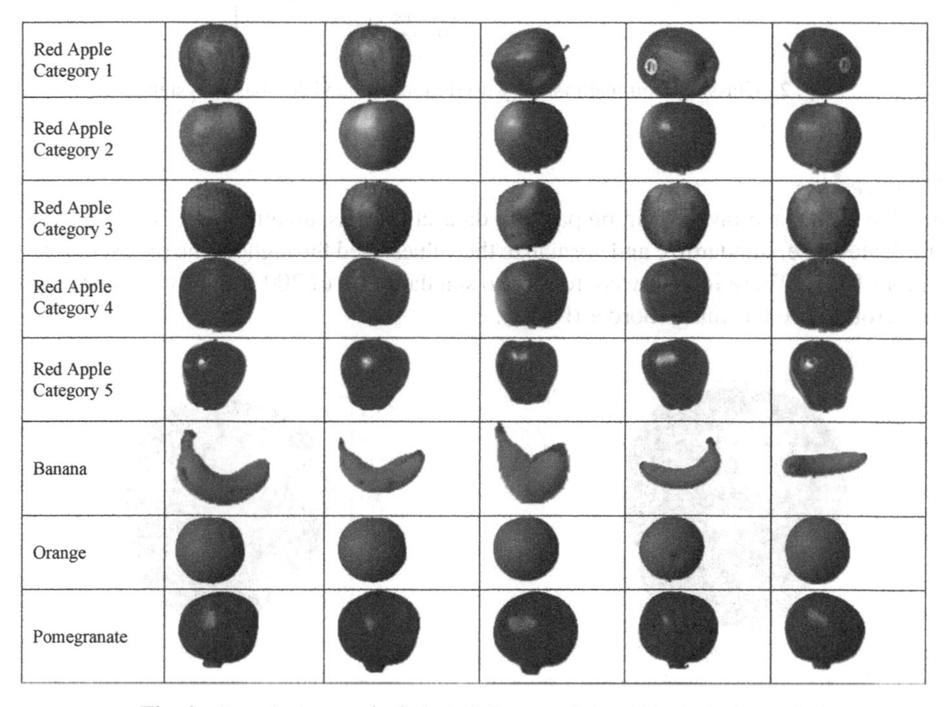


**Fig. 1.** Sample Images in fruit-360 dataset of size 100 × 100 [3, 5–8]

## 4 Traditional Machine Learning Approach

*Methodology Used*

General methodology used in machine vision based system to classify different fruits involves the following steps given in Fig. 2. Various features of the fruit like color, size, height and width of the fruit extracted from the preprocessed image to classify the fruit. Suitable machine learning algorithm was used to build a classifier model. Test data set is used to check the accuracy of the classifier that is pre-trained.

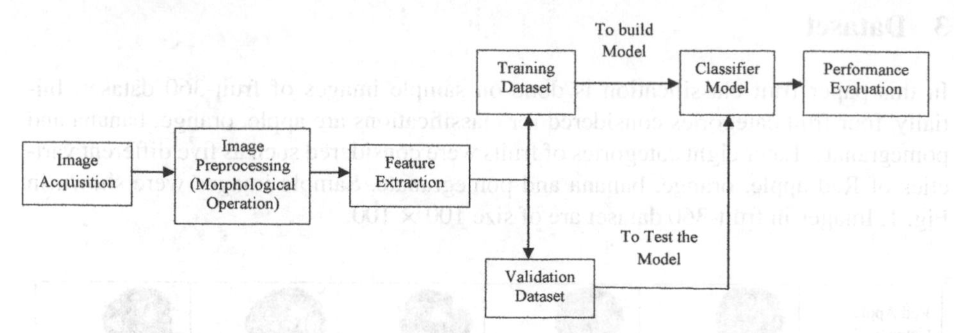


**Fig. 2.** Classification methodology used in machine vision based system

*Preprocessing*

It is the initial step involved in preparing a dataset for classification process. It aims to eliminate noise, standardize and normalize the values used throughout the data set using various filters. These images were resized to standard size of 200 × 200 by padding the background pixel as image border (Fig. 3).

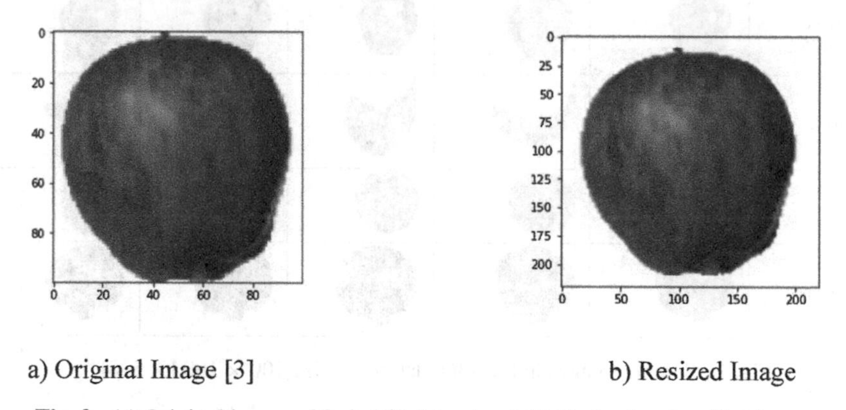


a) Original Image [3] b) Resized Image

**Fig. 3.** (a) Original image of fruit-360 dataset and (b) Resized and padded image

Then the image is converted into gray scale to differentiate fruit from its background (Fig. 4).

$$GrayScalImag0.299 * R + 0.587 * G + 0.114 * B \quad (1)$$

Grayscale image is applied with 'otsu' thresholding. Morphological operations i.e. erosion followed by dilation is applied over the image to smoothen the boundaries of the image after removing noise. Erosion of image A on structuring element B is

$$A \ominus B = \{z \in E | B_z \subseteq A\} \quad (2)$$

Dilation is the process used to enlarge the boundary of the foreground. Dilation of image A on structuring element B is

$$A \oplus B = \bigcup_{b \in B} A_b \quad (3)$$

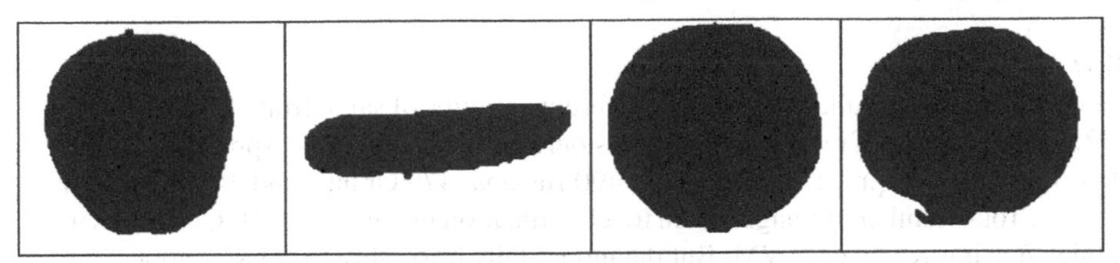

**Fig. 4.** Sample grayscale images after preprocessing

*Feature Extraction*

Feature selection is the important phase in fruit classification and grading system. Classification of fruits mainly depends on the key features. At the same time, too many feature inclusion and also consideration of only few features for classification may degrade the system performance. Some of the features have been considered in our work for fruit classification is color, size, height and width of the fruit. Contour of the preprocessed image is drawn for extracting useful features from the image. Color of the image is extracted by calculating average RGB value per row. Size of the image is calculated through the number of pixels covered by the image. Rectangle is bounded over the image based on the contour area. It is used to find the height and width of the image given in Fig. 5.

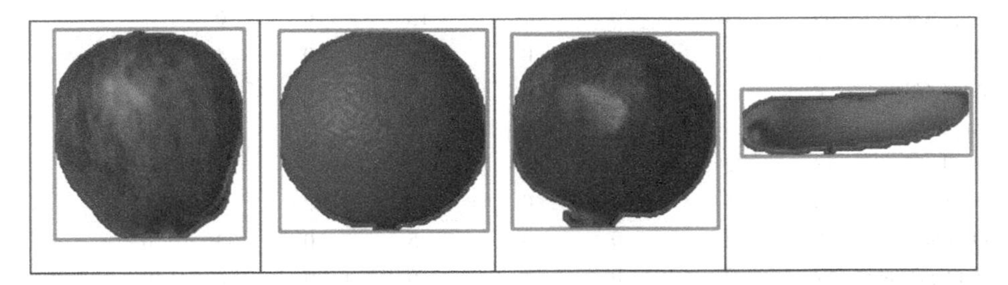

**Fig. 5.** Contour detection and bounding rectangle

*Result of Traditional Machine Learning Algorithm*

*Part I*

In this work, images of fruit-360 dataset have been taken. Initially only four categories of different fruits considered are apple, banana, orange and pomegranate. 492 sample images of red apple, 490 banana, 479 orange and 246 pomegranate images were taken for training. Extracted features of these images were applied over KNN and SVM algorithms. KNN is a lazy learner algorithm which can be used when there is no prior knowledge about distribution of data. It is mainly based on feature similarity. SVM is most probably used classifier in high dimension space. 70% of dataset is used as training data and remaining 30% is considered for testing. Accuracy of the classifier model is 93.85% using KNN and 100% using SVM.

*Part II*
Later classifier was build to recognize different varieties of same fruit. Additionally, five different categories of red apples were considered. It consists of 494 Apple1, 492 Apple2, 919 Apple3, 490 Apple4, 1411 Apple5, 490 Banana, 479 Orange and 246 Pomegranate images. Total number of images taken for classification process is 5,021. Classifiers were modeled using KNN and SVM. But the model fails miserably to classify apple fruit of different categorization. Overall accuracy of the classifier using KNN was 48.63% and using SVM was 60.65%.

Accuracy of the algorithm has been calculated using the values obtained from confusion matrix. True Positive (TP), True Negative (TN), False Positive (FP) and False Negative (FN) values are used to calculate the performance measures. TP represents the correctly identified prediction. TN represents correctly rejected prediction. FP represents incorrectly identified prediction. FN represents incorrectly rejected predictions. Result of applying KNN and SVM algorithm on the later data set is analyzed using confusion matrix given in Tables 1 and 2. Confusion Matrix contains the result of classification with validation dataset.

**Table 1.** Confusion matrix of KNN

| Actual class | Predicted class | | | | | | | |
|---|---|---|---|---|---|---|---|---|
| | AppleC1 | AppleC2 | AppleC3 | AppleC4 | AppleC5 | Banana | Orange | Pomegranate |
| AppleC1 | 109 | 11 | 20 | 1 | 15 | 0 | 0 | 1 |
| AppleC2 | 12 | 99 | 20 | 0 | 18 | 0 | 4 | 0 |
| AppleC3 | 8 | 19 | 60 | 74 | 112 | 0 | 1 | 0 |
| AppleC4 | 4 | 1 | 89 | 23 | 29 | 0 | 0 | 0 |
| AppleC5 | 18 | 32 | 169 | 37 | 117 | 11 | 14 | 0 |
| Banana | 0 | 6 | 0 | 0 | 14 | 122 | 0 | 0 |
| Orange | 0 | 3 | 0 | 0 | 5 | 0 | 153 | 0 |
| Pomegranate | 17 | 2 | 0 | 1 | 6 | 0 | 0 | 50 |

**Table 2.** Confusion matrix of SVM

| Actual class | Predicted class | | | | | | | |
|---|---|---|---|---|---|---|---|---|
| | AppleC1 | AppleC2 | AppleC3 | AppleC4 | AppleC5 | Banana | Orange | Pomegranate |
| AppleC1 | 28 | 0 | 0 | 0 | 129 | 0 | 0 | 0 |
| AppleC2 | 0 | 102 | 0 | 0 | 51 | 0 | 0 | 0 |
| AppleC3 | 2 | 0 | 119 | 34 | 119 | 0 | 0 | 0 |
| AppleC4 | 0 | 0 | 92 | 39 | 15 | 0 | 0 | 0 |
| AppleC5 | 2 | 3 | 116 | 30 | 247 | 0 | 0 | 0 |
| Banana | 0 | 0 | 0 | 0 | 0 | 142 | 0 | 0 |
| Orange | 0 | 0 | 0 | 0 | 0 | 0 | 161 | 0 |
| Pomegranate | 0 | 0 | 0 | 0 | 0 | 0 | 0 | 76 |

**Table 3.** Parameters for classifier performance evaluation

| Fruits | # of samples used | # of training samples | # of test samples | KNN | | | SVM | | |
|---|---|---|---|---|---|---|---|---|---|
| | | | | TP | FP | FN | TP | FP | FN |
| AppleC1 | 494 | 337 | 157 | 109 | 59 | 48 | 28 | 4 | 129 |
| AppleC2 | 492 | 339 | 153 | 99 | 74 | 54 | 102 | 3 | 51 |
| AppleC3 | 919 | 645 | 274 | 60 | 298 | 214 | 119 | 208 | 155 |
| AppleC4 | 490 | 344 | 146 | 23 | 113 | 123 | 39 | 64 | 107 |
| AppleC5 | 1411 | 1013 | 398 | 117 | 199 | 281 | 247 | 314 | 151 |
| Banana | 490 | 348 | 142 | 122 | 11 | 20 | 142 | 0 | 0 |
| Orange | 479 | 318 | 161 | 153 | 19 | 8 | 161 | 0 | 0 |
| Pomegranate | 246 | 170 | 76 | 50 | 1 | 26 | 76 | 0 | 0 |

Accuracy of the algorithm has been calculated using the values obtained from confusion matrix. Since the number of samples is not evenly distributed other performance measures like precision, recall and F1Score were used for comparison. TP, FP and FN values in Table 3 were used to calculate the performance measures. Precision, recall and F1 Score values of the classifiers are given in the Table 4.

$$\text{Accuracy} = \frac{\text{No of samples predicted correctly}}{\text{Total no of Samples}} \tag{4}$$

$$\text{Precision} = \frac{\text{TP}}{(\text{TP} + \text{FP})} \tag{5}$$

$$\text{Recall} = \frac{\text{TP}}{(\text{TP} + \text{FN})} \tag{6}$$

**Table 4.** Classifier comparison using different performance measures

| Fruit | KNN | | | SVM | | |
|---|---|---|---|---|---|---|
| | Precision | Recall | F1 Score | Precision | Recall | F1 Score |
| AppleC1 | 0.65 | 0.69 | 0.67 | 0.88 | 0.18 | 0.30 |
| AppleC2 | 0.57 | 0.65 | 0.61 | 0.97 | 0.67 | 0.79 |
| AppleC3 | 0.17 | 0.22 | 0.19 | 0.36 | 0.43 | 0.40 |
| AppleC4 | 0.17 | 0.16 | 0.16 | 0.38 | 0.27 | 0.31 |
| AppleC5 | 0.39 | 0.29 | 0.33 | 0.44 | 0.62 | 0.52 |
| Banana | 0.92 | 0.86 | 0.89 | 1 | 1 | 1 |
| Orange | 0.89 | 0.95 | 0.92 | 1 | 1 | 1 |
| Pomegranate | 0.98 | 0.66 | 0.79 | 1 | 1 | 1 |
| Avg | 0.50 | 0.49 | 0.49 | 0.66 | 0.61 | 0.60 |

$$\text{F1 Score} = 2 \times \frac{\text{Precision} \times \text{Recall}}{\text{Precision} + \text{Recall}} \tag{7}$$

The result shows that models accuracy and F1 Score is very low for both the classifiers. Even though SVM can able to classify banana, orange and pomegranate correctly it fails to predict different varieties of apple fruit. The reason for the more misclassification is due to the factor that there is only slight difference between RGB values of different apple fruits. The performance measure shows that the classifiers using KNN and SVM are good only in classifying fruits with distinguishing features. To improve the accuracy of the model each pixel of the fruit can be considered.

## 5 Deep Learning Approach

A Simple CNN was constructed with two convolution layer and a fully connected layer. Input images are split into training and validation dataset. The image is given as input to CNN as 3D matrix along with class label value. Class label for each image is assigned using one hot encoding technique. Image is given as inputs to convolution layer were it is represented as 3D matrix with a dimension for width, height and depth.

Hyper parameter values used in convolution layer is kernel or filter of size 5 × 5 and stride 2. Convolution is performed on the input data using kernel and produces feature map. Different kernels were applied to produce different feature map. ReLU is the activation function used in the each convolution layer to convert the data into non-linear representation. Fully connected layer is the output layer were the flatten data is given as input to fit the data into any given classes. Softmax function is used as activation function in the fully connected layer. After training for 5 Epochs the model accuracy

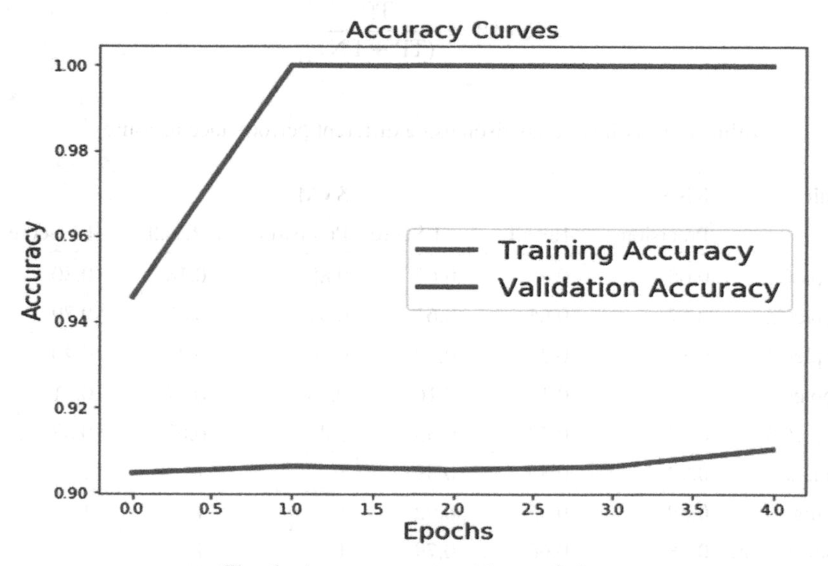


**Fig. 6.** Accuracy curve with over fitting

**Table 5.** Model summary

| Layer (type) | Output shape | Param # |
|---|---|---|
| conv2d_16 (Conv2D) | (None, 30, 30, 32) | 2432 |
| conv2d_17 (Conv2D) | (None, 13, 13, 16) | 12816 |
| flatten_8 (Flatten) | (None, 2704) | 0 |
| dense_15 (Dense) | (None, 32) | 86560 |
| dropout_8 (Dropout) | (None, 32) | 0 |
| dense_16 (Dense) | (None, 8) | 264 |
| Total params: 102,072 | | |
| Trainable params: 102,072 | | |

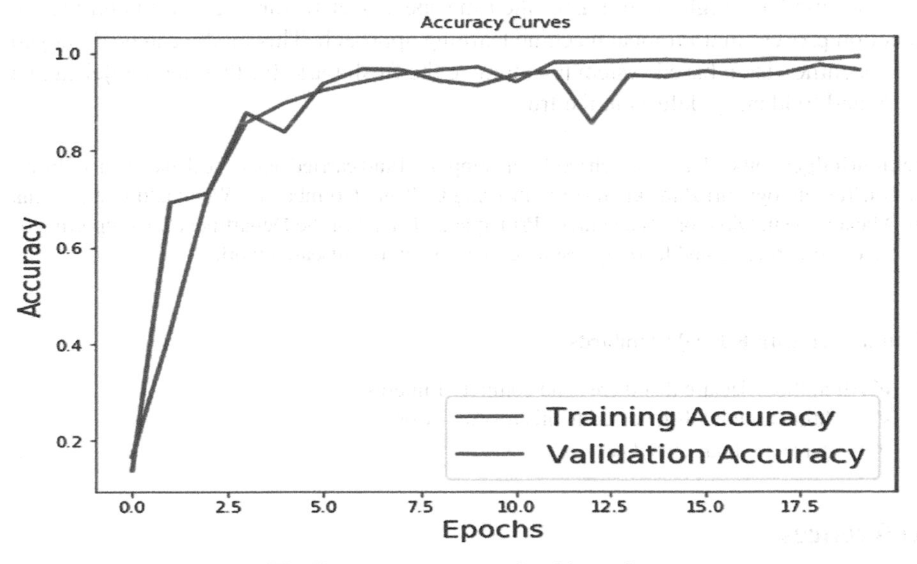


**Fig. 7.** Accuracy curve after 20 epochs

was 90.35% but the accuracy curve in Fig. 6 shows that there is data over fitting i.e. Model predicts the training data well when compared to validation data.

Model was redesigned with dropout layer, which helps to overcome data over fitting. It is used as a network regularization parameter which is added in the fully connected layer. The model summary and accuracy curve after 20 epochs is shown in the Table 5 and Fig. 7 respectively.

Accuracy of the CNN classifier with a validation set after 20 Epoch is 96.49%. Figure 7 shows the accuracy curve of the model with validation dataset has less data over fitting. Total time taken to train the network is 1783 s i.e. 30 min with Intel Core i3 processor and 4 GB RAM System. This time taken includes feature extraction, feature selection and also building of classifier.

## 6 Conclusion

This paper presented a fruit classification system using KNN, SVM and CNN for a machine vision based system. Experiment was carried out with five different categories of apple, banana, orange and pomegranate. Classifier model were constructed using traditional machine learning algorithms KNN and SVM. Deep Learning approach was also applied for classification. Experimental results of fruit classification over the validation dataset show that the CNN model outperforms the other two traditional methods. Comparatively, Deep Learning accuracy is quite high than other machine learning algorithms. Classification accuracy of KNN and SVM was 48.63 and 60.65 respectively. Classification accuracy of these algorithms depends on number of training samples, quality of the images and distinguishing feature selection. The accuracy of fruit Classification system using CNN was 96.49% after 20 Epochs of training. This is because each and every pixel of the image is considered for feature extraction. Even though time taken to build the classifier model is high, it eliminates the time spent over feature extraction and feature selection process in traditional machine learning approach. This model can be deployed in a machine vision based system to automate the fresh park. Further, the model can be re-trained to identify defects in the fruit.

**Acknowledgements.** This research work was supported and carried out at the department of Information Technology, Sri Ramakrishna Engineering College, Coimbatore. We would like to thank our Management, Director (Academics), Principal and Head of the Department for supporting us with the infrastructure and learning resource to carry out the research work.

**Compliance with Ethical Standards**

✓ All authors declare that there is no conflict of interest
✓ No humans/animals involved in this research work.
✓ We have used our own data.

## References

1. Kamilaris, A., Prenafeta-boldú, F.X.: Deep learning in agriculture: a survey. Comput. Electron. Agric. **147**, 70–90 (2018)
2. Dimatira, J.B.U., Dadios, E.P., Culibrina, F., et al.: Application of fuzzy logic in recognition of tomato fruit maturity in smart farming. In: IEEE Region 10 Annual International Conference, Proceedings/TENCON, pp. 2031–2035 (2017)
3. Zhang, Y., Wang, S., Ji, G., Phillips, P.: Fruit classification using computer vision and feedforward neural network. J. Food Eng. **143**, 167–177 (2014)
4. Zhang, Y., Wu, L.: Classification of fruits using computer vision and a multiclass support vector machine. Sensors **12**, 12489–12505 (2012)
5. Srinivasan, K., Porkumaran, K., Sainarayanan, G.: A new approach for human activity analysis through identification of body parts using skin colour segmentation. Int. J. Signal Imaging Syst. Eng. **3**(2), 93–104 (2010)
6. Srinivasan, K., Porkumaran, K., Sainarayanan, G.: Background subtraction techniques for human body segmentation in indoor video surveillance. J. Sci. Ind. Res. **73**, 342–345 (2014)

7. Srinivasan, K., Porkumaran, K., Sainarayanan, G.: Enhanced background subtraction techniques for monocular video applications. Int. J. Image Process. Appl. **1**, 87–93 (2010)
8. Jana, S., Parekh, R.: Shape-based fruit recognition and classification, pp. 184–196. Springer (2017)
9. Karis, M.S., Hidayat, W., Saad, M., et al.: Fruit sorting based on machine vision technique. J. Telecommun. Electron. Comput. Eng. **8**(4), 31–35 (2016)
10. Moallem, P., Serajoddin, A., Pourghassem, H.: Computer vision-based apple grading for golden delicious apples based on surface features. Inf. Process. Agric. **4**(1), 33–40 (2017)
11. Mahendran, R., Gc, J., Alagusundaram, K.: Application of computer vision technique on sorting and grading of fruits and vegetables. J. Food Process. Technol. **10**, 2157–7110 (2012)
12. Zeng, G.: Fruit and vegetables classification system using image saliency and convolutional neural network. In: IEEE Conference, pp. 613–617 (2017)
13. Hou, L., Wu, Q.: Fruit recognition based on convolution neural network. In: International Conference on Natural Computation, Fuzzy Systems and Knowledge Discovery, pp. 18–22 (2016)
14. Khaing, Z.M., Naung, Y., Htut, P.H.: Development of control system for fruit classification based on convolutional neural network. In: IEEE Conference, pp. 1805–1807 (2018)
15. Ma, L.: Deep learning ımplementation using convolutional neural network in mangosteen surface defect detection. In: IEEE International Conference on Control System, Computing and Engineering (ICCSCE 2017), Penang, Malaysia, pp. 24–26 (2017)
16. Tahir, M.W., Zaidi, N.A., Rao, A.A., Blank, R., Vellekoop, M.J., Lang, W.: A fungus spores dataset and a convolutional neural networks based approach for fungus detection. IEEE Trans. Nanobiosci. **17**, 281–290 (2018)

# Supervised and Unsupervised Learning Applied to Crowdfunding

Oscar Iván Torralba Quitian[1(✉)], Jenny Paola Lis-Gutiérrez[2,3], and Amelec Viloria[4]

[1] Universidad Central, Bogotá, Colombia
otorralbaq@ucentral.edu.co
[2] Universidad Nacional de Colombia, Bogotá, Colombia
jplis@unal.edu.co
[3] Fundación Universitaria Konrad Lorenz, Bogotá, Colombia
[4] Universidad de la Costa (CUC), Calle 58 # 55-66, Atlantico Baranquilla, Colombia
aviloria@cuc.edu.co

**Abstract.** This paper aims to establish the participation behavior of residents in the city of Bogotá between 25 and 44 years of age, to finance or seek funding for entrepreneurial projects through crowdfunding? In order to meet the proposed objective, the focus of this research is quantitative, non-experimental and transactional (2017). Through data collection and data analysis, we seek patterns of behavior of the target population. Two machine learning techniques will be used for the analysis: supervised learning (using the learning algorithm of the decision tree) and unsupervised learning (clustering). Among the main findings are that (i) most of the people who would participate as an entrepreneur and donor and entrepreneur simultaneously belong to stratum 3; (ii) Crowdfunding projects based on donations do not have a high interest on the part of Bogotans, but those in which they aspire to recover the investment.

**Keywords:** Supervised learning · Unsupervised learning · Crowdfunding · Bogotá · Decision tree · Clustering · Machine learning

## 1 Introduction

Recently the implementation of collaborative economy scenarios has taken effect, in which as financing options are the loans between peers (P2P lending) and the Crowdfunding [1–3]. Peer-to-peer loans allow people who need access to capital to apply directly to another individual online [4]. Crowdfunding, on the other hand, is the process by which capital is raised for an initiative, project or enterprise through numerous and relatively small financial contributions or investments made by individuals through the Internet [5, 6].

Unlike peer loans, crowdfunding has no obligation to return the money collected for the development of projects and initiatives. Its actions are related to the motivations, incentives and consumption patterns that people must make their contribution. For this reason, this alternative financing mechanism could serve people and organizations that


S. Smys et al. (Eds.): ICCVBIC 2019, AISC 1108, pp. 90–97, 2020.
https://doi.org/10.1007/978-3-030-37218-7_11

wish to obtain financing and do not comply with the requirements demanded by traditional sources of financing.

Also, it should be considered that people who make digital consumption have potential to participate in Crowdfunding, because they use the Internet to make transactions and can access the platforms for this type of funding, to know projects and decide whether they want to participate or not, either as donors or project proponents. In Colombia, this population is between 25 and 44 years old [7].

In this context we propose the following research question: what is the participation behavior of residents in the city of Bogotá between 25 and 44 years of age, to finance or seek funding for entrepreneurial projects through crowdfunding? In order to meet the proposed objective, the focus of this research is quantitative, non-experimental and transactional (2017). Through data collection and data analysis, it is intended to establish patterns of behavior of the target population [2]. Two machine learning techniques will be used for the analysis: supervised learning (using the learning algorithm of the decision tree) and unsupervised learning (clustering).

## 2 Literature Review

According to [8], the collaborative economy refers to the efficient use of goods and services. In this scheme, in order to make use of a good, the consumer does not have to make a purchase of it, but the ownership of the good is temporarily transferred to the consumer for the time he makes use of it. Thanks to the collaborative economy, extra costs are avoided, since the consumer does not make the complete purchase. In this scheme, the use of the good is made to supply a need with the following characteristics: (i) the sale of a good (either first or second hand) is not executed; (ii) between consumers (C2C); (iii) it is not considered as a lease or rent.

For their part, [9] defined collaborative economics as a new business model that coordinates exchange between individuals in a flexible way, separates from traditional markets, breaking down industry categories and maximizing the use of limited resources. The Collaborative Economy arose from a number of technological developments, which simplified the action of sharing tangible and intangible goods and services, through the availability of various information systems on the Internet [9], where users can participate by making their contributions or providing ideas and suggestions through Crowdfunding's web platforms. These platforms aim to increase the number of successful projects [10] by attracting investors and people who want to finance their projects. According to the work done by [11], the platforms are divided into two classes: the first based on investment and the second on donations and retributions. In the latter, investors offer their support in exchange for a product or offer their support for a cause or a combination of the two. [12] stated that collaborative economics refers to businesses seeking financing, platforms and consumers seeking to share resources and services, all within the framework of online applications based on business models. However, one of the forms of collaborative economics is Crowdfunding, which is reviewed below.

In the term collaborative economy, according to [3] are the following products and services: crowdfunding, loans between individuals (P2P), file sharing, free software, distributed computing, some social media, among others. One of the most employed is Crowdfunding, this term is synonymous with Micromatronage, collaborative financing, collective financing, massive financing or Crowdinvesting [13].

In 1989, Luís Von Fanta and Robe Iniesta, members of the Spanish rock band Extremoduro, raised funds to record their album Rock Transgresivo. Their initiative consisted of selling ballots at 1,000 pesetas, people who contributed received the ballot in exchange for their contribution. The total collection was 250,000 pesetas, with the value collected by this public group, their musical work in 1989, later edited and remastered in 1994 [7]. Later, in 1997, the British group Marillion carried out a process like Extremoduro. Announcing to their fans that they wished to make a tour in the United States, stating that in order to carry out the tour they would have to collect USD $60.000. His followers made the necessary monetary contributions to complete the proposed amount and thus managed to make the planned tour. It should be noted that the initiatives of Extremoduro and Marillion were pioneers, but at the time they were not known as Crowdfunding, because the term was proposed until 2006 by Michael Sullivan founder of Fundavlog, a website that contained a gallery of videos related to projects and events [7]. This site sought donations to support bloggers. Since then this practice has been known as Crowdfunding.

[14] identified that three main actors interact in Crowdfunding (i) organizations or people with ideas or entrepreneurial projects; (ii) intermediaries who are responsible for the promotion and publicity of entrepreneurial projects or ideas, in order to collect the resources needed by the speakers of the projects and initiatives to materialize them; (iii) taxpayers or financiers. Crowdfunding is a phenomenon that involves entrepreneurs seeking capital or funds through collective cooperation, which is usually done through web platforms [14]. Entrepreneurs can choose Crowdfunding as a means of financing their projects. In this process, in addition to the entrepreneurs, the following are involved: the Crowdfunding organization or platform that is in charge of promoting the entrepreneurial projects that they wish to materialize, and the people who access these platforms and make their contributions in order to finance the projects proposed by the entrepreneurs [15].

In brief, crowdfunding is a financing process where three main actors interact: (i) the end user who needs to obtain resources to finance his idea or project, (ii) the people who want to finance the initiatives of the end user and (iii) an organization that allows through the promotion and publication of projects to finance the initiatives of the end user. In this way, we identify that Crowdfunding allows to establish a connection between investors and entrepreneurs. Considering that the latter obtain resources from people who wish to contribute to their projects, through specialized platforms).

# 3 Method

## 3.1 The Data

Data collection was done through the development and dissemination of a survey and data collection was random. This instrument was constructed based on previous research on the specific motivations of Crowdfunding business models. The instrument was conceived to obtain relevant information for the research such as: residence in the city of Bogotá, socioeconomic stratum, age and intention to participate. Subsequently, the analysis and interpretation of the data was prepared in order to determine the motivations for investment and financing by means of crowdfunding.

The projection of the population of the city of Bogotá for the year 2017 was 8,080,734, which was composed of 3,912,910 men and 4,167,824 women [16]. 2,505,793 people were within the defined age range, equivalent to 31% of the city's population. Having defined the size of the city's population, which was in the established age range, the sample size was calculated with an error rate of 5% and a confidence level of 95%, in order to obtain an optimal sample quality. The required sample was 385 individuals, however, the sample reached was 394, implying an error of 4.94% and a confidence level of 95.28%.

### 3.2 Instrument

The instrument consists of 10 questions, 8 compulsory questions and 2 with Likert scales. The construction of the instrument was done through Googleforms, to facilitate the dissemination, collection and storage of data. By creating the survey using this application, real-time information on the number of completed surveys can be obtained and the data can be exported to office packages such as Microsoft Excel. This avoids the data entry process manually and allows the diffusion of the instrument through various media such as email, social networks and instant messaging applications such as Whatsapp, Facebook Messenger, Line, Snapchat among others. The instrument was sent to two researchers classified in Colciencias. These experts evaluated and issued suggestions, which were considered for the adjustment and initiation of dissemination of the instrument.

For the pilot test, the instrument was sent to a group of 20 people. The initial means of dissemination was by e-mail, from this group initially only 13 responses were received. Subsequently, the Cronbach Alpha coefficient was calculated to determine the reliability of the instrument before it was definitively sent. The initial method used to calculate the Cronbach alpha coefficient was through the variance of response values and was subsequently calculated using the variance of the total score. The value obtained for the group of data collected by the method of variance of the items in the pilot test was $\alpha = 0.995176$. According to [17] the minimum expected value of the coefficient should be 0.7. Because the value obtained was greater than the threshold, the instrument is reliable. Likewise, the calculation was made using the correlation matrix of the items and the result obtained was $\alpha = 0.96386$ confirming the reliability of the calculation by the first method, considering that between the results there is only a difference of 0.031316.

# 4 Results

The decision tree algorithm was used using the Orange software [18]. This algorithm makes it possible to identify a pattern in the responses obtained from the surveys. In other words, the algorithm analyzes the data for patterns and uses them to define the sequence and conditions for the creation of the classification model. Figure 1 presents the result. According to Fig. 1, a first form of grouping corresponds to the Bogotans who would use Crowdfunding if they did not comply with the conditions of the banks and from the investor's point of view, most of them would participate to recover the investment. Another form of classification is associated with those who do not meet the criteria to obtain resources from non-traditional institutions, but different from the financial sector. In this case, the interest of donors would be related to contributing directly to the project [19].

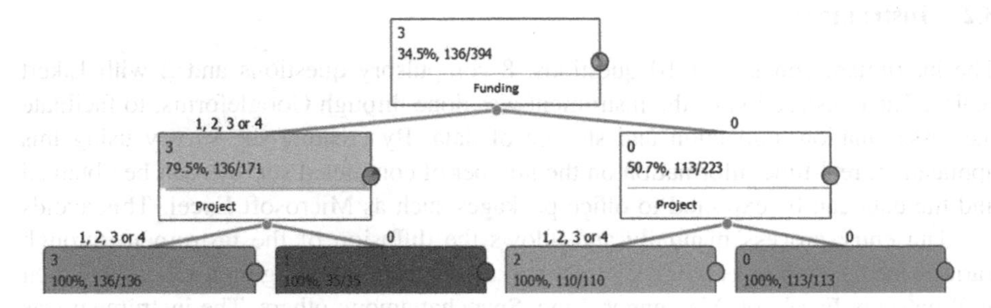


**Fig. 1.** Application of the decision tree algorithm to the database

Now, making the classification for each one of the individuals that answered the survey, we can appreciate that for a level of explanation of the grouping at 95% and considering as a variable to explain if one would be interested or not in participating in crowdfunding activities, 3 groups are identified. The blue group corresponds to the grouping of entrepreneurs, while the red and green groups mix those who would not only be donors or who would participate as donors and entrepreneurs simultaneously (Fig. 2).

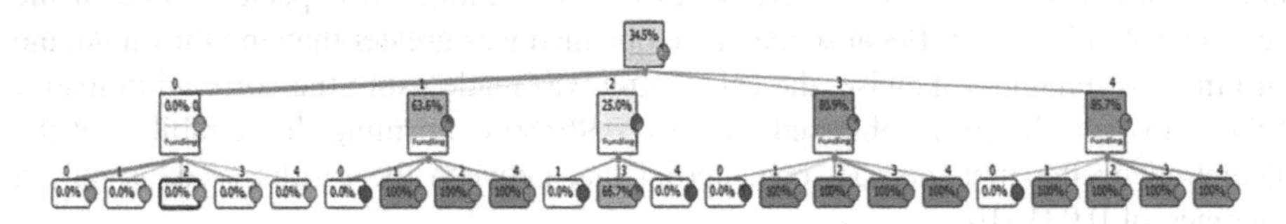

**Fig. 2.** Application of the decision tree algorithm to the database

Figure 3 shows that most of the people who would participate as entrepreneurs and donors and entrepreneurs simultaneously belong to stratum 3, although due to dispersion they were not significant (Fig. 4).

**Linkage:** complete
**Annotation:** 8) Con base en la definición anterior, usted estaría interesado en participar como:
**Prunning:** 10 levels
**Selection:** at 95.0 of height
**Cluster ID in output:** Cluster (as meta variable)

**Fig. 3.** Clustering visualization

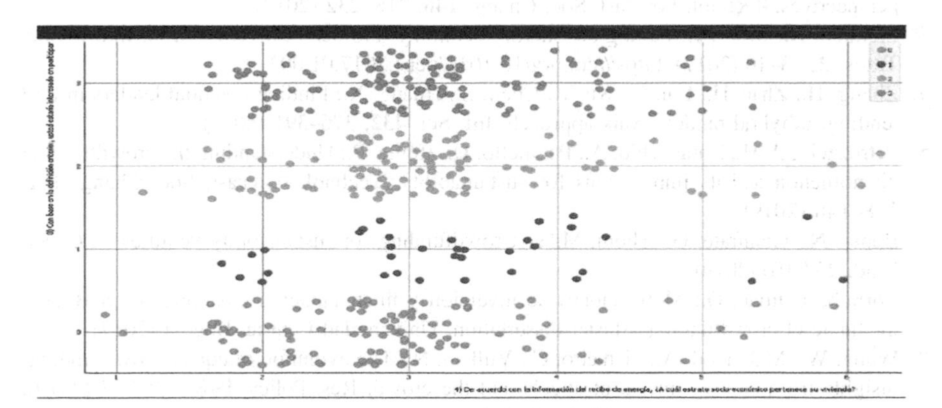

**Fig. 4.** Stratum vs participation in crowdfunding

## 5 Conclusions

From the application of supervised and unsupervised learning methods we found that Crowdfunding projects based on donations do not have a high interest on the part of Bogotans, that is to say that this type of projects would not have the necessary support of people living in the city of Bogota to materialize through Crowdfunding. The roles of investor and entrepreneur have greater acceptance by the population of Bogota, acquiring resources through an alternative means, offers the opportunity to access credit to people who cannot meet the requirements demanded by traditional and alternative funding entities.

For future work, these results could be comparable with other departmental capital cities or other countries. The instrument, considering that it has already been validated, could be used in other researches. Finally, the data collected could be used to estimate multivariate probit or logit models in order to make the findings more statistically robust. Similarly, other supervised learning algorithms could be used to predict participation in crowdfunding mechanisms in Colombia or other countries.

**Compliance with Ethical Standards.**

- All authors declare that there is no conflict of interest
- No humans/animals involved in this research work.
- We have used our own data.

## References

1. Yang, Y., Bi, G., Liu, L.: Profit allocation in investment-based crowdfunding with investors of dynamic entry times. Eur. J. Oper. Res. **280**(1), 323–337 (2019)
2. Bagheri, A., Chitsazan, H., Ebrahimi, A.: Crowdfunding motivations: a focus on donors' perspectives. Technol. Forecast. Soc. Chang. **146**, 218–232 (2019)
3. Frenken, K., Schor, J.: Putting the sharing economy into perspective. Environ. Innov. Soc. Trans. **23**, 3–10 (2017). https://doi.org/10.1016/j.eist.2017.01.003
4. Zhang, H., Zhao, H., Liu, Q., Xu, T., Chen, E., Huang, X.: Finding potential lenders in P2P lending: a hybrid random walk approach. Inf. Sci. **432**, 376–391 (2017)
5. Petruzzelli, A.M., Natalicchio, A., Panniello, U., Roma, P.: Understanding the crowdfunding phenomenon and its implications for sustainability. Technol. Forecast. Soc. Chang. **141**, 138–148 (2019)
6. Bento, N., Gianfrate, G., Thoni, M.H.: Crowdfunding for sustainability ventures. J. Clean. Prod. **237**(10) (2019)
7. Torralba Quitian, O.: Motivaciones de inversión y financiamiento colaborativo en Bogotá mediante el crowdfunding. Master dissertation, Universidad Central, Bogotá (2017)
8. Wang, W., Mahmood, A., Sismeiro, C., Vulkan, N.: The evolution of equity crowdfunding: insights from co-investments of angels and the crowd. Res. Policy **48**(8), 103727 (2019). https://doi.org/10.1016/j.respol.2019.01.003
9. Hamari, J., Sjöklint, M., Ukkonen, A.: The sharing economy: why people participate in collaborative consumption. J. Assoc. Inf. Sci. Technol. **67**(9), 2047–2059 (2016)
10. Gera, J., Kaur, H.: A novel framework to improve the performance of crowdfunding platforms. ICT Express **4**(2), 55–62 (2018)

11. Belleflamme, P., Lambert, T., Schwienbacher, A.: Crowdfunding: tapping the right crowd. Core **32**(2011), 1–37 (2011)
12. Son, S.: Financial innovation - crowdfunding: friend or foe? Procedia Soc. Behav. Sci. **195**, 353–362 (2015)
13. Ordanini, A., Miceli, L., Pizzetti, M., Parasuraman, A.: Crowd-funding: transforming customers into investors through innovative service platforms. J. Serv. Manage. **22**(4), 443–470 (2011)
14. Martínez, R., Palomino, P., Del Pozo, R.: Crowdfunding and social networks in the music industry: implications. Int. Bus. Econ. Res. J. **11**(13), 1471–1477 (2012)
15. Wang, X., Wang, L.: What makes charitable crowdfunding projects successful: a research based on data mining and social capital theory. In: International Conference on Parallel and Distributed Computing: Applications and Technologies, pp. 250–260. Springer, Singapore, August 2018
16. DANE: Proyecciones de Población [database]. DANE, Bogotá (2019)
17. Oviedo, C., Campo, A.: Aproximación al uso del coeficiente alfa de Cronbach. Revista Colombianan de Psiquiatría **34**(4), 572–580 (2005)
18. Demsar, J., Curk, T., Erjavec, A., Gorup, C., Hocevar, T., Milutinovic, M., Mozina, M., Polajnar, M., Toplak, M., Staric, A., Stajdohar, M., Umek, L., Zagar, L., Zbontar, J., Zitnik, M., Zupan, B.: Orange: data mining toolbox in Python. J. Mach. Learn. Res. **14**, 2349–2353 (2013)
19. Lis-Gutiérrez, J.P., Reyna-Niño, H.E., Gaitán-Angulo, M., Viloria, A., Abril, J.E.S.: Hierarchical ascending classification: an application to contraband apprehensions in Colombia (2015–2016). In: International Conference on Data Mining and Big Data, pp. 168–178. Springer, Cham, June 2018

# Contextual Multi-scale Region Convolutional 3D Network for Anomalous Activity Detection in Videos

M. Santhi[✉] and Leya Elizabeth Sunny

Department of Computer Science and Engineering,
Mar Athanasius College of Engineering, Kothamangalam, Kerala, India
santhimuniasamy@gmail.com, leyabejoy81@gmail.com

**Abstract.** The importance of the surveillance camera has been greatly increased as a result of increase in the crime rates. The monitoring of the videos recorded is once again challenging. Thus, there is an incredible interest for intelligent surveillance system for security purposes. The proposed work aims at a anomalous activity detection in video using the contextual multi-scale region convolutional 3D network (CMSRC3D) for. The work is deal with different time scale ranges of activity instances; the temporal feature pyramid is used to represent normal and abnormal activities of different temporal scales. In the temporal pyramid are 3 levels, that is short, medium and long for video processing. For each level of the temporal feature pyramid, an activity proposal detector and an activity classifier are learned to detect the normal and abnormal activities of specific temporal scales. Most importantly, the entire system model at all levels can be trained endwise. The activities detected are classified as the normal and the abnormal using the activity class. In the project CMSRC3D, each scale model uses convolutional feature maps of specific temporal resolution to represent activities within specific temporal scale ranges. More importantly, the whole detector can identify anomalities with in all sequential ranges in a single shot, which makes it very efficient for computation. And also provide if any anomalous activities detected in video, we give alert to avoid future mishappening.

**Keywords:** CMSRC3D · TFP APN · ACN

## 1 Introduction

Over a last few years it has been seen that fast development as well as amazing improvement in real-time video analysis. The aim of video analytics is to find the potential threaten activities with fewer or no human interference. Because, as today crime rate is increased. So the surveillance cameras are placed in malls, educational institute, railway stations, airports etc., for task is to detect the anomalous activities that can affect normal human being's life. And also another important task is to detect the activities in untrimmed video is a challenging problem, since instances of an activity can happen at arbitrary times in arbitrarily long videos. There are several techniques are established to find a solution for an issue and much progress have been made it. In any


S. Smys et al. (Eds.): ICCVBIC 2019, AISC 1108, pp. 98–108, 2020.
https://doi.org/10.1007/978-3-030-37218-7_12

case, how to identify temporal activity boundaries in untrimmed videos precisely is as yet an open inquiry. Motivated by the boundless achievement of R-CNN and Faster RCNN, which is used for object detection, most standing state-of-the-art activity detection methods deal with problem as detection is done by classification.

In most cases, a subset of proposals is generated via sliding window or proposal method and the classifier is used to classify these temporal segments as a specific activity category. Most approaches suffer from the following major drawbacks. (1) When identifying activity instances of numerous temporal scales, they depend on either handmade feature maps, or deep convolutional feature maps like VGG [2], ResNet [3] or C3D [4] with fixed resolution. When use fixed resolution to represent the activity in video frames. It provides an irregularity between the inherent variability in temporal length of activity ranges and the fixed temporal resolution of features, thus, leading lower performance. Because activity of instances in same category, it has a different duration. So if we use a fixed resolution is leading performance degradation of activity detection. So it can clear that it will increase inference time. (2) And also fixed resolution of the convolutional feature maps, contextual information, which has been demonstrated to be very effective in describing activities and boosting activity classification performance is not fully exploited.

In our work, we proposed a CMSC3D [1] for anomalous activity detection. This method is helps to detect the both normal and abnormal activities in videos. To deal with normal and abnormal activities of various temporal scales, we create multi-scale temporal feature pyramid to represent the different duration of video frames. One model is trained on each level of the temporal feature pyramid to detect normal and abnormal activity instances within specific duration ranges in video. Each model has its own convolutional feature maps of different temporal resolutions, which can handle activities of different temporal scales better than using a single scale approach. The key role of the project: an innovative CMSC3D architecture for anomalous activity as well as normal activity detection is proposed, where each scale model is trained to handle activities different scale ranges in video. More importantly the architecture uses a single pass network. Another contribution of this project Contextual information is fused to recognize activities, which can improve detection performance and also if we detecting anomalous activity in video then give alert to avoid mishappening in future.

The major three main contributions of this work: (1) A new multi-scale region convolutional 3D network architecture for anomaly activity detection is proposed, where each scale model is trained to handle activities of a specific temporal scale range. More importantly, the detector can detect activities in videos with a variety temporal scales in a single pass. (2) Contextual information is fused to identify activities, which can increase detection performance. (3) The proposed CMS-RC3D detector achieves state-of-the-art results on our own datasets.

The overview of this project is to detect the normal and abnormal activities using contextual multi-scale region convolutional 3D network detection. First of all, we extract the frames in the videos. After that we give video frames as the input and then perform feature extraction using optical flow algorithm. Depends upon the feature extraction, we created a three levels of pyramid. Then we detect the activity proposal for the pyramid. Finally, we classify activity using deep convolutional neural network, first we give a testing video to the network. The outputs the feature map and region of interests coordinates. Then the output fed in to the RoI pooling, after that it outputs the fixed size feature map for each map. Then we classify normal and abnormal activity is done by feature map of testing video is tested with trained model.

## 2 Related Work

In this area, an ongoing field of work in the anomalous activity detection and normal activity detection are discussed. There are variety of architecture developed for anomalous activity detection. The diverse researchers have utilized distinctive systems to identify anomalous activity in videos.

Buch et al. [5] have approach for time-based recognition of human activities in since a long time ago untrimmed video classifications. It presents Single-Stream Temporal Action Proposals (SST), another successful and effective deep architecture aimed at the creation of temporal action proposals. This system is being able to run constantly in the solitary stream in excess of very large input video sequence, it doesn't want to split input into short overlapping clips or temporal windows for cluster processing. The technical approach of this strategy is to produce temporal action proposals in long uncutted videos. The specialized methodology of this technique are: (i) the design examine the video input precisely one time at numerous timescales without overlapping sliding windows, while it provides an outcome in quick runtime during inference; (ii) it considers and assesses a generous amount of action proposals over densely tested timescales as well as locations, which outcomes in the system creating proposals with high temporal cover through the ground truth activity intervals. Input: The model taking input as long untrimmed video sequence of frames. Visual Encoding: The objective of a module is to calculate a feature representation that summarizes the pictorial content of the input video. Sequence Encoding: The aim of a module is to gather proof across time as the video sequence developments. The thought is that so as to probably deliver great recommendations, the model should most likely give aggregate data until it is certain that a move is making place in the video, at the same time disregarding immaterial foundation of the video. Output: The aim of this module is to deliver confidence scores of numerous proposals at each time step. This module takes as input the hidden representation is determined by the Sequence Encoding layer at period. The disadvantage of this paper is doesn't provide consistent image of activity detection.

Girshick et al. [6] approaches the technique for recognizing objects and localizing them in pictures. It is a standout amongst the most key and testing issues in computer vision. There has been important advancement on this issue in the course of the most recent decade due to great extent to the utilization of low-level image features, for example SIFT and HOG, in refined machine learning works. SIFT and HOG stand in semi-local orientation histograms, a demonstration it might relate generally through difficult cells, the first cortical region in the primate visual path. In any case, it additionally realizes that recognition happen numerous steps downstream, which recommends that there might be hierarchical, multi-stage processes for predicting features that are still more useful for visual recognition. In this paper, it describes an object detection and segmentation system that utilizes multi-layer convolutional networks to process exceedingly discriminative, yet invariant, features. It utilizes these features to categorize an image region, which would be able to output as detected bounding boxes or pixel-level segmentation masks. Dissimilar to classification of image, detection strategy requires limiting (likely various) objects within an image. Unique methodology of the frame detection as a regression issue. This design can be work fit for localizing a single object, but detecting many objects requires complex workarounds or an ad hoc assumption about the number of objects per image. The disadvantage of paper is the method only detect the action in the single image. That is takes image is an input.

Biswas and Babu [7] approaches the technique by using cues from the motion vectors in H.264/AVC compressed domain area. The effort is primarily inspired via an observation that motion vectors (MVs) show distinctive qualities amid anomaly. Seen that H.264 motion vector magnitude holds related information, which could be used to model the usual behavior (UB) efficiently. Like somewhat video compression standard, greater part of compression is accomplished over an eliminating temporal redundancy between neighboring frames, by using Motion Vectors (MVs). MVs contains the move of large scale obstructs among hopeful and reference outlines. Just the movement repaid mistakes alongside MVs are coded, in this way lessening the measure of bits to be coded. In unrefined sense, MVs is viewed as a rough guess of optical flow. Half pixel and quarter pixel motion expectation are in this manner used to enhance precision of square movement, resultant in almost optical stream similar attributes. MVs are frequently utilized for contras tent video examination errands together with video object division, activity acknowledgment, and so forth. In this work, the MVs for identifying irregular occasion.

Anomaly is characterized by the parting from normal features. At that point, abnormal is characterized to be the occasion whose likelihood of event is not exactly a specific limit. The issue is diminished to separating related features, it can distinguish the behaviors between the normal and the abnormal.

## 3 Proposed Work

The proposed work, we use multi-scale region for detecting abnormal activities in videos. The system is also detected normal activities, and it provides good image about the activities. All activities are detected in a single shot and it will also detect what activity has more happened in the video frames. If we change architecture by adding more layers, it will produce an efficient architecture. In this project every process is done in a single pass network. The proposed framework is consisting of five phase. That are, gathering of the inputs, extracting of features, forming a pyramid network utilizing the temporal feature, proposes the activity and classifies the activity.

### 3.1 Network Input

The input of the network is an original video with arbitrary length, which is only restricted by GPU memory. The video is consisting of set of frames. We use a loop to extract key frames from the video frames. It contains categories of activities of normal and abnormal. When we predict all the categories of activities in video.

### 3.2 Feature Extraction Network

For the extraction of feature, shared feature extractor is used which is extracts spatio-temporal features from input videos, it can be any kind of typical architectures. In our project, we used a deep CNN for feature extraction in the training of videos. And also to track an object in the video frame using optical flow algorithm.

Optical flow algorithm, which can be used to characterizes an exact motion of each pixel among two succeeding frames of videos [11]. Using this algorithm, can be used in acquiring the videos motion and the velocity. The optical flow is determined among the two consecutive frames S1 and S2, which is denoted by v = (u, v) where u and v are horizontal and vertical components of optical flow.

Convolutional neural network: the images fed as input are taken in one direction through the CNN to the output. The CNN [12] is comprised of three layers of convolutional, followed by pooling and fully connected layer. The first layer the convolution behaves as a feature extractor learning the complete features of the input. Every neurons in a feature map, constitute weights that are forced to be equal; however, there are cases where the different features map could have different weights even when present in the same convolution layer. So that a numerous of features could be extracted from each image.

### 3.3 Feature Pyramid Network

In unconstrained settings, activities in videos have varying temporal scales. To address this fact, many previous works (e.g. [10]) create different temporal scales of the input and perform several forward passes during inference. Although this strategy makes it possible to detect activities with different temporal scales, it inevitably increases the inference time. So that we create a temporal feature pyramid (TFP) with different temporal strides to detect normal and abnormal activities of different temporal scales. More specifically, a three-level pyramid is created, which is designed to detect short, medium and long duration of normal and abnormal activities simultaneously. Use of this feature pyramid, the system can efficiently detect normal and abnormal activities of all temporal scale range in single pass.

There are three levels of feature pyramid is represented as K = 3. In the existing method [10], to create the temporal feature pyramid, down sampling is applied on the convolutional map to produce two extra level feature. The down sampling is the reduce the size of video frames. So it can be divide the video frame into two or more levels. For the simplicity of the implementation, we use max pooling and 3-D convolution with stride 2 are used for down-sampling. The strides of these levels are considered as 8, 16, 32 respectively. The anchor setting of 8 stride is {1:7} and stride 16 is {4:10}, they contain {8–56} and {64–160} temporal scale ranges. Similarly stride 32 is {6:16} anchor setting and it has {192–516} temporal scale ranges. As compared to [10], each scale model in our architecture only needs to predict activities whose temporal scales lie in a limited range, rather than a single wide range. We assign anchors of specific temporal scales to each level according to training statistics. With this temporal feature pyramid, our detector can efficiently detect activities of all temporal scales in a single forward pass. One of use of this method is, for example if we take running video of 10 s, it may be wrongly detected as walking, but we take 30 s or 50 s of that video, we clearly detected that is running. For this reason, we created temporal feature pyramid for activities representations. Using this temporal feature pyramid can save a substantial amount of computational cost compared to generating multi-scale video segments, which have to pass through the backbone network several times.

### 3.4 Activity Proposal Network (APN)

To find candidate activity segments for the final classification network. To detect activity proposals of different scales more efficient, the single temporal scale feature map is replaced by the multi-scale TFP. We learn an APN on each level of the TFP, a further 3D convolutional filters with kernel size $3 \times 3 \times 3$, which is operates in each level of the TFP to generate feature maps for predicting temporal proposals. In order to detect anomalous and normal activity proposal of different durations, the anchors of multiple scales are pre-defined, in which an anchor denotes a candidate time window associated with a scale and works as a reference segment for activity proposals at each temporal location. As compared to [9], each scale model in our architecture only needs to predict activities whose temporal scales lie in a limited range, rather than a single wide range. We assign anchors of specific temporal scales to each level according to training statistics. As mentioned previously, there are $K = 3$ levels of the feature pyramid. The strides of these levels are {8, 16, 32}, respectively. We use {7, 7, 11} anchors of different temporal scales for the corresponding levels of the temporal feature pyramid. And the temporal scale ranges of each level are 8–56, 64–160, 192−512 frames, respectively. Therefore, our proposal detector can deal with a very wide range of temporal scales in an input video in a single shot. For identify the activities in the video frames, we assign a positive or negative labels to the anchor setting. Already anchor settings are defined in feature pyramid section. We choose a positive label, if the anchor segment is overlap with intersection over union (IoU) higher than 0.7 or it has highest IoU overlap with some ground truth activity. If anchor setting has overlap lower than 0.3 with all ground truth activities. Then we give negative label to the segment.

### 3.5 Activity Classification Network (ACN)

Activity Classification Network (ACN): ACN is a segment based activity detector. Due to the arbitrary length of activity proposals, 3D Region of Interest (3D-RoI) pooling can utilized to extracted the fixed-size features for selected proposal segments. Then, pooled features are fed into fc6 and fc7 layers in the C3D network, trailed by two sibling networks, which output activity categories and refine activity limits by classification and regression. If we are finding an abnormal activity, then give alarm generation will have created to avoid future mishappening [8].

In the existing method [9], ACN is only done on single scale feature map. In our project work, we use temporal feature pyramid, assign activity proposals of different scale range to the levels of the TFP. So we perform 3D-RoI pooling in activity classification network. 3D-RoI pooling is a neural network layer, which is used for object detection. It can achieve a significant speedup of both training and testing. It also maintains a high detection accuracy. The 3D-RoI has region proposal and final classification. In the region proposal, if we give the input image frame, it finds all possible places where objects can be located. The output of this stage should be list of bounding boxes is called region of interest. In the final classification; for every region proposal from the previous stage, decides whether its belongs to target classes or background. Here we use deep convolution network for final classification. This layer takes two inputs: (1) fixed size feature map is obtained from deep convolutional network with several convolutions and maximum pooling layers. (2) Matrix of representing of list of region of interest. The result of 3D-RoI is list of rectangles with different sizes. We can quickly get a list of corresponding feature maps with fixed size. Then this features are fed in to C3D network. Then it outputs the predicted activities and then it is classified as normal and abnormal classes.

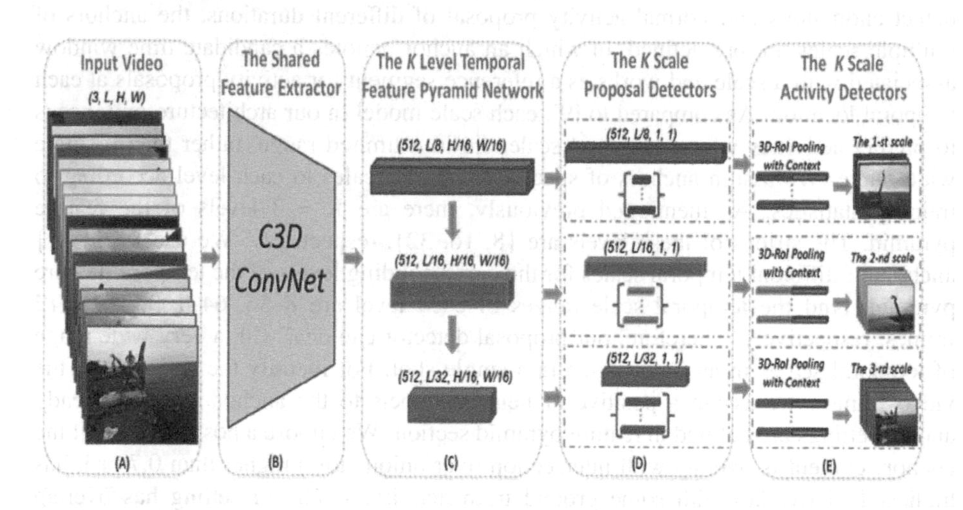


**Fig. 1.** System architecture of CMSRC3D model [4]

In the above Fig. 1(A) Input videos can contain set of activities of different temporal scale ranges, we use loop for extracting the frames in video, then it is given as input to network. In the above figure, input video represented as (3, L, H, W), here L- number of video frames, H- height of video frame and W- width of video frame. (B) The shared feature extractor is used to extracts the spatio-temporal features from input videos, which can be any kind of typical architectures; in this project we use deep convolutional neural architecture. For the feature extraction in video frames, we use optical flow algorithm to find apparent motion of objects in the video frames. Then we generate feature map using optical feature. The feature map is the mapping of features in the different location of the frame. (C) To deal with normal and abnormal activities of different temporal scales, the temporal feature pyramid is created, on which each level of a different temporal resolutions is used to represent activities of short, medium and long temporal scales simultaneously. The strides of feature pyramid are 8, 16 and 32, then anchors also assigned to the pyramid. (D) On each level of the temporal feature pyramid, an activity proposal detector is learned to detect candidate activity segments of frames within specific temporal scale ranges. In this we assign a positive and negative labels to the anchor setting for the detecting activity proposal. (E) Finally activity classification network is uses 3D-RoI pooling, which can extract contextual and non-contextual features for every selected activity proposal from each level of the temporal feature pyramid. At that point, the contextual and non-contextual features are concatenated and fed into the kth specific activity classifier, which outputs activity categories and refines activity segment boundaries.

# 4 Implementation

## 4.1 Data Sets

We create a data set by downloading videos from YouTube and we take some videos from ActivityNet dataset to detect the normal activities. Then we create a data set for abnormal activities, the videos are download from YouTube and also UCSD. The dataset consists of different activity video of normal and abnormal. The normal activities like walking, jumping and handclapping etc., and abnormal activities like attacking person, theft etc. The dataset is divided and used for training, validation and testing phase separately.

## 4.2 Preprocessing

The preprocessing is an important step in process of video. Which is used to convert all videos in one format with use of normalization. In the project, first we perform a frames extraction from thc video and normalized the video frames. With use of the normalization is change all the video types into a common format. In this project we use loop to extracts the key frames from videos.

## 4.3 Training

When training the CMS-RC3D detector, each mini-batch is created from one video, selected uniformly at random from the training dataset. For each frame in a video, which is resized to 172 × 128 (width × height) pixels, and we randomly crop regions of 112 × 112 from each frame of video. To fit each batch into GPU memory, we generate a video buffer of 768 frames. The buffer is generated by sliding from the beginning of the video to the end and from the end of the video to the beginning to rise the amount of training data, which can be seen as a data augmentation strategy. We also employ data augmentation by horizontally flipping all frames in a batch with a probability of 0.5.

The training process is depending upon the architecture and the amount of data. If the amount of data is huge then the data to be extracted will be more then automatically the time will be increased. In this project we use ActivityNet and UCSD data sets for predicting the normal and abnormal activities in video. First we preprocess the video and extract video frames. In the feature extraction phase, we loading the video from the dataset. Then read frame by frame of a video. Convert video frame in to gray image for optical flow calculation. In this phase dilation and filing holes are used. So we create a new graphical image, which is filled with 0 (black) and append computed values to original image. Then we create back image and calculated the optical flow values for video frames. That is, first we create a black frame which is same as that of the current video frame. Then we find frame transpose and convert the image into gray to BGR with points and arrows inside the original frame. As the object moves the arrows will also move to the same direction. That points in which the changes occurred is optical flows of video. The optical flow calculation based on the two video frame, that is compare with current and previous image frame. Then apply some threshold value to find the object in the video frame. Find the region of motion in video frames. Then temporal feature pyramid is created based on the feature of video frames. For example, if we take 30 s of the video, we couldn't predict the activity. Sometimes is lead to wrong prediction. So TFP is used to represent different scale range of video and is useful for predicting activity. Three levels of pyramid are created for activity prediction of normal and abnormal activities. We assign anchors of specific temporal scales to each level according to training statistics. With this temporal feature pyramid, our detector can efficiently detect activities of all temporal scales in a single forward pass. In the implementation, for simplicity, max pooling and 3-D convolution with stride 2 are used for down-sampling.

In order to detect activity proposals of different durations, anchors of multiple temporal scales are pre-defined, in which an anchor denotes a candidate time window associated with a scale and serves as a reference segment for activity proposals at each temporal location. Once we got optical flow values and temporal feature pyramid is created, the optical flow values of the video are stored in the one list. The labels in the another list. The labels are 0, 1, 2 etc., with activity name labels are assigned in the activity proposal network.

Then extract the features from the video using optical flow algorithm, it can give a features as optical value of video. After that we create temporal feature pyramid for the extracted features in different levels. TFP is used to represent different scale range of video and is useful for predicting activity. In the activity proposal detector, we create a proposal for different feature pyramid. Once we have feature values, then perform the mapping between the data and label. The mapping is a manual process in which the data is stored in a list and the corresponding label will be stored in another list. We access then using the index values. Then we create a trained model.

## 4.4 Testing

In the testing phase, testing videos with trained model. We give an input video and the system will extract frame wise. Then the trained model will classify the frame into normal or abnormal type (or type of activity).

## 5 Conclusion

In this project, we proposed a contextual multi-scale region convolutional C3D network for anomaly activity detection in video. In our CMS-RC3D, each scale model uses convolutional feature maps of specific temporal resolution to represent activities within specific temporal scale ranges. More importantly whole detector can detect activities within all temporal ranges in a single shot, which makes it computationally efficient. It uses a convolutional neural networks to extract the features. It detects multi scale temporal activities are detected and its uses single pass network. This method also Detect both normal and abnormal activities, finally give an alert to avoid mishappening when abnormal activity is detected.

**Compliance with Ethical Standards.**

- All authors declare that there is no conflict of interest
- No humans/animals involved in this research work.
- We have used our own data.

## References


1. Bai, Y., Xu, H., Saenko, K., Ghanem, B.: Contextual multi-scale region convolutional 3D network for activity detection. arXiv:1801.09184v1 [cs.CV], 28 January 2018
2. Simonyan, K., Zisserman, A.: Very deep convolutional networks for large-scale image recognition. arXiv preprint arXiv:1409.1556 (2014)
3. He, K., Zhang, X., Ren, S., Sun, J.: Deep residual learning for image recognition. In: Proceedings of the IEEE Conference on Computer Vision and Pattern Recognition, pp. 770–778 (2016)
4. Tran, D., Bourdev, L., Fergus, R., Torresani, L., Paluri, M.: Learning spatiotemporal features with 3D convolutional networks In: Proceedings of the IEEE International Conference on Computer Vision, pp. 4489–4497 (2015)
5. Buch, S., Escorcia, V., Shen, C., Ghanem, B., Niebles, J.C.: SST: single-stream temporal action proposals. In: CVPR (2017)
6. Girshick, R., Donahue, J., Darrell, T., Malik, J.: Region based convolutional networks for accurate object detection and segmentation. IEEE Trans. Pattern Anal. Mach. Intell. **38**(1), 142–158 (2016)
7. Biswas, S., Babu, R.V.: Real time anomaly detection in H.264 compressed videos. In: Proceedings of the 4th National Conference on Computer Vision, Pattern Recognition, Image Processing and Graphics (NCVPRIPG), pp. 1–4, December 2013
8. Chaudharya, S., Khana, M.A., Bhatnagara, C.: Multiple anomalous activity detection in videos. In: 6th International Conference on Smart Computing and Communications, ICSCC 2017, Kurukshetra, India, 7–8 December 2017
9. Xu, H., Das, A., Saenko, K.: R-C3D: region convolutional 3D network for temporal activity detection. CoRR, abs/1703.07814 (2017)
10. Shou, Z., Wang, D., Chang, S.-F.: Temporal action localization in untrimmed videos via multi-stage CNNs. In: Proceedings of the IEEE Conference on Computer Vision and Pattern Recognition, pp. 1049–1058 (2016)

11. Parvathy, R., Thilakan, S., Joy, M., Sameera, K.M.: Anomaly detection using motion patterns computed from optical flow. In: 2013 Third International Conference on Advances in Computing and Communications (2013)
12. Kůrková, V., Manolopoulos, Y., Hammer, B., Iliadis, L., Maglogiannis, I. (eds.): Artificial Neural Networks and Machine Learning ICANN 2018. Springer, Cham (2018)

# Noise Removal in Breast Cancer Using Hybrid De-noising Filter for Mammogram Images

D. Devakumari[1(✉)] and V. Punithavathi[2]

[1] Department of Computer Science, Government Arts College, Coimbatore, Tamil Nadu, India
ramdevshri@gmail.com
[2] Government Arts College, Coimbatore, Tamil Nadu, India
punithavm@gmail.com

**Abstract.** Breast Cancer is a one of the major disease for women in today's world. The aim of this paper is to develop a robust and image pre-processing methods to realize mammogram images features with different dimensions. To categorize an image is to be illustrated or signified by particular features. In this paper, proposed a mammogram image pre-processing extraction process for deriving the image classification. The proposed method present two phases namely, (1) Image Re-sizing: The original MIAS mammogram image database images are resized into predefined sizes; (2) Image pre-processing. The Hybrid De-noising Filter is obtained by Median Filter and Applied Median Filter Integration. It is established out using Hybrid Denoising Filter (HDF) to discard the noise; Different image de-noising algorithms are discussed in literature review. The proposed Hybrid Denoising Filter algorithm performs an important function in image feature selection, segmentation, classification, and investigation. According to the experimental results the Hybrid Denoising Filter algorithm mainly focuses on discarding irrelevant noise and focuses on the execution time using MATLAB R2013a Tool. The proposed Hybrid Denoising Filter has less Root Mean Square Error (RMSE) variation and High Peak Signal Noise Ratio (PSNR) when compared with other de-noising algorithms of Gaussian, Wiener, Median and Applied Median Filters.

**Keywords:** Breast Cancer · Image pre-processing · Hybrid De-noising Filter · MIAS

## 1 Introduction

Image processing is a technique to carry out several operations on an image, in order to obtain an enhanced image or to mine some precious information from it. It is a type of signal processing in which input (contribution) is an image and result could be image or characteristics or attributes linked with that image. Digital Image Processing (DIP) is an ever budding area with a variety of applications. It forms core research area within engineering and computer science regulations too. Digital image processing deals with developing a digital system that perform operations on digital image.

A cancer is a type of disease having source in the abnormal growth of the cells. Breast cancer known as breast disease having staring point is breast tissues and it can


S. Smys et al. (Eds.): ICCVBIC 2019, AISC 1108, pp. 109–119, 2020.
https://doi.org/10.1007/978-3-030-37218-7_13

extend over a large area of the ducts or lobes of their breast. When limiting of the breast cells are not proper, they divide and appear in the form of lumps or tumor that is called breast tumor or cancer. Breast cancer is the one of the second significant common disease which leads to death.

Mammography is the screening program for exposing and identification of breast cancer. This process is utilized for the diagnosis of breast cancer. Mammogram is an X-ray exam which is utilized to expose and identify the advance breast cancer. A mammogram can expose the breast cancer when cancer is very small to notice it. The mammogram having good standard and it can expose 80–90% breast cancer and there is no screening tool which is 100% fruitful.

The most efficient method for identifying breast tumors is the mammography. The small contrast of the tiny tumors to the background, which is occasionally close to the noise, creates that small breast cancer lesions can hardly be distinguished in the mammography [1]. In this perception, an image preprocessing to decrease the noise stage (level) of the image preserving the mammography configurations is a significant item to recover the detection of mammographic attributes or features. De-noising techniques have been based on relevant linear filters as the Wiener filter to the image, but linear methods tends to blur the edge formation of the image. Based on nonlinear filters several De-noising techniques have been established to avoid this problem [2–4].

In this proposed work, a novel Hybrid De-noising Filter (HDF) algorithm for mammographic images is offered. The proposed filter is compared with various image de-noising filters like Wiener, Gaussian, Median, Applied Median filters of the image.

The objective of this work is to develop a novel image de-noising algorithm for mammogram images which are efficient to remove the irrelevant noises. De-noising process of MIAS image database (Mammogram images) is presented using Point Spread Function (PSF) which gives a good De-noising result with high Peak Signal Noise Ratio (PSNR) ratio.

## 2 Related Work

Ponraj et al. [5] has reviewed the previous approaches of preprocessing in mammographic images. The objective of preprocessing is to develop the excellence of the image and make it prepared for further processing by discarding the irrelevant noise and unnecessary portions (parts) in the background of the mammogram. There are different techniques of preprocessing (Adaptive Median Filter, Denoising Using filters, Mean Filter, Adaptive Mean Filter, Histogram Equalization and so on.,) a mammogram image advantages and disadvantages are discussed.

Angayarkanni and Kumar [6] has discussed breast cancer is a serious and become frequent disease that changes thousands of women every year. Premature detection is necessary and dangerous for valuable treatment and patient recovery. The authors presented a clear idea of Extracting features (attributes) from the mammogram image to discover cancer involved region is an important step in breast cancer detection and verification. Several segmentation algorithms were used to segment the mammogram image. Except with the assist of segmentation procedure they could not obtain the clear idea where will the cancer cell with elevated is small density mass.

Monica Jenefer B and Cyrilraj V has designed that "An Efficient Image Processing Methods for Mammogram Breast Cancer Detection method presented a mammogram enhancement can be attained by eliminating the noise and develop the quality of the image using speckle noise elimination and Expectation Maximization (EM) algorithm correspondingly. Unchallengeable standard for Magnetic Resonance Image (MRI) is the attendance of speckle noise is arbitrary and deterministic in an image. Speckle has negative collision on ultrasoundimaging, Radicaldiminishment interestingly resolve may be in charge of the poor flourishing determination of ultrasound as contrasted with MRI. If there should arise an event of medicinal written works, speckle noise is otherwise called texture [7]".

Angayarkanni, Kumar and Arunachalam [8] has discussed a clear idea of extracting attributes from the mammogram image to discover cancer affected region which is a crucial step in breast cancer detection and verification. Combinations of algorithms were used to discover the cancer cell area in mammogram image. The mammogram images are extracted directly from the original gray scale mammogram with steps of image processing algorithms. Mammography has established to be the majority efficient tool for perceiving breast cancer in its initial and most treatable phase, so it persists to be the key imaging modality for breast cancer screening and identification. Additionally, this examination permits the discovery of other pathologies and might suggest the nature such as normal, benign or malignant. The beginning of digital mammography is measured mainly the significant development in breast imaging. Thus the algorithm provides an accuracy of 99.66% and over in numerous cases of images and gives 100% accuracy in the case of excellence image.

Aditya Ravishankar et al., has discussed that "A Survey on Noise Reduction methods in Computed Tomography (CT) scan images presented the detailed imaging of tumor increase inside the lungs and are extensively used. Still assorted noise types like those declared exceeding can be encountered during a CT scan. Elimination of these noises is serious for medical diagnoses and is complete using filters. Filtering is a technique by which they can improve and de-noise the image [9]. They presented a summary of lung cancer and a review on various noises (Wiener Filter, Median Filter, Gaussian Filter, Mean Filter and Contraharmonic Filter) and their removal techniques, including the usage of various filters".

Anshul Pareek and Dr. Shaifali Madan Arora, has discussed about breast cancer is the second most common tumor in the world and more prevalent in the women population, not only a women disease its present in men even. Since the root cause of the disease remains unclear, early detection and diagnosis is the optimal solution to prevent tumor progression and permit a successful medical intervention, save lives and reduce cost. The authors to be processed to eliminate noise such as out of focus and salt and pepper noise, with a set of image processing techniques to enhance or restore the image for meaningful interpretation and diagnosis [10].

Ognjen Magud, Eva Tuba and Nebojsa Bacanin has designed that "Medical Ultrasound Image Speckle Noise Reduction by Adaptive Median Filter algorithm was tested on different ultrasound images and different evaluation metrics counting mean square error, peak signal to ratio, normalized cross correlation, average difference, structural content, maximum difference, normalized absolute error and image enhancement factor were used as measure of the quality of noise removal [11]".

Sulong et al. [12] has presented "Preprocessing digital breast mammograms using adaptive weighted frost filter of preprocessing technique is an essential element of any imaging modalities whose foremost aim is to perform such processes which can bring the image to that quality where it is suitable for further analysis and extraction of significant data. The authors talked about pre-processing which has great significance in mammographic image analysis due to poor quality of mammograms since they are captured at low dose of radiation while the high amount of radiation may expose the patient's health. They concluded that the Adaptive Weighted Frost filter is the best suitable choice for eliminating noise from mammographic images and performs better comparatively. The Frost filter attained better PSNR values, for example in mdb001, the PSNR values produced by modified Frost filter is 34.6 dB".

## 3 Proposed Work

The proposed methodology considers the order to test all the experiments were conducted using Mammographic Image Analysis Society (MIAS) - a benchmarked dataset [12]. The Hybrid De-nosing Filter algorithm successfully eliminates the noise from mammogram images, the resultant mammogram is enhanced and noise free images are obtained which can be used for further processing like segmentation and classification. The overall HDF flow diagram is described in Fig. 1.

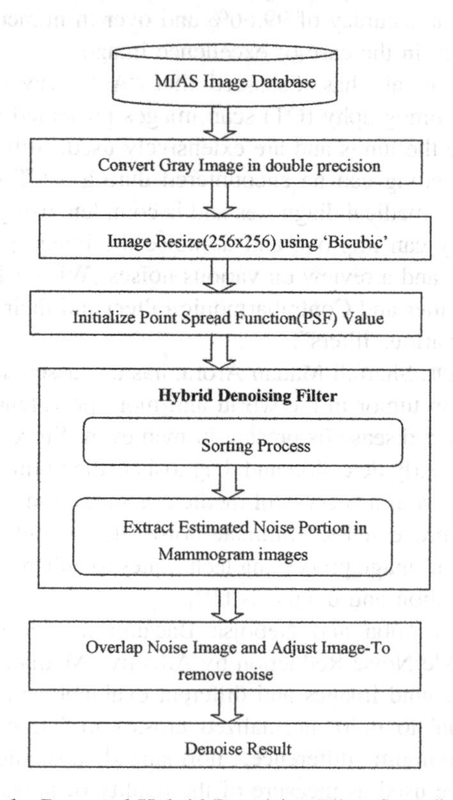


**Fig. 1.** Proposed Hybrid Denoising Filter flow diagram

In Fig. 1, MIAS Image datasets are taken as inputs in MATLAB simulation for Hybrid Denoising Filter algorithm. The first process is converting image from MIAS database into double precision and the image is resized into 256 × 256 using bicubic method. Bicubic method resizes the image without any data loss of the original image. Next step is initializing the Point Spread Function (Optimization method) or kernel value to 3 × 3 for denoising process. Third process is the proposed work called Hybrid Denoising Filter (HDF) method which is applied to extract and remove the noisy portions in an mammogram image. Finally image adjustment procedure is carried out to adjust the image intensity of final Denoise result (HDF result).

### 3.1 Image Preprocessing

Image preprocessing is a technique that involves converting raw image data into reasonable format. Preprocessing method is separated into two stages, specifically; (i) Image Re-sizing conversion (ii) Hybrid De-noising Filter. In this pre-processing phase is an improvement of the mammogram image portion contains unnecessary distortions or improves various image attributes significant for further processing. In this phase, original MIAS mammogram image database of pixels size (1024 × 1024 uint8) is converted into predefined 256 × 256 sizes exclusive of pixel information failure using 'bicubic' interpolation method. After that, MIAS database images should be of the equal dimension and are believed to be connected with indexed images on a common gray color map.

### 3.2 Hybrid De-noising Filter (HDF)

The proposed HDF algorithm successfully eliminates the noise from mammograms, the resultant mammogram is enhanced and noise free images which can be used for further processing like segmentation and classification. In this method an image is taken from MIAS database as input image and image dimensions are taken for processing. The HDF algorithm is used to remove the noise with the help of various attributes such as PSF Value, Sorting, Pixel Portion and Image Adjustment. The formula used is given below:

$$HFD = \sum_{i=1}^{row} \sum_{j=1}^{column} \left\{ \begin{array}{c} (img_{i-1}, img_{j-1}) \\ (img_{i}, img_{j-1}) \\ (img_{i-1}, img_{j}) \end{array} \right. \tag{1}$$

**Algorithm 1:** Hybrid De-noising Filter (HDF)
**Step 1:** Initialize Point Spread Function matrix 3 × 3.
**Step 2:** At each pixel, sort image pixels with average filter differences $A_k(i, j)$ between the image pairs in horizontal and vertical directions.
**Step 3:** Search the sorted image locations according to Eq. (1)
**Step 4:** Calculate the image adjust over the whole image.
**Step 5:** Denoise result image

The proposed algorithm compares various image de-noise filtering algorithms of median, Gaussian, Wiener, Applied Median filter. The comparison results are shown in Fig. 2. By integrating Median and Applied Median Filter, the proposed Hybrid De-noising Filter is found out and shown in Fig. 3.

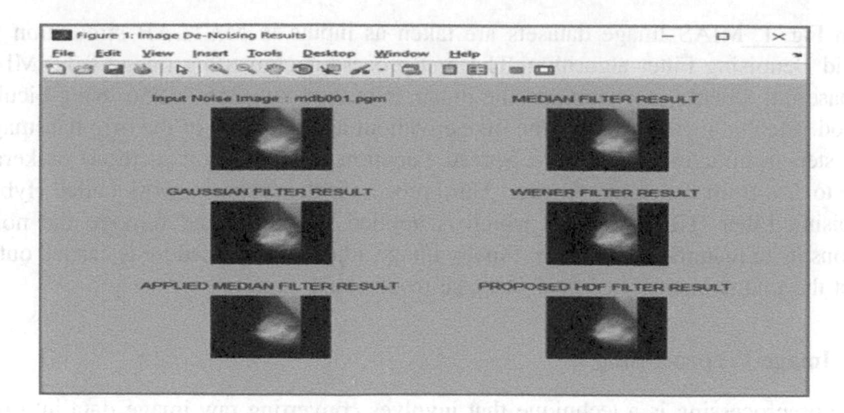


**Fig. 2.** Comparison of various image denoising filters [5]

By comparing original image with de-noised image, the following metrics are calculated. The performance parameters are most significant measure to justify the simulation results. Peak Signal to Noise Ratio (PSNR) and Root Mean Square Error (RMSE) are considered parameters for measuring the quality of images [6]. The Peak Signal to Noise Ratio (PSNR) is measured using Decibel (dB).

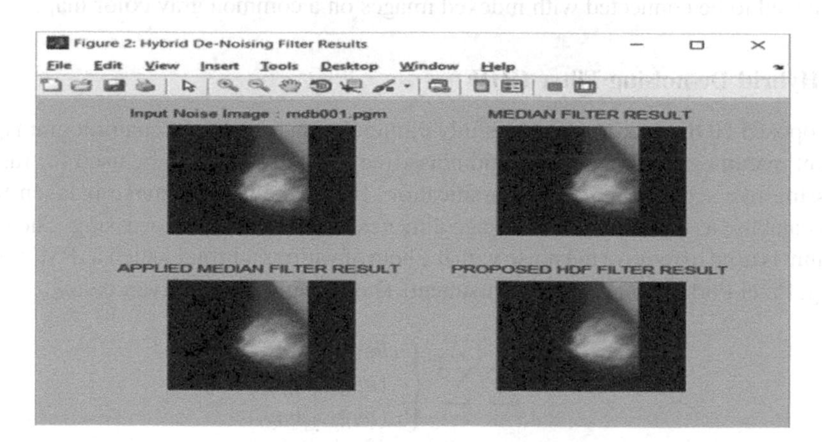


**Fig. 3.** Integration of Median, Applied and proposed Hybrid De-noising Filter

Root Mean Square Error (RMSE): RMSE is the root mean square error between original and denoised image with image dimensions (row × column). It is the cumulative square error between the encoded and the original 322 digitized images and the average RMSE is evaluated using,

$$\text{RMSE} = \text{sqrt}(\text{mean}(\text{input} - \text{denoiseImage})^2) \tag{2}$$

Table 1 shows the comparison of RMSE values with fifteen sample images from MIAS database.

**Table 1.** Comparison of RMSE values with fifteen sample images from MIAS database

| Images | Wiener | Gaussian | Applied Median | Median | Proposed HDF |
|---|---|---|---|---|---|
| mdb001.png | 6.0333 | 6.0338 | 4.2627 | 0.18952 | **0.18516** |
| mdb002.png | 7.0109 | 7.0106 | 4.9552 | 0.26002 | **0.22827** |
| mdb003.png | 7.1967 | 7.1991 | 5.0905 | 0.36252 | **0.33516** |
| mdb004.png | 7.6966 | 7.6995 | 5.441 | 0.1204 | **0.10291** |
| mdb005.png | 7.1716 | 7.1918 | 5.0831 | 0.23076 | **0.2155** |
| mdb013.png | 3.7058 | 7.1527 | 5.0599 | 0.33455 | **0.34706** |
| mdb018.png | 3.7491 | 6.4879 | 4.5876 | 0.25822 | **0.23479** |
| mdb031.png | 4.8945 | 7.3685 | 5.2107 | 0.3925 | **0.40241** |
| mdb033.png | 4.7416 | 6.7688 | 4.7844 | 0.2986 | **0.28849** |
| mdb035.png | 3.7941 | 6.2327 | 4.4043 | 0.26819 | **0.25116** |
| mdb037.png | 4.0218 | 6.0643 | 4.2854 | 0.25533 | **0.22824** |
| mdb044.png | 9.161 | 5.9009 | 4.1725 | 0.26537 | **0.21767** |
| mdb050.png | 5.0736 | 6.9322 | 4.9009 | 0.28034 | **0.26031** |
| mdb052.png | 6.1594 | 6.9221 | 4.8934 | 0.28246 | **0.26386** |
| mdb059.png | 4.8439 | 6.6159 | 4.6771 | 0.26271 | **0.24755** |

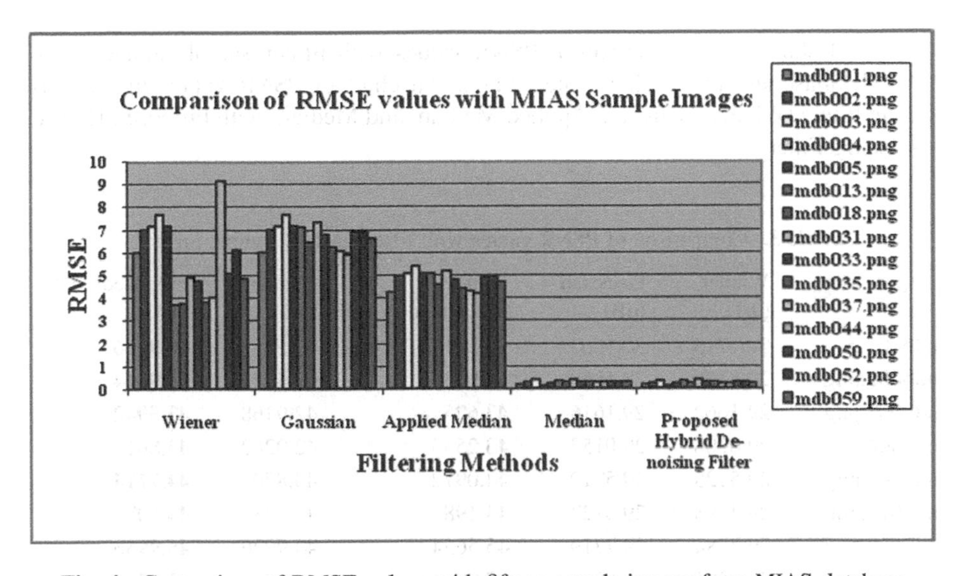


**Fig. 4.** Comparison of RMSE values with fifteen sample images from MIAS database

The comparison chart of RMSE values for various existing filters (Wiener, Gaussian, Applied Median, and Median) with Proposed Hybrid De-noising Filter is shown in Fig. 4.

Table 2 shows the evaluation of RMSE values for various filters with Proposed Hybrid De-noising Filter. The Comparison of Average RMSE Ratio for Overall 322 images from MIAS Database with various Existing and Proposed Filter is shown in Fig. 5.

**Table 2.** Average RMSE ratio for overall 322 images in MIAS database with various filters

| Measure | Wiener | Gaussian | Applied Median | Median | Proposed Hybrid De-noising Filter |
|---|---|---|---|---|---|
| RMSE | 7.2371 | 7.2396 | 5.1187 | 0.277 | **0.2569** |

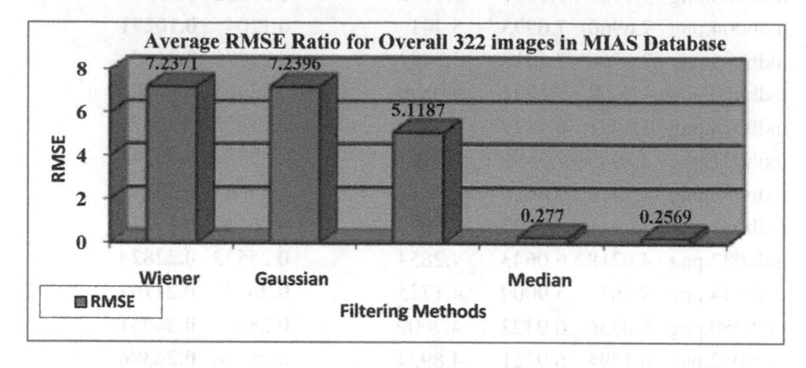


**Fig. 5.** Average RMSE ratio for overall 322 images from MIAS database with existing and proposed filter

Table 3 shows the comparison of PSNR values with fifteen sample images from MIAS database and Fig. 6 shows the comparison chart of PSNR values for various existing filters (Wiener, Gaussian, Applied Median, and Median) with Proposed Hybrid De-noising Filter.

**Table 3.** Comparison of PSNR values with fifteen MIAS sample images

| Images | Wiener (dB) | Gaussian (dB) | Applied Median (dB) | Median (dB) | Proposed HDF (dB) |
|---|---|---|---|---|---|
| mdb001.png | 30.1713 | 30.1663 | 30.1713 | 47.583 | **48.9996** |
| mdb002.png | 29.5596 | 29.5558 | 42.983 | 42.4871 | **42.9914** |
| mdb003.png | 29.1662 | 29.1616 | 43.625 | 42.9168 | **43.5942** |
| mdb004.png | 29.0208 | 29.0157 | 43.2533 | 42.9262 | **43.8612** |
| mdb005.png | 29.5125 | 29.5312 | 44.0972 | 43.8703 | **44.7114** |
| mdb013.png | 29.4193 | 29.4127 | 44.198 | 43.833 | **44.6078** |
| mdb018.png | 29.7784 | 29.7719 | 45.5654 | 44.9026 | **45.9955** |
| mdb031.png | 29.6591 | 29.6531 | 43.9061 | 43.386 | **43.9247** |
| mdb033.png | 29.9694 | 29.9632 | 45.683 | 44.9637 | **45.7363** |
| mdb035.png | 30.1075 | 30.1016 | 46.3653 | 45.4951 | **46.6076** |
| mdb037.png | 30.1055 | 30.1 | 46.2509 | 45.3857 | **46.4589** |
| mdb044.png | 30.0515 | 30.0448 | 45.4693 | 45.2444 | **47.2456** |
| mdb050.png | 29.7054 | 29.7003 | 45.5095 | 45.0399 | **46.1409** |
| mdb052.png | 29.7287 | 29.7239 | 45.2483 | 45.8145 | **45.8853** |
| mdb059.png | 29.8369 | 29.8325 | 45.7528 | 45.1627 | **46.155** |

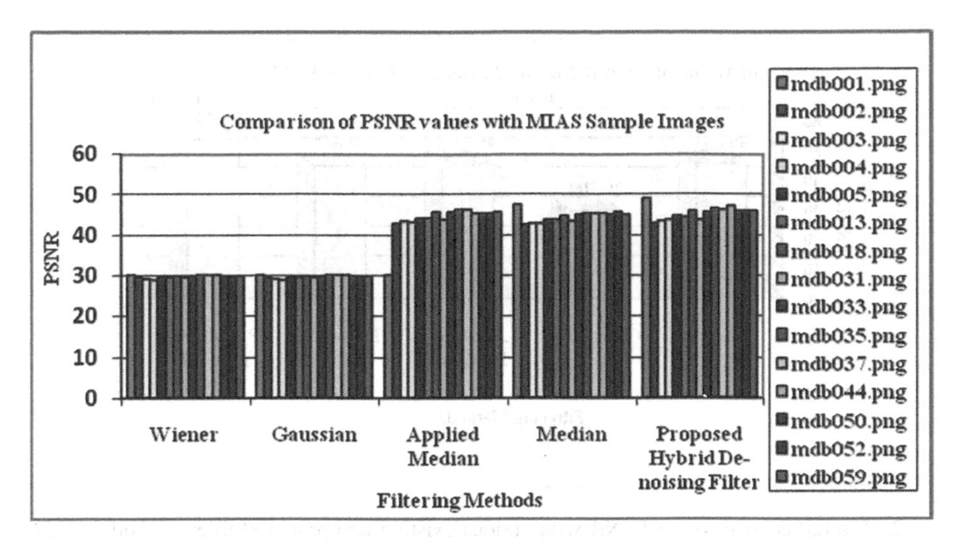


**Fig. 6.** Comparison of PSNR values with fifteen MIAS sample images

Table 4 shows the evaluation of Mammogram image de-noising measures of PSNR on MIAS image database. Overall Comparison of PSNR value for 322 images from MIAS Database is shown in Fig. 7. PSNR is measured using the formula,

$$PSNR = 20log_{10}\left(\frac{MAX_f}{\sqrt{MSE}}\right) \tag{3}$$

The overall comparison of PSNR value for 322 images in MIAS database are executed in order to find out the duration for the noise removal process on a Windows machine with an Intel core I5 processor with 3.20 GHz speed and 8 GB of RAM. The elapsed time is estimated for Applied Median, Median filter methods and proposed HDF as 293.820402, 115 016032 and 114.290015 s. Therefore experimental results proves that HDF is the best method for removing noise with less time when compared to the existing filter methods.

**Table 4.** Average PSNR ratio comparison of overall 322 images in MIAS database with various filtering methods

| Measure | Wiener (dB) | Gaussian (dB) | Applied Median (dB) | Median (dB) | Proposed HDF (dB) |
|---|---|---|---|---|---|
| PSNR | 45.633 | 29.702 | 45.237 | 44.637 | **53.846** |

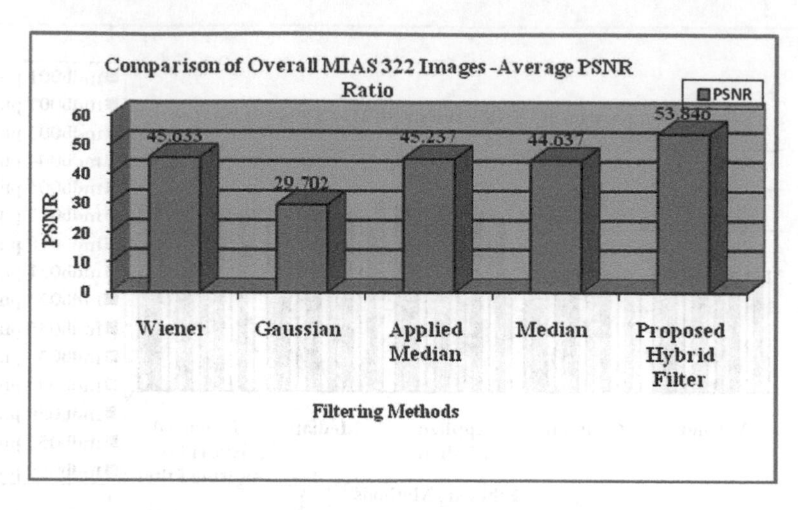


**Fig. 7.** Overall comparison of PSNR with various existing and proposed filter method for 322 images in MIAS database

# 4 Conclusion

In this paper analyzed the advancement of the information and communication technology in the field of mammogram image preprocessing. The proposed method works with two phases namely Image Re-sizing Reduction and Image De-noising. The original MIAS mammogram image database images are resized into predefined sizes and Image De-noising: is carried out using Hybrid De-noising Filter to remove the noise in mammogram images. The proposed Hybrid De-noising Filter algorithm is best eminent method for removal of noises in mammogram images when compared with various filters. The HDF is obtained by Median and Applied Median Filter integration. Comparison of PSNR and RMSE with different De-nosing filter such as Wiener, Gaussian, Median and Applied Median filter with proposed HDF method attains overall mean RMSE value for 322 images is found to be 0.2569 and overall mean PSNR value for 322 images is found to be 53.846 dB which is obtained by using the proposed Hybrid Denoising Filter. Hence it proves that Hybrid De-noising filter have less RMSE value and High PSNR ratio which is the best result when compared with the existing De-noising methods. It also proves that the experimental results of Hybrid De-noising Filter algorithm takes 114.290015 s for executing the overall 322 images. Therefore HDF possess less processing time and it is the best method for noise removal in mammogram images.

**Compliance with Ethical Standards.**

✓ All authors declare that there is no conflict of interest

✓ No humans/animals involved in this research work.

✓ We have used our own data.

## References

1. Dengler, J., Behrens, S., Desaga, J.F.: Segmentation of microcalcifications in mammograms. IEEE Trans. Med. Imaging **12**, 634–642 (1993)
2. Donoho, D., Johnstone, I., Kerkyacharian, G., Picard, D.: Wavelet shrinkage: asymptopia? J. R. Stat. Soc. B **57**, 301–369 (1995)
3. Catté, F., Lions, P., Morel, J., Coll, T.: Image selective smoothing and edge detection by nonlinear diffusion. SIAM Numer. Anal. **29**(1), 182–193 (1992)
4. Hyvärinen, A.: Sparse code shrinkage: denoising of nongaussian data by maximum likelihood estimation. Neural Comput. **11**, 1739–1768 (1999)
5. Ponraj, N., et al.: A survey on the preprocessing techniques of mammogram for the detection of breast cancer. J. Emerg. Trends Comput. Inf. Sci. **2**(12), 2079–8407 (2011)
6. Angayarkanni, N., Kumar, D.: Mammogram breast cancer detection using pre-processing steps of binarization and thinning technique. Pensee J. **76**(5), 32–39 (2014)
7. Monica Jenefer, B., Cyrilraj, V.: An efficient image processing methods for mammogram breast cancer detection. J. Theor. Appl. Inf. Technol. **69**(1), 32–39 (2014)
8. Angayarkanni, N., Kumar, D., Arunachalam, G.: The application of image processing techniques for detection and classification of cancerous tissue in digital mammograms. J. Pharm. Sci. Res. **8**(10), 1179–1183 (2016)
9. Ravishankar, A., Anusha, S., Akshatha, H.K., Raj, A., Jahnavi, S., Madhura, J.: A survey on noise reduction techniques in medical images. In: International Conference on Electronics, Communication and Aerospace Technology (ICECA) (2017)
10. Pareek, A., Arora, S.M.: Breast cancer detection techniques using medical image processing. Int. J. Multi. Educ. Res. **2**(3), 79–82 (2017)
11. Magud, O., Tuba, E., Bacanin, N.: Medical ultrasound image speckle noise reduction by adaptive median filter. WSEAS Trans. Biol. Biomed. **14**, 2224–2902 (2017)
12. Talha, M., Sulong, G.B., Jaffar, A.: Preprocessing digital breast mammograms using adaptive weighted frost filter. Biomed. Res. **27**(4), 1407–1412 (2018)

# Personality Trait with E-Graphologist

Pranoti S. Shete[1](✉) and Anita Thengade[2]

[1] MIT World Peace University, Pune, India
pranotishetelll@gmail.com
[2] MIT College of Engineering, Pune, India
anita.thaengade@mitcoe.edu.in

**Abstract.** Signature analysis helps in analyzing and understanding individual's personality. Graphology is the scientific technique that helps us predict the writer's personality. Different types of strokes and patterns in writer's signature are considered for predicting their personality trait. Social skills, achievements, work habits, temperament, etc. can be predicted by using the writer's signature. It helps us in understanding the person in a better way. As signature is directly related and develops a positive impact on your social life, personal life as well as for your career it is essential to practice correct signature for good results. The main objective here is predicting authors personality trait based on features such as Skewness, Pen Pressure, Aspect Ratio, Margin, and the difference between the first and last letter of the signature. As your signature has a direct impact on any of your assets and career, the proposed system will also provide suggestions for improvement in the signature if needed. This research paper proposes an off-line signature analysis. We have created our own dataset for the analysis purpose. We have also provided them with some questionnaire to check the accuracy of the proposed system.

**Keywords:** Signature analysis · Personality prediction · Graphology · Artificial neural network · Human personality · Personality · Signature · Margin · Aspect Ratio · Difference between the first and last letter of the signature · Pen-pressure

## 1 Introduction

E-Graphologist is emerging as a most demanded profession in this modern world of technologies. Graphology also guides people to derive a proper solution in order to overcome their problems. Manual graphology test is a time consuming process. It involves lot of human intervention and it remains as a more expensive method. Besides, the rate of accuracy of any analysis, it depends on the skills of the graphologist. This system uses image processing techniques to extract the features in your signature. These features tell you about your personality as well as it provides suggestions for improvement if needed.

The characteristics used to predict personality of a writer are Skewness, Pen Pressure, Aspect Ratio, Margin, and difference between first letter and the last letter of the signature. These features are selected based on 5 important traits that are used for predicting personality which is known or memorized as O-C-E-A-N traits.


S. Smys et al. (Eds.): ICCVBIC 2019, AISC 1108, pp. 120–130, 2020.
https://doi.org/10.1007/978-3-030-37218-7_14

Following table shows the personality assigned to each trait and the feature that we are detecting this personality from (Table 1).

**Table 1.** Personality traits and features.

| 5 important traits | Low | High |
|---|---|---|
| Openness | Prefer to be alone *(Skewness)* | Creative, love to be surrounded by people *(Skewness)* |
| Conscientious | Quick reaction and impatience *(Aspect Ratio)* | Self disciplined, Understanding *(Difference between first and last letter of the signature)* |
| Extroversion | Pessimistic *(Skewness)* | Optimistic *(Skewness)* |
| Agreeable | Capricious *(Difference between first and last letter of the signature)* | Reliable, Assertive *(Aspect Ratio)* |
| Neuroticism | Easy going, not over emotional, Forward looking person *(Pen-pressure and Margin)* | Over emotional, Tendency to cling to the past *(Pen-pressure and Margin)* |

- **Openness:** People with high score in openness are the people who love learning new things and getting new experience, learning new things and need to be surrounded by people. People having low score in Openness likes to be alone and not surrounded by the people all the time.
- **Conscientiousness:** Conscientious people are organized with strong sense of duty. They're disciplined, down to earth, understanding and goal oriented. These type of people roam around the world with not just bag-pack but with plans. People who are low in conscientious tends to be more careless, impatient and aggressive.
- **Extraversion:** The extrovert person is also called as social butterfly. Extraverts are sociable and chatty. They are assertive and cheerful. Introverts, on the other hand, need plenty of alone time. People with high extroversion can be classified as optimistic where as people with low extroversion which is also called as introversion can be classified as pessimistic.
- **Agreeableness:** It measures the warmth and kindness of a person. The more agreeable the person is, the more likely he is to be trusting, helpful and compassionate. Disagreeable people on other hand are suspicious of others, cold and not very cooperative.
- **Neuroticism:** It is one of the Big Five personality traits in the study of psychology. High scorers in neuroticism are kind of moody, they just think of the past mistakes and has feelings like worry, anger, frustration, fear, anxiety, etc.

## 2 Related Work

In [1] BPN and structured identification algorithm is used to detect ascending underline and dot structure with the accuracy of 100% and 95% respectively. In [2] personality is predicted using both handwriting and signature. Eight features with multi-structure algorithms and six features with ANN with the accuracy rate of 87–100% and 52–100% respectively were classified. In [3, 5–7] handwriting and signature features like baseline, pen pressure and slants were detected to predict writers personality [3, 5, 7] used ANN for feature extraction whereas [6] used SVM for the same and obtained the accuracy 93.3%. In [3] handwriting features like 't-bar' and loop of 'y' were extracted using template matching algorithm. Shell, dot structure, underline, streak and shape was identified in [5] using ANN and BPN. The main aim of [4] was to show a complete methodology to develop a system for predicting personality of a writer using as a fundamental part of the right people with the features such as slant, zones, pen pressure, connections, spacing, margin, letter size, speed, and clarity to increase the concern of developing more efforts in this area. In [8] two areas were considered for predicting personality in which the first area is to identify the styles n strokes of the capital letters such as 'O', 'T' and 'A' using Learning Vector Quantization(LVQ). The second area was Signature analysis in which features like curved start, end stroke, shell, middle streaks, underline were identified using ANN and extreme margin, streaks disconnected and dot structure were identified with multi structure algorithm. LVQ was also used for training data in second area (signature analysis). For 100 test data the system accuracy was only 43% because while training the data, most of the data was not identified.

In [9] the signature analysis was carried out using WACOM pen tablet (online signature analysis) to identify the features like position of pen in y-axis and x-axis, angle of the pen and pen pressure. Methodologies like Template matching and Hidden Markov Model (HMM) were used to train off-line signature data. In [10] analysis is done on the features such as T-bar, margin, slant and baseline where Polygonalization and Template matching are used for identifying the slant of the baseline and compared input with model signature to identify in which class will that signature belong to. In [11, 12, 15, 16] features like baseline, pen pressure and slant of handwriting was detected using ANN. In [14] loops of 'i' and 'f' were also considered. Whereas in [17] new feature like "unique stroke" of letter 'a', 'd', 'i', 'm', 't' were detected using structured analysis. In [13] the features like page margin, line spacing, slant, zones and direction are extracted. Lex and Yacc software was used for analysis along with context free grammar with nearly 50 rules for accurate prediction. In [14] the effort for developing handwritten Indian Script (Devnagarari) of numeral database are created which consist of 22,556 samples. A multistage method was implemented for recognizing there handwritten data separately as well as in mixed script more accurately. In [18] baseline and pen pressure is detected using orthogonal projection and gray level respectively. These two features separately as well as combined are tested on more than 550 text images.

## 3 System Architecture

- Image is taken from the database for pre-processing purpose. The input image is properly scanned with 300 dpi scanner.
- The signatures are then converted in gray scale to identify the pen-pressure of the writer.
- In noise removal process, single black pixel which is not a part of signature but is plotted on white background and single white pixel which is not a part of signature but is plotted on black background is removed for accurate predictions.
- The features extracted in this paper are Skewness, Pen Pressure, Aspect Ratio, Margin, and the difference between the first and last letter of the signature and the predictions are made based on these features.
- Support Vector Machine (SVM) classifier is used for feature extraction and the personality is predicted based on the strokes of individual's signature (Fig. 1).

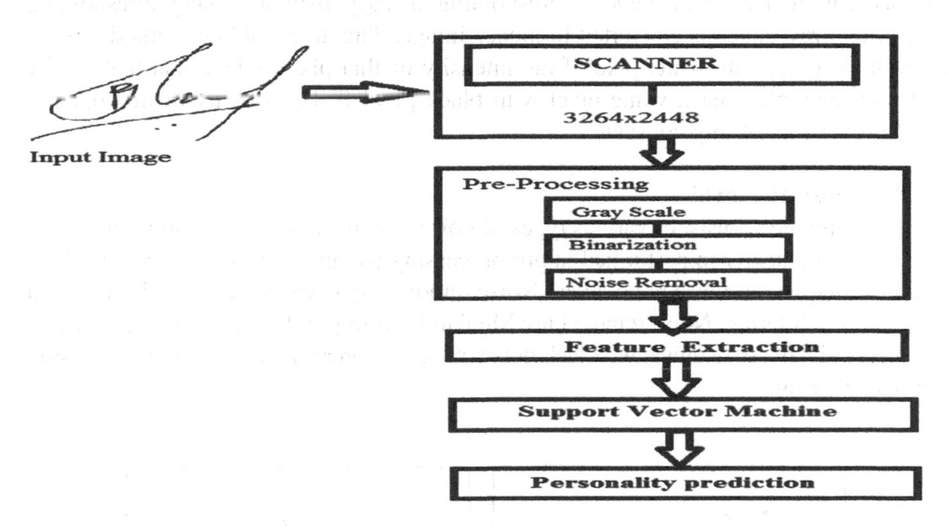


**Fig. 1.** General architecture of personality prediction system.

## 4 Paper Work

This research paper proposes an off-line signature analysis. We have created our own dataset for analysis purpose. We also provided them with some questionnaire to check the accuracy of our system. Image pre-processing is used to remove unwanted noise from the images so that the next step of feature extraction and personality prediction should have higher accuracy.

### 4.1 Scanning the Signature

All the signatures that we are processing need to be scanned properly. The English and Marathi signatures dataset was collected manually, and scanned the signature using 350 dpi resolution scanner. Now the scanned images are preprocessed.

### 4.2 Pre-processing

There are three steps in pre-processing,

#### 4.2.1 Gray Scale

There are three steps in pre-processing, Gray image consist of values from 0–255, where, 0 is black and 255 is white.

#### 4.2.2 Binary Image

To perform image segmentation Thresholding is easy method. Using thresholding technique grey image is converted to binary image. This thresholding method replace each black pixel with white pixel if the intensity of that pixel is less than that of any constant, or replace each white pixel with black pixel if the intensity of that pixel is greater than that of any constant.

#### 4.2.3 Noise Removal

As digital images consist of various types of noise which cause error resulting in wrong output. So the next step performed in pre-processing technique is noise removal. Three main techniques for noise removal are Noise removal by linear filtering, Noise removal by Adoptive filtering, Noise removal my Median Filtering. In Fig. 2 we can see the first image consist of some unwanted black dots which is been removed in this step by using median filtering.

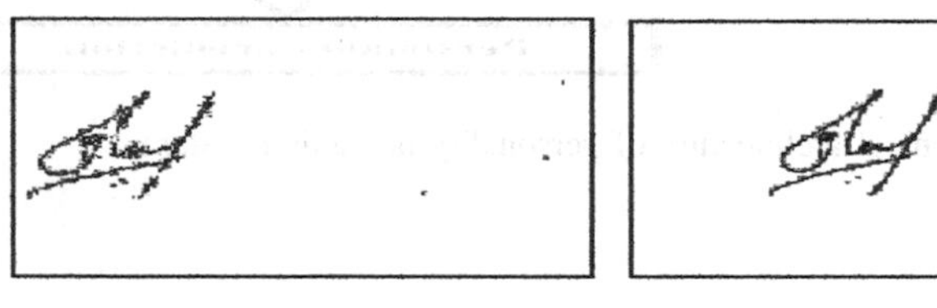


**Fig. 2.** Output of noise removal

## 5 Features

### 5.1 Skewness

Skewness is to find in which direction the writer has signed his signature in downward or upward direction. In this paper to find skewness, the regionprops function is used. It is obtained by computing normalized centered moments. The matrix of Inertia is obtained from the normalized moments of order. With the help of Eigen vectors and Eigen values, the direction of Eigen vector can be found which can then be converted to angle. If the angle is found to be greater than that of 0 than it is classified as upward oriented, and if the angle is found to be less than 0 then the signature is downward oriented.

### 5.2 Pen-Pressure

A pen-pressure is one of important feature in personality prediction. It helps in identifying a person emotional capabilities and decisiveness. In this paper to find pen-pressure we have set threshold to 0.8. If the pen-pressure is more than 0.8 then it is classified as heavy pen pressure or emotional person and if signature is less than 0.8 so the person is not very emotional.

### 5.3 Aspect Ratio

The aspect ratio is the relationship between the height and the width of the signature. Using aspect ration of the writers signature we can predict his personality if he is calm and easy going person or short tempered etc.

### 5.4 Difference Between First and Last Letter of the Signature

A difference between first and last letter indicates if his signature is narrow or broad and based on that the writer will be classified into respective classes.

### 5.5 Margin

Graphologist says the left margin spacc by thc writer should be at least 30% of the length of his signature, if not then the person has a tendency to cling his past which may result in bad out come in his current and future life. We have calculated margin as follows, firstly all the horizontal pixels are calculated and then to calculate length of the signature all the pixels from row 0 to $n - 1$ along with all the pixels of column 0 to $n - 1$ are calculated.

## 6 Features and Their Meaning

The five most important traits for predicting personality is O-C-E-A-N traits. These traits are accepted worldwide. By considering the above traits we are extracting the features such as,

### 6.1 Skewness

The upward oriented signatures indicate that the writer is optimist, active and ambitious, and the signature with downward orientation indicates the writers nature is Pessimistic.

### 6.2 Pen-Pressure

If the writer has heavy pen-pressure that means the writer is too emotional and the person with light pen-pressure indicates the person is less emotional.

### 6.3 Aspect Ratio

People with the long length signature are reliable and assertive, whereas the people with short length signature tends to be impatient and react quickly.

### 6.4 Difference Between First and Last Letter in the Signature

Difference between the first and last letters in your signature tells you about the writer's flexibility. That is if huge amount of difference is identified then the writer is capricious, and flexible enough to accept others work method and if the difference between the signatures is less than the writer is classified as self-disciplined and flexible.

### 6.5 Margin

The margin that we leave while writing while doing signature tells lot about our personality. Signature to the very left indicates the person has a Tendency to cling to the past whereas the signature more to the right indicates that writer is a Forward looking Person.

## 7 System Algorithm

1. if(orientation= upwards) then
2. write "Optimistic"
3. else (orientation= downwards) then
4. write "Pessimistic"
5. end if.
6. if(pressure=heavy) then
7. write "Too emotional"
8. else (pressure=light) then
9. write "Not too emotional"
10. end if.
11. if(length= long) then
12. write "Understanding, Reliable and Assertive"
13. else (length= short) then
14. write "Quick reaction and Impatience"
15. end if.
16. if(difference= huge difference between first and last letter) then

17. write “Capricious”
18. else (difference= less difference between first and last letter) then
19. write “Self-disciplined and flexible.”
20. end if.
21. if(margin= less) then
22. write “Tendency to cling to the past”
23. else (margin= more) then
24. write “Forward looking person”
25. end if.
26. Stop.

## 8 System Input/Output

**Input**

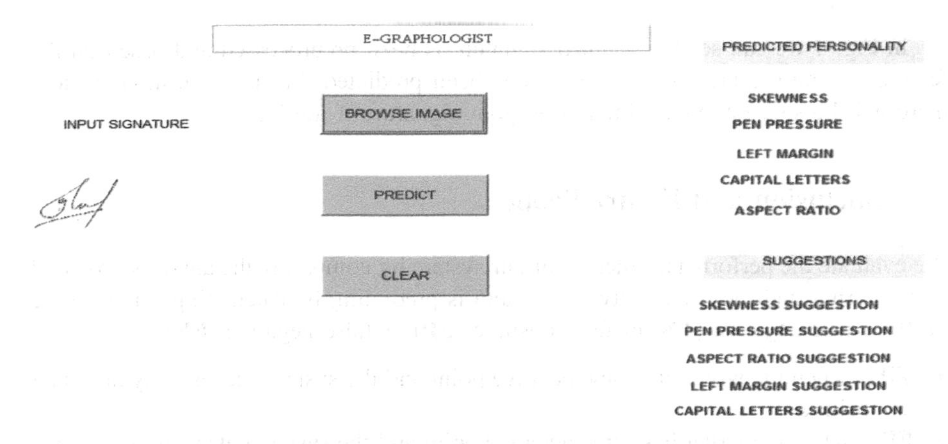


**Fig. 3.** System input image.

Figure 3 is the Graphical User Interface of the E-Graphologist System. Here we need to browse the image from the dataset and then clicking on predict will predict the writers personality. Now there are again two sections on the right hand side, i.e. Predicted personality and suggestions.

**Output**

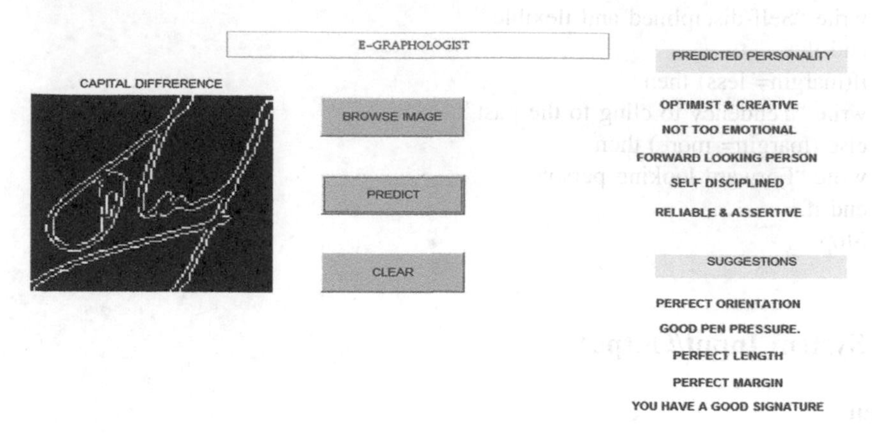


**Fig. 4.** Output image.

In Fig. 4 we can see the scanned signature is been preprocessed and based on the features in the signature the personality is been predicted. In the suggestion section, there will be suggestions for improving your signature if needed.

# 9 Conclusion and Future Scope

We evaluate the performance metrics of our system by comparing the answers provided by the writer and the personality our system is predicting by calculating true positive (**TP**), or true negative (**TN**) or false positive (**FP**) or false negative (**FN**).

- **TP** is when the writer has some positive point and the system successfully detects it as positive.
- **FP** is when the writer has some negative point and the system detects it as positive.
- **FN** is when the writer has some negative point and the system successfully detects it negative.
- **TN** is when the writer has some positive point and the system detects it as negative.

The metrics used to evaluate performance are the standard methods such as accuracy, precision, recall and F score (Fig. 5).

**Precision = TP/TP+FP**

**Recall = TP/TP+FN**

**Accuracy = TP+TN/TP+TN+FP+FN**

**F Score = 2*(Precision * Recall/Precision+ Recall)**

**Fig. 5.** Equations to evaluate performance metrics.

Future target is to develop tool for predicting personality with more number of features. The additional features such as Slant, underline, and character recognition can be taken into consideration.

**Compliance with Ethical Standards**

✓ All authors declare that there is no conflict of interest

✓ No humans/animals involved in this research work.

✓ We have used our own data.

## References


1. Lokhande, V.R., Gawali, B.W.: Analysis of signature for the prediction of personality traits. In: 2017 1st International Conference on Intelligent Systems and Information Management (ICISIM), 5–6 October 2017. IEEE (2017)
2. Djamal, E.C., Darmawati, R., Ramdlan, S.N.: Application image processing to predict personality based on structure of handwriting and signature. In: International Conference on Computer, Control, Informatics and Its Applications (2013)
3. Champa, H.N., AnandaKumar, K.R.: Automated human behavior prediction through handwriting analysis. In: First International Conference on Integrated Intelligent Computing (2010)
4. Varshney, A., Puri, S.: A survey on human personality identification on the basis of handwriting using ANN. In: International Conference on Inventive Systems and Control (ICISC-2017) (2017)
5. Sharma, V., Depti, Er.: Human behavior prediction through handwriting using BPN. Int. J. Adv. Res. Electron. Commun. Eng. (IJARECE) **6**(2), 57–62 (2017)
6. Prasad, S., Singh, V.K., Sapre, A.: Handwriting analysis based on segmentation method for prediction of human personality using support vector machine. Int. J. Comput. Appl. **8**(12), 25–29 (2010). (0975 – 8887)
7. Champa, H.N., AnandaKumar, K.R.: Artificial neural network for human behavior prediction through handwriting analysis. Int. J. Comput. Appl. **2**(2), 36–41 (2010). (0975 – 8887)
8. Djamal, E.C., Ramdlan, S.N., Saputra, J.: Recognition of handwriting based on signature and digit of character using multiple of artificial neural networks in personality identification. In: Information Systems International Conference (ISICO), 2–4 December 2013
9. Faundez-Zanuy, M.: Signature recognition state-of-the-art. IEEE Aerosp. Electron. Syst. Mag. **20**(7), 28–32 (2005)
10. Joshi, P., Agarwal, A., Dhavale, A., Suryavanshi, R., Kodolikar, S.: Handwriting analysis for detection of personality traits using machine learning approach. Int. J. Comput. Appl. **130**(15), 40–45 (2015). (0975 – 8887)
11. Kedar, S., Nair, V., Kulkarni, S.: Personality identification through handwriting analysis: a review. Int. J. Adv. Res. Comput. Sci. Softw. Eng. **5**(1), 548–556 (2015)
12. Grewal, P.K., Prashar, D.: Behavior prediction through handwriting analysis. Int. J. Comput. Sci. Technol. **3**(2), 520–523 (2012)
13. Sheikholeslami, G., Srihari, S.N., Govindaraju, V.: Computer aided graphology. Research Gate, 28 February 2013
14. Bhattacharya, U., Chaudhuri, B.B.: Handwritten numeral databases of Indian Scripts and multistage recognition of mixed numerals. IEEE Trans. Pattern Anal. Mach. Intell. **31**(3), 444–457 (2009)

15. Djamal, E.C., Febriyanti: Identification of speed and unique letter of handwriting using wavelet and neural networks. In: Proceeding of International Conference on Electrical Engineering, Computer Science and Informatics (EECSI 2015), Palembang, Indonesia, 19–20 August 2015
16. Bobade, A.M., Khalsa, N.N., Deshmukh, S.M.: Prediction of human character through automated script analysis. Int. J. Sci. Eng. Res. 5(10), 1157–1161 (2014)
17. https://www.futurepointindia.com/article/en/your-nature-is-revealed-in-your-signature-8792
18. Bala, A., Sahaa, R.: An improved method for handwritten document analysis using segmentation, baseline recognition and writing pressure detection. In: 6th International Conference on Advances in Computing & Communications, ICACC 2016, Cochin, India, 6–8 September 2016

# WOAMSA: Whale Optimization Algorithm for Multiple Sequence Alignment of Protein Sequence

Manish Kumar(✉), Ranjeet Kumar, and R. Nidhya

Department of CSE, MITS, Madanapalle, India
mk9309@gmail.com

**Abstract.** In the past few years, the Multiple Sequence Alignment (MSA) related problems has gained wide attractions of scientists and biologists as it is one of the major tools to find the structural and behavioral nature of Biomolecules. Furthermore, MSA can also be utilized for gene regulation networks, protein structure prediction, homology searches, genomic annotation or functional genomics. In this paper, we purpose a new nature and bio-inspired algorithm, known as the Whale Optimization Algorithm for MSA (WOAMSA). The algorithm works on the principle of bubble-net hunting nature of the whale with the help of objective function we tried to solve the MSA problems of protein sequences. In order to focus on the effectiveness of the presented approach, we used BALiBASE benchmarks dataset. At the last, we have compared the obtained result for WOAMSA with other standard methods mentioned in the literature. After comparison, it was concluded that the presented approach is better (in terms of obtained scores) when compared with other methods available in the considered datasets.

**Keywords:** Proteins · Bioinformatics · Whale Optimization Algorithm · Multiple Sequence Alignment

## 1 Introduction

Alignment of multiple sequence or Multiple Sequence Alignment (MSA) [1] is the alignment of three or more amino/nucleotide acid. Generally, MSA is a method used to determine homology among the given pair of sequences. The most studied branches in bioinformatics, is the sequence similarity. The currently available molecular data is capable enough to teach us about evolution, structure and functions of molecules or biomolecules in the related domains. The motive to study or research on MSA is to have better aligned sequences which are capable of finding biological relationship between related sequences. But, developing optimal MSA is not at all easy. Let us take an example: Here, we will consider large number of sequences (N) as input for MSA with a defined scoring meter for creating an optimal MSA. Although, a very simple requirement is mentioned here for developing a MSA. MSA may require the criteria for input selection along with the comparison model to evaluate the obtained alignment.


S. Smys et al. (Eds.): ICCVBIC 2019, AISC 1108, pp. 131–139, 2020.
https://doi.org/10.1007/978-3-030-37218-7_15

There are various points indicated in the literature [2–10] for alignment problems. First one is the structure of protein sequences, which are too complex. Second one, the newly developed protein or DNA sequences which are chosen by automatics methods contains considerable amount of bugs. But, there are various approaches suggested in the literature for MSA problem such as classical, iterative and progressive algorithms. All these algorithms are either local or global in nature. The global algorithm based on the principle of end to end alignment of sequences. Whereas, local approach first looks for a substring within the available strings and later tries to align it with the reference string [11].

While dealing with local alignment approaches, we often encounter a problem in identifying the region of similarity with the given sequences. Therefore, efforts should be made to resolve the issues related to local alignment. A dynamic programming (DP) [11] approach which generally uses both the technique (local & global) is the Smith–Waterman algorithm [12] and Needleman-Wunsch algorithm [13]. The dynamic based approach can only be used for aligning sequences which are lesser in number. Here, it should be noted that MSA is a combinatorial type of problem and as the numbers of sequences increases the computational efforts becomes probative. In order to know the relationship between sequences, authors (Feng and Dolittle) [14] constructed an evolutionary tree based on Needleman and Wunsch methods.

The progressive alignment algorithm method often faced the problem of getting trapped in local optima. To deal with such kind of problems, it is suggested in the literature [1–14] to use either iterative or stochastic methods [15]. Furthermore, by referring to several literature studies we came to a conclusion that none of the existing or presented algorithms are capable enough to give optimum output with all the standard datasets. As a result, with the help of certain bio inspired or evolutionary algorithms we have developed an iterative method using whale optimization algorithm which can work on some standard datasets available in the literature. However, all the listed methods in the literature [1–15] have shown their effectiveness in aligning the distantly related sequences for almost all the standard datasets. But, some accuracy was lost in aligning such sequences which are situated at a distance apart.

The above paragraphs, clearly indicates that none of the available algorithms or methods can give accurate and meaningful alignment under all possible condition. Above all, progressive alignment method is fast and creative, but if any errors occur in the very first alignment and it gets propagated to related alignments, then it is very hard to correct the errors in the alignment. But, such kind of problems or hurdles are not there for iterative kind of methods. Furthermore, we can say that iterative method is much slower than progressive method and can be used in place where alignment of sequences are most important as compared to computational cost [16].

Algorithm such as the genetic algorithm (GA) [8] which follows the natural selection process, are often applied and implemented for iterative method. This type of algorithms can fit to any type of fitness function and therefore most considered for iterative types of problems. Furthermore, algorithms like GA can help us to achieve different objective functions. Also, evolutionary algorithms like GA or other such kind of related algorithms can give multi-core processor and low cost cluster because they can be parallelized for meeting the current trend scenario.

In this study, we have considered bio inspired algorithms called Whale Optimization Algorithm (WOA) [17], for our experimental analysis. The proposed Whale Optimization Algorithm for Multiple Sequence Alignment (WOAMSA) follows the multi-objective principle in which the nature inspired methodology has also been considered. Here, we generate the initial population randomly and then with the help of objective function we calculated the fitness score for each column of the protein sequences available in standard dataset known as BALiBASE.

The advantages that we have with the bio-inspired algorithms or nature inspired algorithms (such as Biogeographically Based Optimization [18], Gravitational Search Algorithm [19], Ant Colony Optimization [20] etc.) is that they only requires a feasible objective function for evaluating the proposed algorithm or scheme. Since, the above mentioned algorithms are highly implicit parallel techniques; therefore they can be utilized for real time large scale problems for analyzing the biological sequences for alignment problems.

Further in this paper we will proceed as follows. The section that follows tells us about the motivation behind this research work which includes important preliminaries required to understand this research work. After that we will discuss about the proposed scheme that will follow the experimental set up and the comparative results with conclusion.

## 2 Motivation

Sequences which are smaller in length can be align manually but sequences of higher length requires certain algorithm (such as BBO, ACO, WOA etc.) for successful alignment. Dynamic programming which comes under the progressive technique suffers from the problem of early convergence and hence cannot be used for handling MSA related problems. For such kind of problems, we need iterative kind of algorithm such as the Genetic Algorithm.

While going through the literature reviews, we came to know about the importance of protein structure alignment in near future and therefore considered proteins for our experimental analysis. Until now, sequence similarity method is only considered for predicting the functions and evolutionary history of biomolecules. But, literature studies [1–20] suggested that there is a remarkable improvement in the tools for handling issues related to multiple sequence alignment. Furthermore, studies also stated that we need to combine sequence alignment with some of the known protein structure in order to have a close look over the alignment of protein sequences. We can also expect a better alignment, if we utilize the phylogenetic relationship among sequences.

By referring to number of literature studies [1–21], we came to know that there are still huge challenges for aligning protein sequence properly. The first and foremost thing is the less aligned locally conserved region within the sequence. This problem has been considered for aligning any protein sequences, irrespective of their lengths. Second factor is the mismatch of the motifs which may find a disordered region within the alignment. Third and most important problem is the dataset or the alignment error, which we encounter every time we access the protein datasets available around the globe.

Taking the literature survey into consideration, and in order to known the effectiveness of the presented (WOAMSA) approach, a comparative study has been made in

Table 1 with some of the other methods such as BBOGSA [22], GA based approach [23], QBBOMSA [24].

The GA based method presented in [23] tells about the role of genetic operators (crossover and mutation) for aligning the protein sequences in terms of scores. If score is good (approaching towards 1) then we can consider it as good or optimal. Whereas, the BBOGSA approach [22] employed the two new bioinspired algorithms to create optimal alignment for MSA of protein sequence. In this scheme, the author merged two nature inspired algorithm and compared their proposed scheme with some of the standard methods available in the same field.

In the proposed scheme, we have used BALiBASE datasets for protein sequences and with the help of Whale Optimization Algorithm we tried to align the unstructured protein sequences. We first initialize the initial population and then with the help of a fitness function we calculated the fitness score for each column of protein sequence. Based on this score and with the help of WOA we tried to find out the best optimal alignment for protein sequence and compared the same with some of the available methods for MSA.

# 3 Proposed Approach

In this section, we will describe our proposed approach for aligning multiple protein sequences of variable length using Whale Optimization Algorithm (WOA).

The inputs for the WOAMSA is the protein sequences taken from BALiBASE standard dataset. Our experimental analysis followed the Whale Optimization technique in which we first randomly initialized the population of the search agents. And then, the fitness score is calculated for every test case in the dataset by defined objective function. While we are dealing with WOA concept, we need to initialize the current best agent and needs to modify their positions so that they can proceed towards current best search agent. We will continue this process until the optimal solution is achieved. Now, we will demonstrate the steps for aligning protein sequences using Whale Optimization Algorithm. Here, we will discuss the steps involve in WOA and will frame our objective of alignment through WOA.

***Step 1:*** *In this step, Population is initialized randomly, and as said earlier we have to search for the best search randomly. The first or the inception population that we have considered for WOAMSA is:* $t_i$ $(i = 1, 2, \ldots, k)$ *and we have represented the best search agent as* $t*$.

***Step 2: Fitness calculation:*** *To calculate the fitness function we have considered the length of the sequences, where N represents the column and M represent row. The fitness of the chromosome is considered as the sum of the sub-scores obtained from each column. For each column we have consider the chromosomes in* $N * M$ *fashion and tries to calculate the score. For each match, mismatch and Gap we have assumed a score rating of 2, 1 &* $-1$*. Furthermore, we have considered PAM 250 scoring matrix to evaluate the overall score for each column of the protein sequence.*

***Step 3: Encircling prey:*** *Here, in this phase the position of the prey will be predicted by the humpback whale beforehand, and then the whale will try to catch the prey by* *encircling them. Then, the whale will assumes that current solution (prey) is the best it can have and then the search agents update their position based on the position of the current best agent. The objective followed in this study for encircling the whales are as follows:*

$$\vec{R} = \left| \vec{S} . \vec{t^*}(v) - \vec{t}(v) \right| \tag{1}$$

*Where, v represents the current run,* $\vec{t}$ *tells the position of the vector and* $\vec{t^*}$ *represents the position vector for the best solution. Vector coefficient is represented by* $\vec{S}$*. Equation* (1) *represents the currents best position of the agents and Eq.* (2) *will represent the new or updated position:*

$$\vec{t}(a+1) = \vec{t^*}(v) - \vec{M} . \vec{R} \tag{2}$$

*Where,* $\vec{M}$ *represents coefficient vector as mentioned in Eq.* (1)*. Now,* $\vec{S}$ *and* $\vec{M}$ *will be represented by Eqs.* (3) *and* (4)

$$\vec{M} = 2\vec{n} . \vec{r} - \vec{n}$$

$$\vec{S} = 2.\vec{r}$$

*Here, the value of* $\vec{n}$ *is made to fall from 2 to 0 and* $\vec{r}$ *gives the representation of the random vector in [0, 1]. The values of* $\vec{M}$ *and* $\vec{S}$ *are updated or modified in search of best search agent available in the surrounding places.*

***Step 4: Exploitation Phase:*** *This step consists of two important phases. The first one is shrinking encircling phase and the second one is spiral updating position.*

*In the first phase i.e. shrinking encircling phase,* $\vec{M}$ *is set to [−1 to 1] furthermore, current best agent position and initial position will be used to represent the new updated position for the agent.*

*For spiral phase, we can represent the equation for spiral updating as follows:*

$$\vec{t'}(v+1) = \vec{R'} . b^{cq} . cos(2\pi\ell) + \vec{t^*}(v) \tag{3}$$

*Where, c is the constant notation, q has the range between [−1, 1], (.) represents multiplication of one element with other.* $\vec{R'}$ *will be evaluated as follow:*

$$\vec{R'} = \left| \vec{t^*}(v) - \vec{t}(v) \right| \tag{4}$$

*Where, position vector is represented by* $\vec{t}$ *whereas the position vector of the best solution is represented by* $\vec{t^*}$.

*Now, in-between the encircling mechanism and the spiral position, the current position of the search agent will be updated.*

$$\vec{t}(v+1) = \begin{cases} \vec{t^*}(v) - \vec{M} . \vec{R} \; if \; p < 0.5 \\ \vec{R'} . b^{cq} . cos(2\pi\ell) + \vec{t^*}(v) \; if \; p > 0.5 \end{cases} \tag{5}$$

*Here, p is a number which can range between [−1, 1]*

***Step 5:*** ***Exploration Phase:*** *Here in this phase, the randomly chosen agents will be used to update the position of the search agent.*

$$\vec{R} = \left| \vec{S} \cdot \vec{t}_{\mathrm{rand}} - \vec{t} \right|$$

$$\vec{t}(a+1) = \vec{t}_{\mathrm{rand}} - \vec{M} \cdot \vec{R}$$

$\vec{t}_{rand}$ used to represent the random position vector.

***Step 6:*** *Here, we will updated the position of the agent until we find a best solution.*

## 4 Results

The aim of this report is to have a feasible study on the behavior of new nature inspired algorithm such as WOA for solving problems related to alignment of multiple protein sequences. Here, we can evaluate the quality of sequences by observing and comparing its obtained score with that of the standard score mentioned in the BALiBASE datasets. For this experimental analysis, we have considered Whale Optimization Algorithm.

For experimental evaluation and comparison of the developed method, we have compared our results with some other methods which has used similar nature inspired algorithm.

In this case, we have considered BALiBASE standard dataset and within that we have considered ref. 1, 2 & 3 as a reference set for evaluating and comparing our obtained result. It was observed that for few datasets our newly proposed method was not giving optimal result. But for most of the cases, the concept of whale optimization has obtained optimal result. We can say that in the current scenario, most of the nature inspired algorithms such as ACO, GSA, GA, PSO, BBO are very much useful in successful alignment of protein sequences or handling of structural related problems of protein sequences.

With the given table, one can identify the number of reference alignment for which our proposed method has shown better results as well as one can easily find out for which test cases our method needs improvement. Here, the score of 1 means the protein sequence is successfully aligned and 0 means our references alignment does not matches at all with the standard reference alignment available in the BALiBASE dataset. For most of the cases, we have successfully aligned the sequences and for those we are lacking we will look forward to align or to improve our score over alignment.

The bold faced scores indicate the superiority of the results when compared to different methods. Furthermore, unavailability of result in the literature is represented by n/a. In the Table 1, we will find there are many datasets for which we don't have any results to compare. From the Table 1, it can also be concluded that the results presented by us using the whale optimization technique is far better than those presented using similar algorithms or technique for the same application of protein sequence alignment.

Average Score is also calculated for each method against every datasets mentioned in Table 1. It can be seen that when we calculate the overall average score and compare it with the methods considered in this paper, the overall average score of the purposed method is less than all other methods considered in this paper. However, the individual score of WOAMSA is better as compared to other methods considered in the paper. The best score among all are presented/highlighted in bold.

**Table 1.** Obtained results on various references sets (1 to 5)

| Reference no. | SN | ASL | DATASETS | GA based method | QBBOMSA | GA-BBO | WOAMSA |
|---|---|---|---|---|---|---|---|
| Ref. 1 | 3 | 374 | 1ped | n/a | 0.758 | 0.793 | **0.879** |
| | 4 | 220 | 1uky | n/a | 0.574 | 0.878 | **0.954** |
| | 4 | 474 | 2myr | 0.621 | 0.427 | **0.746** | 0.547 |
| | 5 | 276 | kinase | **0.981** | 0.703 | 0.894 | 0.846 |
| Ref. 2 | 23 | 473 | 1lvl | 0.812 | n/a | 0.850 | **0.877** |
| | 18 | 511 | 1pamA | n/a | 0.824 | **0.856** | 0.741 |
| | 15 | 60 | 1ubi | n/a | n/a | 0.988 | **1.000** |
| | 20 | 106 | 1wit | n/a | n/a | **0.788** | 0.784 |
| | 16 | 294 | 2pia | n/a | 0.897 | 0.912 | **0.981** |
| | 15 | 237 | 3grs | 0.793 | n/a | **0.877** | 0.684 |
| Ref. 3 | 23 | 287 | kinase | 0.847 | 0.795 | 0.929 | **0.962** |
| | 19 | 427 | 4enl | 0.845 | n/a | 0.924 | **0.956** |
| | 28 | 396 | 1ajsA | 0.249 | n/a | **0.655** | 0.357 |
| | 22 | 97 | 1ubi | 0.576 | n/a | 0.768 | **0.944** |
| | 19 | 511 | 1pamA | 0.894 | 0.856 | 0.820 | **0.922** |
| | 24 | 220 | 1uky | 0.452 | n/a | **0.944** | 0.571 |
| Ref. 4 | 18 | 468 | Kinase2 | n/a | 0.629 | 0.943 | **0.963** |
| | 6 | 848 | 1dynA | n/a | 0.237 | **0.765** | 0.247 |
| Ref. 5 | 8 | 328 | 2cba | n/a | 0.871 | 0.859 | **0.874** |
| | 15 | 301 | S51 | n/a | 0.869 | **0.983** | 0.674 |
| Average score | | | | 0.707 | 0.703 | 0.858 | 0.698 |

SN: Sequence Number
ASL: Average Sequence Length

## 5 Conclusion

As MSA has become one of the important research area to know the structural and behavioral study of biomolecules therefore, this study present a whale optimization based method to identify the problems related to alignment of protein sequences. Here, a standard reference sets from BALiBASE is considered and with the help of new nature inspired algorithm i.e. Whale optimization Algorithm we tried to align protein sequence. The result so obtained gives the conclusion that our proposed method is much efficient and feasible for alignment of protein sequences as compared to other nature inspired algorithm available in the same field. Thus, after seeing the result outcome we can conclude that our proposed method has successfully aligned unstructured protein sequence from standard dataset and have obtained better score as compared to other methods discussed in this paper for most of the test cases.

**Compliance with Ethical Standards**

✓ All authors declare that there is no conflict of interest
✓ No humans/animals involved in this research work.
✓ We have used our own data.

## References

1. Hamidi, S., Naghibzadeh, M., Sadri, J.: Protein multiple sequence alignment based on secondary structure similarity. In: International Conference on Advances in Computing, Communications and Informatics, pp. 1224–1229 (2013)
2. Eddy, S.R.: Multiple alignment using hidden Markov models. In: Proceeding of the International Conference on Intelligent Systems for Molecular Biology, vol. 3, pp. 114–120 (1995)
3. Peng, Y., Dong, C., Zheng, H.: Research on genetic algorithm based on pyramid model. In: 2nd International Symposium on Intelligence Information Processing and Trusted Computing, pp. 83–86 (2011)
4. Zhu, H., He, Z., Jia, Y.: A novel approach to multiple sequence alignment using multiobjective evolutionary algorithm based on decomposition. IEEE J. Biomed. Health Inform. **20**(2), 717–727 (2016)
5. Wong, W.C., Maurer Stroh, S., Eisenhaber, F.: More than 1,001 problems with protein domain databases: transmembrane regions, signal peptides and the issue of sequence homology. PLoS Comput. Biol. **6**(7), e1000867 (2010)
6. Besharati, A., Jalali, M.: Multiple sequence alignment using biological features classification. In: International Congress on Technology, Communication and Knowledge (ICTCK), Mashhad, pp. 1–5 (2014)
7. Razmara, J., Deris, S.B., Parvizpour, S.: Text-based protein structure modeling for structure comparison. In: International Conference of Soft Computing and Pattern Recognition, pp. 490–496 (2009)
8. Naznin, F., Sarker, R., Essam, D.: Progressive alignment method using genetic algorithm for multiple sequence alignment. IEEE Trans. Evol. Comput. **16**, 615–631 (2012)
9. Aniba, M.R., Poch, O., Thompson, J.D.: Issues in bioinformatics benchmarking: the case study of multiple sequence alignment. Nucleic Acids Res. **38**, 7353–7363 (2010)
10. Katoh, K., Kuma, K., Toh, H., Miyata, T.: MAFFT version 5: improvement in accuracy of multiple sequence alignment. Nucleic Acids Res. **33**, 511–518 (2005)
11. Hong, C., Tewfik, A.H.: Heuristic reusable dynamic programming: efficient updates of local sequence alignment. IEEE/ACM Trans. Comput. Biol. Bioinform. **6**, 570–582 (2009)
12. Fu, H., Xue, D., Zhang, X., Jia, C.: Conserved secondary structure prediction for similar highly group of related RNA sequences. In: Control and Decision Conference, pp. 5158–5163 (2009)

13. Needleman, S.B., Wunsch, C.D.: A general method applicable to the search for similarities in the amino acid sequence of two proteins. J. Mol. Biol. **48**, 443–453 (1970)
14. Feng, D.F., Dolittle, R.F.: Progressive sequence alignment as a prerequisite to correct phylogenetic trees. J. Mol. Evol. **25**, 351–360 (1987)
15. Mohsen, B., Balaji, P., Devavrat, S., Mayank, S.: Iterative scheduling algorithms. In: IEEE INFOCOM Proceedings (2007)
16. Pop, M., Salzberg, S.L.: Bioinformatics challenges of new sequencing technology. Trends Gene. **24**, 142–149 (2008)
17. Mirjalili, S., Lewis, A.: The whale optimization algorithm. Adv. Eng. Softw. **95**, 51–67 (2016)
18. Cheng, G., Lv, C., Yan, S., Xu, L.: A novel hybrid optimization algorithm combined with BBO and PSO. In: 2016 Chinese Control and Decision Conference (CCDC), Yinchuan, pp. 1198–1202 (2016)
19. Ganesan, T., Vasant, P., Elamvazuthi, I., Shaari, K.Z.K.: Multiobjective optimization of green sand mould system using DE and GSA. In: 2012 12th International Conference on Intelligent Systems Design and Applications (ISDA), Kochi, pp. 1012–1016 (2012)
20. Ping, G., Chunbo, X., Yi, C., Jing, L., Yanqing, L.: Adaptive ant colony optimization algorithm. In: 2014 International Conference on Mechatronics and Control (ICMC), Jinzhou, pp. 95–98 (2014)
21. Neshich, G., Togawa, R., Vilella, W., Honig, B.: STING (Sequence to and with In graphics). PDB viewer. Protein Data Bank Quart. Newslett. **85**, 6–7 (1998)
22. Kumar, M., Om, H.: A hybrid bio-inspired algorithm for protein domain problems. In: Shandilya, S., Shandilya, S., Nagar, A. (eds.) Advances in Nature-Inspired Computing and Applications, Chap. 13, pp. 291–311. Springer, Cham (2018)
23. Kumar, M.: An enhanced algorithm for multiple sequence alignment of protein sequences using genetic algorithm. EXCLI J. **14**, 1232–1255 (2015)
24. Zemali, E.A., Boukra, A.: A new hybrid bio-inspired approach to resolve the multiple sequence alignment problem. In: 2016 International Conference on Control, Decision and Information Technologies (CoDIT), St. Julian's, pp. 108–113 (2016)

# Website Information Architecture of Latin American Universities in the Rankings

Carmen Luisa Vásquez[1(✉)], Marisabel Luna-Cardozo[1], Maritza Torres-Samuel[2], Nunziatina Bucci[1], and Amelec Viloria Silva[3(✉)]

[1] Universidad Nacional Experimental Politécnica Antonio José de Sucre, Av. Barquisimeto, Venezuela
{cvasquez,mluna,nbucci}@unexpo.edu.ve

[2] Universidad Centroccidental Lisandro Alvarado, Barquisimeto, Venezuela
mtorres@ucla.edu.ve

[3] Universidad de La Costa (CUC), Calle 58 # 55-66, Atlantico, Baranquilla, Colombia
aviloria@cuc.edu.co

**Abstract.** In the 2019 editions of SIR, QS, ARWU and Webometric university rankings, after the process of evaluating and positioning through academic and research quality criteria established in their methodologies, 22 Latin American university institutions located in Brazil, Mexico, Chile, Argentina, and Colombia coincide in their positions in all the rankings considered for the study. This paper intends to characterize the website main menu design of these universities to describe and enlist the information available to users as an alternative to make their academic and scientific processes visible. Results highlight that 18% of the options of the main menu refer to library, academic life, institutional mail services, language translator, search engine, and information to visitors.

**Keywords:** University Rankings · Latin America · ARWU · SIR scimago · QS · Webometrics

## 1 Introduction

Increasing visibility of academic and scientific processes is a great challenge faced by universities today [1–5], particularly in relation to knowledge generation and the achievement of their researchers [6, 7]. In this sense, university website design, with menus, navigation bars, and frames offer a board to access information on academic processes, teaching activity, research, innovation and extension products, transparency, services such as digital libraries, email and social networks, available to teachers, students, graduates, administrative and support staff, and visitors.

This study characterizes the website main menu of Latin American universities positioned in the SIR, QS, ARWU and Webometrics rankings, 2019 editions, to describe and list the information available to users. Users include scholars and university students, but also visitors in general. For this purpose, the positioning of these


S. Smys et al. (Eds.): ICCVBIC 2019, AISC 1108, pp. 140–147, 2020.
https://doi.org/10.1007/978-3-030-37218-7_16

universities in the respective website of the following four world rankings was reviewed:

- Academic Ranking of World Universities (ARWU), 2019 edition [8].
- Scimago Institutions Rankings (SIR) Latin America, 2019 edition [9].
- The QS World University Rankings, QS Latin America University Rankings, 2019 edition [10].
- Ranking Web of Latin American Universities (Webometric), July 2019 edition [11].

## 2 Visibility and Internationalization of Universities

Internationalization is a key factor for enhancing quality, efficiency, excellence, and competitiveness of universities in a global environment. It is based on the dissemination and communication of knowledge generated by researchers; hence their institutional strategy is linked to policies of excellence and reputation in the global market of knowledge generation, dissemination, and transfer [12]. University institutions, in order to promote and position themselves in the national and international market, and to guarantee its economic viability, require to develop communication and information skills to attract teachers, students, and economic resources [13].

In this way, those responsible for institutional/external relations of Spain universities give 82% of importance to quality and impact of promotion and web marketing instruments, considering it as a quality criterion for the university internationalization [13]. Similarly, UNESCO (2009) states that results of scientific research should be more widely disseminated through ICTs, also supported on free access to scientific documentation [14].

Thus, in an environment of globalized higher education, universities face challenges with respect to the visibility of their academic and scientific processes with the integration of ICT [15], and the governing authorities are aware of the importance of increasing their international profile, strategies and internationalization activities.

One of these challenges is the assessment and positioning of the academic and scientific community through international rankings with different classification criteria according to their quality, which ratifies the importance of scientific and academic web visibility for higher education institutions.

## 3 Web Indicators and the Presence of Universities on Internet

Cybermetrics is an emerging discipline that applies quantitative methods to describe the processes of scientific communication on the Internet, the contents of the site, their interrelationships, and the consumption of that information by users, the structure and use of search tools, the invisible Internet or the particularities of email-based services [15].

The fundamental measurement tool for this purpose is the use of so-called indicators, which can be used in combination with bibliometric equivalents for describing different aspects of the academic and scientific communication processes. There are three main groups of web indicators for cybermetric analysis: descriptive measures that count the number of objects found in each website (pages, average or rich files, link density); visibility and impact measures that count the number and origin of external incoming links, such as the famous Google PageRank algorithm; and popularity measures which take into account the number and characteristics of visits to websites.

### 3.1 Content Indicators

The main indicators are those that describe the volume of content published on the site. They can measure the number and size of computing objects found in each site, but the second data is irrelevant since it depends on factors linked to the format and not to the content.

### 3.2 Visibility and Impact Indicators

Citation analysis techniques can be applied to the description of the global scenario. The measure of visibility is given by the number of external links (from third parties) received by a domain.

### 3.3 Popularity Indicators

Information consumption can be measured by counting the number and describing the characteristics of users and visitors to a site. Relative values could be provided by a popularity ranking such as Alexa [16]. This information, in addition to the global position, can be used in comparative studies.

## 4 Information Architecture (IA)

Information architecture (IA) is the result of organizing, classifying, and structuring the contents of a website so that users can meet their goals with the least possible effort. Its main purpose is to facilitate, as much as possible, the understanding and assimilation of information, as well as the tasks performed by users in a specific information space [17, 18].

A fundamental component in the website's IA is the home page. It is the showcase or the front door of the institution. Although it must share the same style of interior pages, it must show the user where it is and the organization of the website (sections and navigation, layout of menus). It is, therefore, the starting point in the navigation scheme of the site, so it must be clearly and unequivocally organized. The home page is so important that the user must be given access to it with a single click from any interior page. It is recommended to provide a search option. The functions of the home page are as follows:

(a) Show the user's location.
(b) Show news.

(c) Provide access to the information required by the user. Another key element in IA is navigation, so all interfaces must provide the user with answers to three questions: Where am I? Where do I come from? Where can I go? [17].

## 5 Information Architecture (IA) of Latin America Universities Positioned in Sir Csimago, Qs, Arwu and Webometrics Rankings

Among the world rankings that evaluate universities, especially for their scientific and academic quality are [19]:

a. Academic Ranking of World Universities (ARWU) of Shanghai Jiao Tong University [8]. This ranking conceives higher education as equivalent to scientific research, valuing, among other factors, prestige, opinion of peers, research, and the obtaining of Nobel prizes by professors and researchers. It ranks only the first 800 universities recognized by this Ranking.
b. Webometrics Rankings of World Universities of the Cybermetrics Laboratory of the CSIC in Spain [11]. With semi-annual publishing, it considers the productivity and effect of university academic products uploaded to the Internet.
c. SIR Ranking (SCimago Institutions Rankings). Published annually by SCimago Grupo [20], it takes into account variables such as production and research, innovation and social impact, and aspects associated with the institution's position in the global, Ibero-American, Latin American and national contexts, the percentage of the institution's production that has been published in collaboration with foreign institutions, and the impact of research, among others.

22 universities that coincide in the Latin American rankings were identified. As shown in Fig. 1, approximately 55% of them correspond to Brazilian universities, with USP [21] ranking first in three (3) of the four (4) rankings. Universities in Chile and Argentina with 13% of the ranked universities. Finally, Colombia and Mexico, with 10% each.

In order to categorize the information in the main menu options, the information was grouped into 16 items, the nomenclature of which is shown in Table 1.

Figures 2 and 3 show the percentage of information obtained from main menu and its representation in Pareto diagram, respectively. It can be noted that the highest percentage corresponds to Library and Services (18%), followed by Teaching, Extension, and Community with 15% each, and 13% of institutional information.

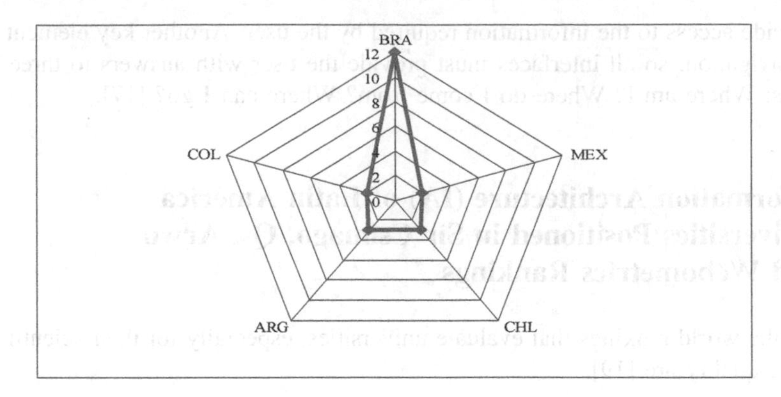


**Fig. 1.** University positions in SIR Csimago, QS, ARWU y Webometrics Rankings in 2019 editions

**Table 1.** Indicators and nomenclature of main menu options of universities positioned in SIR Csimago, QS, ARWU and Webometrics Rankings

| No | NOM | ITEM | CONTAINS |
|---|---|---|---|
| 1 | BEG | Home | Home option |
| 2 | BIB | Library and other services | Library, academic life, email service, language translator, search engine and information to visitors |
| 3 | COM | Communication | Communication media, TV channels, Radio, press, and ICT |
| 4 | E&C | Extension and Community | Engaged organizations, culture, sport, relationship with community and others |
| 5 | STU | Students | Student life |
| 6 | STI | Student admission | Student admission |
| 7 | GRA | Egresados | Information and services to graduates |
| 8 | INS | Institutional | Institutional information, directors, faculties and others |
| 9 | INT | International Cooperation | International Cooperation, foreign visitors and others |
| 10 | RES | Research and Innovation | Research, innovation and related information |
| 11 | MAN | Management | Strategic Planning, Biddings and others |
| 12 | NUM | University figures | Figures on students, teachers and others |
| 13 | NOP | Number of options | Number of options in the main menu |
| 14 | TEA | Teaching | Undergraduate study offers, postgraduate, careers and others |
| 15 | TES | Teachers | Teachers information |
| 16 | TRA | Transparency | Transparency |

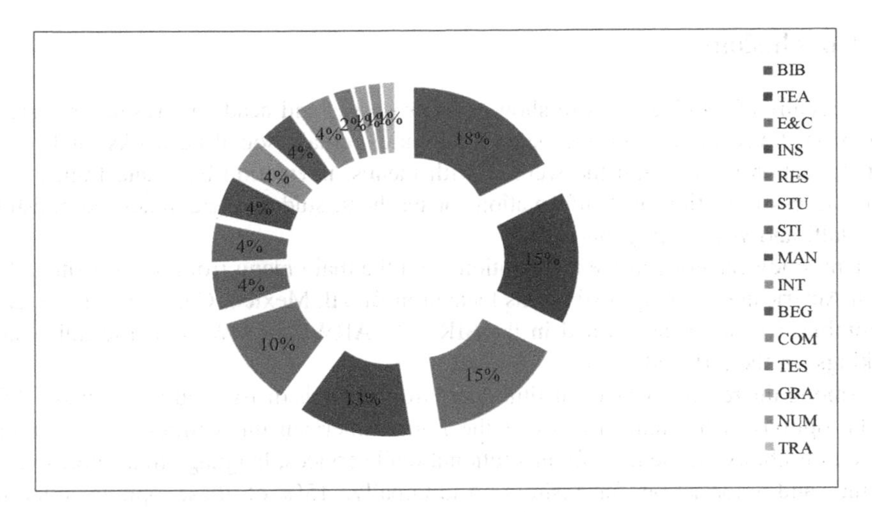


**Fig. 2.** Average distribution in categories according to number of options in main menu of Latin American universities in SIR csimago, QS, ARWU y Webometrics Rankings

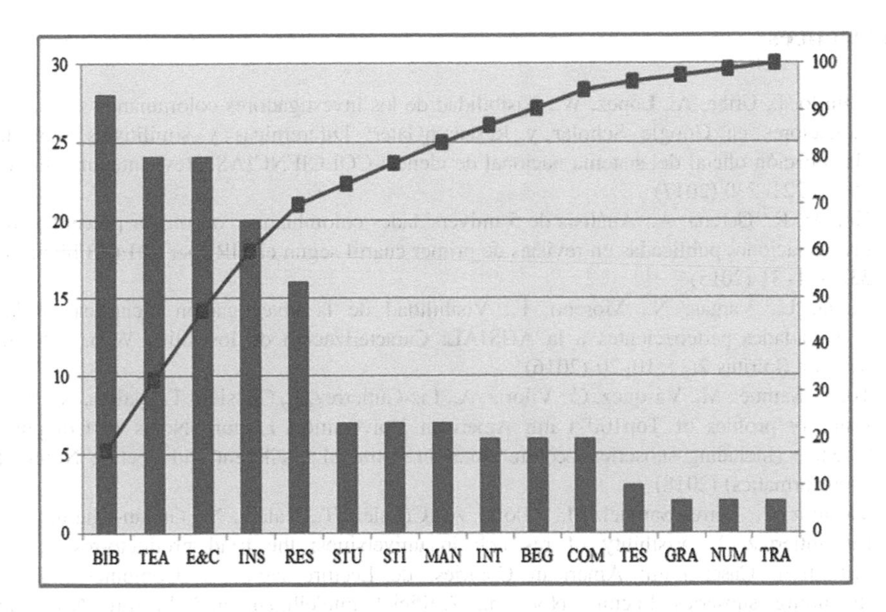


**Fig. 3.** Pareto diagram with information obtained from the number of options in main menu of Latin American universities in SIR Csimago, QS, ARWU y Webometrics Rankings

## 6 Conclusions

An alternative for universities to show their processes and academic results, research, extension, transparency, services such as libraries, email, social networks, and internationalization is to design the website with menus, navigation bars, and frames that allow access to institutional information for teachers, students, graduates, administrative staff, and visitors in general.

This study categorized the information from the main menu from the website of 22 Latin American university institutions located in Brazil, Mexico, Chile, Argentina, and Colombia, which are positioned in the SIR, QS, ARWU and Webometric university rankings, in the 2019 editions.

Among the results, 60% of institutional information is distributed in this way: 18% of the options in the main menus of the Latin American universities contain information on library, academic life, institutional mail services, language translator, search engine, and information for visitors. Additionally, 15% of these options refer to information about teaching in these institutions as well as another 15% about extension and community. In 13% of the options, institutional information stands out.

## References


1. Aguillo, I., Uribe, A., López, W.: Visibilidad de los investigadores colombianos según sus indicadores en Google Scholar y ResearchGater: Diferernicas y similitudes scon la clasificación oficial del sistema nacional de ciencia-COLCIENCIAS. Rev. Interam. Bibliot **40**(3), 221–230 (2017)
2. Gómez, R., Gerena, A.: Análisis de 5 universidades colombianas con mayor porcentaje de investigaciones publicadas en revistas de primer cuartil según el SIR Iber 2014. Bibliotecas **35**(3), 1–31 (2015)
3. Mejía, L., Vargas, N., Moreno, F.: Visibilidad de la investigación científica en las universidades pertenecientes a la AUSJAL: Caracterización de los sitios Web. Salutem Scientia Spiritus **2**(1), 10–20 (2016)
4. Torres-Samuel, M., Vásquez, C., Viloria, A., Lis-Gutierrez, J., Crissien, T., Valera, N.: Web visibility profiles of Top100 Latin American Universities. Lecture Notes in Computer Sciencie (Including subseries Lecture Notes in Artificial Intelligent and Lecture Notes of Bioinformatics) (2018)
5. Vásquez, C., Torres-Samuel, M., Viloria, A., Crissien, T., Valera, N., Gaitán-Angulo, M., Lis-Gutierrez, J.: Visibility of research in universities: the triad producto-researcher-institution. Case: Latin American Contries. de Lecture Notes in Computer Science (Including subseries Lecture Notes in Artificial Intelelligent and Lecture Notes in Bioinformatics) (2018)
6. Vílchez-Román, C., Espíritu-Barrón, E.: Artículos científicos y visibilidad académica: combinación impostergable y oportunidad que debe aprovecharse, Biblios, p. 9 (2009)
7. Henao-Rodríguez, L.-G.J., Gaitán-Angulo, M., Vásquez, C., Torres-Samuel, M., Viloria, A.: Determinants of researachgate (RG) score for the TOP100 of Latina American Universities at Webometrics. Communications in Computer and Informations Science (2019)
8. ARWU: Shanghai Jiao Tong University Ranking (2019). http://www.shanghairanking.com/es/. accessed 01 Sept 2019
9. S.R. Group: SIR IBER 2019. SCImago Lab, Barcelona, España (2019)

10. QSTOPUNIVERSITIES: QS World University Rankings. QS Quacquarelli (2019). https://www.topuniversities.com/qs-world-university-rankings. accessed 01 Sept 2019
11. Webometrics: Web-Webometrics Ranking (2019). http://www.webometrics.info/es/Latin_America_es. accessed 10 Aug 2019
12. de Wit, H., Gacel-Ávila, J., Knobel, M.: Estado del arte de la internacionalización de la educación superior en América Latina. Revista de Educación Superior en América Latina ESAL **2**, 1–4 (2017)
13. Haug, G., Vilalta, J.: La internacionalización de las universidades, una estrategia necesaria: Una reflexión sobre la vigencia de modelos académicos, económicos y culturales en la gestión de la internacionalización universitria, Fundación Europea Sociedad y Educación, Madrid (2011)
14. UNESCO: Conferencia mundial sobre educación superior 2009. Nuevas dinámicas de la educación superior y de la investigación para el cambio social y el desarrollo, UNESCO, Paris, (2009)
15. Aguado-López, E., Rogel-Salazar, R., Becerrill-Garcia, A., Baca-Zapata, G.: Presencia de universidades en la Red: La brecha digital entre Estados Unidos y el restyo del mundo. Revista de Universidad y Sociedad del Conocimeinto **6**(1), 4 (2009)
16. Amazon: Alexa, Amazon (2019). www.alexa.com. accessed 01 Feb 2019
17. Romero, B.: Usabilidad y arquitectura de la informaición. Universidad Oberta de Cataluña, Barcelona, España (2010)
18. Morville, P., Rosenfeld, L.: Information Architecture for the World Wide Web: Desining Large-Scale Web Sites. O'Really, Sebastopol (2006)
19. Torres-Samuel, M., Vásquez, C., Viloria, A., Hernández-Fernandez, L., Portillo-Medina, R.: Analysis of patterns in the university Word Rankings Webometrics, Shangai, QS and SIRScimago: case Latin American. Lecture Notes in Computer Science (Including subseries Lecture Notes in Artificial Intelligent and Lecture Notes in Bioinformatics) (2018)
20. S.I. Rankings: SCImago Institutions Rankings, SCImago Lab & Scopus (2019). https://www.scimagoir.com/. accessed 08 Aug 2019
21. USP: Universidade de Sao Paulo (2019). https://www5.usp.br/. accessed 25 Aug 2019

# Classification of Hybrid Multiscaled Remote Sensing Scene Using Pretrained Convolutional Neural Networks

Sujata Alegavi(✉) and Raghvendra Sedamkar

Thakur College of Engineering and Technology, Thakur Village, Kandivali - (East), Mumbai 400101, Maharashtra, India
{sujata.dubal, rr.sedamkar}@thakureducation.org

**Abstract.** A novel hybrid multiscaled Remote Sensed (RS) image classification method based on spatial-spectral feature extraction using pretrained neural network approach is proposed in this paper. The spectral and spatial features like colour, texture and edge of RS images are extracted by using nonlinear spectral unmixing which is further scaled by using bilinear interpolation. The same RS image in parallel path is first scaled using bilinear interpolation and further spectrally unmixed. These two paths are further fused together using spatial-spectral fusion to give multiscaled RS image which is further given to a pre-trained network for feature extraction. For authentication and discrimination purposes, the proposed approach is evaluated via experiments with five challenging high-resolution remote sensing data sets and two famously used pre-trained network (Alexnet/Caffenet). The experimental results provides classification accuracy of about 98% when classified at multiscale level compared to 83% when classified at single scale level using pretrained convolutional networks.

**Keywords:** Multiscale remote sensed images · Convolutional neural networks · Spatial-Spectral analysis · Pretrained networks

## 1 Introduction

In recent era there is a major boom in the earth observation programs has resulted in easier collection of high resolution imagery for various applications. Most of the research is done in developing a robust feature representation map in different domains. Scene classification at low-level or mid-level works good on some remote sensing scenes but with complex spatial and spectral layouts these techniques also fails for achieving a good classification accuracy. In Previously each CNN network was trained fully for a defined problem [4, 6, 7]. To overcome these aforementioned problems, existing CNN models which are trained on large datasets are transferred to other recognition problems [7, 8]. Training a completely new CNN have not yielded a better classification accuracy compared to using a pretrained network for remote sensing scene classification. In case of few hundreds of training samples training a fully new CNN becomes much more difficult. This transfer learning helps in forming a feature representation. The novelty of this paper is:


S. Smys et al. (Eds.): ICCVBIC 2019, AISC 1108, pp. 148–160, 2020.
https://doi.org/10.1007/978-3-030-37218-7_17

(1) We present a multiscale feature extraction of a RS image in spatial and spectral domain.
(2) We propose the use of a Pretrained network for forming feature discrimination map.

## 2 Related Works

A complete CNN network consists of many such blocks, followed by many fully connected layer and a final classifier layer. For an input data X and a CNN with n computational blocks, the output from the last block is given by [7].

$$\mathrm{fL}(\mathrm{X}) = \sum \mathrm{Xi} * \mathrm{Wi} + \mathrm{bi} \tag{1}$$

where Xi, Wi and bi represent input, weights, and bias respectively. * is the convolution operation. In remote sensing there are many applications developed by exploiting the spatial-spectral information of high resolution RS imagery. As RS images are diverse in nature and there are many discriminative features at different layers, mixed pixel problems are very common in a single RS layer [1]. More Spectral-Spatial information of RS images are applied to CNN to extract the structural information [10–13]. Famous Pretrained networks which have achieved state-of the-art performance can be fine tuned for a given task without altering the parameters of the network [7, 8] fined-tuned networks in [7].

## 3 Proposed Work

Figure 1, shows a diagrammatic representation for hybrid classification of multiscale SAR/Hyperspectral images. As remotely sensed images are diverse in nature, hence while classification many aspects needs to be considered for precise classification. Currently many features of RS images are considered such as edge, colour, texture, etc. but they all are considered in single scale [5]. This is convenient in case of normal images, whereas in case of RS images lot of information can be yielded when the information is extracted at a multiscale level which helps in precise classification of the images. Mapping of similar regions with higher accuracy is possible with extraction of features at multiscale level. The above approach deals with classification of RS images at multiscale level using various different features like, Texture (Contrast, Correlation, Energy & Homogenity), Edge (Edge Histogram) and Wavelets (Daubechies 1 wavelets). The Multiscale analysis is done in two different ways to understand the changes in the classification results and a comprehensive analysis is being done to come up with the best solution proving multiscale analysis to be better as compared to single scale analysis.

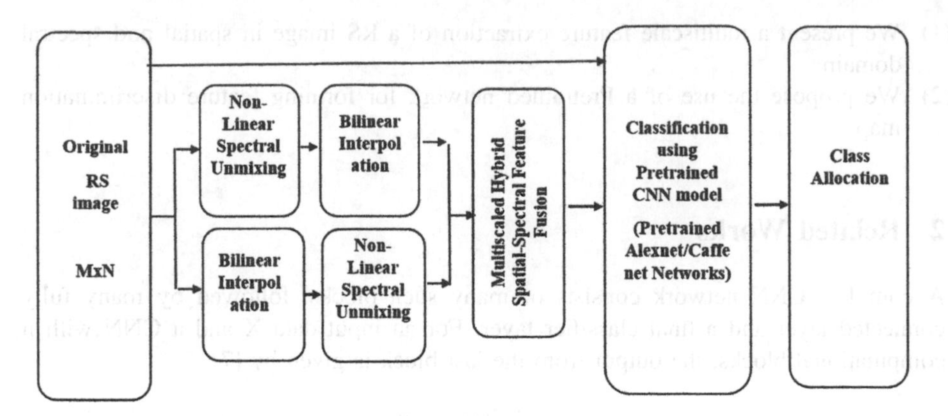


**Fig. 1.** Block diagram for hybrid classification of hybrid multiscale RS images using pretrained convolutional neural network model

1. In the first phase a RS image is directly fed to the pretrained network for classification and classes are determined.
2. In the second phase, low resolution RS image is taken which is processed in two parallel paths.
   a. In the first path, Spectral features of the RS images are derived by non linear spectral unmixing technique assuming that, the end member components are randomly distributed throughout the Instantaneaous Field of View (IFOV) causing multiple scattering effect.

$$f(x, y) = F[M, \alpha(x, y)] + n(x, y) \quad (2)$$

   where, M is number of end members and $\alpha(x, y)$ is the spectral coefficient. Further Bilinear Interpolation is applied to this spectrally unmixed images for achieveing scaling at multiple levels.

$$f(x, y) = \sum \sum a_{ij} x^i y^j \quad (3)$$

   where, a is the coefficient.
   Texture (Contrast, Correlation, Energy & Homogenity), Edge (Edge Histogram) and Wavelets (Daubechies 1 wavelets) are used to extract features from these multiscaled images.
   b. In the second parallel path, Bilinear Interpolation is applied to the RS image for scaling the image at multiple levels using Eq. (3) and then Spectral features of the scaled images are derived by non linear spectral unmixing technique using Eq. (2).
   Texture (Contrast, Correlation, Energy & Homogenity), Edge (Edge Histogram) and Wavelets (Daubechies 1 wavelets) are used to extract features from these multiscaled images.

3. Further these spectral and spatial features are fused together by a weighted parameter using integration.
4. This fused image is further given to the pretrained network (Alexnet/Caffenet) for further feature extraction.
5. Finally class allocation is done by soft max classifier which is used at the end of the fully connected network in the pretrained network.

For getting the spatial values, RS images are first unmixed in the spectral domain to form fractional images further, bilinear interpolation is applied to these fractional images to get the spatial characteristics of these images.

$$X = a1s1 + a2s2 + \ldots. + aMsM + n \quad (4)$$

$$= \Sigma aisi + w = Sa + w \quad (5)$$

Where M is the number of endmembers, S is the matrix of endmembers, and n is an error term for additive noise.

The predicted values in spatial domain are given by [15],

$$Pv_{spa}(kj) = (k = 1, 2, \ldots\ldots.., \; K) \; (j = 1, 2, \ldots.., M) \quad (6)$$

Where Pvspa is the predicted values in spatial domain in x and y direction, which are derived from spectral unmixing followed by bilinear interpolation are selected as spatial terms. Spectral terms are derived by first deriving high resolution RS imagery from low resolution RS imagery using bilinear interpolation. For getting the spectral values, RS images are first scaled using bilinear scaling and further fractional images are generated using spectral unmixing.

The predicted values in spectral domain are given by [15]:

$$Pv_{spe}(kj) = (k = 1, 2, \ldots.., K) \; (j = 1, 2, \ldots\ldots.., M) \quad (7)$$

Further the two predicted values Pvspa(kj) and Pvspe(kj) are integrated using the weight parameter w to obtain the finer multiscale values Pv(kj) [15].

The following equation is used for calculation of multiscaled images,

$$Pv(kj) = w \; Pvspe(kj) + (1 - w)Pvspa(kj) \quad (8)$$

The linear optimization technique is employed to transform the finer multiscale values Pv(kj) into a hard classified land cover map at a multiscale level. Once single RS image is converted to multiscale image we apply this multiscaled image to a pretrained network for further classification.

## 4 Result Analysis and Discussions

The experiments are implemented using MATLAB 2017a, and the platform has X64 based PC, Intel (R) Core (TM), i5 processor – 7400 CPU @ 3.00 GHz, 3001 MHz, 4 cores CPU, 8 GB RAM, NVIDIA Titan XP GPU and Windows 10 Pro operating system. The database consists of scenes from Indian_pines_corrected, JasperRidge2_F198, Jasperridge2_F224, SalinasA_corrected and Urban_F210. Original and Multiscale RS images are applied to pretrained convolutional neural networks. Alexnet and Caffenet Pretrained networks are used. Alexnet is the winner of Imagenet Large Scale Visual Recognition Competition 2012. Architecture of Alexnet [14] and Caffenet [16] is similar to each other only with a difference of multiple GPU used in Alexnet and a single GPU used in Caffenet. These Pretrained networks have eight layers with five convolutional layers and three fully connected layers and the last layer is the soft max classifier. Different tests are carried out by varying different parameters to test the results using Pretrained CNN. Different parameters are calculated and the results precisely shows that, when features of images are derived at multiscale level yields very good recognition rate compared to features of single scale images. Further pretrained CNN network enhances the overall classification results compared to other classification techniques. The results show that, with the use of convolutional neural networks the classification accuracy improves for multiscale RS images compared to original RS images. For class 1 total of 200 images are considered, for class 2 total of 205 images are considered, for class 3 total of 198 images are considered, for class 4 total of 197 images are considered and for class 5 total of 194 images are considered for training and testing purpose.

### 4.1 Results of Training and Testing of Original and Multiscale RS Images on Alexnet Pretrained Network

Phase - I deals with applying of Original and Multiscale RS image to Alexnet network. The following table shows confusion matrix for all the five classes.

**Table 1.** Confusion matrix of original RS image applied to Alexnet pretrained network

Confusion Matrix

| Output Class \ Target Class | 1 | 2 | 3 | 4 | 5 | |
|---|---|---|---|---|---|---|
| 1 | 87<br>20.9% | 1<br>0.2% | 0<br>0.0% | 0<br>0.0% | 1<br>0.2% | 97.8%<br>2.2% |
| 2 | 0<br>0.0% | 97<br>23.3% | 0<br>0.0% | 0<br>0.0% | 0<br>0.0% | 100%<br>0.0% |
| 3 | 13<br>3.1% | 4<br>1.0% | 69<br>16.5% | 14<br>3.4% | 37<br>8.9% | 50.4%<br>49.6% |
| 4 | 0<br>0.0% | 0<br>0.0% | 0<br>0.0% | 37<br>8.9% | 0<br>0.0% | 100%<br>0.0% |
| 5 | 0<br>0.0% | 0<br>0.0% | 0<br>0.0% | 0<br>0.0% | 57<br>13.7% | 100%<br>0.0% |
| | 87.0%<br>13.0% | 95.1%<br>4.9% | 100%<br>0.0% | 72.5%<br>27.5% | 60.0%<br>40.0% | 83.2%<br>16.8% |

Table 1 shows, classification accuracy for each individual class. We can clearly observe that, for class1, class4 and class 5, the classification accuracy is 100%. The overall correct classification is 83.21% whereas the incorrect classification is 16.79%. Total time taken for training per epoch is 312 s. We can clearly see from the confusion matrix that, class 1, 4 and 5 are correctly classified but class 2 and 3 are incorrectly classified using Alexnet network.

**Table 2.** Confusion matrix of hybrid multiscaled RS image applied to Alexnet pretrained network

Confusion Matrix

| Output Class \ Target Class | 1 | 2 | 3 | 4 | 5 | |
|---|---|---|---|---|---|---|
| 1 | 98<br>23.5% | 1<br>0.2% | 0<br>0.0% | 0<br>0.0% | 1<br>0.2% | 98.0%<br>2.0% |
| 2 | 0<br>0.0% | 99<br>23.7% | 0<br>0.0% | 0<br>0.0% | 0<br>0.0% | 100%<br>0.0% |
| 3 | 2<br>0.5% | 2<br>0.5% | 69<br>16.5% | 0<br>0.0% | 1<br>0.2% | 93.2%<br>6.8% |
| 4 | 0<br>0.0% | 0<br>0.0% | 0<br>0.0% | 51<br>12.2% | 0<br>0.0% | 100%<br>0.0% |
| 5 | 0<br>0.0% | 0<br>0.0% | 0<br>0.0% | 0<br>0.0% | 93<br>22.3% | 100%<br>0.0% |
| | 98.0%<br>2.0% | 97.1%<br>2.9% | 100%<br>0.0% | 100%<br>0.0% | 97.9%<br>2.1% | 98.3%<br>1.7% |

Table 2 shows, classification accuracy for each individual class. We can clearly observe that, for class2, class4 and class 5, the classification accuracy is 100%. The overall correct classification is 98.32% whereas the incorrect classification is 1.68%. Total time taken for training per epoch is 312 s. We can clearly see from the confusion matrix that, class 2, 4 and 5 are correctly classified but class 1 and 3 are incorrectly classified using Alexnet network.

### 4.2 Results of Training and Testing of Single Scale and Multiscale Model of RS Images on Caffenet Pretrained Network

Phase - II deals with applying of Original and Multiscale RS image to Caffenet network. The following table shows confusion matrix for all the five classes.

Table 3 shows, classification accuracy for each individual class. We can clearly observe that, for class1, class 2, class4 and class 5, the classification accuracy is 100%. The overall correct classification is 78.65% whereas the incorrect classification is 21.34%. Total time taken for training per epoch is 315 s. We can clearly see from the confusion matrix that, class 1, 2, 4 and 5 are correctly classified but class 3 is incorrectly classified using Caffenet network.

Table 4 shows, classification accuracy for each individual class. We can clearly observe that, for class2, class4 and class 5, the classification accuracy is 100%. The overall correct classification is 98.32% whereas the incorrect classification is 1.68%. Total time taken for training per epoch is 315 s. We can clearly see from the confusion

**Table 3.** Confusion matrix of original RS image applied to Caffenet pretrained network

Confusion Matrix

| Output Class \ Target Class | 1 | 2 | 3 | 4 | 5 | |
|---|---|---|---|---|---|---|
| 1 | 92<br>22.1% | 0<br>0.0% | 0<br>0.0% | 0<br>0.0% | 0<br>0.0% | 100%<br>0.0% |
| 2 | 0<br>0.0% | 87<br>20.9% | 0<br>0.0% | 0<br>0.0% | 0<br>0.0% | 100%<br>0.0% |
| 3 | 8<br>1.9% | 15<br>3.6% | 69<br>16.5% | 36<br>8.6% | 30<br>7.2% | 43.7%<br>56.3% |
| 4 | 0<br>0.0% | 0<br>0.0% | 0<br>0.0% | 15<br>3.6% | 0<br>0.0% | 100%<br>0.0% |
| 5 | 0<br>0.0% | 0<br>0.0% | 0<br>0.0% | 0<br>0.0% | 65<br>15.6% | 100%<br>0.0% |
| | 92.0%<br>8.0% | 85.3%<br>14.7% | 100%<br>0.0% | 29.4%<br>70.6% | 68.4%<br>31.6% | 78.7%<br>21.3% |

**Table 4.** Confusion matrix of hybrid multiscaled RS image applied to Caffenet pretrained network

Confusion Matrix

| Output Class \ Target Class | 1 | 2 | 3 | 4 | 5 | |
|---|---|---|---|---|---|---|
| 1 | 99<br>23.7% | 1<br>0.2% | 0<br>0.0% | 0<br>0.0% | 0<br>0.0% | 99.0%<br>1.0% |
| 2 | 0<br>0.0% | 99<br>23.7% | 0<br>0.0% | 0<br>0.0% | 0<br>0.0% | 100%<br>0.0% |
| 3 | 1<br>0.2% | 2<br>0.5% | 69<br>16.5% | 0<br>0.0% | 3<br>0.7% | 92.0%<br>8.0% |
| 4 | 0<br>0.0% | 0<br>0.0% | 0<br>0.0% | 51<br>12.2% | 0<br>0.0% | 100%<br>0.0% |
| 5 | 0<br>0.0% | 0<br>0.0% | 0<br>0.0% | 0<br>0.0% | 92<br>22.1% | 100%<br>0.0% |
| | 99.0%<br>1.0% | 97.1%<br>2.9% | 100%<br>0.0% | 100%<br>0.0% | 96.8%<br>3.2% | 98.3%<br>1.7% |

matrix that, class 2, 4 and 5 are correctly classified but class 1 and 3 are incorrectly classified using Caffenet network.

**Table 5.** Comparison of classification accuracy of original and Multiscaled RS images using Alexnet and Caffenet pretrained networks

| Classification accuracy | Original image | Multiscaled image |
|---|---|---|
| Alexnet | 83.213 | **98.321** |
| Caffenet | 78.657 | **98.321** |

It can be clearly seen from the above two confusion matrix, that the classification accuracy increases by 15% for hybrid multiscaled RS image as compared to the original image when applied to Alexnet pretrained network and the classification accuracy increases by 20% for hybrid multiscaled RS image as compared to the original image when applied to Caffenet pretrained network (Fig. 2).

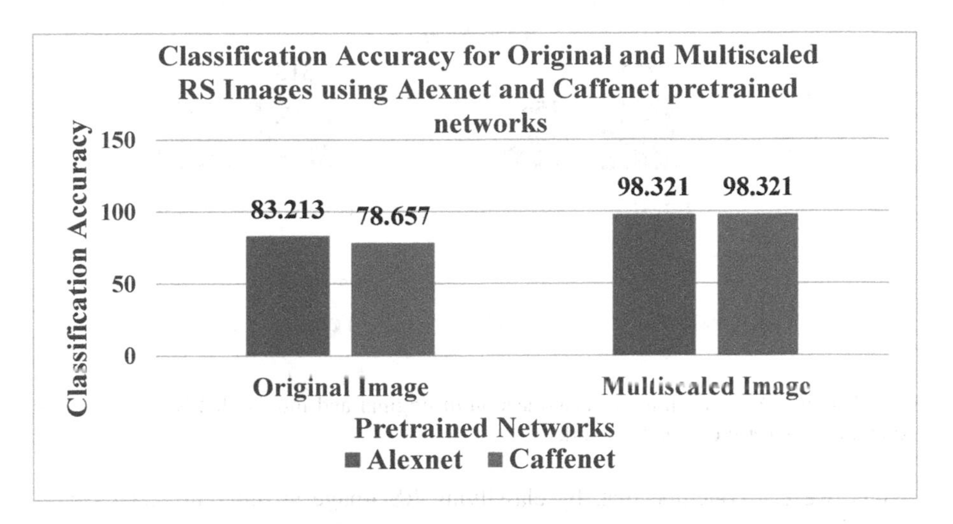


**Fig. 2.** Comparison of classification accuracy of original and multiscaled RS images using Alexnet and Caffenet pretrained networks

The above figure shows comparative analysis of RS images whose features are extracted at different scales. These multiscaled images are trained using fine tuned seven pretrained CNN models in our proposed approach, to implement classification. Besides, we evaluate the performance of original image and multiscaled image on both of the pretrained networks that are widely used for classification. Then, the feature vectors are directly fed into Softmax classifier without dimensionality reduction. Table 5, presents the experimental results on SAR/Hyperspectral datasets obtained by our proposed approach.

**Table 6.** Comparison between training and testing time for classification of Original and Multiscaled RS image using Alexnet and Caffenet pretrained networks

| Time taken | Original RS image | Multiscaled RS image |
|---|---|---|
| Training time per epoch | 312 | 315 |
| Testing time | 156 | 158 |

It, can be clearly seen from table no. 6, that the time required for training and testing of original and multiscaled image is nearly the same, with increased classification accuracy as in case of multiscaled RS image compared to classification accuracy of original RS image (Fig. 3).

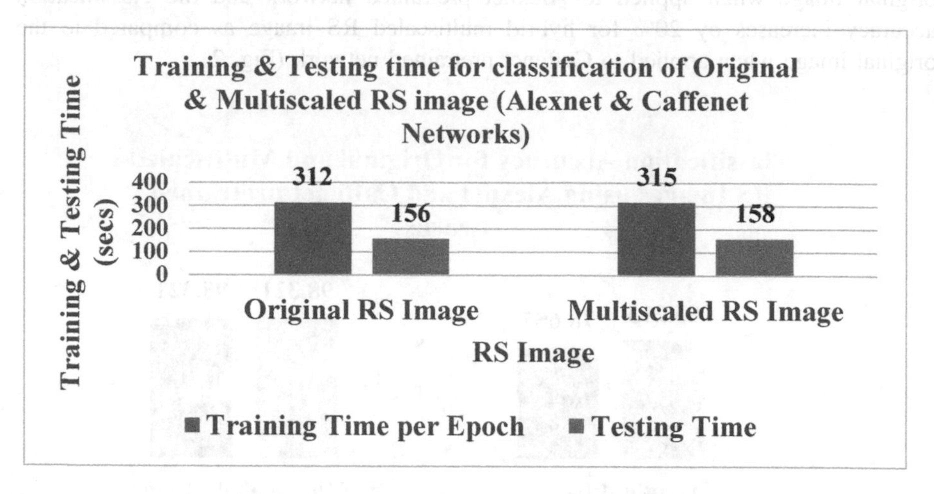


**Fig. 3.** Training & testing time for classification of original and multiscaled RS Images using Alexnet and caffenet pretrained networks

Thus, we can conclude that, by classifying RS image at different scales (multi-scaled) gives us better classification accuracy over single scale with nearly the same training and testing time as that of original RS image.

According to the performances with Alexnet and Caffenet pretrained CNN models, we can easily conclude that images taken at different scales performs well compared to images taken at a single scale. Multiscaled images outperform all the other single scale parameters for the two pretrained network. Overall, the proposed method achieves competitive accuracies (more than 97%) for Alexnet and Caffenet pretrained models (Table 7).

**Table 7.** Performance comparison of the state-of-art methods with our proposed method

| Methods | Classification accuracy % |
|---|---|
| Spatial BOW [19] | 81.19 |
| SIFT + BOW [8] | 75.11 |
| SC + Pooling [20] | 81.67 ± 1.23 |
| SPM [21] | 86.8 |
| PSR [21] | 89.1 |
| Caffenet [22] | 95.02 ± 0.81 |
| Googlenet [22] | 94.31 ± 0.89 |
| GBRCN [23] | 94.53 |
| Caffenet + fine tuning [23] | 95.48 |
| Googlenet + fine tuning [23] | 97.10 |
| **Multiscale + Alexnet (Our Proposed Method)** | **98.32** |
| **Multiscale + Caffenet (Our Proposed Method)** | **98.32** |

We can very well see from the above graph that all CNN based methods have achieved higher classification accuracy compared to the traditional classification methods. The CNN models were trained with 60% of data for training and remaining 40% of data was used for testing. Pretrained networks were used for feature extraction instead of designing a whole new CNN for classification purpose to save on computational complexity while maintain the good results. We can also see from the above graph that, transfer learning was used by fine tuning pretrained caffenet and googlenet networks for classification as reported in [23]. Our proposed method uses transfer learning by fine tuning Alexnet and Caffenet network with input as multiscale features which yields good classification results in our case compared to the existing sate-of-art methods (Fig. 4).

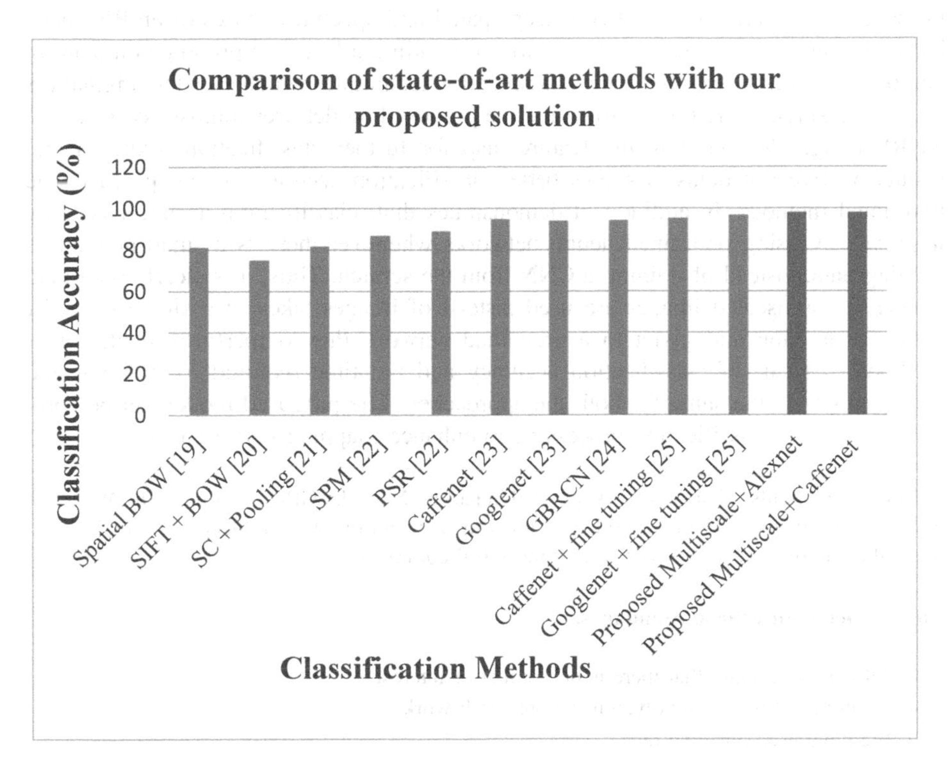


**Fig. 4.** Comparison of the proposed method with the state-of-art methods

A comparison was made between the performances when using different feature vectors at multiscale level for the pretrained networks. Dataset was fine tuned to understand the impact of original data and multiscaled data on the pretrained network (Alexnet & Caffenet). The results were compared after pretraining the data and classification accuracy was calculated for both the pretrained networks. The results suggest that, Multiscaling RS images increases classification accuracy as compared to using original RS image. Moreover using pretrained CNN improves the classification error in turn it also improves the recognition rate when there is a small amount of training data. This is consistent for all data sets. Larger datasets can also be tested successfully using CNN. More number of feature vectors can be trained at different scales. The result are positive, because when there is more training data available for the target data set, more of the variability in the data is captured and features can be trained from scratch to be robust to the variability.

## 5 Conclusion

This paper focuses on use of pretrained networks for feature extraction of remotely sensed images and presents a multiscale model for classification of RS images. The proposed multiscale model extracts deep spatial and spectral features of an RS image before giving it to the Pretrained network for building a feature representation map. In comparison with the existing scene classification techniques which uses the original RS image at a single scale for classification the proposed model uses multiscaled layers of an RS images for building the feature map for further classification. Experimental results on five test datasets shows better classification accuracy as compared to the traditional methods. In addition, it demonstrates that, classification accuracy actually improves by using pretrained neural networks whenever there is a small amount of training data, instead of training a CNN from the scratch. Thus, it is clearly seen that, whenever multiscaled images are used instead of images taken at a single scale for feature extraction and given to a pretrained network they outperform all the other methods in terms of classification accuracy and the time required for training and testing is nearly the same for both the approaches. The proposed model can be combined with post-classification processing to enhance mapping performance.

**Acknowledgements.** This work is supported in part by NVIDIA GPU grant program. We thank NVIDIA for giving us Titan XP GPU as a grant to carry out our work in deep learning. We also thank the anonymous reviewers for their insightful comments.

**Compliance with Ethical Standards.**

✓ All authors declare that there is no conflict of interest
✓ No humans/animals involved in this research work.
✓ We have used our own data.

## References

1. Alegavi, S., Sedamkar, R.R.: Improving classification error for mixed pixels in satellite images using soft thresholding technique. In: Proceedings of IEEE Conference in International Conference on Intelligent Computing and Control Systems (2018)
2. Schmidhuber, J.: Deep learning in neural networks: an overview. Neural Netw. **61**, 85–117 (2015)
3. Zhang, L., Zhang, L., Du, B.: Deep learning for remote sensing data: a technical tutorial on the state of the art. IEEE Geosci. Remote Sens. Mag. **4**(2), 22–40 (2016)

4. Vakalopoulou, M., Karantzalos, K., Komodakis, N., Paragios, N.: Building detection in very high resolution multispectral data with deep learning features. In: Proceedings of IEEE International Geoscience Remote Sensing Symposium (IGARSS), Milan, Italy, July 2015, pp. 1873–1876 (2015)
5. Jiao, L., Tang, X., Hou, B., Wang, S.: SAR images retrieval based on semantic classification and region-based similarity measure for earth observation. IEEE J. Sel. Top. Appl. Earth Obs. Remote Sens. **8**(8), 3876–3891 (2015)
6. Szegedy, C., et al.: Going deeper with convolutions. In: Proceedings of IEEE Conference on Computer Vision and Pattern Recognition (CVPR), Boston, MA, USA, June 2015, pp. 1–9 (2015)
7. Nogueira, K., Penatti, O.A.B., dos Santos, J.A.: Towards better exploiting convolutional neural networks for remote sensing scene classification. Pattern Recognit. **61**, 539–556 (2017)
8. Hu, F., Xia, G.-S., Hu, J., Zhang, L.: Transferring deep convolutional neural networks for the scene classification of high-resolution remote sensing imagery. Remote Sens. **7**(11), 14680–14707 (2015)
9. Penatti, O.A.B., Nogueira, K., dos Santos, J.A.: Do deep features generalize from everyday objects to remote sensing and aerial scenes domains? In: Proceedings of IEEE Computer Society Conference on Computer Vision and Pattern Recognition Workshops (CVPRW), Boston, MA, USA, June 2015, pp. 44–51 (2015)
10. Yue, J., Zhao, W., Mao, S., Liu, H.: Spectral–spatial classification of hyperspectral images using deep convolutional neural networks. Remote Sens. Lett. **6**(6), 468–477 (2015)
11. Zhao, W., Du, S.: Learning multiscale and deep representations for classifying remotely sensed imagery. ISPRS J. Photogramm. Remote Sens. **113**, 155–165 (2016)
12. Makantasis, K., Karantzalos, K., Doulamis, A., Doulamis, N.: Deep supervised learning for hyperspectral data classification through convolutional neural networks. In: Proceedings of International Geoscience Remote Sensing Symposium (IGARSS), Milan, Italy, July 2015, pp. 4959–4962 (2015)
13. Zhao, W., Du, S.: Spectral–spatial feature extraction for hyperspectral image classification: a dimension reduction and deep learning approach. IEEE Trans. Geosci. Remote Sens. **54**(8), 4544–4554 (2016)
14. Krizhevsky, A., Sutskever, I., Hinton, G.E.: ImageNet classification with deep convolutional neural networks. In: Proceedings of 26th Annual Conference on Neural Information Processing System (NIPS), Lake Tahoe, NV, USA, 2012, pp. 1097–1105 (2012)
15. Wang, P., Wang, L., Chanussot, J.: Soft-then-hard subpixel land cover mapping based on spatial-spectral interpolation. IEEE Geosci. Remote Sens. Lett. **13**(12), 1851–1854 (2016)
16. Jia, Y., et al.: Caffe: convolutional architecture for fast feature embedding. In: Proceedings of ACM Conference on Multimedia (MM), Orlando, FL, USA, November 2014, pp. 675–678 (2014)
17. Li, E., Xia, J., Du, P., Lin, C., Samat, A.: Integrating multilayer features of convolutional neural networks for remote sensing scene classification. IEEE Trans. Geosci. Remote Sens. **55**(10), 5653–5665 (2017)
18. Windrim, L., Melkumyan, A., Murphy, R.J., Chlingaryan, A., Ramakrishnan, R.: Pretraining for hyperspectral convolutional neural network classification. IEEE Trans. Geosci. Remote Sens. **56**(5), 2798–2810 (2018)
19. Yang, Y., Newsam, S.: Bag-of-visual-words and spatial extensions for land-use classification. In: Proceedings of 18th ACM SIGSPATIAL International Conference on Advances in Geographic Information System (GIS), San Jose, CA, USA, November 2010, pp. 270–279 (2010)

20. Cheriyadat, A.M.: Unsupervised feature learning for aerial scene classification. IEEE Trans. Geosci. Remote Sens. **52**(1), 439–451 (2014)
21. Chen, S., Tian, Y.: Pyramid of spatial relations for scene-level land use classification. IEEE Trans. Geosci. Remote Sens. **53**(4), 1947–1957 (2015)
22. Xia, G.-S., et al.: AID: a benchmark data set for performance evaluation of aerial scene classification. IEEE Trans. Geosci. Remote Sens. **55**(7), 3965–3981 (2017)
23. Zhang, F., Du, B., Zhang, L.: Scene classification via a gradient boosting random convolutional network framework. IEEE Trans. Geosci. Remote Sens. **54**(3), 1793–1802 (2016)

# Object Detection and Classification Using GPU Acceleration

Shreyank Prabhu, Vishal Khopkar(✉), Swapnil Nivendkar, Omkar Satpute, and Varshapriya Jyotinagar

Veermata Jijabai Technological Institute, Mumbai 400019, India
{svprabhu_b14, vgkhopkar_b14, ssnivendkar_b14, ohsatpute_b14, varshapriyajn}@ce.vjti.ac.in

**Abstract.** In order to speed up the image processing for self-driving cars, we propose a solution for fast vehicle classification using GPU computation. Our solution uses Histogram of Oriented Gradients (HOG) for feature extraction and Support Vector Machines (SVM) for classification. Our algorithm achieves a higher processing rate in frames per second (FPS) by using multi-core GPUs without compromising on its accuracy. The implementation of our GPU programming is in OpenCL, which is a platform independent library. We used a dataset of images of cars and other non-car objects on road to feed it to the classifier.

**Keywords:** Graphics processer unit · GPU · Object detection · Image processing · HOG · OpenCL · Self-driving cars · SVM

## 1 Introduction

Self-driving cars is the buzzword in today's world. It uses multiple cameras to record images in all directions and needs to process them in real time, keeping in mind the speed of the car, which can be as high as 100 km/h especially on access-controlled roads. In order to achieve this, we need an algorithm that can process the images clicked in real time at as much high frames per second as possible with maximum accuracy.

The motivation behind optimizing an object detection algorithm is the massive number of use cases which support local processing of data. With the ongoing trend of using cloud computing there are disadvantages associated with it, such as latency, costs of services, unavailability or failure of servers. This needs efficient processing of the images captured in the cameras of the vehicle – front, left, right and back, in order to achieve this there are several algorithms such as:

- Histogram of Oriented Gradients
- YOLO (You Look Only Once)
- Fast YOLO

Apart from this, algorithms such as histogram of linear binary patterns are also used. However, in order to achieve a good speed-accuracy trade-off we paralleled the histogram of oriented gradients (HOG) [1] algorithm on GPU using OpenCL, on open


S. Smys et al. (Eds.): ICCVBIC 2019, AISC 1108, pp. 161–170, 2020.
https://doi.org/10.1007/978-3-030-37218-7_18

source platform meant for heterogeneous computing i.e. the same algorithm will work on different GPU vendors such as AMD, Nvidia, Intel etc. as well as on CPU. Our code optimizes the object detection algorithm by parallelising it on GPU in order to decrease the time and increase number of frames per second. Our solution increases the current speed by 4–10 times depending on the hardware, without compromising on the accuracy. Paralellising HOG algorithm has provided efficient results in the case of pedestrian detection using CUDA on NVIDIA platforms, which has been shown with results in this paper [2]. Additionally, OpenCL and GPU computing has also been used for face detection and accelerated data mining [3–6]. The following sections give a brief idea on GPU Computing, OpenCL and HOG before explaining the actual approach adopted. Finally, we show a comparative study of different algorithms against ours as our result.

### 1.1 GPU Computing

The Time consuming and computing intensive parts of the code are offloaded to the GPU to accelerate applications running on the CPU. Remaining application can be processed on the CPU which can be linearly computed, the application runs faster because it uses the GPU cores to boost the performance. This is known as hybrid computing.

### 1.2 OpenCL

OpenCL [8] is a framework for writing programs for parallel processing on heterogenous vendor independent platforms like CPUs and GPUs. OpenCL improves the speed and efficiency of wide number of applications in numerous markets such as gaming, scientific and medical software as well as vision processing. OpenCL supports both task and data-based programming models. This language is suited to play an increasingly significant role in emerging graphics that use parallel compute algorithms. Our main purpose of using OpenCL is due to the heterogeneous nature which helps run our algorithm on different kind of systems, making it platform friendly.

### 1.3 Histogram of Oriented Gradients (HOG)

The steps for the algorithm in detail are as follows:

1. Extracting histograms from the image's gradient and magnitude vectors
2. Normalising the consolidated histogram into a full feature vector

The algorithm is fully explained alongside our implementation in Sect. 3.

## 2 Implementation

Our entire code was written in two parts:

1. OpenCL kernel
2. A python code as a driver for OpenCL

The OpenCL kernel code is written in C as it has a store of functions that can be easily translated into GPU understandable code. The python code is the main program that is called, which acts as a driver and involves in creation and retrieval of arrays from the kernel code. It uses PyOpenCL, a computation API [9].

## 2.1 Parallelizing HOG in OpenCL [10]

The histogram of oriented gradients algorithm comprises of different steps. The algorithm has a scope for parallelisation, since the operations it involves are performed on two dimensional arrays (matrices). In traditional sequential processing, the compiler will process each element of the array one by one. This, however, is time consuming, as images are large matrices. For the purpose of this project, our images are 64 × 64 pixels in size. Our kernel code has been implemented using three functions:

1. The first function extracts the magnitude and orientation, and later produces the histogram.
2. The second function creates the array to be normalized from the histogram obtained in step 1 and passes it to the third function as a vector of the entire image.
3. The third function normalizes the array obtained from step 2. This finally forms our feature vector for the given image. We adopted the block normalization approach.

All three functions are explained in detail below:

### 2.1.1 Step 1

At first, the 2D matrix image is retrieved back from the one-dimensional array passed. The derivatives of the image $d_x$ and $d_y$ are found using the classic gradient operators, i.e.

$$d_x = \frac{\partial f(x,y)}{\partial x} = f(x+1,y) - f(x-1,y) \tag{1}$$

$$d_y = \frac{\partial f(x,y)}{\partial y} = f(x,y+1) - f(x,y-1) \tag{2}$$

The next step is to create a 9-bin histogram for every 8 × 8 sub-image. This is also done parallelly, for every pixel, by determining which element of the histogram array the magnitude should be added to [11]. Algorithm 1 gives the list of computing steps.

```
1. Input image (as a pointer to array)
2. Get 2D matrix mat[y][x] = img[w*y + x]
3. Find derivatives dx and dy
4. magnitude_mat[y][x] = sqrt(dx[y][x]^2 + dy[y][x]^2)
5. angle_mat[y][x] = atan(dy[y][x]/dx[y][x])
6. There will be w/8*h/8 histograms. Determine which
   histogram is responsible for the current pixel.
7. Populate the appropriate histogram element with the
   current pixel data.
   curr_ind = angle_mat[y][x]/20
   rem = angle_mat[y][x] % 20
8. At the end, you get a histogram array of
   (w/8)*(h/8)*9 elements
```

**Algorithm 1.** Step 1: Finding the gradients and histogram of the image in HOG

#### 2.1.2 Step 2

In order to normalise the histogram array, we used 16 × 16 block normalisation technique. Step 2 involves creation of the array which will be finally normalised in step 3.

Therefore, the size of the entire vector obtained

$$\begin{aligned} &= \left(\frac{w}{8} - 1\right) \times \left(\frac{h}{8} - 1\right) \times 36 \\ &= \left(\frac{64}{8} - 1\right) \times \left(\frac{64}{8} - 1\right) \times 36 \\ &= 7 \times 7 \times 36 \\ &= 1764 \end{aligned} \tag{3}$$

This is far larger than the size of the total size of the histogram array for the image. This is because during this process, the 16 × 16 blocks overlap with each other and the elements of the histogram array are present in the final array multiple times. Those on the corner blocks are repeated only once; those on the edges (but not corners) are repeated twice; while those in the interior are repeated four times in the final image. Figure 1 gives an illustration and algorithm 2 shows the steps.

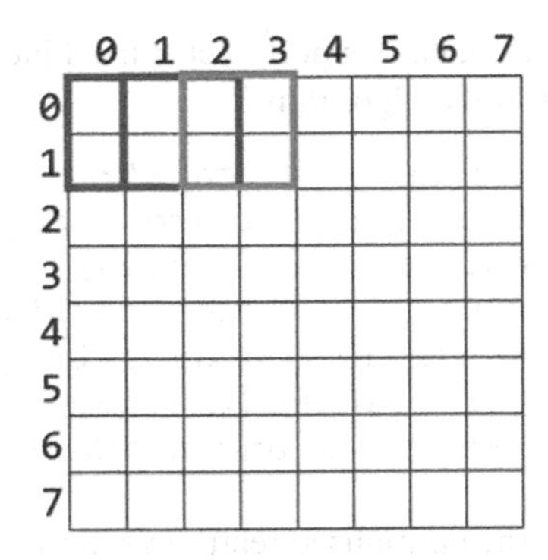


**Fig. 1.** 16 × 16 blocks for histogram calculation

```
1.    Input consolidated vector from step 2
      (tb_norm[(w/8 - 1)(h/8 -1)*36])
2.    Let squared_sum[36] be an array (initialised with
      0s).
3.    normal[i] = tb_norm[i]*tb_norm[i]
4.    sub_number = i/36
5.    squared_sum[i/36] += normal[i]
6.    if(sub_number == 0) normal[i] = 0
      else
      normal[i] =
      tb_norm[i]/(sqrt(squared_sum[sub_number]))
7.    normal array contains the normalised vector for the
      entire image
```

**Algorithm 2.** Finding the normalised feature vector of the image in HOG

### 2.1.3 Step 3

This function takes the input of the consolidated vector obtained from step 2 and normalises it. The detailed steps are shown in Fig. 4. First the square of 36 contiguous elements is calculated and summed up.

There will be (w/8−1) × (h/8−1) such squares, which in our case is 49. Every element is divided by the square root of the corresponding sum of squares to get the normalised vector of the entire image. This vector also measures (w/8−1) × (h/8−1) × 36 elements, which in our case is 1764.

## 2.2 Training the Model

The implementation of the histogram of oriented gradients algorithm stated above gives us the feature vector for the image. In order to classify the image as a vehicle or a non-vehicle, it is essential to apply a classifying algorithm. In this case, we used a supervised learning algorithm, linear support vector machine. The pseudocode of this implementation is as shown in the algorithm 3.

```
for all images 'image' in 'cars':
      features = extractFeatures(image)
      carFeatures.append(features)
for all images 'image' in 'not cars':
      features = extractFeatures(image)
      notCarFeatures.append(features)
train SVM on carFeatures and notCarFeatures
```

**Algorithm 3.** Finding the normalised feature vector of the image in HOG

The SVM training algorithm was not parallelized on OpenCL kernel because it is executed only once during training. Once the SVM is created, while predicting we need to only plot the feature set of the image that we want to predict as an object. This is not time consuming and therefore there is no need to parallelize SVM as well.

The model thus learnt was stored in a file and used to classify images using SVM prediction algorithm [12].

### 2.3 Datasets

The dataset used in this project was a set of images of car and non-car data. The total size of the dataset is 250. Each of the vehicles and non-vehicle images were split into groups of four according to their position in the image. The four groups were namely far, left, right and middle. The samples of each one is shown in Fig. 2.

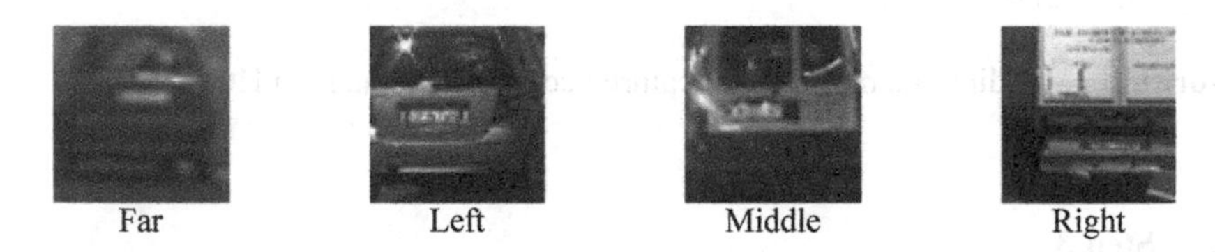


**Fig. 2.** Samples of the dataset: The images are $64 \times 64$ in size and hence appear unclear [7].

On the other hand, the non-vehicle images were not distributed in any groups since they had no criteria such as far, left, etc. under which we could categorize each image.

With this dataset and by randomly splitting the images into test and train data, we observed an accuracy of 95.65% with 111 frames per second on Nvidia GTX 1060 6 GB. The original HOG algorithm run entirely on CPU gives us an accuracy of 99.95% but could only process a complete image in a third or a fourth of a second, making it impossible for real time detection which requires an FPS of at least 27. However, it must also be noted that our algorithm did not detect false negatives. Some non-cars were indeed classified as cars, but no car was classified as non-car. This problem could be attributed to the fact that there can be a lot of images that classify as "not car" therefore making the system learn something as not car was difficult as not all such images could be covered in our training. Therefore, if the system came across any not car image that it has not seen before but has a silhouette of a car then it can confuse it for a car.

Currently, the YOLO (You Look Only Once) algorithm, which gives an accuracy of 58% with 45 frames per second and its enhanced version, Fast YOLO gives an accuracy of 70.7% with 155 frames per second. The Fig. 3 compares the accuracy and FPS of different algorithms.

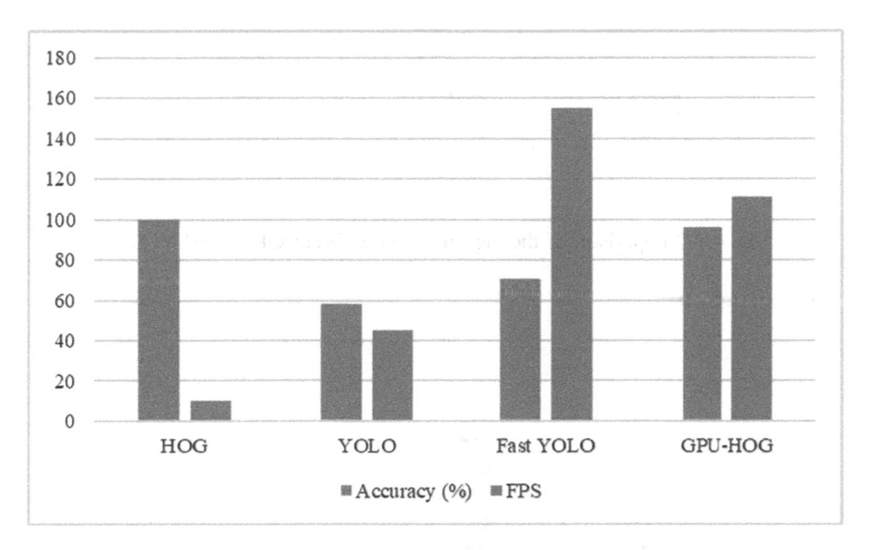


**Fig. 3.** Accuracy and frames per second for different algorithms: The image provides a comparison between the current fast processing algorithms, the traditional object detection in images and the algorithm mentioned above denoted by 'GPU-HOG'.

# 3 Comparison on Different Platforms

We ran the program on different platforms, CPUs, integrated GPUs as well as dedicated GPUs.

Figures 4 and 5 show the details of a comparison between different GPUs and CPUs.

As one can see, all GPUs beat CPUs in processing frames per second because of their higher number of threads for parallel processing. However, some less powerful CPUs and GPUs beat their more powerful counterparts; which could be due to noise influencing our results.

Figure 5 shows the comparison of the algorithm on dedicated and integrated GPUs. As one can see, dedicated GPUs process faster than integrated ones. This is because of several better specifications of the former ones. As an example, Intel HD 620 GPU has 24 pipelines with DDR4 memory. On the other hand, Radeon R7 M445 has 320 pipelines, with GDDR5 memory.

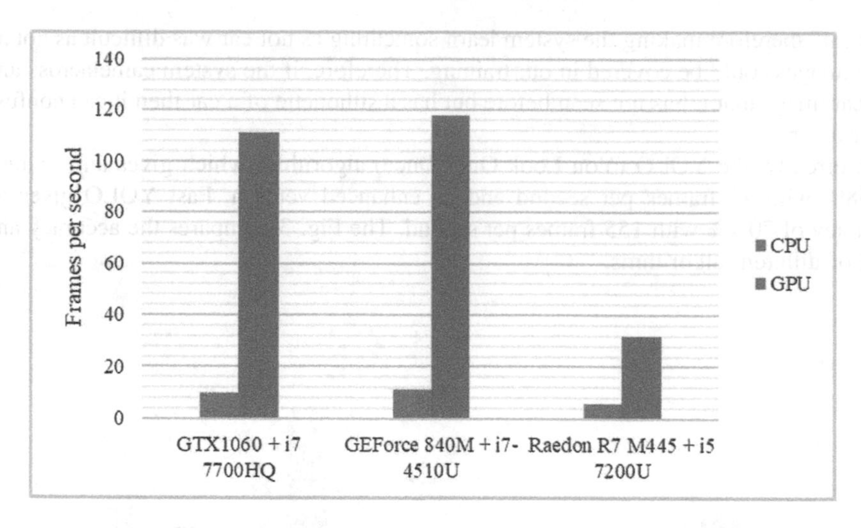


**Fig. 4.** Comparison of the algorithm on different GPUs and CPUs

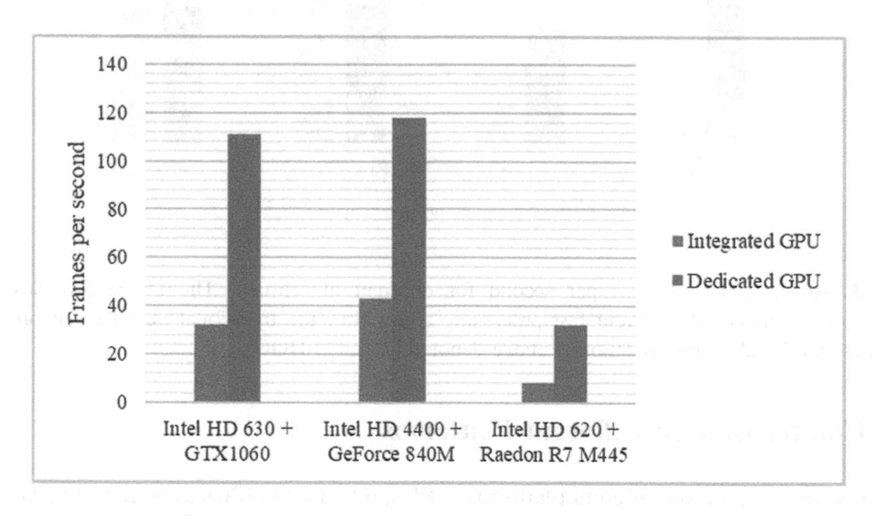


**Fig. 5.** Comparison of integrated and dedicated GPUs

## 4 Conclusion

To summarise, the GPU version of the HOG algorithm was able to retain most of the accuracy of its CPU counterpart, but at the same time it was much faster and gave over 96% accuracy, making the algorithm much more efficient and effective in real world applications. Although there were some issues such as the algorithm the data was being transferred between the CPU and GPU a total of six times, which could be further reduced.

As our testing showed that running the HOG algorithm for the GPU also helps preserve the high accuracy of the HOG, also at the same time it gives an FPS of over 60. This unique ability of the algorithm allows it to outperform both the YOLO and Fast YOLO which are popular algorithms in use currently. Some of the other advantages are that this algorithm is in OpenCL and therefore works on both Nvidia and AMD GPUs, also the hardware used such as the Nvidia 1060 GTX is commodity, which is not that expensive and also the CPU remains free for other instructions as the GPU will be busy in the computations.

Currently, our GPU algorithm is capable of extracting features from an image. As a future work, we would try parallelising an algorithm for detection of objects from an image with a background and detecting them in real-time in a video.

**Acknowledgments.** This project was undertaken for the final year course in B. Tech. Computer Engineering at Veermata Jijabai Technological Institute, Mumbai for the academic year 2017–2018. It was made possible with the guidance of Mrs. Varshapriya J N, Assistant Professor, Dept. of Computer Engineering.

**Compliance with Ethical Standards.**

✓ All authors declare that there is no conflict of interest
✓ No humans/animals involved in this research work.
✓ We have used our own data.

## References

1. Dalal, N., Triggs, B.: Histograms of oriented gradients for human detection. In: International Conference on Computer Vision & Pattern Recognition (CVPR 2005), vol. 1, pp. 886–893. IEEE Computer Society, June 2005
2. Campmany, V., Silva, S., Espinosa, A., Moure, J.C., Vázquez, D., López, A.M.: GPU-based pedestrian detection for autonomous driving. Procedia Comput. Sci. **80**, 2377–2381 (2016)
3. Naik, N., Rathna, G.N.: Robust real time face recognition and tracking on GPU using fusion of RGB and depth image. arXiv preprint arXiv:1504.01883 (2015)
4. Azzopardi, B.: Real time object recognition in videos with a parallel algorithm (Doctoral dissertation, University of Malta) (2016)
5. Cheng, K.M., Lin, C.Y., Chen, Y.C., Su, T.F., Lai, S.H., Lee, J.K.: Design of vehicle detection methods with OpenCl programming on multi-core systems. In: The 11th IEEE Symposium on Embedded Systems for Real-time Multimedia, pp. 88–95. IEEE, October 2013
6. Gurjar, P., Varshapriya, J.N.: Survey for GPU accelerated data mining. Int. J. Eng. Sci. Math. **2**(2), 10 (2013)
7. What Is GPU Computing? https://www.boston.co.uk/info/nvidia-kepler/what-is-gpu-computing.aspx. Accessed 15 June 2019
8. OpenCL Overview - The Khronos Group Inc. https://www.khronos.org/opencl/. Accessed 15 June 2019
9. PyOpenCL. https://mathema.tician.de/software/pyopencl/. Accessed 15 June 2019
10. Histogram of Oriented Gradients||Learn OpenCV. https://www.learnopencv.com/histogram-of-oriented-gradients/. Accessed 15 June 2019

11. Kaeli, D., Mistry, P., Schaa, D., Zhang, D.P.: Examples. In: Heterogenous Computing with OpenCL 2.0 3rd edn. Waltham, Elsevier, pp. 75–79 (2015)
12. Understanding Support Vector Machine algorithm from examples (along with code). https://www.analyticsvidhya.com/blog/2017/09/understaing-support-vector-machine-example-code/. Accessed 15 June 2019

# Cross Modal Retrieval for Different Modalities in Multimedia

T. J. Osheen(✉) and Linda Sara Mathew

Computer Science and Engineering, Mar Athanasius College of Engineering, Kothamangalam, Ernakulam 686666, Kerala, India
osheentharakan@gmail.com, lindasaramathew@gmail.com

**Abstract.** Multimedia data like text, image and video is used widely, along with the development of social media. In order, to obtain the accurate multimedia information rapidly and effectively for a huge amount of sources remains as a challenging task. Cross-modal retrieval tries to break through the modality of different media objects that can be regarded as a unified multimedia retrieval approach. For many real-world applications, cross-modal retrieval is becoming essential from inputting the image to load the connected text documents or considering text to choose the accurate results. Video retrieval depends on semantics that includes characteristics like graphical and notion based video. Because of the combined exploitation of all these methodologies, the cross-modal content framework of multimedia data is effectively conserved, when this data is mapped into the combined subspace. The aim is to group together the text, image and video components of multimedia document which shows the similarity in features and to retrieve the most accurate image, text, or video according to the query given, based on the semantics.

**Keywords:** Cross-modal retrieval · Feature selection · Subspace learning

## 1 Introduction

The fast inclination towards urbanization, in rising markets across the countries worldwide, highly demanding elegant surveillance solutions are used in visual analytic techniques to augment the current digital life scenarios. Many methods were invented to sharply detect and analyze visual objects in an image or video, this facilitate a significant challenge for the researchers. Thus, cross-modal retrieval is becoming essential for many worldwide assistance where the end user perceives as a single entity, such as using image to search the relevant text documents or using text to search the relevant images. Because of the combined exploitation, the fundamental cross-modal content framework of multimedia data is properly preserved when this data is mapped into the joint subspace. Multimedia technology is developing constantly with the exponential increase in internet applications, which makes a terrific growth in the multimedia.

Currently, social media are adopted all over the world, making its development mandate. As the processing speed must be made faster with a high accuracy level. The main features of querying and retrieval process is its speed and the accuracy. The data


S. Smys et al. (Eds.): ICCVBIC 2019, AISC 1108, pp. 171–178, 2020.
https://doi.org/10.1007/978-3-030-37218-7_19

in this case is found as a multi-modal pairs. So, it's a challenging issue to fetch data from heterogeneous types of data that is simply known as cross-modal retrieval. Nowadays, many statistics and machine learning techniques are considered for CMR. To benefit completely from a huge set of multimedia content and to make desirable benefit of quickly deploying multimedia techniques, without external intervention are necessary to confirm a relational connection between each multimedia component to one another.

Depending on the past research works, the content of cross modal knowledge accumulated examination, which could not find a solution for the cross modal complementary relationship measuring issue. Data mining is one area where these methodologies are mainly used. The problems regarding the learning process in characteristics of cross modal data of heterogeneous variety reduces the semantic gap between high level and low level features. Researchers have introduced a different kind of cross modal understanding technologies to search across the possible combinations among media content, increasing the performance of cross modal knowledge accumulation. The main correspondence in which the miniature and acquired video data are essential including real time videos on required concept, inclusion identification, strip track, sport occasions, criminal examination frameworks, and numerous others.

A definitive objective is to empower clients to recover some ideal substance from huge measures of video information in a proficient and semantically significant way. There are essentially three dimensions of video content which are crude video information, low dimension highlights and semantic substance. Second, low-level highlights are portrayed by sound, content, and visual highlights, for example, surface, shading conveyance, shape, movement, and so on. Third, semantic substance contains abnormal state ideas, for example, items and occasions. The initial two dimensions on which content displaying and extraction approaches are based utilize consequently removed information, which speak to the low dimension substance of a video, yet they barely give semantics which are substantially more fitting for clients. Be that as it may, it is extremely hard to remove semantic substance specific from crude video information. This is on the grounds that the video is a fleeting succession of edges without an immediate connection with its semantic substance.

## 2 Background and Motivation

A powerful huge scale multi modular grouping technique is proposed by Wang et al. [1] is to manage the heterogeneous portrayal of extensive scale information. The cross modal bunching utilizes the similitude of the examples in diverse states, and the grouping of various states is utilized to uncover the potential sharing structure among various states. Recently, pertinent specialists have advanced some compelling cross modal grouping strategy. By utilizing non negative framework factorization, [2] scan good grouping technique for various states. A cross-modal grouping technique dependent on CCA is proposed to extend information onto a low-dimensional portion. The proposed model is utilized in [3], to refresh the grouping model by utilizing exchange spread requirements among various states. To develop a classifier dependent on the mutual highlights of the articles found in the different modes, cross modal characterization [4, 5] strategy is used, that is utilized to group the interactive media. The cross modal retrieval technique proposed by Rasiwasia et al. [6] in view of canonical correlation analysis (CCA) to acquire the mutual depiction between various states. To get familiar with a maximally related subspace by CCA, and then figure out how to pick classifiers in the semantic space for cross-modal retrieval, [6] proposed this method. A cross media highlight extraction technique dependent on CCA is proposed to acknowledge cross media characterization.

A 3D object class demonstration is done by paper [7], which depended on the desirable value of posture and the acknowledgment of imperceptible states. A functioning exchange learning strategy is proposed to associate the highlights of various scenes [8]. In view of boosting, a feeble classifier is found out in every mode and the coordinated classifier is delivered by weighted mix. These attributes are huge, heterogeneous and in high measurement. Staggered attributes will without a doubt increment the trouble of recovering, writing and the relationship between the base and top is called semantic hole. In light of the customary substance of cross media learning research, we can't settle the cross media relationship metric issue.

As of late, scientists proposed an assortment of cross modal knowledge accumulation strategy to investigate across the ability of connecting between media information, enhancing the productivity of cross modal studies. Scientists have attempted incredible endeavors in creating different strategies from administered to unsupervised, from direct to non straight, for cross-modal retrieval. Canonical Correlation Analysis is the workhouse in CMR [9] by boosting the connecting relationship of heterogeneous highlights. In view of CCA, different variations are considered to show the mixed modal connection. Afterwards, they build up the methodology with the replacement of KCCA. What's more, a general strategy for utilizing CAA to think about the semantic depiction of web pictures and closely connected writings was introduced. As of late, some advancement has been made in the field of cross media characterization.

## 3 Proposed Work

In this segment, the documentations and problem statement is depicted. At that point, the general target work is planned. At long last, a viable improvement calculation is introduced to tackle the defined enhancement issue.

**A. Notations and Problem Definition**

In this paper, we mostly examine the normal cross-modal retrieval undertakings between picture, content and video: Image recovers Text (I2T) and Text recovers Image (T2I). Let $P_1 = [p_1^1, \ldots, P_n^1] \in \mathbb{R}^{d_1 \times n}$ and $P_2 = [p_1^2, \ldots, P_n^2] \in \mathbb{R}^{d_2 \times n}$ represent the characteristic notion of denoting image and text modalities, respectively. $\mathrm{Q} = [\mathrm{q}_1, \ldots, \mathrm{q}_\mathrm{c}] \in \mathbb{R}^{n \times c}$ represents a content defining matrix, and $i^{th}$ row denotes the content vector similar to $p_i^1$ *and* $p_i^2$. Specifically, if $p_i^1$ *and* $p_i^2$ belong to the $j^{th}$ class, we assign $q_{ij}$ to 1, contrarily it is assigned to 0. It is explained in Fig. 1 below.

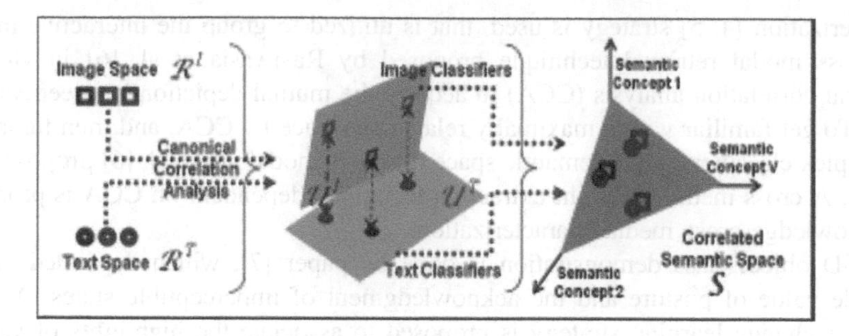


Fig. 1. Modalities mapped to different subspaces

**B. Objective Formulation**

The intention of the work is to acquire both pairs of matrices already mapped $A_m \in \mathbb{R}^{d_1 \times C}$ and $B_m \in \mathbb{R}^{d_1 \times C}$ (m = 1, 2 which depicts 2 varieties of intermediary recovery tasks) for image and content modalities. With the scholarly projection lattices, the multi-modular information can be mapped into a mutual subspace for similitude estimation. The general learning structure is figured as:

$$\min f(A_m, B_m) = L(A_m, B_m) + C(A_m, B_m) + S(A_m, B_m) + \Omega(A_m, B_m) \quad (1)$$

The studies on target work contains 4 sections: which is a straight relapse term which straight forwardly relates the methodology explicit subspace with unequivocal abnormal state semantics. Connection investigation term *C* (*A*, B) is that protects the pairwise semantic consistency of two modalities in the semantic spaces and the element choice term *S* (*A*, B) is that keeps away from over-fitting issue and chooses the discriminative highlights for methodology connection improvement. Ω (*A*, B) is a graph regularization term. The exploited features in hidden complex structure to protect the modality in between and intra-modality based on semantic consistency.

1. Derivation of I2T: While dealing with I2T task, two optimal projection matrices are calculated equivalent to the image and text modalities as: $A_1 \in \mathbb{R}^{d_1 \times C}$ and $B_1 \in \mathbb{R}^{d_2 \times C}$, respectively. The real function of I2T assigned here as follows:

$$\min f(A_1, B_1) = \|P_1^T A_1 - Q\|_F^2 + \|P_1^T A_1 - P_2^T B_1\|_F^2 + \lambda_1 \|A_1\|_{21} + \lambda_2 \|B_1\|_{21} + \lambda_3 \Omega\,(A_1, B_1) \quad (2)$$

$$\min f(A_2, B_2) = \|P_2^T B_2 - Q\|_F^2 + \|P_1^T A_2 - P_2^T B_2\|_F^2 + \lambda_1 \|A_2\|_{21} + \lambda_2 \|B_2\|_{21} + \lambda_3 \Omega\,(A_2, B_2) \quad (3)$$

Here λi (i = 1, 2, 3) represents balancing parameters

**C. Feature Extraction**

Amid the previous couple of years, profound CNN has illustrated a solid capacity for image characterization on some freely accessible datasets, for example, COCO and

Flickr. Some ongoing articles illustrated that the CNN models pretrained on extensive informational collections with information assorted variety, e.g., COCO, can be straightforwardly exchanged to remove CNN visual highlights for different visual acknowledgment errands for example, image grouping and item location. To adjust the parameters pretrained on COCO to the target dataset, images are used from the target dataset to fine tune the CNN. The relationship between two modalities is worked by their mutual ground truth label(s).

i. Extracting Visual Features From Pretrained CNN Model
   Enlivened by Yunchao et al. [8], Wang et al. [4], exhibited the remarkable execution of the CNN visual features from different acknowledgment undertakings, the pretrained CNN model is used commonly to separate CNN visual features for CMR. Specifically, each image is first resized to $256 \times 256$ and given into the CNN. We just use the center patch of the image to deliver the CNN visual highlights. fc6 and fc7 represent the 4096 dimensional highlights of the initial two completely associated layers after the rectified linear units (ReLU) [2].
ii. Extracting Visual Features From Fine-Tuned CNN Model
   Since the classifications (and the quantity of classes) among COCO and the target data set are typically extraordinary, straightforwardly utilizing the pretrained CNN model to extricate visual highlights may not be the best technique. Every image from the objective informational collection is resized into $256 \times 256$ without trimming. The yield of the last fully-connected layer is then encouraged into a c-way delicate max which delivers a likelihood dissemination over c classes. The square of loss function can accomplish a comparable or shockingly better arrangement precision when the quantity of classes of target dataset is less. Nonetheless, with the development of the quantity of classes, cross entropy loss function can achieve a superior characterization result.

**D. Cross Modal Mapping in Subspaces**

Deep learning has been credited to staggering advances in PC vision, normal language handling, and general example understanding. Ongoing disclosures have empowered proficient vector portrayals of both visual and composed upgrades. To probably exchange among visual and content portrayals vigorously remains a test that could yield benefits for recovery applications. Vector portrayals of every methodology are mapped to a Common Vector Space (CVS) utilizing individual inserting systems. These cross-modular subspace learning techniques have the basic work process of learning a projection framework for every methodology to extend the information from various modalities into a typical equivalent subspace in the preparation stage.

**E. Training of Tensor Box**

The Tensor Box venture takes an ".idl" document for preparing. It must contain a line for every one of the picture, with the relative jumping encloses the items are kept in the image. Here some extrapolation from Tensor Board of the preparation stage for the model over 2 M emphases. Every one of the capacities turns out to be level, yet have inadequate outcomes and tops out of the level esteem, this is because of the speculation of the model, that isn't learning diverse covers for each item class, however just a single for every one of them. This brings down the entire precision. This implies the model wasn't adapting any more. Tensor Box gives an all the more clear view on the exactness of the model and furthermore gives the likelihood to watch the test stage on the pictures.

**F. Dataset Preparation**

The datasets for the Cross modular recovery is made from the 3 Datasets speaking to various modalities. Explanations are done, which maps diverse modalities in highlight space. The datasets are:

a. COCO dataset:

   The COCO dataset is an extensive visual database intended for use in visual item acknowledgment, object discovery, division, and subtitling dataset programming research. More than 14 million URLs of pictures have been hand-commented on to show what objects are imagined; in something like one million of the pictures, bouncing boxes are additionally given. The database of explanations of outsider picture URLs is openly accessible specifically.

   COCO has a few highlights:

   i. 91 stuff classes
   ii. Super pixel stuff division
   iii. 1.5 million item occurrences
   iv. 80 object classes
   v. 250,000 individuals with keypoints

b. Flickr Dataset:

   It comprises of 25000 pictures downloaded from the social webpage Flickr through its open API combined with full manual comments. Specifically for the examination network devoted to enhancing picture recovery. It has gathered the client provided picture Flickr labels just as the EXIF metadata and make it accessible in simple to-get to content records. Furthermore, it gives manual picture explanations on the whole accumulation reasonable for an assortment of benchmarks.

c. TRECVID Dataset:

   The primary objective of the TREC Video Retrieval Evaluation (TRECVID) is to advance the substance based examination of and recovery from computerized video by means of open and measurements based assessment. TRECVID is a research center style assessment that endeavors to demonstrate genuine circumstances or noteworthy part undertakings associated with such circumstances. In TRECVID stood up to known-thing seek and semantic ordering frameworks with another arrangement of Internet portrayed by a high level of assorted variety in maker, content, style, creation characteristics, unique gathering gadget/encoding, language, and so on, as is normal in much "Web video". The accumulation additionally has related watchwords and portrayals given by the video contributor.

## 4 Conclusion

With the improvement of the Internet, sight and sound data, for example, picture and video, is progressively advantageous and quicker. Accordingly, how to locate the required sight and sound information on an extensive number of assets rapidly and precisely, has turned into an exploration center in the field of data recovery. A constant web CMR dependent on profound learning is proposed here. The strategy can accomplish high accuracy in picture CMR, utilizing less recovery time. In the meantime, for an expansive database of information, the technique can be utilized to recognize the mistake rate of picture acknowledgment, which is useful for machine learning. The structure joins the preparation of direct projection and the preparation of nonlinear shrouded layers to guarantee that great highlights can be educated. In the field of cross modular recovery, recovery exactness and recovery time utilization is a critical pointer to assess the nature of a calculation.

**Compliance with Ethical Standards**

✓ All authors declare that there is no conflict of interest.
✓ No humans/animals involved in this research work.
✓ We have used our own data.

## References

1. Wang, L., Sun, W., Zhao, Z., Su, F.: Modeling intra- and inter-pair correlation via heterogeneous high-order preserving for cross-modal retrieval. Sig. Process. **131**, 249–260 (2017)
2. Bai, X., Yan, C., Yang, H., Bai, L., Zhou, J., Hancock, E.R.: Adaptive hash retrieval with kernel based similarity (2017)
3. Wang, K., He, R., Wang, W., Wang, L., Tan, T.: Learning coupled feature spaces for crossmodal matching. In: Proceedings of the IEEE International Conference on Computer Vision, pp. 2088–2095 (2013)
4. Wang, K., He, R., Wang, L., Wang, W., Tan, T.: Joint feature selection and subspace learning for cross-modal retrieval. IEEE Trans. Pattern Anal. Mach. Intell. **38**(10), 2010–2023 (2016)
5. Wang, J., He, Y., Kang, C., Xiang, S., Pan, C.: Image-text cross-modal retrieval via modality-specific feature learning. In: Proceedings of the 5th ACM on International Conference on Multimedia Retrieval, pp. 347–354 (2015)
6. Rasiwasia, N., Costa Pereira, J., Coviello, E., Doyle, G., Lanckriet, G.R.G., Levy, R., Vasconcelos, N.: A new approach to cross-modal multimedia retrieval. In: Proceedings of the 18th ACM on Multimedia Conference, pp. 251–260 (2010)
7. Jiang, B., Yang, J., Lv, Z., Tian, K., Meng, Q., Yan, Y.: Internet cross-media retrieval based on deep learning. J. Vis. Commun. Image Retrieval **48**, 356–366 (2017)
8. Wei, Y., Zhao, Y., Lu, C., Wei, S., Liu, L., Zhu, Z., Yan, S.: Cross-modal retrieval with CNN visual features: a new baseline. IEEE Trans. Cybern. **47**, 251–260 (2016)
9. Pereira, J.C., Vasconcelos, N.: Cross-modal domain adaptation for text-based regularization of image semantics in image retrieval systems. Comput. Vis. Image Underst. **124**, 123–135 (2014)

10. He, J., Ma, B., Wang, S., Liu, Y.: Multi-label double-layer learning for cross-modal retrieval. Neuro-computing 123–135 (2017)
11. Hu, X., Yu, Z., Zhou, H., Lv, H., Jiang, Z., Zhou, X.: An adaptive solution for large-scale, cross-video, and real-time visual analytics. In: IEEE International Conference on Multimedia Big Data, pp. 251–260 (2015)
12. Lavrenko, V., Manmatha, R., Jeon, J.: A model for learning the semantics of pictures. In: NIPS (2003)

# Smart Wearable Speaking Aid for Aphonic Personnel

S. Aswin[1(✉)], Ayush Ranjan[1], and K. V. Mahendra Prashanth[2]

[1] Department of Electronics, SJB Institute of Technology, Bangalore, India
saswin1996@gmail.com, ayushranjan445@gmail.com

[2] Department of Electronics and Communication, SJB Institute of Technology, Bangalore, India
kvmprashanth@sjbit.edu.in

**Abstract.** Gesturing is an instinctive form of non-verbal or non-vocal communication in which perceptible bodily actions are used to communicate definite messages in conjunction with or without speech. But, aphonic personnel rely solely on this way of communication. This inhibits them from interacting or communicating freely with other people. This paper deals with a Smart Wearable Speaking aid which translates Indian Sign Language into text and voice in real time based on SVM (Support Vector Machine) Model using Raspberry Pi, Flex Sensors, and Accelerometer. Additionally, LCD is used to display the text corresponding to gestures and equivalent output in the form of voice signal is given to the Bluetooth Speaker. The prototype was designed to identify 5 gestures. During the testing of this device, it was observed that the accuracy of the prototype is 90%.

**Keywords:** Indian Sign Language · SVM model · Raspberry Pi · Speaking aid · Wearable aid

## 1 Introduction

Communication plays a crucial role in conveying an information from one person to another person. They can be a verbal or non-verbal form of communication. Verbal communication relies on conveying the message through Language whereas non-verbal communication (usually termed as gesturing) is an instinctive method of conveying message either in conjunction with or without speech [1]. But people with speech or hearing impairment have gesturing alone as their principal means of communication. This might create a communication barrier between aphonic personnel and normal people.

As per one of the survey report, there are 466 million people around the world who are hearing or speech impaired of which 34 million are children [2]. From the census of 2011, it is observed that there are around 30 million people in India who are aphonic [3]. Around the world, each nation has developed its own variety of sign languages. There are around 300 diverse sign languages in the world [4]. Indian Sign language is commonly used among aphonic people in India. Further, different parts of India adopt different signs.


S. Smys et al. (Eds.): ICCVBIC 2019, AISC 1108, pp. 179–186, 2020.
https://doi.org/10.1007/978-3-030-37218-7_20

However, grammar is common & same throughout the country. Spoken language is not associated to sign languages and has its own grammar structure. Hence it is difficult to standardize the sign language regionally or globally.

Research into the interpretation of sign language gestures in real time is explored progressively in recent decades. There are multiple methods of performing sign language recognition, namely, vision-based and non-vision based approaches. In vision-based approach, a camera is used to capture the gestures and processed to identify the sign. The vision-based method can be further classified into direct and indirect. The direct method processes the captured raw image directly to identify the sign whereas, in the indirect method, the sign is identified based on RGB color space segmented to different colors for each finger using a data glove. In non-vision based approach, sensors are used to analyze the flexing of fingers and identify sign [5–10]. In this paper the non-vision based approach to recognize the sign language is being adopted.

A simple device is designed using Raspberry Pi, Flex sensors, and Accelerometer to function as Smart Wearable Aid. The Flex sensors are used for detecting the Hand gesture and Accelerometer is used for distinguishing the orientation of Palm. The data from the sensors are processed in Raspberry Pi and resulting text is displayed using an LCD screen and voice output is given through a Bluetooth Speaker. Battery with voltage regulator and a current booster is used to power the Raspberry Pi to ensure complete portability.

## 2 Methodology

### 2.1 Prototype Design

The proposed prototype consists of Raspberry Pi 3 (RPi), Flex sensors, MCP3008, Accelerometer, LCD Display, I2C Converter, Bluetooth speaker, and Power Supply (Fig. 1).

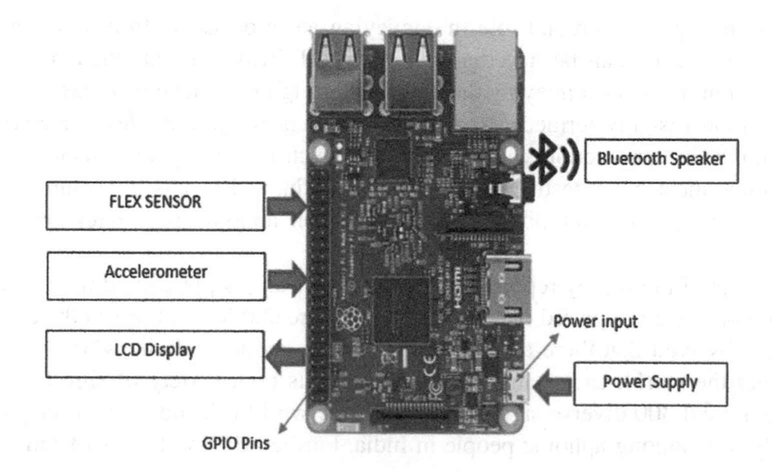


**Fig. 1.** Block diagram of designed prototype

RPi needs 5 V, 2 A to function normally. To make the prototype portable, a 6 V battery is used with IC7805 and a current booster circuit to meet this requirement [11] (Fig. 2). GPIO pins are used to interface the sensors and to power them. Bluetooth 4.1 is used to connect to the Bluetooth Speaker for generating voice output.

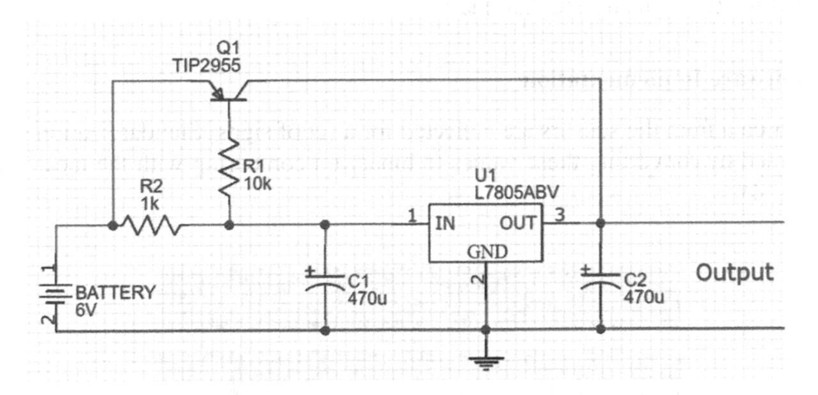


**Fig. 2.** Power supply

Flex sensors works on basis of variable resistor; whose resistance depends on the amount of flexing. Five flex sensors are used for building the prototype. For interfacing RPi with these sensors a voltage divider circuit is used (See Fig. 3). The voltage signal strength at junction increases as the sensor is flexed. Since RPi does not support analog inputs, IC MCP3008 is used to interface the sensors to RPi. MCP3008 is a 10-bit DAC with 8 channels. The device is capable of conversion rates of up to 200 ksps [12]. It is interfaced to RPi using an SPI (Serial Peripheral Interface).

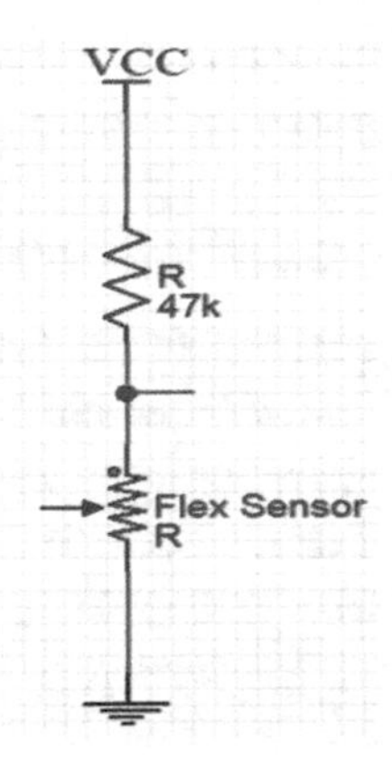


**Fig. 3.** Flex sensor interfaced with voltage divider

Accelerometer ADXL345 is used to identify the orientation of palm. It is an ultralow power, 3-axis accelerometer with 13-bit resolution measuring up to ±16 g. The digital output is formatted as 16-bit twos complement data and are capable of communicating over SPI or $I^2C$ interface [13]. In the prototype, the $I^2C$ digital interface is used to connect the Accelerometer. The LCD is used to display the text output and is interfaced to RPi by using $I^2C$ interface.

## 2.2 Software Implementation

The input data from the sensors are collected for a set of signs. Standardization of data is performed by converting these values to binary on comparing with the mean values [14] (Fig. 4).

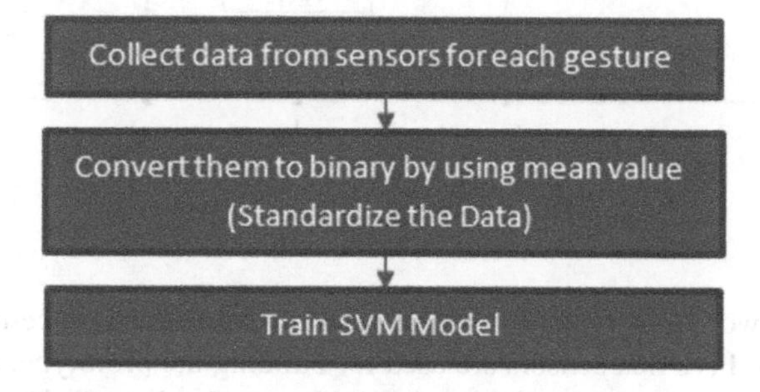


**Fig. 4.** Flowchart for training the SVM model

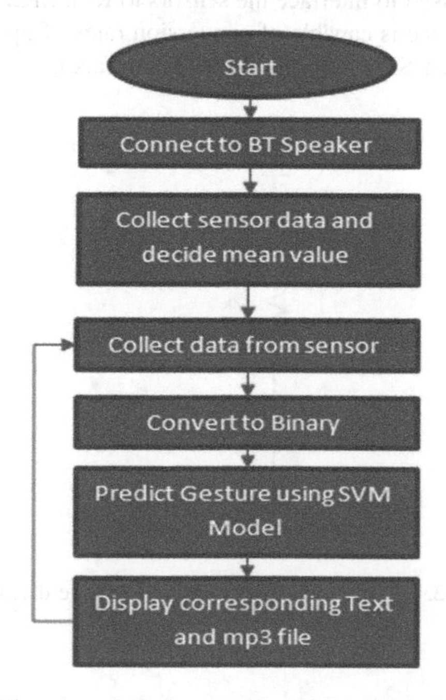


**Fig. 5.** Flowchart depicting working of Raspberry Pi module

The model obtained through various processes as described is then used in the real-time detection of gestures. The program for this is coded in python and is set to run on boot up. After booting up, RPi will automatically get connected to the Bluetooth speaker. Once the speaker is connected, a series of data from the sensors are collected to obtain the mean value. These Mean values are then used to convert the series of future sensor data to binary. The SVM model with the binary sensor values is used to predict the gestures. The corresponding text is displayed in the LCD and the pre-recorded mp3 file of the respective gesture is played through Bluetooth speaker (Fig. 5).

## 3 Results

The python code for real time translation using the pre-trained model is set to run on the startup. Initially, the Bluetooth speaker gets connected to the RPi. After this, we can see the "SIGN LANGUAGE TRANSLATOR" message getting displayed on the screen after a small delay. After this, a message is displayed on LCD to inform the user to make gestures of an open palm and a fist, during which a series of data from sensors are collected to decide the mean and threshold values which are used to convert the future values into binary. After this, the device can be used for the real-time translation of gestures (Figs. 6, 7, 8 and 9).

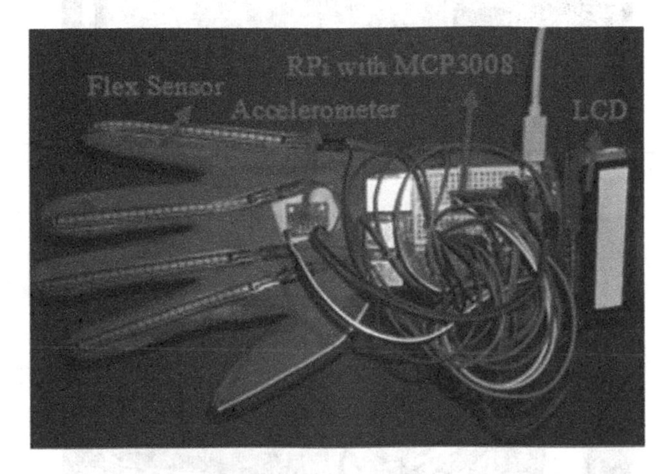


**Fig. 6.** Snapshot of designed prototype

**Fig. 7.** A snapshot illustrating start-up

**Fig. 8.** Displaying instructions to user

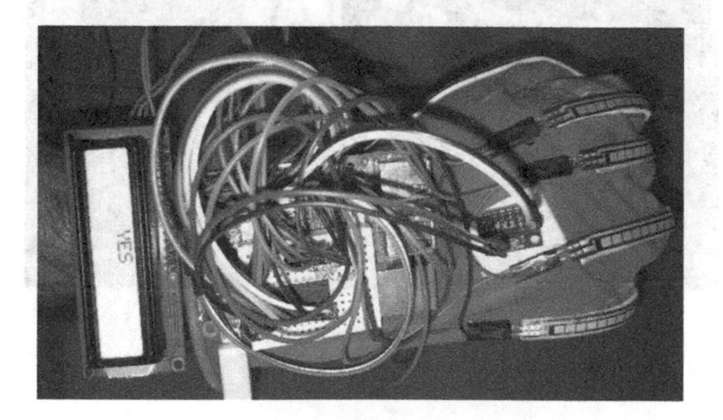


**Fig. 9.** Translator displaying respective text

## 4 Conclusion and Future Scope

A simple cost-effective prototype is designed using Raspberry Pi, Flex sensors, and accelerometer to function as Smart Wearable Aid. The Flex sensors are used for detecting the hand gesture and accelerometer for distinguishing the orientation of Palm. The data from the sensors are processed in Raspberry Pi and resulting text is displayed using an LCD screen and voice output is given through a Bluetooth Speaker. Battery with voltage regulator and a current booster is used to power the Raspberry Pi to ensure complete portability.

From the obtained results it is observed that accuracy of the prototype is 90%. This could be further improvised by training with a large and raw data set rather than converting them to binary. Multiple classification models were compared and tested to select the applicable one for the given application; the real-time implementation could not achieve appreciable accuracy while using other classification models. It is suggested to identify and select a better real-time classification model for this application. As in this perspective of comparison, i.e., within the considered classification models for comparison, SVM was found to obtain better accuracy. The effectiveness of this device could be improved by effectively connecting the sensors and peripherals. Product can also be introduced into the market by devising a single powerful chip to perform all required functions.

**Compliance with Ethical Standards.**

✓ All authors declare that there is no conflict of interest.
✓ No humans/animals involved in this research work.
✓ We have used our own data.

## References


1. Gesture: en.wikipedia.org. Wikipedia. https://en.wikipedia.org/wiki/Gesture
2. Deafness and Hearing Loss: World Health Organization. http://www.who.int/news-room/fact-sheets/detail/deafness-and-hearing-loss
3. Census 2011 India. http://www.censusindia.gov.in/DigitalLibrary/data/Census_2011/Presentation/India/Disability.pptx
4. List of Sign Languages: en.wikipedia.org. Wikipedia. https://en.wikipedia.org/wiki/List_of_sign_languages
5. Mulder, A.: Hand gestures for HCI. In: Technical Report 96-1, vol. Simon Fraster University (1996)
6. Sriram, N., Nithiyanandham, M.: A hand gesture recognition based communication system for silent speakers. In: International Conference on Human Computer Interaction, Chennai, India, August 2013
7. Lee, B.G., Lee, S.M.: Smart wearable hand device for sign language interpretation system with sensors fusion. IEEE Sens. J. **18**(3), 1–2 (2018)
8. Quiapo, C.E.A., Ramos, K.N.M.: Development of a sign language translator using simplified tilt, flex and contact sensor modules. In: 2016 IEEE Region 10 Conference (TENCON), p. 1 (2016)

9. Ekbote, J., Joshi, M.: Indian sign language recognition using ANN and SVM classifiers. In: 2017 International Conference on Innovations in Information Embedded and Communication Systems (ICIIECS), p. 1 (2017)
10. Pawar, S.N., Shinde, D.R., Alai, G.S., Bodke, S.S.: Hand glove to translate sign language. IJSTE – Int. J. Sci. Technol. Eng. **2**(9), 1 (2016)
11. Raspberry Pi 3 Model B. http://www.raspberrypi.org/products/raspberry-pi-3model-b/
12. MCP3008, 10-bit ADC. http://www.farnell.com/datasheets/1599363.pdf
13. ADXL345. http://www.analog.com/en/products/sensors-mems/accelerometers/adxl345.html#product-overview
14. Introduction to Support Vector Machine. https://docs.opencv.org/3.0-beta/doc/tutorials/ml/introduction_to_svm/introduction_to_svm.html

# Data Processing for Direct Marketing Through Big Data

Amelec Viloria[1(✉)], Noel Varela[1], Doyreg Maldonado Pérez[1], and Omar Bonerge Pineda Lezama[2]

[1] Universidad de la Costa (CUC), Calle 58 # 55-66, Barranquilla, Atlantico, Colombia
{aviloria7,nvarela1,dmaldona}@cuc.edu.co

[2] Universidad Tecnológica Centroamericana (UNITEC), San Pedro Sula, Honduras
omarpineda@unitec.edu

**Abstract.** Traditional marketing performs promotion through various channels such as news in newspapers, radio, etc., but those promotions are aimed at all people, whether or not interested in the product or service being promoted. This method usually leads to high expenses and a low response rate by potential customers. That is why, nowadays, because there is a very competitive market, mass marketing is not safe, hence specialists are focusing efforts on direct marketing. This method studies the characteristics, needs and also selects a group of customers as a target for the promotion. Direct marketing uses predictive modeling from customer data, with the aim of selecting the most likely to respond to promotions. This research proposes a platform for the processing of data flows for target customer selection processes and the construction of required predictive response models.

**Keywords:** Data stream · WEKA · MOA · SAMOA · Big Data · Direct marketing

## 1 Introduction

Direct marketing is a data set-oriented process for direct communication with target customers or prospects [1, 2]. But at the same time, marketers face the situation of changing environments. Current datasets constitute computers insert data into each other making environments dynamic and conditioned by restrictions such as limited storage capacity, need for real-time processing, etc. This means that *Big Data* processing approaches [3, 4] are required; Hence, in order to establish and maintain the relationship with clients, specialists have anticipated the need to change the methods of intuitive group selection for more scientific approaches aimed at processing large volumes of data [5], obtaining rapid responses that allow selecting customers who will respond to a new offer of products or services, under a distributed data flow approach [6].

Several researches relate to computational and theoretical aspects of direct marketing, but few efforts have focused on technological aspects needed to apply data mining to the direct marketing process [7–11]. Situation is that gains in complexity


S. Smys et al. (Eds.): ICCVBIC 2019, AISC 1108, pp. 187–192, 2020.
https://doi.org/10.1007/978-3-030-37218-7_21

when distributed data flow environments exist. Researchers need to devote effort and time to implementing data stream simulation environments, both demographic and historical, shopping, because you don't have a platform that handles tasks like that while being scalable, allowing the incorporation of new variants of algorithms for their respective evaluation. Faced with such a problem and the importance of even as a basis for future research in the area of data flow mining for direct marketing; this research proposes a processing platform based on Free *Software* technologies and with a high level of scalability.

## 2 Direct Marketing

Direct marketing is a set of techniques that allows you to create personal communication with each potential buyer, especially segmented (social, economic, geographical, professional, etc.), in order to promote a product, service, idea, and maintain it over time. using for this means or direct contact systems (emailing, telemarketing, couponing, mailboxing, new technologies that offer us "virtual markets", multimedia systems and all the new media that facilitate technological advances online).

Basically it consists in sending individualized communications addressed to a previously selected audience based on certain variables and with which it is sought to have a continuous relationship. The elaboration of your message follows, with adaptations, the same process as that of mass advertising, that is, creativity, production and dissemination, which are located within the corresponding planning.

It is basically solved in two ways:

- personalized mail or mailing, personalized deliveries to your home or workplace, which may include response formulas.
- mailbox and brochures, which are distributed in homes and workplaces without a recipient's address and according mainly to geographical criteria.

Direct marketing is considered essentially part of relational marketing whose strategic objective is to convert any sale or contact with customers into lasting relationships based on the satisfaction of their needs and preferences. To achieve this, apply the techniques of direct marketing and other of its main tools, telemarketing. In this case the contact tries to establish itself through the telephone that, in front of the mail, presents its own advantages and disadvantages. Among the advantages, it should be noted the rapidity in communication, a greater security of contact with the target audience, a lower degree of reticence and the possibility of counterargument and even offering alternatives adapted to each person.

## 3 Methodology

Large volumes of that data are generated every day in many organizations, which can be used to establish and maintain a direct relationship with customers to identify what groups of customers would respond to offers Specific [14].

There are companies such as *Nielsen Datasets Transforming Academic Research in Marketing* (2015), specialized in collecting and maintaining sales data for marketing companies [15]. A special category is databases with demographic information, which provide additional information at the group level, and are useful for determining target groups that have similar properties. Contrary to the external database, company internals provide more relevant and reliable information. Regarding customer behavior.

These attributes (RFMs) are identified as the three most representative variables for marketing modeling. When these variables are used in combination, very accurate predictive results are achieved and are also very useful to support customer relationship management (CRM) [17]. RFMs tend to be 10 or less for a dataset; hence, attribute reduction is not a critical step when using these variables.

In the *Bank Marketing Data Set* the data is related to an advertising campaign of a Portuguese banking institution, aimed at several customers. There are other datasets such as the one proposed by *Nielsen Datasets* (2015) [18] that record people's consumption habits in the United States, which are available to academic researchers. The following table lists the attributes (Characterization of datasets for Direct Marketing), see Table 1.

**Table 1.** Characterization of datasets for Direct Marketing

| Data set | Features | Instances | Attributes |
|---|---|---|---|
| Bank marketing data set | Multivariate | 45478 | 12 |
| Insurance company benchmark (COIL 1000) data set | Multivariate | 9964 | 87 |

# 4 Results

## 4.1 Technology Selection Criteria

There are currently two approaches to Big Data processing. One uses incremental algorithms for processing and the other approach is based on distributed computing processing. Since the nature of the data sources generated by customers' consumption habits are ubiquitous, the article proposes a processing of distributed data flows, capable of solving the tasks of each of the stages of the CRISP-DM methodology. To do this, there are several frameworks and technologies that are very useful, and in some cases indispensable for the evaluation and development of algorithms capable of processing data, both for stationary environments and for data flow environments.

## 4.2 Principles for Technology Selection

As you mentioned above, there are several technologies for data mining. That is why several selection principles were followed for the design of the proposed platform [19]:

1. Suitability: The tools used in the design perform the required function at each stage.
2. Interoperability: easy to integrate and connect the output of certain processes as inputs from others.

3. Homogeneous technology: all the tools used for the design are developed under the same paradigms and programming technologies.
4. Scalable: new algorithms can be incorporated for each of the stages of data processing included in the CRISP-DM methodology, as well as modify existing ones.
5. Free Software: the selected tools comply with the four freedoms of Free Software, thus guaranteeing the deployment of the proposed architecture, such as the reproducibility of the experiments.

### 4.3 Selected Technologies

- WEKA (Waikato Environment for Knowledge Analysis): It is a tool that allows the experimentation of data analysis through the application, analysis and evaluation of the most relevant data analysis techniques, mainly those from machine learning, on any set of user data. Weka also provides access to databases via SQL thanks to the Java Database Connectivity (JDBC) connection and can process the result returned by a query made to the database. You cannot perform multi-relational data mining, but there are applications that can convert a collection of related tables in a database into a single table that can already be processed with Weka.
- MOA: Massive Online Analysis [21] is a framework for implementing and evaluating real-time learning algorithms in data flows. MOA includes a collection of algorithms for classification and grouping tasks. The framework follows the principles of green computing using processing resources efficiently, being a required feature in the data stream model where data arrive at high speeds and algorithms must process them under time and space constraints. The framework integrates with WEKA.
- SAMOA: Scalable Advanced Massive Online Analysis is a platform for mining distributed data flows [22]. Like most data flow processing frameworks, it is written in Java. It can be used as a framework for development or as a library in other software projects. As a framework it allows developers to abstract themselves from tasks related to runtimes, which makes it easier for code to be used with different stream processing engines, making it possible to implement an architecture on different engines like Apache Storm, Apache S4 and Samza. SAMOA achieves seamless integration with MOA and MOA classification and processing algorithms can therefore be used in distributed flow environments. It is a very useful platform in itself for both researchers and deployment solutions in real-world environments.
- Apache Spark: [23, 24] is an open source *framework* built for fast big data processing. Easy to use and sophisticated analysis. Spark has several advantages compared to other Big Data processing *frameworks* that are based on MapReduce such as Hadoop and Storm. This *framework* offers unified functionalities for managing Big Data requirements, including data sets of various in both nature and data source. Spark beats the processing speed of other MapReduce-based *frameworks* with capabilities such as in-memory data storage and real-time processing. It keeps intermediate results in memory instead of writing them to disk.

### 4.4 Proposal

A distributed processing platform for demographic and historical purchasing data streams from various sources is proposed, based on the interoperability of the technologies described above. To do this, a processing cluster was deployed on a set of mid-range desktop computers, running SAMOA configured with multiple processing engines (SPEs) between them Storm and S4. These take care of tasks such as data serialization, which are evaluated in *Scalable Advanced Massive Online Analysis* (SAMOA) [25] using *Massive Online Analysis* (MOA) algorithms [26] and WEKA [27] for the preprocessing and data modeling stages [28].

Several of the algorithms available in the above-mentioned *frameworks* were evaluated, on the *Bank Marketing Data Set* data set managing to identify the target groups and proposing a predictive response model in real time.

## 5 Conclusions

With the development of Computer and Telecommunications Technologies, companies are adopting direct marketing strategies that allow to identify potential customers of their products and/or services, thus avoids large resource expenditures and efforts in mass advertising campaigns. This article described the nature of the data used by some-rhythms for the selection of target customer groups, highlighting updated datasets. In addition, a platform was proposed for the processing of customer data from different sources. For this was evaluated for technologies for environments flows of distributed data, this makes it possible to analyses themselves in real time. Several selection criteria were applied, ensuring interoperability, and the freedoms offered by Free *Software* products. New algorithms requiring processing of data flows. This facilitates tasks such as simulating distributed environments, serializing data, among others that are functional requirements for research related to this field of action.

## References

1. Marston, S., Li, Z., Bandyopadhyay, S., Zhang, J., Ghalsasi, A.: Cloud computing—the business perspective. Decis. Support Syst. **51**(1), 176–189 (2011)
2. Bifet, A., De Francisci Morales, G.: Big data stream learning with Samoa. Recuperado de (2014). https://www.researchgate.net/publication/282303881_Big_data_stream_learning_with_SAMOA
3. Mell, P., Grance, T.: The NIST definition of cloud computing. NIST Special Publication 800–145 (2011)
4. Valarie Zeithaml, A., Parasuraman, A., Berry, L.L.: Total, quality Management services. Diaz de Santos, Bogota (1993)
5. Sitto, K., Presser, M.: Field Guide to Hadoop, pp. 31–33. O'Reilly, California (2015)
6. Sosinsky, B.: Cloud Computing Bible, 3 p. Wiley Publishing Inc., Indiana (2011)
7. Casabayó, M., Agell, N., Sánchez-Hernández, G.: Improved market segmentation by fuzzifying crisp clusters: a case study of the energy market in Spain. Expert Syst. Appl. **41** (3), 1637–1643 (2015)

8. Izquierdo, N.V., Lezama, O.B.P., Dorta, R.G., Viloria, A., Deras, I., Hernández-Fernández, L.: Fuzzy logic applied to the performance evaluation. Honduran coffee sector case. In: Tan, Y., Shi, Y., Tang, Q. (eds.) Advances in Swarm Intelligence, ICSI 2018. Lecture Notes in Computer Science, vol 10942. Springer, Cham (2018)
9. Pineda Lezama, O., Gómez Dorta, R.: Techniques of multivariate statistical analysis: an application for the Honduran banking sector. Innovare: J. Sci. Technol. **5**(2), 61–75 (2017)
10. Viloria, A., Lis-Gutiérrez, J.P., Gaitán-Angulo, M., Godoy, A.R.M., Moreno, G.C., Kamatkar, S.J.: Methodology for the design of a student pattern recognition tool to facilitate the teaching - learning process through knowledge data discovery (Big Data). In: Tan, Y., Shi, Y., Tang, Q. (eds.) Data Mining and Big Data, DMBD 2018. Lecture Notes in Computer Science, vol. 10943. Springer, Cham (2018)
11. Zhu, F., et al.: IBM cloud computing powering a smarter planet. In: Libro Cloud Computing, vol. 5931, pp. 621–625 (2009)
12. Fundación Telefónica: La transformación digital de la industria española. Informe preliminar. Ind. Conectada 4.0, p. 120 (2015)
13. Thames, L., Schaefer, D.: Software defined cloud manufacturing for industry 4.0. Procedía CIRP **52**, 12–17 (2016)
14. Gilchrist, A.: Industry 4.0 the industrial Internet of Things. In: 2016th edited by Bangken, Nonthaburi, Thailand (2016)
15. Schweidel, D.A., Knox, G.: Incorporating direct marketing activity into latent attrition models. Mark. Sci. **31**(3), 471–487 (2013)
16. Setnes, M., Kaymak, U.: Fuzzy modeling of client preference from large data sets: an application to target selection in direct marketing. IEEE Trans. Fuzzy Syst. **9**(1), 153–163 (2001)
17. Thompson, E.B., Heley, F., Oster-Aaland, L., Stastny, S.N., Crawford, E.C.: The Impact of a student-driven social marketing campaign on college student alcohol-related beliefs and behaviors. Soc. Mark. Q. 1514500411471668 (2013)
18. Thorson, E., Moore, J.: Integrated Communication: Synergy of Persuasive Voices. Psychology Press/Lawrence Erlbaum Associates, Inc., Publishers, New Jersey (2013)
19. Lars Adolph, G.: Bundesanstalt: DIN/DKE-Roadmap GERMANS TANDARDIZATION
20. Morales, P.G., España, J.A.A., Zárate, J.E.G., González, C.C.O., Frías, T.E.R.: La Nube Al Servicio De Las Pymes En Dirección a La Industria 4.0. Pist. Educ. **39**(126), 85–98 (2017)
21. Wu, Q., Yan, H.S., Yang, H.B.: A forecasting model based support vector machine and particle swarm optimization. In: 2008 Workshop on Power Electronics and Intelligent Transportation System, pp. 218–222 (2008)
22. Ellingwood, J.: Apache vs Nginx: Practical Considerations. Disponible en (2015). https://www.digitalocean.com/community/tutorials/apache-vs-nginx-practical-considerations
23. Gilbert, S., Lynch, N.: Perspectives on the CAP theorem. Computer **45**(2), 30–36 (2012)
24. Gouda, K., Patro, A., Dwivedi, D., Bhat, N.: Virtualization approaches in cloud computing. Int. J. Comput. Trends Technol. (IJCTT) **12**(4), 161–166 (2014)
25. Hollensen, S.: Marketing Management: A Relationship Approach. Pearson Education, México (2015)
26. James, G., Witten, D., Hastie, T., Tibshirani, R., Hastie, M.T., Suggests, M.: Package 'ISLR' (2013)
27. Kacen, J.J., Hess, J.D., Chiang, W.K.: Bricks or clicks? Consumer attitudes toward traditional stores and online stores. Glob. Econ. Manag. Rev. **18**(1), 11 (2013)
28. Karim, M., Rahman, R.M.: Decision tree and Naïve Bayes algorithm for classification and generation of actionable knowledge for direct marketing. J. Softw. Eng. Appl. **6**, 196–206 (2013). Recuperado de http://file.scirp.org/pdf/JSEA_2013042913162682.pdf

# Brain Computer Interface Based Epilepsy Alert System Using Neural Networks

Sayali M. Dhongade(✉), Nilima R. Kolhare, and Rajendra P. Chaudhari

Government Engineering College, Aurangabad, India
sayalidhongade@gmail.com, nilimageca@gmail.com, rpchaudhari@yahoo.com

**Abstract.** At present, Epilepsy detection is a very time consuming process, and it requires constant clinical attention. In order to overcome this challenge, we have designed an epilepsy alert system, which takes the raw EEG signal, to monitor the blood pressure and the accelerometer application for human fall detection. The raw EEG signal is captured using a Brain Computer Interface module. This design is done using MATLAB GUI. The EEG raw signal is trained by using neural networks and a total of 500 epilepsy patient data is fed to train the network. Accuracy varies with different number of epochs. As the number of epochs increases, the accuracy will also be increased. Fuzzy system is designed to detect the epilepsy with three main parameters.

**Keywords:** Electroencephalograph (EEG) · Epilepsy · Seizure · Brain Computer Interface (BCI) · Windowing · Radial basis network · Fuzzy logic · Epochs

## 1 Introduction

Epilepsy is one of the most common neurological disorders that has affected almost 50 million people all over the world and the count is still increasing [6]. Epilepsy is a disorder of the central nervous systems, where the brain activities increases abruptly. Due to this abnormal condition, there are many unusual sensations [4]. Patients who are mainly diagnosed with epilepsy are children and elderly population, who falls in the age limit of 65–70 years.

### 1.1 Epilepsy

EEG is electrophysiological strategy to record the electrical action that occurs on the surface of the brain. Electroencephalogram (EEG) signals taken from its recording are utilized to exercise the epileptic activities inside the brain [7]. It is commonly a strategy with the electrodes set on the surface of the scalp as shown in Fig. 1. It measures the voltage variations that results from the ionic current present inside the neurons of the brain. A seizure can be described as an abrupt [2] and an uncontrollable electric signals in the brain.


S. Smys et al. (Eds.): ICCVBIC 2019, AISC 1108, pp. 193–203, 2020.
https://doi.org/10.1007/978-3-030-37218-7_22

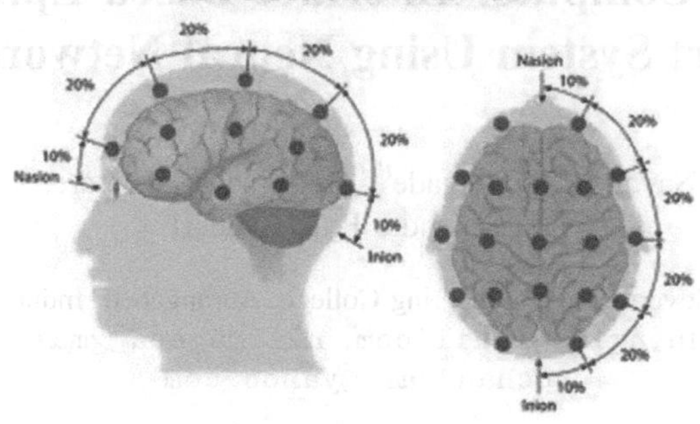


**Fig. 1.** Electrode placement [3]

The paper is organized as follows. Firstly, the work related to BCI, the basic blocks of BCI, and the frequency bands of raw EEG are discussed in Sect. 2. Based on related work being discussed the proposed work is carried out in Sect. 3. Following the proposed work, the Materials required and Methodology is discussed in Sects. 4 and 5. In the next Sect. 6, the Simulation and the results are discussed. Lastly, we summarize the whole project in Conclusion and Future scope.

## 2 Related Work

Brain is one of the most important organs of our body. Brain computer interface has been designed for the user to allow them to control the external devices or computers according to their intentions, these are decoded in binary form from brain activity [9]. The application of BCI has been developed to such a level that we can control the EEG signals for epilepsy patients in seizure control.

The basic blocks of brain computer interface are Pre-processing, Feature extraction, Feature Classification.

1. **Preprocessing:** This is a process which amplifies the EEG signals without disturbing the significant information within it. Signal strengthening guarantees signal quality by improving SNR [11]. EEG sign are examined at 256 Hz which fulfills Nyquist testing theorem. To get high SNR band pass filtering is used to remove the DC bias and high frequency noise.
2. **Feature extraction:** Before classification test and actual BCI, [4] the classifier is firstly trained and then supervised using a classification method [7] which classifies then using feature vectors called "target" and "non target". Different methods are used for feature extraction such as: Discrete wavelet transform [2], independent component analysis, principal component analysis.
3. **Feature classification:** For classification also, various methods are implemented like Linear Discriminant analysis, Stepwise linear discriminant analysis and Support vector machine (Fig. 2).

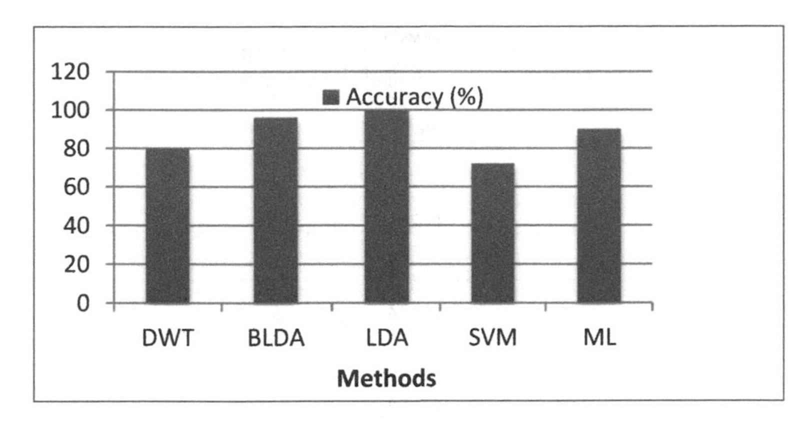


**Fig. 2.** Comparison between different algorithms

**Methods:**

1. Discrete Wavelet Transform (DWT): The [2] performance is 7.7 characters per minute and accuracy is around 80%. Accuracy is greater than 90% for 2.2 characters per minute.
2. Bayesian linear discriminant analysis (BLDA): 95% false positive classification.
3. Linear discriminant analysis: Accuracy is close to 100% [12], and best classification accuracy for disabled body was on an average 100%, varies with the no. of electrodes.
4. Support vector machine (SVM): 96.5% accuracy,
5. Maximum likelihood (ML): Accuracy is 90% with communication rate of 4.2 symbols/min.

The electroencephalogram (EEG) is a multi-channel time series which captures the electrical activity recorded from the scalp [8] from specific locations (electrodes). After spatial analysis then [1], EEG waveform is analyzed in the frequency domain. In which the reason why frequency bands are important is that during some particular tasks it is not the amplitude of whole signal that changes rather the amplitude of particular frequency changes. Using wavelet transform the EEG signal [7] can be classified into following bands shown in the Fig. 3, and the frequencies and period at which they occur is discussed in Table 1.

## 3 Proposed Work

In the proposed system the work is on developing an epilepsy alert system, using EEG signal, accelerometer and blood pressure sensor. In MATLAB we have designed a GUI. The EEG signal is obtained from the Brainsense headband, which takes the raw EEG signal from the headband. The [1] raw EEG signal is then sent to the neural network, for classification. The neural network will then detect if seizure has occurred, if occurred it will check the readings of blood pressure sensor and the accelerometer

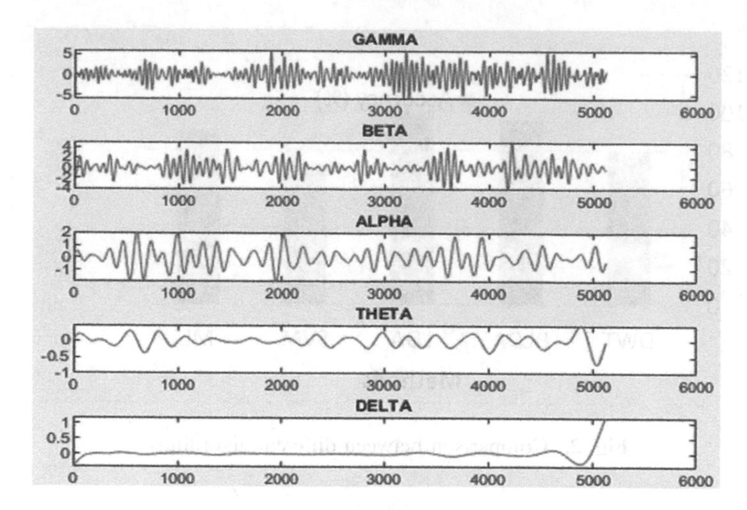


**Fig. 3.** Time domain analysis of raw EEG signal

**Table 1.** Decomposition into various frequencies of raw EEG signal

| Band | Frequency range (Hz) | Period of occurrence |
|---|---|---|
| Delta | 0–3.5 | In profound sleep at any age |
| Theta | 4–7 | In deep relaxed sleep |
| Alpha | 8–13 | Awake and rest mode |
| Beta | 14–30 | Awake and mental activity |
| Gamma | >30 | Highly focused mode during information processing |

which will detect human fall. If the seizure has not occurred it will again check the raw EEG signal till it encounters a seizure. After detecting values from all the three sensors, the values are sent to a fuzzy system, where there are certain rules which are defined to check whether epilepsy is detected or not.

The reason why we are considering all three measures i.e. the EEG signal, blood pressure values and the accelerometer is that sometimes due to excess stress or hypertension our EEG signal is abrupt, which may detect epilepsy and may give false results. So to get true results we are using all sensors. If all the rules in the fuzzy system get satisfied, the epilepsy is detected and the SMS is generated on the cellphone (Fig. 4).

## 4 Materials Required

The hardware required in the epilepsy alert system is the blood pressure sensor, Arduino UNO, Brainsense band and a smartphone. The software used is MATLAB (Fig. 5).

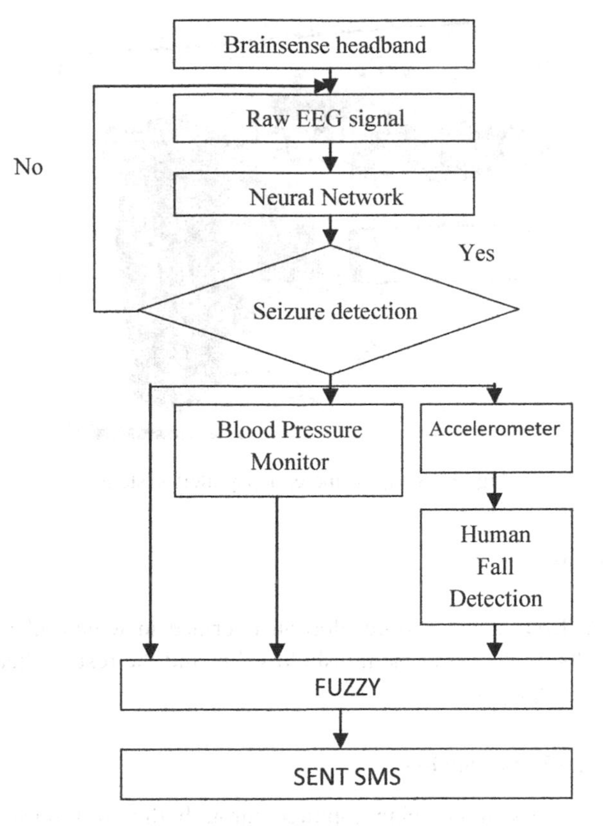


**Fig. 4.** Proposed system

- **Brainsense:**
  This [10] module consists of TGAM 1 Module, dry electrode and Earchip electrode. It is compatible with automatic wireless pairing, Bluetooth 2.1 Class2 is used which has 10 m range. It also possesses Android support.
- **Blood pressure sensor:**
  It reads at 9600 baudrate the blood pressure and heart rate and displays output on the screen. It displays measurement of systolic, diastolic and pulse. It's accuracy clinically. Working voltage is +5 V, 200 mA. It stores 60 group memory measurements.
- **Accelerometer:**
  We use the accelerometer and GPS location from the mobile phone. Human fall is detected by using the traditional human fall detection algorithm [10]. The software also sends the GPS Coordinates from the mobile phone to send alert to the care tracker.

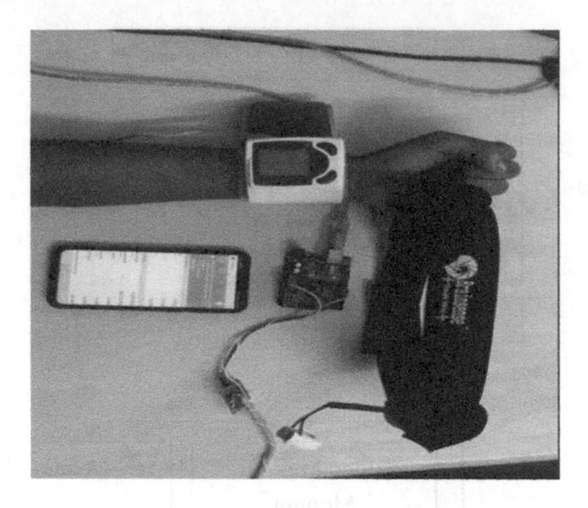

**Fig. 5.** Setup of the epilepsy alert system

## 5 Methodology

The GUI in MATLAB is used to develop an interface in which all the sensors will display their results by checking the threshold value and the result whether epilepsy is detected or not is displayed.

### 5.1 Sampling and Windowing

The EEG signal is first after being captured through the Brainsense is sampled at 256 Hz. A raw EEG signal is sampled and is transformed into discrete values. Due to this transformation Fourier transform can't be taken that is why we take FFT [3]. The frequency domain shows if the clean signal in time domain contains noise, jitter and crosstalk [4]. Since the topologies of time and frequency domain are similar, so end points of time waveform are considered. The FFT is fine when the signal is periodic and the integer number of periods fills the time interval. But when there are non integer periods we may get a truncated waveform which is different from continuous time signal with discontinuous end points [11]. It appears as if one frequency's energy leaks into another. This is called spectral leakage. Windowing is a technique which at the end of the finite sequence reduces discontinuities. This includes the multiplication of time signal with a finite length window which in turn reduces the sharp discontinuities. Hamming window gives wide peak but low side lobes. It does a great job at reducing the nearest side lobes but is poor at other lobes [5] (Fig. 6).

### 5.2 Radial Basis Function Neural Network

Radial Basis Function (RBF) is used for feature classification. It is used to approximate the functions. The neurons are continuously added in the hidden layer of the network till it meets the specified goal of mean square error. The data of 500 epileptic patients are taken to train the network. Where, GOAL is the mean squared error goal which by

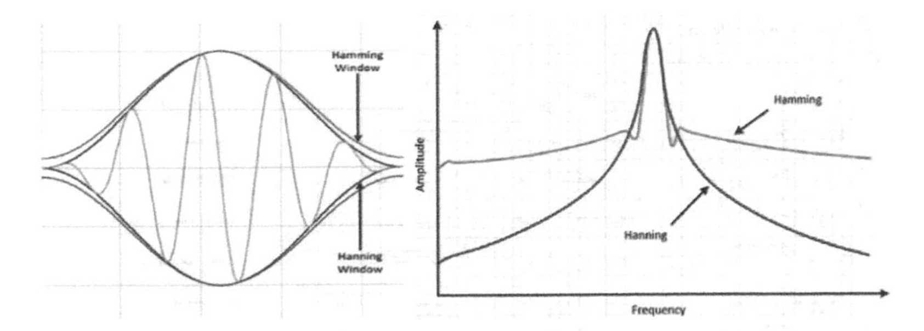


**Fig. 6.** Windowing techniques (Hamming and Hanning) give rise to high peak but nice side lobes.

default is 0.0 .SPREAD, denoted by σ is extent of stretch of the curve of RBF, default = 1.0.

Here for epilepsy detection the ranges and the abruptness are both trained using radial basis function. There are in total 48 hidden layers. The more hidden layers present in the network the more accuracy we get [1].

The two main advantages of the RBF network is that first it keeps mathematics part simple i.e. it uses linear algebra. Second, it keeps computations cheap, i.e. it uses generalized gradient descent algorithm. The approximate function is explained in the equation

$$y(x) = \sum_{i=1}^{N} w_i \, \varphi(\|x - x_i\|)$$

Where y(x) is the approximation function, x is the inputs and w is the weight (Fig. 7).

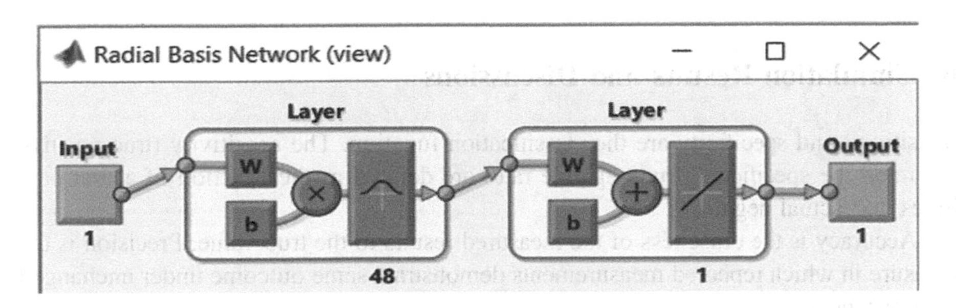


**Fig. 7.** Radial basis function network

### 5.3 Fuzzy Logic

In this project complex system behavior are modeled using simple logic rules using fuzzy designer, we have used Mamdani based fuzzy system, and the fuzzy rules are designed in system as shown in Fig. 8, α is seizure, μ is the blood pressure, € is human fall and d is the output.

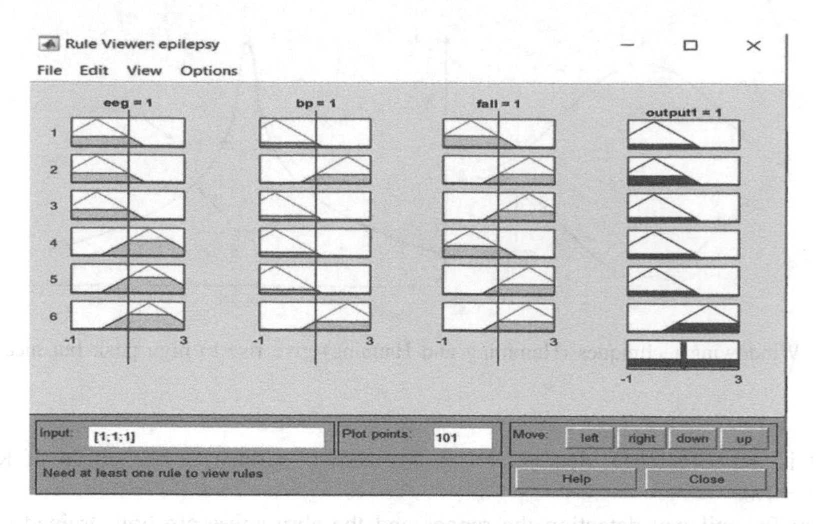


**Fig. 8.** Fuzzy inference system designed.

There are total 4 fuzzy rules that are designed in the system. These are discussed in the Table 2.

**Table 2.** Rules in fuzzy implemented system

| Sr. No. | Rule |
|---|---|
| 1 | If (α is True) and (μ is High) and (€ is False) then (d is False) |
| 2 | If (α is True) and (μ is Low) and (€ is False) then (d is False) |
| 3 | If (α is True) and (μ is Low) and (€ is True) then (d is False) |
| 4 | If (α is True) and (μ is High) and (€ is True) then (d is High) |

## 6 Simulation Results and Discussions

Sensitivity and specificity are the classification function. The sensitivity (true positive rate) and the specificity (true negative rate) are defined as the function of actual positives and actual negatives.

Accuracy is the closeness of the measured results to the true value. Precision is the measure in which repeated measurements demonstrate same outcome under unchanged circumstances.

$$Accuracy = \frac{TP + TN}{TP + TN + FP + FN}, \quad Precision = \frac{TP}{TP + FP}$$

F-score is to calculate the accuracy of the test. It takes into account the precision values and recall values. F1-score has best score at 1 and worst score at 0. Precision is the ratio of related instances to the recovered instances. Recall is the ratio of related instances over the total number of related instances that have been recovered.

$$F = 2.\frac{precision * recall}{precision + recall}$$

Where, $TP$ = true positive, $TN$ = true negative, $FP$ = False positive, $FN$ = False negative.

During neural network training, the radial basis function is trained and a total of 300 epochs are obtained. An epoch is defined as the number of times our learning algorithm is trained. The more the epochs the more accurate results are obtained, Fig. 9 shown below.

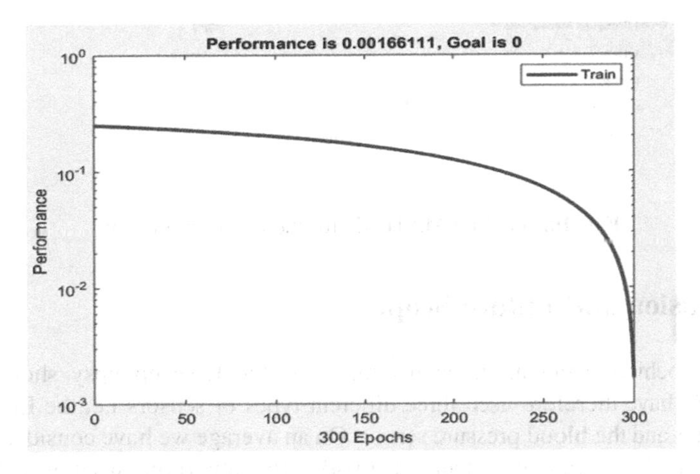


**Fig. 9.** Graph of performance vs. epochs

Parameter analysis at different epochs (Table 3):

**Table 3.** Results of various parameters for given epochs

| Sr. No. | Epochs | Accuracy | F1-score | Precision |
|---|---|---|---|---|
| 1. | 100 | 0.85 | 0.857143 | 0.9 |
| 2. | 150 | 0.85 | 0.842105 | 0.83 |
| 3. | 200 | 0.9 | 0.9 | 0.9 |
| 4. | 300 | 0.92 | 0.92 | 0.95 |

The GUI in MATLAB is designed as shown in Fig. 10.

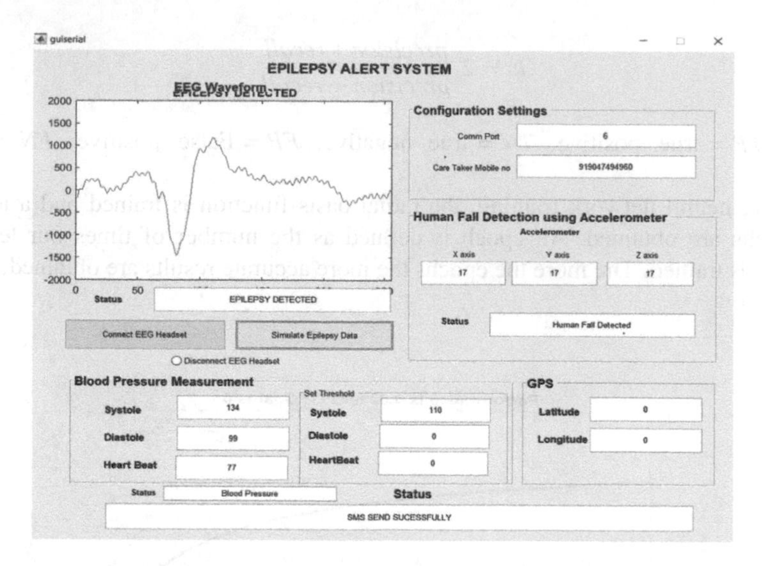


**Fig. 10.** GUI on MATLAB for the proposed system

## 7 Conclusion and Future Scope

The motive behind using all three parameter is that false epilepsy should not be detected. We have therefore used three different types of sensors i.e. the EEG sensor, accelerometer and the blood pressure sensor. On an average we have considered data of almost 500 patients train our neural network. The algorithm used is radial basis function (RBF) network. On the basis of all the three readings of sensors the fuzzy rules finally decide whether epilepsy is detected or not. This system uses single EEG electrode. In future we will be using 8 and 16 channel electrodes to measure the different epilepsy levels.

**Acknowledgments.** This work was supported by the "JJ Plus Hospitals Pvt. Ltd. & NEURON International", Aurangabad for recording EEG and providing data of epileptic patients.

**Compliance with Ethical Standards**

✓ All authors declare that there is no conflict of interest.
✓ No humans/animals involved in this research work.
✓ We have used our own data.

## References

1. Mamun Rashid, Md., Ahmad, M.: Epileptic seizure classification using statistical features of EEG signal. In: International Conference on Electrical, Computer and Communication Engineering (ECCE), Cox's Bazar, Bangladesh, 16–18 February 2017 (2017)
2. Al-Omar, S., Kamali, W.: Classification of EEG signals to detect epilepsy problems. In: 2nd International Conference on Biomedical Engineering (2013)

3. Manjusha, M., Harikumar, R.: Performance analysis of KNN classifier and K-means clustering for robust classification of epilepsy from EEG signals. In: IEEE WiSPNET 2016 Conference (2016)
4. Choubey, H., Pandey, A.: Classification and detection of epilepsy using reduced set of extracted features. In: 5th International Conference on Signal Processing and Integrated Networks (SPIN) (2018)
5. AlSharabi, K., Ibrahim, S., Djemal, R.: A DWT-entropy-ANN based architecture for epilepsy diagnosis using EEG signals. In: 2nd International Conference on Advanced Technologies for Signal and Image Processing - ATSIP, Monastir, Tunisia, 21–24 March 2016 (2016)
6. da Silva, F.H.L.: The impact of EEG/MEG signal processing and modeling in the diagnostic and management of epilepsy. IEEE Rev. Biomed. Eng. **1**, 143–156 (2008)
7. Saker, M., Rihana, S.: Platform for EEG signal processing for motor imagery - application brain computer interface. In: 2nd International Conference on Advances in Biomedical Engineering (2013)
8. Ilyas, M.Z., Saad, P., Ahmad, M.I., Ghani, A.R.I.: Classification of EEG signals for brain-computer interface applications: performance comparison (2016)
9. Hsieh, Z.-H., Chen, C.-K.: A brain-computer interface with real-time independent component analysis for biomedical applications (2012)
10. Lasefr, Z., Ayyalasomayajula, S.S.V.N.R., Elleithy, K.: Epilepsy seizure detection using EEG signals (2017)
11. Wang, L., Xue, W.: Automatic epileptic seizure detection in EEG signals using multi-domain feature extraction and nonlinear analysis. Entropy **19**, 222 (2017). Article in MDPI
12. Gajic, D., Djurovic, Z., Di Gennaro, S., Gustafsson, F.: Classification of EEG signals for detection of epileptic seizures based on wavelets and statistical pattern recognition. Biomed. Eng. Appl. Basis Commun. **26**(2), 1450021 (2014)

# Face Detection Based Automation of Electrical Appliances

S. V. Lokesh[1(✉)], B. Karthikeyan[1], R. Kumar[2], and S. Suresh[3]

[1] School of Electronics Engineering, Vellore Institute of Technology, Vellore, India
lokeshsuganthan@gmail.com, bkarthikeyan@vit.ac.in

[2] Electronics and Instrumentation Engineering, NIT Nagaland, Dimapur, India
rajagopal.kumar@nitnagaland.ac.in

[3] Electronics and Communication Engineering, Sri Indu Institute of Engineering and Technology Sheriguda, Hyderabad, India
write2sureshs@gmail.com

**Abstract.** This paper addresses the problem of automation of electrical appliances with the use of sensors. The proposed method uses Viola-Jones method a real time face detection algorithm for controlling the electrical appliances which are part of our day to day life. A detector is initialized by Viola-Jones based face detection algorithm automatically without the need for any manual intervention. After initializing the detector, the position where the face detected is computed. The region captured by the camera is split into four different regions. Based on the position of the detected face the electrical appliances subjected to that particular region can be controlled.

**Keywords:** Face detection · Viola Jones · Split to four regions virtually

## 1 Introduction

With the growth in the field of electronics, constant efforts have been made to automate various day-to-day tasks. One of these tasks is the automation of electrical appliances. This can be accomplished using GSM which requires the users to check the status of the appliances through mobile phones when they step out, or it can also be done through voice commands [12]. However this is still subjected to errors as it requires human effort and is not completely automated. Another way of controlling electrical devices is through the use of sensors [13] which provides complete automation, however the poor responsiveness of the sensors and complexity of network troubleshooting brings down the efficiency of the entire system.

Due to the drawbacks posed by the already existing methods, this paper proposes automation of electrical appliances by using face detection. The simplicity of the circuit involved, the wide coverage range and a fully automated setup ensures power efficient control of appliances. Face detection is used to compute the size and identify the position of the face in a given input image. The background information are ignored and only the facial feature from the image are detected. From the detected face we can able to predict the age of the person, gender, facial expression etc. in general terms face


S. Smys et al. (Eds.): ICCVBIC 2019, AISC 1108, pp. 204–212, 2020.
https://doi.org/10.1007/978-3-030-37218-7_23

detection is to localize the face in the given input image. Localization helps to know the amount of faces and where the faces are detected in the image.

## 2 Related Work

This section addresses few techniques which are used for computer-vision face detection.

Detection based on skin colour segmentation aims at analysing the visuals in different colour spaces and identifying the face with certain threshold values [14]. The different colour spaces that have been widely used are YCbCr, HIS and binary and some works have proposed the combination of different spaces to threshold the facial features. However this method can be efficient for stationary images but not for real time video streams.

The second section involves feature extraction for object detection using features from SMQT. In order to increase the computation of SNOW classifier, the classifier is divided into multiple stages on which a single classifier is trained to form a weak classifier and these weak classifiers are combined to form a strong classifier in cascade which help in face detection [1].

Neural network consists of perceptron which are artificial neurons are available in multiple layers and are connected to each other. Perceptron network are under unsupervised learning and no kind of programming is required [2]. Perceptron network composes of a threshold function or summation function. Face detection using neural network is determined by checking whether the detected region is face or not at a small window level. It is not necessary to train with non-face images and thus computational speed is also reduced [3]. Face detection using neural network composes of two steps. A part of image of 20 * 20 size is fed to the neural network as input in the first step. The output of the filter determines whether the detected region is face or not a face. In the second step the efficiency of the overall neural network is increased by fusing all the single neural network to find the overlapping detections and this reduces the false detections.

Maximum hyperplane or set of hyper planes in an infinite dimensional space is constructed using a supervised learning model which groups the information in two classes and separation is from two categories is through the hyperplane [4]. The area between hyper plane and closest data points is called the margin. Data points that exists on the margin boundary of the hyper plane are called as support vectors. This method can be applied for object detection then the object can be assumed from positive samples and non-object classes are from the negative samples.

The kernels can be of type polynomial kernels or radial kernels. The main difficulty lies in finding the appropriate kernel for the given problem. The selected kernel is used for the classification of problems and the results produced by them may not be good when compared to the result produced from sample set. They are used in application like detection of pedestrians. Haar wavelets are applied on samples which are positive as well as negative to extract features and they are helpful in discrimination of classes [5].

# 3 Proposed Method

## 3.1 Face Detection

The face detection was carried out through Viola Jones Algorithm. Paul viola and Michael jones proposed the algorithm in 2001 [6, 7]. Since this many works have been done using this algorithm [8–11]. It is mainly used for detecting faces in real time rather than object detection. The algorithm is very robust and also the computation speed is less. The main goal of face detection is to distinguish face from non faces. It is the first step in face recognition. The algorithm comprises of four steps which are:

(1) **Haar Feature Selection:** Haar-like features are used in object recognition. Computation of image intensity (i.e. calculation of RGB pixel values from every pixel made the computation to be very slower and expensive. In a human face the nose bridge region is lighter than the eye region etc. Using Haar basis function the properties of face are compared using Haar features. They are rectangular pixels of dark and white regions and the resultant of each haar feature is calculated by subtracting the black pixel count from the white pixel count (Fig. 1).

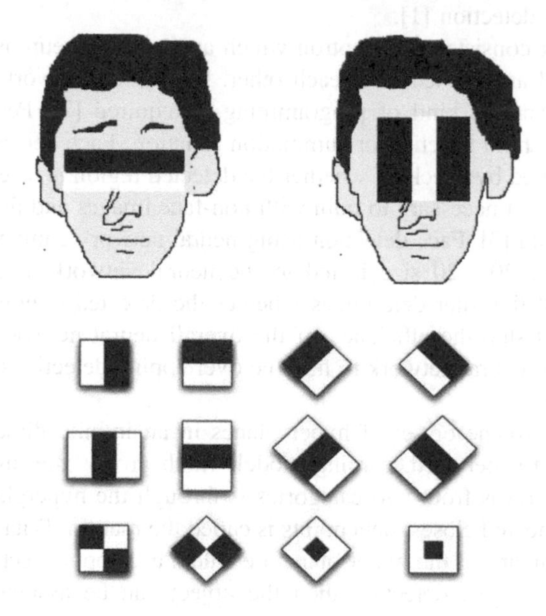

**Fig. 1.** Haar features

(2) **Creating an Integral Image:** For a faster and effective way for calculating the pixel values we go for integral image. If we have to use integral image we go for the conversion of original image into gray scale image. Computation of adjacent pixel values with respect to the pixel value at (x, y) forms the integral image which helps in speeding up the computation. Integral image is the summation of values

pixels above, left and its own pixel value at (x, y). This is done to make the Haar feature compute efficiently (Fig. 2).

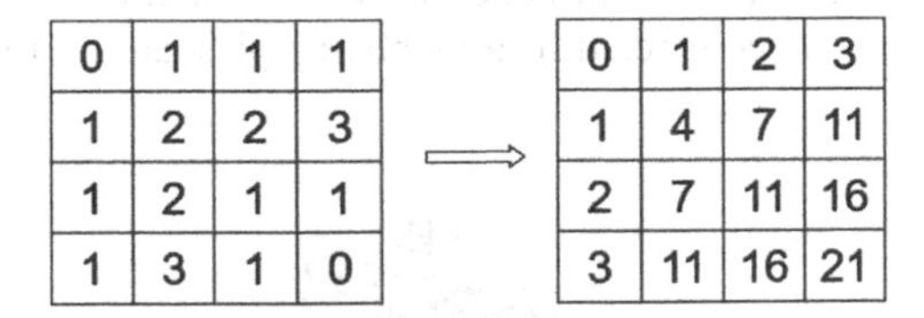


**Fig. 2.** Integral Image

(3) **Adaboost Training:** In order to separate relevant and irrelevant haar like features we go for Adaboost training. Irrelevant features doesn't give any information about the detected image whether it is face not. Relevant features are the features which provide information about face. Relevant features are provided with weights to form a weak classifiers. Adaboost training gives a strong classifiers which we able to say whether the detected image is a face or not a face. These strong classifiers are nothing but a linear combination of all the weighted weak classifiers.

$$f(x) = \alpha_1 f_1(x) + \alpha_2 f_2(x) + \ldots\ldots + \alpha_n f_n(x) \tag{1}$$

(4) **Cascading Classifiers:** Adaboost gives information about the classifiers which will be able say whether the detected image is face or not. Cascade classifiers are used to integrate many classifiers which are useful in neglecting background windows so that more operation are performed on regions where the face is present. Each stage in the classifier consists of distinct sets of face features which is checked step by step for the input video to detect the presence of a face. The stronger classifiers are kept in set 1, so a when the detected image is not a face no need to check for all the classifiers. From the first set of features itself we can be able to predict whether the detected image is face or not a face (Fig. 3).

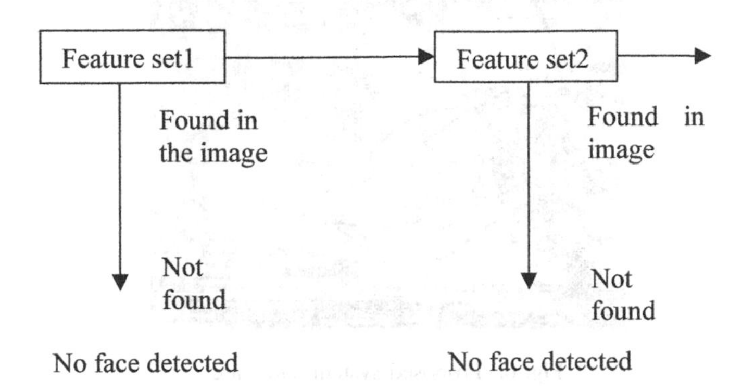


**Fig. 3.** Cascade classifiers

### 3.2 Automation of Appliances

Raspberry pi is a microprocessor which is a small size CPU which runs on Rasbian OS. It supports cameras as peripherals and hence it is the best choice for this application. Up to three cameras can be connected to a raspberry pi board at a time (Fig. 4).

**Fig. 4.** Raspberry Pi model 3B

The automation of appliances happens by controlling them using a programmed Raspberry pi microcontroller to which a web camera is connected to record the real time video (Figs. 5, 6 and 7).

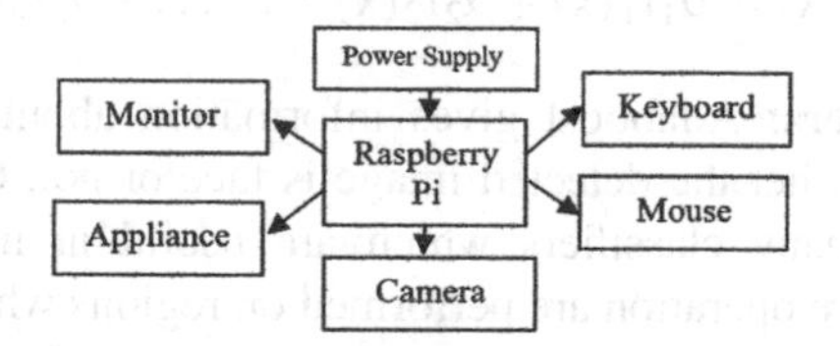


**Fig. 5.** Block scheme of hardware

**Fig. 6.** Proposed system hardware

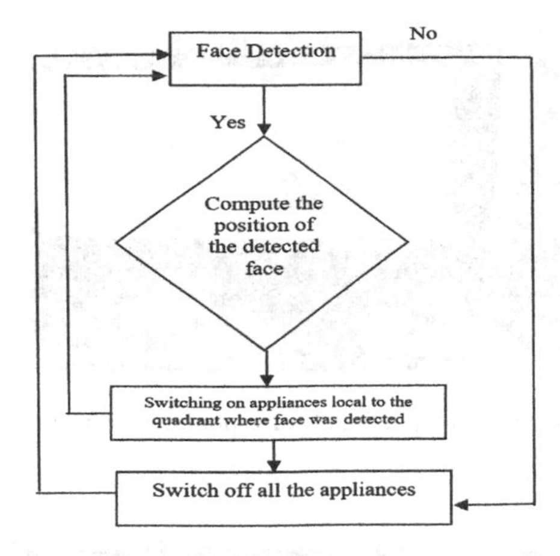


**Fig. 7.** Design flow of hardware

## 4 Results and Discussion

OpenCV is an open source computer vision library. The library is written in Python and runs under Raspbian Operating System. The proposed method with Viola-Jones detection was simulated in Python Idle using OpenCV Python library. The simulations were done on Raspberry Pi 3 Model B which has RAM of 1 GB. This system was tested on real time videos captured in laboratory. Video is captured at a resolution of 640 * 480 using a web camera. The video is split virtually into four regions. Each region has specific appliance which will be automated only if faces are detected in that region (Fig. 8).

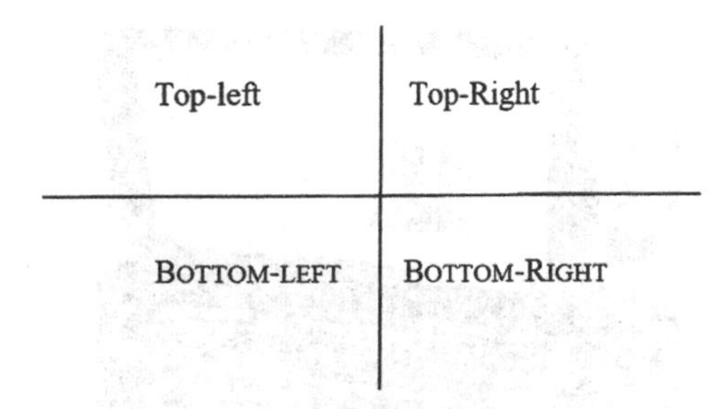


**Fig. 8.** Region split of a video

Based on the region where the face got detected we can be able to say whether is very close or far to the camera. When camera placed at a certain height from the ground, if face gets detected at the top region we can say that the person is far from the camera since the scale factor of the image will be less. If the image gets detected in the bottom region we can say that the person will be closer to the camera since the scale factor will be high when face gets detected in this region (Figs. 9, 10, 11, 12 and 13).

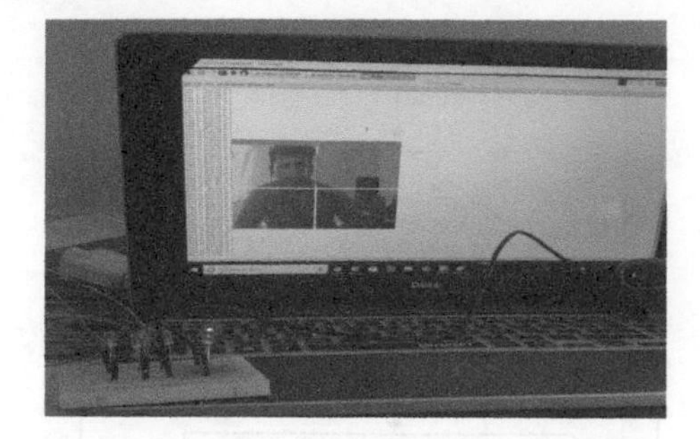

**Fig. 9.** Shows the face detected in top left and corresponding led is turned on and rest of the led is turned off.

**Fig. 10.** Shows the face detected in bottom left and corresponding led is turned on and rest of the led is turned off.

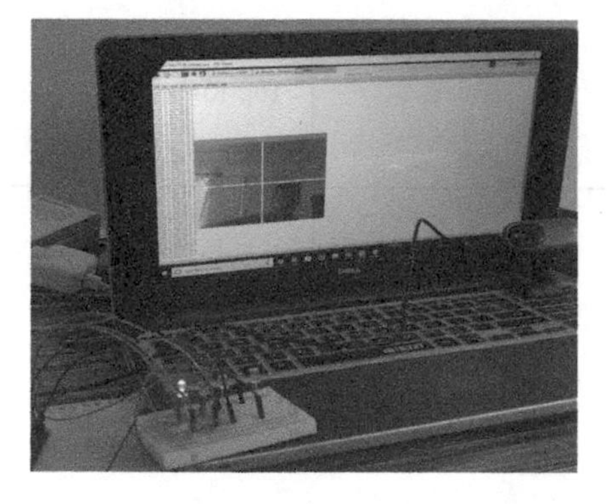

**Fig. 11.** Shows the face detected in bottom right and corresponding led is turned on and rest of the led is turned off.

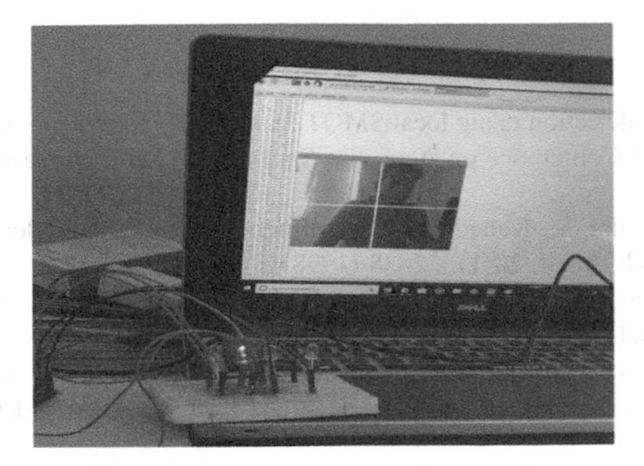

**Fig. 12.** Shows the face detected in top right and corresponding led is turned on and rest of the led is turned off.

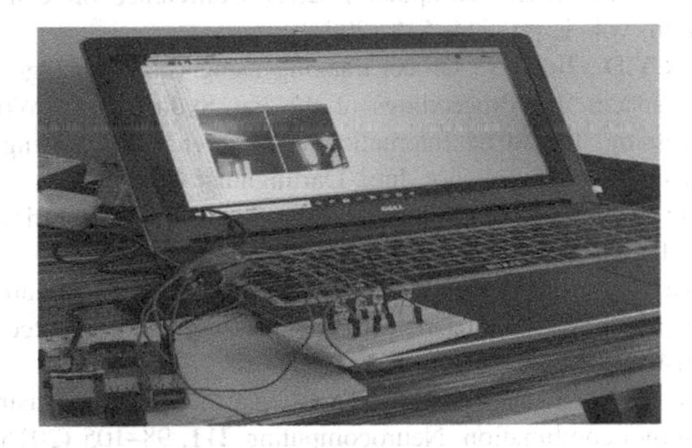

**Fig. 13.** Shows all the led turned off when there is no face detected.

The proposed method will help in efficient use of power. The system on the whole covers the drawbacks like human fatigue in conventional manual control of appliances, small coverage area of sensors. The system will have a disadvantage when there is no sufficient light to detect face. The above discussed idea was prototyped using led s. The work will be extended to controlling the device based on face recognition and also integrating the appliances to the system.

**Compliance with Ethical Standards**

✓ All authors declare that there is no conflict of interest

✓ No humans/animals involved in this research work.

✓ We have used our own data.

## References


1. Claesson, I.: Face detection using local SMQT features and split up snow classifier. In: 2007 IEEE International Conference on Acoustics Speech and Signal Processing - ICASSP 2007 (2007)
2. Rowley, H.A., Baluja, S., Kanade, T.: Neural networks based face detection. IEEE Trans. Pattern Anal. Mach. Intell. **20**(1), 22–38 (1998)
3. Sung, K.K., Poggio, T.: Example-based learning for view-based human face detection. IEEE Trans. Pattern Anal. Mach. Intell. **20**(1), 39–51 (1998)
4. Boser, B., Guyon, I.M., Vapnik, V.: A training algorithm for optimal margin classifiers. In: ACM Workshop on Conference on Computational Learning Theory (COLT), pp. 142–152 (1992)
5. Yilmaz, A., Javed, O., Shah, M.: Object tracking: a survey. ACM Comput. Surv. **38**(4), 45 (2006)
6. Paul, V., Jones, M.J.: Rapid object detection using a boosted cascade of simple features. In: Proceedings of the 2001 IEEE Computer Society Conference on Computer Vision and Pattern Recognition, vol. 1, pp. 511–518 (2001)
7. Allen, J.G., Xu, R.Y.D., Jin, J.S.: Object tracking using CamShift algorithm and multiple quantized feature spaces. In: Proceedings of the Pan-Sydney Area Workshop on Visual Information Processing. In: ACM International Conference Proceeding Series, vol. 100, pp. 3–7. Australian Computer Society, Inc., Darlinghurst (2004)
8. Jain, V., Patel, D.: A GPU based implementation of robust face detection system. Procedia Comput. Sci. **87**, 156–163 (2016)
9. Chatrath, J., Gupta, P., Ahuja, P., Goel, A., Arora, S.M.: Real time human face detection and tracking. In: International Conference on Signal Processing and Integrated Networks (SPIN), pp. 705–710 (2014)
10. Tao, Q.-Q., Zhan, S., Li, X.-H., Kurihara, T.: Robust face detection using local CNN and SVM based on kernel combination. Neurocomputing **211**, 98–105 (2016)
11. Da'san, M., Alqudah, A., Debeir, O.: Face detection using Viola and Jones method and neural networks. In: International Conference on Information and Communication Technology Research, pp. 40–43 (2015)
12. Singh, P., Chotalia, K., Pingale, S., Kadam, S.: Smart GSM based home automation system. Int. Res. J. Eng. Technol. (IRJET) **03**(04), 19838–19843 (2016)
13. D'mello, A., Deshmukh, G., Murudkar, M., Tripathi, G.: Home automation using Raspberry Pi 2. Int. J. Curr. Eng. Technol. **6**(3), 750–754 (2016)
14. Kang, S., Choi, B., Jo, D.: Faces detection method based on skin color modeling. J. Syst. Archit. **64**, 100–109 (2016)

# Facial Recognition with Open Cv

Amit Thakur, Abhishek Prakash(✉), Akshay Kumar Mishra, Adarsh Goldar, and Ayush Sonkar

Bhilai Institute of Technology, Raipur, Raipur, Chhattisgarh, India
abhishekprakash256@gmail.com

**Abstract.** In this digital era, the wide-range implementation of digital authentication techniques has not only impacted the home security systems but also in difference facets of industries. The ever-growing facial recognition technology is now dominated by the digital authentication system with a steady increase in the sophistication of the real time image processing operations, which is more accurate and reliable. The system is more secured by implementing facial recognition as a primary security feature for observing a person in a 3D perspective. In recent times, digital authentication system aims to recognize individuals, payment confirmation, criminal identification etc. In this research work, we have worked with a type of facial recognition system, which is based on "Open-CV" model. Open-CV is a library that is built by intel in 1999. It is a cross platform library, primarily built to work on real time image processing systems, which includes the state-of-the-art computer vision algorithms. The system that we have built can achieve an accuracy of about 85–90% accuracy based on the lighting condition and camera resolution. Based on the database, this system is built by taking still images of the subject, then it trains a model and creates deep neural networks and then for detection it matches the subject in real time and give results based on the match. The system that we have built is quite simple, robust and easy to implement, where it can be used in multiple places for a vast application.

**Keywords:** Digital authentication · Facial recognition · Image processing · Open-CV · Real time image processing

## 1 Introduction

The facial Recognition system is a revolutionary technique that is taken in account in this paper as a system that has been proven to be used in the multiple places as for the authentication system to mass crowd monitoring and much more. As per the development in the artificial system and machine learning system the computers are getting far better than the humans in many of the places in lot of the automation fields humans are being replaced by the human but when we look at the field of the facial recognition the story is quite different as per our latest achievement in [1] still humans are very accurate at the facial recognition but with the evolution in the field of AI and help of Machine Learning and Advancement in the Algorithm has made this gap very close with Deep Face AI now getting an accuracy up to 97.35%. As that we have made an approach with a library that is open CV which was developed by Intel in 1999 [2] for


S. Smys et al. (Eds.): ICCVBIC 2019, AISC 1108, pp. 213–218, 2020.
https://doi.org/10.1007/978-3-030-37218-7_24

regarding the facial recognition and it is used in real time image processing and the library is implemented with python. The implementation of the facial recognition system comes up with its own problem as are: Aging, Pose, Illumination and Emotions these are the factors that we deal and there are physical factors as well like facial hairs, face makeup, face wearables as shown in "Fig. 1". The other kind of problems that we can face regarding the recognition is the hardware problems like with the image capturing device mainly with the camera the quality of images that it captures and better the quality better will be the image and low quality possess the problem regarding the image processing and detection of facial features that may affect the accuracy of the system.

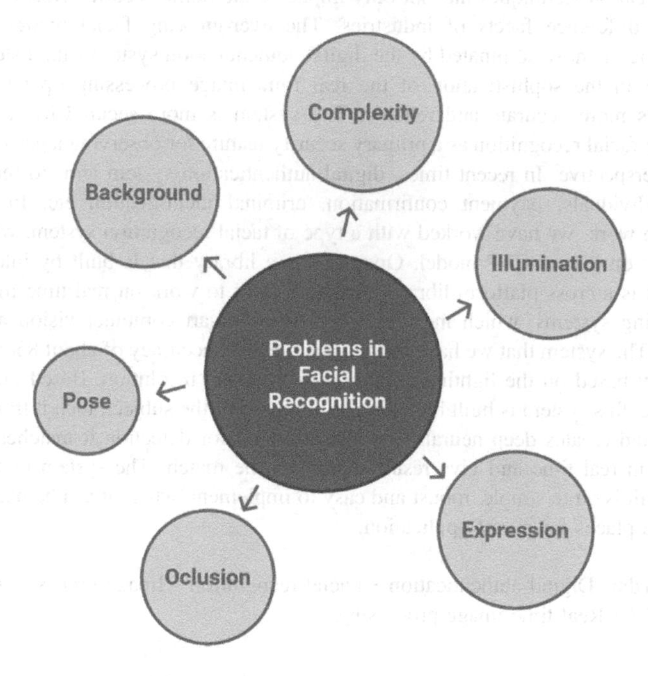


**Fig. 1.** Problems in face recognition

## 2 History

The facial recognition system started in the 1960s by "Woodrow Wilson Bledsoe" known as the father of the facial recognition system that time basis of the system is developed regarding measurement of the certain facial features like: nose, eyes, hairline and mouth. These are then stored in a database (Graphics tablet or on RAND tablet) and when a certain photograph came it is matched with the features the results are given, they provide a base for the mechanism for the system. In 1973 the Japanese [3] researcher Kaneda was able to build first automation system which was able to recognize faces. The system was built on the principles that was developed by the bell labs as first the image is digitized and the points are identified for the recognition and then measurements are taken and stored in the database and 21 points are taken into the consideration.

With evolution of the social media that created a huge database for the images and also these sites use the recognition feature as to identify the individuals for tagging and also a lot of data mining is done by these sites regarding the faces and advanced ML algorithm are used that raised the accuracy of the system. As per the new major breakthrough came in 2017 with the apple's Face Id feature that is major attention seeker with 3d scanning of face and also is used as primary authentication system in the phone and there is constant increase in system as shown in "Fig. 2".

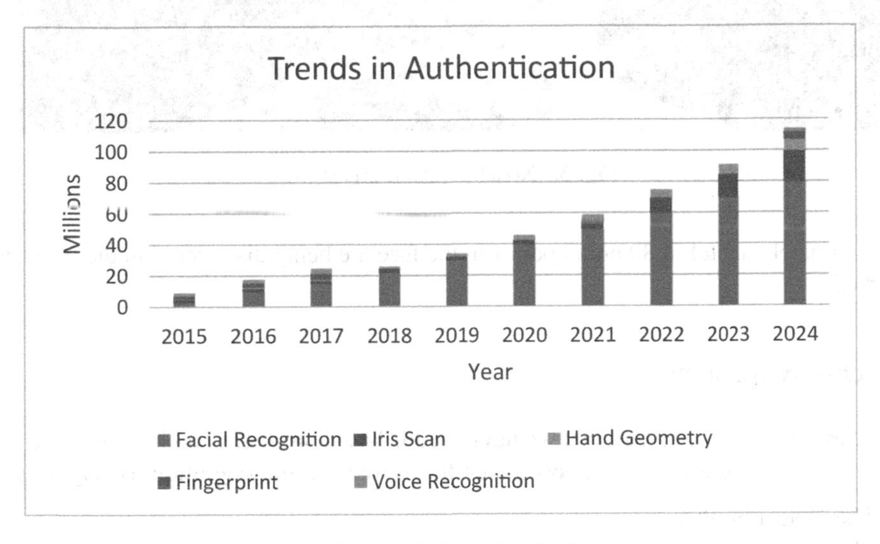


**Fig. 2.** Trends in authentication

## 3 The Nodal Points

The pillars of the facial recognition system are the facial features that helps in distinguishing the face of the individuals the nodal points of the face that define the significant face print of the individuals as these are that landmark differ they are like the: Separation between the two eyes, end length of the nose, bottom distance of the eye socket, shape of the cheek bones, the length jaw line as shown in "Fig. 3". [3] These nodal points total measure makes up the face print that are converted into the numerical code that to be stored in the database. When a face image is seen the nodal points are measured and are compared with the stored data of the face and the matching makes the result. The most significant point about the nodal point is that don't tend to change with the "aging" and also, they are not affected with fat deposition on the face so that creates a validation for the authenticity of the face they nodal points are chosen to be remain constant throughout the age of a person and doesn't change much. Now

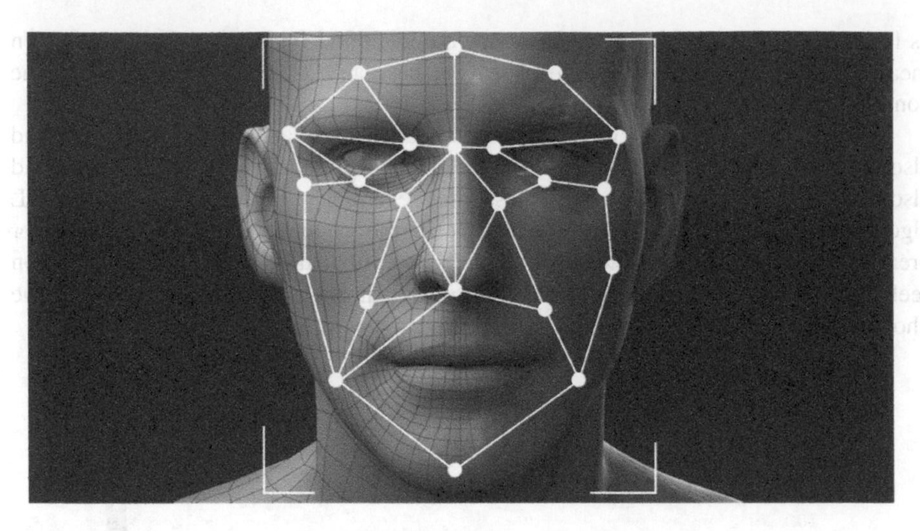

**Fig. 3.** Nodal points in image [3]

approximately a total of 80 nodal points in the face are being discovered in the face and marked.

# 4 Our Approach

The current system approach that we have used is based on stages combine to make the final results, these each stage plays a significant role facial recognition as stages are:

- The Detection Phase
- The Capturing Phase and Database Creation
- Creation of a classifier and adding data
- The Recognition Phase (Fig. 4).

## 4.1 Detection, Capturing and Creating Dataset

The capturing phase is used to take the still images of the subject as the detection phase tells the camera about the presence of a face that is made by "harr cascade" implemented in Open CV which has pre-defined features of the face as frontal and side oriented features like: nose, eyes, mouth, eye socket etc. Then the camera takes 20–30 still images of the subject depend on the camera time allotted in the program and then images are cropped to fit to only face image and background are trimmed out and also grey scale conversion is done.

The implementation of the system is done in Python 3.0 with the support of the Open CV library that implemented the classifier in LBPH [5, 8, 9] and the also the recognition of the face. The Algorithm has these parameters:

- Neighbors: The number of the central point that build around the local central pixel.

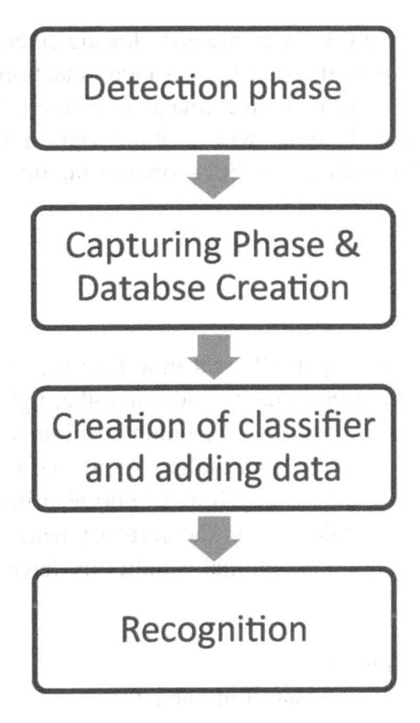


**Fig. 4.** Phases of working

Training of system as each image a certain id is set to it make an input recognition and also in the output. The first step is followed by creating a sliding window on the image.

- The values of the neighbor are set in the 0 to 1 based on higher or lower (Fig. 5).

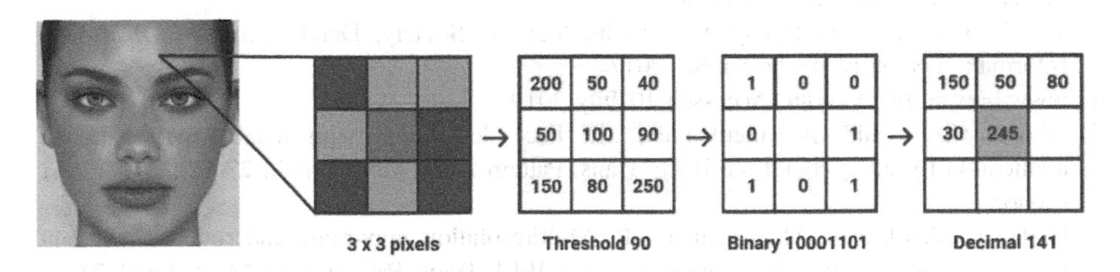


**Fig. 5.** Matrix representation of image [4]

## 4.2 Recognition

The recognition part is done in to two stages:

- The first part is the detection of the face in environment.
- The second part is the histogram conversion of image and the matching with present dataset.

The first part uses the open CV "harr cascade" for the detection of a face in camera that is processed in real time with open CV. If face detection is completed then the second part is performed as the real time image is converted to the histogram, the comparison of the histogram is done with that present in the dataset for this the Euclidean distance is used and the algorithm proposed the image Id which has highest match.

## 5 Conclusion

With the implementation of the open CV real time face recognition system we get an accuracy up to 90–95% varies upon many conditions like: lighting conditions, camera resolution, the training of dataset and also the number of images taken for training. We have used certain conditions varies the results like as shown in Fig:, the varies in number of dataset images are done from 10 to 50 and also in lighting conditions, the camera resolution also plays a major role in the accuracy better resolution camera have shown a higher accuracy rate, the optimal conditions have given us accuracy of 95.34%.

**Compliance with Ethical Standards**

✓ All authors declare that there is no conflict of interest.

✓ No humans/animals involved in this research work.

✓ We have used our own data.

## References


1. https://research.fb.com/publications/deepface-closing-the-gap-to-human-level-performance-in-face-verification/. Accessed 8 July 2019
2. https://en.wikipedia.org/wiki/OpenCV
3. The Impact of Facial Recognition Technology on Society, Derek Benson COMP 116: Information Security, 13 December 2017
4. www.howstuffworks.com. Accessed 10 July 2019
5. Ahonen, T., Hadid, A., Pietikainen, M.: Face description with local binary patterns: application to face recognition. IEEE Trans. Pattern Anal. Mach. Intell. **28**(12), 2037–2041 (2006)
6. Ojala, T., Pietikainen, M., Maenpaa, T.: Multiresolution gray-scale and rotation invariant texture classification with local binary patterns. IEEE Trans. Pattern Anal. Mach. Intell. **24**(7), 971–987 (2002)
7. Ahonen, T., Hadid, A., Pietikäinen, M.: Face recognition with local binary patterns. In: Computer Vision, ECCV 2004, pp. 469–481 (2004)
8. Ahonen, T., Hadid, A., Pietikäinen, M.: Face recognition with local binary patterns. Machine Vision Group, Infotech Oulu, University of Oulu, Finland. http://www.ee.oulu.fi/mvg/
9. LBPH OpenCV. https://docs.opencv.org/2.4/modules/contrib/doc/facerec/facerec_tutorial.html#local-binary-patterns-histograms. Accessed 12 July 2019

# Dynamic Link Prediction in Biomedical Domain

M. V. Anitha(✉) and Linda Sara Mathew

Computer Science and Engineering, Mar Athanasius College of Engineering, Kothamangalam 686666, Kerala, India
Anitha19deepa@gmail.com, lindasaramathew@gmail.com

**Abstract.** Anonymous adverse response to medicines available on the flea market presents a major health threat and bounds exact judgment of the cost/benefits trade-off for drugs. Link prediction is an imperative mission for analyzing networks which also has applications in other domains. Compared with predicting the existence of a link to determine its direction is more difficult. Adverse Drug Reaction (ADR), leading to critical burden on the health of the patients and the system of the health care. In this paper, is a study of the network problem, pointing on evolution of the linkage in the network setting that is dynamic and predicting adverse drug reaction. The four types of node: drugs, adverse reactions, indications and protein targets are structured as a knowledge graph. Using this graph different dynamic network embedding methods and algorithms were developed. This technique performs incredibly well at ordering known reasons for unfavorable responses.

**Keywords:** Link prediction · Dynamic network · Nonnegative matrix factorization

## 1 Introduction

An adverse drug reaction (ADR) can be definite as 'an significantly hurtful or unkind response ensuing from an intercession connected to the use of a medicinal product; adverse effects generally predict threat from future management and license inhibition, or exact action, or variation of the dose routine, or removal of the product. Meanwhile 2012, the description has comprised responses happening as a result of fault, abuse, and to supposed responses to drugs that are unrestricted or being used off-label in addition to the sanctioned use of a therapeutic product in usual dosages.

Although this alteration possibly varies the reporting and observation agreed out by constructions and medicines managers, in clinical practice it should not affect our method to handling ADRs.

Pivotal research commenced in the late 20th and early 21st period in the USA and the UK established that ADRs are a common exhibition in clinical preparation, including as a cause of unscheduled hospital admissions, occurring during hospital admission and establishing after discharge.

Medications that have been especially embroiled in ADR-related emergency clinic confirmations incorporate antiplatelets, anticoagulants, cytotoxics, immunosuppressants,


S. Smys et al. (Eds.): ICCVBIC 2019, AISC 1108, pp. 219–225, 2020.
https://doi.org/10.1007/978-3-030-37218-7_25

diuretics, antidiabetics and anti-microbials. Deadly ADRs, when they happen, are regularly owing to discharge, the most widely recognized presumed cause being an antithrombotic/anticoagulant co-controlled with a non-steroidal mitigating drug (NSAID).

In recent times, various studies examine topological structures and disclose network roles for inclusive considerate the vital charms of composite networks. Link prediction is place onward to resolve the associated difficulties in some investigates and has engrossed more courtesies. Connection forecast prediction specifies the most effective method to exploit the data of edge and system association to foresee the joining probability of each separate farthest point. It very well may be connected in numerous areas for example discovering drug-to-target connections, learning the probable contrivance which drives coauthor ship progression, rebuilding airline networks, applauding friends and espousing purchasing deals, additional [1]. The closeness depend procedure as interrelation premonition is distinct on the convolution pattern and is more appropriate other procedures. The similitude record is generally exhibited to depict the likelihood of judging the missing in addition up and coming connections. This is a reprobate that immense costs are brought about by scarcely mining of attributes from limits in connection forecast.

Seeing the metaphorical closeness dependent on the system development, ordinary strategies for connection forecast can be distanced within three category [2]. With these broad spread of the Internet, informal organizations have turned out to be across the board and individuals can utilize them to develop more extensive impacts.

In addition informal communities, organize design are industriously identified in certifiable frameworks, for example, street traffic systems and neural systems. At the point when fabricate a system form to topologically genuinely exact complex frameworks, missing or lay off connections may unavoidably emerge because of time and cost restrictions in tests led for making the systems. Furthermore, since the system connections might be enthusiastically changing in nature, some potential connections may appear later on. Hence, it is necessary to notice such inconspicuous connections, or future connections, from the topological structure of the current system. This is a mission of connection forecast in complex systems. Connection forecast demonstrates a key part in taking care of numerous true issues. For example, in natural systems like ailment quality systems, protein-protein communication systems, and metabolic systems, joins assign the interface relations between the creature and the illnesses spoken to by the hubs they associate with.

Multifaceted networks are highly dynamic objects they raised very fast over time with the addition of new edges and nodes making their study very hard. A progressively addressable issue is to comprehend the relationship between two explicit hubs. How new communications are formed? What are the factors responsible for the formation of new associations? More specifically, the problem being addressed here is a prediction of likelihood association between two nodes in the near future. This problem is a link-prediction problem.

## 2 Related Works

Two authors advised a standout amongst the most simple connection expectation models that work unmistakably on an interpersonal organization [1–3]. Every apex in the diagram demonstrates an individual and boundary betwixt two peaks suggests the relations between the general population [2, 4, 5]. Assortment of affiliations can be exhibited obviously by allowing parallel edges or by tolerating a proper weighting framework for the edges [6–9].

With the boundless achievement of the Internet [10, 11], social networks have turned out to be prevalent and individuals can utilized them to assembled more extensive associations. Other than informal organizations, arrange structures are continually seen in certifiable framework, for example, road traffic networks and neural networks. At the point when fabricate a system pattern network to topologically surmised complex framework, absents or repetitive connection may unavoidably happen inferable from time and cost confinements in trials led for building the networks. This is an assignment of link expectation in complex networks. Link forecast assumes a key job in taking care of numerous genuine issues [7] (Fig. 1).

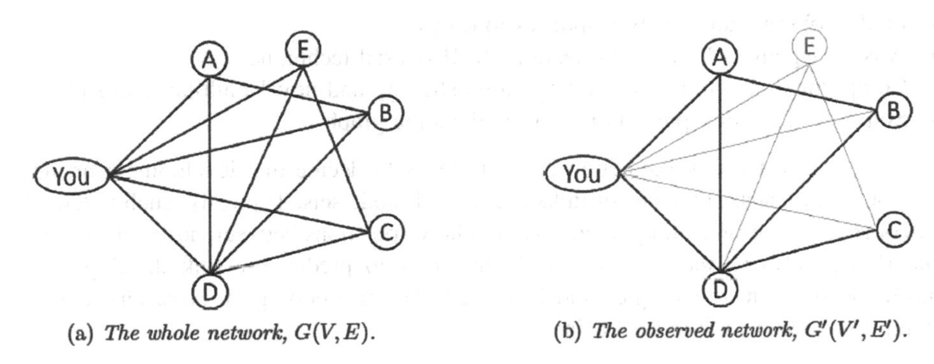


(a) *The whole network,* $G(V, E)$. (b) *The observed network,* $G'(V', E')$.

**Fig. 1.** A case of a whole interpersonal organization, and how it may look to an eyewitness

For example, in a biological network like disease-gene network, protein-protein interaction network, and metabolic systems, link demonstrate the communication relations between the life form and the diseases represented by the hubs they associate with. Complex systems are exceedingly dynamic objects; they developed quickly after some time with the expansion of new edges and hubs making their investigation troublesome. A progressively addressable issue is to comprehend the relationship between two explicit hubs. How new connections are framed? What are the variables in charge of the arrangement of new affiliations? All the more explicitly, the issue being tended to here is the prediction of probability relationship between two hubs sooner rather than later. This issue is the link prediction issue.

## 2.1 Problem Definition

Given problem of recognizing associated ADR from drug interaction, are anticipating in the meantime whether the two medications are associating with one another, and what ADRs will be caused. A system G = <V, E> in all peak e = [u, v] E says to the relationship among u furthermore v that existed at a specific time t(e). Think about numerous collaborations among u as well as v as various fringes, with conceivably unique future-impressions. For multiple times t, t', allow G[t, t'] signify this sub-diagram of G comprising of whole borders among one period seal among t along with t'. Also, it is apparent, the interim [t0, t'0] is preparing interim and [t1, t'1] is the test interim [1]. Presently, the strong definition of the connection expectation issue is, given four distinct occasions t0, t'0, t1, t'1 given G[t0, t'0] contribution to our connection forecast calculation, the yield is a rundown of all plausible future interactions (ends) neither at this time in G[t0, t'0] and all available at G[t1, t'1].

Furthermore, the interpersonal organizations develop on hubs with the inclusion of new people or components in the system, notwithstanding the development through new affiliations. As it isn't feasible for new hubs not present in the preparation interim, future edges for hubs present in the preparation interim are anticipated.

- Predict obscure unfriendly responses to drugs
- A Non-negative Matrix Factorization (NMF)-based technique
- Comparing with Graph embedding based strategy and novel multilayer model
- Solution of the link prediction issue in dynamic graph

To take care of the link prediction issue, it needs to decide the development or disintegration potential outcomes of links between all node sets. Typically, such potential outcomes can be estimated by likenesses or relative positions between node sets. For an underlying network, there are two different ways to predict the link development: similitude based methodologies and learning-based methodologies. Here take anticipating the new/missing/in secret links.

## 2.2 Predicting Unknown ADR

The work process of the expectation algorithm is the medicine that reason a inured ADR ought to share convinced highlights for all intents and purpose that are identified with the mechanism(s) by which they cause the ADR, for instance, protein targets, transporters or substance highlights. One-way to deal recognize those highlights is by a basic test. The advancement evaluation test act a shortsighted method to emulate a personal thinking procedure. Such as, since part of obsessed ADR is intrigue, the realized explanation abide probably going toward likewise source queasiness, yet the demand not call such accordingly some remedy this can create sickness keep a reason for ADR of intrigue since realize best medications can basis queasiness. As it were sickness is certainly not a particular component of the avowed reasons for an ADR also effect disregard it [1].

This outcome may be accomplished in strategy by examining for the enhancement of every component (for example likewise causing queasiness or focusing on a particular target) as entire the familiar reasons for an ADR of intrigue versus every single other medication. Highlights that are observed to be fundamentally advanced for the noted reasons for an ADR are utilized in ideal prescient model, different highlights are excluded.

The arrangement of highlights can attention of as a "meta-medicate" along just advanced highlights of the obvious reasons for ADR.

Any medication would presently able to be rack up for its closeness to this portrait, along with await the noted reasons for this ADR to grade moderately profoundly. Several medication this isn't presently avowed to orginate the ADR however such has high comparability to enhanced "meta tranquilize" structure is anticipated to-be a promising new reason. This procedure is rehashed for each ADR to produce new forecasts. The highlights delivered against the system (related to the enhancement test) can be utilized for order by any standard calculations. The execution of the expectation calculation was surveyed usual three different ways:

- Capacity to accurately order the popular reasons for various ADR
- Strength to guss (supplant) peak erased from the chart
- Talent to infer ADRs not current in the map but rather saw in EHRs (Fig. 2).

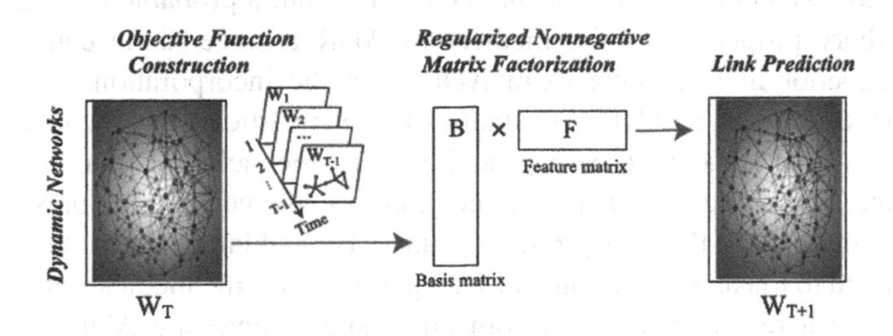


**Fig. 2.** Review of the NMF calculation. It comprises of three segments: (A) the target work development strategy successfully consolidates the fleeting systems as a regularizer, (B) the improvement methodology processes the refresh rules for the NMF calculation, and (C) the connection expectation the method acquires the fleeting connections at time T+1 dependent on the element lattice F. (Source: Physica A 496 (2018). P. 125).

# 3 Methodology

Thereby yielding better execution. The time and basic data influence the technique to accomplish forecast outcomes that are progressively precise [13].

The highest quality level in our test is developed by the information accessible in an outer database DrugBank. The DrugBank database is an extraordinary bioinformatics and cheminformatics asset that joins point by point medicate (for example substance, pharmacological and pharmaceutical) information with thorough medication target (for example succession, structure, and pathway) data. We look for all the 23 medications to recognize whether a medication is accounted for to associate with some other medications utilizing the Interax Interaction Search 3 motor given by DrugBank.

Before going to prediction construct a graph or network (heterogeneous network). It contain different nodes and their relationship. Many of the existing work is focusing on single drug reaction but here concentrate on two or more drugs.

## 4 Conclusion

To tackle the link prediction issue, it needs to decide the arrangement or disintegration conceivable outcomes of connections between all hub sets. Generally, such conceivable outcomes can be estimated by similitudes or relative positions between hub sets. For an underlying system, there are two different ways to foresee the connection development: comparability based methodologies and learning-based methodologies. Here take anticipating the new/missing/in secret links.

A critical qualification between the current ways to deal with ADR expectation is whether the forecasts are produced for medication like atoms as of now being developed, or for medications that have experienced clinical preliminaries. From a demonstrating point of view, the key contrast between these two circumstances is the accessibility of a symptom profile for the medication (but a probable fragmented one). Natural auxiliary properties of the medication ADR arrange alone can accomplish shockingly superior in foreseeing extra ADRs, yet the incorporation of extra information enhances execution. ADR expectations for lead particles have would in general spotlight on a synthetic element, perhaps at the same time including quality articulation profiles or medication targets. This examination is centered around foreseeing extra ADRs for medications in the post-promoting stage. By and large, the strategy exhibited could be utilized to make expectations for lead particles, yet the medication information diagram utilized herein is restricted to objective, signs, including ADRs. Absence in either compound highlights implies within each next to no information staying in the present learning chart for a lead particle.

## References

1. Bean, D., Wu, H., Dzahini, O., Broadbent, M., Stewart, R.J., Dobson, R.J.B.: www.nature.com/scientificreports
2. Liben-Nowell, D., Kleinberg, J.: The link-prediction problem for social networks. J. Am. Soc. Inf. Sci. Technol. **58**(7), 1019–1031 (2007)
3. Leskovec, J., Huttenlocher, D., Kleinberg, J.: Predicting positive and negative links in online social networks. In: International Conference Worldwide Web, pp. 641–650 (2010)
4. Al Hasan, M., Chaoji, V., Salem, S., Zaki, M.: Link prediction using supervised learning. In: Proceedings of SDM Workshop of Link Analysis, Counterterrorism and Security (2006)
5. Bilgic, M., Namata, G.M., Getoor, L.: Combining collective classification and link prediction. In: Proceedings of the Workshop on Mining Graphs and Complex Structures at ICDM Conference (2007)
6. Wang, C., Satuluri, V., Parthasarathy, S.: Local probabilistic models for link prediction. In: Proceedings of International Conference on Data Mining, ICDM 2007 (2007)

7. Doppa, J.R., Yu, J., Tadepalli, P., Getoor, L.: Chance-constrained programs for link prediction. In: Proceedings of Workshop on Analyzing Networks and Learning with Graphs at NIPS Conference (2009)
8. Taskar, B., Wong, M.F., Abbeel, P., Koller, D.: Link prediction in relational data. In: Proceedings of Neural Information Processing Systems, NIPS 2003 (2003)
9. Popescul, A., Ungar, L.H.: Statistical relational learning for link prediction. In: Proceedings of Workshop on Learning Statistical Models from Relational Data at IJCAI Conference (2003)
10. Popescul, A., Ungar, L.H.: Structural logistic regression for link analysis. In: Proceedings of Workshop on Multi-relational Data Mining at KDD Conference (2003)
11. Sarukkai, R.R.: Link prediction and path analysis using Markov chain. In: Proceedings of the Ninth World Wide Web Conference, WWW 2000, pp. 377–386 (2000)
12. De Verdière, É.C.: Algorithms for embedded graphs. MPRI (2013–2014)
13. Ma, X., Sun, P., Wang, Y.: Graph regularized nonnegative matrix factorization for temporal link prediction in dynamic networks. Sci. Direct **496**, 121–136 (2018). www.elsevier.com/locate/physa

# Engineering Teaching: Simulation, Industry 4.0 and Big Data

Jesús Silva[1(✉)], Mercedes Gaitán[2], Noel Varela[3], and Omar Bonerge Pineda Lezama[4]

[1] Universidad Peruana de Ciencias Aplicadas, Lima, Peru
jesussilvaUPC@gmail.com
[2] Corporación Universitaria Empresarial de Salamanca – CUES, Barranquilla, Colombia
m_gaitan689@cues.edu.co
[3] Universidad de la Costa (CUC), Calle 58 # 55-66, Atlantico, Barranquilla, Colombia
nvarela1@cuc.edu.co
[4] Universidad Tecnológica Centroamericana (UNITEC), San Pedro Sula, Honduras
omarpineda@unitec.edu

**Abstract.** In the educational field, there is a paradigm shift, where the focus is on the student and his active role in the training process, and where there is a turn that involves moving from content teaching to the training of competences. This necessarily requires higher education institutions to articulate innovation processes that involve the entire academic community. On the other hand, in the current context of technological development and innovation, companies, particularly manufacturing companies, are committed to reviewing and adapting their processes to what has been called Industry 4.0, a circumstance that entails the need to require new professional profiles that have competencies not only technological, but fundamentally those that will allow them to be competitive in a world where technology is renewed at an ever-increasing speed. The work presents implementation of innovation strategies in the teaching methodology from the integration of simulation software under the educational format of Problem-Based Teaching, which on the one hand aims to develop various competencies increase levels of motivation and satisfaction.

**Keywords:** Innovation · Simulation · Teaching · Engineering · Big data

## 1 Introduction

There is a coincidence in the academic fields of the need for a change in traditional teaching methods, however, it is not easy to innovate in the pedagogical models established in the different institutions [1]. It is noted that trends in education promote change from a passive learning approach, to an active one in which the participation and interaction of students and teachers is what is really significant [2]. These trends are highly positive in the training of engineers, because of the role they play, it is not enough that they have knowledge of basic and technological sciences, but it is essential


S. Smys et al. (Eds.): ICCVBIC 2019, AISC 1108, pp. 226–232, 2020.
https://doi.org/10.1007/978-3-030-37218-7_26

that during their training they achieve develop capacities to apply such knowledge in practice.

In the field of engineering teaching, the paradigm of competency-based training and the application of teaching and evaluation methods that contribute to achieving this achievement are installed.

On the other hand, from the productive sector, the concept of industry 4.0 appears, as a new stage of technological development and manufacturing processes, it is a relatively new concept that refers to the introduction of technologies in the manufacturing industry [3, 4]. The concept emerged in 2011, in Germany, to refer to a government economic policy sustained in high-tech strategies [5], such as automation, process digitization and the use of technologies in the field of electronics and applied to manufacturing [6]. These changes are geared towards the personalization of production, the provision of services, high value-added businesses, as well as towards interaction capabilities and the exchange of information between people and machines [7].

The conceptualizations that exist to date on the so-called Industry 4.0 are recent, however it can be defined as those that are physical and devices with sensors and software that work in the network and allow to predict, control and better plan business and organizational outcomes [8]. Therefore, there is an interest in the emergence of an industry that is based on the development of systems, the Internet of Things (IoT) and the internet of people and services [9], along with additive manufacturing, 3D printing, simulation, reverse engineering, big data and analytics, artificial intelligence and virtual reality among others. It is not new that discrete event simulation has great potential to contribute to the performance of engineers and managers of strategic and operational planning. These are techniques that improve the ability to analyze complex processes and facilitate decision-making [10]. Following specialists in the field, the simulation seeks to describe the actual behavior of a system or part of it, also contributes to the construction of theories and hypotheses allowing the observation of impacts and risks [11]. It also allows you to test new physical devices or systems, without compromising resources or work positions. The simulation time can also be sectioned, expanded or compressed, allowing the analysis of tasks, long-term scenarios or complex phenomena [13].

This is how, in manufacturing processes, simulation has been increasingly used in order to identify risks and impacts to make preventive decisions in the sector [14].

In short, it can be said that to talk about Industry 4.0, is to address an innovative approach to products and processes that arise from smart factories, that integrate into work networks and foster new forms of collaboration. In this context, the concept of intelligent manufacturing arises that implies the possibility of digitally representing every aspect of the process with support in design and assisted manufacturing software (CAD, CAM), product lifecycle management systems (PLM), and use of simulation software (CATIA, DELMIA, among others) [15].

As noted, simulation is one of the pillars of change in the productive areas. While the contribution of simulation technology to education is undeniable, it is noted that there are some restrictions in this regard. On the one hand, the need to explore and experiment with the various tools available on the market (commercial or open source), and fundamentally develop case studies. It is a complex activity, which requires intense interaction between specialists in the subject on the one hand and those with experience in the management of the software, on the other. As well as simulation techniques in the field of industry, they have an important added value, in terms of education the use of simulation software allows students to acquire skills in the management of systems very similar to those that in their work environment [16]. In the same vein [17] they consider that the use of simulation is particularly useful, from the perspective of the teaching and learning process, since, through these techniques, the student is able to operate models in a real system, to carry out experiences with assess the consequences and limitations of its application.

## 2 Materials and Methods

The case that is presented-Robotic Simulation-, has been developed by the Software Delmia Quest, by the technical team of the Laboratory of Simulation and Computational Modeling, of the University of Mumbai in India in 2017. It has been applied in the subject Logical Processes and has been integrated into the Bank of Business Cases that the university has. It aims to perform movements of an anthropomorphic robot and in turn save points in space, so that they can then be used by the robot. For the practice a robot brand KUKA, model KR6 R900 of 6 degrees of freedom is used. It is equipped with a pneumatic gripper for the clamping of parts (see Fig. 1).

**Fig. 1.** Robot brand KUKA, model KR6 R900 6-degree freedom.

To achieve this objective, several steps were considered: (a) system definition, (b) model formulation, (c) variable identification, (d) data collection, (e) model validation, (f) implementation and interpretation.

"Digital Enterprise Lean Manufacturing Interactive Application" (DELMIA) is a technology solution [18], focused on digital production and manufacturing. It is a tool that supports the engineering and manufacturing areas of companies, by virtue of which through it you can plan and orient actions towards continuous improvement, starting with the generation of 3D simulations with which it is obtained in different possible scenarios. Based on the information it provides, at the industry level, it supports critical decision-making on production lines and therefore helps ensure line efficiency. In the DELMIA environment, different software plug-ins can be differentiated, such as (Table 1).

**Table 1.** DELMIA add-ons. Source: DELMIA Products and Services.

| Complement | Use | Advantages |
|---|---|---|
| Delmia Quest [19] | Discrete event simulator focused on generating 3D scenarios that provide certainty for production line engineers | - Detection of bottlenecks<br>- Productivity assessment<br>- Obtaining statistics<br>- Validation of resource utilization and material handling systems with various mixtures of products |
| Delmia Robotics [20] | Production cell simulator with Virtual Robots to recreate plant operation behaviors | - Simulation of robots of various brands<br>- Simulation of product handling with robots<br>Simulation of spot welding<br>- Simulation of arc welding |
| Delmia Ergonomics [21–26] | Workstation simulator where company staff are involved | It seeks to meet two main objectives:<br>- Ergonomic evaluation where anthropometric characteristics can be established to obtain results of the type RULA, NIOSH and SNOOK and CIRIELLO<br>- Evaluation of the workspace, where the activities of the operators in their environment are generated Work |

Students receive a guide describing the case in advance, a simplified instruction manual for operating the software and the slogan to be developed (Fig. 1). At the end of the practice, and before retiring from the Laboratory they are asked to complete a survey, the structure of which includes three axes: Pre-class initiative in the laboratory; Development of activity in the laboratory and Development of practical work. The results obtained are systematically treated on a regular basis.

## 3 Results

Taking into account that the interest of the project, is focused on evaluating the experience of such innovation in the teaching and learning process, at the end of the laboratory activity, a survey is distributed that seeks to gather information that practice towards continuous improvement of it. The survey, since 2017, was applied to 4171 students according to the following distribution: 2017 (1452 students); 2018 (1587 students); 1st Quarter 2019 (1132 students). With respect to the profile of respondents, the average age of the total number of students is slightly higher than 27 years (28.55). In terms of sex, 83.2% are men and 18.3% are women, while 1.7% do not specify it.

In the surveys it is possible to identify some questions that are related to the understanding, use and application of the software seen in the laboratory, since they have been selected based on the interest that is pursued in this work. Including:

1. Existence of training instance in the use and scope of the software used: affirmative answers to the question is in the range [87.33%; 94.12%] with a confidence of 95%. No practices were identified in which negative answers to this question will be found.
2. Teacher's concern to clarify doubts about the use of the software: The percentage of affirmative answers to the question is in the range [96.88%; 99.11%] with a confidence of 95%.
3. Adequacy of time to develop the practice: The information received through the surveys allowed, to adapt the practice, so that, without losing its essence, the concepts are fixed and the programming competence acquired to the extent expected. This possibility of readjusting the practices based on the results obtained in the survey favored, the indicator, which in this year (2019), stands at 96.66%. The percentage of affirmative responses for this variable is in the range [91.14%; 98.60%] with a confidence of 95%.
4. Level of understanding of the class.

This variable was measured with a Likert scale ([1–6] being 1 Unsatisfied and 5 Very satisfied): all quarters are above 4.26 on average.

## 4 Conclusions

Based on the results obtained it can be observed that the innovation proposal in the teaching strategy implemented through the Laboratory of Simulation and Computational Modeling has had a very good acceptance by the students. It is an experience that on the one hand motivates them, and on the other they value the tool as an approach to the world of work, through open problems, in which to solve them it is necessary to interact with colleagues.

They also value the dedication and concern of teachers to approach and provide additional explanations, aimed at solving the problem that is posed to them; it is also noted that the previous instance of induction to the software is important and necessary in its training process. In relation to the level of understanding of the class and satisfaction with the software we also find a satisfactory level of response. In all cases it is observed that the values of the responses have improved over time.

Improvements in the teaching process would be determining an improvement in overall assessments over the experience [27]. One factor that is repeated is that the time spent on the case is insufficient for students from two points of view: (a) on the one hand in relation to the practice itself. In this sense, work has been done to readjust its scope so that the case can be resolved in a shorter time. (b) However, the greatest point is in relation to the meager time they consider to be devoted to this type of activity during the race. In this case, it is important to note that the restrictions that the institutions have, both buildings and equipment, which are very rigid factors and therefore their modification requires long-term planning.

Also, in relation to the software used, another limitation is the number of licenses available to the Academic Unit, taking into account that it is facing proprietary software. These are aspects, which are being worked on in order to expand the coverage of subjects that are involved by developing cases to address different topics, as well as providing the Laboratory with a technical critical mass trained, for the structuring of cases that are proposed from the various chairs. If we analyze the experience longitudinally, it is observed that there has been a positive trend in the perception of students and this is taken as a signal so that from the institutions, work continues on the analysis of the scopes and characteristics to implement a sustained process towards the continuous improvement of the ongoing educational innovation that is being developed.

## References


1. Marston, S., Li, Z., Bandyopadhyay, S., Zhang, J., Ghalsas, A.: Cloud computing—the business perspective. Decis. Support Syst. **51**(1), 176–189 (2011)
2. Armbrust, M., Fox, A., Griffith, R., Joseph, A.D., Katz, R., Konwinski, A., Lee, G., Patterson, D., Rabkin, A., Stoica, I., Zaharia, M.: A view of cloud computing. Commun. ACM **53**(4), 50–58 (2010)
3. Mell, G.: The NIST definition of cloud computing. NIST Special Publication 800-145 (2011)
4. Valarie Zeithaml, A., Parasuraman, A., Berry, L.L.: Total, quality management services. Diaz de Santos, Bogota (1993)
5. Sommer, L.: Industrial revolution—Industry 4.0: are German manufacturing SMEs the first victims of this revolution? J. Ind. Eng. Manag. **8**, 1512–1532 (2015)
6. Tao, F., Zuo, Y., Xu, L., Zhang, L.: IoT-based intelligent perception and access of manufacturing resource toward cloud manufacturing. IEEE Trans. Ind. Inf. **10**(2), 1547–1557 (2014)
7. UNESCO: Engineering: Issues Challenges and Opportunities for Development (2010)
8. Yajma, K., Hayakawa, Y., Kashiwaba, Y., Takahshi, A., Oiguchi, S.: Construction of active learning environment by the student project. Procedia Comput. Sci. **96**, 1489–1496 (2016). https://doi.org/10.1016/j.procs.2016.08.195
9. Izquierdo, N.V., Lezama, O.B.P., Dorta, R.G., Viloria A., Deras, I., Hernández-Fernández, L.: Fuzzy logic applied to the performance evaluation. Honduran coffee sector case. In: Tan, Y., Shi, Y., Tang, Q. (eds.) Advances in Swarm Intelligence, ICSI 2018. LNCS, vol. 10942. Springer, Cham (2018)
10. Amelec, V.: Increased efficiency in a company of development of technological solutions in the areas commercial and of consultancy. Adv. Sci. Lett. **21**(5), 1406–1408 (2015)
11. Viloria, A., Lis-Gutiérrez, J.P. Gaitán-Angulo, M., Godoy, A.R.M., Moreno, G.C., Kamatkar S.J.: Methodology for the design of a student pattern recognition tool to facilitate the teaching - learning process through knowledge data discovery (big data). In: Tan, Y., Shi, Y., Tang, Q. (eds.) Data Mining and Big Data, DMBD 2018. LNCS, vol. 10943. Springer, Cham (2018)
12. May, R.J.O.: Diseño del programa de un curso de formación de profesores del nivel superior. Mérida de Yucatán, UADY, p. 99 (2013)

13. Medina Cervantes, J., Villafuerte Díaz, R., Juarez Rivera, V., Mejía Sánchez, E.: Simulación en tiempo real de un proceso de selección y almacenamiento de piezas. Revista Iberoamericana de Producción Académica y Gestión Educativa (2014). http://www.pag.org.mx/index.php/PAG/article/view/399. ISSN 2007-8412
14. Minnaard, C., Minnaard, V.: Sumatoria de Creatividad + Brainstorming la Vida Cotidiana: Gestar problemas de Investigación/Summary of Creativity + Brainstorming + Everyday Life: Gestating Research Problems. Revista Internacional de Aprendizaje en la Educación Superior, España (2017)
15. Thames, L., Schaefer, D.: Software defined cloud manufacturing for industry 4.0. Procedía CIRP **52**, 12–17 (2016)
16. Gilchrist, A.: Industry 4.0 the Industrial Internet of Things, 2016th edn. Bangken, Nonthaburi (2016)
17. Arango Serna, M., Zapata Cortes, J.: Mejoramiento de procesos de manufactura utilizando Kanban; Manufacturing process improvement using the Kanban. Revista Ingenierías Universidad de Medellín 14(27), 2248–4094, 1692–3324 (2015)
18. Banks, J.: Introduction to simulation. In: Proceedings of the Simulation Conference, pp. 7–13 (1999)
19. Cataldi, Z.: Metodología de diseño, desarrollo y evaluación de software educativo. UNLP, p. 74 (2000)
20. Chen, F., Deng, P., Wan, J., Zhang, D., Vasilakos, A.V., Rong, X.: Data mining for the internet of things: literature review and challenges. Int. J. Distrib. Sens. Netw. **11**, 103–146 (2014)
21. Gilberth, S., Lynch, N.: Perspectives on the CAP theorem. Comput. IEEE Comput. Soc. **45**, 30–36 (2012)
22. Gouda, K., Patro, A., Dwivedi, D., Bhat, N.: Virtualization approaches in cloud computing. Int. J. Comput. Trends Technol. (IJCTT) **12**, 161–166 (2014)
23. Platform Industrie 4.0: Digitization of Industrie – Platform Industrie 4.0, pp. 6–9, April 2016
24. Schrecker, S., et al.: Industrial internet of things volume G4: security framework. Industrial Internet Consortium, pp. 1–173 (2016)
25. An Industrial Internet Consortium and Plattform Industrie 4.0: Architecture alignment and interoperability (2017)
26. OpenStack: Introduction to OpenStack, chap. 2. Brief Overview. http://docs.openstack.org/training-guides/content/module001-ch002-brief-overview.html
27. OpenStack: Introduction to OpenStack, chap. 4. OpenStack Architecture. http://docs.openstack.org/training-guides/content/module001-ch004-openstack-architecture.html

# Association Rule Data Mining in Agriculture – A Review

N. Vignesh(✉) and D. C. Vinutha

Department of Information Science and Engineering,
Vidyavardhaka College of Engineering, Mysuru, Karnataka, India
vigneshviggu97@gmail.com

**Abstract.** Agriculture is one of the most important occupations carried out by centuries - old people. Agriculture is our nation's backbone occupation. This is actually the occupation that fulfills the people's food needs. In this paper, we examine certain data mining techniques (Association rule) in the field of agriculture. Some of these techniques are explained, such as Apriori algorithm, K-nearest neighbor and K-means and the implementation of these techniques is represented in this field. In this field the effectiveness of the Associative rule is successful. In view of the details of agricultural soils and other data sets, this paper represents the role of data mining (association rule). This paper therefore represents the various algorithms and data mining techniques in agriculture.

**Keywords:** Agriculture · Data mining · Apriori · K-means · K-nearest neighbor

## 1 Introduction

Agriculture contributes nearly 10–15% of the GDP to India's economy [1]. The dataset is enormous in agriculture. Due to population growth, the demand for food supply is growing day by day. Therefore, scientists, trainers, researchers and governments always work to increase food supplies. For this purpose, the various techniques are applied. We discuss the conversion of an enormous dataset into knowledge. However, there is one major problem with the selection of appropriate data mining techniques and algorithms [2].

Here the data set can be converted into knowledge of ancient techniques and even current trends in agriculture [3]. As we can see, one example. i.e. Knowledge of crops like when and the harvest period should be plant. Even in the field of losses that can occur or the profit of the crop. It is very important in this field to know how the soil reacts with which fertilizers and to know the strength of the soil are important aspects. In this field, data mining techniques are merely statistics in which they can be properly explained [4]. In agriculture, the data can be of two types which are fundamental and technical in the unique way [3]. Which can be transformed into knowledge and profits beyond normal.

The association rule is very useful where the data is enormous, which makes it very appropriate in this area. Where there is no such end to the data, so that we can finish our analysis, because new techniques or patterns are always evaluated. The Data mining is


S. Smys et al. (Eds.): ICCVBIC 2019, AISC 1108, pp. 233–239, 2020.
https://doi.org/10.1007/978-3-030-37218-7_27

therefore used fully to restore the data that is evolving and to keep the old data together. Future forecasts are very important in the field of agriculture [2]. Because this field depends entirely on the future. Data mining for agriculture has many techniques, but here we use few techniques. Like the Apriori algorithm, K-neighbor, K-means that are more than enough to explain the purpose of this research paper.

## 2 Background of the Work

The mining of data is the process of discovering the way things are in large data sets, including statistics, machine learning and database systems. It also adds new techniques discovered in the data set previously saved. But we concentrate only on statistics for this paper. The main goal of data mining is to separate the knowledge from the available data and then customize it into a simple, comprehensible substance and also use it for prediction methods [1].

## 3 Data Mining Techniques

The explanation of the techniques used here is briefly given below. The data processing techniques and can be divided into two types i.e. Clustering and classification: This is mainly based on the grouping of unknown samples acquired from classified samples. These are training sets with which classification techniques are trained. In fact, K-nearest neighbor is referred to as a training set because it learns nothing but classifies the given data into an ideal group. Because it uses a training group every time the classification needs to be done [6]. If classification is not available or if no previous data are available, the clustering techniques are used i.e. K-means where unknown data can also be converted into clusters. And it is the most used techniques for this operation. This technique is used in many fields and suits the agricultural sector very well because unknown samples are always evolving frequently. This paper demonstrates efficiently the different mining techniques and algorithms that can be used in agriculture. Since it contains the large data set. It is therefore a difficult task to process, so that we use some efficient techniques, such as the A priori algorithm, the nearest neighbors, the k-means, the bi-clusters. The details of the above techniques are explained below.

### 3.1 Apriori Algorithm

Apriori is a bottom - up approach. The Apriori algorithm is an impact for mining for frequent data sets over the Boolean association rule and is intended to work on transactions. It works in this field on soil, crops, income transactions, profit and losses. The pre-processing of numerical data sets is much needed in the association rule, which increases efficiency (also known as a training set). After the pre-processing of these data, they are used to compare the attributes of days a week to weeks a month. The values can be derived from the minimum price, the maximum price and even the moderate price of the trendy crops on the market. Because of its combination explosion, it is necessary to determine the relationship between these sessions in a day with a minimum of more than 10. Once the frequent pattern has been obtained, it is easy to create an association rule with a trust greater than or equal to 20.

Apriori algorithm is a mining technique that is mainly used to find the pattern that is very common. In this paper, we store the data set on crops and soil activities and the activities that took place during daytime sessions. Let us consider crops which trend in one crop during the day's session let the size of the crop be k. Let the crop be fk and the work be wk. Here Apriori searches and scans the data set and derives the frequent trends in a marketing day size of 1 on the particular day.

1. Generate wk + 1, work performed on the frequent trend in a day for certain seeds and the market size will be k + 1 and the specific market size will be k [2].
2. Varify the data set and support for the work performed on a day-to-day basis for all types of crops on a specific market [2].
3. Combining the day-to-day work for a market crop that meets the minimum requirements for saturation on the market

Apriori Algorithm does work on two steps: i.e. it creates wk + 1 from fk

1. *Join Step:* Generate Rk + 1, which is the day-to-day trend for crops of market size f k + 1. Which is done by uniting the two part-to-day trend for crops of market size K
2. *Prune Step:* Examine the trends of the day-to-day morning, noon and evening session for all crops of market size k in Rk + 1 which will create the regular trend for crops of market size k in Rk + 1.

This step will take count of the frequent trends that occurred recently or in certain period of time as we specified earlier above the three sessions in a day morning, noon and evening. The crop it will be taken as 'k' as the market size. Then the frequent trend that occurred on the same plot that will be taken as 'k + 1'. The frequent trend that occurred in particular time in particular sessions of the day will be taken for count.

Examine the data which has been collected previously from the frequent trend that occurred before. The trend may occur in any of the three sessions. Compute the contribution provided by each trend that occurred in any particular time in a day. That will be available in the particular market.

The final step is that adding all the trends that occurred before in particular time i.e., in particular session of the particular day for the specific crop in the specific market. Which the support given by those trends to the demand that is required by Fk + 1. Thus if the minimum demand must be satisfied if not the steps must be carried on until we get the result.

### 3.2 K-nearest Neighbor

K-nearest neighbor is one of the methods used in data mining classifications and regression. It is also referred to as a lazy learning algorithm and also as a non-parametric based technique [5]. The bottom line assumption in the KNN algorithm is that it should have the same sample data set with the same classifications. It is mainly an experimental theoretical assumption and does not match the practical data set. In this case, the parameter k counts the number of samples known to classify the unknown. And the classifications repeatedly used in the nearest neighboring sample are allocated to the unknown samples used The results of the KNN are helpful in the practical world [6, 7]. The KNN method will provide us with a very classic and simple classification rule and it is expensive to proceed, as we said the bottom line is to consider unknown samples and the distance from all the known samples nearby must be calculated and this calculation is cost-effective [3]. Of course, the KNN method uses the training set but does not extract it.

```
for all Un-Sample(i)    ||Unknown samples
        for all sample(j)   ||Known samples
          Compute the distance between Un-Samples(i) and Samples(j)
            end for
              search the K smallest distance
              locate the correlate samples Sample(ji...jk)
              assign Un-Sample(i) to the group which appears more often
            end for [1].
```

### 3.3 K-Means Algorithm

It is a clustering method for vector quantification originally from signal processing, which is most commonly used in data mining clustering analysis. The system is given with a data set that is unknown in class. The main aim of the K-means algorithm is to find the same data set in a cluster. The measurement is provided with efficient distance. Assign the value to each sample at random in the clusters and calculate the center for each cluster. If clusters are unstable, the distance between the center and the sample should be calculated. The closer sample to the center must be found. And the center for the changed cluster should be reputed [6–8].

```
assign values to the object sample to one of the K clusters S(j)
        find the center c(j) for all the sample S(j)
        while (cluster unstable)
            for each Sample(i)
                  find out the distance from center c(j) to Sample(i)
                  find j* in c(j)* is closer Sample(i)
                  Assign Sample(i) to the cluster S(j)*
                  Re-compute center
            end for
        end while [1].
```

Many k-means algorithms have been created and even tried to implement in the field of data mining [6]. However, K-means does not allow empty clusters. Many types have been created, some of which are independent parameters where the k value is assigned by the execution time (Table 1).

**Table 1.** Comparison between the different association rule data mining algorithm in agriculture

| Parameters | Apriori | k-NN | k-Means |
|---|---|---|---|
| Rain prediction | Yes | No | No |
| Disease prediction | Yes | No | No |
| Pattern of crops | No | Yes | Low changes |
| Trends in sessions that are count | No | No | Yes |
| Specifying market crops | Yes | No | No |
| Dependency of crops | No | Yes | No |
| Specifying sessions in a particular day | Yes | No | No |
| Soil pattern | No | Yes | Yes |
| Nutrients in soil | No | Yes | Yes |

# 4 Application of Data Mining Techniques in Agriculture

Agriculture is a giant dataset and very difficult to understand. But this agriculture has been followed for ages that meet the basic needs of human food. I'm planning this field was started long before, but that doesn't mean it's already improved. If a system works properly, it doesn't mean that it doesn't have any improvements anymore [9, 10]. Therefore, this paper discusses the few techniques that can be implemented in the field of agriculture. We discuss crops, soil activities and much more. The problem can occur and the problems can arise in future forecasts and so on.

## 4.1 Application of Apriori Algorithm

The Apriori algorithm is used to find the pattern that is often used. We can frequently take the pattern of cultivation that people used to perform. The pattern has to be decided according to the soil wealth. That's how the Apriori algorithm helps. The activity pattern of the three-shift crop should be taken into account. Apriori helps you to search and scan the data set, which is very frequent and time consuming. It estimates the crop activity and the market value, which will also help in future forecasting.

## 4.2 Application of k-Nearest Neighbors

In every field the technology has been greatly improved. We can implement many things in agriculture in this paper when we discuss soil and fertility. Using satellite sensors to observe land and soil activities is very easy to carry out in this field, researchers use means and k-NN techniques that are most commonly very efficient in either way. Sensors can observe and provide enormous amounts of data set at least for a long time, which is much needed in the field of agriculture [6, 7].

## 4.3 Application of k-Means Algorithm

This algorithm is used to determine the fertility of the soil. To find the fertility, this uses AHP to determine soil nutrient values. Then the k-means algorithm is added to continue. The optimum efficiency indicates the classification that improves the clustering algorithm's efficiency. The weighted k-means clustering was more accurate than the unweighted clustering, and more efficient. The evolution of the changes that can occur in the soil must be studied in advance and with which fertilizer the soil works well. Using these methods, the soil nutrient value will improve after years. The result is that improved clustering is an efficient method for determining soil fertility [6, 7].

## 5 Conclusion

Agriculture is one of the major occupations of Indian peoples. And in developing countries such as India, technology must be implemented in this field. Technology implementation will adversely improve the field. And it will also help farmers in a variety of ways, where the term intelligent work is used instead of hard work. Data mining plays its important role in this area if it occasionally finds difficulties and helps them to make decisions. We discuss soil and crop activities in this paper. Therefore, the characteristics of the soil can be effectively studied instead of using some old techniques, we will use modern techniques to study these activities. And the process of growing and forecasting the future for growing can be carried out efficiently. Therefore, revenue can also be improved and the losses that can occur can also be avoided.

**Acknowledgements.** The authors express gratitude towards the assistance provided by Accendere Knowledge Management Services Pvt. Ltd. In preparing the manuscripts. We also thank our mentors and faculty members who guided us throughout the research and helped us in achieving desired results.

**Compliance with Ethical Standards**
✓ All authors declare that there is no conflict of interest
✓ No humans/animals involved in this research work.
✓ We have used our own data.

## References

1. Jaganathan, P., Vinothini, S., Backialakshmi, P.: A study of data mining techniques to agriculture. Int. J. Res. Inf. Technol. **2**, 306–313 (2014)
2. Thakkar, R.G., Kayasth, M., Desai, H.: Rule based and association rule mining on agriculture dataset. Int. J. Innov. Res. Comput. Commun. Eng. **2**, 6381–6384 (2014)
3. Khan, F., Singh, D.: Association rule mining in the field of agriculture: a. Int. J. Sci. Res. Publ. **4**, 1–4 (2014)
4. Nasrin Fathima, G., Geetha, R.: Agriculture crop pattern using data mining techniques. Int. J. Adv. Res. Comput. Sci. Softw. Eng. **4**, 781–786 (2014)
5. Ramesh Babu, P., Rajesh Reddy, M.: An analysis of agricultural soils by using data mining techniques. Int. J. Eng. Sci. Comput. **7**, 15167–15177 (2017)
6. Mucherino, A., Papajorgji, P., Pardalos, P.M.: A survey of data mining techniques applied to agriculture. Oper. Res. **9**, 121–140 (2009)

7. Patel, H., Patel, D.: A brief survey of data mining techniques applied to agricultural data. Int. J. Comput. Appl. **95**, 6–8 (2014)
8. Ramesh, V., Ramar, K., Babu, S.: Parallel K-means algorithm on agricultural database. Int. J. Comput. Sci. **10**, 710–713 (2013)
9. Bharatha Krishnan, K., Suganya, S.S.: Application of data mining in agriculture. Int. J. Res. Comput. Appl. Robot. **5**, 18–21 (2017)
10. Ayesha, N.: A study of data mining tools and techniques to agriculture with applications. Spec. Issue Publ. Int. J. Trend Res. Dev. 1–4 (2017)

# Human Face Recognition Using Local Binary Pattern Algorithm - Real Time Validation

P. Shubha(✉) and M. Meenakshi

Department of Electronics and Instrumentation Engineering,
Dr. Ambedkar Institute of Technology, Bangalore 560056, India
shubhajagi@gmail.com, meenakshi_mbhat@yahoo.com

**Abstract.** A real time face recognition using LBP algorithm and image processing techniques are proposed. Face image is represented by utilizing information about shape and texture. In order to represent the face effectively, area of the face is split into minute sections, then histograms of Local Binary Pattern (LBP) are extorted which are then united into a single histogram. Secondly, the recognition is carried out on computed feature space using nearest neighbor classifier. The developed algorithm is validated in real time by developing a prototype model using Raspberry Pi single board computer and also in simulation mode using MATLAB software. The above obtained results match with each other. On comparing both the results, recognition time taken by the prototype model is more than that of the simulation results because of hardware limitations. The real time experimental results demonstrated that the face recognition rate of LBP algorithm is 89%.

**Keywords:** Local Binary Pattern (LBP) · Matlab · Histogram · Classifier · Raspberry Pi · Real time face recognition

## 1 Introduction

Face recognition is a popular image processing application. Using this method a person can be identified and verified with the database. The process identification involves face recognition, which compares unknown face image with the large set of face image database. The obtained value is the measure of the similarity between two images. The face image is recognized when the obtained value is equal to or greater than the threshold value. The recognition process includes the following three stages. Firstly, Face representation carries out the process of modeling a face. Secondly, feature extraction stage includes extraction of unique and useful features of a face. Lastly, classification stage compares the face image with database images.

Applications of face recognition includes in various fields such as banking, airports, passport office, government offices, security systems and many more. There are two phases of face recognition. The first phase includes training of the faces which are saved in the database and second phase is the verification stage to obtain the exact image of the face.

Face recognition technique is considered as a challenging method due to variation in size, color, pose variations, illumination variations, and face rotation. To overcome


S. Smys et al. (Eds.): ICCVBIC 2019, AISC 1108, pp. 240–246, 2020.
https://doi.org/10.1007/978-3-030-37218-7_28

the above problem Local Binary Pattern algorithm (LBP) is proposed. The paper is organized in the following steps: First, Existing methods of face recognition is given in Sect. 2. Next, in Sects. 3 and 4 the proposed method and face recognition algorithm are explained. Sections 5 and 6 gives the working principle and the results of the experiment. Finally Sect. 7 is referred in order to draw conclusions.

## 2 Existing Techniques of Face Recognition

The various methods are available for face recognition which includes: Principal Component Analysis (PCA) and Linear Discriminant Analysis (LDA).

In PCA algorithm [2] the number of variables can be reduced statistically. The relevant features are extracted from the face. In the training set, the obtained images are combined as weighted eigen-vectors and eigenfaces. The covariance matrix of images in training database provides the eigen-vectors known as basic function and the corresponding weight is determined by choosing a set of relevant eigenfaces. Values of eigen are increased to obtain higher face recognition.

Classification of images according to a group of features is carried out by LDA [2, 3]. Between class scatter matrix (SB) and within class scatter matrix (SW) of all the classes of all samples are defined. The differences between classes are found for recognition. The above methods cannot produce complete information of face due to the loss of information and they are sensitive to noisy image and lighting conditions in the images [4]. PCA algorithm fails to recognize the identical twin faces.

## 3 Proposed Method

In order to overcome the above problems, local binary pattern algorithm is proposed [5]. The algorithm is less sensitive to illumination variations and scaling variations. LBP are very efficient approach for feature extraction. It encodes differences between pixel intensities in a local neighborhood of a pixel. Tiny patterns from a face image are identified by applying the local binary pattern operator [6].

Face recognition system includes the following modules. Firstly, image acquisition module in which the details of the image are gathered and these samples are taken for analysis. Secondly, feature extraction module extracts only significant features from the image. Lastly, classification and training classifier database module classifies the obtained samples.

### 3.1 Face Detection

Face detection is represented by a block diagram as given in Fig. 1. A face can be detected by using Haar cascade classifier from OpenCV. Detection of various organs on the face is carried out by Haar cascade classifier which operates based on AdaBoost algorithm. The detected faces are stored in a FR100.

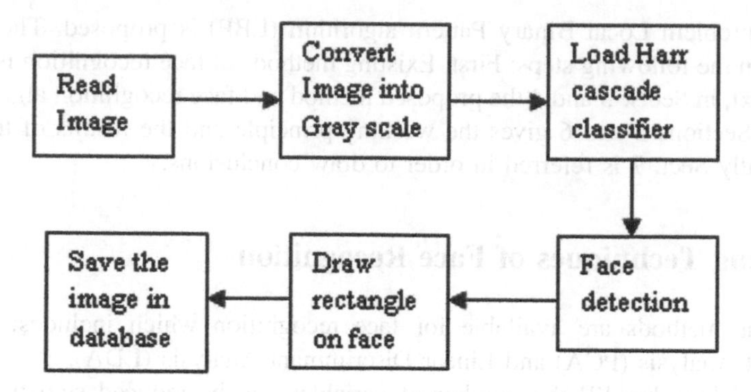


**Fig. 1.** Block diagram representing detection of a face

### 3.2 Face Recognition

The technique of face recognition is depicted in Fig. 2. The calculation of the binary pattern of a face is carried out by converting it into a gray scale image [7]. The LBP operator is not fixed with the radius and neighborhood. Therefore the algorithm can be applied to the images of different size and texture. The face description is represented through feature vector. The histograms for the region are calculated. The united histogram represents an overall description of the face [8].

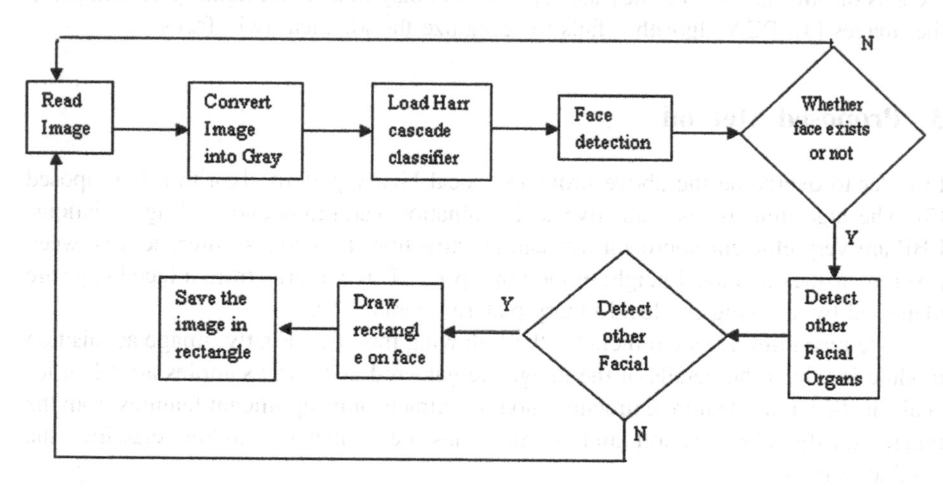


**Fig. 2.** The block diagram representation of the face recognition technique.

### 3.3 Database

Database FR100 is created to recognize a human face. The database stores 100 images of each face with different facial expressions and postures in an image. During acquisition process database images are converted to gray scale images for feature extraction. Then the images are normalized for better recognition results. Each image undergoes a process of normalization to eradicate noise and to align it in proper position.

## 4 Block Diagram and Working Principle of Hardware Setup

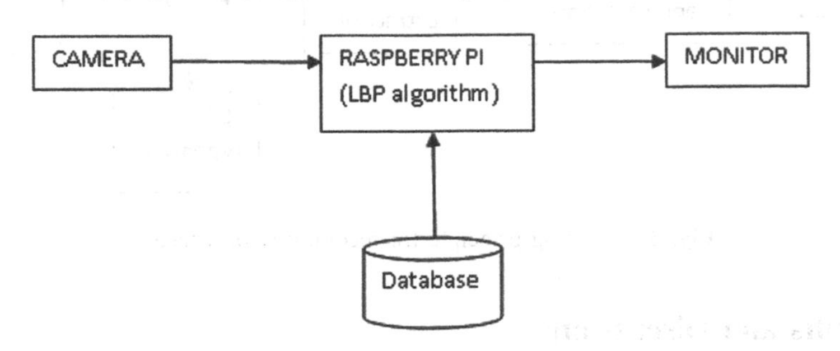


**Fig. 3.** Block diagram representation of hardware setup

The hardware setup is shown in Fig. 3 whereas working principle of the face recognition system is depicted by means of a block diagram as shown in Fig. 4. The hardware module consists of Raspberry pi single board computer powered by a micro USB socket with 5 V, 2.5 A. The LBP algorithm is made to run in python IDE. The image is captured by a camera. The obtained color image is converted into grayscale image for further processes. The required facial region is cropped and is divided into several blocks. The unique ID is assigned to each image in the database as a result a database is created. Then the extraction of texture features of LBP from the test image is carried out. A face is recognized by using Harr cascade classifier and recognizer [9, 10]. Database image is used by the classifier to compare it with the test image. Face recognition is successful only when the test image matches with the images of the database [11].

## 5 Face Recognition Algorithm

Local binary algorithm is applied to perform face recognition. Proposed algorithm is as described below:

1. Inputting of the face image.
2. Dividing the face image into blocks
3. Calculating histogram for each block
4. Combining LBP histogram into single histogram
5. Comparing the face image with the database image for further processing
6. Successful recognition of the face only when the test image matches with the images of the database.

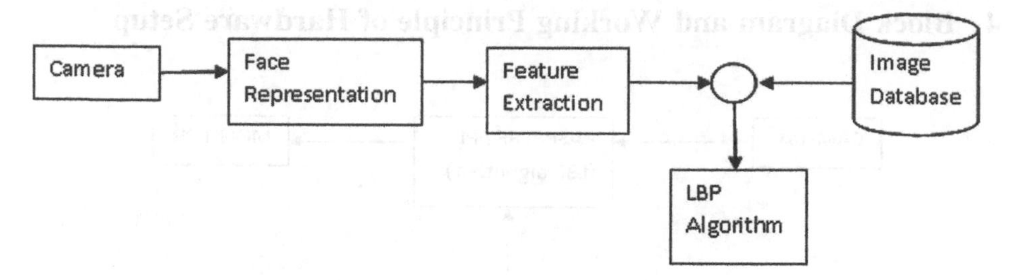


**Fig. 4.** Working principle for recognition of a face

## 6 Results and Discussions

Algorithm is tested for its performance using different test images in simulation environment and in hardware setup.

Before applying an algorithm, required face region is cropped as depicted in Fig. 5 (a). Input image and LBP Image are illustrated in Fig. 5(b). The recognition result is represented in Fig. 5(c). The hardware setup result is described in Fig. 6.

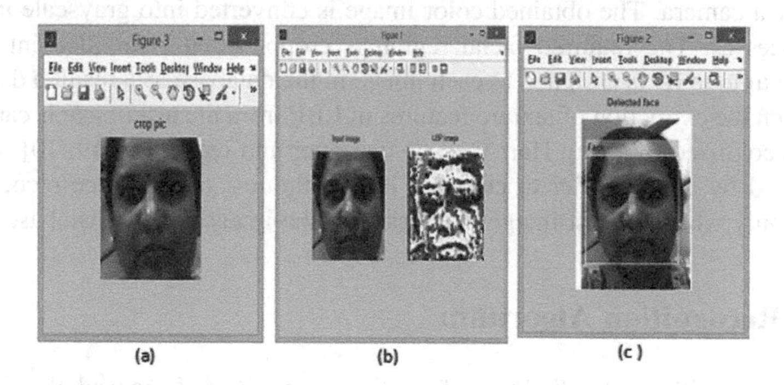


**Fig. 5.** Simulation results of LBP algorithm (a) Cropped image, (b) Original image and LBP image, (c) Detected image

The overall rate of recognition of an LBP algorithm is 89% which is observed in Table 1. The experiment is also extended for images with varying positions of the face. From Table 2, it can be concluded that rate of recognition of a face facing the front is greater as compared to the other positions of the face.

The results of Table 3 depicts that the recognition rate is high in LBP algorithm when compared to the other face recognition algorithms such as PCA, 2D-PCA and LDA. This is represented in Fig. 7.

**Fig. 6.** Face detected from experimental setup

**Table 1.** Rate of recognition of FR100 database

| No. of images in the database | No. of input faces tested | Recognized face | Un recognized face | Recognition rate |
|---|---|---|---|---|
| 5500 | 55 | 49 | 06 | 89 |

**Table 2.** Recognition rate of FR100 database at various position of face

| Left | Front | Right |
|---|---|---|
| 58 | 89 | 45 |

**Table 3.** Face Recognition rate of standard face recognition algorithm [12]

| Standard face recognition methods | Recognition rate (%) |
|---|---|
| LBP | 89 |
| PCA | 64 |
| 2D-PCA | 63.1 |
| LDA | 55 |

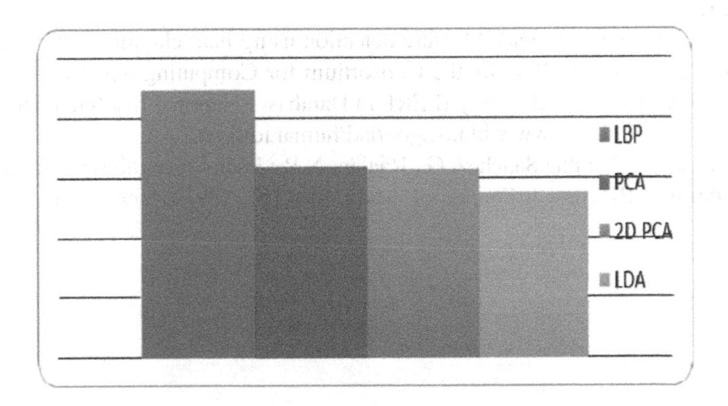


**Fig. 7.** Comparison of various algorithms

## 7 Conclusion

The paper presented a simple and efficient way of face recognition using LBP algorithm. The design and validation of real time face recognition is proposed. The proposed system is developed by using LBP algorithm and techniques of image processing. MATLAB software is used to validate the results in offline mode whereas experimental setup is used to validate the results in real time. These results match with each other. The rate of recognition of a face by LBP algorithm is obtained as 89% by which we can prove that above algorithm is better when compared to other face recognition algorithms. The same work can be extended to an autonomous robot.

## References

1. Singh, S., Kaur, A., Taqdir, A.: Face recognition technique using local binary pattern method. Int. J. Adv. Res. Comput. Commun. Eng. **4**(3), 165–168 (2015)
2. Shubha, P., Meenakshi, M.: Design and development of image processing algorithms-A comparison. In: 1st International Conferences on Advances in Science & Technology ICAST-2018 in Association with IET, India, Mumbai, 6th–7th April 2018 (2018). ISSN 0973-6107
3. Riyanka: Face recognition based on principal component analysis and linear discriminant analysis. IJAREEIE **4**(8) (2015)
4. Gonzalez and Woods: Digital Image Processing. 3rd edn. Prentice Hall, Pearson Education, Upper Saddle River (2008)
5. Bruunelli, R., Poggio, T.: Face recognition: features versus templates. IEEE Trans. Pattern Anal. Mach. Intell. **15**(10), 1042–1052 (1993)
6. Khanale, P.B.: Recognition of marathi numerals using artificial neural network. J. Artif. Intell. **3**(3), 135–140 (2010)
7. Rahim, A., Hossain, N., Wahid, T., Azam, S.: Face recognition using local binary patterns. Glob. J. Comput. Sci. Technol. Graph. Vis. **13**(4), 1–9 (2013)
8. Huang, D., Shan, C., Ardabilian, M., Wang, Y., Chen, L.: Local binary patterns and its application to facial image analysis: a survey. IEEE Trans. Syst. Man Cybern. Part C Appl. Rev. **41**(6), 765–781 (2011)
9. Gudipathi, V.K., Barman, O.R., Gaffoor, M., Harshagandha, A., Abuzneid, A.: Efficient facial expression recognition using adaboost and Harr cascade classifiers. In: IEEE 2016 Annual Connecticut Conference on Industrial Electronics, Technology & Automation (CT-IETA) (2016)
10. Wilson, P.I., Fernandez, J.: Facial feature detection using haar classifiers. JCSC **21**(4), 127–133 (2006). Copyright © 2006 by the Consortium for Computing Sciences in Colleges
11. The Facial Recognition Technology (FERET) Database: National Institute of Standards and Technology (2003). http://www.itl.nist.gov/iad/humanid/feret/
12. Vishnu Priya, T.S., Vinitha Sanchez, G., Raajan, N.R.: Facial recognition system using local binary patterns (LBP). Int. J. Pure Appl. Math. **119**(15), 1895–1899 (2018)

# Recreation of 3D Models of Objects Using MAV and Skanect

S. Srividhya[1(✉)], S. Prakash[2], and K. Elangovan[3]

[1] Department of ISE, BNM Institute of Technology, Bangalore, India
s.srividhyaa@gmail.com
[2] Department of CSE, Dr. Ambedkar Institute of Technology, Bangalore, India
prakash.hospet@gmail.com
[3] Er. Perumal Manimekalai College of Engineering, Hosur, India
drelangovank@gmail.com

**Abstract.** Mapping of static objects in an indoor environment is a process to obtain the description of that environment. Simultaneous localization and mapping (SLAM) is an active area of research in mobile robotics and localization technology. To accomplish enhanced performance with low cost sensors, Microsoft Kinect has been mounted on a developed aerial platform in association with Skanect. Skanect transforms the structure sensor Kinect into a less expensive 3D scanner that can create 3D meshes. An experimental result shows how the proposed approach is able to produce reliable 3D reconstruction from the Kinect data.

**Keywords:** SLAM · Quadrotor · Skanect · RANSAC · SURF

## 1 Introduction

The main important application of mobile robots is to arrive at and discover terrains which are unreachable or considered unsafe to humans. Such surroundings are commonly stumble upon in search and rescue scenarios where prior information of the surroundings is unknown but necessary before any rescue procedure can be organized. A small mobile robot furnished with a proper sensor bundle can be viewed as the best guide in such situation. The robot is estimated to navigate itself through the place and produce maps of the surroundings which can be used by human rescuers for navigation and localization of victims. In spite of the wealth of investigation in planar robot mapping, the restrictions of traditional 2D feature maps in providing constructive and comprehensible information in these situations have thrust the increase in research efforts towards the generation of richer and more descriptive 3D maps instead [1].

Pioneering work carried out by the research team of Prof. Vijay Kumar, director – GRASP Laboratory, University of Pennsylvania has almost pushed the MAV technology to its thresholds of applicability especially in collaborative autonomous missions by swarm of nano quad copters. One of the striking features of his work is the demonstration of 3D image reconstruction and live transmission by integration of versatile visual/motion sensors in MAVs which has triggered the imagination of this research work.


S. Smys et al. (Eds.): ICCVBIC 2019, AISC 1108, pp. 247–256, 2020.
https://doi.org/10.1007/978-3-030-37218-7_29

In this article, the development of a quadrotor with a payload capacity of up to 500 gm and integrating the readily available sensor hardware Kinect® to a MAV platform and creating the 3D view of targeted objects in an indoor environment using Skanect has been shown and explained. It's not affordable to capture a color 3D model of an object, a human or a room. Skanect will transform the Structure Sensor, Microsoft Kinect camera into a less expensive 3D scanner, which will create 3D mesh out of real landscape in a few minutes. The functionalities of sub systems like 3D scanner, camera, microchip etc. would be individually studied with intent to remove the extraneous elements and reduce the system weight [2].

## 2 Goals of the Study

It's a challenging task to map the internal environment of an environment. In that capacity, it is compulsory to assess the necessities of this task, as:

- Device cost
- Practical application of the method
- Computational cost
- Easy to use and exploitation of the attained results.

The following parameters can be considered to accomplish these necessities:

- Less expensive 3D sensing equipment
- Less expensive computer

### 2.1 Principles of 2D/3D Reconstruction

#### 2.1.1 Sensing

Different types of sensors will be used by SLAM to obtain information with probably different errors. To manage with metric preconception and with noisy measures it is important to have statistical independence. Those type of optical sensing equipment's may be 3D or 2D cameras. There has been an extreme research in VSLAM using mainly visual sensors, due to the increase use of cameras in mobile phones. As a compliment to unpredictable wireless measures, the recent methods pertain quasi-optical wireless ranging for multi-lateration (RTLS) or multi-angulations in conjunction with SLAM

SLAM has been introduced to human pedestrians' by means of shoes accumulated with inertial measurement unit as the major sensor in order to avoid walls. This method is known as FootSLAM which is used to make the floor plans of buildings automatically which can be used by an indoor localization system.

#### 2.1.2 Positioning

The outcome from sensing will provide the algorithms for localization or positioning.

According to geometrical plan, the sensing should include at least one iteration and (n + 1) estimating equations for n dimensional problem. It should also have some additional information like a priori knowledge about the adjustment of results or the coordinates of the systems with rotation and mirroring.

#### 2.1.3 Modeling

Mapping can be used in 2D exhibiting and corresponding illustration or in 3D forming and 2D depiction. To improve the estimation of sensing under intrinsic condition and noise, the kinematics of the robot is inculcated as an element in the model itself. From various sensors and partial error models the dynamic model has been balanced. It finally encompasses a sharp virtual description as a map with the position and direction of the robot. The final representation of these model can be represented by mapping and the map should be either descriptive or conceptual design of the model.

### 2.2 Need for Custom Platforms

We have proposed to utilize a stripped down Kinect Sensor custom fitted onto a vibration proof chassis and mounted on top of a quadrotor. Kinect is a structure sensor by Microsoft for Xbox 360 and Xbox One video game consoles and Microsoft Windows PCs. A 3D scanning evoke can be used to measure the 3D outline of an object using anticipated pattern of the light and a camera. It works on the principle by foretelling a narrow band of light onto a 3D structured plane producing a line of illumination that shows distorted from other point of view than that of the projector and can be used for an accurate geometric restoration of the surface outline.

## 3 Development of the Mobile Platform

### 3.1 Selection of Mobile Platform

The quadrotor has unbounded applications and it is also an emerging Micro Aerial Vehicle (MAV). Recent quadrotors are surfacing into miniature and agile vehicles. The emerging research in the area of miniature vehicles allows quadrotor to communicate smartly with other autonomous vehicles, to discover unknown environments or surroundings and to move in dense environment with high speed and with a prior knowledge. These advancements will let the quadrotors to accomplish the missions like surveillance, search and rescue operations. Even though it is rotor based, it is different from the helicopters and predominantly makes it attractive as an MAV. It is highly reliable and manoeuvrable with the four rotors. It can also be used in applications like multi-craft communication, environment exploration and manoeuvrability to increase the feasibility of quadrotors. A quadrotor is a kind of rotorcraft that has two pairs of counter-rotating, fix-pitched blades for lift. The quadrotor propellers are connected to four individual motors using the fixed-pitched blades. This leads to the non usage of control pitch. These motors are then associated in an 'X' arrangement. The battery and microcontrollers are placed near to the center of the craft in order to power and control the rotors. The altitude and attitude of the craft can be changed by shifting the speed of individual rotors. Figure 1 represents the layout of these components and shows the various changes in rotor to adjust the state of the craft. Using these basic designs it is easy to design vehicles that are much smaller than the conventional rotorcrafts. It is also important to consider the payload and size based upon the requirements of the task [3].

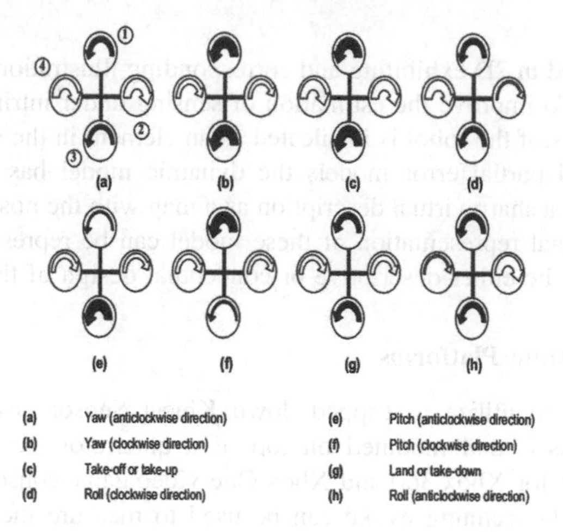


**Fig. 1.** Control scheme of quadrotor design

Due to its intrinsic stability, the quadrotor design provides tremendous flight characteristics. As each pair of rotor blades are spinning in opposite direction the torques are terminated and keeps the quadrotor flying straight and accurate. The propellers rotating in anti clockwise direction increase the efficiency and flight time. And no extra propel is required to balance the unwanted rotation (Table 1).

### 3.1.1 Mobile Aerial Platform Control and Dynamics

**Reference Frames.** This segment explains the different frames for the references and coordinate systems that is utilized in demonstrating the position and direction of the platform. It is essential to use numerous diverse coordinate systems for the subsequent causes:

i. The coordinate frame is substituted with Newton's equations of motion appended to the aerial platform.
ii. The body frame is applied with aerodynamic forces and torques.
iii. Accelerometers and rate gyros are the on board sensors from which the values of the body frames will be measured.
iv. The inertial frame contains the information's about hover points and flight trajectories.
v. There are several coordinated systems available for a quadrotor. The frames of the co-ordinator are discussed below.

### 3.1.2 Forces and Moments

Due to the gravity and the four propellers the force and moments are generated.

**Table 1.** Selected quadrotor technical specifications are listed below (DragonFly)

| Standard configuration<br>Dimensions<br>Width: 64.5 cm (25.4 in.)<br>Length: 64.5 cm (25.4 in.)<br>Top Diameter: 78.5 cm (30.9 in.)<br>Height: 21 cm (8.3 in.) | Without Rotors<br>Dimensions<br>Width: 36 cm (14.1 in.)<br>Length: 36 cm (14.1 in.)<br>Top Diameter: 51 cm (20 in.)<br>Height: 21 cm (8.3 in.) |
|---|---|
| Without rotor blades or landing gear<br>Dimensions<br>Width 36 cm (14 in.)<br>Length: 36 cm (14 in.)<br>Top Diameter: 51 cm (3 in.)<br>Height: 7.5 cm (3 in.) | Weight & Payload<br>Helicopter Weight (with battery): 680 g (24 oz)<br>Payload Capability: 250 g (8.8 oz)<br>Maximum Gross Take-Off Weight: 980 g (33 oz) |
| Rotor Blades<br>Four Counter Rotating Blades<br>Rotor Blade Material: Molded Carbon Fiber<br>Electric Motors | Brushless Motors: 4<br>Configuration: Direct Drive (One Motor per Rotor)<br>Ball Bearing: 2 per Motor<br>Rotor Mounting Points: Integrated<br>Voltage: 14.8 V |
| Rechargeable Helicopter Battery<br>Cell Chemistry: Lithium Polymer<br>Voltage: 14.8 V<br>Capacity: 2700 mAh<br>Connectors: Integrated Balance and Power<br>Recharge Time: Approx. 30 min (after typical flight) | Landing Gear<br>Installed Height. 17 cm (7 in.)<br>Stance Width: 26 cm (12 in.)<br>Skid Length: 31 cm (12 in.) |
| Landing Gear Material: Molded Carbon Fiber<br>Materials<br>Carbon Fiber<br>Glass Filled Injected Nylon<br>Aluminum A Stainless Steel Fasteners RoHS Compliant | |

The whole force acting on the quadrotor is represented by:

$$F = Ff + Fr + Fb + Fl$$

Figure 2 shows the top angle representation of the quadrotor. An rising force F and a torque T is generated by every motor. The front and rear motors and the right and left motors spin in clockwise direction and anti clockwise direction respectively (Fig. 3).

Due to Newton's 3rd law, a yawing torque will be produced by the drag of the propellers on the body of the quadrotor. The way of the torque and motion of the propeller will be in the opposite direction to each other. Consequently the total yawing torque is represented by Eq. (1)

$$T\varphi = Tr + Tl - Tf - Tb \tag{1}$$

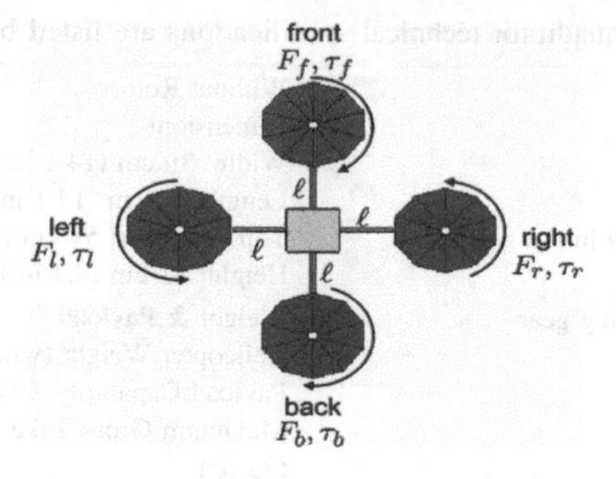


**Fig. 2.** Top angle view of the quadrotor

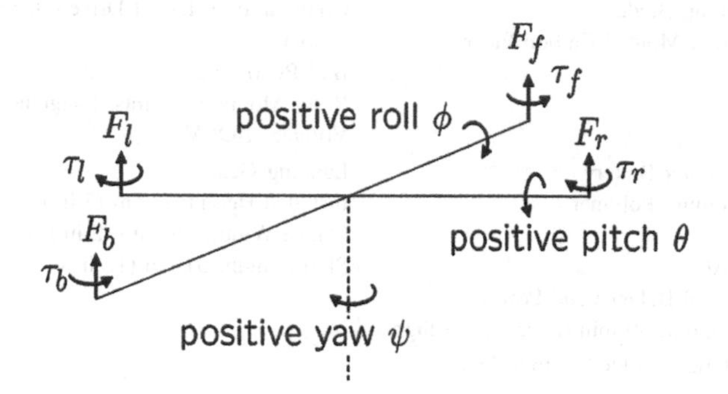


**Fig. 3.** Forces and Torques in the Quadrotor

The force and torque of each motor can be represented as:

$$F* = k1\delta* \tag{2}$$

$$T* = k2\delta* \tag{3}$$

Where, k1 and k2 are constants that has to be identified practically, δ* is the command signal of the motor and τ represents f, r, b, and l.

Matrix form represents the forces and torques on the quadrotor

$$\begin{matrix} F \\ \tau\varphi \\ \tau\theta \\ \tau\varphi \end{matrix} = \begin{matrix} k1 & k1 & k1 & k1 \\ 0 & -lk1 & 0 & lk1 \\ lk1 & 0 & -lk1 & 0 \\ -k2 & k2 & -k2 & k2 \end{matrix} = M \begin{matrix} \delta f \\ \delta r \\ \delta b \\ \delta l \end{matrix} \tag{4}$$

Forces and torques are the control strategies resulted in succeeding sections. The real motors commands can be represented as:

$$\begin{matrix} \delta f \\ \delta r \\ \delta b \\ \delta l \end{matrix} = M^{-1} \begin{matrix} F \\ \tau\varphi \\ \tau\theta \\ \tau\varphi \end{matrix} \tag{5}$$

The PWM commands are necessary to be between 0 and 1. In accumulation to the motor force, gravitational force will also be applied on the quadrotor. The gravity force performing on the center of mass in the vehicle frame Fv is given by

$$\begin{matrix} \delta f \\ \delta r \\ \delta b \\ \delta l \end{matrix} = M^{-1} \begin{matrix} F \\ \tau\varphi \\ \tau\theta \\ \tau\varphi \end{matrix} \tag{6}$$

It should be changed to the body frame to give

$$f_g^v = \begin{matrix} 0 \\ 0 \\ mg \end{matrix}$$

$$f_g^b = R_v^b \begin{matrix} 0 \\ 0 \\ mg \end{matrix} \tag{7}$$

$$= \begin{matrix} -mg\, sin\theta \\ mg\, cos\theta\, sin\varphi \\ mg\, cos\theta\, cos\varphi \end{matrix}$$

The Kinect sensor will be fitted on the developed MAV to extract the 3D view of an environment using Skanect as shown in Fig. 4 [5]:

**Fig. 4.** Quadrotor Fitted with a Kinect Sensor

## 4 Methodology

This part deals with the method used to create the 3D mapping of an object in a room using the MAV in association with Skanect [6]. Up to 30 frames per second of dense 3D information can be obtained in a scene. If the Structure Sensor is moved roughly, Kinect can capture a full set of point of views and we will get a 3D mesh in real time [7, 8].

The configuration details are:

To sense: Microsoft Kinect v1
To localize: RGDBSLAM algorithm
To characterize: 3D point cloud

RGBDSLAM [9] is a SLAM solution committed for RGB-D cameras. Using a hand held Kinect camera the 3D models of objects and interior views can be captured easily. The key points are detected using the visual features that in turn will be used in feature matching step. SURF), SIFT, or ORB [10, 11]. The obtained solution will be key points are used in determining the points extracted, the results are key-points, localized in 2D and a précis of their surrounding area. 3D transformation of successive frames is enumerated using the A RANSAC (Random Sampling and Consensus). It takes more processing power to calculate the key points. So General Processing on GPU is used to calculate the key points with less processing time. In order to reduce the errors the resulting graph for the for Pose graphs on Manifolds is optimized with hierarchical optimization (Table 2).

The Fig. 5 a, b, c shows the depth image of the objects such as chair, monitor and helmet.

**Table 2.** Number of points and their storage size

| Place | No. of points (millions) | Storage size (MB) |
|---|---|---|
| Office | 17.0 | 271 |
| Corridor | 12.2 | 202 |
| Lab | 18.1 | 298 |
| Chair | 5 | 101 |
| Monitor | 3.6 | 88 |
| Helmet | 2.4 | 71 |

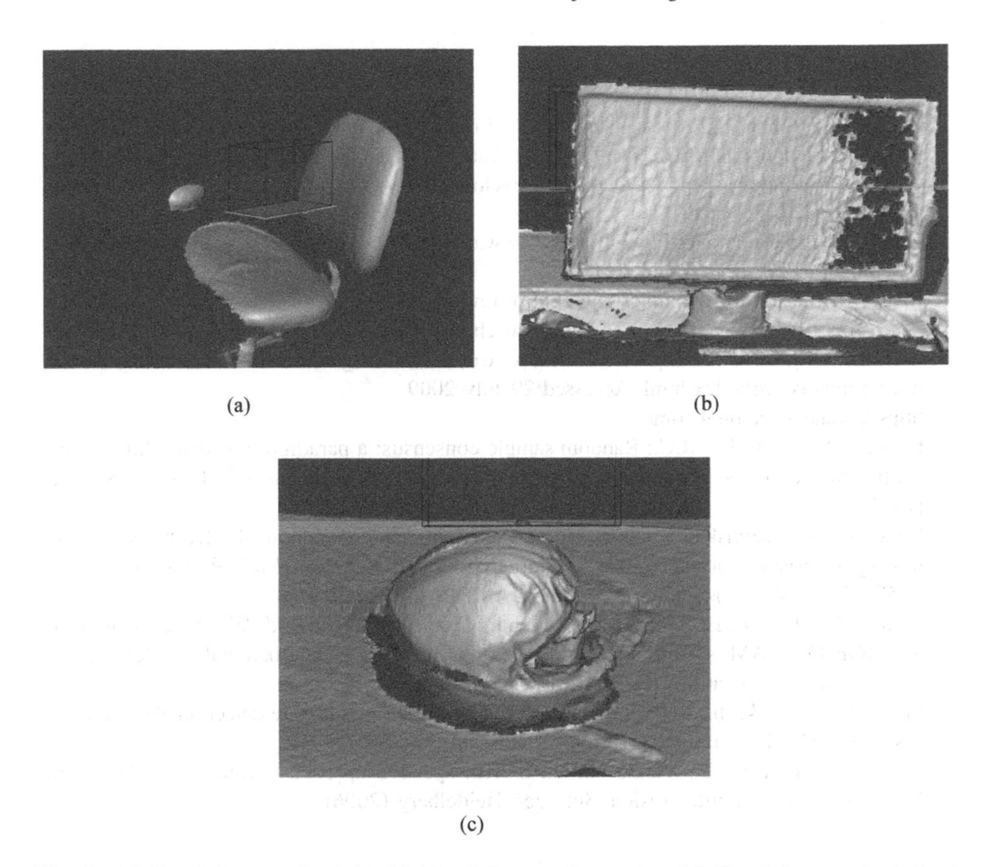

Fig. 5. (a). Depth image of a chair (b). Depth image of a monitor (c). Depth image of a helmet

## 5 Conclusion

In this paper an approach for creating the 3D view of objects in indoor environment using robotics method is presented. In this paper we have shown an illustration of actual use of a Kinect sensor fitted on a MAV platform to plot the interior environment. A less expensive sensor Kinect in association with Skanect emerges to be a good combination to execute this assignment. The accomplished research showed an essential aspect that has to be dealt with. In perspective of memory, the information acquired from the Kinect and RGBDSLAM are huge. Using high end graphics card with adequate dispensation power the point clouds can be generated.

**Compliance with Ethical Standards.**

✓ All authors declare that there is no conflict of interest
✓ No humans/animals involved in this research work.
✓ We have used our own data.

## References

1. Zhou, W., Jaime, V., Dissanayake, G.: Information efficient 3D visual SLAM for unstructured domains. IEEE Trans. Robot. Autom. **24**, 1078–1087 (2008)
2. Kinect for Windows: (n.d.). Microsoft Developer Network. http://msdn.microsoft.com/en-us/library/jj131033.aspx
3. DragonFly: (n.d.). DragonFly X4. http://www.draganfly.com/uav-helicopter/draganflyer-x4/specifications/. Accessed Sept 2013
4. Beard, R.: Quadrotor dynamics and control rev 0.1. (2008)
5. "gumstix: finally! - a very small linux machine". gumstix.org. 8 April 2004. Archived from the original on 8 April 2004. http://web.archive.org/web/20040408235922, http://www.gumstix.org/index.html. Accessed 29 July 2009
6. https://skanect.occipital.com/
7. Fischler, M.A., Bolles, R.C.: Random sample consensus: a paradigm for model fitting with applications to image analysis and automated cartography. Commun. ACM **24**(6), 381–395 (1981)
8. Genevois, T., Zielinska, T.: A simple and efficient implementation of EKF-based SLAM relying on laser scanner in complex indoor environment. J. Autom. Mob. Robot. Intell. Syst. **8**, 58–67 (2014). https://doi.org/10.14313/JAMRIS_2-2014/20
9. Endres, F., Hess, J., Engelhard, N., Sturm, J., Cremers, D., Burgard, W.: An evaluation of the RGB-D SLAM system. In: Proceedings of the IEEE International Conference on Robotics and Automation (ICRA) (2012)
10. Srividhya, S., Prakash, S.: Performance evaluation of various feature detection algorithms in VSLAM. PARIPEX Indian J. Res. **6**(2), 386–388 (2017)
11. Bay, H., Tuytelaars, T., Van Gool, L.: SURF: speeded up robust features. In: European Conference on Computer Vision. Springer, Heidelberg (2006)

# Inferring Machine Learning Based Parameter Estimation for Telecom Churn Prediction

J. Pamina(✉), J. Beschi Raja, S. Sam Peter, S. Soundarya, S. Sathya Bama, and M. S. Sruthi

Sri Krishna College of Technology, Coimbatore, India
jpamina8@gmail.com, beskiraja@gmail.com

**Abstract.** Customer churn is an important issue and major concern for many companies. This trend is more noticeable in Telecom field. Telecom operators requires an essential proactive method to prevent customer churn. The existing works fails to adopt best feature selection for designing model. This works contributes on developing churn prediction model, which helps telecom operators to identify the customers who are about to churn. The significance for the recall evaluation measure, which actually solves the real-time business problem is highlighted. The prominent goal of this churn analysis is to perform binary classification with customer records and figure out who are likely to cancelled in the future. Fifteen machine learning methods with different parameters are employed. The performance of the model is evaluated by various measures like Accuracy, Recall, Precision, F-score, ROC and AUC. Our aim in this work is to produce highest recall value which has a direct impact on real-world business problems. Based on experimental analysis, we observed that Decision Tree model 3 outperforms all other models.

**Keywords:** Telecom · Machine learning · Parameters · Churn prediction · Recall

## 1 Introduction

Due to wide range of global development, Information Technology has remarked as high increase in different sorts of service providers which finally cause to great rivalry among them. Tackling customer churn and satisfying the customers are most common challenges for them [1, 2, 22]. Due to high impact of revenues of the companies, predicting factors that maximize customer churn is necessary to reduce customer churn. Churn is defined as when a customer terminates the service from one service providers and switches to another for some reasons. This activity impacts more on company revenues [22]. So, it's always better to design effective churn model to predict specific customer cluster who are likely to churn and acknowledge their fidelity before cancel the subscription. The current research trends are emerging in analysing customers and developing patterns from it, especially in Telecom filed. Churn prediction is a major factor because it shows direct effect on profit and revenues of the company [3, 4]. The structure of this article is as follows: In Sect. 1, describes related works related to churn prediction using ML Techniques. Section 2, reveals about the pre-processing using


S. Smys et al. (Eds.): ICCVBIC 2019, AISC 1108, pp. 257–267, 2020.
https://doi.org/10.1007/978-3-030-37218-7_30

PCA [7–10]. Section 3, describes various machine learning techniques with different parameters. In Sect. 4, we discussed about the cross-validation and it's for accumulating results. Section 5, presents the description of dataset and its feature. Section 6, presents summary and results of 15 ML algorithms. Finally, Sect. 7, concludes this paper.

## 2 Pre-processing

The pre-processing work includes converting object data types to numeric type. The dataset contains two different kinds of customers: those who left the telecom service and those who stayed [12–15]. The average of both the groups are depicted in Fig. 1. The Table 1 shows the statistics of customers those who left vs who stayed. It is observed that on average those who left had higher monthly Charges and lower tenure. It also noticed that only 7032 observations are for Total Charges. so, we filled NAN values with median values. After converting the dummy variables, we noticed that impact of churn is very less in gender and phone service.

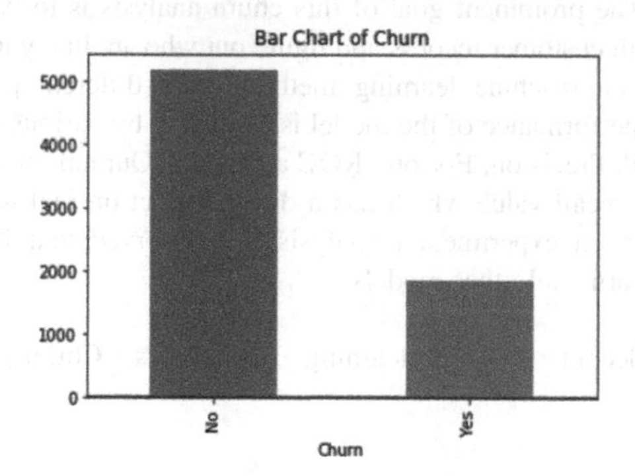


**Fig. 1.** Availability of churn

**Table 1.** Statistics on churners

| Churn | Senior citizen | Tenure | Monthly charges |
|---|---|---|---|
| NO | 0.128721 | 37.569965 | 61.265124 |
| YES | 0.254682 | 17.979133 | 74.441332 |

### 2.1 Principal Component Analysis (PCA)

Principal component analysis is used for reduction of dimensions which involves zeroing out the smallest components and leads in less-dimensional projection of the data that cause maximal data variance preservation. In PCA, relationship is computed by finding principal axes list in the data and describe the dataset. It is noticed that first feature interprets 21% approximately of variance within dataset.

PCA removes least principal axis information and leaving the high variance data components. The fractional variance that cut out is evaluated roughly by measuring how much information is removed during dimensionality reduction. We chose to add all features in designing our models. Figure 2 denotes proportional to points spread to the line formation. The train and test data features are standardized. This will transform features by scaling to given range between zero and one.

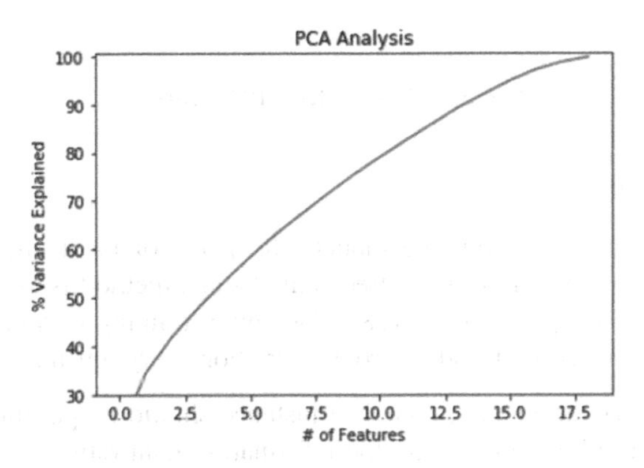


**Fig. 2.** PCA analysis

# 3 Machine Learning Models

## 3.1 KNN Models

We have created three KNN models by changing weights and nature algorithms. KNN algorithm is transparent, simple, consistent, straight forward, easy implement when data is distributed with no prior knowledge [17]. If $k_x > k_y \forall x \neq y$ then put $f$ in class $x$. Three types of KNN models are created by changing the parameters and algorithms. The K range is set between 20 to 50. The main goal is to increase Recall (Churn = yes). We set K value as 27 for all models since this point creates highest AUC value. First, KNN model was created by setting algorithm as auto and weights are set as uniform. Here 'uniform' denotes uniform weights. The data points at every neighbourhood are equally weighted. Second model was designed by setting algorithm as auto and weights parameter as distance. Here 'distance' denotes weight points by the inverse of their distance. In this position, closer neighbours' points will experience higher impact than neighbours far away. Third KNN was created by setting algorithm as Kd_tree and weights as uniform. Kd trees is a binary tree that constructs trees very fast by recursively partition the parameter with data axes. The evaluation measures such as confusion matrix, cross-validation, recall, accuracy, precision, AUC ROC and F1 scores are computed using packages. After evaluating the measures for three models, the third model with kd algorithm gives the optimal scores for all evaluation metrics. We have selected third KNN model as the best KNN from our experiments (Fig. 3).

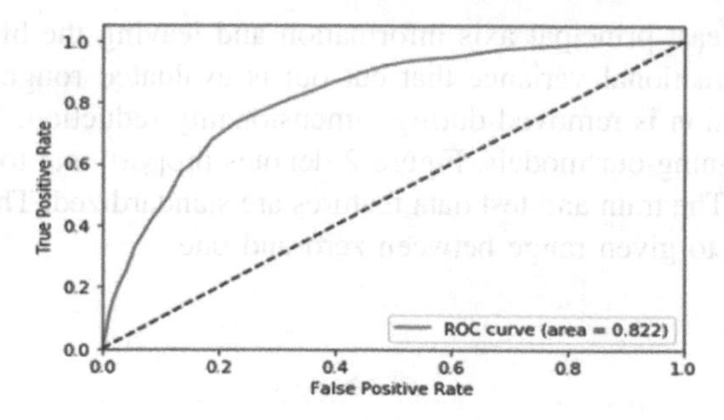


**Fig. 3.** KNN model 3 ROC curve

### 3.2 Decision Tree

Next, we created three decision tree models. It is one of the methods for decision making. The common idea in decision tree is multistage method is to divide up a more complex decision of simple decision [18]. Cast every training instance on the root of the decision tree. Flag current node to root node. For every attribute

- Based on the attribute value, every data instance should be partitioned.
- Based on the partitioning calculate the information gain ratio.

The feature that contributes to the highest information gain ratio should be identified and set this as the splitting criterion. The current node will be flagged as the leaf and returned if higher information gain ratio is 0. According to the prominent feature divide, all instances based on the attribute value Represent every partition as a child node of the current node. For each child node:If the child node mark it as a leaf and return else mark the child node as the current node and go back to process of attribute partition.

We chose Max_depth parameter which would increase the recall metric. We notice that the model overfits for greater values. The tree correctly detects the train data. Still, it declines to predict the new data. Based on this, we chose Max_depth value as 5 to decrease overfitting. When testing the model with Test data, we observed that the accuracy is 78.2% with Recall of 0.60 and the value exceeds all three KNN models. Next, we designed Decision Tree model 2 using Min_samples_leaf parameter. We selected min_samples_split = 0.225 and this produced the highest AUC score for the Test and train data. For Decision model 3 selected min_samples_split = K which represents minimum sample number. Decision tree model 3 provides less accuracy value when compared to DT 1 and DT2 but gives the increased recall of 0.71 (Fig. 4).

### 3.3 Random Forest Model

Random forest is integration of decision trees such that each decision tree relays on values of vector independently and with distribution for all decision trees [19]. Random Forest 1 model was created by using Max_Depth parameter. The default parameter is making all the leaves prune and expanded for this data which leads to overfitting. We observed the when max_depth increases, train curve approach 1.0 while test curve began to level off at 7 (Fig. 5).

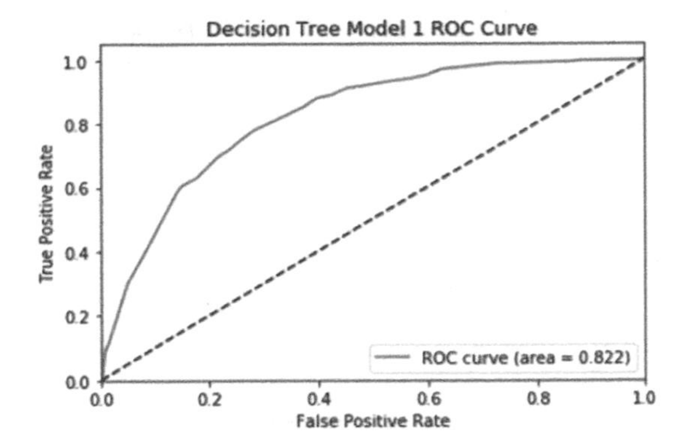


**Fig. 4.** Decision tree model 1 ROC curve

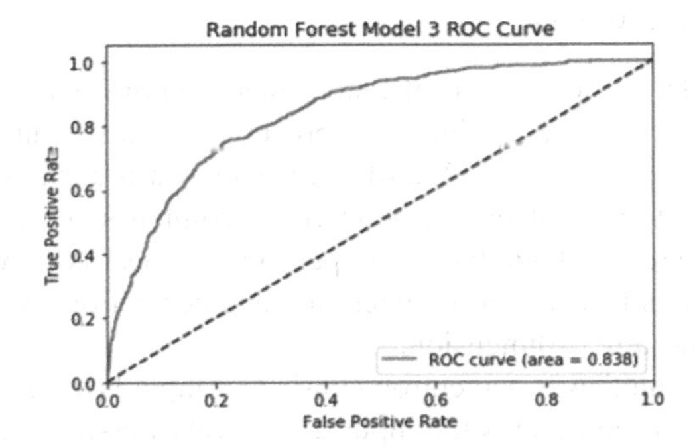


**Fig. 5.** Random forest 1 ROC curve

### 3.4 Stochastic Gradient Descent Model

Stochastic Gradient Descent is a technique for optimization of objective function. The term "stochastic" refers to randomly elected samples to predict the gradients [20]. In stochastic gradient descent, the exact gradient of $G(x)GG$ is imprecise by a gradient at a sole example:

$$x := x - \eta \nabla G_i(x)$$

In this work, three Stochastic Gradient Descent Models were created. The first SGD model was designed by setting loss = 'modified_huber' which is a smooth loss which helps to handle outliers and probability estimates. For second model we fixed loss = 'log' which yields logistic regression classifier. The third model was designed by fixing loss = 'log' and penalty = 'l1' which helps for feature selection. From the experiment, we selected second model as best since it gives highest recall value among all models (Fig. 6).

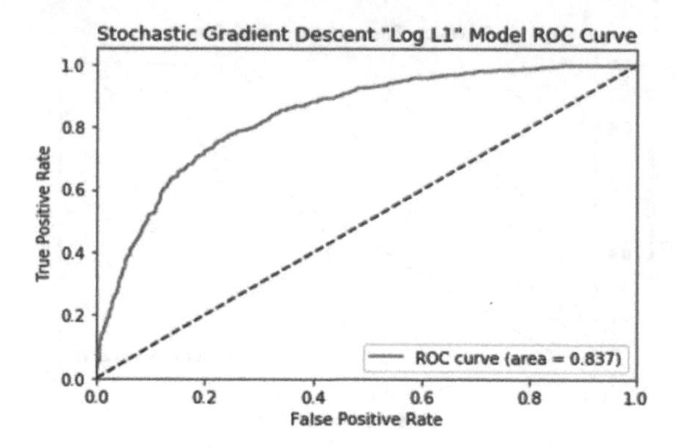


**Fig. 6.** Stochastic gradient 2 ROC curve

## 3.5 Neural Network Models

Neural network model is a collection of units called neurons which are connected to each other. The neuron receives a signal for processing and weights can be added or reduced according the problem [21]. Multilayer perceptron is a classifier considered as a part of feed-forward artificial neural network. It combines several perceptron's to create a non-linear decision boundary. The perceptron consists of weights, bias, the summation processor and an activation function. Each perceptron provides a non-linear mapping function into a new dimension.

Let us consider a multilayer perceptron consists of input layer, hidden layer and an output layer. Input layer represents the input data, it takes node's weights and bias to the next layer. Intermediate layer or hidden layer which present next to the input layer performs activation function to make sense of the given input.

Node's in hidden layer performs activation function i.e. sigmoid (Logistic) function, it is used to predict the probability of anything that ranges between 0 and 1. If anything ranges from −1 to 1 then we can use hyperbolic tangent function.

Sigmoid function,

$$f(x) = \frac{1}{1 + e^{-x_i}}$$

Finally, nodes in output layer performs loss function i.e SoftMax function. To find the exact class of the input given by the user.

SoftMax function,

$$f(x_i) = \frac{e^{x_i}}{\sum_{k=1}^{N} e^{x_k}}$$

Where N in the output layer corresponds to the number of classes. Multi-layer perceptron model was used for designed Neural Network models. The first MLP model was designed using hidden layer = (20,8). The activation function is set to 'tanh' and solver = 'adam'. second model was designed using 'relu' activation function with 15 hidden layers. Last model was created by changing 'solver = sgd' with 30 hidden layers (Fig. 7).

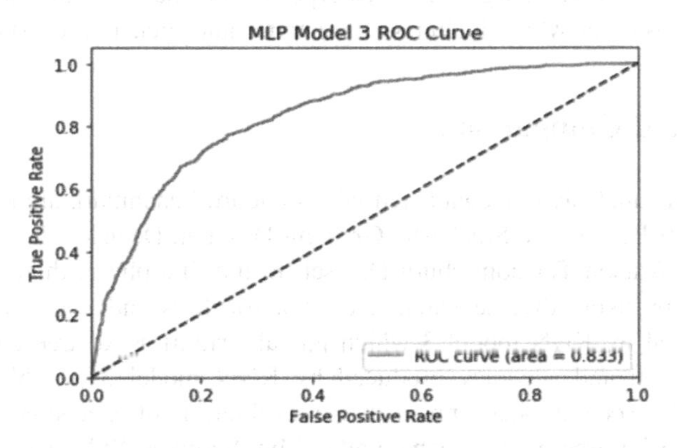


**Fig. 7.** MLP model-2 ROC curve

# 4 Cross Validation

The models in paper are tested by 10-fold cross validation for accumulating best results. The k-fold CV otherwise called rotation estimation. Cross validation is used to assess the results of a statistical analysis on an independent dataset [15]. It is sometimes called out of sampling which is mainly used to analyse the performances of various predictive models. The whole dataset S is divided randomly into K folds or subsets S1, S2, S3…SK of equal size approximately. This process evaluates the accuracy of overall number of correct classifications. The estimation of cross validation is evaluated by random number which depends on the splits into subsets (folds). The formula for cross validation is mentioned below

$$acc_{cv} = \frac{1}{n} \sum_{(v_i, y_i) \in S} \delta\left(L\left(S \setminus S_{(i)}, c_i\right), bi\right)$$

where, S(i) be th test dataset that contains Xi = uci; bii, is estimation of cross validation. using 10-fold cross validation the errors and variance are minimum and provides best results.

## 5 Dataset

The experiments are carried out with a standard benchmark IBM Watson Telecom churn dataset. The dataset contains the following information about services that each customer has opted multiple lines, device protection, internet, online security, tech support and streaming TV and movies. The second type of information is about customer account information which contains contract, paperless billing, payment method, monthly charges and total charges. The last type of information contains demographic content of the customer which includes age, gender and their marital status.

## 6 Results and Comparison

The experimental analysis was conducted with standard benchmark algorithms such as KNN, Multilayer Perceptron, Stochastic Gradient Descent, Decision Tree and Random Forest on IBM Watson Telecom churn Dataset. In the first phase, three KNN models were designed by using diverse parameters in algorithms and weights. The highest value is produced by KNN model 3 which has algorithm as kd tree and weights as distance. the recall and accuracy produced by KNN model 3 is 0.57 and of 0.78 respectively. The second phase presents the evaluation of Decision Tree models. Among all, Decision tree model 3 which adds Min_Samples_Split parameter receives the highest recall and accuracy of 0.71 and 0.73 respectively. The third phase contains the creation of three Random Forest models using various parameters in that, Random forest models 2 and 3 receives the recall of 0.49 and accuracy of 0.79. The next phase presents the creation and analysis of Stochastic Gradient Descent models. In that, all three SGD models provides the highest accuracy of 0.79 but SGD model 2 (Loss = 'log') yields the highest recall of 0.54. The final session is composed of creating ANN models. We used Multi-layer perceptron for designing ANN models. The second and third models provides the highest accuracy of 0.79 but the third models (solver = 'sgd') achieves the highest recall value of 0.53.

After the experimental analysis of 15 models, we selected the decision tree model-3 as the best model for real-time business problem. Decision Tree model-3 secured highest recall value compared to other models. The values and experimental analysis for 15 models are shown in Table 2. So, this model has been selected for feature selection to find the impact of churn.

**Table 2.** Experimental results and summary of 15 ML models

| No | Method | Parameters | Accuracy | AUC | Recall | CV score | Precision | F1 score |
|---|---|---|---|---|---|---|---|---|
| 1 | KNN-1 | Algorithm = 'Auto' Weights = 'Uniform' | 0.78 | 0.80 | 0.52 | 0.76 | 0.59 | 0.55 |
| 2 | KNM-2 | Algorithm = 'Auto' Weights = 'Distance' | 0.76 | 0.79 | 0.47 | 0.75 | 0.56 | 0.51 |
| 3 | KNN-3 | Algorithm = 'K d_Tree' Weights = 'Uniform' | 0.78 | 0.82 | 0.57 | 0.77 | 0.60 | 0.58 |
| 4 | DT-1 | Max_Depth | 0.78 | 0.82 | 0.60 | 0.78 | 058 | 0.59 |
| 5 | DT-2 | Min_Samples_Leaf | 0.74 | 0.75 | 0.61 | 0.75 | 0.51 | 0.55 |
| 6 | DT-3 | Min_Sample_Split | 0.73 | 0.50 | **0.71** | 0.73 | 0.50 | **0.71** |
| 7 | RF-1 | Max_Depth | **0.79** | **0.83** | 0.48 | 0.79 | **0.65** | 0.55 |
| 8 | RF-2 | Estimators, Max_Depth | **0.79** | **0.83** | 0.49 | 0.79 | **0.65** | 0.56 |
| 9 | RF-3 | Estimators, Max_Depth, Min_Samples_Leaf | **0.79** | **0.83** | 0.49 | 0.79 | **0.65** | 0.56 |
| 10 | SGD-1 | Modified_Huber | **0.79** | **0.83** | 0.52 | 0.79 | 0.64 | 0.57 |
| 11 | SGD-2 | Loss = 'Log' | **0.79** | **0.83** | 0.54 | **0.80** | 0.63 | 0.59 |
| 12 | SGD-3 | Penalty = 'L1' | **0.79** | **0.83** | 0.52 | 0.79 | 0.64 | 0.57 |
| 13 | MLP-1 | Solver = 'adam' | **0.79** | 0.82 | 0.49 | 0.78 | 0.64 | 0.55 |
| 14 | MLP-2 | Solver = 'relu' | 0.78 | 0.81 | 0.50 | 0.78 | 0.60 | 0.55 |
| 15 | MLP-3 | Solver = 'sgd' | **0.79** | **0.83** | 0.53 | 0.79 | 0.63 | 0.58 |

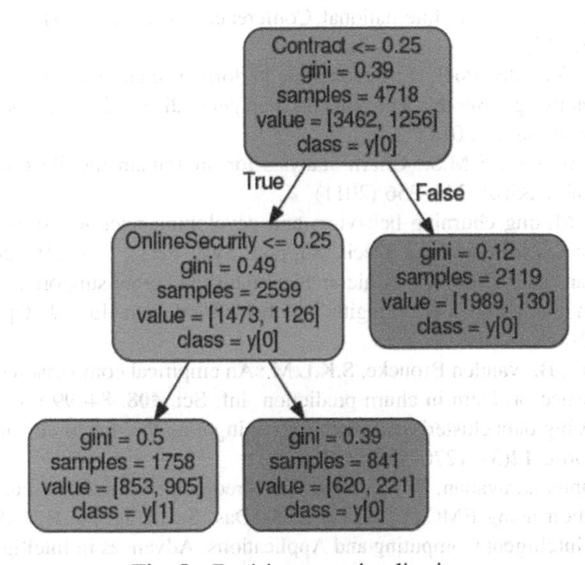


**Fig. 8.** Decision tree visualization

## 7 Conclusion

This work concentrates on designing an efficient churn prediction model that helps telecom sector to predict the customers who are going to churn. Principal Component Analysis is employed for pre-processing and normalization of data. A comparative experimental analysis of 15 models with different parameters were employed and carried out for churn dataset. Recall metric makes a strong impact of real-time business to reduce the churn. Decision Tree model-3 secured highest recall value and hence it is used for feature selection which impact the churn of customers. From Fig. 8. the variables “contract” and “online security” are prominent factors impacting churn. Online security refers to whether the person hires online security or not. Contact refers the term and duration of a person. The extension of this work is intended toward make a comparative analysis of with many machine learning methods with large volume of records for other different applications.

**Compliance With Ethical Standards.**

✓ All authors declare that there is no conflict of interest
✓ No humans/animals involved in this research work.
✓ We have used our own data.

## References


1. Verbeke, W., et al.: Building comprehensible customer churn prediction models with advanced rule induction techniques. Expert Syst. Appl. **38**(3), 2354–2364 (2011)
2. Zhao, L., et al.: K-local mamum margin feature extraction algorithm for churn prediction in telecom. Clust. Comput. **20**(2), 1401–1409 (2017)
3. Idris, A., Khan, A., Lee, Y.S.: Genetic programming and adaboosting based churn prediction for telecom. In: 2012 IEEE International Conference on Systems, Man, and Cybernetics (SMC). IEEE (2012)
4. De Bock, K.W., Van den Poel, D.: Reconciling performance and interpretability in customer churn prediction using ensemble learning based on generalized additive models. Expert Syst. Appl. **39**(8), 6816–6826 (2012)
5. Keramati, A., Ardabili, S.M.S.: Churn analysis for an Iranian mobile operator. Telecommunications Policy **35**(4), 344–356 (2011)
6. Lee, H., et al.: Mining churning behaviors and developing retention strategies based on a partial least squares (PLS) model. Decis. Support. Syst. **52**(1), 207–216 (2011)
7. Chen, Z.-Y., Fan, Z.-P., Sun, M.: A hierarchical multiple kernel support vector machine for customer churn prediction using longitudinal behavioral data. Eur. J. Oper. Res. **223**(2), 461–472 (2012)
8. Zhu, B., Baesens, B., vanden Broucke, S.K.L.M.: An empirical comparison of techniques for the class imbalance problem in churn prediction. Inf. Sci. **408**, 84–99 (2017)
9. Bi, W., et al.: A big data clustering algorithm for mitigating the risk of customer churn. IEEE Trans. Ind. Inform. **12**(3), 1270–1281 (2016)
10. Babu, S., Ananthanarayanan, N.R.: Enhanced prediction model for customer churn in telecommunication using EMOTE. In: Dash, S., Das, S., Panigrahi, B. (eds.) International Conference on Intelligent Computing and Applications. Advances in Intelligent Systems and Computing, vol. 632 (2018)

11. Qi, J., et al.: ADTreesLogit model for customer churn prediction. Ann. Oper. Res. **168**(1), 247 (2009)
12. Karahoca, A., Karahoca, D.: GSM churn management by using fuzzy c-means clustering and adaptive neuro fuzzy inference system. Expert Syst. Appl. **38**(3), 1814–1822 (2011)
13. Adris, A., Iftikhar, A., ur Rehman, Z.: Intelligent churn prediction for telecom using GP-AdaBoost learning and PSO undersampling. Clust. Comput., 1–15 (2017)
14. Vijaya, J., Sivasankar, E.: An efficient system for customer churn prediction through particle swarm optimization-based feature selection model with simulated annealing. Clust. Comput., 1–12 (2017)
15. Kohavi, R.: A study of cross-validation and bootstrap for accuracy estimation and model selection. Ijcai **14**(2), 1137–1145 (1995)
16. García, S., Luengo, J., Herrera, F.: Data Preprocessing in Data Mining. Springer, New York (2015)
17. Adeniyi, D.A., Wei, Z., Yongquan, Y.: Automated web usage data mining and recommendation system using K-Nearest Neighbor (KNN) classification method. Appl. Comput. Inform. **12**(1), 90–108 (2016)
18. Safavian, S.R., Landgrebe, D.: A survey of decision tree classifier methodology. IEEE Trans. Syst. Man Cybern. **21**(3), 660–674 (1991)
19. Breiman, L.: Random forests. Mach. Learn. **45**(1), 5–32 (2001)
20. Mei, S.: A mean field view of the landscape of two-layer neural networks. Proc. Natl. Acad. Sci. **115**(33), E7665–E7671 (2018). https://doi.org/10.1073/pnas.1806579115. PMC 6099898. PMID 30054315
21. Zhang, G.P.: Time series forecasting using a hybrid ARIMA and neural network model. Neurocomputing **50**, 159–175 (2003)
22. Pamina, J., et al.: An effective classifier for predicting churn in telecommunication. J. Adv. Res. Dyn. Control Syst. **11**, 221–229 (2019)

# Applying a Business Intelligence System in a Big Data Context: Production Companies

Jesús Silva[1(✉)], Mercedes Gaitán[2], Noel Varela[3], Doyreg Maldonado Pérez[3], and Omar Bonerge Pineda Lezama[4]

[1] Universidad Peruana de Ciencias Aplicadas, Lima, Peru
jesussilvaUPC@gmail.com
[2] Corporación Universitaria Empresarial de Salamanca – CUES, Baranquilla, Colombia
m_gaitan689@cues.edu.co
[3] Universidad de la Costa (CUC), Calle 58 # 55-66, Atlantico, Baranquilla, Colombia
{nvarela1,dmaldona}@cuc.edu.co
[4] Universidad Tecnológica Centroamericana (UNITEC), San Pedro Sula, Honduras
omarpineda@unitec.edu

**Abstract.** Industry 4.0 promotes automation through computer systems of the manufacturing industry and its objective is the Smart *Factory.* Its development is considered a key factor in the strategic positioning not only of companies, but of regions, countries and continents in the short, medium and long term. Thus, it is no surprise that governments such as the United States and the European Commission are already taking this into consideration in the development of their industrial policies. This article presents a case of the implementation of a BI system in an industrial food environment with Big Data characteristics in which information from various sources is combined to provide information that improves the decision-making of the controls.

**Keywords:** Industry 4.0 · Big Data · Business Intelligence (BI) · Food industry · ERP systems

## 1 Introduction

Industry 4.0 envisages within the same paradigm the latest advances in *hardware* and *software* at the service of industry such as the Internet of Things (with machines collaborating with other machines) [1], Systems cyber-physicists (in which the human being interacts with the machine naturally and safely), the *'doer'* culture or *maker-culture* (with customization devices such as 3D printers), simulation systems (such as Augmented Reality, Simulation of Productive Processes for Predicting Behaviors of Environments and Productive Assets) or Big Data Technology for the processing, processing and representation of large volumes of data [2, 3].

*Big Data* can be simply defined as a system that allows the collection and interpretation of datasets that because of their large volume it is not possible to process them with conventional tools of capture, storage, management and Analysis. Big Data technologies are designed to extract value from large volumes of data, from different types


S. Smys et al. (Eds.): ICCVBIC 2019, AISC 1108, pp. 268–274, 2020.
https://doi.org/10.1007/978-3-030-37218-7_31

of sources, quickly and cost-effectively [4–6]. The objective is to support the decision-making process based on information by managing the five "V" [7]: Volume, Variety, Speed, Veracity and Value. A Big Data system can be structured into three layers: (1) the infrastructure layer (such as a computer storage and distribution service), (2) the compute layer (with middleware that includes tools for data integration and management, along with the programming model) and, (3) the application-visualization layer (with business applications, web services, multimedia, etc.).

Big Data has begun to be the present and will hold the future of employment in the world. Thus, several sources refer to the employment opportunity that exists around Big Data.

In addition to Big Data, BI systems are part of Industry 4.0, and include applications, infrastructure, tools, and best practices that enable access and analysis of information to improve both the decision-making process and its impact [8–10]. In [11] they define BI systems as the set of techniques and tools for transforming large amounts of raw data into consistent and useful visual information for business analysis in seconds. Big Data and BI systems are changing businesses and industry [12], although their degree of implementation is not, at the moment, as high as you would expect [13, 14].

## 2 Materials and Methods

The present implementation of a BI system in a context of Big Data has been developed in a multinational production sector with approximately 5800 employees spread over 6 production plants that invoiced during 2015 1254.2 Million, thanks to selling about one trillion product units on several million customer order lines. The multinational handles information regarding different business areas of at least 5 exercises in its decision-making processes, thus giving an idea of the amount of data being handled. For example, for the quality of the product it manufactures, each unit must be able to be plotted unitally in front of any potential quality incidence, extracting from which raw materials and production processes the same is obtained in order to establish possible correlations, statistical analysis, etc. In reference to customer service, all of the above line items have different service policies and penalties that must be able to be analyzed unitally in the event of any potential service failure that the customer wishes to claim. These examples are just some of the examples that show that the multinational needs to comply and meet the premises that make up the 5 "Vs" of big data mentioned above.

### 2.1 Methods

The multinational has deployed its BI tool (Qlikview®) for data analysis both centrally (for purchasing, logistics, sales, etc.) and in a decentralized manner (to meet the demand for plant information). To do this, the method applied has been that of the Quick Prototype or Quick Prototyping, which, as explained in [15], consists of acting before disposing of the results and being persuaded by the results to refine them. The Rapid Prototype consists of the concurrent performance of the design and development stages of a product so that the project team can test it experimentally, observe it and even feel it as soon as possible in order to improve it.

The implementation methodology followed a cyclical process of selection and development of the area (purchases, production, etc.) to be implemented, to subsequently concurrently perform requirements, conceptual designs, functional designs and prototypes which users tested for validation or tailoring. The tests were performed without integrating the development made with the rest of the company's systems (i.e., with experimental data and in a specific hardware and software test environment, not interacting with the other developments made so far). If necessary the development would be adapted until its acceptance, and at that time it was tested in a test environment that already integrated the programing res- to the other areas of the company. Once the solution was completed, it was closed and moved to a maintenance phase while the IT team approached another area of the company.

The deployment of the company's BI system includes, broadly speaking, the development of 5 online analytical processing (or On-Line Analytical Processing, OLAP) cubes and Supply Chain and Corporate. These OLAP cubes collect up-to-date information from the servers to copy and compress to a BI data server that will then be used by the BI engine itself (located on another server) for near-instant querying and viewing. Each of these areas supports decision-making on issues as diverse as the management of purchase contracts, the management of products blocked by quality, claims regarding low service fees, etc.

## 2.2 Materials

The deployment shown in the following sections of the article was performed based on the use of a number of servers used to (a) support ERPs (Enterprise Resource Planning or Resource Planning systems (b) additional application and documentary servers, (c) the BI tool management server, and (d) the BI data management tool server.

Thus, the set of case BI cubes is mainly fed on plant ERPs and marketers, which were initially found on all types of ERP platforms (SAP R3, Baan, Navision and AX) and which are in a migration process to AX2012. This information is complemented by different types of files found on other types of platforms (e.g. Sharepoint) and applications of various kinds (e.g. SAP BPC). Additionally, group ERPs interact one-way or two-way with hardware and applications of various kinds, such as barcode scanning guns, invoice scanning systems, etc. [16, 17].

## 2.3 Features

Once the model is built, the BI tool periodically reads the data and visualizes the cross-data that is created appropriate. In this case, the tool is used to be able to filter and obtain any data that is in the filter area of the BI screen. The tool displays as a result (in the middle of the same screen) all kinds of information requested, such as:

(a) who has produced a pallet that has a number of SSCCs (Serial Shipper Container Code, identification number of the logistics unit produced, in this case the pallet) in particular, together with when it has been drawn up, (b) which workers have produced more or less units in a period, or (c) as shown in Fig. 1, what product quantities have been produced in a period on a line both in total (at the bottom) as well as production by reference (at the top of the image).

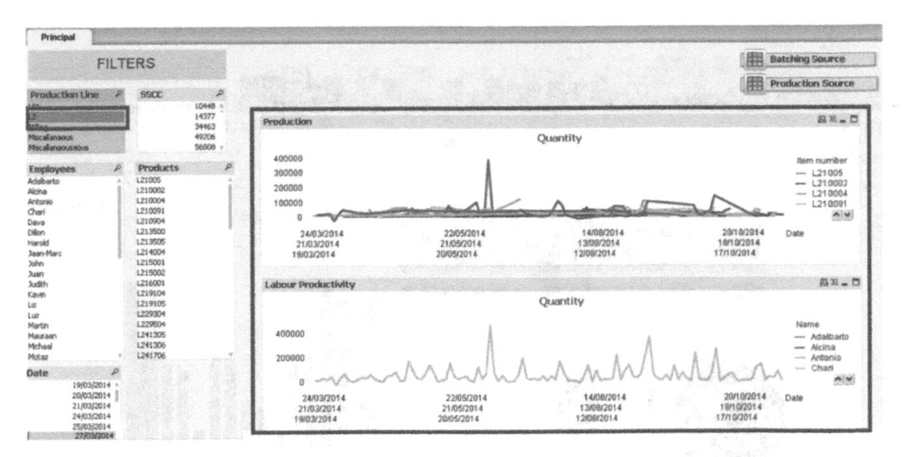


**Fig. 1.** Chart of units produced on a production line per anonymized period

The advantages of having the system implemented are obvious: prior to the implementation of it, and in reference to what is shown in Fig. 1, the visualization of this type of information was not graphical, nor as easy to filter as with the BI tool. With regard to the unit traceability of finished products to investigate who had made each product in the face of incidents, in the past it was 100% manual and bond-based (neither agile nor operational) while currently such information you get it instantly.

## 3 Results

The Supply Chain cube is shown in Fig. 2. This screen is used to track the rate of service provided on time and quantity (On Time In Full, OTIF) per plant, customer, reference, etc. in the period studied showing also an evo- tion of the finished product stock both on heeldays as in Tons (chart at the bottom right). Finally, Fig. 3 shows a screenshot of part of the corporate BI cube, which tracks the group's 75 strategic improvement projects: they track to see the extent to which the expected improvement ('Target Annual Savings (K') is being achieved through current progress ('Actual Annual Savings (K€)').

As part of the Spanish continuous market, the multinational must make its results public on a quarterly basis. As noted in the company's release of results on a quarterly basis, improvements made in the area of operations (production and *Supply Chain)*have favored an improvement of the business of both the previous year and this year. Specifically, the letter justifies the results obtained on the basis of "the improvement in direct production costs, mainly in personnel costs and transport costs, together with the improvement of other operating expenses". It is understood that BI has served to support decision support and has therefore influenced the outcome. It is worth noting that, for example, and as shown in the (top center) of Fig. 2 the OTIF service fee has improved by 6.42% compared the current year to the previous year, even if the costs associated with transport have been reduced.

To the quantitative results obtained can be added another series of advantages obtained as a result of the implementation carried out, since the system has served,

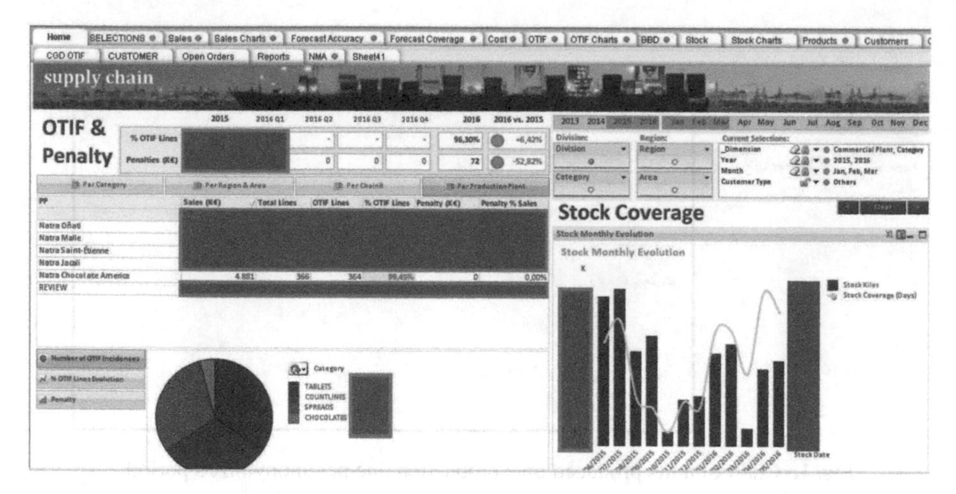


Fig. 2. Supply chain BI cube screen

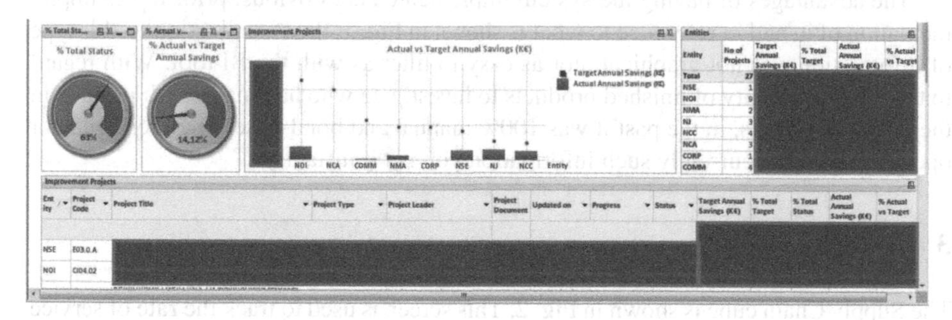


Fig. 3. Corporate BI cube screen

among others, to: (a) discriminate customers based on business criteria based on data, (b) prioritize customers when a limited amount of resources have been available and it was not possible to serve everyone in the short term, (c) act agilely in the face of fluctuations in market availability in relation to certain raw materials, (d) improve monitoring of products that could become obsolete, etc.

## 4 Conclusions

This article presents the potential of Big Data and BI systems within the framework of Industry 4.0 indicating that, however, the literature does not find a significant number of success stories or good practices of its application in the area of or manufacturing industry and specifying that no such cases have been found concerning the food industry. The following is the deployment and implementation of BI in an industrial food group that can be considered as a case of big data. This deployment sets out in detail the development of the solution and the functionalities obtained for the specific case of a production management control system and shows in its overall capacity part of the

5 cubes created in the group. Subsequently, the results and advantages obtained through the implementation of the system are presented, among which are an improvement in the overall turnover, emphasizing an increase in the rate of service to customers and a reduction in production costs and Transport. At the pedagogical level it is worth noting that the developed case has served to develop pedagogical examples in the degrees and Master of Engineering in the classes that the authors teach. To do this, the free version of the software applied in the company shown (Qlikview®) has been used, offering students modified versions of the data handled in reality and facing them the challenge of building their own cubes. The results have been encouraging, as in some cases students have suggested to teachers improvements that the company has considered appropriate to implement.

As for future lines, these are oriented to the analysis of how much is necessary the adequacy of the processes and information systems of the different parts of the company used as a support, for example, a common ERP system in all its areas so that it is in training obtained is fully confrontational, as the use of different processes and platforms makes it difficult to compare between different plants and delegations.

**Compliance with Ethical Standards**

✓ All authors declare that there is no conflict of interest.
✓ No humans/animals involved in this research work.
✓ We have used our own data.

## References

1. Blanchet, M., Rinn, T., Von Thaden, G., Georges, D.T.: Industry 4.0 The new industrial revolution How Europe will succeed. Rol. Berger Strateg. Consult. 1–24 (2014)
2. Bauer, W., Schlund, S., Ganschar, O., Marrenbach, D.: Industrie 4.0
3. Volkswirtschaftliches Potenzial für Deutschland. Bitkom, Fraunhofer Inst., pp. 1–46 (2014)
4. E. y P. Comisión Europea: Foro de políticas estratégicas sobre el emprendimiento digital, Dirección General de Mercado Interno, Industria. Report: Digital Transformation of European Industry and Enterprises (2015)
5. Luftman, J., Ben-Zvi, T.: Key issues for IT executives 2009: difficult economy's impact on IT. MIS Q. Exec. **9**(1), 203–213 (2009)
6. Mazuin, E., Yusof, M., Othman, M.S., Omar, Y., Rizal, A.: The study on the application of business intelligence in manufacturing: a review **10**(1), 317–324 (2013)
7. Ur-Rahman, N.: Textual data mining for next generation intelligent decision making in industrial environment: a survey. Eur. Sci. J. **11**(24), 1857–7881 (2015)
8. Fitriana, R., Djatna, T.: Progress in Business Intelligence System research: a literature review. Int. J. Basic Appl. Sci. IJBAS-IJENS **11**(03), 96–105 (2011)
9. Fitriana, R., Djatna, T.: Business intelligence design for decision support dairy agro industry medium scaled enterprise. Int. J. Eng. Technol. **12**(05), 1–9 (2012)
10. Rashid Al-Azmi, A.-A.: Data text and web mining for business intelligence: a survey. Int. J. Data Min. Knowl. Manag. Process. **3**(2), 1–21 (2013)
11. Izquierdo, N.V., Lezama, O.B.P., Dorta, R.G., Viloria, A., Deras, I., Hernández-Fernández, L.: Fuzzy logic applied to the performance evaluation. Honduran coffee sector case. In: Tan, Y., Shi, Y., Tang, Q. (eds.) Advances in Swarm Intelligence, ICSI 2018. Lecture Notes in Computer Science, vol. 10942. Springer, Cham (2018)

12. Amelec, V.: Increased efficiency in a company of development of technological solutions in the areas commercial and of consultancy. Adv. Sci. Lett. **21**(5), 1406–1408 (2015)
13. Viloria, A., Lis-Gutiérrez, J.P., Gaitán-Angulo, M., Godoy, A.R.M., Moreno, G.C., Kamatkar, S.J.: Methodology for the design of a student pattern recognition tool to facilitate the teaching - learning process through knowledge data discovery (big data). In: Tan, Y., Shi, Y., Tang, Q. (eds.) Data Mining and Big Data, DMBD 2018. Lecture Notes in Computer Science, vol. 10943. Springer, Cham (2018)
14. May, R.J.O.: Diseño del programa de un curso de formación de profesores del nivel superior, p. 99. UADY, Mérida de Yucatán (2013)
15. Medina Cervantes, J., Villafuerte Díaz, R., Juarez Rivera, V., Mejía Sánchez, E.: Simulación en tiempo real de un proceso de selección y almacenamiento de piezas. Revista Iberoamericana de Producción Académica y Gestión Educativa (2014). ISSN 2007 – 8412. http://www.pag.org.mx/index.php/PAG/article/view/399
16. Minnaard, C., Minnaard, V.: Sumatoria de Creatividad + Brainstorming + la Vida Cotidiana: Gestar problemas de Investigación / Summary of Creativity + Brainstorming + EverydayLife: GestatingResearchProblems. Revista Internacional de Aprendizaje en la Educación Superior. España (2017)
17. Thames, L., Schaefer, D.: Software defined cloud manufacturing for industry 4.0. Procedía CIRP **52**, 12–17 (2016)

# Facial Expression Recognition Using Supervised Learning

V. B. Suneeta(✉), P. Purushottam, K. Prashantkumar, S. Sachin, and M. Supreet

Department of ECE, BVBCET, Hubballi 580031, India
suneeta_vb@bvb.edu

**Abstract.** Facial Expression Recognition (FER) draws much attention in present research discussions. The paper presents a relative analysis of recognition systems for facial expression. Facial expression recognition is generally carried out in three stages such as detection of face, extraction of features and expressions' classification. The proposed work focuses on a face detection and extraction method is presented based on the Haar cascade features. The classifier is trained by using many positive and negative images. The features are extracted from it. For standard Haar feature images like convolutional kernel are used. Next, Fisher face classifier a supervised learning method is applied on COHN-KANADE database, to have a facial expression classifying system with eight possible classes (seven basic emotions along with neutral). A recognition rate of 65% in COHN-KANADE database is achieved.

**Keywords:** Facial expression recognition · Supervised learning · Fisher face · Haar cascade · Feature extraction · Classification · PCA

## 1 Introduction

FER is important in interpersonal development of relationships. The intentions or situations can be anticipated by interpreting facial expressions in the social environment, allows people to respond appropriately. Human facial expressions can reveal feelings and emotions that control and monitor the behavior among themselves. The fast growing technology leads to humans and robots communication where robots are instructed through facial expressions. FER of humans by machines has been a very emerging area of research. Following are the major issues identified in FER: expressions' variations, fast processing and product specific applications. Presently FER has become an essential research field due to its large range of applications and efforts are towards increasing FER accuracy, by using the facial expression database involving large span of ages. Universally there are eight basic expressions i.e., disgust, sadness, fear, surprise, happiness, anger, contempt and neutral. These expressions can be found in all societies and cultures. For any FER system, following are the necessary steps: (A) Face Detection (B) Features Extraction (C) Facial Expressions Classification.


S. Smys et al. (Eds.): ICCVBIC 2019, AISC 1108, pp. 275–285, 2020.
https://doi.org/10.1007/978-3-030-37218-7_32

Authors carried out an extensive work on FER techniques using various feature extraction methods. Various techniques for face and FER have been presented. However each of them have their own limitations and advantages. The Eigenfaces and Fisherface algorithms, that are built most recent 2D PCA, and Linear Discriminant Analysis (LDA) are few examples of holistic methods. Various authors studied these methods but, local descriptor have caught attention because of their robustness to illumination and posed variations. Some of the other approaches are discussed further in the fore coming sections.

Authors in paper [1] discussed about Automated Facial Expression using features of salient patches. This paper has presented facial expression recognition for six universal expressions. Authors in [2] discussed about FER that deals about facial expression for aging. Here Gabor and log Gabor filters are employed for extracting facial features. The use of log Gabor filter shows superior results when compared to Gabor filter. But the processing time in log Gabor filter is greater than that of Gabor filter. The accuracy of implementation using Gabor filter or log Gabor filter with SVM classifier is greater when compared to natural human prediction on standard as well as synthetic database. Authors in [3] discussed about Facial Expression Recognition which deals about aging with two databases such as lifespan and faces collected in psychology society.

They have detailed explanation about the reasons of aging and their influence on FER.

Authors in [4] discussed about FER using feed forward neural networks. Cohn Kanade dataset is used for recognition tests. It has given a better rate of recognition as compared to conventional linear PCA and histogram equalization. Authors in [5] presented their discussion about the working of the most recent face recognition algorithms. It was inferred from results that the latest algorithms are 10 times better than the face recognition algorithms of year 2002 and 100 times better than those of 1995. Some of the algorithms were capable of leaving behind human participants in recognizing faces. Also these algorithms were capable of identifying even identical twins. Authors in [6] discussed about Vytautas Perlibakas (2004) using Principal Component Analysis and Wavelet Packet Decomposition which permits the usage of PCA based face recognition involving enormous number of training images and are faster training than PCA method.

Authors in [7] have introduced a face recognition algorithm and classification, depending on linear subspace projection. The subspace is established using neighboring component analysis (NCA) which uses dimensionality reduction introduced recently. Authors in [8] discussed about features fusion method using Fisher Linear Discriminant (FLD). The technique extracts features by employing Two-Dimensional PCA (2DPCA) and Gabor Wavelets, and then fuses their features that are extracted with FLD respectively. From literature review during 1980's typical classification techniques were used for facial expression recognition. In recent years extraction of features like eyes, mouth etc. are developed using neural networks classifier.

The organization of the paper is as follows. The Sect. 2 explores the proposed methodology and Sect. 3 highlights the comparison of different methodologies used and Sect. 4 elaborates on the results and their analysis. The paper is concluded in Sect. 5.

The contributions of the paper are to,

- Develop a framework to define the expression detection problem as machine learning problem.
- Propose a framework for feature extraction and training using the dataset.

## 2 Proposed Framework for Feature Extraction

In this section the architecture of facial expression recognition is being summarized, where each stage enhances the application of image data set with a brief discussion on them.

### 2.1 Face Detection

Face detection is a technology which recognizes human faces in images. To obtain a proper classification of expressions building an efficient algorithm is required. From the survey it is clear that fisher face classifier for classification and Haar Cascade technique for feature extraction yield good results. The data set is first sorted and images with different expressions are put into separate baskets. Then these sorted images are re-scaled into same size. After this the image sets are manually labeled as happy, sad, etc. into eight different labels as shown in Fig. 1.

### 2.2 Features Extraction

When the face is detected, key features like eyes, eyebrows, lips etc are extracted from the facial image. Extraction of features is related to dimensionality reduction. Feature extraction starts from initial stage of the given data and it constructs derived features to interpret informative and non redundant features. Feature extraction using Haar Cascade algorithm is applied to these labeled set of images as shown in Fig. 2. Calculation speed of Haar like features is high over other feature extraction techniques. Use of integral images provides the easy calculation of Haar-like features of any size in constant time.

Every feature type can specify the availability or non availability of certain characteristics in the image, such as edges or modifications in texture. Feature extraction algorithm yields specific features for each eight different classes. Through this feature matrix $X$ is obtained. This, along with label $Y$ is sent to fisher-face classifier algorithm to obtain hypothesis $h_\theta(X)$ for each class.

### 2.3 Facial Expressions Classification

Last step of the automatic FER system is the classification that can be implemented either by attempting recognition or by interpretation as shown in Fig. 4. After the extraction of these important features, the expression is categorized by making the comparison between the image with the images in the training dataset using fisher face classifier. Classification is a general process related to categorization in which objects/images are identified, differentiated and understood. Classification of expressions using fisher-face classifier gives high success rate than Eigen-face classifier and bag of features technique. LDA is used in

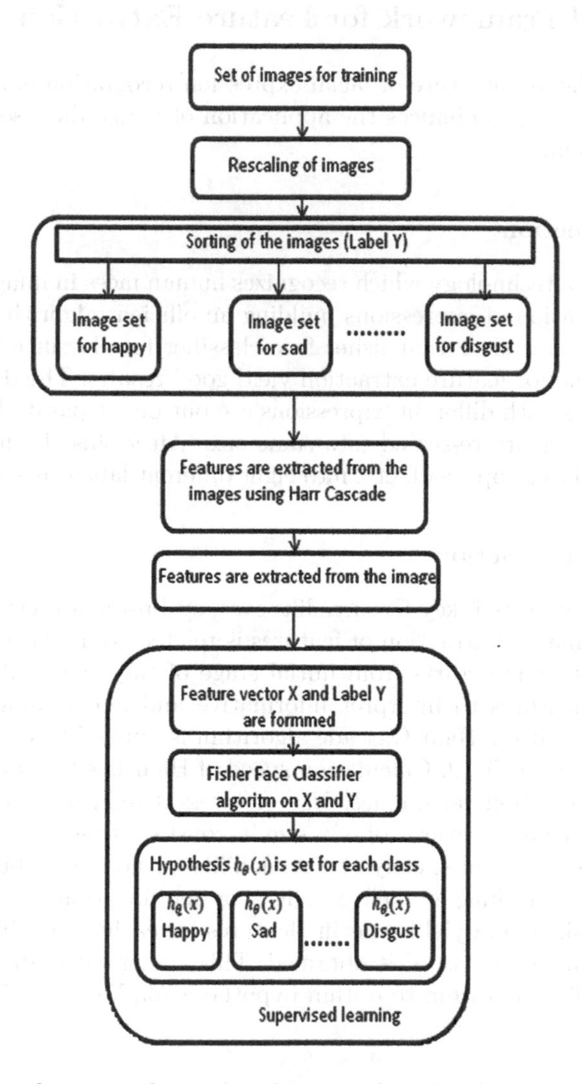


**Fig. 1.** A framework to train the machine learning algorithm for seven facial expressions using the image data set as training data set.

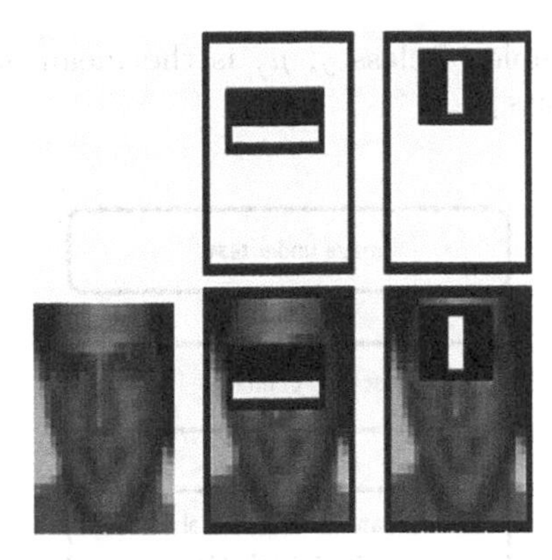

**Fig. 2.** A framework to train and test the machine learning algorithm for eight facial expressions using the Haar cascade features to get extracted features [3].

fisher-face classification to reduce the dimension there by number of features required to classify face expressions is reduced.

To determine the Fisher-faces, the data in every category is assumed to be Normally distributed ($ND$), the multivariate $ND$ as $N_i(\mu_i, V_i)$, with mean $\mu_i$ and covariance matrix $V_i$, and its pdf is given by,

$$f_i(x|\mu_i, V_i) \tag{1}$$

In the C class problem, we have $N_i(\mu_i, V_i)$, with $i = 1....C$. Given these NDs and their class prior probabilities $P_i$, the classification of a test sample $x$ is given by comparing the log-likelihoods of,

$$f_i(x|\mu_i, V_i)P_i \tag{2}$$

for all $i$, i.e.,

$$\arg\min_{1\leq i\leq C} d_i(\mathbf{x}) \tag{3}$$

where,

$$d_i(\mathbf{x}) = (\mathbf{x} - \mu_i)^T \Sigma_i^{-1}(\mathbf{x} - \mu_i) + ln|\Sigma_i| - 2\ln P_i \tag{4}$$

However, in the case where all the covariance matrices are identical, $V_i = V, \forall(i)$, the quadratic parts of $d_i$ cancel out, providing linear classifiers. These classifiers are known as linear discriminant bases. Assume that $C = 2$ and that the classes are homoscedastic normals (Fig. 3).

It ensures that number of wrongly classified samples in the original space of $p$ dimensions and in this subspace of just one dimension are the same. This can be verified easily. In general, there will be more than 2-classes. In such case, there will be reformulation of the above stated problem as that of decreasing within-class differences and increasing between-class distances. Within class differences can be determined using the within-class scatter matrix given by,

$$\mathbf{S}_w = \sum_{j=1}^{C}\sum_{i=1}^{n_j}(\mathbf{x}_{ij} - \mu_j)(\mathbf{x}_{ij} - \mu_j)^T \tag{5}$$

where $x_{ij}$ is the $i^{th}$ sample of class $j$, $\mu_j$ is the mean of class $j$, and $n_j$ the number of samples in class $j$.

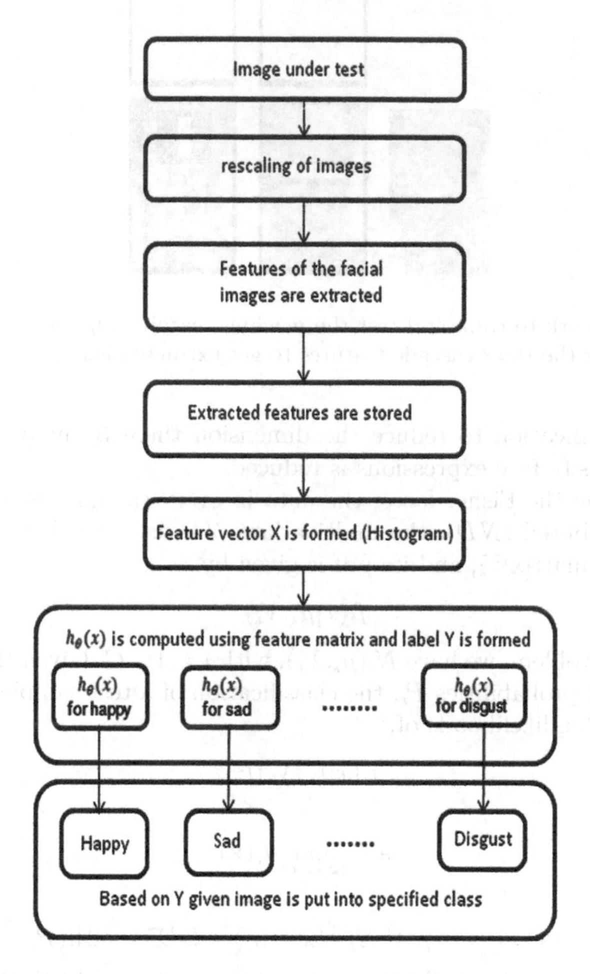


**Fig. 3.** A framework to test the machine learning algorithm for seven facial expressions using the image data set as testing set.

## 3 Comparison of Different Methods

Since correlation techniques are computationally expensive and need large quantity of storage, it is essential to follow dimensionality reduction schemes.

### 3.1 Eigen Face Mathematical Equations

(1) N sample images $x_1, x_2, x_3, .......x_N$

(2) C classes $X_1, X_2, X_3, ..........X_c$
(3) New feature vector

$$y_k \in R^m y_k = W^T x_k \tag{6}$$

where k = 1,2,...N and W matrix with orthonormal columns.

### 3.2 Fisher Face Mathematical Equations

(1) Between class scatter matrix

$$\mathbf{S}_b = \sum_{j=1}^{C} (\mu_j - \mu)(\mu_j - \mu)^T \tag{7}$$

where $\mu$ represents the mean of all classes.
(2) Within class scatter matrix

$$\mathbf{S}_w = \sum_{j=1}^{C} \sum_{i=1}^{n_j} (\mathbf{x}_{ij} - \mu_j)(\mathbf{x}_{ij} - \mu_j)^T \tag{8}$$

In case of Eigenface recognition scatter being increased is not only due to between-class scatter but, also due to the within-class scatter. But in Fisherface, ratio of the determinant of the between-class scatter matrix of the projected samples to that of within-class scatter matrix projected samples is maximized. Hence Fisherface is preferred.

## 4 Results and Discussions

Face expressions are classified into respective classes using the Fisherface classifier algorithm. Results and optimization technique are discussed below in three sub-sections. A. Face detection and Feature extraction. B. Classification. C. Performance variation on test size. comparison of fisherface and eigen face classifier algorithm is done in this section and conclusion is drawn.

### 4.1 Face Detection and Feature Extraction

Face detection with Feature extraction is the first step performed in FER system. In this step faces of different expressions are detected from the image data-set and then all important features of faces with different expressions are extracted from the image data set using Haar cascade features. Detected faces with different expressions and corresponding features of face are as follows.

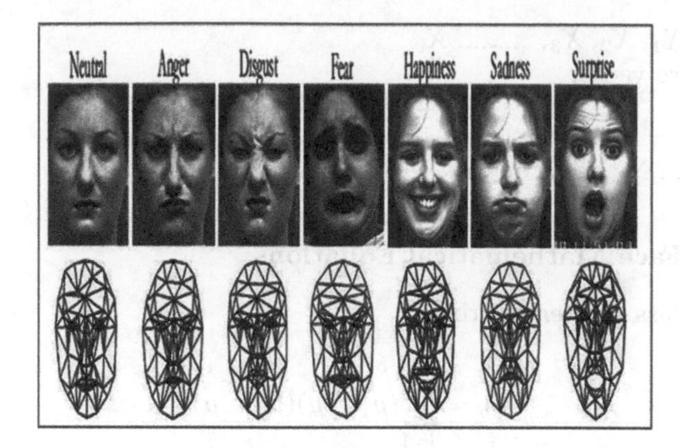


**Fig. 4.** Figure depicts the six different facial expressions of human [7].

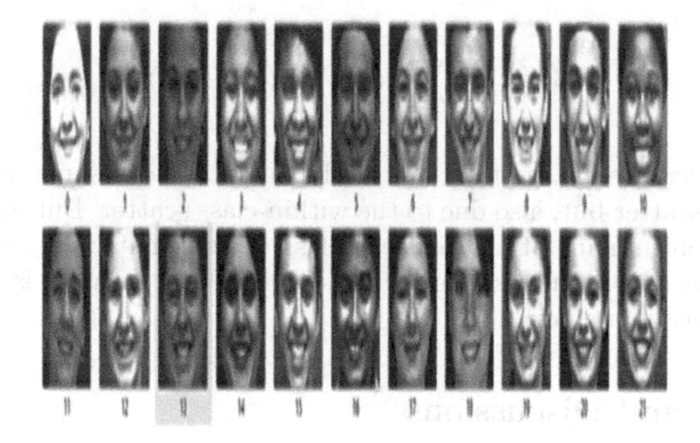

**Fig. 5.** Figure shows the images of faces with happy expressions classified into class HAPPY.

### 4.2 Classification

Classification of images is being generalized using feature vector of projected test image, set hypothesis of each expression are computed and the images are put into respective class as shown in Figs. 5 and 6, depending on the label $Y$ found in this step. Fisherface classifier gives more success rate than Eigen face classifier. In case of Eigen face only in between class scatter matrix is used where as in Fisherface both in between class scatter matrix and within class scatter matrix are used which contribute more success rate than eigenface.

### 4.3 Performance Variation on Test Size

Success rate defines the number of images rightly classified among test image set. It is expressed in terms of percentage. Here the dependency of success rate

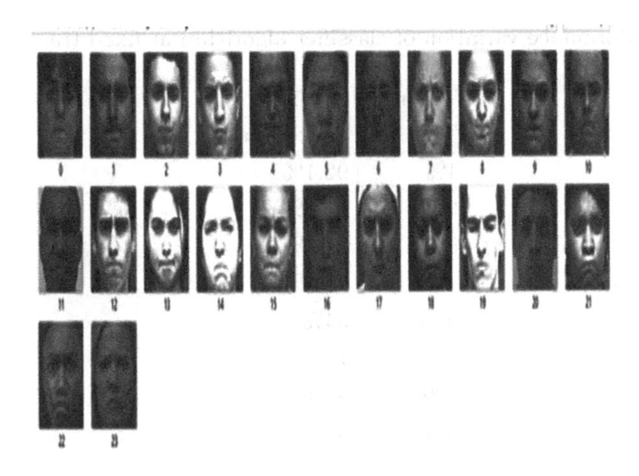

**Fig. 6.** Figure shows the images of faces with anger expressions classified into class ANGER [6]

on test size variation is discussed. In general for a fixed training data set as test size increases, success rate decreases initially and after reaching certain test size, success rate increases with increase in test size.

**Table 1.** Performance variation of classifier algorithm at fixed train size = 95%

| Train size = 95% | |
|---|---|
| Test size | Success rate |
| 1% | 96.2% |
| 3% | 96% |
| 5% | 93.5% |
| 10% | 60% |
| 20% | 82.5% |
| 30% | 88.9% |
| 40% | 90.6% |
| 50% | 93.7% |

The increasing success rate in the later stage is due to repeated occurrence of images in both test and train set of data. During the initial stage upto 10% of test size, system would get hardly few images for testing. Thus the rate of rightly classifying the images is more. A bias towards classes which have number of instances of standard classifier algorithms behavior is the reason for it's behavior. The tendency is to predict only the majority class data. The minority class features are treated as noise and are often neglected. Thus it leads to a high probability of misclassification of these minority classes in comparison with the

**Table 2.** Performance variation of classifier algorithm at fixed train size = 85%

| Train size = 85% | |
|---|---|
| Test size | Success rate |
| 1% | 92.1% |
| 3% | 91.5% |
| 5% | 89.4% |
| 10% | 58.5% |
| 20% | 64% |
| 30% | 75.6% |
| 40% | 81% |
| 50% | 84% |

**Table 3.** Performance variation of classifier algorithm at fixed train size = 75%

| Train size = 75% | |
|---|---|
| Test size | Success rate |
| 1% | 89.3% |
| 3% | 88% |
| 5% | 85.6% |
| 10% | 56.7% |
| 20% | 60% |
| 30% | 65% |
| 40% | 73.6% |
| 50% | 78.1% |

majority classes. By observing Tables 1, 2 and 3 the system has good mixture of both training and test data in Table 2. Hence data with train size of 85% is preferred.

## 5 Conclusion

An image processing and classification method has been implemented in which face images are used to train a dual classifier predictor that predicts the seven basic human emotions given a test image. The predictor is relatively successful at predicting test data from the same dataset used to train the classifiers. However, the predictor is consistently poor at detecting the expression associated with contempt. This is likely due to a combination of lacking training and test images that clearly exhibit contempt, poor pre-training labellings of data, and the intrinsic difficulty at identifying contempt.

**Compliance with Ethical Standards.** All authors declare that there is no conflict of interest No humans/animals involved in this research work. We have used our own data.

## References


1. Fu, Y., Guo, G.-D., Huang, T.S.: Age synthesis and estimation via faces: a survey. IEEE Trans. Pattern Anal. Mach. Intell. **32**(11), 1955–1976 (2010)
2. Flynn, P.J., Scruggs, T., Bowyer, K.W., Worek, W., Phillips, P.J.: Preliminary face recognition grand challenge results. In: Proceedings of the 7th International Conference on Automatic Face and Gesture Recognition, pp. 15–24, April 2016
3. Perlibakas, V.: Distance measures for PCA-based face recognition. Pattern Recogn. Lett. **25**(6), 711–724 (2004)
4. Butalia, M.A., Ingle, M., Kulkarni, P.: Facial expression recognition for security. Int. J. Mod. Eng. Res. (IJMER) **2**, 1449–1453 (2012)
5. Zhang, Z., Zhang, J.: A new real-time eye tracking for driver fatigue detection. In: 6th IEEE International Conference on Telecommunications Proceedings, pp. 181–188, April 2006
6. Guo, Y., Tian, Y., Gao, X., Zhang, X.: Micro-expression recognition based on local binary patterns from three orthogonal planes and nearest neighbor method. In: International Joint Conference on Neural Networks (IJCNN), pp. 3473–3479 (2014)
7. Tian, Y.L., Kanade, T., Cohn, J.F.: Facial expression recognition. In: Li, S., Jain, A. (eds.) Handbook of Face Recognition, pp. 487–519. Springer, London (2011)
8. Pawaskar, S., Budihal, S.V.: Real-time vehicle-type categorization and character extraction from the license plates. In: International Conference on Cognetive Informatics and Soft Computing-2017. Advances in Intelligent Systems and Computing a Springer book chapter, VBIT, Hyderabad, pp. 557–565, July 2017
9. Wu, Y.W., Liu, W., Wang, J.B.: Application of emotional recognition in intelligent tutoring system. In: First IEEE International Workshop on Knowledge Discovery and Data Mining, pp. 449–452 (2008)
10. Youssif, A., Asker, W.A.A.: Automatic facial expression recognition system based on geometric and appearance features. Comput. Inf. J. **4**(2), 115–124 (2011)
11. Gomathi, V., Ramar, K., Santhiyaku Jeevakumar, A.: A neuro fuzzy approach for facial expression recognition using LBP histograms. Int. J. Comput. Theory Eng. **2**(2), 245–249 (2010)
12. Hawari, K., Ghazali, B., Ma, J., Xiao, R.: An innovative face detection based on skin color segmentation. Int. J. Comput. Appl. **34**(2), 6–10 (2011)

# Optimized Machine Learning Models for Diagnosis and Prediction of Hypothyroidism and Hyperthyroidism

Manoj Challa(✉)

Department of Computer Science and Engineering, CMR Institute of Technology, Bengaluru, India
manojreddi@gmail.com

**Abstract.** Thyroid is a major complaint happens due to absence of thyroid hormone among the human beings. The diagnosis report of thyroid test composed of the T3 (Triiodothyronine, T3-RIA) by radioimmunoassay, T4 (Thyroxine), FT4 (FT4, Free Thyroxin), FTI (Free Thyroxine Index, FTI, T7), TSH (Thyroid Stimulating Hormone). Manual evaluation of these factors is tedious, so machine learning approaches will be used for this diagnosis and prediction of hypothyroidism and hyperthyroidism from the larger datasets. In this proposed research, the optimized machine learning classification approaches Decision Tree, Random Forest, Bayes Classification and K- nearest neighbor methods are used for predicting the thyroid disorder. The performance measure of all these approaches are estimated and compared with R Programming Tool.

**Keywords:** Thyroid · Hypothyroidism · Hyperthyroidism · Diagnosis · R

## 1 Introduction

The thyroid gland is located below the Adam's apple as shown in the Fig. 1 wrapped around the trachea (windpipe). A thin area of tissue in the gland's middle, known as the isthmus, joins the two thyroid lobes on each side. When the thyroid produces either too much or too little of these hormones it causes the gland to work wrongly, leading to disorders like hyperthyroidism and hypothyroidism. The Risk factors of hypothyroidism and hyperthyroidism are depicted in the Table 1.


S. Smys et al. (Eds.): ICCVBIC 2019, AISC 1108, pp. 286–293, 2020.
https://doi.org/10.1007/978-3-030-37218-7_33

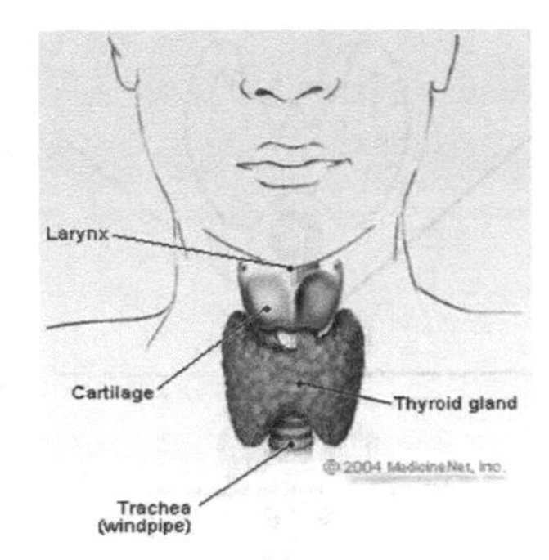


**Fig. 1.** Thyroid Disorder [6]

**Table 1.** Hypothyroidism and hyperthyroidism risk factors

| Hypothyroidism | Hyperthyroidism |
|---|---|
| Age and gender | Gender |
| Preexisting condition | Family or personal history of autoimmune disorders |
| Pituitary gland disorder | Vitamin D and selenium deficiencies |
| Pregnancy | Current pregnancy |

### 1.1 Factors that Affect Thyroid Function

Figure 2 depicts the routine factors that rise the possibility of thyroid complaints. Some routine factors rise the thyroid complaints are listed below.

- Smoking.
- Psychological stress.
- Injury to the thyroid.
- High amounts of lithium and iodine usage.

### 1.2 Thyroid Function Tests

Thyroid function diagnosis composed of the following parameters

(1) T3 - T3 (Triiodothyronine, T3-RIA) by radioimmunoassay
(2) T4 - T4 (Thyroxine)
(3) FT4 - T4 (FT4, Free Thyroxin)
(4) FTI - FTI (Free Thyroxine Index, FTI, T7)
(5) TSH - TSH (Thyroid Stimulating Hormone)

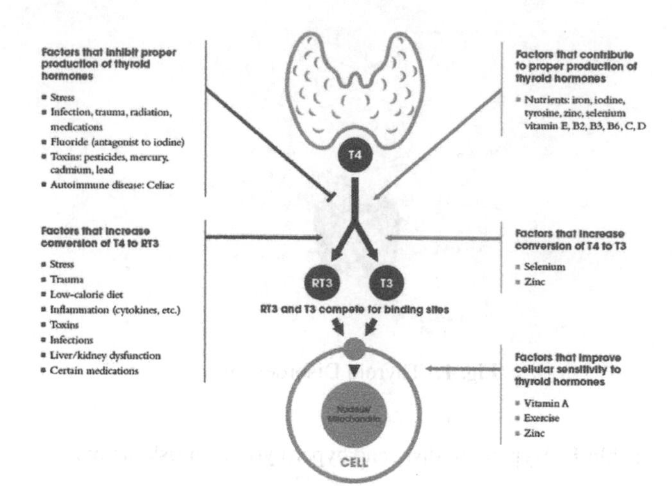


**Fig. 2.** Factors that affect Thyroid Function [5]

## 2 Proposed Classification Models

**2.1 Decision tree** is one of the very widespread machine learning approach. Decision Tree is a Supervised Machine Learning algorithm which resembles like an inverted tree. Each node in the tree represents a predictor variable (feature), the link between the nodes denotes a Decision and each leaf node denotes an outcome (response variable).
**2.2 Random Forest (RF)** is an classification approach that comprises of several decision trees and outputs the class that is the mode of the classes output by individual trees. Random Forests are frequently used when there are very large training datasets and a very large number of input variables. A random forest model is normally made up of tens or hundreds of decision trees.
**2.3 Bayesian classification** is basically based on Bayes' Theorem. This Bayesian classification approach can predict the probability of class membership. For example, the probability that a given tuple belongs to a particular class. Classification algorithms have found a simple Bayesian classifier known as naïve Bayes classifier to be comparable in performance with decision tree and selected neural network classifiers. High amount of accuracy and speed will be achieved through Bayesian Classifiers.
**2.4 K-NN (Nearest-Neighbor Method)** Nearest-neighbor classifiers approach is based on learning by analogy. It compares a given test dataset with training dataset that are similar to it.

## 3 Implementation

The dataset is extracted from the UCI Machine Learning Repository. The dataset contains 3163 instances. In this 151 data comes under hypothyroid and 3012 data is negative cases. The attributes are shown in the table below (Table 2):

**Table 2.** Hypothyroidism and hyperthyroidism attributes

| S. No | Attribute name | Possible values | S. No | Attribute name | Possible values |
|---|---|---|---|---|---|
| 1 | AGE | NUMERIC | 15 | HYPOPITUITARY | F, T |
| 2 | SEX | M (0), F (1) | 16 | PSYCH | F, T |
| 3 | ON THYROXINE | F, T | 17 | TSH MEASURED | F, T |
| 4 | QUERY ON THYROXINE | F, T | 18 | TSH | NUMERIC |
| 5 | ANTITHYROID MEDICATION | F, T | 19 | T3 MEASURED | F, T |
| 6 | SICK | F, T | 20 | T3 | NUMERIC |
| 7 | PREGNANT | F, T | 21 | TT4 MEASURED | F, T |
| 8 | THYROID SURGERY | F, T | 22 | TT4 | NUMERIC |
| 9 | I131 TREATMENT | F, T | 23 | T4U MEASURED | F, T |
| 10 | QUERY HYPOTHYROID | F, T | 24 | T4U | NUMERIC |
| 11 | QUERY HYPERTHYROID | F, T | 25 | FTI MEASURED | F, T |
| 12 | LITHIUM | F, T | 26 | FTI | NUMERIC |
| 13 | GOITRE | F, T | 27 | TBG MEASURED | F, T |
| 14 | TUMOR | F, T | 28 | TBG | NUMERIC |

The following Table 3 depicts the vital attributes that plays a major role in both hypothyroidism and hyperthyroidism disorder

**Table 3.** Mean, Median, Standard Deviation, Range, Skewness and Kurtosis

| S. No | Target Attribute Name | Min Value | Max Value | Mean | Median | Standard Deviation | Range | Skewness | Kurtosis |
|---|---|---|---|---|---|---|---|---|---|
| 1 | AGE | 1 | 98 | 51.15 | 54.00 | 19.29 | 97 | −0.16 | −0.96 |
| 18 | TSH | 0 | 530 | 5.92 | 0.70 | 23.90 | 530 | 10.21 | 153.65 |
| 20 | T3 | 0 | 10.20 | 1.94 | 1.80 | 1.00 | 10.20 | 2.12 | 10.25 |
| 22 | TT4 | 2 | 450 | 108.85 | 104 | 45.49 | 448 | 1.51 | 6.94 |
| 24 | T4U | 0 | 2.21 | 0.98 | 0.96 | 0.23 | 2.21 | 0.92 | 3.41 |
| 26 | FTI | 0 | 881 | 115.40 | 107 | 60.24 | 881 | 5.10 | 47.10 |
| 28 | TBG | 0 | 122 | 31.28 | 28.00 | 19.22 | 122 | 2.82 | 8.96 |

## 4 Results and Discussion

The Naïve Bayes, Decision Tree, KNN and SVM classification approaches implementation is done using R programming tool with imported thyroid datasets from UCI machine learning repository. The results are estimated and evaluated, and it is clearly visible that Decision Tree Approach shows the high accuracy for the prediction of Hyperthyroidism when compared to the other classification algorithms Naïve Bayes, KNN and SVM (Fig. 3 and Table 4).

**Table 4.** Performance evaluation for hyperthyroidism

| | | Recall | Precision | Sensitivity | Specificity | F-Measure |
|---|---|---|---|---|---|---|
| Hyperthyroidism | Naïve Bayes | 0.444 | 0.400 | 0.444 | 0.976 | 0.421 |
| | Decision Tree | 0.778 | 1 | 0.778 | 1 | 0.875 |
| | KNN | 0.680 | 0.960 | 0.100 | 0.100 | 0.960 |
| | SVM | 0.060 | 0.870 | 0.110 | 0.810 | 0.880 |

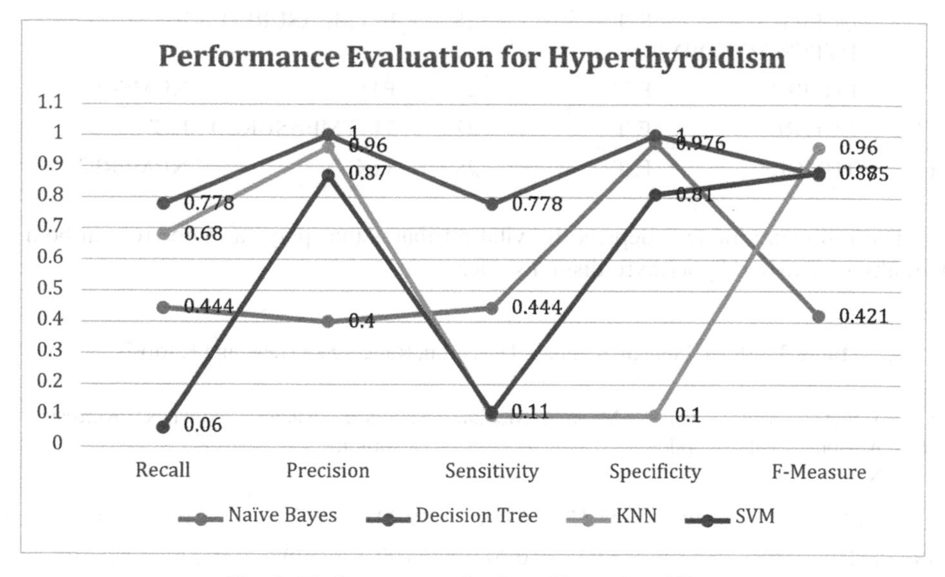


**Fig. 3.** Performance evaluation of hyoerthyroidism

The Naïve Bayes, Decision Tree, KNN and SVM classification approaches implementation is done using R programming tool with imported thyroid datasets from UCI machine learning repository. The results are estimated and evaluated, its clearly visible that Decision Tree Approach shows the high accuracy for the prediction of Hypothyroidism when compared to the other classification algorithms Naïve Bayes, KNN and SVM (Figs. 4, 5 and Tables 5, 6).

**Table 5.** Performance evaluation for hypothyroidism

| | | Recall | Precision | Sensitivity | Specificity | F-Measure |
|---|---|---|---|---|---|---|
| Hypothyroidism | Naïve Bayes | 0.500 | 0.455 | 0.500 | 0.976 | 0.476 |
| | Decision Tree | 0.900 | 0.600 | 0.900 | 0.976 | 0.720 |
| | KNN | 0.680 | 0.960 | 0.100 | 0.100 | 0.960 |
| | SVM | 0.060 | 0.870 | 0.110 | 0.810 | 0.880 |

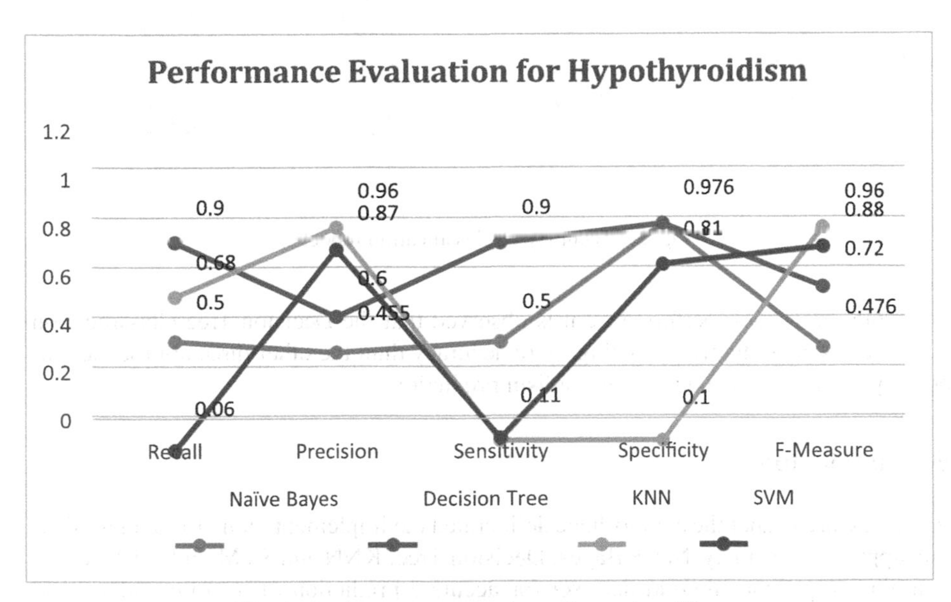


**Fig. 4.** Performance evaluation of hyperthyroidism

## 4.1 Accuracy of Classification Methods

The accuracy of the classification approaches of the thyroid dataset based on the classification models Naïve Bayes, Decision Tree, KNN and SVM algorithms is depicted in

**Table 6.** Accuracy of classification models

| Models | Accuracy |
|---|---|
| Naïve Bayes | 91.60 |
| Decision Tree | 96.90 |
| KNN | 95.10 |
| SVM | 96.00 |

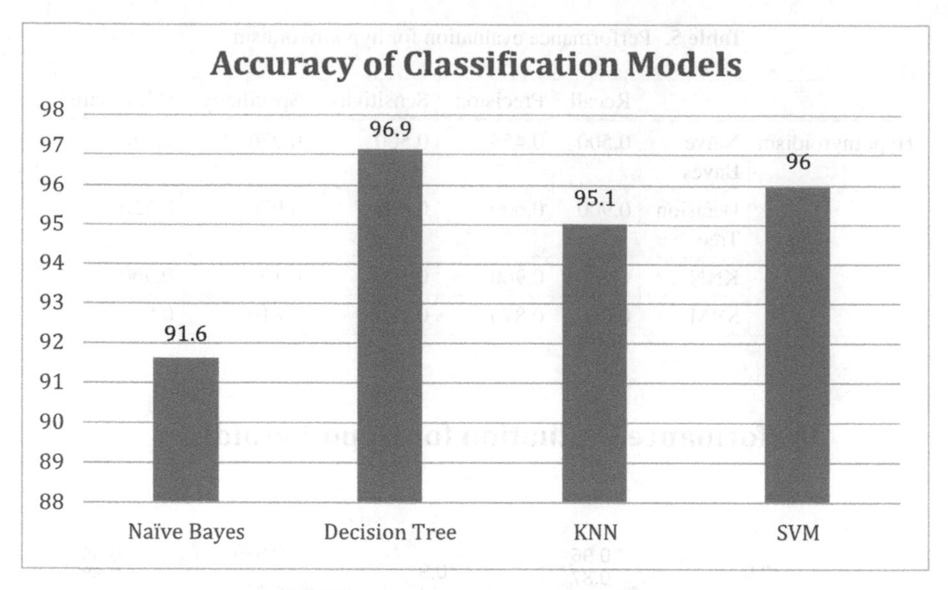


**Fig. 5.** Accuracy of classification models

the Table. From their performance it is observed that the Decision Tree Classification Approach is consistently producing more accuracy than the other three approaches for both hypothyroidism and hyperthyroidism prediction.

## 5 Conclusion

In this research paper the authors have deliberated the implementation of four classification approaches namely, Naïve Bayes, Decision Tree, KNN and SVM on UCI Machine Learning Repository thyroid data set for accurate prediction of hypothyroidism and hyperthyroidism disorder. The best classification approach was the decision tree approach in all the effectuated implementations. The future work will focus on the diagnosis and prediction of risk factors that affect the thyroid diseases and on testing more machine learning approaches for the classification of vital human diseases.

**Compliance with Ethical Standards**

✓ All authors declare that there is no conflict of interest.
✓ No humans/animals involved in this research work.
✓ We have used our own data.

## References

1. Bologna, G.: A model for single and multiple knowledge based networks. Artif. Intell. Med. **28**, 141–163 (2003)
2. Chang, W.W., Yeh, W.C., Huang, P.C.: A hybrid immune-estimation distribution of algorithm for mining thyroid gland data. Artif. Intell. Med. **41**, 57–67 (2007)

3. Chen, H.L., Yang, B., Wang, G., Liu, J., Chen, Y.D., Liu, D.Y.: A three stage expert system based on support vector machines for thyroid disease diagnosis. J. Med. Syst. **36**, 1953–1963 (2012)
4. Dogantekin, E., Dogantekin, A., Avci, D.: An automatic diagnosis system based on thyroid gland. ADSTG. Expert Syst. Appl. **37**, 6368–6372 (2010)
5. Dogantekin, E., Dogantekin, A., Avci, D.: An expert system based on Generalized Discriminant analysis and Wavelet Support Vector Machine for diagnosis of thyroid disease. Expert Syst. Appl. **38**(1), 146–150 (2011)
6. Duch, W., Adamezak, R., Grabezewski, K.: A new methodology of extraction, optimization and application of crisp and fuzzy logic rules. IEEE Trans. Neural Netw. **12**, 277–306 (2001)
7. Temurtas, F.: A comparative study on thyroid disease diagnosis using neural networks. Expert Syst. Appl. **36**, 944–949 (2009)
8. Jiawei, H., Micheline, K., Jian, P.: Data Mining Concepts and Techniques. Morgan Kaufmann, Tokyo (2006)
9. Keles, A., Keles, A.: ESTDD: expert system for thyroid disease diagnosis. Expert Syst. Appl. **34**, 242–246 (2008)
10. Keleş, A., Keleş, A.: ESTDD: expert system for thyroid diseases diagnosis. Expert Syst. Appl. **34**(1), 242–246 (2008)
11. Kodaz, H., Ozsen, S., Arslan, A., Gunes, S., Arslan, A., Gunes, S.: Medical application of information gain based artificial immune recognition system (AIRS): diagnosis of thyroid disease. Expert Syst. Appl. **36**, 3086–3092 (2009)
12. Kumari, M., Godara, S.: Comparative study of data mining classification methods in cardiovascular disease prediction 1. IJCST **2** (2011). ISSN 2229-4333
13. Lavarello, R.J., Ridgway, B., Sarwate, S., Oelze, M.L.: Imaging of follicular variant papillary thyroid carcinoma in a rodent model using spectral-based quantitative ultrasound techniques. In: IEEE 10th International Symposium on Biomedical Imaging (ISBI), pp. 732–735 (2013)
14. Li, L.N., Ouyang, J.H., Chen, H.L., Liu, D.Y.: A computer aided diagnosis system for thyroid disease using extreme learning machine. J. Med. Syst. **36**, 3327–3337 (2012)
15. Liu, D.Y., Chen, H.L., Yang, B., Lv, X.E., Li, L.N., Liu, J.: Design of an enhanced fuzzy k-nearest neighbour classifier based computer aided diagnostic system for thyroid disease. J. Med. Syst. **36**, 3243–4354 (2012)
16. Ng, S.K., McLachlan, G.J.: Extension of mixture-of-experts networks for binary classification of hierarchical data. Artif. Intell. Med. **41**, 57–67 (2007)
17. Ozyilmaz, L., Yildirim, T.: Diagnosis of thyroid disease using artificial neural network methods. In: Proceedings of ICONIP 2002 9th International Conference on Neural Information Processing, Orchid Country Club, Singapore, pp. 2033–2036 (2002)
18. Pasi, L.: Similarity classifier applied to medical data sets. In: 10 sivua. Fuzziness in Finland 2004. International Conference on Soft Computing, Helsinki, Finland & Gulf of Finland & Tallinn, Estonia (2004)
19. Polat, K., Şahan, S., Güneş, S.: A novel hybrid method based on artificial immune recognition system (AIRS) with fuzzy weighted pre-processing for thyroid disease diagnosis. Expert Syst. Appl. **32**(4), 1141–1147 (2007)
20. Parimala, R., Nallaswamy, R.: A study of spam e-mail classification using feature selection package. Global J. Comput. Sci. Technol. **11**, 44–54 (2011)
21. Godara, S., Singh, R.: Evaluation of predictive machine learning techniques as expert systems in medical diagnosis. Indian J. Sci. Technol. **910** (2016)
22. Savelonas, M.A., Iakovidis, D.K., Legakis, I., Maroulis, D.: Active contours guided by echogenicity and texture for delineation of thyroid nodules in ultrasound images. IEEE Trans. Inf. Technol. Biomed. **13**(4), 519–527 (2009)

# Plastic Waste Profiling Using Deep Learning

Mahadevan Narayanan, Apurva Mhatre, Ajun Nair(✉), Akilesh Panicker, and Ashutosh Dhondkar

Department of Computer Engineering, SIES Graduate School of Technology, Navi Mumbai, India
{mahadevan.narayanan16,apurva.mhatre16,ajun.nair16, akilesh.panicker16,ashutosh.dhondkar16}@siesgst.ac.in

**Abstract.** India is the second populated country across the globe, with a rough population of 1.35 billion. Around 5.6 million tonnes of plastic wastes are generated every year, which is approximately about 15,342 tonnes per day [1]. India produces more plastic than its recycling limit. Multi-National companies like Frito Lay, PepsiCo are accountable for most of the waste generated. The Government of India has decided to make India plastic-free by 2022, though the rules and regulations pertaining to reach the goal are not robust enough to inflict any change but the citizens can make this change. We will explore a potential solution using Machine Learning, Image processing and Object detection to assist the smooth profiling of the waste in our locality.

**Keywords:** Deep Learning · Object detection · Image processing · Plastic waste · Convolutional Neural Network · TensorFlow Object Detection API · Faster RCNN

## 1 Introduction

The plastic pollution crisis in India is at its apex and the issue is India doesn't have any efficient waste management system, hence the country produces more waste than it can recycle. Half of the waste generated comes from major metropolitan cities like Mumbai, Delhi and Kolkata. Majority of the waste generated consists of packaged plastic consumables which include chips biscuits, chocolates and bottled drinks. The companies responsible for manufacturing these are in a way responsible for the current circumstances. EPR (Extended Producer Responsibility) is a law which was issued by the government which states that it is the responsibility of the producer to manage and recycle the post-consumer goods [2]. Most of the multinational companies don't take responsibility for the waste they produce. Food Manufacturing Companies are the major contributors in the game. Out of the total plastic waste generated, most of the plastic waste is generated by Nestle, PepsiCo and Coca-Cola and many such multinational companies [3]. This waste is the root cause of all the underlying plastic pollution caused in India. High time is during monsoon when this waste gets accumulated in the sewage system causing the water level to rise and hence causing floods. We have developed a


S. Smys et al. (Eds.): ICCVBIC 2019, AISC 1108, pp. 294–303, 2020.
https://doi.org/10.1007/978-3-030-37218-7_34

solution to restrain this problem by empowering the citizens to help contribute to control plastic pollution. Another perspective for this project can be for the multinational companies to help them set up collection points for plastic waste strategically to incur lower costs.

## 2 Survey of Object Detection Methods

Object detection is a common term in computer vision for classification and localization of objects in an image. Object detection in recent times is largely based on use of convolutional neural networks (CNN). Most deep learning models use CNN at some point in their architecture. Some of the models that are trending today are Faster R-CNN, YOLO (You Only Look Once) and Single Shot Detector (SSD) [9] Some of these models gives an accuracy greater than a human eye can identify an object. It is difficult to have a fair comparison among different models for object detection and there is no single answer on which model is the best amongst all. For real-life applications like plastic waste brand detection, we had to make choices to balance accuracy, speed, object size, number of proposals and input image resolution. Besides these many other choices impact the performance like data augmentation, feature extractors, training data, number of proposals, strides, etc [7].

YOLO (You Only Look Once) works on a single convolutional network on the whole input image to predict bounding boxes with confidence scores for each class simultaneously. This approach is simple and hence the YOLO model is extremely fast (compared to Faster R-CNN and SSD). Single Shot Detector (SSD) also has a lightning speed but less than YOLO as it takes only one shot of the image to detect multiple objects within an image. Methods such as YOLO or SSD work very fast, but this amounts to a decrease in accuracy. On the other hand, Faster R-CNN achieve higher accuracy but is a bit more

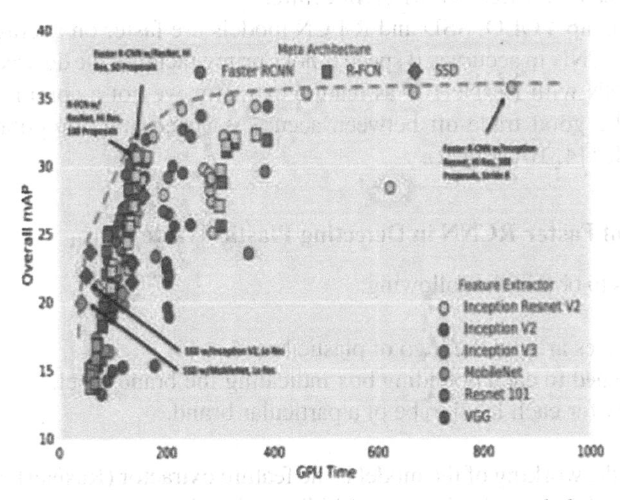


**Fig. 1.** Accuracy vs Time, with colours denoting feature extractor and shapes denoting meta-architecture. Each (feature extractor, meta-architecture) pair can correspond to more than one points due to changing strides, input sizes, etc.

expensive to run. R-FCN (Region based Fully Convolutional Network) uses position-sensitive score maps which helps to increase processing speed while maintaining good accuracy as Faster R-CNN. R-FCN and Faster R-CNN demonstrate a good advantage in terms of accuracy if real-time speed is not desired by the application. SSD works well for real-time processing (Fig. 1).

The next important factor is the selection of a good feature extractor. Inception v2 and Mobile net feature extractors are most accurate and are the fastest respectively. ResNet falls somewhere in between them. But if the number of proposals in ResNet were reduced then the accuracy improved.

Another important factor is how the model performs on different sizes of objects. Not surprisingly, all models give pretty good results on large objects. But SSD models with any architecture typically has a below par performance on small objects. Although it is on par with R-FCN and Faster RCNN to detect large objects. In our use case we want to label plastic brands that we find lying around in streets or public spaces. Hence our model needs to have a good accuracy in detecting small plastic wrappers as well.

It has been observed by other authors that input resolution can significantly impact object detection accuracy. Higher the resolution better is the detection performed on small objects. Resolution does not impact to a certain extent for large objects.

For R-FCN and Faster R-CNN, we can alter the number of proposals that is given by the region proposal network. Number of proposals is the number of regions in the feature map (output of the feature extractor CNN) identified by the region proposal network which can be an object. Observations were made such that when the number of proposals was set to 100, the speed and accuracy for Faster R-CNN with ResNet becomes comparable to that of R-FCN models (which uses 300 proposals).

Memory analysis in this experiment showed that more powerful feature extractor used up more memory. Faster RCNN with inception ResNet architecture used way more memory space than with ResNet101 architecture.

Hence to sum up YOLO, SSD and R-FCN models are faster on average but cannot beat the Faster R-CNN in accuracy if speed is not a prime factor in the desired application. And Faster RCNN with ResNet101 as feature extractor we got a commendable speed which provided a good trade-off between accuracy, speed and computational power which was needed [4, 10] (Fig. 2).

### 2.1 Working of Faster RCNN in Detecting Plastic Waste

Our final goal is to obtain the following:

1. Bounding boxes around the logo of plastic brand.
2. A label assigned to each bounding box indicating the brand label.
3. A probability for each label to be of a particular brand.

First step in the working of the model is the feature extractor (ResNet). Images given as input to the model are in the form of multidimensional arrays (also known as tensors) having Height × Width × Depth, which are passed through a pre-trained CNN. After going down a few layers an intermediate layer is reached called a convolutional feature map which will be used in the next part.

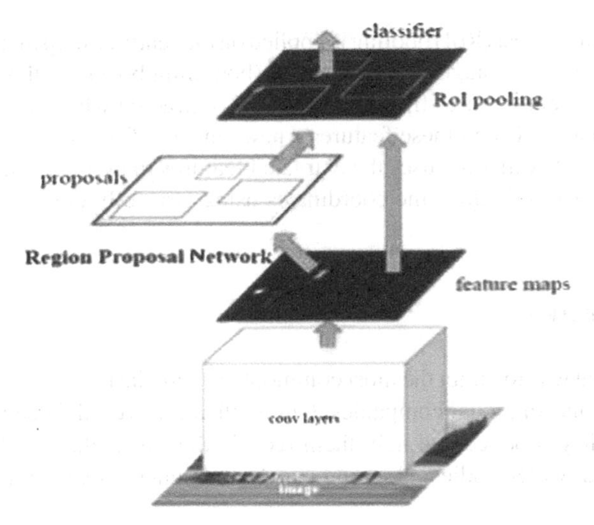


**Fig. 2.** Faster R-CNN architecture

Every convolutional layer in CNN creates some abstractions based on information from the previous layer. The first layers of CNN usually learn edges or borders, the second layer identifies patterns in edges and as the layer deepens it starts to earn more complex feature that humans cannot visualize. Hence a convolutional feature map is created which has information about the spatial dimensions which is much smaller compared to original image, but has greater depth. In our case ResNet is used as a feature extractor.

Next in the architecture is a Region Proposal Network (RPN). Using the convolutional feature map that feature extractor module computed in the earlier stage, RPN locates some regions in the original image (bounding boxes), which may contain some objects. Number of these regions are already defined. These boxes may or may not contain the whole object. One of the hardest issues with Deep Learning (DL) used for object detection is generation of a list of variable length bounding boxes that is when RPN proposes the existence of an object at a particular spot the bounding box is not accurate. The variable-length problem is solved in the RPN by using anchors. Anchors are a set of many bounding boxes that are placed throughout the image having varying sizes and are used as reference while predicting location of objects.

The RPN takes in all the anchors (reference boxes) and gives a set of good proposals (where the object is located). The RPN computes two outputs which is used in the next step. First is the probability that the object within the anchor is some relevant object. The RPN does not take into account what class the object belongs to. The only focus is that it does look like an object. The RPN thus gives two scores to every anchor, the score of it(anchor) being background (not an object) and the score of it being foreground (an actual object). The second output is the change in bounding box for adjusting the anchors to better locate the object in the image [11].

The final score of each anchor is used to compute object coordinates and thus a good set of proposals for objects in the image is obtained.

Then region of interest (RoI) pooling is applied on the features map that was generated by the CNN in the earlier stage and the anchors (bounding boxes) with relevant objects by RPN. Then relevant features from feature map is extracted which would correspond to the relevant objects. Using these features a new tensor is formed.

At last, the R-CNN module uses the extracted features to assign a class to the object in the bounding box and adjust the coordinates using two fully connected layers as in CNN.

## 3 Data Collection

Here the main goal was to target the most commonly found plastic waste on streets which are produced by multinational companies. Our application would force them to reduce the plastic that they produce and help them recycle the same plastic which otherwise ends up in streets, water bodies, clogging canals and causing every form of pollution there is.

Some plastics are burnt which is hazardous to the atmosphere. Many online resources and surveys were also considered. Lays, Kurkure, Sprite, Vimal and Pepsi were the brands most commonly found on Indian streets and public places [3]. Frito-Lay, Inc is a subsidiary of PepsiCo (Pepsi manufacturers) that manufactures and sells snack foods known as Lays. This company also manufactures Kurkure. Sprite is manufactured by The Coca Cola company. All these companies have a huge market share in India. These are also some of the most plastic pollution companies in the world as well as in India.

We web scraped images of all the above brands by using a simple python script. These images needed to be filtered out and we ended up with approximately 100 images per brand. But the problem with these images were that it was a full resolution clear image most of which have a white background. For building a robust model the training data should have crumbled plastic waste, torn or dis-coloured logos, and distorted images with different background where people may find waste plastic.

To create a good dataset the only source for collecting data was to click images of thrash, dustbins, plastics lying around in streets, near railway stations, dumping ground and other public places. More than 200 images of each brand were collected. The application was designed keeping in mind citizens as end users. Hence, the training data almost covered all the possible places from where citizens could click images. A total of more than 1800 images were collected.

## 4 Data Augmentation

Data augmentation is a powerful technique that allows to significantly increase the diversity and quantity of data available for training as well as testing without actually having to collecting any new data. Data augmentation were performed on the data that we manually collected from various places. Following techniques were performed using some simple python script.

**Flip**
Flipping images horizontally and vertically.

**Rotation**
It involves simple rotation of the image by 90°.

**Cropping**
Some portion of the image is cropped and then resized to the size of original image also known as random cropping.

**Gaussian Noise**
Over-fitting of a deep learning model is most likely to happen when neural network tries to learn patterns that occur more frequently. A method to solve this is the salt and pepper noise, which is random black and white pixels spread through the image. Adding these grains to images will decrease the chances of over- fitting. These data augmentation techniques increase size of the relevant dataset 4–5 times the original number of dataset.

**Scale**
The training images are scaled outward (Zoom in) or inward (Zoom out) to get a diversified data. For scaling outwards most image frameworks crop out some part from the new image but end image is of size equal to the original image.

## 5 Implementation

### 5.1 Object Detection with Customized Model

TensorFlow Object Detection API can be used with a variety pre-trained model. In this work, a Faster Rcnn model with Resnet101 (faster_rcnn_resnet101_coco) was chosen. The model had been trained using MSCOCO Dataset which consists of 2.5 million labelled object instances in 328,000 images, containing 91 object types.

All the dataset was labelled using LabelImg tool. LabelImg is a simple tool to annotate objects and xml file is created for each image comprising of annotations coordinates. Training of the model was performed with 85% of the total dataset. Another 10% was used for testing and 5% data was used for validating the results.

The pre-trained Faster Rcnn model (Faster Rcnn resnet101) was fine-tuned for our dataset using manually labelled images. A provided configuration file (faster-rcnn_resnet101.config) was used as a basis for the model configuration (after testing different configuration settings, the default values for configuration parameters were used). The provided checkpoint file for faster_rcnn_resnet101_coco was used as a starting point for the finetuning process. Training the model in took about 48 h with Nvidia 940Mx GPU. The total loss value was reduced rapidly as training was started from the pre-trained checkpoint file [6, 8].

## 5.2 User Interface

**Image input**

A desktop application written in python was developed where there are options to upload an image to detect plastic item brands in those images. These images then get stored in an archive folder from where results can be viewed. One can send images which will be loaded into a folder where it is passed on to the neural network to perform plastic waste detection. Location (of the device) from where the image was taken is also recorded along with date and time. The results of object detection will be same as input image but with a bounding box surrounding the logo of the plastic waste item found in that image. A confidence percentage adjoining the box will also be visible which will be greater than 60 as the threshold is set at 60%. This output image will be stored in the archive folder. But the output that gets stored in the database is in the form of number of instances (count) of a plastic waste identified in the image along with date, time and location.

**Video Input**

Input can also be given in video format. A separate option is listed in the menu for uploading a video. The video format is converted into images via a simple python script. To avoid duplication and speed up the processing one frame is captured among ten consecutive frames from the video. For example, consider that the video file has 20 frames per second, only 2 frames will be considered and these two frames (images) will be sent as an image input to the neural network. Output of each frame will be a frame with a bounding box surrounding the plastic brand logo. Count of each brand in each frame will be recorded. Unless there is an increase in the count of plastic waste for a certain brand in consecutive frames the resultant count of that plastic brand will not be updated. If a decrease in count for a certain brand is found in subsequent frame then the previous greater value will still be the final count of that brand and that count will be updated in the database. Duplicate entries will be avoided in this manner. The output frames are then collated to generate a video. This video gets stored in another archive folder.

**Plastic Waste Analysis and Output Visualization**

As soon as the output is generated whether as an image or video the count of plastic waste from that image or video is updated in the database by using SQL queries. This database has following columns- image name, date, time, location and count of each plastic brand in that image. For easier visualization the count of each plastic brand recorded in database over a certain period of days and location can be viewed as a line graph with an option to select a particular location and date to get a more precise visualization. A cumulative count for a particular location can be visualized by choosing a bar graph.

For further analysis the database for a specific location and days can be exported in csv format from the application directly to email inbox (Figs. 3, 4 and 5).

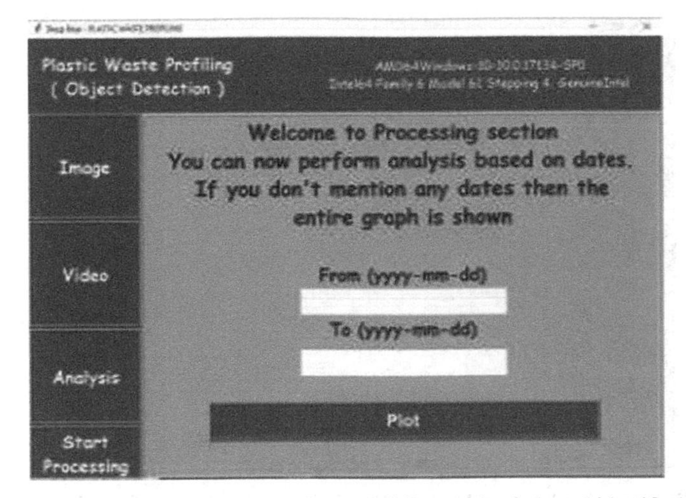


**Fig. 3.** Interface to visualize the stored database of different plastic brand identified in between a span of the dates entered by the analyst.

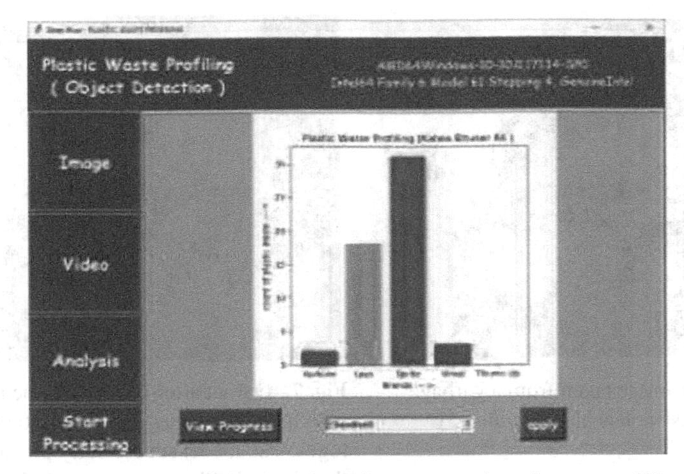


**Fig. 4.** Visualization using bar graph where X - axis represents brand names and Y- axis represents count of plastic waste generated at a particular location.

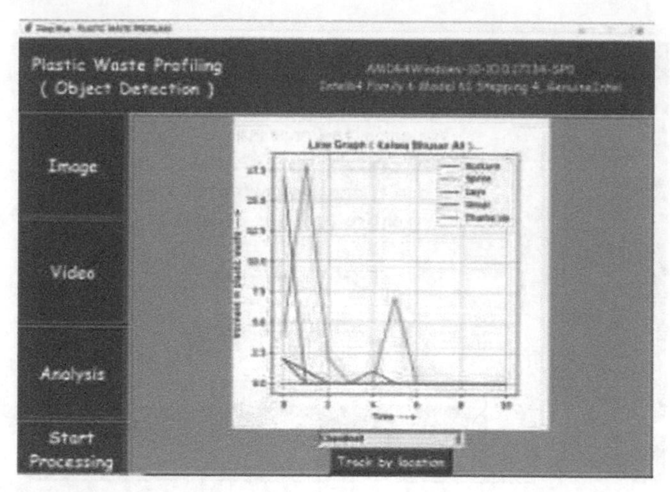


Fig. 5. Visualization using line graph where X-axis represents a span of number of days entered by the user and Y- axis represents count of plastic waste obtained in that span. Colour of the line indicates the different brands of plastic waste.

Fig. 6. Input image taken from a garbage dump that has various plastic items [5]

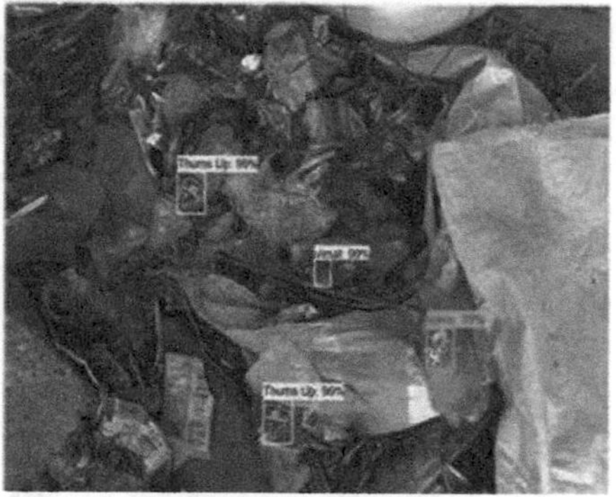

Fig. 7. Output after processing the image with bounding boxes around plastic brand logos. The model successfully predicted two thumbs up bottles, a wrapper of vimal and one sprite bottle.

## 6 Results

The accuracy of the model on validation data provided a good accuracy. Around 150 images in the validation dataset were tested and the results were fairly accurate. The model performed accurate detection in almost all cases. Some images where the plastic was torn in pieces or where the logos were covered with dust are the images where the model failed. Very few images were present in the training data. Otherwise the model did a good prediction on test images taken in garbage dumps, streets and gardens. The model provided accurate results 90% of times from the test dataset. There is a scope for improvement in accuracy by adding more training images by crowd sourcing the data. More brands of plastic can be added to make this application practically viable (Figs. 6 and 7).

## 7 Conclusion

The goal of this project was to create a profile of companies that causes plastic waste pollution. The audience are the citizens who send the images of plastic waste lying around in public spaces. Creating a profile of such waste will help to curb multinational companies who are causing pollution. Government authorities can take action for those companies that not abiding the law.

This deep learning model can be trained to detect many brands of plastic, which in turn increases the scope of applications. The solution was implemented as a proof of concept and we can further expand this solution to a larger scale where we can significantly reduce plastic pollution in our country and throughout the world.

**Compliance with Ethical Standards**

✓ All authors declare that there is no conflict of interest.
✓ No humans/animals involved in this research work.
✓ We have used our own data.

## References


1. https://india.mongabay.com/2018/04/india-to-galvanise-greater-action-against-plastic-waste-on-world-environment-day/
2. https://cn.wikipedia.org/wiki/Extended_producer_responsibility
3. https://www.downtoearth.org.in/news/africa/coca-cola-pepsico-nestl-worst-plastic-polluters-in-global-cleanups-brand-audits-61834
4. Huang, J., Rathod, V., Sun, C., Zhu, M., Korattikara, A., Fathi, A., Fischer, I., Wojna, Z., Song, Y., Guadarrama, S.: Speed/accuracy trade-offs for modern convolutional object detectors. In: 2017 IEEE Conference on Computer Vision and Pattern Recognition (CVPR) (2017)
5. Ren, S., He, K., Girshick, R., Sun, J.: Faster R-CNN: towards real-time object detection with region proposal networks. IEEE Trans. Pattern Anal. Mach. Intell. **39**(6), 2–5 (2017)
6. https://medium.com/@14prakash/understanding-and-implementing-architectures-of-resnet-and-resnext-for-state-of-the-art-image-cf51669e1624
7. https://github.com/tensorflow/models/blob/master/research/object_detection/g3doc/detection_model_zoo.md
8. https://medium.com/@sidereal/cnns-architectures-lenet-alexnet-vgg-googlenet-resnet-and-more-666091488df5
9. Ren, S., He, K., Girshick, R., Zhang, X., Sun, J.: Object detection networks on convolutional feature maps. IEEE Trans. Pattern Anal. Mach. Intell. **39**(7), 1476–1481 (2017)
10. https://mc.ai/object-detection-speed-and-accuracy-comparison-faster-r-cnn-r-fcn-ssd-and-yolo/
11. Tang, C., Feng, Y., Yang, X., Zheng, C., Zhou, Y.: The object detection based on deep learning. In: 2017 4th International Conference on Information Science and Control Engineering (ICISCE), Changsha, pp. 723–728 (2017)

# Leaf Disease Detection Using Image Processing and Artificial Intelligence – A Survey

H. Parikshith(✉), S. M. Naga Rajath, and S. P. Pavan Kumar

Department of Computer Science and Engineering, Vidyavardhaka College of Engineering, Mysuru, Karnataka, India
parikshithh@gmail.com, nagarajath98@gmail.com, pavanalina@vvce.ac.in

**Abstract.** In today's world, crop diseases are one of the main threats to crop production and also to food safety. Disease detection using traditional methods that are not so accurate. Current phenotyping methods for plant disease are predominantly visual and are therefore slow and sensitive to human error and variation. Accuracy can be achieved using technologies such as artificial intelligence, IoT, algorithm based on rules, machine learning regression techniques, image processing, transfer learning, hyper-spectral imagery, leaf extraction and segmentation.

**Keywords:** Leaf disease · Artificial intelligence · Image processing · Crop

## 1 Introduction

Crop diseases remain a major threat to the world's food supply. Due to advances in technology, new methods have been developed for the early detection and diagnosis of plant diseases. These methods are mainly based on deep learning techniques or models that include convolutionary neural networks (CNN), transfer learning, support vector machines (SVM) and genetic algorithms for hyper-spectral band selection.

A Convolutionary Neural Network (CNN or ConvNet) is a special type of multi - layer neural network that recognizes visual patterns directly from pixel images with minimal preprocessing. The CNNs are inspired by the natural biological process that is similar to the visual cortex of animals. AlexNet and GoogleNet are the basis of the CNN model. AlexNet is the name given to Alex Krizhevsky's CNN model, containing a total of eight layers; the first five layers are convolutionary layers and the last three are fully connected layers. AlexNet uses the function of non- saturated ReLu activation.

Transfer learning is a popular machine learning method that uses pre - trained models as a starting point for computer vision and the processing of natural languages. Transfer learning and imaging techniques are used in the detection of diseases. Image recognition with transfer learning prevents the complex process of extracting features from images to train models.

Another approach used to address the problem of disease detection in plants is the use of genetic algorithms as an optimizer and as a classifier of vector machines. These methods include the acquisition of hyper-spectral images for classification and


S. Smys et al. (Eds.): ICCVBIC 2019, AISC 1108, pp. 304–311, 2020.
https://doi.org/10.1007/978-3-030-37218-7_35

the use of the charcoal rot protocol where disease progression was manually evaluated by measuring the length of soybean external, internal and dead tissue lesions. It's a difficult problem to segment leaves from natural images. It's a difficult problem to segment leaves from natural images. To overcome this problem, a unique method, including leaf extraction, segmentation, leaf structure refining and morphological processing, was developed and developed. The refining of the leaf structure consists of primary and secondary vein detection.

## 2 Literature Review

Sarangdhar et al. [1] have proposed a system that uses vector- based regression to identify cotton leaf diseases and that also uses Gabor filter, median filter and an android application with Raspberry pi integration to display and control the disease.

Morco et al. [2] have developed a mobile application called e- Rice that provides farmers with information to identify rice problems using algorithms based on rules to detect and prescribe possible control methods.

Dutta et al. [3] proposed an approach to detect salad leaf disease using Hyper Spectral sensing based on machine learning to acquire different vegetation spectrum profiles and have used portable high - resolution ASD spectroradiometers. And used PCA, LDA classifiers to classify salad leaves affected by disease.

Deep learning and robotics have been used by Durmus et al. [4] to detect leaf diseases in tomato plants and the deep learning network architectures used by AlexNet and SqueezeNet. The diseases they examined had physical changes in the leaves and RGB cameras showed these changes.

Ashourloo et al. [5] submitted a paper on the detection of Leaf Rust disease using Machine Learning regression techniques and investigated PLSR, v- SVR and GPR regression methods for wheat leaf rust disease detection.

Mohanty et al. [6] used deep learning to detect plant disease. They used the neural network of convolution to identify crop species of 14 types and diseases of 26 types. They aim to develop smartphone-assisted disease diagnostics in order to make it better for farmers.

Ramcharan et al. [7] have submitted a paper to detect cassava disease caused by viruses using transfer learning that is efficient, fast, affordable and easy to deploy on mobile devices.

Nagasubramanian et al. [8] submitted a paper on the identification of carbohydrate disease in soybean by selecting hyper-spectral band. They showed that selected bands gave more information than RGB images and that this can be used in multispectral cameras to identify the disease remotely.

Anantrasirichai et al. [9] proposed a system to operate the venous leaves system in conjunction with other low- level features. They have developed primary secondary vein detection and removal of the protrusion clamp to refine the extracted leaf.

Ramcharan et al. [10] presented a paper on the mobile deep learning model for the detection of plant diseases. Each disease category was tested with two serious symptoms, mild and pronounced.

**Table 1.** Comparison of different leaf disease detection using image processing and artificial intelligence

| Ref. No | Approach | Methods | Datasets | Result | Advantages | Recommendation |
|---|---|---|---|---|---|---|
| #1 | Regression system based on the support vector machine to identify and classify five cotton- leaf diseases | Machine based regression system, Support Vector Machine, Gabor Filter, Median Filter, Raspberry pi Android app | – | The android app is designed to display disease and sensor information together with the relay's ON/OFF. The app also handles the movement of the entire system from one location to another so that the farmer can move the system to check the soil and control the relay condition | The current system gives 83.26% accuracy for disease detection | – |
| #2 | Application to help farmers identify problems in rice plants and provide practical advice on detecting and diagnosing diseases in rice plants and prescribing possible control options | Rule-Based algorithm as a classification method to generate rules | | The developed mobile application system provides information on how to determine the possible causes of diseases such as bacteria, fungi or viruses and information on rice plants with symptom-based malnutrition. And information about rice plants that suffer from symptoms-based malnutrition | Application reacts well with its functionality | Including rice plant varieties in the country and including insects and nematodes as factors affecting rice plants to broaden the scope of diagnosis of diseases and to calculate the cost of controlling rice crop problems may be included to help farmers decide which type of fertilizer or chemicals to *use* |

(*continued*)

**Table 1.** *(continued)*

| Ref. No | Approach | Methods | Datasets | Result | Advantages | Recommendation |
|---|---|---|---|---|---|---|
| #3 | The machine- based approach together with a high- resolution spectroradiometer | Machine learning algorithms and Hyper Spectral sensing, portable high resolution ASD FieldSpec Spectroradiometer<br>[1] Principal Component Analysis (PCA), Multi-Statistics Feature ranking and Linear Discriminant Analysis (IDA)<br>[2] Evaluating the impact of training sample size on the results<br>[3] companions between the performances of SVIs and machine learning techniques | – | First, an early evaluation of the effectiveness of the use of a hyper-spectroradiometer in the detection of salad leaf disease Secondly, the spectral data collected from field experiments could be used to formulate a problem of multi-classification based on the linear separation of classes | Machine learning based classification paradigm based on PCA Multi-Statistics and LDA data is a highly effective methodology for the rapid detection of salad leaf diseases | To capture a variety of salad leaves and related diseased leaves to test the long- term effectiveness of this newly developed detection technology in the context of real- life farming |
| #4 | Using deep learning and robotics for the detection of tomato leaves | Deep learning, precision fanning, deep learning network architectures were AlexNet and SqueezeNet | PlantVillage dataset | Trained models are tested on the GPU" validation set. AlexNet did a little bit better than SqueezeNet | SqueezeNet is a good mobile deep learning candidate because of its lightweight and low computing requirements | Leaf extraction study for the completion of the system from the complex background |

*(continued)*

**Table 1.** *(continued)*

| Ref. No | Approach | Methods | Datasets | Result | Advantages | Recommendation |
|---|---|---|---|---|---|---|
| #5 | Machine learning techniques have been used to estimate vegetation parameters and detect diseases | Machine Learning, partial least square regression (PLSR), v support vector regression(v-SVR), and Gaussian process regression (GPR) methods for wheat leaf rust disease detection, spectroradio meter in the electromagneticregion of 350 to 2500 nm | Spectral dateset, ground truth dataset | The experiment was conducted under controlled conditions to study the various effects of symptoms on the leaf reflection | This study shows that the training data set of GPR results in greater precision than other implemented methods | In order to be used in the field, PLSR, _-, GPR must be tested on different sensors and different types of wheat |
| #6 | Applying deep CNN AlexNet and GoogleNet classification problems | [1] Dataset description: Analysis of dataset [2] Measurement of performance: Run experiments across training dataset | ImageNet dataset, PlantVillage dataset | Former AlexNet and GoogleNet were better equipped to train publicly available data set images | – | A more diverse set of training data is needed to improve accuracy. Since many diseases do not occur on the upper side of the leaves, images from many different perspectives should be obtained |
| #7 | Application of deep CNN transmission learning for cassava image data set using the latest version of the initial model based on GoogleNet | Cassava leaf images were captured and screened for co- infections to limit the number of images with multiple diseases | Cassava dataset, Leaflet cassava dataset | The overall accuracy of the original casavva data set ranged from 73% to 91%. The leaf let cassava data set the accuracy ranged from 80 to 93% | – | The aim is to develop mobile applications that allow Technicians to monitor:the prevalence of disease quickly |

*(continued)*

**Table 1.** (*continued*)

| Ref. No | Approach | Methods | Datasets | Result | Advantages | Recommendation |
|---|---|---|---|---|---|---|
| #8 | Used genetic algorithm as an optimizer and classifier for support vector machines | Used hyperspectral image acquisition where hyperspectral line scanning imager was used for imaging and also used charcoal rot rating protocol | – | A binary classification of healthy and infected samples achieved 97% accuracy and a F1 score of 0.97 for the infected class | Field inoculations of various soybean genotypes are imaged using a multispectral camera with the selected wavebands from the GASVM model for the early identification of carbohydrate disease to understand the resistance of specific genotypes to the disease. This study provides an efficient method for selecting the most valuable wavebands from hyperspectral data for early detection of disease | Use field inoculations and assessments to increase this technology |
| #9 | The leaf isolation closest to the camera as this is not obscured by other leaves and lesioma segmentation is further applied to the leaf segment | [1] Leaf extraction<br>[2] Leaf structure refinement<br>[3] Segmentation<br>[4] Morphological processing | – | The proposed method is tested with 100 random leaf images with an average accuracy and a reminder of 0.92 and 0.90 | – | Develop an automatic lesion segmentation and classification of the leaf disease |
| #10 | Use the model on a mobile app and test its performance on mobile images and videos of 720 leaflets with disease | Convolutional neural network and SSD(Single Shot Detection) algorithm as object detection algorithm | COCO [Common Objects in context) dataset.<br>Cassava leaf dataset | The CNN(Convolutional Neural Network) model achieves accuracy of 80.6 ± 4.10% for symptomatic leaves and 43.2 ± 20.4% for mild symptomatic leaves | | |

## 3 Comparison of Different Leaf Disease Detection Using Image Processing and Artificial Intelligence

The Table 1, gives us a picturesque idea about the methodologies used by various authors in the field of leaf disease detection using image processing and artificial intelligence. It also gives the list of recommendations which we thought, could have been implemented in the system in future.

## 4 Conclusion

Food losses from pathogens caused by crop infections are persistent problems in agriculture throughout the world for centuries. In order to prevent and detect these plant diseases early, this paper describes disease detection methods, such as training a CNN model based on AlexNet on the plant image data set. Other methods include vector regression support technique, transfer learning, a genetic algorithm consisting of hyperspectral band selection and morphological leaf processing used to train the network. The present system, which includes the above techniques used for the detection of diseases in plants, proves its efficiency in the detection and control of leaf disease by improving crop production for farmers.

**Acknowledgements.** The authors express gratitude towards the assistance provided by Accendere Knowledge Management Services Pvt. Ltd. In preparing the manuscripts. We also thank our mentors and faculty members who guided us throughout the research and helped us in achieving desired results.

**Compliance with Ethical Standards**

✓ All authors declare that there is no conflict of interest.
✓ No humans/animals involved in this research work.
✓ We have used our own data.

## References

1. Sarangdhar, A.A., Pawar, P.V.R., Blight, A.B.: Machine learning regression technique for using IoT. In: International Conference on Electronics, Communication and Aerospace Technology, pp. 449–454 (2017)
2. Morco, R.C., Bonilla, J.A., Corpuz, M.J.S., Angeles, J.M.: e-RICE: an expert system using rule-based algorithm to detect, diagnose, and prescribe control options for rice plant diseases in the Philippines. In: CSAI 2017, pp. 49–54 (2017)
3. Dutta, R., Smith, D., Shu, Y., Liu, Q., Doust, P., Heidrich, S.: Salad leaf disease detection using machine learning based hyper spectral sensing. In: IEEE SENSORS 2014 Proceedings, pp. 511–514 (2014)
4. Durmu, H., Olcay, E., Mürvet, K.Õ.Õ.: Disease detection on the leaves of the tomato plants by using deep learning. In: 6th International Conference on Agro-Geoinformatics (2017)
5. Ashourloo, D., Aghighi, H., Matkan, A.A., Mobasheri, M.R., Rad, A.M.: An investigation into machine learning regression techniques for the leaf rust disease detection using hyperspectral measurement. IEEE J. Sel. Top. Appl. Earth Obs. Remote Sens. **9**, 4344–4351 (2016)

6. Mohanty, S.P., Hughes, D.P., Salathé, M.: Using deep learning for image-based plant disease detection. Front. Plant Sci. **7**, 1419 (2016)
7. Ramcharan, A., Baranowski, K., Mccloskey, P., Legg, J., Hughes, D., Hughes, D.: Using transfer learning for ımage-based cassava disease detection. Frontiers (Boulder) 1–10 (2017)
8. Nagasubramanian, K., Jones, S., Sarkar, S., Singh, A.K.: Hyperspectral band selection using genetic algorithm and support vector machines for early identification of charcoal rot disease in soybean. Arxiv. 1–20 (2017)
9. Anantrasirichai, N., Hannuna, S., Canagarajah, N.: Automatic leaf extraction from outdoor images. Arxiv. 1–13 (2017)
10. Ramcharan, A., Mccloskey, P., Baranowski, K., Mbilinyi, N., Mrisho, L.: Assessing a mobile-based deep learning model for plant disease surveillance. Arxiv (2018)

# Deep Learning Model for Image Classification

Sudarshana Tamuly(✉), C. Jyotsna, and J. Amudha

Department of Computer Science and Engineering, Amrita School of Engineering, Bengaluru, Amrita Vishwa Vidyapeetham, Bengaluru, India
sudarshanatamuly@gmail.com,
{c_jyotsna, j_amudha}@blr.amrita.edu

**Abstract.** Starting from images captured on mobile phones, advertisements popping up on our internet browser to e-retail sites displaying flashy dresses for sale, every day we are dealing with a large quantity of visual information and sometimes, finding a particular image or visual content might become a very tedious task to deal with. By classifying the images into different logical categories, our quest to find an appropriate image becomes much easier. Image classification is generally done with the help of computer vision, eye tracking and ways as such. What we intend to implement in classifying images is the use of deep learning for classifying images into pleasant and unpleasant categories. We proposed the use of deep learning in image classification because deep learning can give us a deeper understanding as to how a subject reacts to a certain visual stimuli when exposed to it.

**Keywords:** Image classification · Computer vision · Eye tracking · Heat map · Deep learning

## 1 Introduction

Image classification refers to the process of categorizing images logically and methodically into sub groups. This grouping is done depending upon the traits exhibited by the images, with the images sharing common traits being stacked into one subgroup. This indexing of images into logical groups is very helpful because by doing this, maintaining of databases which deals with images becomes much easier. Companies such as giant e-commerce sites, websites delivering wallpapers and likewise benefit greatly from image classification. Till now, computer vision and eye tracking were the notable methods used for the classification of images. However, we propose the use of deep learning for the classification of images. Deep learning employs different machine learning algorithms to get a clearer picture of how a subject is reacting to a given visual stimuli. It helps us get a deeper understanding of the various subtle interactions going on between the subject and the visual stimuli displayed which is something we can't accurately monitor had we just used computer vision or eye tracking. Extraction of multiple features from each layer of the neural network of an image is possible with the help of deep learning algorithms. One popular deep learning algorithm which goes by the name of Convolutional Neural Networks (CNN) is extensively used because it provides high image classification accuracy. Following a


S. Smys et al. (Eds.): ICCVBIC 2019, AISC 1108, pp. 312–320, 2020.
https://doi.org/10.1007/978-3-030-37218-7_36

hierarchical model, CNN is laid out like neurons which are built on a network resulting in the formation of a layer which is fully connected. The main objective of this paper was to develop a deep learning model for classifying images using VGG-16 architecture. The system overview of this architecture is depicted in Sect. 3 including the stimuli used and the methodology adopted in Sects. 3.1 and 3.2 respectively. Section 4 contained results from the tests which were conducted using the VGG-16 architecture. Three image datasets were considered for these tests, namely: IAPS, NUSEF and MIT-300 and the results recorded were compared with those obtained by using only computer vision and eye tracking for image classification.

With the help of deep learning and tools associated with it, this paper tries to develop a method to automatically categorize pictures as pleasant, non-pleasant and neutral. For doing this, the first challenge will be to categorize which eye feature gets triggered/activated upon exposure to different scenes or images. Different images will result in the eye having a different response to it. Tools such as TensorFlow, Keras and the VGG-16 architecture were used for analyzing and classifying the test images according to the three categories which are pleasant, unpleasant and neutral.

## 2 Related Works

Image classification can be done by the application of deep learning algorithm such as Convolutional Neural Network (CNN) [1–3]. CNN was tested on UC Merced Land Use Dataset and SUN database [1], and in turn each dataset was tested for different number of iterations. In total 10 experiments were carried out: face vs. Non-face image classification, dense residential vs. Forest, dense residential vs. Agricultural, constructed regions vs. Green regions, agriculture vs. Forest, building vs. Dense residential, classification from large aerial images, garden scene vs. Street scene, road scene classification and war scene classification. Classification accuracies were found to be very high in the experiments where the two classes were very distinct. In experiments such as building vs. dense residential, agricultural vs. forest, the classification accuracies were found to be low because the set of images were similar. So, fairly good classification accuracy was achieved using the CNN. A miniature CNN was built for functioning well on a CPU [2]. A subset of Kaggle Dog- Cat Dataset was used to train the network. Here an image classifier which identifies and separates the images of dog and cat was build based on CNN. For training and testing, 2000 and 600 images were used respectively. The whole dataset was trained for 50 iterations. Tensorflow was used to classify the input images as dog or cat. The classification accuracy was found to be 75%. So comparatively Convolutional Neural Network worked well on CPU. MNIST and CIFAR-10 datasets were trained using CNN for image classification [3]. Different optimization algorithm and methods of learning rate were analyzed. Errors can be reduced to a greater extent with the help of Stochastic Gradient Descent (SGD) by increasing the number of iterations. The results after applying CNN on the MNIST dataset was not that high compared to existing systems, but it imposes less computational cost and structure is very simple.

Convolutional neural network (CNN) have proven to be particularly useful in tasks involving the use of computer vision like detection, classification and segmentation of objects and likewise. Extensive use of CNN is seen amongst medical researchers now-a-days to diagnose diseases such as Alzheimer's. For the classification of the brain MRI scans, a transfer learning based mathematical model PFSECTLis used [4]. ImageNet database trained VGG-16 architecture (which is used in the CNN structure) is used as the feature extractor.

To predict the amount of ripening of dates, CNN was used [5]. The architecture used was based on VGG-16 and the CNN model also had a classifier block which contained max-pooling, dropout, batch normalization, and dense layers. The image dataset with which the model was trained contained four categories of dates namely Rutab, Tamar, Khalal and defective dates. The CNN model achieved 96.98% classification accuracy overall. The conclusion drawn was that classification methods employing CNN performed much better when compared to the traditional methods of classification.

A system to detect vehicles just with aerial images and not with the intervention of a human operator was proposed [6]. Feature extraction capabilities of VGG-16 and Alexnet neural network models was compared with classical feature extraction techniques like Histogram of Oriented Gradients and Singular Value Decomposition. The problem was converted to a binary class by segregating all the vehicles as one class and the rest as non-vehicles. On the VEDAI dataset that was used, features extracted using neural network achieved a classification accuracy of 99%.

To improve the accuracy of classification when classifying art collections, fine art objects and items likewise, a two stage approach to image classification was proposed [7]. Six different CNNs namely ResNet-50, VGG-16, VGG-19, GoogleNet, AlexNet and Inceptionv3 were pre-trained and used as the first stage classifiers and for the second stage, a shallow neural network was used. Individual patches were the factor which influenced the categorization of images in the first stage while the second stage further categorized the images based on their artistic style. Here, instead of using images, the second stage uses probability vectors and is trained independently on the first stage making it very effective in classifying images as well as compensating for any mistakes done in the first stage while classifying images. Three datasets were used in the experiment performed and in all three, the proposed method showed impressive results in classifying images. Also, eye tracking technology was employed in various other fields such as in the development of applications, evaluating stress levels among others [15–18].

## 3 System Overview

A deep learning network in the form of the VGG-16 architecture was employed for the classification of images. This is because by using deep learning architectures, a higher degree of accuracy in the classification of images can be obtained and also it has been known to work better with large datasets. One of the wider known techniques employed for the classification of images is known as Convolutional Neural Networks (CNN). The overview of the system is represented in Fig. 1.

VGG-16 neural network which has convolution layers of size 3*3, max pooling layers of size 2*2 and fully connected layers at end which make up a total of 16 layers

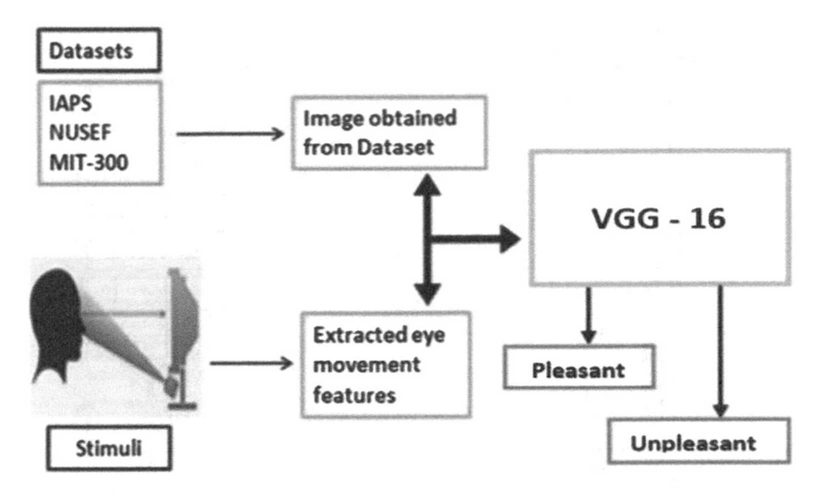


**Fig. 1.** System overview

was the deep learning architecture used here. RGB image of size 224*224*3 is the form of input here. A representation of the various layers of VGG-16 architecture is shown in Fig. 2.

To identify visual patterns directly from pixel images with very little preprocessing, Convolutional Neural Networks is used. It contains two layers (input and output) including several other layers which are concealed. These hidden layers comprise of pooling, convolutional and normalization layers, RELU layer i.e. activation function and fully connected layers. During the training period, an RGB image of fixed size ($224 \times 224$) is used as the input to the ConvNets. For preprocessing, subtraction of only the mean RGB value is done from each of the pixels from the computed training set. Filters with receptive field size of $3 \times 3$ are used in the stack of convolutional layers through which images are passed. To decrease the spatial size of the representations progressively is the work of the pooling layer. The reduction is done to bring down the computation and parameter amount in the network. Also, the operation of a pooling layer on each feature map is independent. Out of max pooling and average pooling, it is seen that max pooling is more preferred.

Three layers which are Fully-Connected (FC) follow a stack of convolutional layers and its arrangement is same in all networks. 4096 channels are present on both the first two channels, 1000 channels (one for each class) is present on the third since it performs 1000-way ILSVRC classification. The last layer is the soft-max layer. The two classes into which the images are classified into are as follows: pleasant and unpleasant. After the splitting of the images into training and testing, 1484 images considered as training samples and 637 images as test samples.

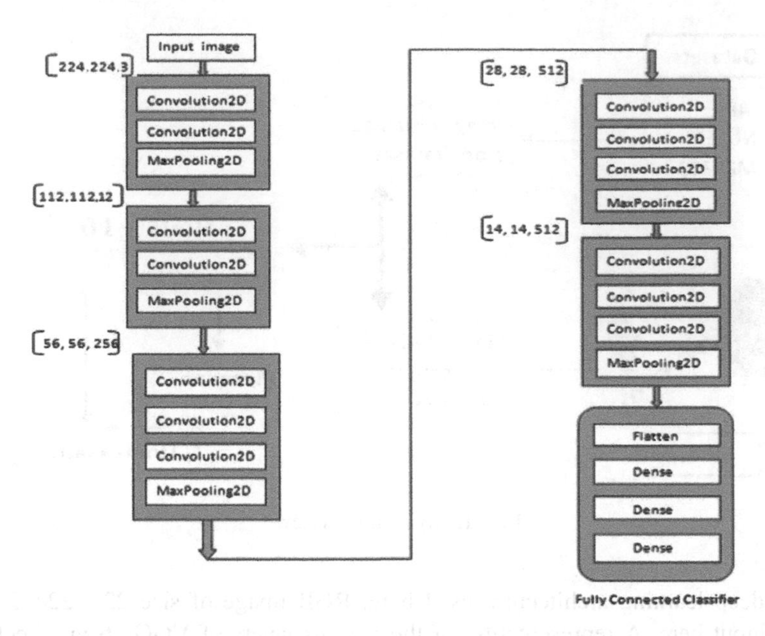


**Fig. 2.** Layout of VGG-16 architecture.

### 3.1 Stimuli

The stimuli included images from IAPS, MIT-300 and NUSEF datasets. IAPS which stands for "International Affective Picture System" is a picture database which is widely used in psychological research. The IAPS dataset has total 1037 images out of which 614 images were pleasant and 423 images were unpleasant. NUSEF dataset consists images of various orientations, scale, illumination among others other factors. It has 329 pleasant and 102 unpleasant images. MIT300 which contains 300 natural indoor and outdoor images is used as a benchmark test set as because it was the first data set with held-out human eye movements. So a total number of 2121 images were used as the stimuli containing 1484 pleasant and 637 unpleasant images.

### 3.2 Methodology

A deep learning network is used for the image categorization. For the classification of images, a deep learning network can be employed. This is because by deep learning architectures, a higher degree of accuracy in the classification of images can be obtained and also it has been known to work better with large datasets. One of the wider known techniques employed for the classification of images is known as Convolutional Neural Networks (CNN).To identify visual patterns directly from pixel images with very little preprocessing, Convolutional Neural Networks is used which is a kind of multi-layered neural network. It contains two layers (input and output) including several other layers which are concealed. These hidden layers comprise of pooling layers. (224 $\times$ 224) is used as the input to the ConvNets. For preprocessing, subtraction of only the mean RGB value is done from each of the pixels from the computed training set. Filters with receptive field size of 3 $\times$ 3 are used in the stack of convolutional layers through which images are passed. The two classes into which the images are classified into are as follows: pleasant and unpleasant. After the splitting of the images into training and testing, 1484 images considered as training samples and 637 images as test samples.

From the eye tracking technology, heat maps and focus maps were extracted. Similar eye features were considered for image classification using machine learning techniques [12]. The focus map as shown in Fig. 3(b) displays the areas which are given less visual attention by the user and thus gives us an idea about the regions which were not seen by the user. Heat maps as shown in Fig. 3(b) is another representation where evaluation is done based upon the frequency of focus of the user's gaze on different regions of an image. The regions which have been gazed more are known as "hot zones". For eye tracking studies, heat maps are one of the widely used visualization technique.

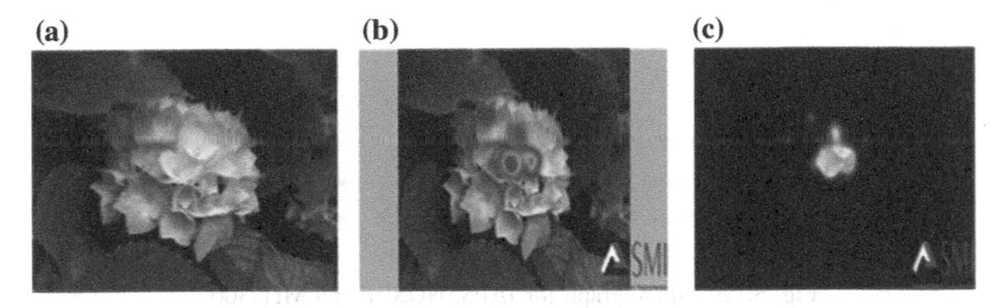


**Fig. 3.** (a) Pleasant image [4]. (b) Heat map, (c) Focus map

# 4 Result and Analysis

A deep learning architecture was applied on the IAPS dataset, the accuracy and the loss was calculated. The architecture used was a VGG-16 architecture. The training accuracy achieved after applying on the 1037 images was 61% and the testing accuracy was 54%. As seen in Fig. 4, the accuracy was calculated for both training and testing samples across 10 epochs.

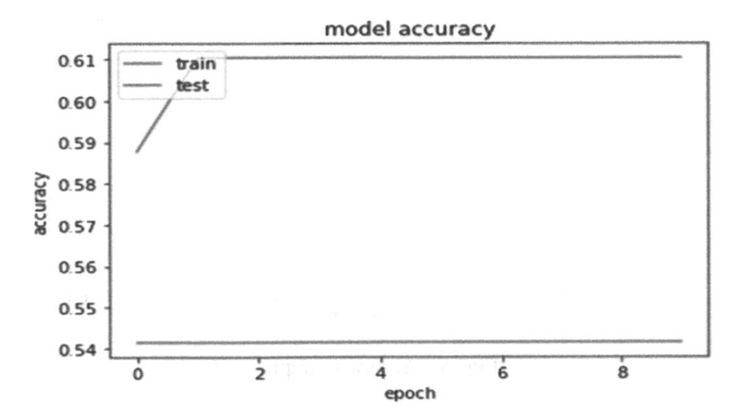


**Fig. 4.** Accuracy graph for IAPS dataset

Again the architecture was applied on the IAPS, MIT-300 and NUSEF datasets, the accuracy and the loss was calculated. The architecture used was a VGG-16 architecture. The training accuracy achieved after applying on the 2121 images was 71% and the testing accuracy was 65%. As seen in Fig. 5, the accuracy was calculated for both training and testing samples across 5 epochs. Training accuracy is the accuracy we get on training data and test accuracy is the accuracy of a model on images which are new and not seen earlier while training, while.

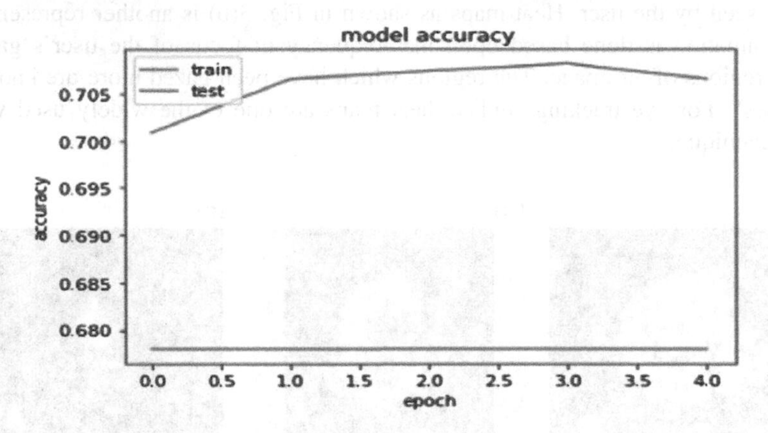


**Fig. 5.** Accuracy graph for IAPS, NUSEF and MIT-300

After the extraction of the focus maps and heat maps from the 20 images and for 11 participants; out of 220 images, 110 were pleasant and 110 were unpleasant. On these 220 images deep learning architecture was applied and the corresponding accuracy and loss graphs were achieved. The training accuracy was found to be 88% and testing accuracy was found to be 80% as shown in Fig. 6.

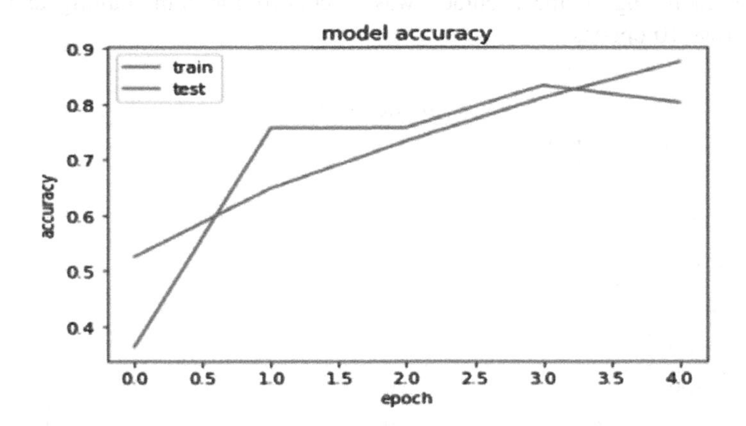


**Fig. 6.** Accuracy graph

The proposed model of VGG-16 architecture was applied on the IAPS dataset. And the training accuracy achieved after applying on IAPS dataset of 1037 images was 61% and the testing accuracy was 54%. Again it was applied on the IAPS, NUSEF and MIT-300 datasets and the achieved training accuracy was 71% and the testing accuracy was 65%. Lastly, the training accuracy achieved after applying on 20 images from IAPS dataset and its corresponding heat and focus maps of 220 images was 88% and the testing accuracy was 80%.

The future enhancement in this project can be determining the derived parameters and using those, the classification of the images can be done in an efficient way. Derived parameters from the eye features like saccade slope and saliency map can be considered to enhance the chances for obtaining a better classification.

## 5 Conclusion and Future Scope

It can be concluded from the various studies done till now that deep learning algorithms can be used for the better classification of images. A model was proposed with VGG-16 architecture and applied on the IAPS dataset and the training accuracy achieved after applying on IAPS dataset of 1037 images was 61% and the testing accuracy was 54%. Again it was applied on the IAPS, NUSEF and MIT-300 datasets and the achieved training accuracy was 71% and the testing accuracy was 65%. Lastly, the training accuracy achieved after applying on 20 images from IAPS dataset and its corresponding heat and focus maps of 220 images was 88% and the testing accuracy was 80%.

The future enhancement in this project can be determining the derived parameters and using those, the classification of the images can be done in an efficient way. Derived parameters from the eye features like saccade slope and saliency map can be considered to enhance the chances for obtaining a better classification.

**Compliance with Ethical Standards.**

✓ All authors declare that there is no conflict of interest.
✓ No humans/animals involved in this research work.
✓ We have used our own data.

## References

1. Jaswal, D., Vishvanathan, S., Kp, S.: Image classification using convolutional neural networks. Int. J. Adv. Res. Technol. **3**(6), 1661–1668 (2014)
2. Shanmukhi, M., Lakshmi Durga, K., Mounika, M., Keerthana, K.: Convolutional neural network for supervised image classification. Int. J. Pure Appl. Math. **119**(14), 77–83 (2018)
3. Guo, T., Dong, J., Li, H., Gao, Y.: Simple convolutional neural network on image classification. In: 2017 IEEE 2nd International Conference on Big Data Analysis (ICBDA), Beijing, pp. 721–724 (2017)
4. Jain, R., Jain, N., Aggarwal, A., Jude Hemanth, D.: Convolutional neural network based Alzheimer's disease classification from magnetic resonance brain images. Cogn. Syst. Res. **57**, 147–159 (2019)

5. Nasiri, A., Taheri-Garavand, A., Zhang, Y.-D.: Image-based deep learning automated sorting of date fruit. Postharvest Biol. Technol. **153**, 133–141 (2018)
6. Mohan, V.S., Sowmya, V., Soman, K.P.: Deep neural networks as feature extractors for classification of vehicles in aerial imagery. In: 2018 5th International Conference on Signal Processing and Integrated Networks (SPIN), Noida, pp. 105–110 (2018)
7. Sandoval, C., Pirogova, E., Lech, M.: Two-stage deep learning approach to the classification of fine-art paintings. IEEE Access **7**, 41770–41781 (2019)
8. Krizhevsky, A., Sutskever, I., Hinton, G.E.: ImageNet classification with deep convolutional neural networks. In: Proceedings of the NIPS, pp. 1097–1105 (2012)
9. Zhou, B., Lapedriza, A., Xiao, J., Torralba, A., Oliva, A.: Learning deep features for scene recognition using places database. In: Proceedings of the NIPS, Montreal, QC, Canada, pp. 487–495 (2014)
10. Rosenblum, M., Yacoob, Y., Davis, L.S.: Human expression recognition from motion using a radial basis function network architecture. IEEE Trans. Neural Netw. **7**, 1121–1138 (1996)
11. Xin, M., Wang, Y.: Research on image classification model based on deep convolution neural network. EURASIP J. Image Video Process. (2019)
12. Tamuly, S., Jyotsna, C., Amudha, J.: Tracking eye movements to predict the valence of a scene. In: 10th International Conference on Computing, Communication and Networking Technologies (ICCCNT) (2019, presented)
13. Cetinic, E., Lipic, T., Grgic, S.: Fine-tuning convolutional neural networks for fine art classification. Expert Syst. Appl. **114**, 107–118 (2018)
14. Chu, W.-T., Wu, Y.-L.: Image style classification based on learnt deep correlation features. IEEE Trans. Multimed. **20**(9), 2491–2502 (2018)
15. Venugopal, D., Amudha, J., Jyotsna, C.: Developing an application using eye tracker. In: IEEE International Conference on Recent Trends in Electronics, Information & Communication Technology (RTEICT). IEEE (2016)
16. Jyotsna, C., Amudha, J.: Eye gaze as an indicator for stress level analysis in students. In: International Conference on Advances in Computing, Communications and Informatics (ICACCI), pp. 1588–1593. IEEE (2018)
17. Pavani, M.L., Prakash, A.B., Koushik, M.S., Amudha, J., Jyotsna, C.: Navigation through eye-tracking for human–computer interface. In: Information and Communication Technology for Intelligent Systems, pp. 575–586. Springer, Singapore (2019)
18. Uday, S., Jyotsna, C., Amudha, J.: Detection of stress using wearable sensors in IoT platform. In: Second International Conference on Inventive Communication and Computational Technologies (ICICCT). IEEE (2018)

# Optimized Machine Learning Approach for the Prediction of Diabetes-Mellitus

Manoj Challa[1(✉)] and R. Chinnaiyan[2]

[1] Department of Computer Science and Engineering, CMR Institute of Technology, Bengaluru, India
manoj.c@cmrit.ac.in

[2] Department of Information Science and Engineering, CMR Institute of Technology, Bengaluru, India
vijayachinns@gmail.com

**Abstract.** Diabetes is one of the most common disorders in this modern society. In general, Diabetes-mellitus refers to the metabolic disorder by means of malfunction in insulin secretion and action. The proposed optimized machine learning models both decision tree and random forest models presented in this paper must predict the diabetes mellitus based the factors like BP, BMI and GL. The results build from the data sets are more precise, crisp and can be applied for health care sectors. This proposed model is more suitable for optimized decision making in health care environment.

**Keywords:** Diabetes · Metabolic disorder · Insulin · Malfunction · Machine learning · Health care

## 1 Introduction

In recent times, Machine Learning methods are often used in diabetes prediction. Decision tree is one of the machine learning methods in health care sector, which has the highest accuracy of classification. Random forest generates many decision trees. In this research paper optimized decision tree and random forest (RF) machine learning methods are adopted for predicting diabetes mellitus.

## 2 Relate Works

In this section we will be carrying out a systematic literature review of existing literatures related to diabetes prediction. According Kumari and Singh [4] clearly explained the hi-tech approaches for prediction of diabetes Mellitus. They emphasized the role of data mining and how effectively it helps to predict that disease. The role of software like MATLAB and neural networks used to build the model to detect the whether the person is suffering from diabetics or not.

As per Sankaranarayanan and Pramananda Perumal [6] mentioned that the patient primary datasets can be processed for predicting the diabetes. They also referred that rule classification and decision tree being used as primary method for their prediction model.


S. Smys et al. (Eds.): ICCVBIC 2019, AISC 1108, pp. 321–328, 2020.
https://doi.org/10.1007/978-3-030-37218-7_37

# 3 Architecture

The below diagram (Fig. 1) explains the high-level framework of building and processing data model for diabetes.

- Data Source
- Data Munging
- Data Processing
- Data Modeling

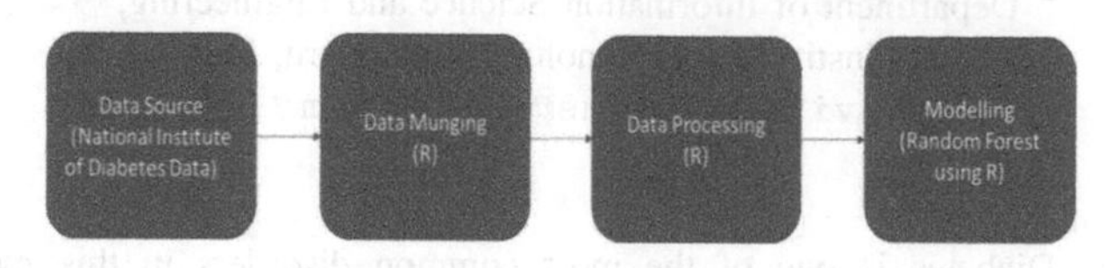


**Fig. 1.** High level framework

## 3.1 Data Source

The data may be in either direct database or flat file which can feed the data analytic tool.

## 3.2 Data Munging

The process of cleansing the data prior to analysis is called data munging. It helps to remove the anomalies and data quality errors before we building the model.

## 3.3 Data Processing

In this process we need to normalize and do label encoding (transform the data from non-numerical or categorical to numerical labels).

## 3.4 Data Modeling

This process is an integration of mining and probability for classifying the labels based on pre-determined labels. The models are composed of predictors, which the attributes which may influence future outputs. Once the dataset is procured, the applicable forecasters and numerical models will be formulated.

# 4 Proposed Methodology

## 4.1 Decision Tree

Decision Tree is one of the very wide spread machine learning approach. Decision Tree is a Supervised Machine Learning algorithm which resembles like an inverted tree. Every node in tree represents a predictor variable (feature), link between the nodes denotes a Decision and each leaf node denotes an outcome (response variable).

**Optimized Decision-Tree Algorithm**
**INPUT:** Diabetes Database
**OUTPUT:** Optimized Decision Tree for Diabetes Prediction
**Decision-Tree:**

**1:** Diabetes dataset is initially processed using R tool
**2:** The Missing values are replaced and Normalization has been done.
**3:** The Processed diabetes data is moved via FS where the unnecessary attributes are removed.
**4:** The final processed dataset is uploaded in R studio
**5:** The Decision Tree algorithm is executed
**6:** Model Creation is done.

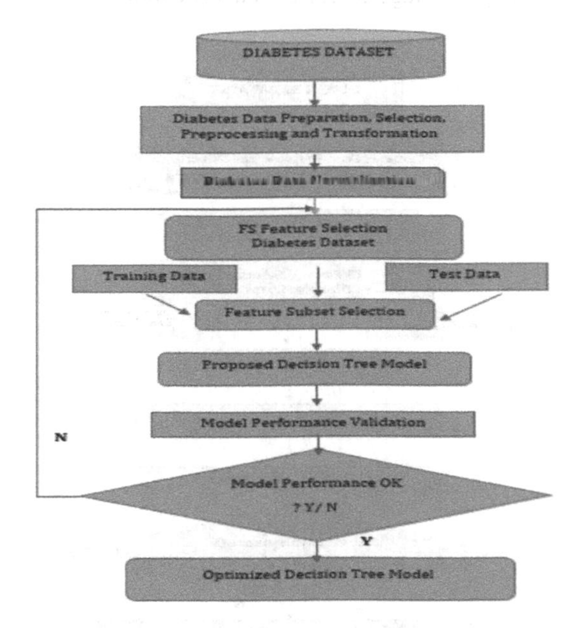


**Fig. 2.** Flowchart for optimized decision tree

## 4.2 Random Forest

Random Forest is classification approach that comprises of several decision trees. Random Forests can be used when the problem is having enormous volume of training datasets and input variables.

**Optimized Random Forest Algorithm**
**INPUT:** Diabetes Database
**OUTPUT:** Optimized Random Forest for Diabetes Prediction
**Random Forest (RF)**

**1:** Diabetes dataset mine fresh sample set with Bootstrap technique repetitive for N times
**2:** Construct a decision-tree with results from step 1.
**3:** Repeat steps 1, 2 and outputs in many trees comprise a random-forest.
**4:** Assume each tree from forest, vote for $X_i$.
**5:** Compute average votes for each class and class who is having more number of votes is classification label for X.
**6:** Accuracy of RF is percentage of correct classification

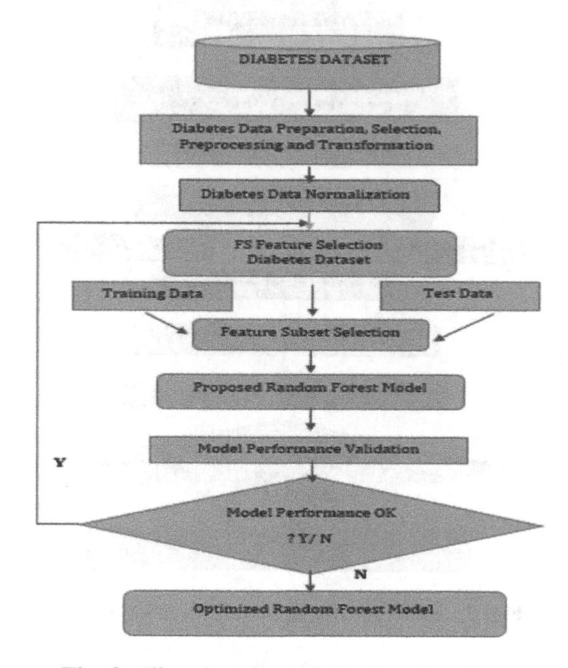


**Fig. 3.** Flowchart for optimized random forest

## 4.3 Data Set

The Pima Indians Diabetes Data set is imported from Machine Learning Repository and R Programming Tool is used for Implementation of Classification. The number of patients (n) in the database is 768, each with 9 attributes as depicted in Table 1.

**Table 1.** Pima Indian Diabetic data set attributes

| Attribute |
|---|
| Pregnancies |
| Glucose |
| BloodPressure |
| SkinThickness |
| Insulin |
| BMI |
| DiabetesPedigree |
| Age |
| Outcome |

(See Table 2).

**Table 2.** Min, Max, Mean and Median of PID dataset

| Attribute | Minimum | Maximum | Mean | Median |
|---|---|---|---|---|
| 1 | 0 | 17 | 3.845 | 3.000 |
| 2 | 0 | 199 | 120.9 | 117.0 |
| 3 | 0 | 122 | 69.11 | 72.00 |
| 4 | 0 | 99 | 20.54 | 23.00 |
| 5 | 0 | 84 | 7 | 3 |
| 6 | 0 | 67.10 | 31.99 | 32.00 |
| 7 | 0.780 | 2.4200 | 0.4719 | 0.3725 |
| 8 | 21 | 81 | 33.24 | 29.00 |

# 5 Results and Discussion

The classifiers performances are assessed by accuracy, sensitivity, specificity and were calculated by following formulas

$$Accuracy = TP + TN/TP + TN + FP + FN.$$

$$Sensitivity = TP/TP + FN.$$

$$Specificity = TN/TN + \mathrm{FP}$$

Confusion matrix performance of various machine learning classification algorithms on the Pima dataset is shown in the Table 3 and are depicted in Fig. 2.

**Table 3.** Confusion Matrix of PID dataset

| Confusion matrix | | Predicted class | |
|---|---|---|---|
| | | Positive | Negative |
| Decision tree | Positive | 242 | 26 |
| | Negative | 131 | 369 |
| SVM | Positive | 239 | 29 |
| | Negative | 142 | 358 |
| KNN | Positive | 208 | 60 |
| | Negative | 60 | 387 |
| Radom forest | Positive | 417 | 83 |
| | Negative | 110 | 158 |

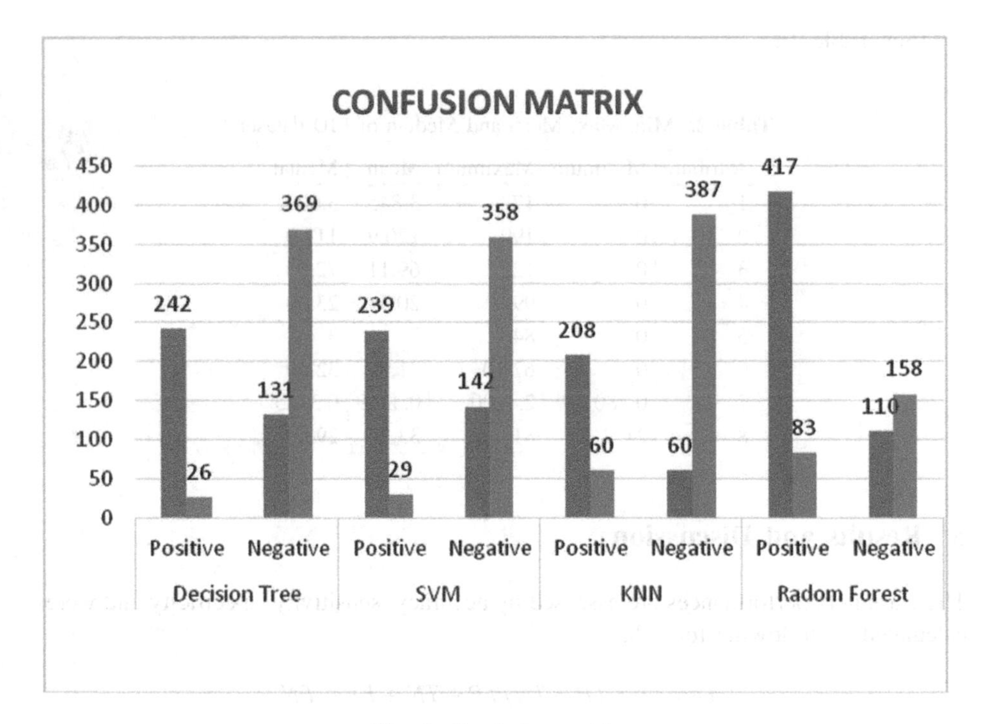


**Fig. 4.** Confusion matrix

The performance measures of various machine learning classification algorithms on the Pima dataset is shown in the Table 4 and are depicted in Figs. 3 and 4.

**Table 4.** Accuracy, Sensitivity and Specificity of PID dataset

| Classification algorithms | Accuracy | Sensitivity | Specificity |
|---|---|---|---|
| Decision tree | 78.25 | 65.31 | 80.00 |
| SVM | 77.73 | 43.75 | 77.33 |
| KNN | 77.47 | 64.10 | 78.90 |
| Random forest | 77.90 | 71.3 | 79.90 |

(See Figs. 5 and 6).

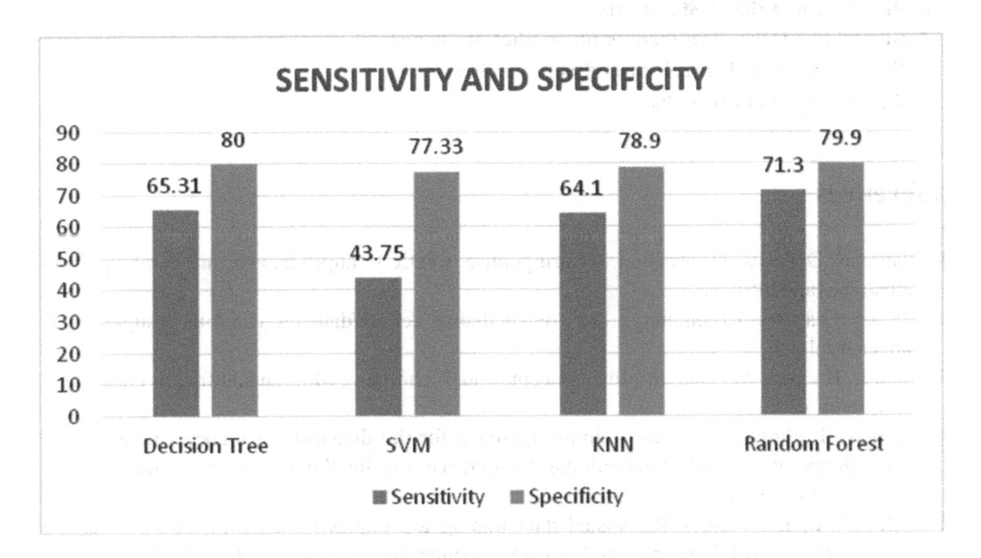


**Fig. 5.** Sensitivity and specificity

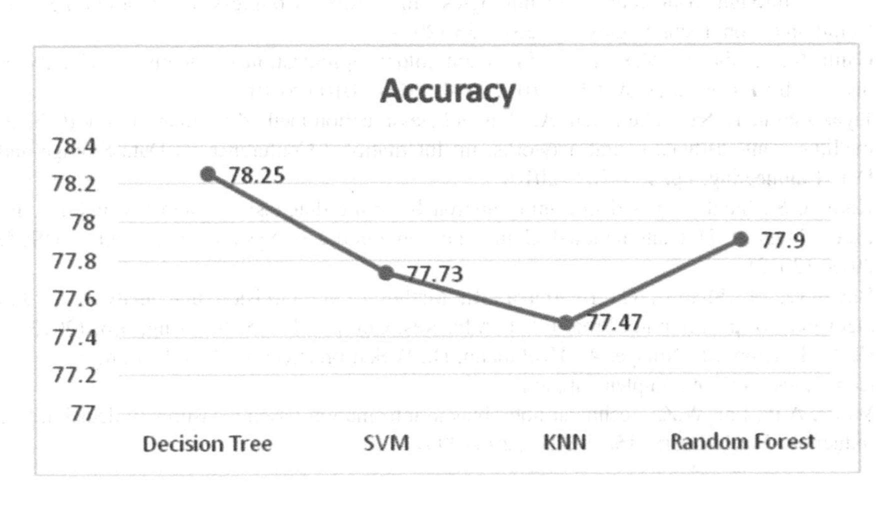


**Fig. 6.** Accuracy of classification approaches

## 6 Future Scope and Conclusion

The Accuracy for classification models Decision Tree, SVM, KNN and Random Forest were done and it is found that **78.25** for Decision Tree, **77.73** for SVM, **77.47** for KNN and **77.90** for Random Forest. Thus the Decision Tree and Random Forest Classification Models were chosen as the best models for predicting the occurrence of diabetes in PIMA database. The benefit of this optimized machine learning model is suitable for patients with diabetes mellitus and it can be applied for all the fields of health care environments.

**Compliance with Ethical Standards**

✓ All authors declare that there is no conflict of interest.
✓ No humans/animals involved in this research work.
✓ We have used our own data.

## References


1. National Diabetes Information Clearinghouse (NDIC). http://diabetes.niddk.nih.gov/dm/pubs/type1and2/#signs
2. Global Diabetes Community. http://www.diabetes.co.uk/diabetes_care/blood-sugar-level-ranges.html
3. Han, J., Kamber, M.: Data Mining Concepts and Techniques. Morgan Kaufmann Publishers, Burlington (2001)
4. Kumari, S., Singh, A.: A data mining approach for the diagnosis of diabetes mellitus. In: Proceedings of Seventh International Conference on Intelligent Systems and Control, pp. 373–375 (2013)
5. Velu, C.M., Kashwan, K.R.: Visual data mining techniques for classification of diabetic patients. In: 3rd IEEE International Advance Computing Conference (IACC) (2013)
6. Sankaranarayanan, S., Pramananda Perumal, T.: Predictive approach for diabetes mellitus disease through data mining technologies. In: World Congress on Computing and Communication Technologies, pp. 231–233 (2014)
7. Ganji, M.F., Abadeh, M.S.: Using fuzzy ant colony optimization for diagnosis of diabetes disease. In: Proceedings of ICEE 2010, 11–13 May 2010 (2010)
8. Jayalakshmi, T., Santhakumaran, A.: A novel classification method for diagnosis of diabetes mellitus using artificial neural networks. In: International Conference on Data Storage and Data Engineering, pp. 159–163 (2010)
9. Kumari, S., Singh, A.: A data mining approach for the diagnosis of diabetes mellitus. In: Proceedings of 71st International Conference on Intelligent Systems and Control (ISCO 2013) (2013)
10. Bhargava, N., Sharma, G., Bhargava, R., Mathuria, M.: Decision tree analysis on J48 algorithm for data mining. Proc. Int. J. Adv. Res. Comput. Sci. Softw. Eng. **3**(6) (2013)
11. Feld, M., Kipp, M., Ndiaye, A., Heckmann, D.: Weka: practical machine learning tools and techniques with Java implementations
12. White, A.P., Liu, W.Z.: Technical note: bias in information-based measures in decision tree induction. Mach. Learn. **15**(3), 321–329 (1994)

# Performance Analysis of Differential Evolution Algorithm Variants in Solving Image Segmentation

V. SandhyaSree(✉) and S. Thangavelu

Department of Computer Science and Engineering, Amrita School of Engineering, Coimbatore, Amrita Vishwa Vidyapeetham, Coimbatore, India
sandhiyasree23@gmail.com, s_thangavel@cb.amrita.edu

**Abstract.** Image segmentation is an activity of dividing an image into multiple segments. Thresholding is a typical step for analyzing image, recognizing the pattern, and computer vision. Threshold value can be calculated using histogram as well as using Gaussian mixture model. but those threshold values are not the exact solution to do the image segmentation. To overcome this problem and to find the exact threshold value, differential evolution algorithm is applied. Differential evolution is considered to be meta-heuristic search and useful in solving optimization problems. DE algorithms can be applied to process Image Segmentation by viewing it as an optimization problem. In this paper, Different Differential evolution (DE) algorithms are used to perform the image segmentation and their performance is compared in solving image segmentation. Both 2 class and 3-class segmentation is applied and the algorithm performance is analyzed. Experimental results shows that DE/best/1/bin algorithm out performs than the other variants of DE algorithms

**Keywords:** Image segmentation · Gaussian mixture model · Differential evolution · Mutation strategies

## 1 Introduction

Image segmentation is an activity of subdividing an image into the constituent parts. The reason for subdividing an image is to detect an object which will be useful for the highlevel process. Image segmentation is important study in pattern recognition and computer vision that is useful in understanding the image. The tonal distribution of a digital image can be represented graphically by Image histogram [1]. Thresholding is the fundamental task to achieve segmented image. Determination of threshold point is found to minimize the average segmentation error. Thresholding can be either Bi-level or Multi-level [2]. In Bi-level thresholding, segmentation of an image is performed based on the selected limit value. Multilevel thresholding isolates the given image into different regions and determines more than one threshold value [3]. Gaussian mixture model is a clustering algorithm that is used to group the same pixels and each group is considered as a class [4].

In recent years, Differential Evolution algorithms are useful in solving problems in various domains and many attempts made to improve the algorithm performance [5, 6].


S. Smys et al. (Eds.): ICCVBIC 2019, AISC 1108, pp. 329–337, 2020.
https://doi.org/10.1007/978-3-030-37218-7_38

Differential Evolution is a population centric, efficient and robust searching method. In this paper segmenting an image is performed using DE algorithm. It is selected for the faster convergence towards the threshold point. Different DE algorithms are used to perform the image segmentation and the performance of those algorithms in image segmentation process is analyzed in this paper. Segmenting the foreground and background objects in an image is achieved through bi-level thresholding. Traditional approaches and intelligent approaches have some problems [7] when expanding their schemes to multi-level thresholding. The DE approach resolves these problems in finding the threshold values.

DE algorithm completely depends on the mutation operation. DE has faster convergence rate when compared to other evolutionary algorithm. Binomial and Exponential are the two crossover methods of DE. Binomial and Exponential can be called as uniform method and two point modulo respectively [8, 9]. Performance of the binomial method is based on uniformly distributed variables and randomly generated numbers between 0 and 1 and this should be less than or equal to fixed predefined parameter called Crossover rate (*Cr*). In this case the mutant vector content copied into trial vector. Otherwise, the target vector content is copied into trial vector. This scheme has been used in this paper Eq. (1.6). Exponential crossover, copy the content of mutant vector into trial verctor as long as the generated random number is less than the $C_r$. If it is not so, thereafter the rest of the contents from target vector is copied into trial vector [10]. Parameter that ends up with the critical problem while minimizing the squared and nature of equations are non-linear. Since analytical solution is not available, So DE algorithm uses numerical method which has grade information based on iterative approach. Nevertheless accepting the gradient descent search method has the problem to fall into local minima, but the initial values play a major role in finding the final solution. Evolutionary based algorithm provides gratification performance in solving the various problems of image processing. Hence finding the suitable parameters and mapping them to their respective threshold values is achieved through DE algorithm. DE algorithms are used for solving variety of applicants in image processing [11], Data mining etc.

The rest of the sections are described as follows. Various concepts of Image segmentation related work is described in Sect. 2 followed by the proposed work at Sect. 3. Results and Analysis are discussed in Sect. 4 and finally Sect. 5 conclude the work.

## 2 Related Work

There are various segmentation techniques in image processing [12, 13]. Those techniques are accommodated into two categories such as Contextual technique and Non-contextual technique. In contextual technique includes different types viz. Simple thresholding, Adaptive thresholding and Colour thresholding. Non-contextual technique types includes Pixel segmentation, Region segmentation, Edge segmentation and Model segmentation. Region segmentation again has two types [14] such as Region growing, Region splitting and Merging. Using machine learning, Segmentation can be achieved by supervised and unsupervised algorithm where K-means clustering algorithm is widely used [15].

Image segmentation falls under two approaches. Discontinuity based and Similarity based. Discontinuity approach is based on intensity level and gray level. This approach is mainly interested in identification of isolated points, lines, and edges and all these together is called as masking and this depends on the coefficient values which is given and based on that decision is taken where operation is performed. In Similarity based approach grouping is done for those pixels which are similar [16]. Thresholding is considered as pre-processing step for image segmentation. Automatic thresholding is an iterative process, where image will be divided into two groups, the first is based on intensity values which are similar to each other and other will be in different group [17]. Based on these groups mean, standard deviation, and variance is calculated.

To attain optimal threshold, Probability Distribution Function (PDF) model has fit to histogram data using mean-squared error approach.

$$h(g) = \frac{n_g}{N}, h(g) \geq 0, N = \sum\nolimits_{g=0}^{L-1} n_g \quad \text{and} \quad \sum\nolimits_{g=0}^{L-1} h(g) = 1 \tag{1.1}$$

In image processing the tonal distribution of the image has $L$ gray levels $[0 \ldots L-1]$. Considering the histogram as a PDF. Normalization of histogram is applied to each pixels in image. The histogram h(g) can be contained in a mix of Gaussian mixture model in this way as defined by [18]

$$p(x) = \sum_{i=1}^{K} P_i \cdot p_{i(x)} = \sum_{i=1}^{K} \frac{P_i}{\sqrt{2\pi}\sigma_i} exp\left[\frac{-(x-\mu_i)^2}{2\sigma_i^2}\right] \tag{1.2}$$

In Eq. (1.2), $K$ represents the total number of classes in the image, $i$ = 1, …, $k$, the priori probability $P_i$ of class $i$, the PDF of gray-level random variable $x$, in class $i$ is $p_i$(x). $\mu_i$ is the mean and $\sigma_i$ is the standard deviation of the $i^{th}$ PDF. In addition, $\sum_{i=1}^{k} P_i = 1$ must be contended.

The estimation of the three parameters $(P_i, \mu_i, \sigma_i)$, $i$ = 1, …, K classes is done by using Mean square error. The mean square error between $p(x_j)$, the Gaussian mixture model function and $h(x_j)$, the histogram function of the experimental values is defined in Eq. (1.3). [19]

$$E = \frac{1}{n}\sum_{i=1}^{n} [p(x_j) - h(x_j)]^2 + \omega \cdot \left| \left(\sum_{i=1}^{K} P_i\right) - 1 \right| \tag{1.3}$$

Here $\omega$ is penalty related with the constraint $\sum_{i=1}^{k} P_i = 1$. For every candidate of a population, the above formula is applied, to calculate the fitness value.

DE is an evolutionary algorithm. It starts with an initial population of size $Np$ (vectors). The dimensional vector values are distributed randomly between the two boundaries, lower initial predefined parameter bound $x_{j,low}$ and the upper initial predefined parameter bound $x_{j,high}$.

$$\begin{aligned} x_{j,i,t} &= x_{j,low+rand(0,1)} \cdot (x_{j,high} - x_{j,low}); \\ j &= 1,2,\ldots D;\ i = 1,2,\ldots,N_p;\ t = 0. \end{aligned} \tag{1.4}$$

Considering $t$ as the index of a generation, the parameter and population indexes as $j$, $i$ respectively, the generation of mutant vector $v_{i,t}$ is achieved by Eq. (1.5).

$$\begin{aligned} V_{i,t} &= X_{best,t} + F \cdot (X_{r1,t} - X_{r2,t}); \\ r1,\ r2 &\in \{1,2,\ldots,N_p\}. \end{aligned} \tag{1.5}$$

In Eq. (1.5), $x_{best,t}$ is the best candidate of current population and $F$ is the scaling control parameter, usually between 0 and 1. Considering the index vectors that have been selected randomly, namely $r1$, $r2$ these should be distinct from each other and also from population index (specifically, $r1 \neq r2 \neq i$).

Next, the trial vector $u_{j,i,t}$ is developed from the target vector elements $x_{j,i,t}$ and the donor vector elements $v_{j,i,t}$. Hence the mixture of target vector and donor vector is a trial vector shown in Eq. (1.6).

$$u_{j,i,t} = \begin{cases} v_{j,i,t}, & if\ rand(0,1) \leq C_r\ or\ j = j_{rand}, \\ x_{j,i,t}, & otherwise \end{cases} \tag{1.6}$$

The contents of donor vector enter into the trial vector $u_{j,i,t}$ with the probability $C_r$ $(0 < C_r < 1)$. Here $C_r$ is the crossover constant, $j_{rand}$ is a random integer from $(1, 2, \ldots, D)$. $j_{rand}$ ensure that $v_{j,i,t} \neq x_{j,i,t}$. The $u_{j,i,t}$ take over the donor vector parameter with random index $j_{rand}$ to make sure that the $u_{j,i,t}$ differs by at least one parameter from the vector to be compared to $x_{j,i,t}$ [20].

The target vector $x_{j,i,t}$ and the trial vector $u_{j,i,t}$ are compared and the one with the lowest function value is admitted to the next generation as shown in Eq. (1.7).

$$X_{i,t+1} = \begin{cases} U_{i,t}, & if\ f(U_{i,t}) \leq f(X_{i,t}), \\ X_{i,t}, & otherwise \end{cases} \tag{1.7}$$

Representing $f$ as the cost function and if the $f$ of $u_{j,i,t}$ is less or equal to the $f$ of $x_{j,i,t}$, then $u_{j,i,t}$ replaces $x_{j,i,t}$ in next generation. $x_{j,i,t}$ remains in the population of next generation otherwise.

Threshold value is calculated by accepting the data classes sorted by their mean values i.e. $\mu_1 < \mu_2 < \ldots\ldots\ldots < \mu_k$. The overall probability error for the 2 adjacent Gaussian mixture model function is as follows [21]

$$E(T_i) = P_{i+1} \cdot E_1(T_i) + P_i \cdot E_2(T_i), \quad i = 1,2,\ldots,K-1 \tag{1.8}$$

Considering Eqs. (1.9) and (1.10)

$$E_1(T_i) = \int_{-\infty}^{T_i} p_{i+1}(x)dx \tag{1.9}$$

$$E_2(T_i) = \int_{T_i}^{\infty} p_{i(x)dx}, \tag{1.10}$$

Probability of misclassified pixels in the $(i + 1)^{th}$ class to the $i^{th}$ class is given as $E_1(T_i)$ and the probability of erroneusly classified pixels in the $i^{th}$ class to the $(i + 1)^{th}$ class is given as $E_2(T_i)$. The threshold value between the $i^{th}$ and the $(i + 1)^{th}$ classes is specified as $T_i$. One $T_i$ is selected in such a way that $E(T_i)$ is minimal. $P_j'$s are the prior probabilities within the combined PDF. Differentiating $E(T_i)$ with respect to $T_i$, Eq. (1.11) provide the $T_i$, the optimum threshold value.

$$\mathrm{A}T_i^2 + BT_i + C = 0 \tag{1.11}$$

where

$$\begin{aligned} \mathrm{A} &= \sigma_i^2 - \sigma_{i+1}^2 \\ B &= 2 \cdot \left(\mu_i \sigma_{i+1}^2 - \mu_{i+1} \sigma_i^2\right) \\ C &= \left(\sigma_i \mu_{i+1}\right)^2 - \left(\sigma_{i+1} \mu_i\right)^2 + 2 \cdot \left(\sigma_i \sigma_{i+1}\right)^2 \cdot \ln\left(\frac{\sigma_{i+1} P_i}{\sigma_i P_{i+1}}\right) \end{aligned} \tag{1.12}$$

The quadratic equation has two solutions, only one is viable (positive and is within the mean bound).

## 3 Proposed Work

Essentially, the problem of image segmentation can be attained by finding the threshold value using the DE algorithm. Initially find foreground and background values using histogram. Histogram outputs only two classes. To rectify that, the priori probability, mean and standard deviation is calculated using Gaussian mixture model from Eq. (1.2) [finds out the value for each pixel (0 to 256)]. By using Gaussian mixture model multiple classes has been found and the output of the Gaussian mixture model does not give the exact threshold value for every iteration. To overcome this differential evolution algorithm is used. To find the exact threshold value for each class, limit bound is set within the class. Fix the bound values by using the groups which is isolated in Gaussian mixture model cluster based on the intensity distribution. Histogram normalization is calculated by the Eq. (1.1). Mean square error is calculated by taking the difference of gaussian function and histogram. This is considered as the fitness function that is shown in Eq. (1.3).

DE initially begins by initializing the population with $Np$ candidate and $D$ as the number of classes, in which the values are randomly distributed between lower bound and upper bound by using the Eq. (1.4). For each candidate of the population the gaussian mixture is applied to cluster the image pixels and the histogram is generated.

A child is generated for every member of the population by following the DE algorithm steps (viz. Mutation and Crossover). The better one among the parent and child moves to the next generation. This process is repeated for 10 generations and the best two candidates from the last two generations are the leading candidates. The whole steps are repeated for 30 times and the average of all best candidates are used to find the optimum threshold value to do the image segmentation.

The various mutation strategies [22] used in this work to do the comparative analysis is listed below.

"DE/rand/1": $v_i^{(t)} = X_{R_1}^{i(t)} + F(X_{R_2}i^{(t)} - X_{R_3}i^{(t)})$.

"DE/best/1": $v_i^{(t)} = x_{best^{(t)}} + F(X_{R_1}i^{(t)} - X_{R_2}i^{(t)})$.

"DE/current-to-best/1" : $v_i^{(t)} = x_i^{(t)} + F(x_{best^{(t)}} - x_i^{(t)}) + F(X_{R_1}i^{(t)} - X_{R_2}i^{(t)})$.

"DE/best/2": $v_i^{(t)} = x_{best^{(t)}} + F(X_{R_1}i^{(t)} - X_{R_2}i^{(t)}) + F(X_{R_3}i^{(t)} - X_{R_4}i^{(t)})$.

"DE/rand/2": $v_i^{(t)} = X_{R_1}i^{(t)} + F(X_{R_2}i^{(t)} - X_{R_3}i^{(t)}) + F(X_{R_4}i^{(t)} - X_{R_5}i^{(t)})$.

## 4 Results and Analysis

The performance of the presented algorithms are estimated and compared with each other. Considering the image "The Cameraman" shown in the Fig. 1(a). The goal is to perform the image segmentation using different types of mutation strategies of DE algorithm. In this process of DE, population size is fixed as 90 (*Np*) individuals, and every candidate holds 6 dimensions for class 2 and 9 dimensions for class 3, as shown in Eq. (1.13).

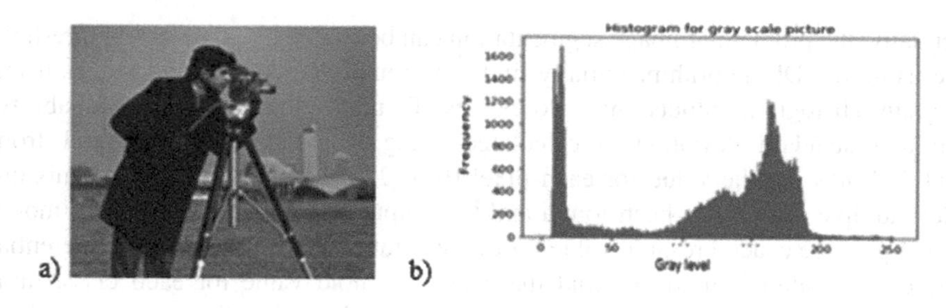


**Fig. 1.** (a) Image of "The Cameraman" [7] and (b) its concurrent histogram

$$I_N = \{P_1^N, \sigma_1^N, \mu_1^N, P_2^N, \sigma_2^N, \mu_2^N, P_3^N, \sigma_3^N, \mu_3^N\} \tag{1.13}$$

The parameters ($P$, $\mu$, $\sigma$) are initialized in random where $\mu$ must be between 0 and 255, the DE algorithm parameters are set with the values as: $\omega = 1.5$, $F = 0.25$, $Cr = 0.8$

After several iterations, the Gaussian mixture model provide the output a shown in Fig. 2(a) and (b), by using the Eq. (1.2). From this $P$, $\mu$, $\sigma$ is calculated for each pixel but the exact threshold value is not obtained hence the output calculated from this equation cannot be considered as the exact solution. Using the DE with Gaussian mixture model improve the performance.

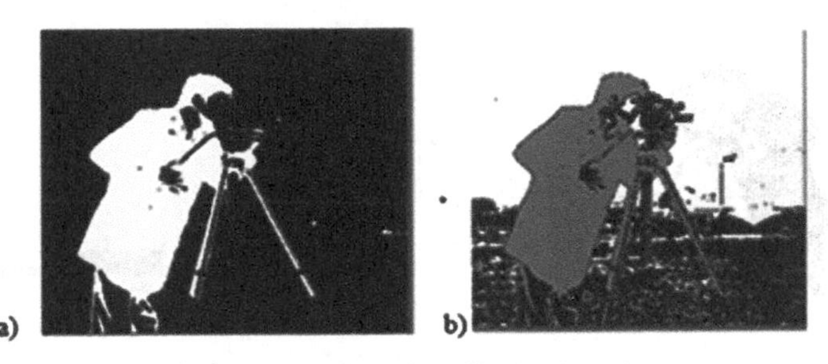


**Fig. 2.** (a) segmented image using Gaussian mixture model for 2 classes (b) segmented image using Gaussian mixture model for 3 classes

Table 1 shows the results of image segmentation for the five different DE algorithmic variants. Both 2 class and 3 class segmentation were done using the algorithms. It is clear from the results shown in Table 1, that DE/best/1 performs well in both 2 class and 3 class image segmentation with less error value obtained from Eq. (1.8). Their respective threshold values obtained to do the image segmentation were also shown in the Table 1. The resultant image is shown in Fig. 3 for both 2 class and 3 class segmentations along with their historgram. The performance/ outcome can be improved further by mixing the algorithms in a distributed environment by exchanging the best values between them [23].

**Table 1.** Comparative analysis of all mutation strategies

| Mutation algorithms | Error values (class 2) | Error values (class 3) | Threshold value (class 2) | Threshold value (class 3) |
|---|---|---|---|---|
| DE/rand/1 | 0.00093 | 0.0011, 0.0008 | 30 | 25,140 |
| DE/best/1 | 0.00050 | 0.0003, 0.0011 | 33 | 26,135 |
| DE/current-to-best/1 | 0.00077 | 0.0004, 0.0002 | 32 | 26,137 |
| DE/best/2 | 0.00077 | 0.0009, 0.0002 | 38 | 30,139 |
| DE/rand/2 | 0.00054 | 0.0007, 0.0003 | 34 | 26,142 |

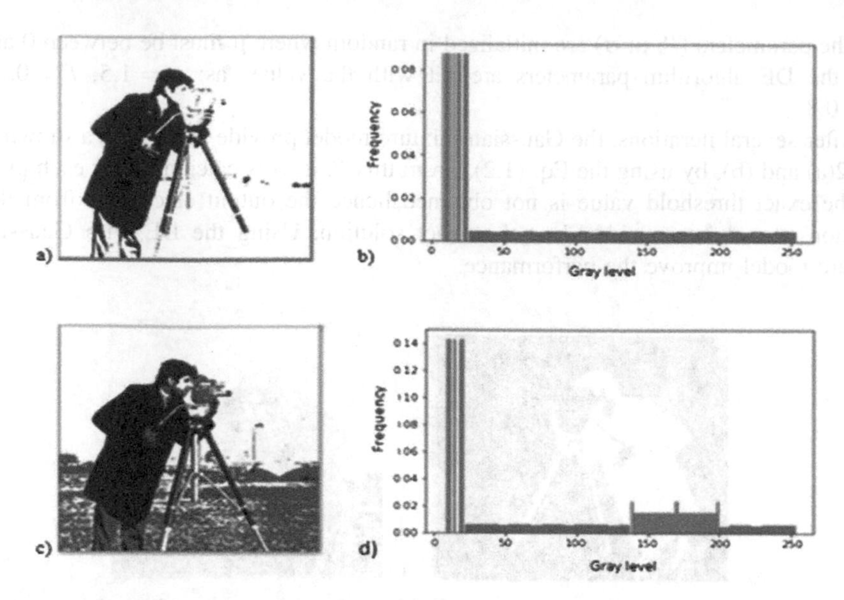


**Fig. 3.** (a) segmented image by using DE/best/1 for 2 classes, (b) its concurrent histogram and (c) segmented image using DE/best/1 for 3 classes, (d) its concurrent histogram

## 5 Conclusion

An effective method for segmentation of image using Differential evolution is employed in this paper. The method involves different mutation strategies. Essentially, histogram is used for 2 class problems and gaussian mixture model cluster with DE algorithm is used for multi class problem. Results shows that DE/best/1 algorithms out performs than other algorithms.

**Compliance with Ethical Standards.**

✓ All authors declare that there is no conflict of interest.
✓ No humans/animals involved in this research work.
✓ We have used our own data.

## References

1. Kaur, H., Sohi, N.: A study for applications of histogram in image enhancement. Int. J. Eng. Sci. **6**, 59–63 (2017)
2. Kotte, S., Kumar, P.R., Injeti, S.K.: An efficient approach for optimal multilevel thresholding selection for gray scale images based on improved differential search algorithm (2016)
3. Chen, B., Zeng, W., Lin, Y., Zhong, Q.: An enhanced differential evolution based algorithm with simulated annealing for solving multiobjective optimization problems (2014)
4. Farnoosh, R., Yari, G., Zarpak, B.: Image segmentation using Gaussian mixture model. Int. J. Eng. Sci. **19**, 29–32 (2008)

5. Tang, L., Dong, Y., Liu, J.: Differential evolution with an Individual-dependent mechanism. IEEE Trans. Evol. Comput. **19**(4), 560–574 (2015)
6. Huang, Z., Chen, Y.: An improved differential evolution algorithm based on adaptive parameter. J. Control Sci. Eng. **2013**, 5 (2013). Article ID 462706
7. Cuevas, E., Zaldívar, D., Perez-Cisneros, M.A.: Image Segmentation Based on Differential Evolution Optimization, pp. 9–21. Springer International Publishing, Switzerland (2016)
8. Tvrdik, J.: Adaptive differential evolution and exponential crossover. IEEE (2008)
9. Weber, M., Neri, F.: Contiguous Binomial Crossover in Differential Evolution. Springer-Verlag, Heidelberg (2012)
10. Das, S., Mullick, S.S., Suganthan, P.N.: Recent advances in differential evolution–an updated survey. Swarm Evol. Comput. **27**, 1–30 (2016)
11. Haritha, K.C., Thangavelu, S.: Multi-focus region-based image fusion using differential evolution algorithm variants. In: Computational Vision and Biomechanics. LNCS, vol. 28, pp.579–592. Springer, Netherlands (2018)
12. Suganya, M., Menaka, M.: Various segmentation techniques in image processing: a survey. Int. J. Innov. Res. Comput. Commun. Eng. **2**(1), 1048–1052 (2014)
13. Kaur, A., Kaur, N.: Image segmentation techniques. Int. Res. J. Eng. Technol. **02**(02), 944–947 (2015)
14. Zaitouna, N.M., Aqelb, M.J.: Survey on image segmentation techniques. In: International Conference on Communication Management and Information Technology. Elsevier (2015)
15. Choudhary, R., Gupta, R.: Recent trends and techniques in image enhancement using DE – a survey. Int. J. Adv. Res. Comput. Sci. **7**(4), 106–112 (2017)
16. Kaur, B., Kaur, P.: A comparitive study of image segmentation techniques. Int. J. Comput. Sci. Eng. **3**(12), 50–56 (2015)
17. Rahnamayan, S., Tizhoosh, H.R., Salama, M.M.: Image thresholding using differential evolution. In: Proceedings of International Conference on Image Processing, Computer Vision and Pattern Recognition, Las Vegas, USA, pp. 244–249 (2006)
18. Osuna-Enciso, V., Cuevas, E., Sossa, H.: A Comparison of nature inspired algorithms for multi-thresholding image segmentation. Expert Syst. Appl. **40**(4), 1213–1219 (2013)
19. Ochoa-Monitel, R., Carrasco Aguliar, M.A., Sanchez-Lopez, C.: Image segmentation by using differential evolution with constraints handling. IEEE (2017)
20. Storn, R., Price, K.: Differential evolution – a simple and efficient heuristic for global optimization over continuous spaces. J. Glob. Optim. **11**, 341–359 (1997)
21. Ali, M., Siarry, P., Pant, M.: Multi-level image thresholding based on hybrid DE algorithm. Application of medical images. Springer-Verlag Gmbh, Germany (2017)
22. Leon, M., Xiong, N.: Investigation of mutation strategies in differential evolution for solving global optimization problems, vol. 8467, pp. 372–383. Springer International Publishing, Switzerland (2014)
23. Thangavelu, S., ShanmugaVelayutham, C.: An investigation on mixing heterogeneous differential evolution variants in a distributed framework. Int. J. Bio-Inspired Comput. **7**(5), 307–320 (2015)

# Person Authentication Using EEG Signal that Uses Chirplet and SVM

Sonal Suhas Shinde(✉) and Swati S. Kamthekar

Department of Electronics and Telecommunication Engineering, Pillai HOC College of Engineering and Technology, Rasayani, India
Sonalshinde42@gmail.com, Swati.kamthekar@gmail.com

**Abstract.** The brain wave signal which is result of neurons activity basically used for authentication due to its benefits over traditional biometric system. However lots of work has been done in brainwave based authentication, also numerous pre-processing, numbers of features were extracted and different classification methods have been investigated for authentication system. This study focuses on EEG (Electroencephalography) signal authentication with excellent precision. This paper consist of identification and authentication of human based on the EEG signal, here the database is used which consist of EEG signal of 29 subject and five trail of each subject. The transform named as Chirplet is used for extraction of the feature and support vector machine is used to improve the efficiency of those EEG-based recognition systems. The paper's goal is to achieve excellent classification rates and consequently assist create a secure and reliable bio-metric scheme using brain wave signal as a basis. This study achieves 99% precision.

**Keywords:** Brain wave · Biometric authentication · Neurons · EEG electrodes · Wavelet transform · Alpha · Beta · Gama · Theta · Delta

## 1 Introduction

The billions of neurons are found in human brain that are accountable for electrical charge when brain operates the sum of all these charges generates an electric field which is in μV range, this field is measured using sensors named as electrode, changes in thinking will alter the interconnections of neurons that generate alterations in brain waves pattern. Now a day's person recognition system is very popular researcher topic.

Old authentication systems provide authentication for example passwords, pin, codes etc. However some biometrics mostly like fingerprints and retina are also used for authentication purposes, brain wave based authentication is new one in authentication history, in this the EEG signal is used to confirm the personal identity. Biometrics systems which are based on fingerprints, retina, and face etc. are suffers from attacks which degrade its performance, compared to the existing biometric, EEG offers better robustness and safety, as it is very difficult to replicate. EEG-based technique of authentication includes extraction and pattern recognition of information preprocessing features, where data collection is carried out with the help of Electrodes, collected data has EEG signal as well as noise usually artifacts so in preprocessing noise is removed.


S. Smys et al. (Eds.): ICCVBIC 2019, AISC 1108, pp. 338–344, 2020.
https://doi.org/10.1007/978-3-030-37218-7_39

EEG signal has gamma, beta, alpha, theta and delta signals from which characteristics are obtained and compared with the pattern recognition database.

This research organized as follows Sect. 2 as related work. General architecture of EEG based authentication system is discussed in Sect. 3. Proposed system is discussed in Sect. 4. Section 5 deals with the result where as proposed system conclude in Sect. 6.

## 2 Related Work

Eye closed and eye open condition were used for authentication process where signal stability was major issue. Performance is good by considering EC (Eye Closed) condition only [1].

This paper proposed analysis of EEG signal with the help of power spectral density as a feature which further used k-nearest neighbors algorithm classifier for classification purpose. Comparing alpha band (8–13 Hz), beta band (13–30 Hz), mixed bands such as alpha and beta (8–30 Hz) and combining delta, alpha, beta and gamma bands (4–40 Hz) compares for the precision outcomes. Overall accuracy in terms of percentage was obtained above 80%.

This paper concludes that highest accuracy obtained by combination of theta, alpha, beta, and gamma which provide highest accuracy about 89.21% [2]. Furthermore, it is common to get brain waves in the research area under relaxed or eye-closed conditions. However, evaluating the verification performance using such brain waves is not realistic [5].

Convolution neural network (CNN) is used to dig the characteristic for separating persons [6]. A wavelet based feature extraction technique is suggested to analyze the Potentials evoked by visual and auditory EEG signal [7]. In the proposed system, sample entropy features obtained from 64 channel EEG signal whereas frequency bands were evaluated for identification [8]. Changes in alpha action (8–13 Hz) during eyes open and closed were collected for each channel as features this results into alpha power increases when subject close their eyes which should be improved with more advanced authentication algorithms in future studies [9]. In proposed approach, the user is requested to visualize a pre decided number and respective EEG signal is captured. This study particularly focuses on the combined frequency bands like Alpha and Beta through all EEG channels, the maximum accuracy obtained was 96.97% [10]. Paper seeks to explore the best set of number of electrode for human authentication using VEP (Visual Evoked Potention) extraction techniques such as cross-correlation, coherence and amplitude mean used for classification purposes [11].

## 3 General Architecture of EEG Based Authentication

Figure 1 shows general architecture of EEG based authentication system where EEG signal is captured with the help of Brain computer Interface (BCI) usually known as electrodes which are placed on the scalp, raw EEG signal consist noise and artifacts which are need to be process in preprocessing stage this artifacts are removed using filter. In feature extraction section different features like mean, power spectral density, energy etc. are extracted from the filtered signal and further they are classified with the help of classifier. Finally classified output is compared with data base which is already stored in system.

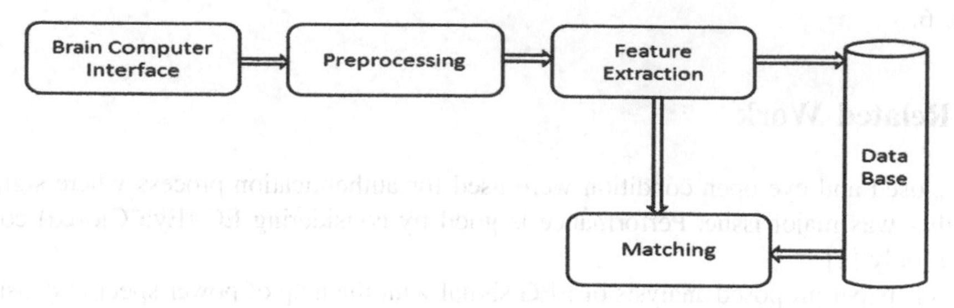


**Fig. 1.** General architecture of EEG based authentication system

# 4 Proposed System

## 4.1 Signal Acquisition

**Dataset Description**

The dataset is downloaded from UCI Machine Learning Repository; this database was obtained by performing the experiments using a commercial device for 29 subjects. The experiment was conducted with the highest ethical norms.

The tests are visual experiments, so that the most interesting signals are the O1 and O2 electrodes. Each subject performed tests and respective EEG signal were capture with the help of 16 electrodes (Fig. 2).

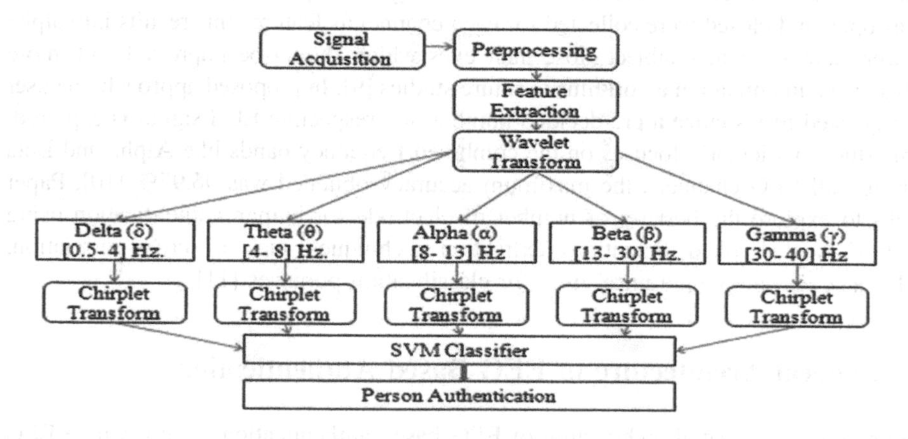


**Fig. 2.** Proposed system.

**Attribute Information**

Total 16 attributes are available, the last 14 of which are the signals from the electrodes and the first two are the time domain and interpolated signal.

### 4.2 Preprocessing

As EEG signals recorded non-invasively, the signal is affected by noise which is removed with the help of low pass filter with sampling frequency 173 Hz. Filtering provide noise free EEG signal having range 0–80 Hz. Further Notch filter is used to get EEG signal having the highest frequency 50 Hz.

### 4.3 Feature Extraction

EEG signal pattern changes with change in thought which changes neuron activity. Wavelet transform is capable to provide time - frequency information simultaneously. EEG signal has five basic component Gamma ($\gamma$) – [30–40] Hz, Beta ($\beta$) – [13–30] Hz, Alpha ($\alpha$) – [8–13] Hz, Theta ($\theta$) – [4–8] Hz and Delta ($\delta$) – [0.5–4] Hz. In this work the wavelet transform is used to separate EEG signal into five distinct frequency bands. The decomposition levels are chosen as 4. As a result, the EEG signals are broken down into D1–D4 as detail coefficient and A4 as a final approximation. The idea of a stationary signal cannot always consider, practically non stationary signal should be considered [12]. Non stationary is much nearer to the existence, such as the EEG signal, signal of rotary equipment during nonstationary events, human speech signals, etc. In non-stationary event, the general time frequency representation of signal is inefficient to describe non-stationary signals and the word instantaneous frequency (IF) plays a crucial role in characterized time-varying signals. As recent advanced time frequency analysis (TFA) technologies can provide more accuracy, however their limitations cannot be omitted by the time frequency analysis method should have following capabilities 1. Well-representing a linear multi-component signal, 2. Depending on the model of mathematics and the original time frequency analysis (TFA) technique, 3. Enabling for the mathematical model. The general form of Fourier transformation and short time transformation is Chirplet transform, which provide flexible frequency time window, linear chirplet transformation which provides greater time frequency representation and exhibit the best technique to characterize nonlinear information as well as being robust against noise [3]. This research uses Chirplet transform to extract the features from the alpha, beta, gama, theta and delta frequency band where frequency components are analysed for particular time interval.

### 4.4 Classification

Extracted features are applied to the classifier. Support vector machine (SVM) is a classifier which is less complex, nonlinear conditions are well treated in SVM. In this study total 725 features are extracted (29 subjects $\times$ 5 trail $\times$ 5 frequency bands) to classify the data, proposed study uses SVM classifier, SVM's primary function is to convert input information into high dimensional space which results into the separation of data by providing hyper plane between different types of data. Hence the largest margin hyperplane is selected which consider as a decision boundary to differentiate different type of data.

## 5 Result and Discussion

The database which obtained by Machine Learning Repository consist VEP signal of 29 subject, which is filter out with sampling frequency 173 Hz, it has been observed noise is neglected and produce EEG signal which is further decomposed into five bands (Fig. 3).

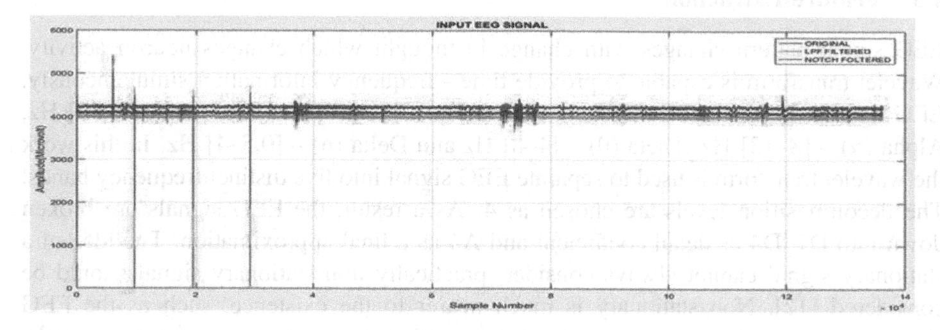


**Fig. 3.** Input signal.

It is very essential to select the no. of decomposition stage in wavelet Transform. The concentrations are selected so that the wavelet coefficients retain those components of the signal which are used to classify the signal. In this research, as the EEG signals have no helpful frequency elements above 50 Hz, the number of concentrations of decomposition was selected as 4. As a result, the EEG signals broken down into D1–D4 details and A4, a final approximation, as shown in Fig. 4.

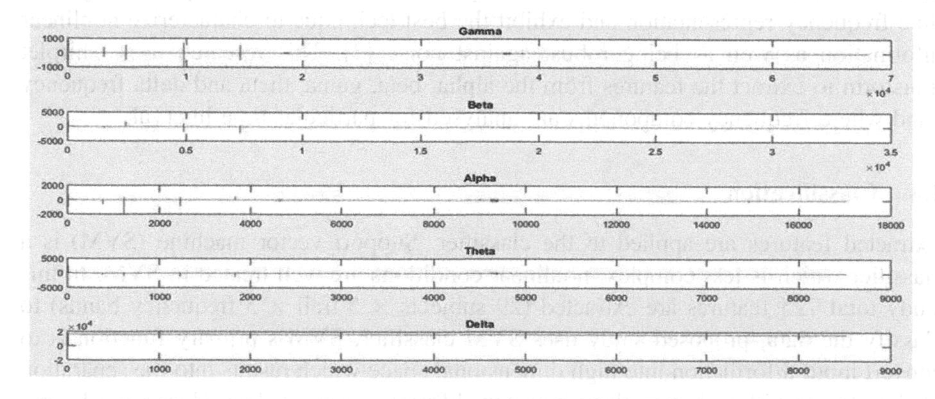


**Fig. 4.** Decomposition of signal

Further decomposed signal classified with the help of classifier. The proposed study use Support vector machine as a classifier. Due to their precision and capacity to handle a big amount of data, SVM is used for information classification.

Without much additional computational effort, nonlinear limits can be used in SVM. In addition, SVM efficiency is highly competitive with other techniques, by using SVM this research achieve 99% accuracy (Table 1).

**Table 1.** Camparision with existing work

| Existing method | Accuracy (%) |
|---|---|
| 7 subjects, EC condition, 7 features, NN [2] | 95 |
| 26 subjects, music included potential, 60 features, SVM [3] | 92.73 |
| 4 subjects, VEP, P300 classification [4] | 89 |
| Serial visual presentation (RSVP), CNN [6] | 91.27 |
| 109 subjects, sample entropy feature, cross-validation framework [8] | 98.31 |
| 101 subjects, VEP, Fuzzy-Rough Nearest Neighbour (FRNN) [11] | 91 |
| Proposed system: 29 subject+5 trails, VEP, LCT, SVM | 99 |

## 6 Conclusion

Old biometric system such as Finger print, retina scan and PIN lock are suffer from different attacks, this study proposed authentication of person using EEG signal which uses the physiological and behavioral features of individual in order to authentication of individual. The precision output is very nice, which authenticates the individual known and dismisses the unknown. This paper de-scribes the technique for improving the effectiveness of current EEG-based person identification technologies and thereby encouraging the future development of EEG-based person recognition for security control.

**Compliance with Ethical Standards**

✓ All authors declare that there is no conflict of interest.

✓ No humans/animals involved in this research work.

✓ We have used our own data.

## References

1. Dmitriev, N.A., Belik, D.V., Zinevskaya, M.S.: Bioengineering system used to restore the patient's brain fields activity after stroke (2016)
2. Lee, C., Kang, J.H., Kim, S.P.: Feature selection using mutual information for EEG-based biometrics (2016)
3. Lin, Y.P., Wang, C.H., Wu, T.L., Jeng, S.K., Chen, J.H.: Support vector machine for EEG signal classification during listening to emotional music (2008)
4. Moonwon, Y., Kaongoen, N., Jo, S.: P300-BCI-based authentication system (2016)

5. Nakanishi, I., Yoshikawa, T.: Brain waves as unconscious biometrics towards continuous authentication. In: 2015 International Symposium on Intelligent Signal Processing and Communication Systems (ISPACS), 9–12 November 2015
6. Wu, Q., Zeng, Y., Lin, Z., Wang, X.: Real-time EEG-based person authentication system using face rapid serial visual presentation. In: 8th International IEEE EMBS Conference on Neural Engineering Shanghai, China, 25–28 May 2017
7. Sundararajan, A., Pons, A., Sarwat, A.I.: A generic framework for EEG-based biometric authentication. In: 12th International Conference on Information Technology - New Generations (2015)
8. Thomas, K.P., Vinod, A.P.: Biometric identification of persons using sample entropy features of EEG during rest state. In: IEEE International Conference on Systems, Man, and Cybernetics, 9–12 October 2016
9. Toshiaki, K.A., Mahajan, R., Marks, T.K., Wang, Y.: High-accuracy user identification using EEG biometrics (2016)
10. Virgílio, B., Paula, L., Ferreira, A., Figueiredo, N.: Advances in EEG-based brain-computer interfaces for control and biometry (2008)
11. Siaw, H., Yun, H.C., Yin, F.L.: Identifying visual evoked potential (VEP) electrodes setting for person authentication. Int. J. Adv. Soft Comput. Appl. 7(3), 85–99 (2015)
12. Gang, U., Yiqui, Z.: Mechanical systems and signal processing (2015)

# Artificial Neural Network Hardware Implementation: Recent Trends and Applications

Jagrati Gupta[1(✉)] and Deepali Koppad[2]

[1] CMR Institute of Technology, Bengaluru, India
jagrati.g@cmrit.ac.in
[2] Ramaiah Institute of Technology, Bengaluru, India
deepali.koppad@gmail.com

**Abstract.** The human brain far exceeds contemporary supercomputers in tasks such as pattern recognition while consuming only a millionth of the power. To bridge this gap, recently proposed neural architectures such as Spiking Neural Network Architecture (SpiNNaker) use over a million ARM cores to mimic brain's biological structure and behaviour. Such alternate computing architectures will exhibit massive parallelism and fault tolerance in addition to being energy efficient with respect to traditional Von-Neuman architecture. This leads to the emergence of neuromorphic hardware that can exceed performance in Artificial Intelligence (AI) driven remote monitoring embedded devices, Robotics, Biomedical devices and Imaging systems etc. with respect to traditional computing devices. In this review paper, we focus on the research conducted in the field of neuromorphic hardware development and the applications. This paper consists of the survey of various training algorithms utilized in hardware design of neuromorphic system. Further, we survey existing neuromorphic circuits and devices and finally discuss their applications.

**Keywords:** Artificial neural network · Neuromorphic · Hardware implementation · Memristor · Spintronics · Brain machine interface

## 1 Introduction

Neuromorphic engineering is the field of engineering where function of artificial unit called neuron made up of various analog and digital structures imitates biological neuron. Human brain has billions of neuron cells which process the information and works as a processor. In biological neuron, interaction between two neurons takes place at the junction called as synapse. Recently a lot of effort has been made for modeling these synapses using latest discovered memristors [1, 2]. Memristor can be used as synapses and thus utilized to shrink the area and complexity of neuromorphic circuits [3]. These Artificial Neuron Networks operate similar to the brain in terms of intelligence and while processing any information. In fact, human intelligence keeps evolving because of past experience. Similarly in artificial neural network, there is requirement of learning and training algorithm so that it can mimic human intelligence. The need arise as there is requirement of computing and processing large amount of data. In this


S. Smys et al. (Eds.): ICCVBIC 2019, AISC 1108, pp. 345–354, 2020.
https://doi.org/10.1007/978-3-030-37218-7_40

review paper, the reference is basically to the research conducted in the domain of hardware implementation of neuromorphic systems. A lot of advancement and work has happened in this field in a decade in order to deal with complexity of calculations as power budget of human brain is far better than best supercomputers [4].

## 2 Training Algorithms

In this section, study of some of the training algorithms has been presented which is widely used in implementation of neural network hardware.

### 2.1 Backpropagation (BP)

The BP targets for the minimum value of error using the method of gradient descent. Weights combination that reduces the error function is accepted as solution of the training problem. Neural networks are categorized as Static and Dynamic Networks. Static networks contains no feedback elements and have no delays where as in dynamic networks, the output is dependent on the current input to the network and also on the present or past inputs, outputs. A general framework for describing dynamic neural networks has been proposed in [5]. Backpropagation algorithms basically contains three steps (i) Forward Propagation (ii) Backward Propagation (iii) Calculation of updated weight values. In backpropagation, the weight values are wij pointing from i to j node. The information fed into the subnetwork in the forward propagation was xi wij, here xi refers to buffered output of unit i. BP step calculates the derivative of E, i.e., $\frac{\partial E}{\partial x_i w_{ij}}$. We finally have,

$$\frac{\partial E}{\partial w_{ij}} = x_i \frac{\partial E}{\partial x_i w_{ij}} \tag{1}$$

Error that is backpropagated at the j-th node is δj, now the partial derivative of E with respect to wij is given by xi δj. Now, wij is recalculated by adding error term, $\Delta wij = -\gamma.xi.\delta j$ hence, this transforms training of neural network. Prior work proposes a hardware architecture to implement both recall and training for Multilayer Feed Forward Spiking Neural Networks [6]. In Spiking Neural Networks information among the neurons is passed via spikes or pulses. Learning process can be categorized as off and on-the chip learning. Off chip learning is slower and thus speed up using on chip implementation of training hardware. Recently, analog backpropagation learning hardware has been proposed for various architectures based on memristors [7]. One of the applications of backpropagation is Sensor interpreter based on neural network where backpropagation is utilized for training the data generated from the CHEMFET (chemically field-effect transistor) sensor [8]. Implementing backpropagation using analog circuits is still a problem so recently efforts have been made to design analog circuits for realizing backpropagation algorithm that can be used with memristive crossbar arrays in neural networks [9]. On chip training is faster so recent work has been presented for on-chip back-propagation training algorithm to implement FPGA based 2 × 2 × 1 neural network architecture [10]. A new memristor crossbar architecture has been proposed where the backpropagation algorithm has been used to train a 3-bit parity circuit [11]. Efforts have been made to develop mathematical models of various non idealities that are found in crossbar implementations such as neuron and source resistance, device difference and analysis of their effects on fully connected network and convolutional neural network which are trained with backpropagation algorithm [12]. Thus, prior works have utilized backpropagation for training various neural networks as it is simple, fast and easy to program.

## 2.2 Spike Timing Dependent Plasticity (STDP)

Biologically, STDP is a phenomenon where the spikes timings directly cause variation in the strength of synapse [13]. This is an emerging area of research to work upon and is being implemented in neuromorphic networks. This learning algorithm can be modeled as below [14]:

$$K(\Delta t) = \{S_{+}\exp(\Delta t/\Delta\tau_{+}) \quad \text{if } \Delta t < 0 \\ - S_{-}\exp(-\Delta t/\Delta\tau_{-}) \quad \text{if } \Delta t \geq 0\} \tag{2}$$

where $K(\Delta t)$ determines the synaptic modification. Here, $\tau_{+}$ and $\tau_{-}$ depicts the ranges of pre-to-postsynaptic interspike intervals. It simply implies that a neuron embedded in a neural network can point out which neighboring neurons are worth listening to by potentiating those inputs that predict its own spiking activity [15]. STDP has been implemented on neuromorphic hardware like SpiNNaker. A pre-synaptic scheme and event-driven model has been used to modify the weights only when pre-synaptic spike appears [16]. STDP has been applied for various applications recently in Dynamic Vision sensor cameras for visual feature training [17]. A new method has been proposed to design causal and acausal weight updation using only forward lookup access of the synaptic connectivity table that is memory-efficient [18].

STDP weight update rules has been shown in Fig. 1 where the leftmost block shows acausal updates produced by new presynaptic input events. Middle block shows processing of causal updates just before acausal updates. Rightmost block depicts that causal updates are produced due to completion events.

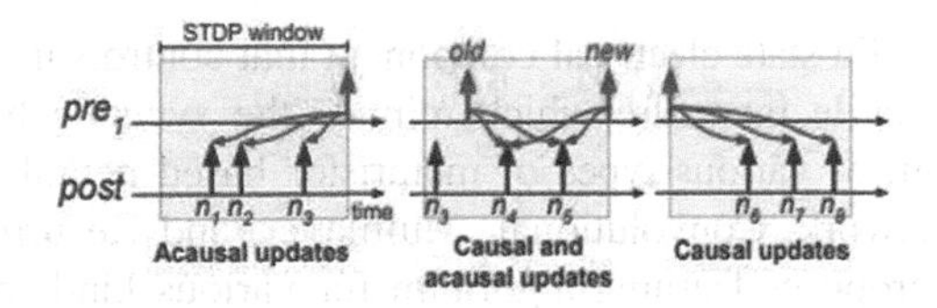


**Fig. 1.** STDP weight update rule [18].

### 2.3 Spike Driven Synaptic Plasticity (SDSP)

This training rule has advantage over Spike Time Dependent Plasticity (STDP) as it can be utilized for implementation of condensed digital synapses. Recent research proposes SDSP-based 3-bit asynchronous programmable synapse [19] with an area of 25 $\mu m^2$. Synaptic efficacy J(t) is defined using analog internal variable X(t), given as the variation in post-synaptic depolarization, or caused by one pre-synaptic spike [20]. Some research also describes circuit implementation of various plasticity rules which includes SDSP [21]. Spike driven threshold training algorithm has been proposed for mixed memristor based systems [22]. Table 1 compares various training algorithms for neural network hardware implementation.

**Table 1.** Comparison of training algorithms used for implementing neural network hardware.

| Ref. | Training algorithm | Advantage/Accuracy | Hardware | Mode | Applications |
|---|---|---|---|---|---|
| [7] | BP | MNIST-93%<br>YALE-78.9% | Memristive Crossbar Arrays | Non Spiking | Near Sensor Processing |
| [8] | BP | R factor = 0.9106 | CHEMFET (Sensor) | Non Spiking | Classification of target ion concentration in mixed solution |
| [17] | STDP | Small delay (9.9 μs) for online learning | FPGA | Spiking | Dynamic Vision Sensor Cameras |
| [18] | STDP | Reconfigurable synaptic connectivity | FPGA | Spiking | – |
| [19] | SDSP | Area of 25 $\mu m^2$ | FDSOI CMOS Process | Spiking | Digital synapse Implementation |

## 3 Neuromorphic Circuits and Devices

### 3.1 Memristors

It is a bi- terminal solid state electrical component that controls the flow of current in a circuit and non volatile in nature which mimics the synaptic behaviour. Prior text discusses the design of various types of memristor based neural architecture such as Spiking Neural Network, Convolutional, Multilayer, and Recurrent Neural Network [23]. Other work proposes learning algorithm for various kinds of memristive architecture [6]. Full circuit level design using Crossbar memristive structures can be simulated for Convolutional Neural Networks [24]. Recently much research focuses on device level implementations and fabrication of memristor crossbar arrays [25], MoS2/Graphene based memristors to mimic artificial neuron [26]. Other research shows that memristor is mainly used for implementation of synapses [27]. Crossbar arrays of memristors has disadvantage of leakage of data and thus recently synapses has been developed using nanoporous memristor synapses where there is possibility of self mending [28]. Usage of memristor for implementation of circuits has shortcomings also since fabrication process is still in development phase [29].

### 3.2 Spintronics Devices

These are nanoelectronic devices which have been implanted with magnetic impurities and materials that allows the usage of spin of electrons and charge associated with it required for transfer of information. Recently spin-orbit torque (SOT) device has been used which shows resistive switching and thus can act as an artificial synapse in neuromorphic systems [30]. Recent research work proposes compound spintronics synapse made up of multiple Magnetic Tunnel Junctions (MTJ) which has the advantage to attain stable multiple resistance states and focuses on the structure of All Spin Artificial Neural Network [31]. The most recent research shows development of spintronic devices which can be used as synapse in neural network and thus reviewed on spintronics based synapse [32]. A neural computing system has been proposed in a work which models neural transfer function for MTJ using its device physics [33]. Problems that occur with nanoscale devices are unstability and noise so experimental research is carried out to prove that nanoscale MTJ's oscillator can help to achieve good accuracy [34]. Some papers review application of spintronics device in Bio-inspired computing [35] and various spintronics structures opening the path for implementation of neurons and synapses which are efficient in area [36].

### 3.3 Floating Gate MOS Devices

Motivation to use Floating Gate MOS (FGMOS) in implementation of neural networks is low power dissipation, slow memory discharge due to charge retention property, use of standard fabrication process and compatibility with CMOS technology. Prior work includes implementation of an Operational Transconductance Amplifier (OTA) using Multiple Input FGMOS which can be used as programmable neuron activation function [37], design and implementation of Neural network using FGMOS [38], design of an analog synapses using complementary Floating gate MOSFETs [39] and excitatory synapse circuit using quasi floating MOS Transistor [40]. Non-Linear Neural Networks has been implemented using FGMOS based neurons which is basically used for image processing application [41]. Several other research discuss Convolutional Neural Network and its hardware implementation using FGMOS at an architecture level [42]. Whereas at circuit level Multiple input FGMOS is used in differential pair to be used in cellular neural network which helps in reducing power consumption [43]. For pattern recognition application hardware for generalized n-input neuron circuit is implemented which utilizes FGMOS based neuron model [44]. Table 2 compares neuromorphic devices and circuits.

**Table 2.** Comparison of neuromorphic devices and circuits.

| Ref. | Device | Mode | Level | Applications | Remarks |
|---|---|---|---|---|---|
| [26] | Memristor | Spiking | Device | Can be used for Random Number generation and hardware security | Novelty: Fabrication of device by forming V-$MoS_2$ on a CVD graphene monolayer |
| [27] | Memristor | Non-Spiking | Circuit | Pattern Recognition | Proposed CMOS memristor synapses |
| [28] | Memristor | Spiking | Device | Pattern Recognition | Fabricated Crossbar array, Accuracy of 89.08% after 15 epochs |
| [34] | Spintronics (MTJ) | Spiking | Circuit | Spoken digit Recognition | Experiment conducted to recognize digits independent of speaker |
| [44] | Floating Gate MOS | Non-Spiking | Circuit | Pattern Recognition | Designed FGMOS based neuron model |

# 4 Applications

ANN hardware is being implemented that can be used for various applications. Neural network applications includes biomedical applications, pattern recognition, robotics and brain machine interface etc.

## 4.1 Biomedical

Prior work proposed a neural network which can configure itself automatically called as (ARANN) architecture to deal with unexpected errors similar to self healing and self recovery concept in nervous system of humans [45]. Another work proposes FPGA implementation based hardware that can be used for detection of nematodes in sheep flocks [46]. Other application includes hardware implementation of convolutional neural network for seizure detection [47] and network-on-chip design for real cerebellum [48].

## 4.2 Signal Processing and Classification/Pattern Recognition

Prior research work demonstrates development of Pattern recognition system for detection of sulphate reducing bacteria [49]. Vibroacoustic signal classifier was designed and tested for FPGA implemented neural network [50]. A lot of work has been done that focuses on application called as Pattern Recognition [23, 25, 27, 28, 30, 31, 33, 40, 44]. Some work also shows that developed design can be used for spoken digit recognition [34]. Another application is for image processing where the designed circuit is meant for gray scale image processing [41].

### 4.3 Robotics

Hardware based implementation of an artificial neural network has been validated for a task for controlling locomotion in a quadrupled walking robot [43]. Other work demonstrates a low power silicon spiking neuron for real time controlling of articulated robot [51], and self repairing robot vehicle using FPGA [52].

### 4.4 Brain Machine Interface

Some recent research focuses on proposal of Hardware processing unit for brain machine interface using implementation on FPGA [53] while another application is development of Brain machine interface controller in real time [54]. Recently, a Brain Machine Interface system has been implemented on FPGA for simulating mental actions [55].

## 5 Conclusion

In this review paper, we focussed on the research work done in the field of hardware implementation of neuromorphic systems. We introduced the concept of neuromorphic in Sect. 1. Further sections showed survey of various training algorithms. Recent literature shows that much research is being carried out on neural systems applying the variants of backpropagation and synaptic plasticity rules. Further, neuromorphic circuits and devices section focused on various devices that can be used for hardware implementation of neural network. A lot of work has been done using memristive devices and thus leading to application in low power, adaptive and compact neuromorphic systems. Systems using FGMOS has very good compatibility with CMOS circuits. Research on implementation of artificial neural network using spintronics device is in pace. Further, these designs and their hardware implementation are being used for validating the model of neural network but many more applications are yet to come in the future in the field of robotics, Biomedical, Signal processing, Healthcare system, Brain Machine Interface and Image processing.

## References


1. Hu, M., Li, H., Chen, Y., Wu, Q., Rose, G.S., Linderman, R.W.: Memristor crossbar-based neuromorphic computing system: a case study. IEEE Trans. Neural Netw. Learn. Syst. **25**, 1864–1878 (2014)
2. Liu, C., Hu, M., Strachan, J.P., Li, H.: Rescuing memristor-based neuromorphic design with high defects. In: IEEE Design Automation Conference, June 2017
3. Nafea, S.F., Dessouki, A.A., El-Rabaie, S., El-Sayed.: Memristor Overview up to 2015. Menoufia J. Electron. Eng. Res. (MJEER), 79–106 (2015)
4. Camilleri, P., Giulioni, M., Dante, V., Badoni, D., Indiveri, G., Michaelis, B.: A neuromorphic a VLSI network chip with configurable plastic synapses. In: International Conference of HIS, September 2007

5. Jesus, O.D., Hagan, M.T.: Backpropagation algorithms for a broad class of dynamic networks. IEEE Trans. Neural Netw. **18**, 14–27 (2017)
6. Nuno-Maganda, M.A., Arias-Estrada, M., Torres-Huitzil, C., Girau, B.: Hardware implementation of spiking neural network classifiers based on backpropagation-based learning algorithms. In: IEEE International Joint Conference on Neural Networks, June 2009
7. Krestinskaya, O., Salama, K.N., James, A.P.: Learning in memristive neural network architectures using analog backpropagation circuits. IEEE Trans. Circuits Syst. I **66**(2), 719–732 (2019)
8. Aziz, N.A., Latif, M.A.K.A., Abdullah, W.F.H., Tahir, N.M., Zolkapli, M.: Hardware implementation of backpropagation algorithm based on CHEMFET sensor selectivity. In: IEEE International Conference on Control System, Computing and Engineering, January 2014
9. Krestinskaya, O., Salama, K.N., James, A.P.: Analog backpropagation learning circuits for memristive crossbar neural networks. In: IEEE International Symposium on Circuits and Systems (ISCAS), May 2018
10. Vo, H.M.: Implementing the on-chip backpropagation learning algorithm on FPGA architecture. In: IEEE International Conference on System Science and Engineering (ICSSE), July 2017
11. Vo, H.M.: Training on-chip hardware with two series memristor based backpropagation algorithm. In: IEEE International Conference on Communications and Electronics (ICCE), July 2018
12. Chakraborty, I., Roy, D., Roy, K.: Technology aware training in memristive neuromorphic systems for nonideal synaptic crossbars. IEEE Trans. Emerg. Top. Comput. Intell. **2**(5), 335–344 (2018)
13. Shouval, H.Z., Wang, S.S.-H., Wittenberg, G.M.: Spike timing dependent plasticity: a consequence of more fundamental learning rules. Front. Comput. Neurosci. **4**, 19 (2010)
14. Song, S., Miller, K.D., Abbott, L.F.: Competitive Hebbian learning through spike-timing-dependent synaptic plasticity. Nat. Neurosci. **3**(9), 919–926 (2000)
15. Markram, H., Gerstner, W., Sjostrom, P.J.: Spike-timing-dependent plasticity: a comprehensive overview. Front. Synaptic Neurosci. **4**(2) (2012)
16. Jin, X., Rast, A., Galluppi, F., Davies, S., Furber, S.: Implementing spike-timing-dependent plasticity on SpiNNaker neuromorphic hardware. In: IEEE World Congress on Computational Intelligence (WCCI), July 2010
17. Yousefzadeh, A., Masquelier, T., Serrano-Gotarredona, T., Linares-Barranco, B.: Hardware implementation of convolutional STDP for on-line visual feature learning. In: IEEE International Symposium on Circuits and Systems (ISCAS), May 2017
18. Pedroni, B.U., Sheik, S., Joshi, S., Detorakis, G., Paul, S., Augustine, C., Neftci, E., Cauwenberghs, G.: Forward table-based presynaptic event-triggered spike-timing-dependent plasticity. In: IEEE Biomedical Circuits and Systems Conference (BioCAS), October 2016
19. Frenkel, C., Indiveri, G., Legat, J.-D., Bol, D.: A fully-synthesized 20-gate digital spike-based synapse with embedded online learning. In: IEEE International Symposium on circuits and systems (ISCAS), May 2017
20. Fusi, S., Annunziato, M., Badoni, D., Salamon, A., Amit, D.J.: Spike-driven synaptic plasticity: theory, simulation, VLSI implementation. J. Neural Comput. **12**(10), 2227–2258 (2000)
21. Azghadi, M.R., Iannella, N., Al-Sarawi, S.F., Indiveri, G., Abbott, D.: Spike-based synaptic plasticity in silicon: design, implementation, application, and challenges. Proc. IEEE **102**, 717–737 (2014)

22. Covi, E., George, R., Frascaroli, J., Brivio, S., Mayr, C., Mostafa, H., Indiveri, G., Spiga, S.: Spike-driven threshold-based learning with memristive synapses and neuromorphic silicon neurons. J. Phys. D Appl. Phys. **51**(34), 344003 (2018)
23. Zhang, Y., Wang, X., Friedman, E.G.: Memristor-based circuit design for multilayer neural networks. IEEE Trans. Circuits Syst.–I **65**(2), 677–686 (2018)
24. Alom, M.Z., Taha, T.M., Yakopcic, C.: Memristor crossbar deep network ımplementation based on a convolutional neural network. In: International Joint Conference on Neural Networks (IJCNN), July 2016
25. Bayat, F.M., Prezioso, M., Chakrabarti, B., Nili, H., Kataeva, I., Strukov, D.: Implementation of multilayer perceptron network with highly uniform passive memristive crossbar circuits. Nat. Commun. **9**, 2331 (2018)
26. Krishnaprasad, A., Choudhary, N., Das, S., Kalita, H., Dev, D., Ding, Y., Tetard, L., Chung, H.-S., Jung, Y., Roy, T.: Artificial neuron using vertical MoS2/Graphene threshold switching memristors. Sci. Rep. **9**, 53 (2019)
27. Rosenthal, E., Greshnikov, S., Soudry, D., Kvatinsky, S.: A fully analog memristor-based neural network with online gradient training. In: IEEE International Symposium on Circuits and Systems (ISCAS), May 2016
28. Choi, S., Jang, S., Moon, J.H., Kim, J.C., Jeong, H.Y., Jang, P., Lee, K.J., Wang, G.: A self-rectifying TaOy/nanoporous TaOx memristor synaptic array for learning and energy-efficient neuromorphic systems. NPG Asia Mater. **10**, 1097–1106 (2018)
29. Chen, Y., Li, H., Yan, B.: Challenges of memristor based neuromorphic computing system. Sci. China Inf. Sci. **61**, 060425 (2018)
30. Fukami, S., Borders, W.A., Kurenkov, A., Zhang, C., DuttaGupta, S., Ohno, H.: Use of analog spintronics device in performing neuro-morphic computing functions. In: IEEE Berkeley Symposium on Energy Efficient Electronic Systems and Steep Transistor Workshop (E3S), October 2017
31. Zhang, D., Zeng, L., Cao, K., Wang, M., Peng, S., Zhang, Y., Zhang, Y., Klein, J.-O., Wang, Y., Zhao, W.: All spin artificial neural networks based on compound spintronic synapse and neuron. IEEE Trans. Biomed. Circuits Syst. **10**(4), 828–836 (2016)
32. Fukami, S., Ohno, H.: Perspective: spintronic synapse for artificial neural network. J. Appl. Phys. **124**(15), 151904 (2018)
33. Sengupta, A., Parsa, M., Han, B., Roy, K.: Probabilistic deep spiking neural systems enabled by magnetic tunnel junction. IEEE Trans. Electron Device **63**(7), 2963–2970 (2016)
34. Torrejon, J., Riou, M., Araujo, F.A., Tsunegi, S., Khalsa, G., Querlioz, D., Bortolotti, P., Cros, V., Yakushiji, K., Fukushima, A., Kubota, H., Yuasa, S., Stiles, M.D., Grollier, J.: Neuromorphic computing with nanoscale spintronic oscillators. Nat. Lett. **547**, 428–431 (2017)
35. Grollier, J., Querlioz, D., Stiles, M.D.: Spintronic nano-devices for bio-inspired computing. Proc. IEEE **104**, 2024–2039 (2016)
36. Sengupta, A., Yogendra, K., Roy, K.: Spintronic devices for ultra-low power neuromorphic computation. In: IEEE International Symposium on Circuits and Systems (ISCAS), May 2016
37. Babu, V.S., Rose Katharine, A.A., Baiju, M.R.: Adaptive neuron activation function with FGMOS based operational transconductance amplifier. In: IEEE Computer Society Annual Symposium on VLSI, May 2016
38. Keles, F., Yildirim, T.: Low voltage low power neuron circuit design based on subthreshold FGMOS transistors and XOR implementation. In: International Workshop on Symbolic and Numerical Methods, Modeling and Applications to Circuit Design (SM2ACD) (2010)
39. Sridhar, R., Kim, S., Shin, Y.-C., Bogineni, N.C.: Programmable Analog Synapse and Neural Networks Incorporating Same, United States (1994)

40. Fernandez, D., Villar, G., Vidal, E., Alarcon, E., Cosp, J., Madrenas, J.: Mismatch-tolerant CMOS oscillator and excitatory synapse for bioinspired image segmentation. In: IEEE International Symposium of Circuits and Systems, May 2005
41. Flak, J., Laihot, M., Halonen, K.: Binary cellular neural/nonlinear network with programmable floating-gate neurons. In: IEEE International Workshop on Cellular Neural Networks and their Applications, May 2005
42. Lu, D.D., Liang, F.-X., Wang, Y.-C., Zeng, H.-K.: NVMLearn: a simulation platform for non-volatile-memory-based deep learning hardware. In: IEEE International Conference on Applied System Innovation (ICASI), May 2017
43. Nakada, K., Asai, T., Amemiya, Y.: Analog CMOS implementation of a CNN-based locomotion controller with floating-gate devices. IEEE Trans. Circuits Syst.–I **52**(6), 1095–1103 (2005)
44. Kele, F., Yldrm, T.: Pattern recognition using N-input neuron circuits based on floating gate MOS transistors. In: IEEE EUROCON, May 2009
45. Jin, Z. Cheng, A.C.: A self-healing autonomous neural network hardware for trustworthy biomedical systems. In: IEEE International Conference on Field Programmable Technology, December 2011
46. Rahnamaei, A., Pariz, N., Akbarimajd, A.: FPGA implementation of an ANN for detection of anthelmintics resistant nematodes in sheep flocks. In: IEEE Conference on Industrial Electronics and Applications (ICIEA), May 2009
47. Heller, S., Hugle, M., Nematollahi, I., Manzouri, F., Dumpelmann, M., Schulze-Bonhage, A., Boedecker, J., Woias, P.: Hardware implementation of a performance and energy-optimized convolutional neural network for seizure detection. In: IEEE Annual International Conference of IEEE Engineering in Medicine and Biology Society (EMBC), July 2018
48. Luo, J., Coapes, G., Mak, T., Yamazaki, T., Tin, C., Degenaar, P.: Real-time simulation of passage-of-time encoding in cerebellum using a scalable FPGA-based system. IEEE Trans. Biomed. Circuits Syst. **10**(3), 742–753 (2016)
49. Tan, E.T., Halim, Z.A.: Development of an artificial neural network system for sulphate-reducing bacteria detection by using model-based design technique. In: IEEE Asia Pacific Conference on Circuits and Systems, December 2012
50. Dabrowski, D., Jamro, E., Cioch, W.: Hardware implementation of artificial neural networks for vibroacoustic signals classification. Acta Physica Polonica Ser. A **118**(1), 41–44 (2010)
51. Menon, S., Fok, S., Neckar, A., Khatib, O., Boahen, K.: Controlling articulated robots in task-space with spiking silicon neurons. In: IEEE International Conference on Biomedical Robotics and Biomechatronics RAS/EMBS, August 2014
52. Liu, J., Harkin, J., McDaid, L., Halliday, D.M., Tyrrell, A.M., Timmis, J.: Self-repairing mobile robotic car using astrocyte neuron networks. In: IEEE International Joint Conference on Neural Networks (IJCNN) (2016)
53. Wang, D., Hao, Y., Zhu, X., Zhao, T., Wang, Y., Chen, Y., Chen, W., Zheng, X.: FPGA implementation of hardware processing modules as coprocessors in brain-machine interfaces. In: IEEE Annual International Conference of Engineering in Medicine and Biology Society (EMBS) (2011)
54. Kocaturk, M., Gulcur, H.O., Canbeyli, R.: Toward building hybrid biological/in silico neural networks for motor neuroprosthetic control. Front. Neurorobotics **9**(8), 496 (2015)
55. Malekmohammadi, A., Mohammadzade, H., Chamanzar, A., Shabany, M., Ghojogh, B.: An efficient hardware implementation for a motor imagery brain computer interface system. Scientia Iranica **26**, 72–94 (2019)

# A Computer Vision Based Fall Detection Technique for Home Surveillance

Katamneni Vinaya Sree(✉) and G. Jeyakumar

Department of Computer Science and Engineering, Amrita School of Engineering, Coimbatore, Amrita Vishwa Vidyapeetham, Coimbatore, India
cb.en.p2cse17011@cb.amrita.students.edu, g_jeyakumar@cb.amrita.edu

**Abstract.** In this modern era where the population and life expectancy are continuously increasing, the demand for an advanced healthcare system is increasing at an unprecedented rate. This paper presents a novel and cost-effective fall detection system for home surveillance which uses a surveillance video to detect the fall. The advantage of the proposed system is that it doesn't need the person to carry or wear a device. The proposed system uses background subtraction to detect the moving object and marks it with a bounding box. Furthermore, few rules are based on the measures extracted from the bounding box and contours around the moving object. These rules are used with the transitions of a finite state machine (*FSM*) to detect the fall. It is done using the posture and shape analysis with two measures viz height and speed of falling. An alarm is sent when the fall is confirmed. The proposed approach is tested on three datasets *URD*, *FDD* and *multicam*. The obtained results show that proposed system works with an average accuracy of 97.16% and excels the previous approaches.

**Keywords:** Fall detection · Home surveillance · Finite state machine · Health care

## 1 Introduction

The part of the population consisting of elderly people is growing day by day. This increase in the population demands improvement in health care. [1] From 2011 to 2028 the deaths due to fall are expected to increase by 138% [2]. One of the major care that can be taken about elderly people is their injuries due to fall. Based on the WHO statistics accidental falls have an effect of 35% to 42% on elderly people each year.

People experience frequent falls as they are infirm and weak from age-related biological challenges. According to United States international organization statistics almost 40% of the old people are currently living independently. The increase in this aged population is estimated quite a pair of billions by 2050. For such growing population, some assistance is needed to lead an independent life. Such assistance, if unmanned, can make their life easier and even surveys say that these elderly people are ready to welcome the concept of such surveillance for their improved security and


S. Smys et al. (Eds.): ICCVBIC 2019, AISC 1108, pp. 355–363, 2020.
https://doi.org/10.1007/978-3-030-37218-7_41

safety. So, the amount spent for these injuries will increase to 70% over a period from 2018 to 2035.

When developing a fall detection system, one should understand "when people fall" and in "what kind of scenarios" it happens. So "when people fall" is vital to understand while we design the sample cases as it impacts the performance of the designed system.

In [3] they state reasons for the above two questions, their study, contains more than 75,000 falls from a hospital, they provide proper information to support the reasons derived for the above raised questions. Age, activities, type of floor they walk and footwear are the factors that are analyzed. In resident's room 62% of the falls took place and 22% of the falls in common areas. But the data did not reveal how much time the resident has spent in each of these places. They state that an elderly person falls once or twice a year [7]. One important data observed in this study is most of the falls occur in the morning and there is 60% more risk from these falls as the caretakers are unavailable. And the other reason might be due to the medication that is used by them last night as most of them are sedative to give them good sleep.

Studies related to fall can be divided as fall prevention, fall detection/identification and injury reduction. It is difficult to prevent people from fall as there will be very little time to react. But a solution can be drawn similar to a biker wearing a helmet to protect him/her in case of fall [4]. As we don't have chances to prevent the falls all that can be done is identification/detecting a fall accurately and try to reduce the loss that happens.

The rest of the paper is arranged as follows Sect. 2 discusses the previous approaches for fall detection. Section 3 introduces the design and implementation of the proposed approach. Next, Sect. 4 summarizes the results of the experiment and compares with the previous works. Conclusion and future work are briefed out in Sect. 5.

## 2 Related Works

There are large verities of products that are available for detecting the fall. But almost all of them are designed to send some alerts after the fall occurred [5]. This section discusses three systems which are available to give a brief view to these kinds of systems. The most commonly used product is from Sweden which is designed as a wearable device for arm with a button to give alarm. When person falls they are required to press the alarm button so that they get help from the nearest base station available, but the drawback of this method is if the person is not able to press the button then the base station can't be alarmed about the fall. Another company from *USA* has designed one system which is almost same as the above discussed product but instead of a security alarm they used an accelerometer [6]. Advantage of this product is that it detects the falls automatically, by analyzing whether the user is moving after the fall or not based on which it decides to give off the alarm. But the disadvantage of this is its limited range; it works up to 180 m only from the base station.

A Swedish company also has a similar type of product but the only difference is that it works on the user's mobile phone. So that it makes the user comfortable to leave the mobile in home with the system still functioning, it detects the person's location and his actions using global positioning system (*GPS*).

A company from Sweden has a similar one as above, with little difference as the system runs on a cellphone used by the user. This helps the user to have his/her own lifestyle as carrying or having a cellphone is very common now-a-days due to advancement in technology. But this system cannot detect the fall if the person is not carrying his/her cellphone. And additionally it keeps track of the *GPS* of the person using it.

The working environment for the above discussed solutions differs. The solution presented in this paper has specific range only due to positioning of the camera. So this solution will not work on its own.

The study carried out gave better understanding on how existing solutions for the fall detection problem work. This gave the novel idea used in the proposed system presented in this paper. A brief summary of the studied methodologies is discussed below and all the referred works are mentioned in the references at the end.

The most common techniques use videos along with accelerometers. Some approaches [13, 14] and [15] use different type of sensing devices and detects the abnormal motions which relate to falls in daily life. But the drawback with sensors is it needs lot of data to determine normal behavior. To overcome these issues a vision based approach is used, in the proposed system, to monitor and detect the falls. This system tracks the moving person continuously from surveillance video and based on the parameters it detects the fall.

On taking into view all the studied methods and approaches [8, 11] and [16] to solve the fall detection problem a novel approach with low cost based on vision computing is proposed. The design and implementation is discussed in the next section. The proposed method uses the video from a surveillance camera and detects occurrence of fall. The method is tested on three different datasets and the results are discussed.

## 3 Design and Implementation

Figure 1 shows the proposed system architecture. Initially the surveillance video is taken and is divided into frames. For each of these frames background subtraction is applied and preprocessing is carried out. Next all the required features are extracted. Then, based on the features, conditions are framed and a finite state machine is formulated to decide the fall.

The vital task in human fall detection is to detect human object accurately. The proposed system employs background subtraction and contour finding for this tracking. The system initially obtains the moving objects in the video using background subtraction and then does foreground segmentation, then eliminates if any noise is present using noise elimination such as morphological operations. Whenever a moving object is identified, the system continuously tracks the object. In the proposed system the object of interest is human being. The moving human being is continuously tracked and the detection is represented using a bounding box around the moving object. From [12, 17] and [18], studied how these moving objects can be detected in the video and formulated the procedure to detect and extract the required features from the bounding box.

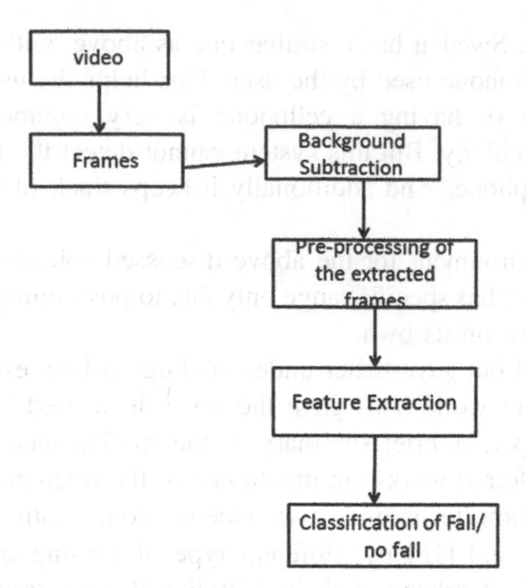


**Fig. 1.** Architecture of proposed system

**Background Subtraction -** As video is the input to the proposed system, the input frames of the video are extracted and a background subtraction algorithm is applied in the proposed method. This system applies background subtraction *MOG2* subtractor from *OpenCv* and employed a Gaussian mixture model so that it will have less variance in differentiating the background and foreground pixels. Based on the output, the pixels are partitioned as foreground and background and marked.

**Pre-Processing of the Extracted Frames -** A morphological model is employed to remove any noise such as fills of holes in the moving objects. In the previous step major noise are removed by background subtraction. So the region of interest is identified which is a moving object in foreground.

**Feature Extraction -** After the extraction of required frames with all the preprocessing done, the system extracts contours and all the required measures from the bounding box. The bounding box is a rectangle drawn around the moving object. The features considered are *aspect ratio*, *horizontal and vertical gradients*, *fall angle*, *acceleration*, *height* and *falling speed*. The usage of the measures for fall detection is explained below.

- ***Aspect Ratio*** - Aspect ratio is the measure that is largely affected when a person posture is changed from standing to any other abnormal position. So, this measure is taken into consideration in this approach.
- ***Horizontal and vertical gradients*** - The measure that is mostly affected when a person falls is their horizontal and vertical gradients in simple terms it is *X* and *Y* axes values so this measure is considered.
- ***Fall Angle*** - Fall angle is the measure calculated from the centroid of the falling object with respect to the horizontal gradient. So when the fall angle is less than 45° it can be observed that there is major chance that the person ha fell down. But, this

measure alone cannot determine the fall as sometimes when a person is sitting or bending the fall angle will be less than 45°.

- ***Change in height*** - Whenever there is a change in the posture the height of the human body varies so considering this is worthy, it is calculated as a difference of the mean height of human body and the height of the detected human at that time frame.

Based on the features extracted a rule based state machine is designed to capture and analyze the bounding box values continuously. The fall detection system is designed as a three state finite machine and framed 8 rules in total to decide the state transition. Figure 2 describes the finite state machine and the states used: *Fall_detection*, *Posture _changed* and *Fall_confirmation*.

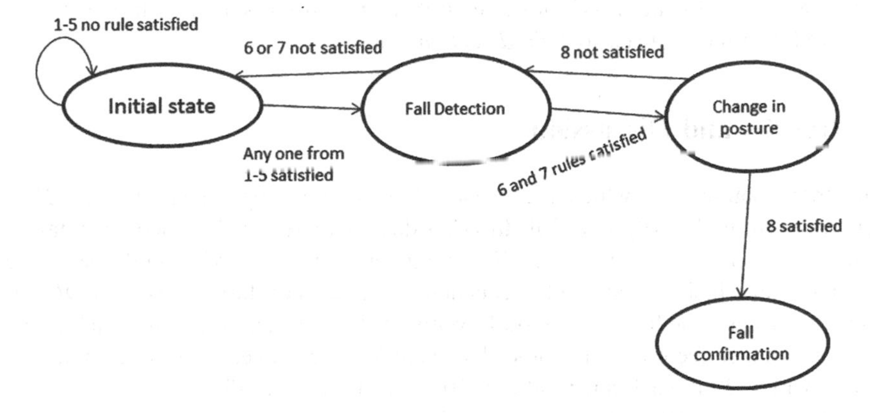


**Fig. 2.** FSM for detection of fall

**State Transition -** The rules that are framed for the *FSM* are as follows

1. ***Rule1*** - Let theta be the fall angle measure from the centroid coordinates ($C_1$,$C_2$) which is the center of mass coordinate of the object. This rule is satisfied if the theta value is less than 45°.
2. ***Rule2*** - Acceleration is one of the good factor that can be considered when detecting fall. When a person falls the acceleration will be lower than 130 $cm^2$ [7].
3. ***Rule3*** - Height is also a good factor as the height of the person will be decreasing during a fall so a threshold of 75 cm is considered and this factor is considered true when the detected person height is less that 75 cm.
4. ***Rule4*** - This rule is to check the posture of the person and is based on the height measure of the person detected. In the earlier works [10], they differentiated the standing and lying postures using a bounding box, whereas this system considers ratio of the height of the person in frame at time *t* to the mean height of the person. This threshold ranges between 0.4 and 0.45. So if the measure is less than 0.45 it is considered as fall otherwise as lying posture.
5. ***Rule5*** - The other measure for posture check is falling speed which is concerned a scenario like a person can simply lay down without falling. To identify this, the

proposed system measures the speed at which the fall occurs. This is calculated as the time difference of person standing or sitting to the time when the person detected to lay down. If the time difference is less than a threshold *T* it is considered as fall otherwise lying posture.
6. ***Rule6*** - This rule is to confirm the fall and generate alarm. In this rule, the system checks for any movement after a fall. If there is little movement or no movement it is considered as fall otherwise no fall.

At the beginning of the video, the *FSM* of the system is in an initial state. If any of the rules from rule 1 to rule 5 is satisfied, the *FSM* moves to *Fall_detection* state. From the *Fall_detection* state, the system analyzes to find any change in posture using rules 6 and 7. If there is no change i.e., 6 and 7 rules are not satisfied the *FSM* moves to the initial state. Else, if satisfied it moves to *posture_changed* state. Then checks for rule 8, if rule 8 is satisfied the *FSM* moves to *Fall_confirmation* state and trigger that a fall occurred else moves back to *Fall_detection* state.

## 4 Results and Discussion

The three datasets on which the proposed method is applied are *URD, FDD* and *Multicam* from [18, 19] and [20]. In *URD* dataset there are 70 videos in total out of which 30 are fall events and other 40 are regular events. In *multicam* dataset there are 24 videos in which 22 have fall events and rest 2 has non-fall events. In *FDD* dataset there are videos for fall and non-fall events in different environments such as office, lecture room, coffee room and home. There are 30 videos in each of these environments out of which 20 in each have fall and 10 in each are non-fall events.

Table 1 represents the results obtained on applying the proposed approach on these datasets. Here TP represents the number of fall events identified as fall events and *TN* represents the number of non-fall events identified as non-fall events. *FP* represents number of non-fall events identified as fall and *FN* represents number of fall events identified as non-fall.

**Table 1.** Recognition results

| Dataset | No. of events | TP | FP | TN | FN |
|---|---|---|---|---|---|
| *URD* | 70 | 30 | 2 | 38 | 0 |
| Multicam | 24 | 21 | 0 | 2 | 1 |
| *FDD* Office | 30 | 15 | 2 | 13 | 0 |
| *FDD* Lecture room | 30 | 14 | 0 | 15 | 1 |
| *FDD* Coffee room | 30 | 15 | 0 | 15 | 0 |
| *FDD* Home | 30 | 15 | 0 | 15 | 0 |

In all the above five sets three measures of evaluation are calculated and Table 2 represents those results:

**Table 2.** Performance of the system

| Dataset | No. of events | Accuracy | Sensitivity | Specificity |
|---|---|---|---|---|
| *URD* | 70 | 97.14 | 100 | 95 |
| Multicam | 24 | 95.83 | 95.45 | 100 |
| *FDD* Office | 30 | 93.33 | 100 | 86.67 |
| *FDD* Lecture room | 30 | 96.67 | 93.33 | 100 |
| *FDD* Coffee room | 30 | 100 | 100 | 100 |
| *FDD* Home | 30 | 100 | 100 | 100 |

The above results show that the proposed approach can detect a fall with an average accuracy of 97.16% and with sensitivity of 98.13% and a specificity of 96.45%. Some of the frames of fall detection are as shown in Fig. 3.

**Fig. 3.** Sample screenshots of fall detection

From the results it is proven that the proposed system gave an accuracy of 97.16% in detecting the fall even in different environments and with different camera positions. This approach excels in performance when compared with the existing approaches [9] and [10].

## 5 Conclusion and Future Work

The advantage of the proposed method is its high accuracy in detecting the fall even with any environmental change or camera positions. This approach doesn't need the person to carry any wearable device; it can solely be done based on a surveillance camera. Moreover, this is a cost-effective system as it doesn't need any special devices to achieve the objective.

Future enhancement of this system can be employing some deep learning or any optimization techniques so as to speed up the computation. An addition of an alert system specific to person and the venue for necessary immediate action would be a better enhancement of this system.

**Compliance with Ethical Standards**

- All authors declare that there is no conflict of interest.
- No humans/animals involved in this research work.
- We have used our own data.

## References

1. Den aldrandebefolknigen. Karin Modig (2013). http://ki.se/imm/den-aldrande-befolkningen
2. Büchele, G., Becker, C., Cameron, I.D., Köning, H.-H., Robinovitch, S., Rapp, K.: Epidemiology of falls in residential aged care: analysis of more than 75,000 falls from residents of Bavarian nursing homes. JAMDA **15**(8), 559–563 (2014)
3. Hövding. Hövding den nyacyklehjälmen (2015). http://www.hovding.se/
4. Tao, J., Turjo, M., Wong, M., Wang, M., Tan, Y.: Fall incident detection for intelligent video surveillance. In: Fifth International Conference on Information, Communication and Signal Processing (2005)
5. Luo, S., Hu, Q.: A dynamic motion pattern analysis approach to fall detection. In: IEEE International Workshop in Biomedical Circuit and Systems (2004)
6. Rougier, C., Meunier, J.: Demo: fall detection using 3D head trajectory extracted from a single camera video sequence. J. Telemed. Telecare **11**(4), 7–9 (2018)
7. Vishwakarma, V., Mandal, C., Sural, S.: Automatic detection of human fall in video. In: Pattern Recognition and Machine Intelligence, PReMI. Lecture Notes in Computer Science, vol. 4815. Springer, Heidelberg (2007)
8. Nasution, A.H., Emmanuel, S.: Intelligent video surveillance for monitoring elderly in home environments. In: Proceedings of the IEEE 9th International Workshop on Multimedia Signal Processing (MMSP 2007), Crete, Greece, pp. 203–206, October (2007)
9. Rougier, C., Meunier, J., St-Arnaud, A., Rousseau, J.: Fall detection from human shape and motion history using video surveillance. In: Proceedings of the 21st International Conference on Advanced Information Networking and Applications Workshops (AINAW 2007), pp. 875–880, May (2007)
10. Meunier, J., St-Arnaud, A., Cleveert, D., Unterthiner, T., Povysiil, G., Hochreiter, S.: Rectified factor networks for biclustering of omics data. Bioinformatics **33**(14), i59–i66 (2017)
11. Sreelakshmi, S., Vijai, A., Senthil Kumar, T.: Detection and segmentation of cluttered objects from texture cluttered scene. In: Proceedings of the International Conference on Soft Computing Systems, vol. 398, pp. 249–257. Springer (2016)
12. Broadley, R.W., Klenk, J., Thies, S.B., Kenney, L.P.J., Granat, M.H.: Methods for the real-world evaluation of fall detection technology: a scoping review. Sensors (Basel) **18**(7), 2060 (2018)
13. Xu, T., Zhou, Y., Zhu, J.: New advances and challenges of fall detection systems: a survey. Appl. Sci. **8**, 418 (2018). https://doi.org/10.3390/app8030418
14. Birku, Y., Agrawal, H.: Survey on fall detection systems. Int. J. Pure Appl. Math. **118**(18), 2537–2543 (2018)
15. Mastorakis, G., Makris, D.: Fall detection system using kinect's infrared sensor. J. RealTime Image Process. **9**(4), 635–646 (2014)

16. Krishna Kumar, P., Parameswaran, L.: A hybrid method for object identification and event detection in video. In: National Conference on Computer Vision, Pattern Recognition, Image Processing and Graphics (NCVPRIPG), Jodhpur, India, pp. 1–4. IEEE Explore (2013)
17. Houacine, A., Zerrouki, N.: Combined curvelets and hidden Markov models for human fall detection. In: Multimedia Tools and Applications, pp. 1–20 (2017)
18. Auvinet, E., Rougier, C., Meunier, J., St-Arnaud, A., Rousseau, J.: Multiple cameras fall dataset. Technical report 1350, DIRO - Université de Montréal, July (2010)
19. Charfi, I., Mitéran, J., Dubois, J., Atriand, M., Tourki, R.: Optimised spatio-temporal descriptors for real-time fall detection: comparison of SVM and Adaboost based classification. J. Electron. Imaging (JEI) **22**(4), 17 (2013)
20. Kwolek, B., Kepski, M.: Human fall detection on embedded platform using depth maps and wireless accelerometer. Comput. Methods Programs Biomed. **117**(3), 489–501 (2014). ISSN 0169-2607

# Prediction of Diabetes Mellitus Type-2 Using Machine Learning

S. Apoorva(✉), K. Aditya S(✉), P. Snigdha,
P. Darshini, and H. A. Sanjay

Nitte Meenakshi Institute of Technology, Bangalore, India
apoorvaapps68@gmail.com, snigdhapriya1805@gmail.com,
darshini1913@gmail.com,
{adityashastry.k, sanjay.ha}@nmit.ac.in

**Abstract.** Around 400 million people suffer from diabetes around the world. Diabetes prediction is challenging as it involves complex interactions or inter-dependencies between various human organs like eye, kidney, heart, etc. The machine learning (ML) algorithms provide an efficient way of predicting the diabetes. The objective of this work is to build a system using ML techniques for the accurate forecast of diabetes in a patient. The decision tree (DT) algorithms are well suited for this. In this work, we have applied the DT algorithm to forecast type 2 diabetes mellitus (T2DM). Extensive experiments were performed on the Pima Indian Diabetes Dataset (PIDD) obtained from the UCI machine learning repository. Based on the results, we observed that the decision tree was able to forecast accurately when compared to the SVM algorithm on the diabetes data.

**Keywords:** Prediction · Machine learning · Decision tree · Diabetes

## 1 Introduction

Diabetes mellitus can harm several parts of the human body. The pancreas can be affected due to diabetes and hence adequate insulin will not be created in the body. Diabetes is of two types [1]:

- TYPE -1: In type-1 diabetes, the pancreatic cells that produce insulin are destroyed. Hence, insulin injections need to be given to the patients affected by type-1 along with restricted nutritional diets and regular blood tests.
- TYPE 2: In diabetes type 2, the numerous body organs turn out to be resistant to insulin, and this increases the insulin demand. On this fact, pancreas doesn't produce the required amount of insulin.

The patients suffering from type-2 diabetes need to follow a strict diet, routine exercise along with regular blood glucose monitoring. The main reasons for this are obesity and absence of physical exercise [2]. The diabetics suffering from type-2 have Pre-Diabetes condition which is a disorder in which the levels of glucose in blood are complex than ordinary but not more than the diabetic patient [3].

Machine learning (ML) represents a field of artificial intelligence (AI) which enables systems to improve by experience by learning automatically devoid of being


S. Smys et al. (Eds.): ICCVBIC 2019, AISC 1108, pp. 364–370, 2020.
https://doi.org/10.1007/978-3-030-37218-7_42

programmed explicitly [4]. The ML techniques emphasize on program development that are capable of self-learning from data [5].

Here we are using different ML Algorithms which are powerful in both classification and predicting. This innovative projected study trails the different machine learning techniques to predict diabetes Mellitus at primary stage to save human life [6]. Such algorithms are SVM, Decision Tree to predict and raise the prediction accuracy and performance [7]. In the historical 30 years, with increasing number of diabetics, people have started to understand that this chronic disease has deeply wedged every family and everyone's daily life. In direction to research the high-risk group of DMs, we need to operate progressive information technology. Therefore, machine learning algorithm is suitable method for us. Diabetes sustaining glycemic control remains a challenge [8].

Due to its endlessly increasing existence, numerous are adversely affected by diabetes mellitus. Most of the diabetic patients are unaware of the pre-diabetes risks which lead to the actual disease [9]. In this work, our main objective is to build a model centered on machine learning methods for forecasting T2DM. The core difficulties that we are trying to resolve are to advance the accurateness of the estimate model, and to make the model adaptive to more than one dataset. The main tasks related to our work are gathering of data, Pre-processing of data and development of prediction model. The benchmark datasets related to diabetes may be obtained from standard websites such as UCI, KEEL, and Kaggle etc.

The main contributions of this work are emphasized below:

- Identification of patients suffering from T2DM.
- Presented a Decision Tree model for forecasting diabetes existing in pregnant women.
- Studied the analytical model using structures that are associated to the diabetes type 2 risk aspects.

The structuring of the paper is as follows. Section 2 describes the literature survey done on the different methods and techniques used to analyse the characteristics of diabetes type 2 and decide if it is a diabetes or not. Section 3 describes the Pima Indian dataset along with its input and output attributes. Section 4 describes the system design and demonstrates step by step procedure of the proposed system. The experimental results are demonstrated in Sect. 5. This is followed by the conclusion and future work.

## 2 Related Work

This section deliberates on some of the recent work carried out for diabetes prediction using ML algorithms and probabilistic approaches. In [2], the authors used two classification algorithms namely decision tree and Support Vector Machine (SVM) to predict which algorithm generates accurate results. In the work [3, 4] WEKA tool was used for executing the algorithms. The experiments were accomplished on the PIDD from UCI [5] having 768 medical records & 8 attributes of female patients. The datasets are given to the WEKA software to predict if the person has diabetes or not using the classification algorithms. Results are achieved by means of internal cross-validation of

10-folds. Accuracy, F-Measure, Recall, Precision, and ROC (Receiver Operating Curve) measures were utilized for model evaluation. It was observed by the authors that the decision tree demonstrated better accuracy than the SVM.

In [6], the authors utilized decision tree classifier and Bayesian network to predict if the person is diabetic or non-diabetic. Here the author is using two phases one is training and the other one is testing phases. In training phase individually record is related with a class label. Training process groups the data into classes 0 and 1. In testing Phase, the testing of algorithm is done on unseen dataset. Experiments were performed on PIDD of UCI [6]. In pre-processing they are changing the data into distinct values depending upon the series of each attribute. Class variable are classifying into two classes (0 or 1) where '1' indicates a positive test for diabetes and '0' indicates a negative test. Training dataset is the input. On validating test records, decision tree performed better than Bayesian network.

In [7], the authors utilized Decision tree classifier and "J48" algorithm existing in the WEKA software to predict if the person is diabetic or non-diabetic. A web based health center diabetes control system located in Tabriz was utilized as the data source. The features such as gender, systolic, age, BP, past family history of diabetes, and BMI were considered for experimentation. Among 60,010 records, 32,044 records were considered as they matched the 6 input feature values in the diagnosis field. The rest of the fields were all omitted. They have implemented 10-fold cross validation method to evaluate the model and calculate the Precision, Recall and Accuracy of the model. After the pre-processing of data and its subsequent preparation, a whole of 22,398 accounts were used for developing the decision tree.

## 3 Dataset Description

The dataset is taken for prediction is collected from "Pima Indians Diabetes Database" of UCI [5]. This dataset was acquired from the National Institute on diseases related to kidney, diabetes and digestive systems. The datasets are used to forecast the occurrence of diabetes in the patient based on certain parameters present in the dataset. It consists of 768 records of patient data. 80% of the data is taken for training and the remaining 20% is taken for testing. The output attribute to be predicted is a variable which is diagnostic and binary that indicates the occurrence of diabetes symptoms in the patient according to the World Health Organization. This standard states that diabetes occurs in a patient if the plasma glucose level during post load is at least 200 mg/dl. The pima diabetes dataset consists of 9 input attributes which are: Number of times pregnant, concentration of plasma glucose, blood pressure (Diastolic) in terms of mm Hg, thickness of triceps skin fold in mm, serum insulin measured over a 2 h period, Body mass index (in kg/m$^2$), function of the diabetes pedigree, age in years. The output attribute represents a binary class which is 1 if the person has diabetes and 0 otherwise.

The diabetes dataset is split into two parts i.e. test and train data sets.

- test.csv – 206
- train.csv – 562

Training Data: It is used to train an algorithm. The training data comprises of both input data and the equivalent output. Based on data, algorithm like decision tree, support vector mechanism can be applied to produce accurate decisions.

Test Data: Using test data we determine the result accuracy. It compares whether the outcome matches with the train data or not and if it matches then how much is the accuracy.

## 4 Proposed Work

This section demonstrates the stepwise procedure followed to build the system. In this work, we have made use of the decision tree. A DT represents a support tool for making decisions by utilizing tree-like structure which demonstrates a series of decisions and their possible outcomes. They are supervised learning algorithms and can be utilized for solving the problems related to regression and classification. The chief aim of DT algorithm is to achieve accurate classification through least number of decisions. The overall design of the diabetes prediction system is depicted in Fig. 1.

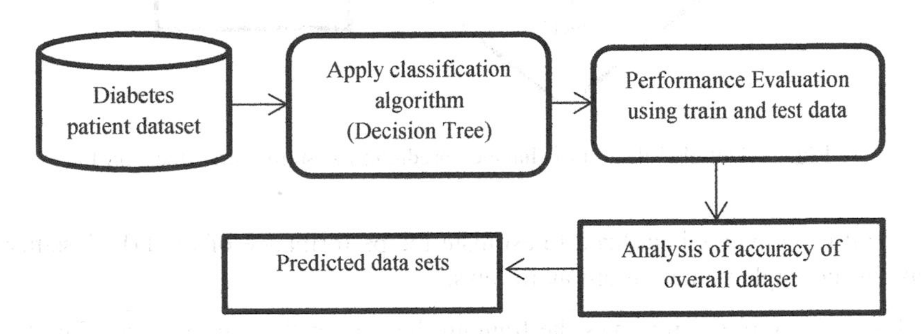


**Fig. 1.** Diabetes prediction analysis model

In this diabetes prediction analysis model, the diabetes dataset is collected from PIMA Indian Diabetes Dataset [5] for predicting diabetes. Then, the algorithm is used for the prediction by taking decision on the basis feature selection method which evaluates taking Glucose and BMI as its highest scorer in feature importance. After that, both the attributes are judged taking decision and performance evaluation is done. We used confusion matrix for the analysis of the performance which takes actual values and predicted values. Based on the confusion matrix technique, the complete accuracy is calculated for the prediction. Finally, there is an outcome file which presents the outcome predicting diabetes (i.e. 0 or 1). Here, 0 means person is not having diabetes and 1 means a person is having diabetes. The detailed design is depicted in Fig. 2 below.

The above given flowchart, starts with taking datasets on which feature selection is performed. Based on feature selection, we calculate feature importance score for all the attributes and the maximum scorer is considered for the further decision. Decision tree classifier divides the attributes in Healthy and Diabetic and finally predicts the diabetes.

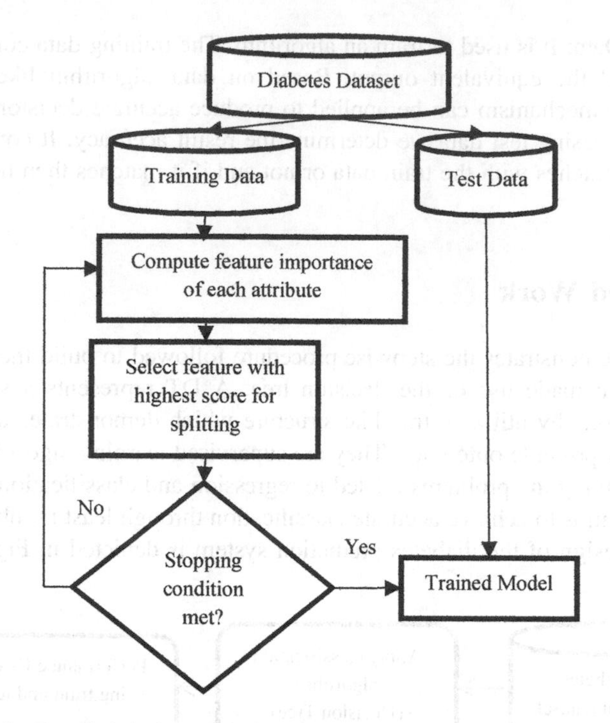


**Fig. 2.** Detailed design of diabetes prediction system using decision tree

The confusion matrix is utilized to estimate the performance of the DT classifier. The steps in the implementation are as follows:

- The training phase involves the training the DT on the training data. The training dataset comprises of examples that are used for learning and is able to fit the input attributes of the PIDD dataset. Most techniques that are used to search through training data are for realistic relationships tend to over fit the data, meaning that they can recognize and exploit obvious relations in the training data that do not embrace in general.
- Feature importance provides a score for every data attribute. A high score indicates that the feature is significant with respect to the output attribute. In this work, we observed that BMI and glucose features were possessing higher feature importance scores when compared to other features. Feature importance is computed using Gini index shown in Eq.-1

$$GI(T) = 1 - \sum_{i=1}^{n} p_j^2 \quad (1)$$

Where, T is a training data set containing samples from n classes, n is the number of classes and pj is the comparative frequency of class j in T.
- The testing phase consists of evaluating the DT model on the test dataset. The test dataset possesses similar distribution of samples as the training records but is independent of it. A better fit of the training samples as opposed to the test instances

leads to over-fitting. A test set is represents a set of instances which is used to evaluate the classifier performance.

Cross-validation (CV) is a technique of resampling which is used to assess ML models on less number of data samples. The parameter 'k' in the CV procedure represents the number of data splits. Due to this, the method is often referred to as the k-fold cross-validation. Test data is the 20% of the whole data and remaining is the train data. Train data is further split into sub-parts by applying the 5-cross validation procedure.

## 5 Experimental Results

This section describes the experimental results and comparison between decision tree and SVM for diabetes prediction. The experiments were conducted using SPYDER on windows 10 platform. Figure 3 depicts the comparison between SVM and DT classifiers for diabetes prediction.

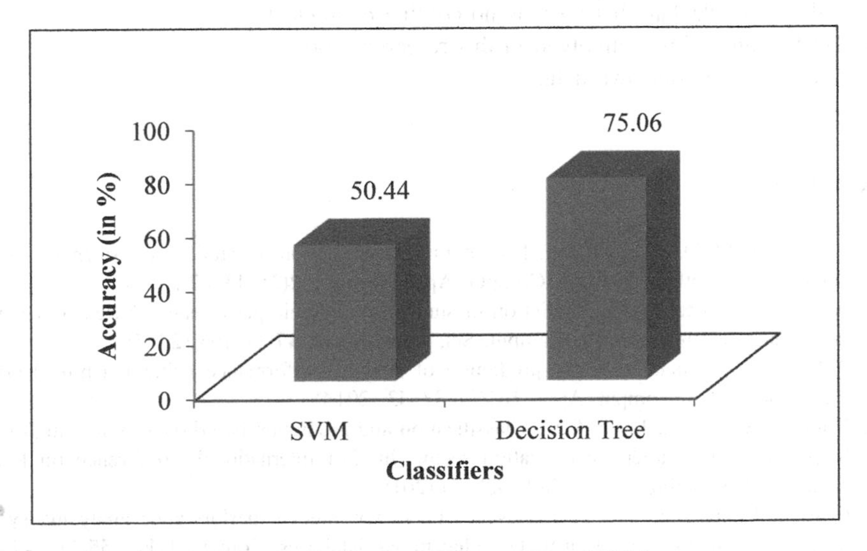


**Fig. 3.** Comparison between DT and SVM

The above graph represents the prediction of SVM and DT. Here, the comparison is done between both the algorithms. The accuracy of SVM is 50.44% and that of DT is 75.06%. Since the accuracy is high in case of DT, DT algorithm is better than SVM for the prediction of diabetes.

## 6 Conclusions and Future Scope

In conclusion, we would like to state that this work deals with the forecast of T2DM using ML algorithms. A model is developed with the intention to predict whether the person is having diabetes or not. Our work applies the DT classifier to detect diabetes. The DT algorithm has been used in many areas due to their ability to solve diverse problems, ease of use, and the model-free property. The main objective is to construct a diabetes prediction model that forecasts the output value based on various input variables. In DT algorithm, every leaf signifies a target attribute value while the input variable values are characterized by the path from root to leaf. Our predicting system based on decision tree is able to learn and detect whether the person is having diabetes or not. The DT classifier was shown to have an improvement of 24.62% over the SVM classifier with regards to diabetes prediction. In future, we plan to provide a cloud based platform for generalized prediction of diseases.

**Compliance with Ethical Standards**

- All authors declare that there is no conflict of interest.
- No humans/animals involved in this research work.
- We have used our own data.

## References

1. Ahmed, A.B.E.D., Elaraby, I.S.: Data mining: a prediction for student's performance using classification method. World J. Comput. Appl. Technol. **2**(2), 43–47 (2014)
2. Altaher, A., BaRukab, O.: Prediction of student's academic performance based on adaptive neuro-fuzzy inference. Int. J. Comput. Sci. Netw. Secur. **17**(1), 165 (2017)
3. Acharya, A., Sinha, D.: Early prediction of student performance using machine learning techniques. Int. J. Comput. Appl. **107**(1), 37–43 (2014)
4. Kaur, P., Singh, M., Josan, G.S.: Classification and prediction based data mining algorithms to predict slow learners in education sector. In: 3rd International Conference on Recent Trends in Computing 2015 (ICRTC-2015) (2015)
5. Guruler, H., Istanbullu, A., Karahasan, M.: A new student performance analysing system using knowledge discovery in higher educational databases. Comput. Educ. **55**(1), 247–254 (2010). https://doi.org/10.1016/j.compedu.2010.01.010. ISSN 0360-1315
6. Vandamme, J.-P., Meskens, N., Superby, J.-F.: Predicting academic performance by data mining methods. Educ. Econ. **15**(4), 405–419 (2007). https://doi.org/10.1080/09645290701409939
7. Abuteir, M., El-Halees, A.: Mining educational data to ımprove students' performance: a case study. Int. J. Inf. Commun. Technol. Res. **2**, 140–146 (2012)
8. Baradwaj, B.K., Pal, S.: Mining educational data to analyze students performance. Int. J. Adv. Comput. Sci. Appl. **2**(6), 63–69 (2011)
9. Shanmuga Priya, K., Senthil Kumar, A.V.: Improving the student's performance using educational data mining. Int. J. Adv. Netw. Appl. **04**(04), 1680–1685 (2013)

# Vision Based Deep Learning Approach for Dynamic Indian Sign Language Recognition in Healthcare

Aditya P. Uchil[1(✉)], Smriti Jha[2], and B. G. Sudha[3]

[1] The International School Bangalore (TISB), Bangalore, India
adityapuchil@gmail.com
[2] LNM Institute of Information Technology, Jaipur, India
smritijha369@gmail.com
[3] Xelerate, Bangalore, India
sudhagrr@gmail.com

**Abstract.** Healthcare for the deaf and dumb can be significantly improved with a Sign Language Recognition (SLR) system, which is capable of recognizing the medical terms. This paper is an effort to build such a system. SLR can be modelled as a video classification problem and here we have used a vision based deep learning approach. The dynamic nature of sign language poses additional challenge to the classification. This work explores the use of OpenPose with convolutional neural network (CNN) to recognize the sign language video sequences of 20 medical terms in Indian Sign Language (ISL) which are dynamic in nature. All the videos are recorded using common smartphone camera. The results show that even without the use of recurrent neural networks (RNN) to model the temporal information, the combined system works well with 85% accuracy. This eliminates the need for specialized camera with depth sensors or wearables. All the training and testing was done on CPU with the support of Google Colab environment.

**Keywords:** Indian sign language recognition · Convolutional neural network · OpenPose · Video classification · Pose estimation · Gesture recognition · Medical terms · Healthcare

## 1 Introduction

India's National Association of the Deaf estimates that 1.8 to 7 million people in India are deaf [1]. To assist the needs of the deaf and dumb community worldwide different sign language dictionaries have been developed with American Sign Language (ASL) being the most popular and common of all. A lot of work has been done for ASL recognition so that deaf people can easily communicate with others and others can easily interpret their signs. For Indian Sign language, recently researchers have started developing novel approaches which can handle static and dynamic gestures.

Signers have to communicate their conditions routinely at health care setups (hospitals). Signs of words used for conveying in health care settings comprise of a continuous sequence of gestures and are hence termed as dynamic gestures. A gesture


S. Smys et al. (Eds.): ICCVBIC 2019, AISC 1108, pp. 371–383, 2020.
https://doi.org/10.1007/978-3-030-37218-7_43

recognition system that recognises the dynamic gestures and classifies the sequence of gestures (video of sign) needs to be developed. As sensor hand gloves or motion tracking devices are expensive and not common in rural health care outlets, we have focused on developing a vision based solution for sign language recognition. Our work is the first attempt to recognize medical terms in ISL. This will involve video of the signer captured from a regular smartphone camera as input. This paper is organised as follows: The next section lists the related work done on Indian and other sign language recognition followed by the proposed methodology, results and discussion.

## 2 Related Works

*ISL*

Ansari and Harit [2] achieved an accuracy of 90.68% in recognizing 16 alphabets of ISL. Using Microsoft Kinect camera, they recorded images of 140 static signs and devised a combination strategy using SIFT (Scale-Invariant Feature Transform), SURF (Speeded Up Robust Features), SURF neural network, Viewpoint Feature Histogram (VFH) neural network that gave highest accuracy. Tewari and Srivastava [3] used 2D-Discrete Cosine Transformation (2D-DCT) and Kohonen Self Organizing Feature Map neural network for image compression and recognition of patterns in static signs. Their dataset consisted of static signs of ISL numerals (0 to 5) and 26 alphabets captured on a 16.0 MP digital camera (Samsung DV300F). Implemented in MATLAB R2010a, their recognition rate was 80% for 1000 training epochs with least time to process. Their approach did not take non-manual features (body, arm, facial expression) into account and included only hand gesture information making it suitable for static signs only. Singha and Das [4] worked on live video recognition of ISL. They considered 24 different alphabets for 20 people (total of 480 images) in the video sequences and attained a success rate of 96.25%. Eigen values and Eigen vector were considered as features. For classification they used Euclidean distance weighed on Eigen values. Kishore and Kumar [5] used a wavelet based video segmentation technique wherein a fuzzy inference system was trained on features obtained from Discrete wavelet transformation and Elliptical Fourier descriptors. To reduce feature vector size, they further processed the descriptor data using Principle component analysis (PCA). For the 80 signs dataset they achieved 96% accuracy. Tripathi et al. [6] used gradient based key frame extraction procedure to recognize a mixture of dynamic and static gestures. They created a database of ten sentences using a Canon EOS camera with 18 mega pixels, 29 frames per second. Feature extraction using Orientation Histogram with Principal Component Analysis was followed by classification based on Correlation and Euclidean distance. Dour and Sharma [7] acquired the input from an i-ball C8.0 web camera at a resolution of 1024 $\times$ 768 pixels and used adapted version of fuzzy inference systems (Sugeno-type and Adaptive-Neuro Fuzzy) to recognize the gestures. 26 signs from 5 different volunteers and 100 samples of each gesture were used for training purpose. So a total of around 2611 training samples was obtained. The classification of the input sign gesture was done by a voting scheme according to the clusters formed. The system was able to recognize for all alphabets. Most of the misclassified samples correspond to the gestures that are similar to each other like letters E and F. Rao and Kishore [8] used a smart phone front camera (5 M pixel) to generate a database of 18 signs in continuous sign language by 10 different signers. Minimum Distance (MD) and Artificial Neural Network (ANN) classifiers were used to train the feature space. The average word matching score (WMS) of this selfie recognition system was around 85.58% for MD and 90% for ANN with minute difference of 0.3 s in classification times. They found that ANN outperformed MDC by an upward 5% of WMS on changing the train and test data.

***Other Sign Languages***

Camgoz et al. [9] approached the problem as a neural machine translation task. The dataset presented "RWTH-PHOENIX-Weather 2014T" covered a vocabulary of 2887 different German words translated from 1066 different signs by 9 signers. Using sequence-to-sequence (seq2seq) learning method, the system learnt the spatio-temporal features, language model of the signs and their mapping to spoken language. They used different tokenization methods like state-of-the-art RNN-HMM (Hidden Markov Model) hybrids and tried to model conditional probability using attention-based encoder-decoders with CNNs. Konstantinidis et al. [10] took the LSA64 dataset of Argentinian sign language consisting of 3200 RGB videos of 64 different signs by 10 subjects. For uniform video length, they processed the gesture video sequences to compose of 48 frames each. They extracted features using a pretrained ImageNet VGG-19 network. This network up toconv4_4 was used for hand skeleton detection and initial 10 layers of this network for detecting body skeleton. They concluded that non manual features like body skeletal joints contain important information that when combined with hand skeletal joints information gave a 3.25% increase in recognition rate. Their proposed methodology classified both one-handed and two-handed signs of the LSA64 dataset and outperformed all variants by around 65% irrespective of the type of classifier or features they employed. Mahmoud et al. [11] extracted local features from each frame of the sign videos and created a Bag of features (BoFs) using clustering to recognize 23 words. Each image was represented by a histogram of visual words to generate a code book. They had further designed a two-phase classification model wherein the BoFs was used to identify postures and a Bag of Postures (BoPs) was used to recognize signs. Hassan et al. [12] used Polhemus G4 motion tracker and a camera to generate two datasets of 40 Arabic sentences. Features were extracted using window-based statistical features and 2D-DCT transformation and classified using three approaches: modified KNN (MKNN) and two different HMM toolkits (RASR and GT2 K). Polhemus G4 motion tracker measured hand position and orientation whereas the DG5-V Hand data gloves measured the hand position and configuration. They found that inspite of more precision in sensor-based data, vision-based data achieved greater accuracy.
Most of the approaches listed above either used

- Specialized camera like Microsoft Kinect which measures the depth information or motion tracker camera or
- Wearable like hand gloves with sensors and
- Extensive Image processing to extract and handcraft the features
- CNN + RNN for temporal information which is computationally expensive
- Static signs

Our proposed method models the problem based on computer vision and deep learning thus representing the sign language recognition as that of sequence of image classification. InceptionV3 model and OpenPose [13] are used to classify the image frames individually. A voting scheme is then applied which classifies the video based on the maximum vote for a particular category.

The following are the contributions towards this paper:

- Created the medical dataset comprising of 80 videos of 19 dynamic and 1 static sign for medical terms each performed by two different volunteers.
- Studied the effect of using OpenPose with CNN to improve the accuracy of recognition of medical signs directly from RGB videos without specialized hardware or hand gloves and extensive image processing or computationally intensive RNN.

Table 1 shows the list of medical terms used for our study. Except the word 'chest', all the other words are dynamic in nature.

**Table 1.** List of medical terms comprising our dataset

| | |
|---|---|
| 1. Allergy | 11. Common Cold |
| 2. Old | 12. Cough |
| 3. Heart Attack | 13. Diabetes |
| 4. Chest | 14. Ear |
| 5. Exercise | 15. Eye |
| 6. Bandage | 16. Chicken Pox |
| 7. Throat | 17. Forehead |
| 8. Blood pressure | 18. Injection |
| 9. Pain | 19. Mumps |
| 10. Choking | 20. Blood |

## 3 Proposed Methodology

The sign language recognition system can be modelled as a video classification problem and can be built using different approaches as mentioned in [14]. Both spatial and temporal information are used in most of the approaches for the classification to give the desired output. But the use of RNN for storing temporal information is a bottleneck as it is more computationally expensive and requires more storage. The two articles [15] and [16] support the same and it is proved that CNN is all that we need. Also researchers have now shifted focus to attention-based network and multi-task learning which formed the basis of OpenPose architecture. OpenPose [13] has proven its ability to identify human poses with face and hand key points which are very crucial for sign language recognition. Hence we have leveraged the power of CNN and OpenPose for recognising dynamic signs.

**Algorithm: Training and Testing using OpenPose and CNN**
Input: Sign Video
Output: Label for the Sign Recognised
**Training**

1. For each of the classes in training set
2. For each of the videos in class folders do
   (a) Pass the RGB video to OpenPose framework to get the video with stick figure embedded.
   (b) Extract the frames from OpenPose videos and write the OpenPose frames to the images folder under the class folder;
   (c) Extract frames for each of the RGB videos and add it to the images folder
   (d) Feed the Inception V3 model with all the images in all the class folders.
   (e) Save the model built

**Testing**

1. Repeat steps 1 to 2 (b) of the training set for the test set
2. Feed the learnt CNN model with all the images in the test set.
3. Get the prediction of each of the frames for a particular video as a list
4. Use majority voting scheme on the list above to get the class prediction of the video.

Advantages of our proposed method:

- Specialized hardware like Kinect cameras that capture depth information is not required. A common smartphone camera is sufficient.
- All the models are trained and tested on CPU and hence no GPU is required. OpenPose is run online on Colab environment with GPU enabled and hence does not require GPU system.
- Previous works focus mainly on signs limited to the wrist whereas our system takes both manual (hand) and non-manual features (arm, body, face) of the sign into account. This is useful for recognition of sign language wherein complexity of the signs is greater.
- Removed the feature engineering part.

**Experimental Setup**
The sign language recognition system was implemented on an Intel Corei5 CPU (1.80 GHz × 2) with 8 GB RAM. The system ran macOS Mojave (version 10.14.3). Lenovo-Intel Corei3 – UCPU with 4 GB RAM running windows7 was also used. The programming language used for implementing the system was Python. OpenPose library was used for keypoint detection from the sign language videos. OpenCV was used for handling data and pre-processing. The recognition system took 10 min on the MacOS and 18 min on the windows system to complete training on the sign language dataset.

### 3.1 Data Acquisition

The sign videos are recorded in a controlled environment by a 12 MP smartphone front camera having frame resolution of 640 × 480 pixels captured at the rate of 30 fps. In the process initially we collected 20 signs of medical terms from Indian sign language. The dataset has been selected from the ISL general, technical and medical dictionaries. Indian Sign Language Research and Training Institute (ISLRTC), New Delhi, [17] has been taken as the standard source for Indian sign language. So, all the videos are performed based on observation from the ISLRTC YouTube channel. The ISLRTC videos were lengthier of the order of 30 s for every word on an average which is very long compared to other sign languages. For example, the Argentinian sign language [18] has words with video lengths of the order of 1 or 2 s. Hence we took only that portion of the video which essentially means the sign performed. So we have fixed 3 s for the signs. Inorder to standardise and simplify processing, the signs are performed with a uniform white background at a constant distance of 1 m right in front of the camera. The signers wore black coloured clothes with sleeves to make sure that only the hands and face part are visible. Each sign is repeated 2–3 times depending on the complexity of the gesture. Due to different recording environments the lighting conditions vary slightly from subject to subject. The dataset hence forth referred to as 'ISL Medical Dataset', consists of 1 static and 19 dynamic signs handpicked from ISL. All volunteers were non-native to sign language. The videos were recorded for a period of three seconds consistently for all the signs. The frame rate of the videos captured was 30 fps. Another Research Institute extensively working on building ISL dictionary – the Ramakrishna Mission Vivekananda Educational and Research Institute (RKMVERI), Coimbatore Campus has their ISL videos listed in [19]. The ISL videos given by RKMVERI were also used as reference by the signers.

### 3.2 Data Pre-processing

#### (i) Preparation of training set

The data pre-processing stage is also pictorially represented in Fig. 1. The video acquired above was fed to OpenPose. OpenPose requires GPU for processing. Hence Google Colab environment which is freely available for academic use [20] was used to get the pose estimation of hands and face keypoints. Thus we get an equivalent video with the keypoints marked in the video. Each video consisted of 90 frames at a frame rate of 30 fps. This key points video was fed to OpenCV for frame extraction to extract every 5$^{th}$ frame. Thus we get a total of 18 frames. Usually the first and last few fractions of a second do not contain useful information. The frames are then cropped on the sides to ensure that the signer is prominent in all the frames and also to reduce the dimensionality for training. Hence the first two and last two frames were omitted from the training set. Thus a total of 14 image frames for each video sequence comprise our training set. Thus for 20 classes, there were 1120 image frames.

To study the effect of using OpenPose with CNN, the training set was prepared in two phases. In the first phase, the original videos without OpenPose pose estimations formed our training set. In the second phase, the training frames were extracted from

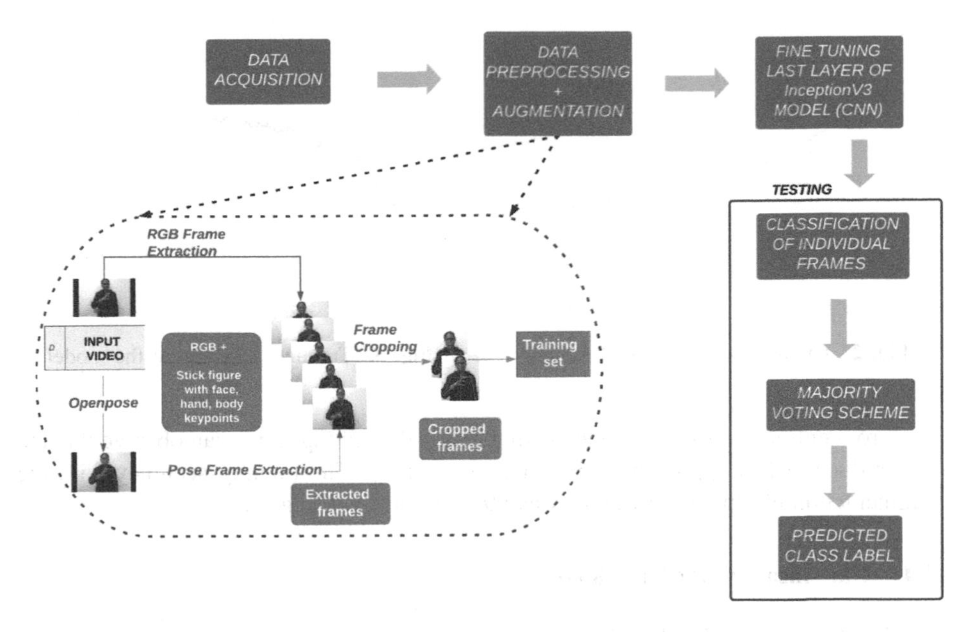


**Fig. 1.** Overall architecture of the proposed CNN+OpenPose system

the output of OpenPose pose estimation. Testing was done after each of these training phases. The results are shown later in the results section.

(ii) **Preparation of test set**

To test the performance of our gesture recognition approach we collected videos of each sign from the official YouTube channel of ISLRTC, New Delhi (Indian Sign Language Research and Training Centre). The signs performed by signers involved repetition of the same gesture multiple times in a video to clearly explain how the sign is performed. The videos also involved the signer explaining the context of the sign so each sign language reference video was observed to be around 30–40 s which was too lengthy for our model to detect features and recognise the gestures. Consequently, we clipped the videos to display the signer performing the key gesture once or twice at maximum and obtained our shortened test videos which were fed for frame extraction.

### 3.3 Fine Tuning Inception V3 Model

The frames extracted from the two phases of pre-processing with and without OpenPose were fed to the Inception V3 model with all other layers frozen except for the top layer. Hence fine tuning of the model was done using our training data for all the twenty categories. The number of training steps was 500 with a learning rate of 0.01. The cross entropy results and accuracy of the training set is shown is Figs. 2 and 3 with OpenPose frames.

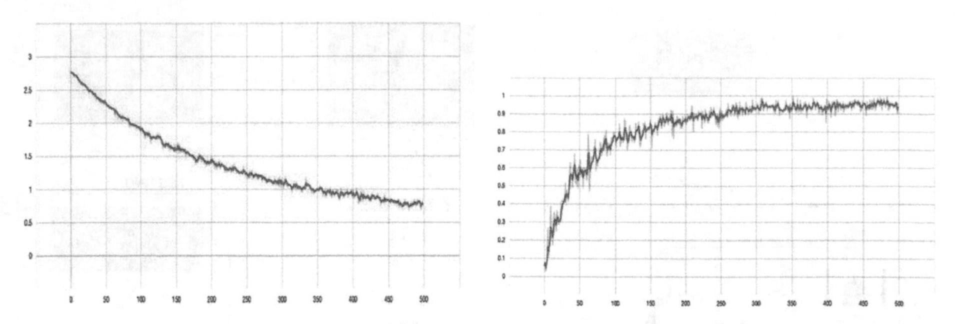

**Fig. 2.** Cross entropy results of training

**Fig. 3.** Training accuracy of the model

Cross entropy was used for calculating loss and from Fig. 2 we can observe that the loss reduced when approaching 500 steps in training. The training accuracy with the augmentation of OpenPose frames was 99% as shown in Fig. 3.

## 3.4 Classification of Sign Video

(i) **Prediction of individual frames**

The test frames from ISLRTC videos were fed to the trained/learnt model to classify each frame individually. The probability score of each frame is given as output. This is done by using the features learnt from the CNN model. The features are

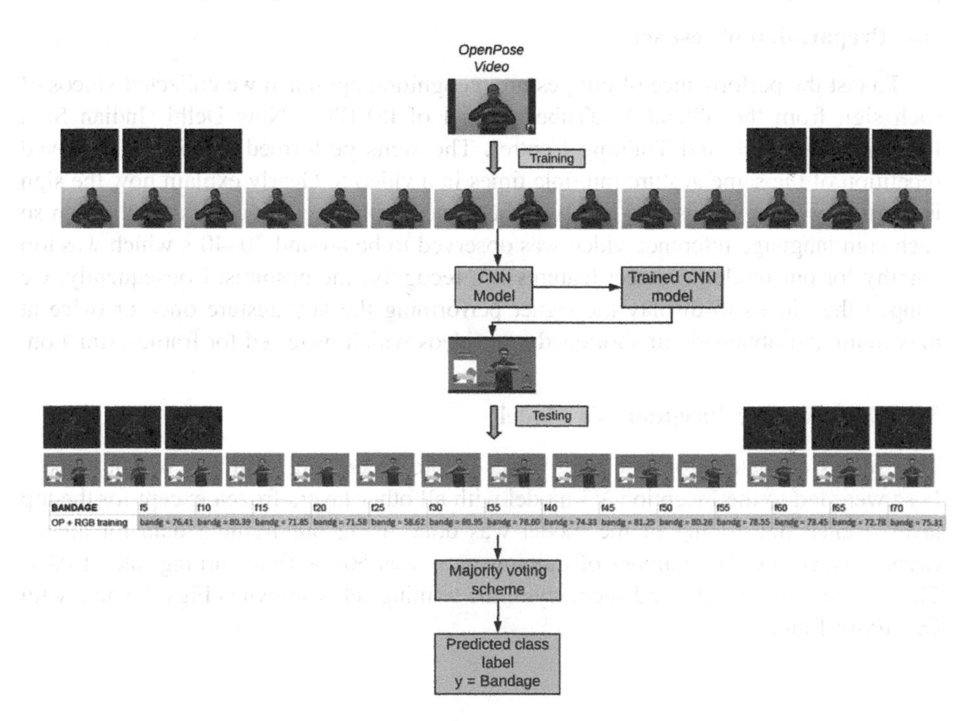


**Fig. 4.** Data flow pipeline for prediction of class label

represented in Fig. 4 as stick figures from OpenPose embedded in the RGB frames. Thus these feature embedded frames give a body part mapping which is very useful in matching the frame to a corresponding frame with similar pose estimation in training set. The same is explained in Fig. 4.

(ii) **Prediction of sign video**

The predictions of each of the 14 frames are given as input to a majority voting scheme which give the predicted label of the video sequence based on the probability scores of individual frames. The formula for majority voting is given in Eq. (1).

$$y = \text{mode}\{C(X_1), C(X_2), C(X_3), \ldots, C(X_{14})\} \quad (1)$$

where 'y' is the class label of the sign video
X1, X2, ... X14 represent the frames extracted for a single sign video and
C(X1), C(X2), ... C(X14) represent the class label predicted for each frame.

The Fig. 4 shows an example of bandage class where the OpenPose video is fed to the model for testing. The individual frames are extracted and classified individually. Both OpenPose and RGB frames are combined for training. As it can be seen, all the frames are classified as 'bandage' class with a probability score of 0.75 on an average. Hence, the test video or the sign is recognised as 'bandage' with high confidence.

## 4 Results and Discussion

The test results given in Table 2 show that without OpenPose key points, the CNN fails to recognise the class correctly. But with OpenPose, the results are amazing, given the large difference in probability score as seen below for the allergy class from 23.09 for old (wrong class) to allergy (correct class) with 87.81 for frame f5.

**Table 2.** Improvement in results with and without OpenPose trained samples

| ALLERGY | f5 | f10 | f15 | f20 | f25 | f50 | f55 | f60 | f65 | f70 |
|---|---|---|---|---|---|---|---|---|---|---|
| w/o OP | old = 23.09 | allergy = 22.13 | old = 24.67 | bandage = 21.50 | allergy = 17.86 | old = 23.22 | old = 20.85 | old = 23.66 | allergy = 20.82 | old = 18.92 |
| w OP | allergy = 87.81 | allergy = 85.55 | allergy = 78.90 | allergy = 87.20 | allergy = 84.16 | allergy = 79.91 | allergy = 76.45 | allergy = 80.22 | allergy = 79.56 | allergy = 83.26 |
| | | | | | | | | | | |
| **PAIN** | f5 | f10 | f15 | f20 | f25 | f50 | f55 | f60 | f65 | f70 |
| w/o OP | allergy = 58.16 | allergy = 46.36 | allergy = 29.79 | allergy = 50.07 | allergy = 41.23 | allergy = 36.05 | allergy = 31.99 | allergy = 32.35 | allergy = 30.07 | allergy = 35.85 |
| w OP | pain = 58.09 | bp = 26.28 | pain = 33.51 | pain = 43.35 | pain = 19.14 | pain = 51.99 | pain = 55.80 | pain = 49.75 | pain = 51.83 | pain = 56.67 |
| | | | | | | | | | | |
| **CHOKING** | f5 | f10 | f15 | f20 | f25 | f50 | f55 | f60 | f65 | f70 |
| w/o OP | chest = 27.57 | chok = 28.89 | chok = 28.62 | chok = 25.78 | chok = 21.73 | chok = 22.65 | chok = 27.41 | chok = 28.74 | chok = 27.34 | chok = 28.32 |
| w OP | chok = 58.88 | chok = 47.41 | chok = 70.34 | chok = 38.78 | chok = 41.98 | chok = 57.84 | chok = 46.44 | chok = 50.91 | chok = 53.65 | chok = 48.65 |
| | | | | | | | | | | |
| | (dark) | misclassified relevant frame | | | | | | | | |
| | (light) | classified correctly | | | | | | | | |

Table 3 shows the test results with exclusive RGB and exclusive OpenPose frames and the combined set of both RGB and OpenPose frames for training. It can be readily seen that when only RGB images are taken for building the model, the features from the CNN on RGB images were not able to model the data accurately and it fails on 12 of the 20 classes. The training set with only OpenPose images did a decent job of classifying more that 50% of the frames correctly. For example, for the Blood Pressure sign the model trained with only OpenPose images classified 10 frames correctly of the total 14 frames in test images. The misclassification reduced further and when both the OpenPose and RGB images were given for training. Thus it was able to classify 13 out of 14 frames correctly as shown (OP + RGB training) below in Table 3.

**Table 3.** Pure RGB vs Pure OpenPose vs Hybrid results comparison

| BLOOD PRESSURE | f5 | f10 | f15 | f20 | f25 | f50 | f55 | f60 | f65 | f70 |
|---|---|---|---|---|---|---|---|---|---|---|
| purely RGB training | h attack = 40.85 | h attack = 46.79 | h attack = 47.78 | h attack = 53.58 | h attack = 55.24 | h attack = 47.45 | h attack = 46.33 | h attack = 49.65 | bp = 28.03 | h attack = 43.69 |
| purely OP training | pain = 32.08 | chest = 38.38 | bp = 29.49 | allergy = 26.50 | bp = 54.95 | bp= 35.41 | bp=31.05 | allergy = 44.87 | bp= 36.87 | bp=43.41 |
| OP + RGB training | bp = 25.88 | bp = 28.50 | bp = 29.22 | bp. = 24.28 | bp = 48.49 | bp = 35.29 | bp = 39.40 | old = 30.24 | bp= 44.02 | bp=43.62 |

Thus our final training set had a total of 2240 frames with the augmentation of OpenPose frames. The inception v3 model pretrained on our RGB image data used for retraining the model. The test results were positive with five more classes correctly identified. Table 4 shows the test results of the top 6 classes for which it performed extremely well. As it can be seen, for most of these classes, except for two or three frames, our model labelled all the frames correctly with increased probability.

**Table 4.** Successful cases (top 6)

| | f5 | f10 | f15 | f20 | f25 | f50 | f55 | f60 | f65 | f70 |
|---|---|---|---|---|---|---|---|---|---|---|
| CHEST | ch = 40.05 | ch = 58.18 | ch = 51.95 | pain = 45.31 | pain = 39.81 | ch = 58.83 | ch = 66.61 | ch = 56.07 | ch = 43.59 | ch = 43.59 |
| | f5 | f10 | f15 | f20 | f25 | f50 | f55 | f60 | f65 | f70 |
| ALLERGY | al = 94.05 | al = 92.49 | al = 91.73 | al = 84.24 | al = 92.91 | al = 81.79 | al = 85.44 | al = 85.96 | al = 85.96 | al = 84.87 |
| | f5 | f10 | f15 | f20 | f25 | f50 | f55 | f60 | f65 | f70 |
| EXERCISE | ex = 88.06 | ex = 78.26 | ex = 57.76 | ex = 80.98 | ex = 73.56 | ex = 78.79 | ex = 69.24 | ex = 69.39 | ex = 81.79 | ex = 80.96 |
| | f5 | f10 | f15 | f20 | f25 | f50 | f55 | f60 | f65 | f70 |
| BANDAGE | bandg = 76.41 | bandg = 80.39 | bandg = 71.85 | bandg = 71.58 | bandg = 58.62 | bandg = 59.02 | bandg = 81.99 | bandg = 76.87 | bandg = 75.89 | bandg = 74.78 |
| | f5 | f10 | f15 | f20 | f25 | f50 | f55 | f60 | f65 | f70 |
| OLD | old = 61.90 | old = 98.09 | old = 94.67 | old = 93.44 | old = 96.26 | old = 93.11 | old = 95.34 | old = 90.02 | old = 93.78 | old = 91.88 |
| | f5 | f10 | f15 | f20 | f25 | f50 | f55 | f60 | f65 | f70 |
| CHOKING | chest = 27.57 | chok = 28.89 | chok = 28.62 | chok = 25.70 | chok = 21.73 | chok=26.82 | chok = 22.22 | chok = 35.28 | chok = 19.16 | chok = 19.16 |
| | f5 | f10 | f15 | f20 | f25 | f50 | f55 | f60 | f65 | f70 |
| COUGH | bp = 24.21 | cough = 29.57 | cough = 38.61 | cough = 32.53 | cough =31.67 | ex = 22.32 | ex = 21.83 | cough = 25.73 | cough = 32.12 | cough=27.35 |

Table 5 shows the failure cases when the system failed. This can be attributed to the absence of key points for the ear, forehead and neck. For example, the 'EAR' class gets confused with the 'Exercise' class because the hands go up and there are no key points for the ear. OpenPose frames does not prove to be useful for such cases and hence the result. Similarly, the 'Forehead' class gets confused with exercise class for the same reason. The 'Mumps' class has one hand near the shoulders similar to the 'Pain' class and hence most of the predictions come out to be 'Pain'.

**Table 5.** Failure cases (Misclassified classes)

| | f5 | f10 | f15 | f20 | f25 | f50 | f55 | f60 | f65 | f70 |
|---|---|---|---|---|---|---|---|---|---|---|
| EAR | ex = 45.21 | ex = 54.80 | ex = 51.93 | ex = 53.97 | ex = 52.62 | ex = 51.89 | ex = 52.23 | ex = 50.67 | ex = 61.32 | ex = 51.06 |
| | f5 | f10 | f15 | f20 | f25 | f50 | f55 | f60 | f65 | f70 |
| FOREHEAD | ex = 27.18 | ex = 30.47 | foreh = 16.09 | ex = 28.51 | ex = 29.67 | ex = 29.55 | ex = 26.33 | ex = 28.01 | ex = 29.02 | ex = 31.66 |
| | f5 | f10 | f15 | f20 | f25 | f50 | f55 | f60 | f65 | f70 |
| MUMPS | pain = 25.09 | inj = 26.70 | pain = 28.34 | mumps= 19.25 | pain = 27.02 | inj = 26.05 | pain = 26.41 | inj = 25.09 | pain = 27.10 | inj = 28.86 |

The test results show that of the 20 medical sign videos, 17 were classified correctly and the above three classes were misclassified. Thus our system has an overall accuracy of 85%. The confusion matrix for the test results is shown in Fig. 5 below.

The misclassification rate can be reduced if more keypoints are included as features. Also training set can be increased to give more variations in the data.

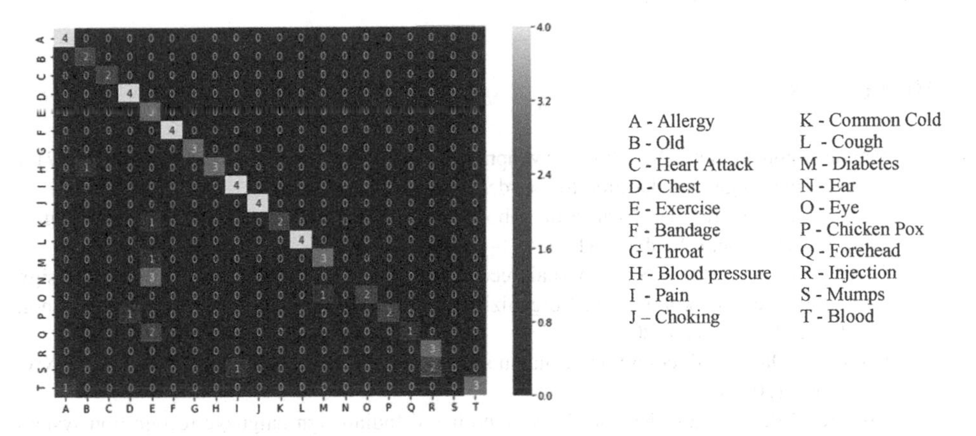


**Fig. 5.** Test Accuracy of the proposed system shown as confusion matrix

## Limitations

- Video classification is not performed in real time.
- Our system depends on OpenPose for improved accuracy.

## Future Enhancements

Currently our work involves classification of 20 medical terms (19 dynamic and 1 static). The training set can be increased with more videos performed by different signers. The Bag of visual features model can be used to classify each of the patches separately for every frame to capture the local information along with a global classifier. A mobile application can be developed to classify the video in real-time.

## 5 Conclusion

Real time recognition of Indian sign Language in a healthcare setting is important and crucial for the improvement in quality of life for the deaf and dumb. But to realise this in a resource limited clinic is a challenge. We have addressed this challenge with minimal system requirements so that the computational and storage requirements are not a bottleneck. If this system could be made available in a smartphone, the applications are endless from condition monitoring to assisting the disabled. We have shown that OpenPose with CNN is all we need for a sign language recognition system.

**Compliance with Ethical Standards.**

✓ All authors declare that there is no conflict of interest
✓ No humans/animals involved in this research work.
✓ We have used our own data.

## References


1. National Public Radio (US). https://www.npr.org/sections/goatsandsoda/2018/01/14/575921716/a-mom-fights-to-get-an-education-for-her-deaf-daughters
2. Ansari, Z.A., Gaurav, H.: Nearest neighbour classification of Indian sign language gestures using kinect camera. Sadhana **41**(2), 161–182 (2016)
3. Tewari, D., Srivastava, S.: A visual recognition of static hand gestures in indian sign language based on kohonen self-organizing map algorithm. Int. J. Eng. Adv. Technol. (IJEAT) **2**(2), 165–170 (2012)
4. Singha, J., Das, K.: Recognition of Indian sign language in live video. arXiv preprint: arXiv: 1306.1301 (2013)
5. Kishore, P.V.V., Rajesh Kumar, P.: A video based Indian sign language recognition system (INSLR) using wavelet transform and fuzzy logic. Int. J. Eng. Technol. **4**(5), 537 (2012)
6. Tripathi, K., Nandi, B.N.G.C.: Continuous Indian sign language gesture recognition and sentence formation. Procedia Comput. Sci. **54**, 523–531 (2015)
7. Dour, S., Sharma, M.M.: Review of literature for the development of Indian sign language recognition system. Int. J. Comput. Appl. **132**(5), 27–34 (2015)
8. Rao, G.A., Kishore, P.V.V.: Selfie video based continuous Indian sign language recognition system. Ain Shams Eng. J. **9**(4), 1929–1939 (2018)
9. Cihan Camgoz, N., Hadfield, S., Koller, O., Ney, H., Bowden, R.: Neural sign language translation. In: Proceedings of the IEEE Conference on Computer Vision and Pattern Recognition, pp. 7784–7793 (2018)
10. Konstantinidis, D., Dimitropoulos, K., Daras, P.: Sign language recognition based on hand and body skeletal data. In: 2018-3DTV-Conference: The True Vision-Capture, Transmission and Display of 3D Video (3DTV-CON), pp. 1–4. IEEE (2018)
11. Zaki, M.M., Shaheen, S.I.: Sign language recognition using a combination of new vision based features. Pattern Recogn. Lett. **32**(4), 572–577 (2011)
12. El-Soud, A.M., Hassan, A.E., Kandil, M.S., Shohieb, S.M.: A proposed web based framework e-learning and dictionary system for deaf Arab students. In: IJECS2828, 106401 (2010)

13. Cao, Z., Hidalgo, G., Simon, T., Wei, S., Sheikh, Y.: OpenPose: real time multi-person 2D pose estimation using Part Affinity Fields. arXiv preprint: arXiv:1812.08008 (2018)
14. Five video classification methods implemented in Keras and Tensorflow. https://blog.coast.ai/five-video-classification-methods-implemented-in-keras-and-tensorflow-99cad29cc0b5
15. Chen, Q., Wu, R.: CNN is all you need. arXiv preprint: arXiv:1712.09662 (2017)
16. Karpathy, A., Toderici, G., Shetty, S., Leung, T., Sukthankar, R., Li, F.-F.: Large-scale video classification with convolutional neural networks. In: Proceedings of the IEEE Conference on Computer Vision and Pattern Recognition, pp. 1725–1732 (2014)
17. Youtube ISLRTC, New Delhi. https://www.youtube.com/channel/UC3AcGIlqVI4nJWCwHgHFXtg/videos?disable_polymer=1
18. Ronchetti, F., Quiroga, F., Estrebou, C.E., Lanzarini, L.C., Rosete, A.: LSA64: an Argentinian sign language dataset. In: XXII CongresoArgentino de Ciencias de la Computación (CACIC 2016) (2016)
19. Indian Sign Language, FDMSE, RKMVERI, CBE (Deemed University). http://www.indiansignlanguage.org/indian-sign-language/
20. Openpose Colab Jupyter notebook. https://colab.research.google.com/github/tugstugi/dl-colabnotebooks/blob/master/notebooks/OpenPose.ipynb

# A Two-Phase Image Classification Approach with Very Less Data

B. N. Deepankan(✉) and Ritu Agarwal

Department of Information Technology,
Delhi Technological University, Delhi, India
{bn.deepankan, ritu.jeea}@gmail.com

**Abstract.** Usually, neural networks training requires large datasets. In this paper, we have proposed a two-phase learning approach that differentiates the multiple image datasets through transfer learning via pre-trained Convolutional Neural Network. Initially, images are automatically segmented with the Fully connected network to allow localization of the subject through creating a mask around it. Then, the features are extracted through a pre-trained CNN in order to train a dense network for classification. Later, those saved weights of the dense network are loaded to train the model once again by fine-tuning it with a pre-trained CNN. Finally, we have evaluated our proposed method on a well-known dog breed dataset, and bird species dataset. The experimental results outclass the earlier methods and achieve an accuracy of 95% to 97% for classifying these datasets.

**Keywords:** Batch normalization · Deep learning · Image segmentation · Transfer learning

## 1 Introduction

Image classification means classifying the subjects using images. It is performed on a daily basis for differentiating the traffic signals or in choosing apple from oranges. Although the process seems to be an easy task for an individual, but the machine cannot classify on the go. However, with the latest advancements in deep learning techniques, training the machine for accurate classification is made possible.

Though learning a machine for efficient classification can be conducted, it will require a huge dataset to do so. However, for some problems sufficient data are not available to train the model precisely. In those situations, data augmentation methods such as rotation, scaling and inverting the available images on the dataset will increase the classification accuracy. Although, excessive training can lead to the overfitting of data, which can cause some inaccurate classifications.

Unlike simple image classification between cats from dogs, differentiating between similar objects remains as a difficult task due to the variety of matching shapes, colour, texture and appearance. Furthermore, images of dog or birds usually contain similar surrounding such as grass, leaves etc. For my experiment, I have considered a dataset containing 8,351 images of 133 dog breeds [1], another dataset with 6,033 images of 200 bird species [2]. These datasets contain images from similar background moreover,


S. Smys et al. (Eds.): ICCVBIC 2019, AISC 1108, pp. 384–394, 2020.
https://doi.org/10.1007/978-3-030-37218-7_44

each type is comparable in shape, size, and light. However, manual classification is possible by a trained professional but it can be a tedious task for a pedestrian. Hence, learning a machine for performing such classification can reduce the hassle involved.

There are many techniques through which features can be extracted from the image which can be used for classification purposes. It involves scale-invariant feature transform (SIFT) [3], speeded up robust features [4] etc. However, it cannot achieve higher accuracy of classification due to the wide range of images belonging to similar classes as well as overhead complexity involved. This occurs because the features extracted from these techniques are similar due to comparable shapes and background of images which can become a cumbersome task to segregate the classes.

In this work, we have used the latest advancement in the deep learning methods to tackle the image classification problem with very fewer data. In our approach, we have performed localization of the subject through a fully connected network which removes the background through minimum bounding box around the object. Dataset of segmented images is used for extracting features from pre-trained CNN which trains the dense network for differentiating between classes. Later, those saved weights are loaded to the same dense network which is kept on top of the convolutional base of pre-trained CNN for further training.

Our proposed method is divided into two phases: phase one consist of feature extraction through the pre-trained network and training the dense network. The second phase involves loading those saved weights fine-tuning it with pre-trained CNN. We have trained our network twice to converge it towards for better classification. Moreover, we have given methodologies to overcome the problem for overfitting while converging it rapidly. The experimental results show that the proposed technique outclasses the earlier methods and achieves an accuracy of 95% for bird species dataset while 97% for dog breed dataset.

The rest of the paper is structured as follows: Sect. 2 gives a brief background about the recent work on image classification and segmentation approaches. Section 3 will provide a detailed explanation of our approach. The experimental results of our approach are shown in Sect. 4. We then conclude our work in Sect. 5.

## 2 Related Work

In the following section, we will give a brief description of the image segmentation and classification applications. We then present related background work which addresses the image classification and image segmentation task.

### 2.1 Image Segmentation

There are a variety of ways has been proposed for image segmentation which can perform binary segmentation: 0 for background and 1 for the object region(s). One of which is YOLO [5] (You only look once) algorithm, for object classification and localization using bounding boxes. It consists of 24 layers of a single CNN network which is followed by a dense network at the end. However, to reduce the training period we can opt for fast YOLO (9 layers of a single CNN network instead of 24).

Though the data availability, there may be a possibility of fewer training samples to train the model effectively for segmentation. U-Net [6] is one of the methods to train with fewer annotated labels, but they can only process 2D images. V-Net [7] can perform 3D segmentation based on volumetric, fully convolutional, neural network. Furthermore, various other methods are proposed which uses CNN to do the object localization process such as Mask-RCNN [8]. Overall, all of the mentioned methods use the concept of the fully convolutional network (FCN) for semantic segmentation.

### 2.2 Image Classification

There are several methods have been proposed to classify dog breed and bird species datasets. The majority of the proposed techniques have used machine learning based methods. Like the work in [9] which coarsely classify dog breeds in 5 groups and for each group principal component analysis (PCA) is used as a feature extractor. The extracted features are compared with the feature template which was derived while classifying the dog images into groups. The image is classified as the breed which gives the minimum distance between two feature vectors.

In [10], they model the classification of a dog as a manifold which uses facial shapes and their associated geometry. They have applied Grassmann manifold to represent the geometry of different dog breeds. After feature extraction, SVM [11] categorizes the features to the corresponding category. An improved dog breed classification is proposed [12] from click-through logs and pre-trained models. In this approach, more dog images are mined through the web with dog-related keywords for gathering more information. After that, important information is derived using the pre-trained model (DCNN) by keeping related neurons in the last layers. Later, those features are classified through a dense neural network.

To classify among different bird species several methods have been proposed, in [13] the authors have extracted textural content of spectrogram images. These features are extracted through three different techniques of digital image processing: Gabor features [14], Local Binary Pattern (LBP) [15] and Local Phase Quantization (LPQ) [16]. After that, SVM classifier is used to classify these textural content and the results were taken as 10 fold-cross validation.

Though the basic feature extraction methods can provide valuable information for categorizing the images. Although, it cannot achieve higher classification accuracy in comparison with deep learning techniques. However, hybrid of techniques were used to classify between bird species as shown in work [17] using decision tree and SVM classifier. In addition, the classification of bird species is performed through bird voices [18] using CNN.

## 3 Proposed Approach

We have conducted a two-phase experiment in which the best weights of the dense network are saved in the first phase. Later, in the second phase, those saved weights are loaded to the dense network for a fine-tuned network with pre-trained CNN.

### 3.1 Pre-trained CNN

Pre-trained networks consist of trained weights according to a particular data set. If a model is built from scratch, then each synapse is initially assigned with random weights. However, after hours of training through a training dataset, it may converge in order to perform the desired task. After that, the accuracy can be measured through testing via blind dataset. Later those weights are saved for different problems through fine-tuning which can reduce the overhead involved in training a model from scratch.

In our approach, we have used the VGG 16 network [19] which is a pre-trained convolutional neural network trained on the ImageNet database. It consists of 5 blocks of the convolutional network each with max-pooling layers which are followed by a dense network, as visualized in Fig. 3. It has a total of 14 million trainable parameters with weights assigned as per the ImageNet database.

### 3.2 First Phase of Image Classification

The first phase follows the process as shown in Fig. 1 in which the bird image dataset is passed to the Fully Connected Network [20] for localization of the object through creating a mask around it. This operation is performed to remove the background features which can cause the network to learn unwanted features. Ultimately, it can cause the model to classify inaccurately. The segmented image dataset is given to the VGG-16 model for feature extraction which extracts bottleneck features.

Later, those significant bottleneck features are used to train the dense network. The Dense network is a two-layer perceptron, the first Dense layer has 256 feature maps with Relu activation function. Batch normalization [21] is conducted on the data presented by the first layer of the dense network to normalize the data such that the mean of the data becomes 0 and the standard deviation equals to 1. Basically, training a neural network through normalized inputs helps it to learn. The idea is that, instead of just normalizing the inputs to the network, we normalize the inputs to layers within the network. The normalized output is transferred to the second dense layer with 20 feature maps with softmax activation function for determining the probability of the bird species. Also, 50% of the network weights are dropped for each step so the model doesn't succumb to the overfitting problem.

The architecture is similar for Dog breed classification, however, there is a slight difference in the dense network implemented for categorizing the dog breeds as shown in Fig. 2. It consists of a single layer dense network with 133 feature maps with softmax activation function. In dog breed classification we haven't used Batch normalization because it didn't provide a rise in classification accuracy.

Furthermore, those extracted features are used to train the dense network for image classification until it converges or it doesn't improve for five epochs. The best weights associated with the dense network is saved for fine-tuning purposes on the second phase of the experiment.

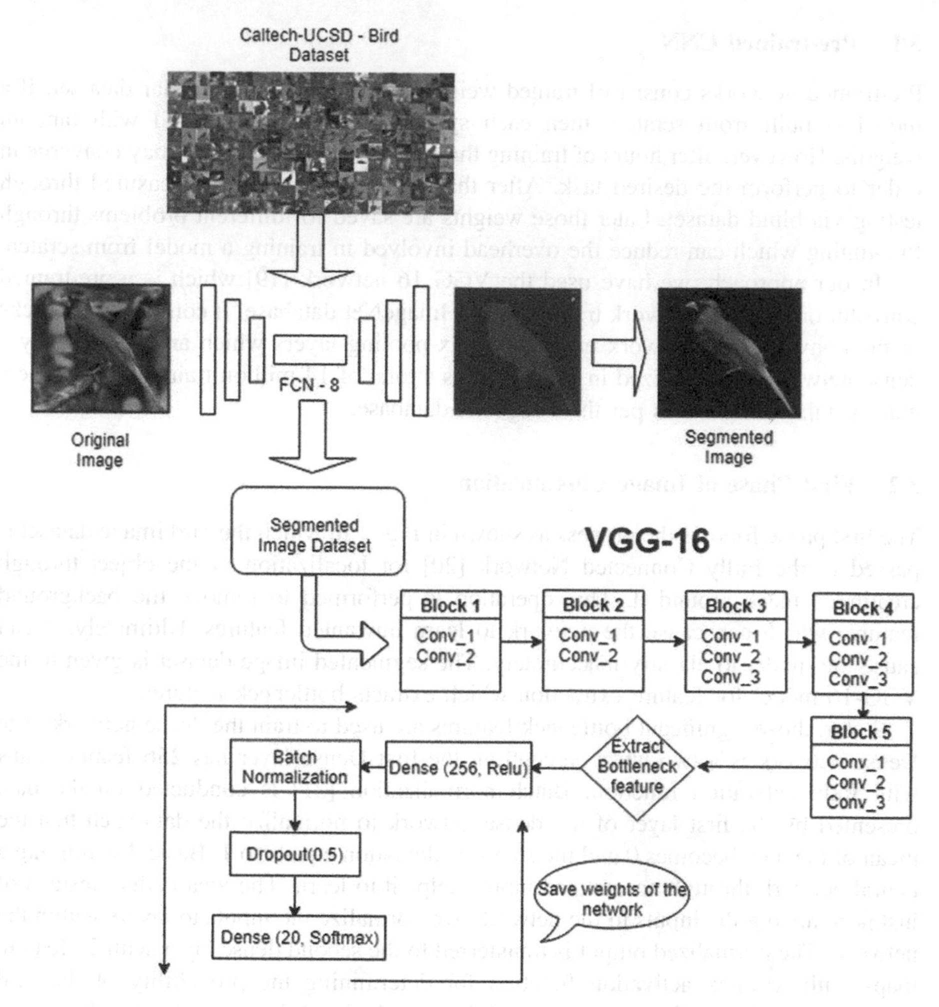


**Fig. 1.** First phase of proposed approach for bird classification [7, 9]

### 3.3 Second Phase of Image Classification

In the second phase of bird classification as shown in Fig. 3, the dense layer is loaded with the best-saved weights from the first phase of the experiment. Also, the last layer of the VGG 16 network remained trainable while the other layers were frozen. Since the whole network has a large entropic capacity and could lead to the problem of overfitting. The features learned by the lower blocks are more abstract than the top-layers of the convolutional block. So, it is sensible to keep the last layers of the model trainable while keeping top layers untrainable. Finally, I have trained the network once again by rescaling the image through data augmentation.

For dog classification, it is the same architecture as shown above but the dense network is according to Fig. 2.

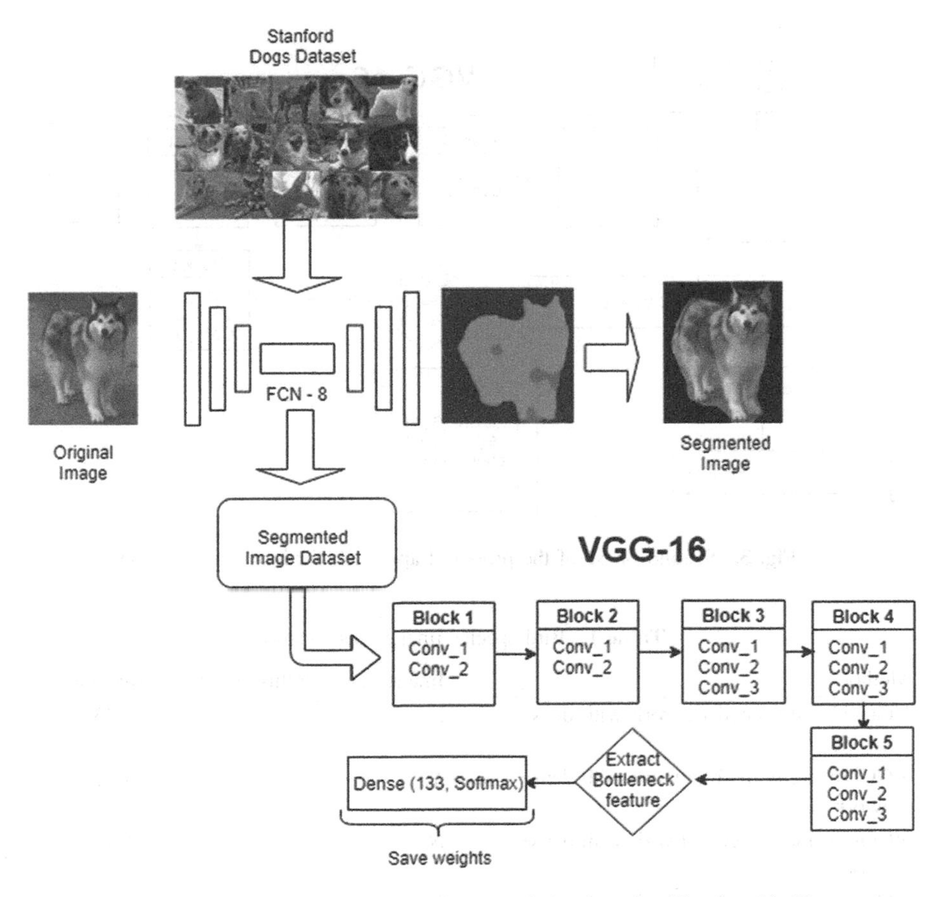


**Fig. 2.** First phase of proposed approach for dog classification [6, 8]

## 4 Experimental Results

We have conducted the experimental results on two datasets. First, we have conducted the operation on 20 different bird species, where we have a total of 1115 images in which train set consist of 789 images while validation set contains 267 images and 59 images in the test set. However, for each type of species, there are around 49 train images which are very much less number of images to train our model efficiently. Table 1 shows the difference between various methods in terms of accuracy, with or without image segmentation for the bird image dataset. However, the dense network is the same as shown in Fig. 1 for each method.

Table 2 provides a comparison between the methods using batch normalization on segmented image dataset. The methods are shown below has the same neural network model discussed above, although the batch normalization layer is skipped in some to get the perspective of using that layer.

Second, The data set used for dog classification consists of 133 different dog breeds of the total 8351 images. The train set consist of 6680 images while the validation set

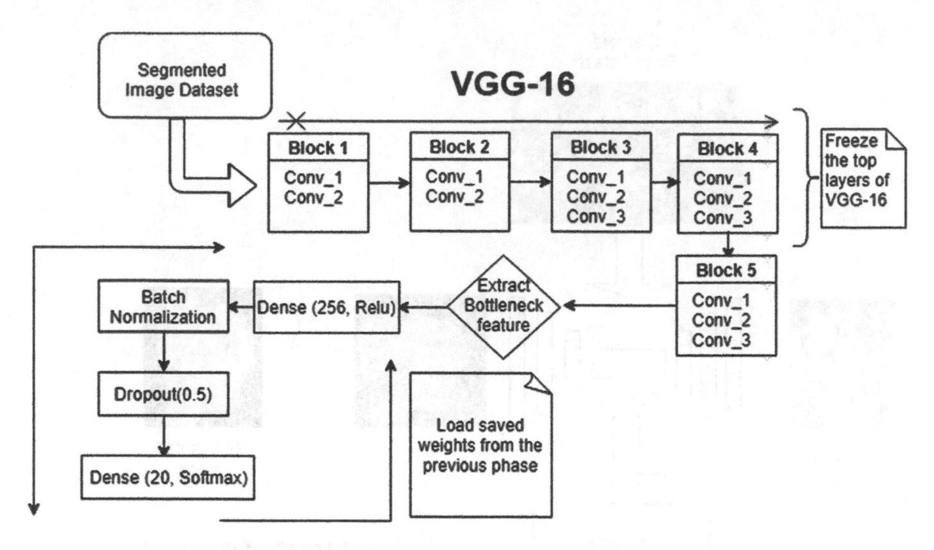


**Fig. 3.** Second phase of the proposed approach for bird classification

**Table 1.** Bird species image classification

| Method | Image segmentation (Y/N) | Accuracy (%) |
|---|---|---|
| VGG-16 pre-trained network with dense network | N | 49.23 |
| VGG-16 pre-trained network with dense network | Y | 54.62 |
| VGG-19 pre-trained network with dense network | N | 44.43 |
| VGG-19 pre-trained network with dense network | Y | 46.15 |
| Proposed Two-phase method | Y | 94.89 |

**Table 2.** The difference between using Batch Normalization for Bird classification

| Method | Batch normalization (Y/N) | Accuracy (%) |
|---|---|---|
| VGG-16 pre-trained network with dense network | N | 50.00 |
| VGG-19 pre-trained network with dense network | N | 42.31 |
| Proposed Two-phase method | N | 89.84 |

consist of 835 images and the test set contains 836 images. Table 3 shows a comparison between various methods with or without image segmentation.

**Table 3.** Dog breed Image classification

| Method | Image segmentation (Y/N) | Accuracy (%) |
|---|---|---|
| VGG-16 pre-trained network with dense network | N | 74.61 |
| VGG-16 pre-trained network with dense network | Y | 86.34 |
| ResNet50 pre-trained network with dense network | N | 82.46 |
| ResNet50 pre-trained network with dense network | Y | 88.95 |
| Proposed Two-phase method | Y | 97.12 |

Table 4 shows the difference between not using the Batch normalization layer on a segmented image dataset for dog breed classification.

**Table 4.** The difference between using Batch Normalization for Dog classification

| Method | Batch normalization (Y/N) | Accuracy (%) |
|---|---|---|
| VGG-16 pre-trained network with dense network | N | 77.27 |
| ResNet50 pre-trained network with dense network | N | 84.26 |
| Proposed Two-phase method | N | 93.87 |

The bird species model with a batch size of 10 is trained for 25 epochs and enabled with early stopping after 5 epochs. If the model doesn't improve for 5 epochs then the model stops training to avoid the overfitting problem. The dog breed model with a batch size of 20 is trained for 20 epochs while has a similar criteria of early stopping. Some of the experimental results of the proposed method are shown in Figs. 4 and 5.

## 5 Conclusion

Classification between similar images is a difficult task for an algorithm. However, by using our proposed deep learning technique, we have achieved the classification accuracy of 94.89% for 20 various types of bird species, while we got an accuracy of 97.12% for classifying between 133 different types of dog breeds.

Although, many other methods use hand-crafted features, the proposed deep-learning approach learns the unique features through the convolutional neural network. Convolutional image classification techniques use a plethora of features extracted from the images with the aim of improving classification performance [12]. In this paper, I

**Fig. 4.** Some of the experimental results for Bird species classification [5]

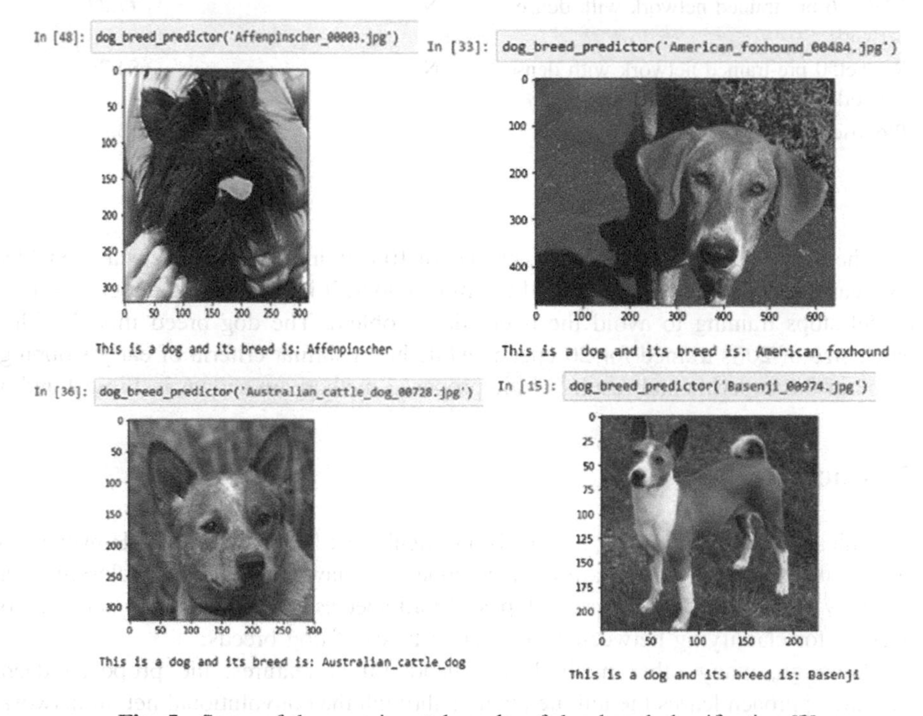


**Fig. 5.** Some of the experimental results of dog breed classification [8]

have discussed the methods involved in fine-tuning a pre-trained neural network to efficiently classify the model to categorize the images with higher accuracy.

Also, describing an image segmentation technique to remove the background from the classifying subject. The experimental results performed on bird dataset and dog dataset work significantly better than other models to classify among different species and breeds which are similar in shape, luminance and having a complex background. The proposed methodology also states a novel method to train the model proficiently, making it robust against erroneous images consisting of multiple subjects.

Furthermore, I have used very little data to train the model to give high accuracy as shown on experimental results. The proposed two-phase algorithm trained twice with the same dataset through data augmentation technique which transforms the image to prevent overfitting. The CNN also learns to the rotation variant information for categorizing the images efficiently.

The proposed algorithm can be extended in classifying between medical images due to the scarcity of sufficient images in training a neural network efficiently. In addition, the applicability of my algorithm on several kinds of classification problems which are similar to the bird and dog classification.

**Compliance with Ethical Standards.**

✓ All authors declare that there is no conflict of interest

✓ No humans/animals involved in this research work.

✓ We have used our own data.

## References


1. Khosla, A., Jayadevaprakash, N., Yao, B., Li, F.-F.: Novel dataset for fine-grained image categorization. In: First Workshop on Fine-Grained Visual Categorization (FGVC), IEEE Conference on Computer Vision and Pattern Recognition (CVPR) (2011)
2. Welinder, P., Branson, S., Mita, T., Wah, C., Schroff, F., Belongie, S., Perona, P.: Caltech-UCSD Birds 200. California Institute of Technology. CNS-TR-2010-001 (2010)
3. Geng, C., Jiang, X.: SIFT features for face recognition. In: 2009 2nd IEEE International Conference on Computer Science and Information Technology, Beijing, pp. 598–602 (2009)
4. Zhou, D., Hu, D.: A robust object tracking algorithm based on SURF. In: 2013 International Conference on Wireless Communications and Signal Processing, Hangzhou, pp. 1–5 (2013)
5. Redmon, J., Divvala, S., Girshick, R., Farhadi, A.: You only look once: unified, real-time object detection. In: 2016 IEEE Conference on Computer Vision and Pattern Recognition (CVPR), Las Vegas, NV, pp. 779–788 (2016)
6. Medical Image Computing and Computer-Assisted Intervention (MICCAI). LNCS, vol. 9351, pp. 234–241. Springer, Cham (2015). arXiv:1505.04597
7. Milletari, F., Navab, N., Ahmadi, S.: V-Net: fully convolutional neural networks for volumetric medical image segmentation. In: 2016 Fourth International Conference on 3D Vision (3DV), Stanford, CA, pp. 565–571 (2016)
8. He, K., Gkioxari, G., Dollar, P., Girshick, R.: Mask R-CNN. IEEE Trans. Pattern Anal. Mach. Intell.

9. Chanvichitkul, M., Kumhom, P., Chamnongthai, K.: Face recognition based dog breed classification using coarse-to-fine concept and PCA. In: 2007 Asia-Pacific Conference on Communications, Bangkok, pp. 25–29 (2007)
10. Wang, X., Ly, V., Sorensen, S., Kambhamettu, C.: Dog breed classification via landmarks. In: 2014 IEEE International Conference on Image Processing (ICIP), Paris, pp. 5237–5241 (2014)
11. Hearst, M.A., Dumais, S.T., Osuna, E., Platt, J., Scholkopf, B.: Support vector machines. IEEE Intell. Syst. Appl. **13**(4), 18–28 (1998)
12. Xie, G., Yang, K., Bai, Y., Shang, M., Rui, Y., Lai, J.: Improve dog recognition by mining more information from both click-through logs and pre-trained models. In: 2016 IEEE International Conference on Multimedia & Expo Workshops (ICMEW), Seattle, WA, pp. 1–4 (2016)
13. Lucio, D.R., Maldonado, Y., da Costa, G.: Bird species classification using spectrograms. In: 2015 Latin American Computing Conference (CLEI), Arequipa, pp. 1–11 (2015)
14. Kamarainen, J.: Gabor features in image analysis. In: 2012 3rd International Conference on Image Processing Theory, Tools and Applications (IPTA), Istanbul, pp. 13–14 (2012)
15. Meena, K., Suruliandi, A.: Local binary patterns and its variants for face recognition. In: 2011 International Conference on Recent Trends in Information Technology (ICRTIT), Chennai, Tamil Nadu, pp. 782–786 (2011)
16. Zhai, Y., Gan, J., Zeng, J., Xu, Y.: Disguised face recognition via local phase quantization plus geometry coverage. In: 2013 IEEE International Conference on Acoustics, Speech and Signal Processing, Vancouver, BC, pp. 2332–2336 (2013)
17. Qiao, B., Zhou, Z., Yang, H., Cao, J.: Bird species recognition based on SVM classifier and decision tree. In: 2017 First International Conference on Electronics Instrumentation & Information Systems (EIIS), Harbin, pp. 1–4 (2017)
18. Narasimhan, R., Fern, X.Z., Raich, R.: Simultaneous segmentation and classification of bird song using CNN. In: 2017 IEEE International Conference on Acoustics, Speech and Signal Processing (ICASSP), New Orleans, LA, pp. 146–150 (2017)
19. Simonyan, K., Zisserman, A.: Very deep convolutional networks for large-scale image recognition. arXiv:1409.1556
20. Long, J., Shelhamer, E., Darrell, T.: Fully convolutional networks for semantic segmentation. In: 2015 IEEE Conference on Computer Vision and Pattern Recognition (CVPR), Boston, MA, pp. 3431–3440 (2015)
21. Ioffe, S., Szegedy, C.: Batch normalization: accelerating deep network training by reducing internal covariate shift. In: ICML (2015)

# A Survey on Various Shadow Detection and Removal Methods

P. C. Nikkil Kumar(✉) and P. Malathi

Department of Computer Science and Engineering, Amrita School of Engineering, Amrita Vishwa Vidyapeetham, Coimbatore, India
nikkilkumar26@gmail.com, p_malathy@cb.amrita.edu

**Abstract.** Shadows plays an inevitable part in an image and also the major source of hindrance in Computer Vision analysis. Shadow detection is the performance enhancement process that increases the accuracy of the Computer Vision algorithms like Object Tracking, Object Recognition, Image Segmentation, Surveillance, Scene Analysis, Stereo, Tracking, etc. Shadows limit the stability of these algorithms, and hence detecting shadows and its elimination are profound pre-processing techniques for improving execution of Vision algorithms efficiently. To label the challenges under various environmental conditions, researches have been carried out to develop various algorithms and techniques for shadow detection. This paper objective is to bring comparative analysis of shadow detection techniques with their pros and cons.

**Keywords:** Shadow · Shadow detection · Image enhancement · Image processing

## 1 Introduction

Shadows are the repeatedly occurring disturbances to many computer vision algorithms which decreases its efficiency in problem solving and detection of object mapping. So in order to overcome this effect many shadow detection algorithms were introduced. Different algorithms works best for different conditions like Indoor and Outdoor lightening. This paper has discussed different algorithms their pros and cons and their efficiency to use in compressive conditions. However detecting false shadows and scattered shadows is quite demanding task. Generally shadows are formed when a light source is blocked from the source light. Cast and self-shadows are two types of shadows. Shadows occurred by illumination of objects by direct light are called cast shadows. Penumbra and Umbra are the two types of cast shadows. Detection of shadows is a complex task in Penumbra and easy in Umbra. Shadows occurred which objects are not exposed to direct light are self-shadows. In this paper focuses on interpreting 13 different Shadow detection and Removal techniques with their main motive and pros and cons. These shadow detection methods are very accurate in various conditions like Outdoor and Indoor.


S. Smys et al. (Eds.): ICCVBIC 2019, AISC 1108, pp. 395–401, 2020.
https://doi.org/10.1007/978-3-030-37218-7_45

## 2 Shadow Detection and Removal Methods

This paper explains comparative analysis of shadow detection techniques with their pros and cons. The comparative analysis of various Shadow detection and Removal methods are shown in Table 1.

From Table 1, it shows that the comparative results of different shadow detection and removal, it is clear that for different methods and algorithm produces different results and output.

For Illumination recovery method [1] a novel approached has been used for single images and aerial images, the input image is decomposed to overlay patches with respect to shadow distribution. Illumination recovery operator is build using correspondence between shadow patch and illuminator patch. It is simple and effective and can produce rich texture image as output.

In Multiple ConvNets [2] an automatic framework is used to detect and remove shadow in real world scenes, it has ability to learn features at super-pixel level. For shadow extraction Bayesian formulation is used. Here a removal framework is used to remove original recovered image from shadow matte.

For Automatic soft shadow detection method [3] Merging the histogram splitting of two data peaks with resultant image to achieve smooth detection results. Main advantage is that it can rebuild even objects in shadow frame. This method can reinstate even objects in High-resolution shadow images.

For 3D intensity surface modeling [4] smoothing filter is used to erase texture information using a technique called image decomposition. 3D modeling helps to obtain the proper intensity of the surface and illumination in shadow areas.

In Kernel Least-Squares Support Vector Machine (LSSVM) [5] main operation is predicting the label of each region and LSSVM technique helps to separate non-shadow regions and shadow regions. Markov Random Field (MRF) is a framework useful for performance enhancement.

For ESRT model [6] first step is to convert image to HSV and 26 parameters were taken. Shadows are removed using Morphological processing, it has better detection rate and accuracy.

For Local Color Constancy computation [7] the illumination match is done using random field model which is contained in color shade descriptors and it also produces coherent shadow fields. It uses anisotropic diffusion for each image pixel in shadow to determine illuminant narrowly.

Tricolor Attenuation Model [8] it has ability to extract shadows for complex outdoor scenes but for one image at a time. This method is more useful in video sequences and complex outdoor images.

Near-Infrared Information [9] normally built-in sensitivity of digital camera sensors are higher, this method uses it detect shadows automatically. Moreover sensitivity is low in near-infrared (NIR).

Ratio Edge [10] geometric heuristics and Intensity constraint are used. These constraints are used to maximize the execution. This method used to detect moving shadows actively. It is very effective and useful in detecting moving shadows.

**Table 1.** Comparative analysis of various shadow detection and method

| Sr. no | Method | Key idea | Advantages | Disadvantages | Author |
|---|---|---|---|---|---|
| 1 | Illumination recovering optimization method | Constructing an optimized illumination recovery operator | Shadow image is processed to high quality texture enriched with non-uniform shadows | It cannot effectively process video sequences as processing an image | Ling Zhang, Qing Zhang, and Chunxia Xiao |
| 2 | Convolutional deep neural networks (ConvNets) | Features are trained at higher level using super-pixel and these trained model is send to conditional random model to produce a polished mask of shadow | For extracting shadow matte and removing shadows Bayesian formulation is widely used | Not very effective for outdoor conditions | Salman H. Khan, Mohammed Bennamoun |
| 3 | Automatic soft shadow detection method | Merging the splitting histogram of two data peaks with resultant image achieves smooth detection results | The ability to restore uniform objects in shadow areas | This method includes two level processes for shadow removal | Ye Zhang, Shu Tian, Yiming Yan, and Xinyuan Miao |
| 4 | 3D intensity surface modeling | Image intensity surface with local smoothness Pattern is retained by edge-preserving filter | Shadow region is obtained by Surface intensity illumination technique and 3D modeling technique | It can deal only with cast shadows | Kai He, Rui Zhen, Jiaxing Yan, and Yunfeng Ge |
| 5 | Kernel Least-Squares Support Vector Machine (LSSVM) | To envision the identification of each region, LSSVM is used for dividing shadow and non-shadow regions | Markov Random Field (MRF) framework is used to improve the performance of region classifier | Less accurate to normal pixels than core pixels of shadow regions | Tomas F. Yago Vicente, Minh Hoai, and Dimitris Samaras |
| 6 | ESRT (enhanced streaming | 1. HSV image with 26 Parameters are taken for measurement of the image | It is better in terms of accuracy and detection rate | Processing stage of algorithm is large | Vicky Nair, Parimala Geetha Kosal Ram, |

(*continued*)

**Table 1.** (*continued*)

| Sr. no | Method | Key idea | Advantages | Disadvantages | Author |
|---|---|---|---|---|---|
| | random tree) model | 2. The algorithm with trained dataset is used for segmentation<br>3. Removal of shadow is performed through morphological processing | when compared with Bayesian classifier | | Sundaravadivelu Sundararaman |
| 7 | Local Color Constancy Computation | Color shade descriptors are used in condition random field model. This model is able to find the illumination pairs and coherent shadow regions | An improved local color constancy is used for computations<br>This has capability to find illuminant at each pixel image in shadow | It doesn't work with images having even shadow surfaces | Xingsheng Yuan, Marc Ebner, Zhengzhi Wang |
| 8 | Tricolor Attenuation Model | A model is formed by using image formation theory and tricolor attenuation model | It extracts shadows even for image with Complex outdoor scenes | Operation can only be used to single image at a time | Jiandong Tian, Jing Sun, and Yandong Tang |
| 9 | Near-Infrared Information | Building a shadow map using pixel image based on visibility | Automatically detect shadows with maximum accuracy and by using sensitivity of digital camera sensors | Simultaneously capturing visible and NIR information is not possible by current photography | Dominic Rüfenacht, Clément Fredembach, Sabine Süsstrunk |
| 10 | Ratio Edge | A novel method has been introduced for moving cast shadows detection | Intensity constraint and geometric heuristics are used to enhance the performance | It can only detect cast shadows | Wei Zhang, Xiang Zhong Fang, Xiaokang K. Yang |
| 11 | Stereo photography | Algorithm involves foreground segmentation, shadow mask, detection of a shadow candidate area, and deshadow | Solves shadow problems caused by flash photography | Can detect only for flash shadows images | Sang Jae Nam, and Nasser Kehtarnavaz |

(*continued*)

**Table 1.** *(continued)*

| Sr. no | Method | Key idea | Advantages | Disadvantages | Author |
|---|---|---|---|---|---|
| 12 | Object segmentation | Moving Cast shadow detection is introduced and merged into segmentation algorithm | Detection of cast shadows is through integrating the temporal regions of background objects | Mainly effective for video sequences | Jurgen Stauder, Roland Mech, and Jorn Ostermann |
| 13 | Color and Edge | Shadow and background uses same color tune value with different intensity values | Well founded technique for colored images | It takes more time for computation. Fails when shadow and background intensity is same | Maryam Golchin et al. |

Stereo photography [11] this technique reduces the shadows occurring during flash photography. It can detect only for flash shadow images.

Object segmentation [12] detecting the cast shadows using temporal integration of covered background regions. The change detection method is employed in detecting object mask from shadow background.

Color Based [13] Shadow and background uses same color tune value with different intensity values. Well founded technique for colored images. It takes more time for computation. It fails when shadow and background intensity is same.

## 3 Conclusion

This paper aim at different comparative analyses of above mentioned shadow detection and removal techniques. There are 13 shadow detection techniques are discussed in this paper. For each technique different analysis of method and their corresponding pros and cons are discussed. It can be conclude after final analysis that there is no such the best Algorithm. So the choice of foremost algorithm is based on dataset feature and Conditions like Outdoor and Indoor sequence.

**Compliance with Ethical Standards**

✓ All authors declare that there is no conflict of interest
✓ No humans/animals involved in this research work.
✓ We have used our own data.

## References

1. Zhang, L., Zhang, Q., Xiao, C.: Shadow remover: image shadow removal based on illumination recovering optimization. IEEE Trans. Image Process. **24**(11), 4623–4636 (2015)
2. Khan, S.H., Bennamoun, M., Sohel, F., Togneri, R.: Automatic shadow detection and removal from a single image. IEEE Trans. Pattern Anal. Mach. Intell. **38**(3), 431–446 (2016)
3. Su, N., Zhang, Y., Tian, S., Yan, Y., Miao, X.: Shadow detection and removal for occluded object information recovery in urban high-resolution panchromatic satellite images. IEEE J. Sel. Top. Appl. Earth Obs. Remote Sens. **9**(6), 2568–2582 (2016)
4. He, K., Zhen, R., Yan, J., Ge, Y.: Single-image shadow removal using 3D intensity surface modeling. IEEE Trans. Image Process. **26**(12), 6046–6060 (2017)
5. Vicente, T.F.Y., Hoai, M., Samaras, D.: Leave-one-out kernel optimization for shadow detection and removal. IEEE Trans. Pattern Anal. Mach. Intell. **40**(3), 682–695 (2018)
6. Nair, V., Kosal Ram, P.G., Sundararaman, S.: Shadow detection and removal from images using machine learning and morphological operations. J. Eng. **2019**(1), 11–18 (2019)
7. Yuan, X., Ebner, M., Wang, Z.: Single-image shadow detection and removal using local colour constancy computation. IET Image Process. **9**(2), 118–126 (2015). https://doi.org/10.1049/iet-ipr.2014.0242
8. Tian, J., Sun, J., Tang, Y.: Tricolor attenuation model for shadow detection. IEEE Trans. Image Process. **18**(10), 2355–2363 (2009)

9. Rüfenacht, D., Fredembach, C., Süsstrunk, S.: Automatic and accurate shadow detection using near-infrared information. IEEE Trans. Pattern Anal. Mach. Intell. **36**(8), 1672–1678 (2014)
10. Zhang, W., Fang, X.Z., Yang, X.K., Wu, Q.M.J.: Moving cast shadows detection using ratio edge. IEEE Trans. Multimed. **9**(6), 1202–1214 (2017)
11. Nam, S.J., Kehtarnavaz, N.: Flash shadow detection and removal in stereo photography. IEEE Trans. Consum. Electron. **58**(2), 205–211 (2012)
12. Stander, J., Mech, R., Ostermann, J.: Detection of moving cast shadows for object segmentation. IEEE Trans. Multimed. **1**(1), 65–76 (1999). https://doi.org/10.1109/6046.748172
13. Golchin, M., Khalid, F., Abdullah, L., Davarpanah, S.H.: Shadow detection using color and edge information. J. Comput. Sci. **9**, 1575–1588 (2013). https://doi.org/10.3844/jcssp.2013.1575.1588

# Medical Image Registration Using Landmark Registration Technique and Fusion

R. Revathy, S. Venkata Achyuth Kumar(✉), V. Vijay Bhaskar Reddy, and V. Bhavana

Department of Electronics and Communication Engineering, Amrita School of Engineering, Bengaluru, Amrita Vishwa Vidyapeetham, Bengaluru, India
kachyuth007@gmail.com, bhavanapyarilal@gmail.com

**Abstract.** The process of image registration and fusion is used to obtain a resultant image with better characteristics. The concept of image registration and fusion can be used in various domains such as remote sensing, satellite imaging, medical imaging, etc. But, as the smallest of errors can have great consequences in medical images, applying these concepts in medical imaging will make a great difference. Aligning images spatially into one single framework is attained through image registration. There are various methods or techniques to implement image registration. Landmark registration, a semi-automatic registration method is used where images are spatially aligned by selecting control points from both the images and applying geometric transformation that helps in translation, rotation, scaling and shearing. During fusion process intermixing of traits from source images is performed using Discrete Wavelet Transform and Bicubic interpolation. The resultant image helps the doctors for better clinical diagnosis.

**Keywords:** Image fusion · Image registration · Landmark registration · Discrete wavelet transform · Bicubic interpolation

## 1 Introduction

A set of finite values called pixels are used to represent a 2D image and is known as a digital image. Devices like computers are used to manipulate these images, known as Digital Image Processing. Several challenges are faced during monitoring and analysing the medical images. Medical imaging has different modalities and it is broadly classified into two categories namely anatomical and functional imaging. Anatomical imaging such as MRI (Magnetic Resonance Imaging) gives anatomic structure of the body, which helps to determine the bone structures, internal abnormalities, etc. Functional imaging such as PET (Positron Emission Tomography) gives physiological information which helps to analyse the change of metabolism, blood flow, etc. During PET or MRI scan, distortions in signal can arise due to several factors such as variation in breathing levels of the patient that can have impacts on the scans.

To eradicate errors as well as to enhance the features of medical images, the following process is implemented. The first step in the process of integration is bringing different modalities into a spatial alignment called image registration [7].


S. Smys et al. (Eds.): ICCVBIC 2019, AISC 1108, pp. 402–412, 2020.
https://doi.org/10.1007/978-3-030-37218-7_46

Image registration is followed by image fusion [1] which is used to display the integrated data from multiple input images. In this work functional image (PET) and anatomical image (MRI). Image registration is the process of finding best transformation method and aligning the images as per the region or structure of interest. In this work, landmark based approach is used for image registration. Two sets of points have been loaded; one as fixed and the other as moving. Moving image points are compared to the fixed image points and are adjusted relatively to bring them to similar orientation in spatial domain. Then, the converted images are used for multimodal image fusion.

When fusion is performed, the characteristics from more than two images are combined to produce a single image frame with better information [9]. Image fusion is technique to qualitatively and quantitatively increase the image quality of imaging features which makes fusion of medical image more efficient. Medical image fusion contains a broad range of techniques to address medical issues reflected through images of human body, organs, and cells. In this work, image fusion is performed through 4 level Discrete Wavelet Transform (DWT) [1] and Bi-cubic interpolation [3]. DWT decomposition [10] is performed to acquire lower and higher frequency coefficients. Bicubic interpolation is an advancement of cubic interpolation [3] in which the weighted average of the nearest sixteen neighbours are taken and the new pixel value is obtained.

Fusion rules [6] have been used to fuse the corresponding coefficients at each level and which gives new coefficients. Inverse discrete wavelet transform (IDWT) is then performed on the new coefficients which is followed by color conversion that results in a final fused image.

## 2 Impact of Image Registration on Clinical Diagnosis

Image registration has got a huge impact on medical image diagnosis. During diagnosis various medical images provide different information such as any issue to deal with bone is found using CT scan and issues related to the structure of tissue which include blood vessels and muscles is found using MRI whereas the ultrasound image deals with problems regarding luminal structures and organs. When medical images are in form of a digital images it is easier for doctors to analyse such images and thus digital image processing and development has a great significance in getting a better image that make medical diagnosis easier. Registration is important as during the process of medical diagnosis there may be a need of correlating two or more images, which may be in different spatial framework, in order to bring them in the same spatial orientation image processing is used. This ensure more efficiency and precision during the diagnosis.

## 3 Literature Review

Different transform techniques such as Discrete Wavelet Transform, Discrete Cosine Transform have been used for wavelet decomposition for fusing medical images. Discrete wavelet Transform gives better results [6] than Discrete Cosine Transform in terms of performance metrics. Fusion rules like mean, maximum, minimum, based on contrast, based on energy for corresponding detailed and approximate coefficients. For approximation and detail coefficients, mean and maximum rules respectively provides better enhanced image.

Table 1 shows that pros and cons of different image registration tools or graphical user interface tools.

**Table 1.** Pros and Cons of image registration tools

| Tool | Pros | Cons |
|---|---|---|
| OsiriX | OsiriX is compatible for both Desktop i.e. OsiriX MD and iPhone and iPad i.e. OsiriX HD Platforms. OsiriX is one of the best medical Image viewer or GUI uses 64 bit computing and multi-threading for better performance | OsiriX is compatible only with the latest Mac OS available and it is not compatible with Windows and Linux. A minimum of 6 GB RAM is required for the processing and a restriction of 128 bit registers is put for all operations that are complex |
| FreeSurfer | FreeSurfer is one of the open source software for better analysis and performance of MRI images of a human brain. It works on both Linux and Mac OS | An additional third party software has to be installed for the working of Free Surfer. As it is not fully automated human interaction is necessary and only few images formats a supported |
| ANTS | ANTS which stands for Advanced Normalization Tools are helpful in extracting information from the complex data sets which includes imaging | Though it is very handy and functional, it doesn't produces equivalent results. ANTS data servers render unpredicted databases that are bound by RAM limits, for large user loads the performance is significantly degraded |
| Mango | Mango is a non-commercial software for viewing, analysing and editing the volumetric medical images. It is compatible with MacOS, Windows and Linux. Mango, Imago, Papaya are some user interface software's works on desktop, iPhone or iPad and browser respectively | Imago in particular requires an iPad with iOS 5.0 or higher, whereas papaya is a medical viewer developed using Java script and it is an online tool |

# 4 Proposed Technique

## 4.1 Pre-processing

Now a day, the most widely used images are digital images, and sensors such as CCD (Charge Coupled Device) is used to obtain these images. The principle behind medical imaging is when a light source (ex. X-Ray source) provides illumination, the sensors arranged opposite to the source collects the energy that passes through the human body. This is the basis for medical computer axial tomography (CAT). MRI and PET which are the source images also comes under CAT principle. The images thus acquired may not be of the same size, thus image scaling which is the process of resizing (i.e. altering the dimension of the image) is used to bring it to a desirable size. These images then undergo filtering which includes smoothing, sharpening, edges detection and denoising for better image enhancement [3]. Histogram helps in analysing and finding the noise is present in an image.

### 4.2 Image Registration

Image Registration [7] is a crucial step in medical imaging because it helps to align different medical images that are taken at different places or different situations or different angles or from different sensors. Image registration helps physicians to analyse the changes or growth of disease in single image instead of multiple images of a patient at different stages. Image Registration is widely used for many anatomical organs of the human body like brain, retina, chest/lungs etc. Out of the two source images, one will be designated as fixed image or reference image or target image and other as moving image or sense image.

Figure 1 represents the block diagram of image registration. At a time, one source image act as reference image and the other source image acts as moving image. The Landmark registration contains of four steps. The first step is the resizing (to make both the images to be of same size) [8], after pre-processing the identification of features from both the fixed and source images. Second, the user needs to select the points corresponding from both the moving and fixed images and then pair the feature points from both the images. Third step is the crucial step for selecting the transformation model like affine or similarity, transformation calculation and finally application of transformation to one of the image set to align to the other image.

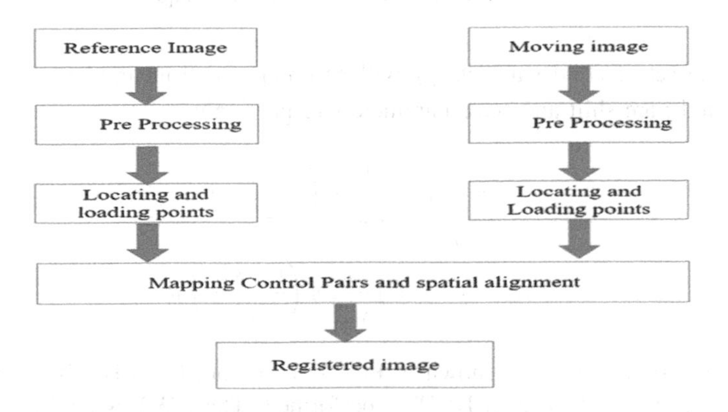


**Fig. 1.** Registration methodology

### 4.3 Image Fusion

The images obtained after landmark registration were passed through the proposed fusion process as shown in Fig. 2. Initially the images were gone through segmentation [1] to obtain low and high frequency components by passing through low pass and high pass filters respectively. DWT [7] provides flexible multi resolution analysis for the image and preserves the important information. In Discrete Wavelet Transform, the mother wavelet is broken down into smaller blocks called wavelets. A four level DWT decomposition as shown in Fig. 4 has been applied on low and high frequency components individually. The decomposition can be done using different wavelet filters. Some of them are Haar, Daubechies, and Coiflet. A Comparison is made using different wavelet filters such as coiflet Daubechies and Haar. Haar has been used for further process. The DWT coefficients that are obtained are used in further stages.

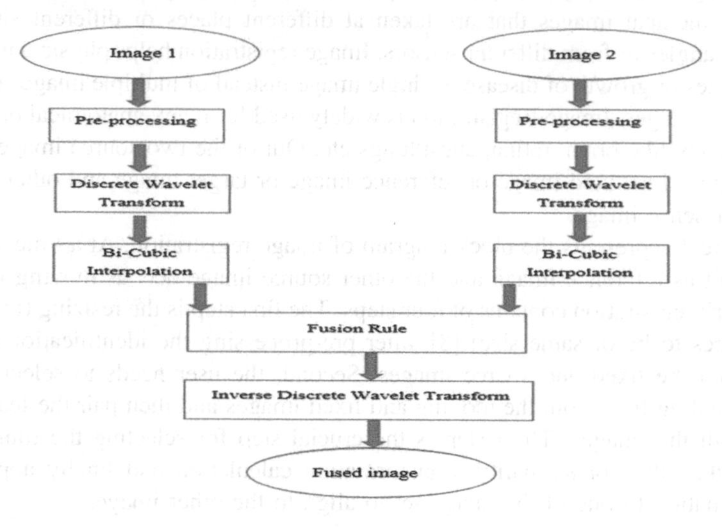


**Fig. 2.** Proposed image fusion technique

Discrete set of child wavelets $\gamma_{jk}$ will be computed from mother wavelet $\varphi_{j,k}(t)$, where k and j are shift and scale parameters respectively.

$$\varphi_{j,k}(t) = \frac{1}{\sqrt{2^j}} \varphi\left(\frac{t - k2^j}{2^j}\right) \tag{1}$$

$$\gamma_{jk} = \int_{-\infty}^{\infty} x(t) \frac{1}{\sqrt{2^j}} \varphi\left(\frac{t - k2^j}{2^j}\right) dt \tag{2}$$

The 4 level DWT process is shown in Fig. 3. To obtain LL, LH, HL, and HH coefficients of an image first level DWT is performed. The DWT is applied for the low frequency coefficients as it has the maximum amount of information. So, DWT is applied for the LL coefficient obtained from first level DWT, then LL2, LH2, HL2, and HH2 coefficients are obtained. The process of decomposition continues for two more times to obtain 4 level DWT coefficients. The DWT coefficients that are obtained after 4 level DWT [1] are interpolated using Bi-Cubic Interpolation [3], where $a_{ij}$ represents sixteen nearest neighbour coefficients. Fusion rules like Maximum and Mean fusion rules [2] are applied. Maximum fusion rule is applied for the higher frequency coefficients for each level of decomposition. Every pixel value of low frequency coefficients of both the images are compared and the resultant pixel value will be the maximum of the two pixel values. The Mean fusion rule is applied for the lower frequency coefficients for each level of decomposition. Every pixel value of low frequency coefficients of both the images are compared and the result pixel value will be the average of the two pixel values. By comparing different fusion rules, it is concluded that maximum fusion rule can preserve the edges and applying the mean fusion rule can preserve the approximations.

$$g(x,y) = \sum_{i=0}^{3}\sum_{j=0}^{3} a_{ij}x^i y^j \quad (3)$$

The corresponding coefficients of 4 level DWT are fused based on the Mean fusion rule and the other corresponding coefficients are fused based on the Maximum fusion rule [6].

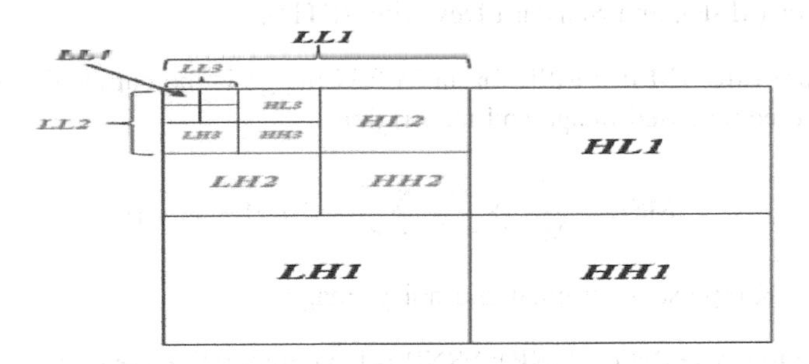


**Fig. 3.** Four level discrete wavelet transform

The Mean fusion rule is used for lower coefficients because they will have maximum details and maximum fusion rule for higher coefficients because it contains vertical, horizontal and diagonal edges [5]. Then inverse DWT is applied to the fused coefficients for the reconstruction of the image [7]. The final image obtained after this process is the fused image that has the characteristics of both PET and MRI source images.

## 5 Algorithm for the Proposed Innovation

1. Two images, MRI image as moving and PET image as reference image are considered for registration to obtain MRI registered on PET.
2. The images are preprocessed [11, 12] by resizing, smoothening, sharpening and filtering to obtain the high and low frequency components.
3. Landmark registration can be achieved by locating and loading the points of interest for the images.

4. The image thus formed by mapping the corresponding points is the registered image using landmark registration.
5. To obtain PET registered on MRI, the landmark registration is performed with the moving image as PET and the reference image as MRI image.
6. The two registered images after landmark registration are processed through proposed fusion methodology.
7. The registered images are performed with the 4 level DWT and followed by Bi-Cubic interpolation.
8. The obtained coefficients are fused using the fusion techniques.
9. IDWT is performed on fused coefficients to obtain the final fused image.

## 6 Results and Analysis

The proposed methodology has been implemented and verified on various performance metrics like Mutual Information (MI), Peak signal to Noise Ratio (PSNR), Mean Square Error (MSE), and Standard Deviation (STD).

Mean Square Error (MSE): MSE for an M * N image is the cumulative squared error between the compressed image and the original

$$\mathrm{MSE} = \frac{1}{\mathrm{MN}} \sum\nolimits_{\mathrm{i=0}}^{\mathrm{M-1}} \sum\nolimits_{\mathrm{j=0}}^{\mathrm{N-1}} [\mathrm{I(i,j)} - \mathrm{K(i,j)}]^2 \tag{4}$$

where I and K represents original and noisy image.

Peak Signal to Noise Ratio (PSNR): PSNR is the comparative measure between highest possible signal power and the power of noise due to corruption in the signal.

$$PSNR = 10.\log_{10}\left(\frac{MAX_I^2}{MSE}\right) \tag{5}$$

where $MAX_I$ represents the highest possible pixel value of the image.

Entropy: Entropy is the measure of information content in an image, which is the average of the uncertainty in information source.

$$H = -\sum\nolimits_{i=1}^{L} P_i \log(P_i) \tag{6}$$

where $P_i$ represents the probability of occurrence of pixel value in an image and L represents the number of gray levels of an image.

Standard Deviation (STD): STD is a measure that is used to calculate the amount of variation in a data set.

$$STD = \sqrt{\frac{1}{MN-1}\sum\nolimits_{i=1}^{M}\sum\nolimits_{j=1}^{N}(x(i,j) - x_{mean})^2)} \quad (7)$$

where x (i, j) is the pixel value of an image and $x_{mean}$ is the mean of all pixel values in an M * N image.

The images taken are MRI image (Fig. 4) with the spatial resolution of 256 * 256 * 3 and the PET image (Fig. 5) with the spatial resolution of 256 * 256 * 3.

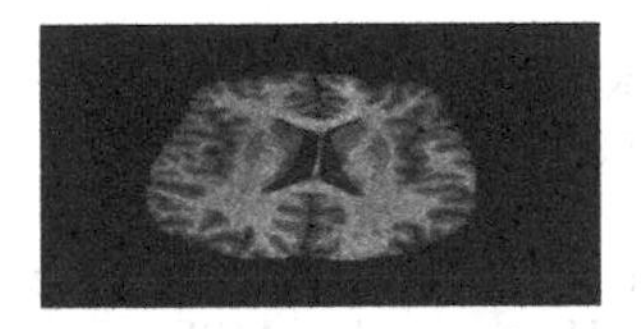

**Fig. 4.** MRI image [2]

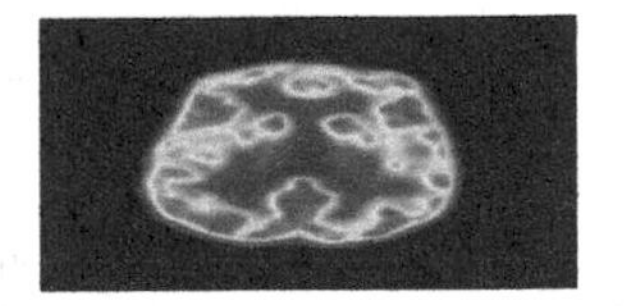

**Fig. 5.** PET image [2]

Figure 6 is the fused image using only proposed fusion Technique and Fig. 7 represents fused image using both proposed registration and fusion methodology.

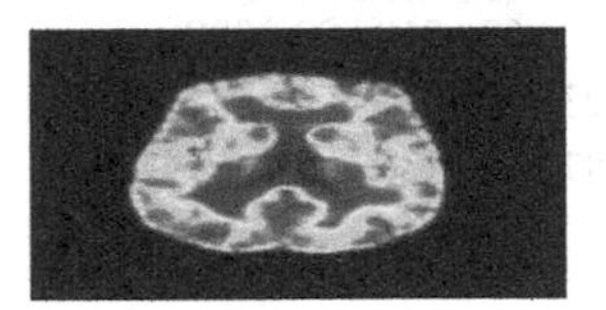

**Fig. 6.** Fused image using only fusion methodology [3]

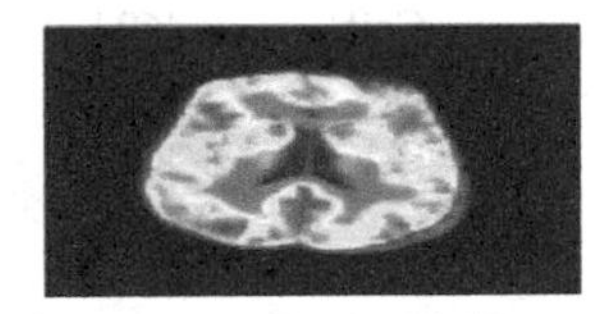

**Fig. 7.** Fused image using registration and fusion [4]

The images taken are MRI image (Fig. 8) with the spatial resolution of 300 * 300 * 3 and the PET image (Fig. 9) with the spatial resolution of 300 * 300 * 3. Both the images has been scaled to 256 * 256 * 3 for better analysis.

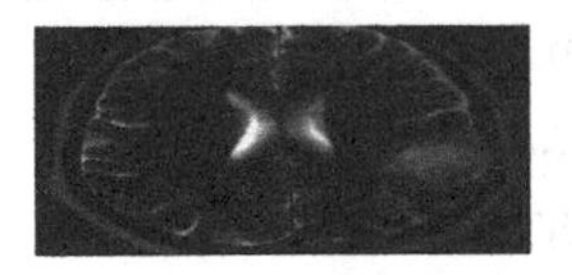

**Fig. 8.** MRI image

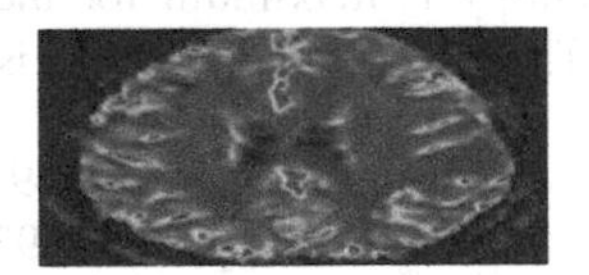

**Fig. 9.** PET image

Figure 10 is the fused image using only proposed fusion Technique and Fig. 11 represents fused image using both proposed registration and fusion methodology.

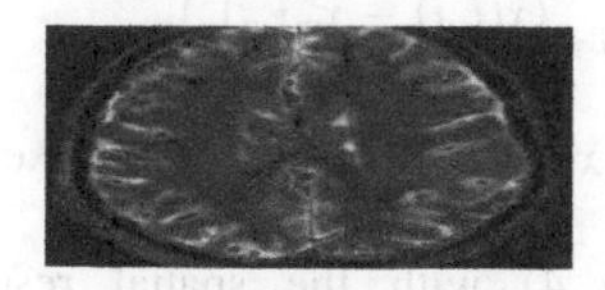

**Fig. 10.** Fused image using only fusion methodology

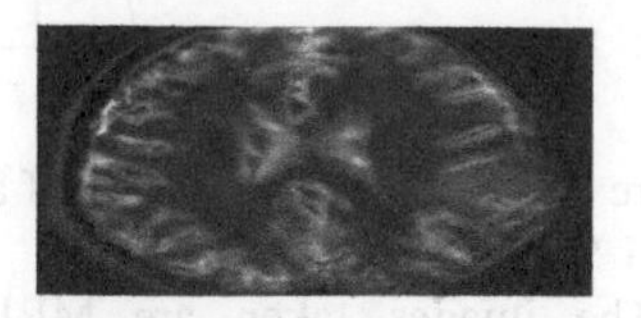

**Fig. 11.** Fused image using registration and fusion

$$\text{Input MRI entropy} = 2.6981 \quad \text{MRI STD} = 78.5347$$
$$\text{Input PET entropy} = 2.4391 \quad \text{PET STD} = 83.1191$$

From Table 2 i.e. fusion without registration, when compared with the results obtained the output entropy is maximum for the Haar wavelet, MSE is minimum, PSNR is maximum, STD is maximum and M.I is minimum for the Haar wavelet.

**Table 2.** Performance indices for the fused image for dataset 1

| Wavelets | Output Entropy | M.I | M.S.E | PSNR | STD |
|---|---|---|---|---|---|
| Coif1 | 3.4624 | 1.6748 | 0.0110 | 67.6510 | 89.3098 |
| Db2 | 3.3254 | 1.8118 | 0.0093 | 67.9810 | 89.9247 |
| Haar | 3.5017 | 1.6355 | 0.0087 | 68.7356 | 90.2457 |

**Table 3.** Performance indices of fusion with registration for data set 1

| Wavelets | Output Entropy | M.I | M.S.E | PSNR | STD |
|---|---|---|---|---|---|
| Coif1 | 3.6904 | 1.4468 | 0.0108 | 67.7965 | 92.7447 |
| Db2 | 3.7924 | 1.3465 | 0.0084 | 68.8880 | 93.0524 |
| Haar | 3.8140 | 1.3232 | 0.0070 | 69.6798 | 93.7143 |

From Table 3 i.e. fusion with registration, when compared with the results obtained the output entropy is maximum for the Haar wavelet, MSE is minimum, PSNR is maximum, STD is maximum and M.I is minimum for the Haar wavelet.

$$\text{MRI entropy} = 6.4919 \quad \text{MRI STD} = 45.1057$$
$$\text{PET entropy} = 6.3934 \quad \text{PET STD} = 70.972$$

**Table 4.** Performance indices for the fused image for data set 2

| Wavelets | Output entropy | M.I | M.S.E | PSNR | STD |
|---|---|---|---|---|---|
| Coif1 | 6.8512 | 5.98118 | 0.0078 | 69.3217 | 53.3037 |
| Db2 | 6.8485 | 6.0368 | 0.0077 | 69.2659 | 53.9885 |
| Haar | 6.8611 | 6.0242 | 0.0074 | 69.4384 | 55.6311 |

From Table 4 fusion without registration, when compared with the results obtained the output entropy is maximum for the Haar wavelet, MSE is minimum, PSNR is maximum, STD is maximum and M.I is minimum for the Haar wavelet.

**Table 5.** Fusion with registration results on different wavelet filters

| Wavelets | Output entropy | M.I | M.S.E | PSNR | STD |
|---|---|---|---|---|---|
| Coif1 | 7.1452 | 5.7401 | 0.0075 | 69.3801 | 51.3893 |
| Db2 | 7.0906 | 5.7947 | 0.0076 | 69.3226 | 50.3267 |
| Haar | 7.2833 | 5.6020 | 0.0069 | 69.7488 | 49.3889 |

From Table 5 i.e. fusion with registration, when compared with the results obtained the output entropy is maximum for the Haar wavelet, MSE is minimum, PSNR is maximum, and M.I is minimum for the Haar wavelet.

## 7 Conclusion

After applying the proposed methodology on the acquired MRI and PET images and comparing the performance evaluation factors such as M.I, M.S.E, PSNR and STD values, it is concluded that this methodology gives better results in terms of higher entropy, effective mutual information, lower Mean Square Error, higher Peak Signal to Noise Ratio and higher Standard Deviation. The resultant image proved to be the one with more information and less error, if implemented in real time would help doctors in clinical diagnosis immensely in analysing the diseases more precisely.

**Compliance with Ethical Standards**

✓ All authors declare that there is no conflict of interest
✓ No humans/animals involved in this research work.
✓ We have used our own data.

## References


1. Bhavana, V., Krishnappa, H.K.: Multi-modality medical image fusion using discrete wavelet transform. Procedia Comput. Sci. **70**, 625–631 (2015). Open Access
2. Madhuri, G., Hima Bindu, C.H.: Performance evaluation of multi-focus image fusion techniques. In: 2015 International Conference on Computing and Network Communications (CoCoNet 2015), Trivandrum, India, 16–19 December 2015
3. Sekar, K., Duraisamy, V., Remimol, A.M.: An approach of image scaling using DWT and bicubic interpolation. In: Proceedings of the IEEE International Conference on Green Computing, Communication and Electrical Engineering, ICGCCEE 2014 (2014). https://doi.org/10.1109/ICGCCEE.2014.6922406

4. Haribabu, M., Hima Bindu, C.H.: Visibility based multi modal medical image fusion with DWT. In: IEEE International Conference on Power, Control, Signals and Instrumentation Engineering (ICPCSI 2017) (2017)
5. Haribabu, M., Satya Prasad, K.: Multimodal medical image fusion of MRI and PET using wavelet transform. In: Proceedings of the 2012 International Conference on Advances in Mobile Networks, Communication and Its Applications, MN, pp. 127–130 (2012). https://doi.org/10.1109/MNCApps.2012.33
6. Jha, N., Saxena, A.K., Shrivastava, A., Manoria, M.: A review on various image fusion algorithms. In: International Conference on Recent Innovations is Signal Processing and Embedded Systems (RISE 2017), 27–29 October 2017
7. Shakir, H., Talha Ahsan, S., Faisal, N.: Multimodal medical ımage registration using discrete wavelet transform and Gaussian pyramids. In: 2015 IEEE International Conference on Imaging Systems and Techniques (IST) (2015)
8. El-Gamal, F.E.-Z.A., Elmogy, M., Atwan, A.: Current trends in medical image registration and fusion. Egypt. Inform. J. **17**(1), 99–124 (2016)
9. Krishna Chaitanya, Ch., Sangamitra Reddy, G., Bhavana, V., Sai Chaitanya Varma, G.: PET and MRI medical image fusion using STDCT and STSVD. In: International Conference on Computer Communication and Informatics (ICCCI 2017), Coimbatore, India, 5–7 January 2017
10. Bhavana, V., Krıshnappa, H.K.: Fusion of MRI and PET images using DWT and adaptive histogram equalization. In: 2016 International Conference on Communication and Signal Processing, ICCSP 2016, Melmaruvathur, Tamilnadu (2016)
11. Radhakrishna, A., Divya, M.: Survey of data fusion and tumor segmentation techniques using brain images. Int. J. Appl. Eng. Res. **10**, 20559–20570 (2015)
12. Corso, J.J., Sharon, E., Yuille, A.: Multilevel segmentation and integrated bayesian model classification with an application to brain tumor segmentation. In: MICCAI 2006 (2006)

# Capsule Network for Plant Disease and Plant Species Classification

R. Vimal Kurup(✉), M. A. Anupama, R. Vinayakumar, V. Sowmya, and K. P. Soman

Center for Computational Engineering and Networking (CEN), Amrita School of Engineering, Coimbatore, Amrita Vishwa Vidyapeetham, Coimbatore, India
vimal.kurup10@gmail.com, anupamamanikandan94@gmail.com, vinayakumarr77@gmail.com, sowmiamrita@gmail.com

**Abstract.** In deep learning perspective, convolutional neural network (CNN) forms the backbone of image processing. For reducing the drawbacks and also to get better performance than conventional neural network, the new architecture of CNN known as, capsulenet is implemented. In this paper, we analyze capsulenet for two datasets, which are based on plants. In the modern world, most of the diseases are contaminating due to the lack of hygienic food. One of the main reasons for this is, diseases affecting crop species. So, the first model is built for plant disease diagnosis using the images of plant leaves. The corresponding dataset consists of 54,306 images of 14 plant species. The proposed architecture with capsulenet gives an accuracy around 94%. The second task is plant leaves classification. This dataset consists of 2,997 images of 11 plants. The prediction model with capsulenet gives an accuracy around 85%. In the recent years, the use of mobile phones is rapidly increasing. Here for both the models, the images of plant leaves are taken using mobile phone cameras. So, this method can be extended to various plants and can be adopted in large scale manner.

## 1 Introduction

In deep learning, one powerful tool for image processing applications is convolutional neural network (CNN). For getting better results, several architectures of CNN have been developed. But still, conventional CNN have lot of drawbacks. It needs a large dataset for building a model. Also, orientation and spatial relationships between the components of image are not very important to CNN. As a solution to this problem, a new architecture called capsulenet is introduced [13]. Capsules forms the capsulenet for information extraction from images, similar to human brain. In this work, two cases namely, plant image diagnosis and plant leaves classification are taken into account to prove the significance of capsulenet.


S. Smys et al. (Eds.): ICCVBIC 2019, AISC 1108, pp. 413–421, 2020.
https://doi.org/10.1007/978-3-030-37218-7_47

One of the basic needs of human is food. Modern farming techniques have the ability to produce food for 7.6 billion people across the world. Although enough food is available, we are still facing lot of diseases. One of the main reasons is due to the problems associated with these food materials. Surveys showing that more than 50% of plants are affected with pests and diseases [23]. Even, high-tech farming also faces similar type of issues. While considering plants, there are so many diseases associated with each plant. Plant diseases are caused by different factors such as climate changes, lack of organic fertilizers etc. Here the main challenge is to find the particular disease affecting particular plant. Majority of farming is done by small scale farmers and they are also unaware of different types of plant diseases.

Various methods are adopted to detect plant diseases. In earlier days, rather than finding the diseases, small scale farmers used several pesticides and chemical fertilizers, which cause severe after effects. So, disease predictions have much importance in recent days. Nowadays, the use of internet and mobile phones are drastically increased. This can be utilized to get a solution for our problem. Smartphones have much computational power and high-resolution cameras. Here, we implemented an image recognition system for plant disease diagnosis.

For disease diagnosis, we propose a deep learning model using capsulenet. The dataset under consideration for plant disease diagnosis is taken from plant village project (Hughes and Salathé 2015). It consists of 54,306 images of 14 plant species with 38 classes. The performance of the given model is measured on the basis of the exact prediction of crop-disease pair. Capsulenet also works well on small datasets. It is proved by the second case. ie, Plant leaves classification. The dataset is taken from leafsnap and it consists of 11 classes [14].

In past few years, computer vision and image recognition tasks have advanced tremendously. So, convolutional neural network is the best thing to be chosen to perform a task like plant disease diagnosis. The neural network consists of a number of layers that map input to output. The challenge is to create functions for layers to correctly map the input to output. The parameters of the network are effectively tuned during the training process to get high accuracy.

## 2 Related Works

The objective of the present work is to build a capsulenet model for two applications. Plant disease diagnosis can be done in several ways. Several plants are affected by different types of diseases. There are different symptoms for each of the diseases. Plant disease diagnoses based on different symptoms are described in [1]. Now-a-days, several approaches are available to predict plant disease. An image processing based technique, which is also used as smart phone application is discussed in [2]. Rather than a stand-alone application, cloud access for large database can provide more accurate results.

In [3], diseases are determined based on micro-organisms. They implemented an image processing technique based on Convolutional Neural Networks (CNN). For performing the deep CNN training, they used Caffe as the deep learning framework, which is developed by Berkiey Vision Learning Centre. Another approach is described in [4], based on real time polymerase chain reaction and DNA analysis. For the plant disease diagnosis there are certain types of non-destructive techniques such as, spectroscopic and image processing [5].

There is a real time Polymerase Chain Reaction based approach, which does molecular tests, that facilitates rapid diagnostics [6]. The digital image processing techniques, which is used to detect, quantify and classify plant diseases using neural networks are done in [7]. Along with Gabor filter, image segmentation is also used to feed the images as input to the Convolutional Neural Networks to get better accuracy [8].

For the image diagnostics in corn only, a deep learning based image processing technique is used in [9]. The same approach is used in rice plants and is followed in [10]. In [11], diseases are predicted across rice varieties and they have also provided the user interface component. There is an approach developed to predict diseases in fruits [12]. Image processing techniques and stereo microscope are used to measure size and shape variations in fruits and thereby predict the diseases.

Apart from capsulenet, plant leaves classifications are also done in different ways. In [15], the leafsnap dataset is analyzed using support vector machine (SVM) and CNN. Plant leaves can be also classified with Artificial Neural Network (ANN) and Euclidean classifier [16].

Arupriya et al. proposed an approach, which consists of three phases such as: pre-processing, grayscale transformation, boundary enhancement and feature extraction [17]. In [18], to find the similarity between shapes during matching process, a probabilistic curve evolution method with particle filters is used. Pooja [20] did x-ray classification using support vector machine and Grand Unified Regularized Least Squares and obtained good results. Swapna [20] did cardiac arrhythmia diagnosis using convolutional neural network, recurrent neural network and LSTM and hybrid cnn and obtained good results which shows the significance of neural networks.

## 3 Proposed Architecture

The proposed architecture is unique for the datasets related to Plants. Further extension can be done with increasing parameters so that we can solve problems related Plants. The proposed architecture can be visualized as shown in Fig. 1.

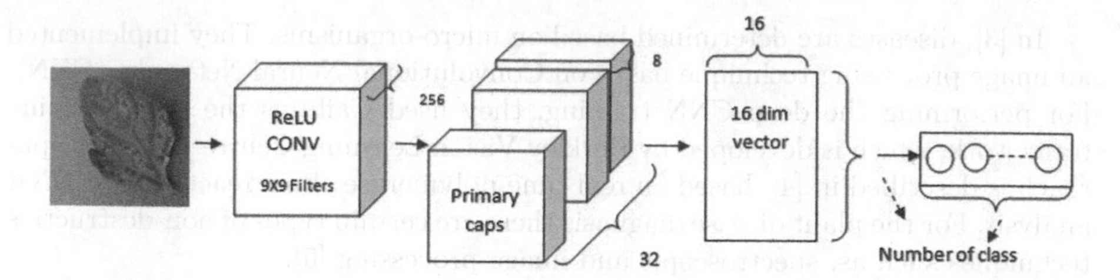


**Fig. 1.** Proposed architecture of capsule network for plant leaves classification and plant disease diagnosis [2].

## 4 Data Description

In this work, the analysis of capsulenet is done on two datasets. One is plant disease diagnosis and the other is plant leaf classification. In both the cases, the images are down scaled while applying to the network. The size at which the image is down scaled is $28 \times 28$.

The dataset for plant leaf prediction is taken from leafsnap [14]. It consists of images taken from species found in Northeastern United States. From the huge dataset, we analyze 11 classes, which contain 2,977 images. Table 4 shows number of images of each class (80% Train and 20% Test data).

In plant disease diagnosis, we analyze 54,306 images of 14 different crop species. It consists of 38 classes, which consists of 26 diseases and 12 healthy cases. Table 1 shows number of images of each class (80% Train and 20% Test data). Each label is a crop disease pair. The data was taken from the project PlantVillage (Hughes and Salathé 2015). Before applying to the network, various transformations are applied to the input images including random cropping, flipping, normalizing etc.

We need a very large dataset for this type of neural network application. Until very recently such data sets are unavailable, solution to this problem is a difficult task. So, Plant Village Project (Hughes and Salathé 2015) collected thousands of images of disease affected leaves as well as healthy and has made them openly and freely available.

Table 4 shows number of images of each class (80% Train and 20% Test data).

## 5 Experimental Results and Observations

We evaluated the applicability of capsule network on plant village dataset and leafsnap dataset. The model is executed in GPU enabled TensorFlow background [22].

Based on the proposed architecture, the capsule network model is evaluated for both the tasks: plant disease diagnosis as well as plant leaves classification.

Tables 2 and 3 shows the configuration details of capsulenet for plant disease diagnosis and Plant leaves classification respectively.

**Table 1.** Configuration details of capsulenet for plant disease diagnosis.

| Layer (type) | Output shape | Param # | Connected to |
|---|---|---|---|
| input1 (InputLayer) | (None, 28, 28, 3) | 0 | |
| conv1 (Conv2D) | (None, 20,20,256) | 62464 | input1[0][0] |
| primarycap conv2d (Conv2D) | (None,6,6,256) | 5308672 | conv1[0][0] |
| primarycap reshape (Reshape) | (None,1152,8) | 0 | primarycap conv2d[0][0] |
| primarycap squash (Lambda) | (None,1152,8) | 0 | primarycap reshape[0][0] |
| plantcaps (CapsuleLayer) | (None, 38,16) | 5603328 | primarycap squash[0][0] |
| input 2 (InputLayer) | (None, 38) | 0 | |
| mask 1 (Mask) | (None, 608) | 0 | plantcaps[0][0], input 2[0][0] |
| capsnet (Length) | (None, 38) | 0 | plantcaps[0][0] |
| decoder (Sequential) | (None, 28, 28, 3) | 3247920 | mask 1[0][0] |

**Table 2.** Configuration details of capsulenet for plant leaves classification.

| Layer (type) | Output shape | Param # | Connected to |
|---|---|---|---|
| input1 (InputLayer) | (None, 28, 28, 3 | 0 | |
| conv1 (Conv2D) | (None, 20,20,256) | 62464 | input1[0][0] |
| primarycap conv2d (Conv2D) | (None,6,6,256) | 5308672 | conv1[0][0] |
| primarycap reshape (Reshape) | (None,1152,8) | 0 | primarycap conv2d[0][0] |
| primarycap squash (Lambda) | (None,1152,8) | 0 | primarycap reshape[0][0] |
| plantcaps (CapsuleLayer | (None, 11,16) | 1622016 | primarycap squash[0][0] |
| input 2 (InputLayer) | (None, 11) | 0 | |
| mask 1 (Mask) | (None, 176) | 0 | plantcaps[0][0], input 2[0][0] |
| capsnet (Length) | (None, 11) | 0 | plantcaps[0][0] |
| decoder (Sequential) | (None, 28, 28, 3) | 3026736 | mask 1[0][0] |

In this work we analysed the Plant disease diagnosis and Plant Leaves classification using capsulenet.

The training set is choosen as 80% of entire dataset for both the cases and validation accuracy is calculated for monitoring the training accuracy. Then based on the parameters obtained the accuracy is calculated for test data. Figure 2 shows the performance of capsnet model for training and validation datasets for plant disease diagnosis and plant leaves classification respectively. The graph indicates that plant disease diagnosis have less error compared to plant leaves classification, since the dataset is small for plant leaves classification task.

For plant disease diagnosis we got 94.5% of accuracy and for Plant Leaves Classification we got 85% accuracy.

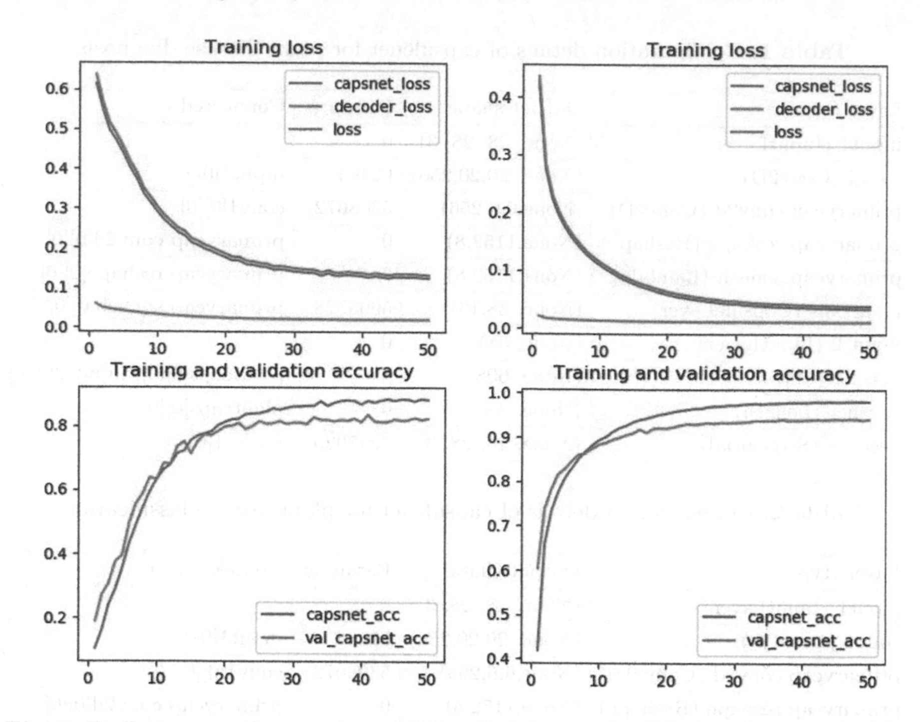


**Fig. 2.** Performance of capsnet model for Training and validation datasets in Plant leaves classification and Plant disease diagnosis.

**Table 3.** Train-test split and accuracy assessment parameters obtained for plant leaves classification using proposed work.

| Sl No. | Classes | Train | Test | Precision | Recall | F1 score | AUC |
|---|---|---|---|---|---|---|---|
| 1 | Acer negundo | 188 | 41 | 0.87 | 0.94 | 0.91 | 0.99 |
| 2 | Acer rebrum | 248 | 49 | 0.64 | 0.79 | 0.71 | 0.94 |
| 3 | Asiminatriloba | 207 | 42 | 0.88 | 0.97 | 0.92 | 0.99 |
| 4 | Broussonettiapapyrifera | 242 | 53 | 0.97 | 0.76 | 0.85 | 0.98 |
| 5 | Catalpa bignonioides | 180 | 37 | 0.92 | 0.92 | 0.92 | 0.99 |
| 6 | Cercis Canadensis | 179 | 37 | 0.77 | 0.93 | 0.84 | 0.99 |
| 7 | Chionanthusvirginicus | 180 | 37 | 0.89 | 0.78 | 0.83 | 0.99 |
| 8 | Ilex opaca | 206 | 39 | 0.63 | 0.73 | 0.68 | 0.96 |
| 9 | Liriodendron tulpifera | 200 | 38 | 0.72 | 0.76 | 0.74 | 0.96 |
| 10 | Maclurapomifera | 378 | 72 | 0.91 | 0.81 | 0.86 | 1.00 |
| 11 | Ulmusrubra | 275 | 54 | 0.76 | 0.76 | 0.76 | 0.96 |

**Table 4.** Train-test split and accuracy assessment parameters obtained for plant disease diagnosis using proposed work.

| Sl No. | Classes | Train | Test | Precision | Recall | F1 score | AUC |
|---|---|---|---|---|---|---|---|
| 1 | Apple scab | 504 | 126 | 0.80 | 0.89 | 0.85 | 1.00 |
| 2 | Apple Black rot | 496 | 125 | 0.92 | 0.93 | 0.93 | 1.00 |
| 3 | Apple cedar | 220 | 55 | 0.55 | 0.91 | 0.68 | 0.98 |
| 4 | Apple healthy | 1316 | 329 | 0.91 | 0.92 | 0.91 | 0.99 |
| 5 | Blueberry healthy | 1202 | 300 | 0.98 | 0.92 | 0.95 | 1.00 |
| 6 | Cherry healthy | 684 | 170 | 0.99 | 0.97 | 0.98 | 1.00 |
| 7 | Cherry powdery mildew | 842 | 210 | 0.95 | 0.91 | 0.93 | 1.00 |
| 8 | Corn cercospora grey spot | 410 | 103 | 0.71 | 0.88 | 0.78 | 0.99 |
| 9 | Corn common rust | 953 | 239 | 1.00 | 1.00 | 1.00 | 1.00 |
| 10 | Corn healthy | 929 | 233 | 1.00 | 0.98 | 0.99 | 1.00 |
| 11 | Corn northern leaf blight | 788 | 197 | 0.91 | 0.89 | 0.90 | 1.00 |
| 12 | Grape black rot | 944 | 236 | 0.88 | 0.92 | 0.90 | 1.00 |
| 13 | Grape esca | 1107 | 276 | 0.93 | 0.91 | 0.92 | 1.00 |
| 14 | Grape healthy | 339 | 84 | 0.99 | 0.97 | 0.98 | 1.00 |
| 15 | Grape leaf blight | 861 | 215 | 0.97 | 0.96 | 0.97 | 1.00 |
| 16 | Orange haunglongbing | 4405 | 1102 | 0.98 | 0.97 | 0.98 | 1.00 |
| 17 | Peach bacterial spot | 1838 | 459 | 0.94 | 0.94 | 0.94 | 1.00 |
| 18 | Peach healthy | 288 | 72 | 0.96 | 0.91 | 0.93 | 1.00 |
| 19 | Pepper bell bacterial spot | 797 | 200 | 0.91 | 0.85 | 0.88 | 0.99 |
| 20 | Pepper bell healthy | 1183 | 295 | 0.96 | 0.92 | 0.94 | 1.00 |
| 21 | Potato early blight | 800 | 200 | 0.95 | 0.90 | 0.92 | 1.00 |
| 22 | Potato healthy | 121 | 31 | 0.61 | 0.83 | 0.70 | 0.99 |
| 23 | Potato late blight | 800 | 200 | 0.84 | 0.86 | 0.85 | 1.00 |
| 24 | Raspberry healthy | 297 | 74 | 0.92 | 0.86 | 0.89 | 1.00 |
| 25 | Soybean healthy | 4072 | 1018 | 0.98 | 0.99 | 0.98 | 1.00 |
| 26 | Squash powdery mildew | 1468 | 367 | 0.98 | 0.98 | 0.98 | 1.00 |
| 27 | Strawberry healthy | 364 | 92 | 0.95 | 0.95 | 0.95 | 1.00 |
| 28 | Strawberry leaf scorch | 887 | 222 | 0.94 | 0.96 | 0.95 | 1.00 |
| 29 | Tomato Bacterial spot | 1702 | 425 | 0.95 | 0.94 | 0.94 | 1.00 |
| 30 | Tomato Early blight | 800 | 200 | 0.71 | 0.80 | 0.76 | 0.99 |
| 31 | Tomato healthy | 1273 | 318 | 0.98 | 0.98 | 0.98 | 1.00 |
| 32 | Tomato Late blight | 1527 | 382 | 0.83 | 0.86 | 0.84 | 0.99 |
| 33 | Tomato Leaf Mold | 761 | 191 | 0.91 | 0.93 | 0.92 | 1.00 |
| 34 | Tomato Septoria leaf spot | 1417 | 354 | 0.90 | 0.92 | 0.91 | 1.00 |
| 35 | Tomato Two-spotted spider mite | 1341 | 335 | 0.92 | 0.92 | 0.92 | 1.00 |
| 36 | Tomato Target Spot | 1123 | 281 | 0.93 | 0.87 | 0.90 | 1.00 |
| 37 | Tomato mosaic virus | 299 | 74 | 0.89 | 0.92 | 0.90 | 1.00 |
| 38 | Tomato Yellow Leaf Curl Virus | 4286 | 1071 | 0.99 | 0.98 | 0.99 | 1.00 |

## 6 Conclusion

This paper analyzes the performance of capsulenet for two tasks related to plants. We chose capsulenet rather than conventional convolutional neural architectures, since CNN have some drawbacks. Capsules present in capsulenet are similar to human brain in capturing the relevant information. The two tasks considered in the present work are plant disease diagnosis, which resulted in accuracy around 94.5%. To evaluate the effectiveness of capsulenet model, the same model is executed in one more task called, plant leaves classification, which has a small dataset and the proposed capsulenet architecture resulted around accuracy of 85%.

Since large computational power is required, we done experiments on this small number of images. But, capsulenet is capable of giving better accuracies for small datasets also. For last few years, the uses of mobile phones are drastically increased; mobile phone images can be used to use such models in real-time. So, in future, it is easier to increase the number of images in the dataset and can re-train the same model.

**Compliance with Ethical Standards.** All authors declare that there is no conflict of interest No humans/animals involved in this research work. We have used our own data.

## References


1. Riley, M.B., Williamson, M.R., Maloy, O.: Plant disease diagnosis. The Plant Health Instructor. https://doi.org/10.1094/PHI-I-2002-1021-01.2002
2. Georgakopoulou, K., Spathis, C., Petrellis, N., Birbas, A.: A capacitive to digital converter with automatic range adaptation. IEEE Trans. Instrum. Meas. **65**(2), 336–345 (2011)
3. Sankaran, S., Mishra, A., Eshani, R., Davis, C.: A review of advanced techniques for detecting plant diseases. Comput. Electron. Agric. **72**(1), 1–13 (2010)
4. Schaad, N.W., Frederick, R.D.: Real time PCR and its application for rapid plant disease diagnostics. Can. J. Plant Pathol. **24**(3), 250–258 (2002)
5. Purcell, D.E., O'Shea, M.G., Johnson, R.A., Kokot, S.: Near infrared spectroscopy for the prediction of disease rating for Fiji leaf gall in sugarcane clones. Appl. Spectro. **63**(4), 450–457 (2009)
6. Schaad, N.W., Frederick, R.D., Shaw, J., Schneider, W.L., Hickson, R., Petrillo, M.D., Luster, D.G.: Advances in molecular-based diagnostics in meeting crop biosecurity and phytosanitary issues. Ann. Rev. Phytopathol. **41**, 305–24 (2003)
7. Barbedo, G.C.A.: Digital image processing techniques for detecting quantifying and classifying plant diseases. Springer Plus **2**, 660 (2013)
8. Kulkarni, A., Patil, A.: Applying image processing technique to detect plant diseases. Int. J. Mod. Eng. Res. **2**(5), 3361–3364 (2012)
9. Lai, C., Ming, B., Li, S.K., Wang, K.R., Xie, R.Z., Gao, S.J.: An image-based diagnostic expert system for corn diseases. Agric. Sci. China **9**(8), 1221–1229 (2010)
10. Sarma, S.K., Singh, K.R., Singh, A.: An expert system for diagnosis of diseases in rice plant. Int. J. Artif. Intell. **1**(1), 26–31 (2010)

11. Abu-Naser, S.S., Kashkash, K.A., Fayyad, M.: Developing an expert system for plant disease diagnosis. J. Artif. Intell. **1**(2), 78–85 (2008). Asian Network for Scientific Information
12. Mix, C., Picó, F.X., Ouborg, N.J.: A comparison of stereomicroscope and image analysis for quantifying fruit traits. SEED Technology, vol. 25, no. 1 (2003)
13. Sabour, S., Frosst, N., Hinton, G.E.: Dynamic routing between capsules. In: Neural Information Processing Systems (2017)
14. Kumar, N., Belhumeur, P.N., Biswas, A., Jacobs, D.W., John Kress, W., Lopez, I.C., Soares, J.V.B.: Leafsnap: a computer vision system for automatic plant species identification. In: Proceedings of the 12th European Conference on Computer Vision (2012)
15. Sulc, M., Matas, J.: Fine-grained recognition of plants from images. Plant Methods **13**, 115 (2017)
16. Satti, V., Satya, A., Sharma, S.: An automatic leaf recognition system for plant identification using machine vision technology. Int. J. Eng. Sci. Technol. **5**, 874 (2013)
17. Arun Priya, C., Balasaravanan, T., Selvadoss Thanamani, A.: An efficient leaf recognition algorithm for plant classification using support vector machine. In: Proceedings of the International Conference on Pattern Recognition, Informatics and Medical Engineering, pp. 428–432 (2012)
18. Valliammal, N., Geethalakshmi, S.N.: Automatic recognition system using preferential image segmentation for leaf and flower images. Comput. Sci. Eng.: Int. J. **1**, 13–25 (2011)
19. Kekre, H.B., Mishra, D., Narula, S., Shah, V.: Color feature extraction for CBIR. Int. J. Eng. Sci. Technol. **3**, 8357–8365 (2011)
20. Pooja, A., Mamtha, R., Sowmya, V., Soman, K.P.: X-ray image classification based on tumor using GURLS and LIBSVM, pp. 0521–0524 (2016). https://doi.org/10.1109/ICCSP.2016.7754192
21. Swapna, G., Soman, K.P., Vinayakumar, R.: Automated detection of cardiac arrhythmia using deep learning techniques. Procedia Comput. Sci. **132**, 1192–1201 (2018). https://doi.org/10.1016/j.procs.2018.05.034
22. Abadi, M., Barham, P., Chen, J., Chen, Z., Davis, A., Dean, J., Devin, M., Ghemawat, S., Irving, G., Isard, M., Kudlur, M.: TensorFlow: a system for large-scale machine learning, vol. 16, pp. 265–283 (2016)
23. Harvey, C.A., Rakotobe, Z.L., Rao, N.S., Dave, R., Razafimahatratra, H., Rabarijohn, R.H., et al.: Extreme vulnerability of smallholder farmers to agricultural risks and climate change in madagascar. Philos. Trans. R. Soc. Lond. B Biol. Sci. **369**, 20130089 (2014). https://doi.org/10.1098/rstb.2013.008

# *RumorDetect*: Detection of Rumors in Twitter Using Convolutional Deep Tweet Learning Approach

N. G. Bhuvaneswari Amma[1(✉)] and S. Selvakumar[1,2]

[1] National Institute of Technology, Tiruchirappalli 620 015, Tamil Nadu, India
ngbhuvaneswariamma@gmail.com

[2] Indian Institute of Information Technology, Una, Himachal Pradesh, India
ssk@nitt.edu, director@iiitu.ac.in

**Abstract.** Nowadays social media is a common platform to exchange ideas, news, and opinions as the usage of social media sites is increasing exponentially. Twitter is one such micro-blogging site and most of the early update tweets are unverified at the time of posting leading to rumors. The spread of rumors in certain situations make the people panic. Therefore, early detection of rumors in Twitter is needed and recently deep learning approaches have been used for rumor detection. The lacuna in the existing rumor detection systems is the curse of dimensionality problem in the extracted features of Twitter tweets which leads to high detection time. In this paper, the issue of dimensionality is addressed and a solution is proposed to overcome the same. The detection time could be reduced if the relevant features are only considered for rumor detection. This is captured by the proposed approach which extracts the features based on tweet, reduces the dimension of tweet features using convolutional neural network, and learns using fully connected deep network. Experiments were conducted on events in Twitter PHEME dataset and it is evident that the proposed convolutional deep tweet learning approach yields promising results with less detection time compared to the conventional deep learning approach.

**Keywords:** Convolutional neural network · Deep learning · Feature extraction · Rumor detection · Social media · Twitter

## 1 Introduction

Online social media is a platform that generates massive amount of data during breaking news [11]. Twitter is one of the social media platforms in which people post information about stock markets, natural hazards, general opinion, etc. But it is hard to trust the information being spread in the social media virally as the users simply click Twitter retweet without verifying the information. This unverified information is termed as *Rumor* and it originates from one or more sources and spread virally from node to node in the social media network. The rumor may


S. Smys et al. (Eds.): ICCVBIC 2019, AISC 1108, pp. 422–430, 2020.
https://doi.org/10.1007/978-3-030-37218-7_48

be true, false, or remain unresolved. The detection of rumors is crucial in crisis situations and therefore, rumor detection system is needed to stop the spread of rumors at the earliest [8]. The rumor detection systems discriminate the rumor from non-rumor by considering it as a binary classification problem. The binary classification techniques are categorized into statistical approach and machine learning approach. The statistical approach requires low computational cost but it needs proper assumptions and inability to fix the assumptions leads to low false positive rate. The machine learning approach requires no prior assumptions and detects the rumor with high accuracy [1].

The recent buzzword in the literature is deep learning which is the subset of machine learning. The deep learning approaches automatically find the relationship in the data and learn the features with different levels of abstractions. The most extensively used deep approach is Convolutional Neural Network (CNN). The CNN extracts the features and compresses the extracted features to make it suitable for learning. Generally, the CNN consists of Convolution Layer (CL), Pooling Layer (PL), Dropout Layer (DOL), and fully connected layer. The features extracted from the tweets play a major role in classification problems. Therefore, feature vector formation is an important step to improve the performance of the rumor detection systems [7]. The following propositions are the key contributions of this paper:

1. Extraction of salient features to detect rumors.
2. A Convolutional Deep Tweet Learning network structure for training the Twitter rumor detection system.
3. *RumorDetect* approach to classify the tweet as rumor or non-rumor.

The rest of the paper is organized as follows: Sect. 2 discusses feature extraction in Twitter rumors and deep learning based rumor detection. Section 3 describes the proposed *RumorDetect* approach. Section 4 discusses the experimental results of the proposed approach. The conclusion with the directions for further research is discussed in Sect. 5.

## 2 Related Works

### 2.1 Feature Extraction in Twitter Rumors

Generally Twitter rumor is the collection of tweets about an event that spread through Twitter. The tweets are similar assertions about the event [8]. The rumors are thousands of tweets with same meaning but the order of words in the tweet is different. The features play a major role in machine learning based classification problems. Therefore, extracting the salient features from the Twitter tweets is an essential step in rumor detection [4]. It is observed from the literature that the features are categorized into content, multimedia, propagation, and topic based features [5].

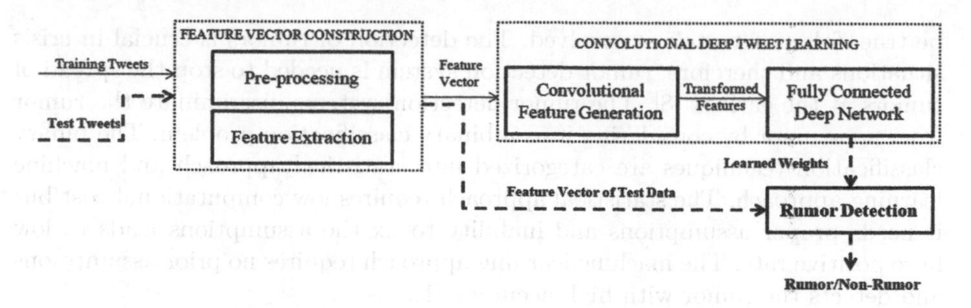


**Fig. 1.** Architecture of proposed *RumorDetect* approach

### 2.2 Deep Learning Based Rumor Detection

The surveys of existing rumor detection systems in social media are given in [9,10]. One category is based on rumors known apriori. In this type of rumor detection, a set of pre-defined rumors are fed to the classifier and it classifies the new tweets as either rumor or non-rumor. This approach is suitable for long standing rumors but not suitable for breaking news as unseen rumors emerge and keywords linked to the rumor are yet to be detected [12]. Another category is based on learning the generalizable patterns that help to identify rumors during emerging events [8]. Deep learning based rumor detection system learns the tweet stream automatically by finding the relationship in the tweets. The deep learning techniques such as autoencoder, CNN, long short term memory, etc., adapted to rapidly changing twitter data streams motivated us to propose a convolutional deep tweet learning classifier which extracts the features with different levels of representation and learns automatically the extracted features to classify the Twitter data stream [2,6].

## 3 Proposed *RumorDetect* Approach

The architecture of proposed *RumorDetect* approach is shown in Fig. 1. This approach consists of training tweets and testing tweets phases. The training tweets phase comprises of Feature Vector Construction and Convolutional Deep Tweet Learning modules. The testing tweets phase uses Rumor Detection module to classify the given tweet as either rumor or non-rumor. The flows of training and testing tweets are represented using solid and dashed lines respectively. The tweets in the dataset consist of source tweet and reactions for each Twitter data.

### 3.1 Feature Vector Construction

The feature vector construction phase includes pre-processing and feature extraction sub modules. The pre-processing sub module comprises of tokenization, parts of speech tagger, stop word removal, stemming, bag of words creation,

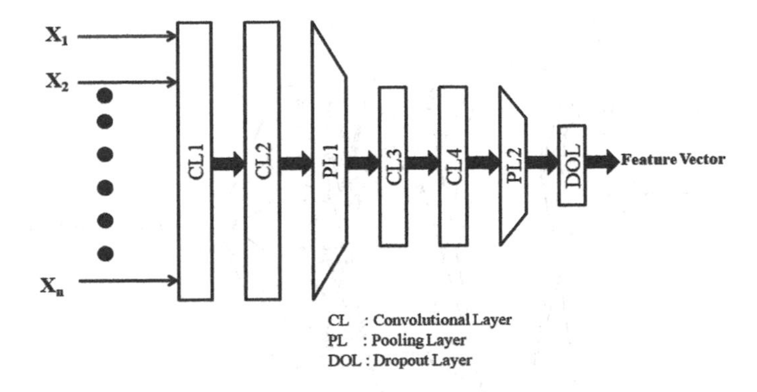


**Fig. 2.** Proposed convolutional feature generation network structure

and term frequency-inverse tweet frequency (tf-itf) calculation [9]. Tokenization is the process of reading the tweets and output tokenized words. The parts of speech tagging is the process of reading the tokenized words and output parts of speech tagged words. Stop word removal is the process of removing less important and meaningless words such as articles and prepositions like a, the, is, etc. Stemming is the process of finding all root forms of the input word. Bag of words creation is the process of taking into account the words and their frequency of occurrence in the tweet disregarding the semantic relationship in the tweet. The tf-itf calculation is the process of computing the weights for each token in the tweet to evaluate how important a word is to a tweet. The feature extraction plays a major role in rumor detection. The extracted features are represented as vector for further learning. The extracted features from the tweets are as follows: Tweet count, Listed count, Retweet count, Favorite count, Account age, Followers count, Hashtag count, Friends count, URL count, Character count, Digit count, and Word count.

### 3.2 Convolutional Deep Tweet Learning

The convolutional deep tweet learning module comprises of two sub modules, viz., Convolutional Feature Generation (CFG) based pre-training and Fully Connected Deep Network (FCDN) based training.

The proposed CFG network compresses and reduces the size of the feature vector. The structure of the CFG network is depicted in Fig. 2. It consists of CL, PL, and DOL. The normalized training tweets are given as input to the CL. The CL [3] produces the transformed convolution as follows:

$$TrCon_i^l = X_{D_i^l} \times F \tag{1}$$

where $X_D$ is the normalized tweet vector and $F$ is the filter. The reason for using two levels of CL before PL is to generate high level features to compress the generated feature vector so as to reduce the feature vector size.

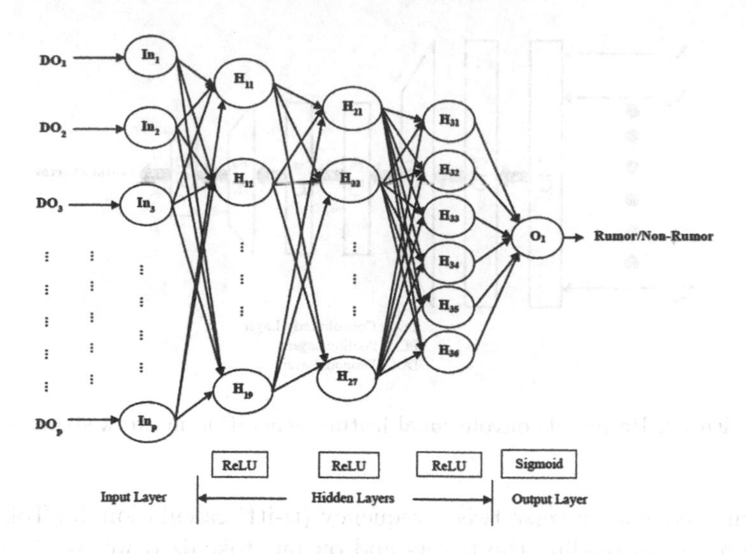


**Fig. 3.** Structure of proposed fully connected deep network

The PL compresses and reduces the size of the feature vector so as to overcome the issue of overfitting. The proposed approach uses average pooling with the size of 2 using (2).

$$TrPoo(TrCol_{ij}^{l}, TrCol_{ik}^{l}) = avg(TrCol_{ij}^{l}, TrCol_{ik}^{l}) \tag{2}$$

The subsequent CLs and PL are computed using (1) and (2) respectively. The application of filter with feature vector creates co-dependency among each other leading to overfitting. In order to overcome this issue, DOL is incorporated in the CFG network. The nodes are dropped out with the probability, $p$, so that the reduced network is passed to the FCDN for further training.

FCDN trains the feature vector and produces learned weights for rumor detection. Figure 3 depicts the proposed FCDN with $q-9-7-6-1$ structure. The intuition behind choosing this structure is to learn the feature vector with different abstractions. The computation in the hidden layer is performed as follows:

$$H_{1i} = \sum_{i=1}^{p} In_{ij} \times W_{ij}^{H1} \tag{3}$$

where $In_{ij}$ is the DOL output which is passed to the input layer of FCDN and $W_{ij}^{H1}$ is the weights between input and hidden layer. The Rectified Linear Unit (ReLU) and sigmoid functions are used in the hidden and output layers respectively. The ReLU is computed as follows:

$$f\left(H_{1i}\right) = max\left(H_{1i}, 0\right) \tag{4}$$

The sigmoid activation function is used in the output layer and computed as follows:

$$f\left(I_{j}^{O}\right) = 1/\left(1 + e^{-I_{j}^{O}}\right) \tag{5}$$

**Algorithm 1.** *RumorDetect* Algorithm

**Input:** Test Tweet, learned weights
**Output:** Rumor/Non-Rumor
1: **for** each record **do**
2: Compute transformed convolution using (1)
3: Compute transformed pooling using (2)
4: Compute sum of product of DOL output and weights
5: Compute ReLU activation function
6: Compute output using (5)
7: **if** $output \leq 0$ **then**
8: Return $Non - Rumor$
9: **else**
10: Return *Rumor*
11: **end if**
12: **end for**

**Table 1.** Statistics of datasets

| Breaking news | Training | | Testing | |
|---|---|---|---|---|
| | Rumors | Non-rumors | Rumors | Non-rumors |
| FU | 275 | 973 | 183 | 648 |
| OS | 170 | 515 | 114 | 344 |
| SS | 143 | 139 | 95 | 92 |
| CHS | 282 | 252 | 188 | 168 |
| GPC | 313 | 419 | 209 | 280 |

In order to train the model, Root Mean Square Error (RMSE) is computed using the class label and the computed output of each tweet in the training dataset [3]. The tweet learning process continues till the RMSE reaches a pre-defined threshold which is suitable for rumor detection.

### 3.3 Rumor Detection

The Rumor Detection module is similar to that of CFG pre-training and FCDN training without RMSE computation. The learned weights obtained from tweet learning are used for detection of rumor. Algorithm 1 depicts the *RumorDetect* Algorithm. The test Twitter tweet is given as input to the Rumor Detection module to detect whether the given tweet is rumor or non-rumor.

## 4 Experimental Results

The proposed approach was implemented in Python on Intel(R) Core(TM) i7-6700 CPU @ 3.40 GHz processor and 16 GB RAM under Windows 10. PHEME dataset was used for experimentation and it consists of Twitter tweets posted

**Table 2.** Overall performance of proposed *RumorDetect* approach

| Breaking news | TN | FP | FN | TP | Precision | Recall | F-measure | Accuracy | ER |
|---|---|---|---|---|---|---|---|---|---|
| FU | 607 | 41 | 31 | 152 | 78.76 | 83.06 | 80.85 | 91.34 | 8.66 |
| OS | 312 | 32 | 20 | 94 | 74.60 | 82.46 | 78.33 | 88.65 | 11.35 |
| SS | 74 | 18 | 13 | 82 | 82.00 | 86.32 | 84.10 | 83.42 | 16.58 |
| CHS | 140 | 28 | 25 | 163 | 85.34 | 86.70 | 86.02 | 85.11 | 14.89 |
| GPC | 251 | 29 | 33 | 176 | 85.85 | 84.21 | 85.02 | 87.32 | 12.63 |

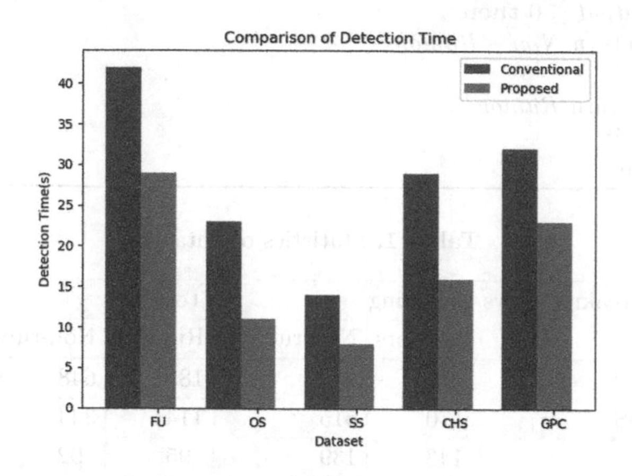


**Fig. 4.** Rumor detection approaches versus detection time

during breaking news [11]. The following are the five events that attracted the media with huge interest: Ferguson Unrest (FU), Ottawa Shooting (OS), Sydney Siege (SS), Charlie Hebdo Shooting (CHS), and Germanwings Plane Crash (GPC). Table 1 shows the training and testing events in the PHEME dataset. The proposed approach is evaluated using True Negative, False Positive, False Negative, and True Positive performance measures. The performance metrics, such as, Precision, Recall, F-measure, Accuracy, and Error Rate (ER) are computed for the proposed approach and tabulated in Table 2. It is observed from the table that all the events in the PHEME dataset yield significant results. Figure 4 depicts the comparison of detection time taken by the proposed approach and conventional deep learning approach. The structure of conventional deep learning classifier used for experimentation consists of two levels of CL and PL with FCDN with $q-9-7-6-1$ structure. It is seen that the detection time of proposed approach is less than the existing deep learning approach for all the events in the dataset as the proposed approach uses DOL after CL and PL.

## 5 Conclusion

In this paper, *RumorDetect* approach is proposed to detect rumors in Twitter tweets. The objective is to overcome the curse of dimensionality problem from the extracted features of Twitter tweets, thereby reducing the detection time of the proposed *RumorDetect* system. The features were extracted from the tweets to generate feature vector. The network was trained using convolutional deep tweet learning approach and the *RumorDetect* system classifies the unseen tweet as either rumor or non-rumor. The performance of the proposed *RumorDetect* approach was evaluated using benchmark Twitter PHEME dataset. The proposed approach achieves significant improvement in performance compared to the conventional deep learning classifier. However, the extension of this work could be the online detection of rumors in Twitter to further automate the system.

**Compliance with Ethical Standards.** All authors declare that there is no conflict of interest No humans/animals involved in this research work. We have used our own data.

## References

1. Ahmed, F., Abulaish, M.: A generic statistical approach for spam detection in online social networks. Comput. Commun. **36**(10–11), 1120–1129 (2013)
2. Ajao, O., Bhowmik, D., Zargari, S.: Fake news identification on Twitter with hybrid CNN and RNN models. In: Proceedings of the 9th International Conference on Social Media and Society, pp. 226–230. ACM (2018)
3. Amma, N.G.B., Subramanian, S.: VCDeepFL: vector convolutional deep feature learning approach for identification of known and unknown denial of service attacks. In: TENCON 2018-2018 IEEE Region 10 Conference, pp. 0640–0645. IEEE (2018)
4. Chen, C., Wang, Y., Zhang, J., Xiang, Y., Zhou, W., Min, G.: Statistical features-based real-time detection of drifted Twitter spam. IEEE Trans. Inf. Forensics Secur. **12**(4), 914–925 (2017)
5. Liang, G., He, W., Xu, C., Chen, L., Zeng, J.: Rumor identification in microblogging systems based on users behavior. IEEE Trans. Comput. Soc. Syst. **2**(3), 99–108 (2015)
6. Liu, Y., Jin, X., Shen, H.: Towards early identification of online rumors based on long short-term memory networks. Inf. Process. Manag. **56**(4), 1457–1467 (2019)
7. Vishwarupe, V., Bedekar, M., Pande, M., Hiwale, A.: Intelligent Twitter spam detection: a hybrid approach. In: Smart Trends in Systems, Security and Sustainability, pp. 189–197. Springer (2018)
8. Vosoughi, S., Mohsenvand, M., Roy, D.: Rumor gauge: predicting the veracity of rumors on Twitter. ACM Trans. Knowl. Discov. Data (TKDD) **11**(4), 50 (2017)
9. Wu, T., Wen, S., Xiang, Y., Zhou, W.: Twitter spam detection: survey of new approaches and comparative study. Comput. Secur. **76**, 265–284 (2018)
10. Zubiaga, A., Aker, A., Bontcheva, K., Liakata, M., Procter, R.: Detection and resolution of rumours in social media: a survey. ACM Comput. Surv. (CSUR) **51**(2), 32 (2018)

11. Zubiaga, A., Liakata, M., Procter, R.: Learning reporting dynamics during breaking news for rumour detection in social media. arXiv preprint arXiv:1610.07363 (2016)
12. Zubiaga, A., Spina, D., Martinez, R., Fresno, V.: Real-time classification of Twitter trends. J. Assoc. Inf. Sci. Technol. **66**(3), 462–473 (2015)

# Analyzing Underwater Videos for Fish Detection, Counting and Classification

G. Durga Lakshmi(✉) and K. Raghesh Krishnan

Department of Computer Science and Engineering, Amrita School of Engineering, Coimbatore, Amrita Vishwa Vidhyapeetham, Coimbatore, India
cb.en.p2cse17010@cb.students.amrita.edu,
k_raghesh@cb.amrita.edu

**Abstract.** Underwater video processing is a valuable tool for analyzing the presence and behaviour of fishes in underwater. Video based analysis of fishes finds its use in aquaculture, fisheries and protection of fishes in oceans. This paper proposes a system to detect, count, track and classify the fishes in underwater videos. In the proposed approach, two systems are developed, a counting system, which uses gaussian mixture model (for foreground detection), morphological operations (for denoising), blob analysis (for counting) and kalman filtering (for tracking), and a classification system, which uses bag of features approach that is used to classify the fishes. In the bag of feature approach, surf features are extracted from the images to obtain feature descriptors. k-mean clustering is applied on the feature descriptors, to obtain visual vocabulary. The test features are input to the MSVM classifier, which uses visual vocabulary to classify the images. The proposed system achieves an average accuracy of 90% in counting and 88.9% in classification, respectively.

**Keywords:** Gaussian mixture model · Kalman filtering · Morphological operations · Blob analysis · Bag of features · Visual vocubulary · SURF features · K-means clustering · FAST (Features from Accelerated Segment Test)

## 1 Introduction

Counting and classification of fishes is an essential tool for fisheries and oceanogra phers, to analyze the presence and behavior of fishes. In practice, oceanographers generally verify the underwater conditions using nets (to collect and test the fishes) or human observation. Manual approaches are time consuming, expensive and less accurate in analyzing the state and behavior of fishes. The process also involves an enormous risk of killing or injuring the fishes and damaging the surroundings. In order to overcome these issues, the oceanographers have started exploring alternate approaches to marine life observation. One such approach is to acquire underwater fish videos and examine them using computer vision techniques, which split the problem into fish counting and classification modules, for observation. Most of the traditional computer vision based approaches to underwater fish analysis in the literature make use of manual counting. In order to increase the speed of the counting process and to


S. Smys et al. (Eds.): ICCVBIC 2019, AISC 1108, pp. 431–441, 2020.
https://doi.org/10.1007/978-3-030-37218-7_49

reduce the human intervention, this work proposes an automatic approach to fish counting and classification.

## 2 Literature Survey

Boudhane Mohcine and Benayad Nsiri, devised a useful method to pre-process and detect fishes in underwater images [1]. Poisson-Gauss theory is used for de-noising, as it accurately removes the noise in the images. A mean-shift algorithm is used to segment the foreground, following which statistical estimation is applied to every segment, to combine the segments into objects. The quality of the reconstructed image is measured using MSE and PSNR. The results estimated using log likelihood ratio test show a reduction in the number of false regions.

Spampinato et al. propose a method for recognition of fish species in underwater videos using a sparse low-rank matrix decomposition, to extract the foreground from the image [2]. The features of the fishes are extracted using a Deep Architecture, which consists of two convolution layers, a feature pooling layer and a non-linear layer. In above layers PCA, binary hashing and block wise histograms used, respectively. In order to collect the information which does not vary with respect to large poses, spatial pyramid pooling algorithm used. The result obtained is given as an input to a linear SVM classifier. The deep network model is trained for 65000 iterations, to achieve an accuracy of 98.64%.

Prabowo et al. propose a system to provide fish and video related information, such as the number of fishes present in the video, the quality of the video and the dominant color of the video [3]. The work uses a gaussian mixture model followed by a moving average detection algorithm, to detect the moving objects in the video. Statistical moments of the gray level histogram are used in the analysis of color and texture, for fish detection and mean shift algorithm is used for fish tracking. The accuracies of detection and counting of fishes are reported as 84% and 85%, respectively.

Palazzo et al. propose an approach to detect moving objects in underwater videos in [4]. The RGB video is converted into gray scale video normal distribution is applied to distinguish the background pixels and objects. The average sensitivity obtained for two test videos is 83% and the precision value is 67.5%. Rathi et al., devise an algorithm to detect fish in marine videos [5]. The covariance-based model is used to collaborate color and texture features. Kernel Density Estimation is used to progress the detection of fish. Mark R. Shortis et al. propose a new approach to classify the fishes in images in [6]. Otsu's thresholding, erosion and dilation are applied to pre-process the images, following which convolutional neural networks are used to classify the fish species. The work reports an accuracy of 96.29%.

Jing Hu discusses an algorithm to classify 6 different species of fishes in images [7]. RGB, HSV color features, statistical texture features, such as gray scale histogram, GLCM features and wavelet-based features are used to differentiate the fish species. These features are used to train and test the LIBMSVM and DAGMSVM and it was observed that the DAGMSVM took less time than LIBMSVM. Fabric et al. propose a systematic approach to count fishes in videos acquired using dynamic camera [8].

Laplacian of the Gaussian (LoG) is used to produce color histograms, which differentiate the background from fish. Following this, Zernike moments are extracted to analyse the shape of the fishes. Blob detector and connected component labelling are used in the counting of fishes.

Prabhakar and Kumar use global contrast correction, to produce dual intensity images, to enhance the underwater images [9]. Contrast enhanced images are produced from the dual intensity images and processed, to enhance the features. To improve the saturation and brightness, color correction is applied on the the contrast enhanced image. Mark R in [10], proposes an automated fish detection system in a video. In this work, PCA is used to model the fish shape. Haar classifier is to detect the fishes in the video. Christopher Richard Wren et al. in [11] conducted experiments in order to detect the fishes in underwater images using histogram-based background subtraction, region growing, edge detection, which proves to be useful when the camera is static.

Khanfar et al. in [12] propose an algorithm that differentiates background and foreground, which simplifies object detection in every frame. However, there are few drawbacks in this approach. The algorithm fails in scenarios where the background is also in motion or the objects to be detected are in the same color as that of the background. The second problem is fixed by incorporating texture and color features in the model. The algorithm is evaluated on underwater videos and is found to produce great and reliable performance.

Fiona H Evans et al. propose an approach based on Expectation-maximization (EM) algorithm, for fish detection [13]. In this approach, a particular threshold intensity is chosen from the intensity values of the fish images and the pixels which have their intensity higher than the chosen threshold are assumed to be the pixels of fishes. The pixel values are defined in terms of x and y sets. The fish image is modelled as gaussian mixture model, considering shape of each fish as a multivariant gaussian. After this, parameters of gaussian mixture model, along with the quantity of fishes are calculated, using EM algorithm. Neetika Singhal, Kalaichelvi et al. propose an image classification algorithm using bag of visual words model [14]. The image features are extracted using FAST algorithm and converted into feature descriptors. K-means clustering is applied on the feature descriptors, to produce a bag of visual words. These words are input to a linear SVM and report an accuracy of 90.8%.

Inspired by the gaussain mixture model, kalman filter and bag of feature approach, this paper attempts to implement an effective system, to solve the problem of underwater fish counting and classification using videos.

## 3 Proposed Approach

This work proposes two systems, as shown in Figs. 1 and 2, to analyze the underwater fish behavior. The first system, namely the fish counting system, is used to *count* the number of fishes present in underwater videos and the second fish classification system, is used to *classify* the fish species. In the first system, the initial step is to convert the video into frames and a gaussian mixture model is applied for foreground detection. Denoising is performed using morphological operators to eliminate the unnecessary noise. Connected component labelling is used to detect the fishes. Finally, Kalman filter is applied, to count and track the fishes.

The Second system used to classify the fish species in videos. Proposed architecture of the fish classification system is shown in Fig. 2. This system uses bag of feature approach, In which the dataset is segregated into training and testing sets for classification. SURF features are extracted from the training images and feature descriptors are framed. These feature descriptors are input to the k-means clustering algorithm, which outputs the visual vocabulary. The test images are given to the trained Multi Class Support Vector Machine (MSVM) classifier, which uses visual vocabulary to classify the test images.

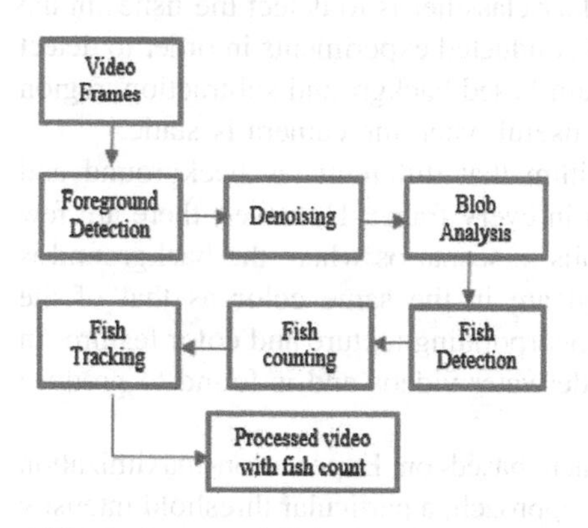


**Fig. 1.** Proposed Fish clounting system

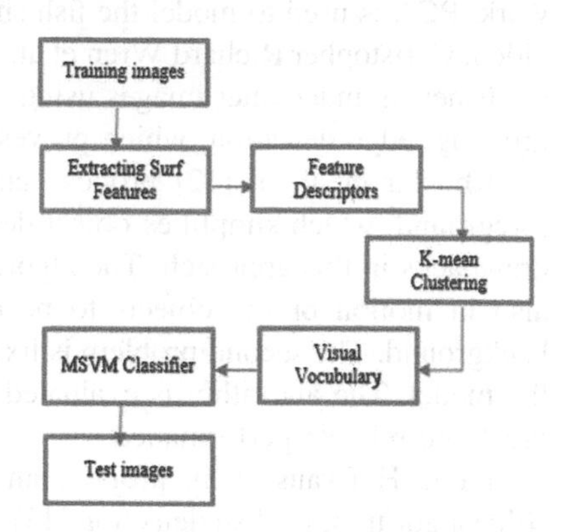


**Fig. 2.** Proposed classification system

## 3.1 Fish Counting System

### 3.1.1 Foreground Detection Using GMM

Background subtraction refers to the task of separating the foreground from the videos for further processing. In Gaussian mixture model, the background pixels are modelled as normal distribution and the variables of the distributions such as mean and variance are determined. The low probability pixels are treated as object and the high probability pixels are considered as the backgound. In GMM, each pixel in every frame is modelled as a Gaussian distribution. Initially, each frame is split in the RGB color space based on its intensity.

Therefore, the intensity distinction is assumed to have a uniform variance. The covariance matrix is calculated as:

$$\sum\nolimits_{i,t} \sigma_{i,t}^2 I \qquad (1)$$

If a gaussian value is greater than the threshold value estimated, it is considered as background and if a gaussian is less than the threshold estimated, it is considered as foreground.

$$B = argmin_b(\sum_{i=1}^{b} \omega_{i,t} > T). \tag{2}$$

Whenever a picture element matches one of the K gaussians, the values of ω, μ and σ are modified as shown below:

$$\omega_{i,t+1} = (1 - \alpha)\omega_{i,t} + \alpha \tag{3}$$

where,

$$\rho = \alpha \times \eta\left(X_t, \mu_i, \sum_i\right) \tag{4}$$

In cases where the pixels do not match with any of the K gaussians, ω is modified as.

$$\omega_{j,t+1} = (1 - \alpha)\omega_{j,t} \tag{5}$$

#### 3.1.2 Denoising

In this work, traditional morphological opening and closing are applied, to remove the noise in foreground.

**Morphological opening.** The opening operation includes performing erosion followed by dilation, with the same structuring element. By performing this morphological operation, noise can be removed from the frame. For the problem at hand, morphological opening removes the noise in the final step.

**Morphological closing.** The closing operation includes performing dilation followed by erosion, with the same structuring element. By performing this morphological operation, holes in object can be filled, to avoid misclassification.

#### 3.1.3 Blob Analysis

Connected components of the foreground pixels are treated as moving objects in the video. On the obtained denoised foreground, blob analysis is applied, to find the connected components, and compute their properties like area, centroid and bounding box. Based on these properties the objects are identified as blobs. These blobs are used in the sub-sequent step of kalman filtering to track the moving objects.

#### 3.1.4 Kalman Filter

Kalman filtering is a popular optimal state estimation algorithm. This work uses Kalman filter, for object tracking, namely the fishes. It mainly focusses on three factors namely, predicting the future positions of object, decreasing mis-detections due to noise, providing multiple object tracks.

The Kalman filter generates mean and covariance of the object state with respect to time. Initially apriori estimate $\hat{x}_{k+1|k}$, for the state is built. When the state of the object changes, posteriori estimate $\hat{x}_{k|k}$ is calculated using the filter. In discrete Kalman filter, this step performed for every specific time period. To obtain the general Kalman filter, a discrete linear system is considered as shown below:

$$w_k = N(0, Q_k) \tag{6}$$

$$v_k = N(0, R_k) \tag{7}$$

$$E\begin{bmatrix} w_k w_k^T & w_k v_k^T \\ v_k w_k^T & v_k v_k^T \end{bmatrix} = \begin{bmatrix} Q_k & 0 \\ 0 & R_k \end{bmatrix} \tag{8}$$

The initial values of state and covariance error are given by:

$$x_0 = E[x_0] \tag{9}$$

$$P_0 = E[(x_0 - \hat{x}_0)(x_0 - \hat{x}_0)] \tag{10}$$

where, $P_0$ is the initial value of the estimation error covariance $P_k$. The apriori estimate can be stated as:

$$\hat{x}_{k+1|k} = A_k \hat{x}_{k|k} + B_k u_k \tag{11}$$

$$P_{k+1|k} = A_k P_{k|k} A_k^T + K_k Q_k K_k^T \tag{12}$$

when the new measurement is available the estimate is filtered:

$$K_F = P_{k|k-1} C_k^T \left(C_k P_{k|k-1} C_k^T + C_k\right)^{-1} \tag{13}$$

$$\hat{x}_{k|k} = \hat{x}_{k|k-1} + K_F\left(y_k - C_k \hat{x}_{k|k-1}\right) \tag{14}$$

$$P_{k|k} = [I - K_F C_k] P_{k|k-1} \tag{15}$$

The Kalman filter calculates the state $x_k$ accurately given $y_k$, the outcome. Since this filter can be used to estimate approximate sampling interval without affecting accuracy in calculating the state, it can be used to calculate the state accurately even if the sample has missing or incorrect data.

## 3.2 Fish Classification System

### 3.2.1 Bag of Features Approach

Bag of features approach is inspired from bag of words concept in natural language processing. As there is no concept of words in computer vision the above-mentioned bag of features are generated based on surf key points, which are extracted using surf detector. These surf feature descriptors are input to the k-means clustering algorithm, which generates the visual vocabulary. This visual vocabulary is used to train the Multi Class Support Vector Machine (MSVM) classifier, to classify the objects.

## 4 Results

The Video dataset used in this work was collected from fish clef data centre [18], which consists of 20 underwater videos with various backgrounds and climatic conditions. The image dataset contains 21346 images of 11 different fish species.

### 4.1 Fish Counting

The proposed approach of counting of fishes is applied on the dataset. In every video, the first 150 frames are used for training the gaussain mixture model and 150 random frames are chosen from the testing video and the fishes are detected in those frames using trained GMM foreground detector.

Foreground detection is achieved through background subtraction, which results in binary frames with the foreground isolated as shown in Fig. 3 (ii). The underwater videos are generally prone to several artifacts, which appear in the form of noise in the background subtracted videos, which lead to the error in the detection of foreground and background pixels. This noise should be minimized to prevent its effect on the subsequent phase of fish detection, in object frames, in order to achieve accurate results. Noise removal is achieved by applying morphological opening and closing with struture elements $3 \times 3$, $5 \times 5$ rectangles respectively. The results of noise removal are shown in Fig. 3(iii).

Connected component analysis is performed on the denoised frames, in order to detect the objects as shown in Fig. 3(iii). Following this, blob analysis is performed to draw the detected fish as shown in Fig. 3(iv). Subsequently Kalman filtering is applied to track and predict the position of the fishes. Figure 3(v) shows the counting of fishes using using Kalman filter.

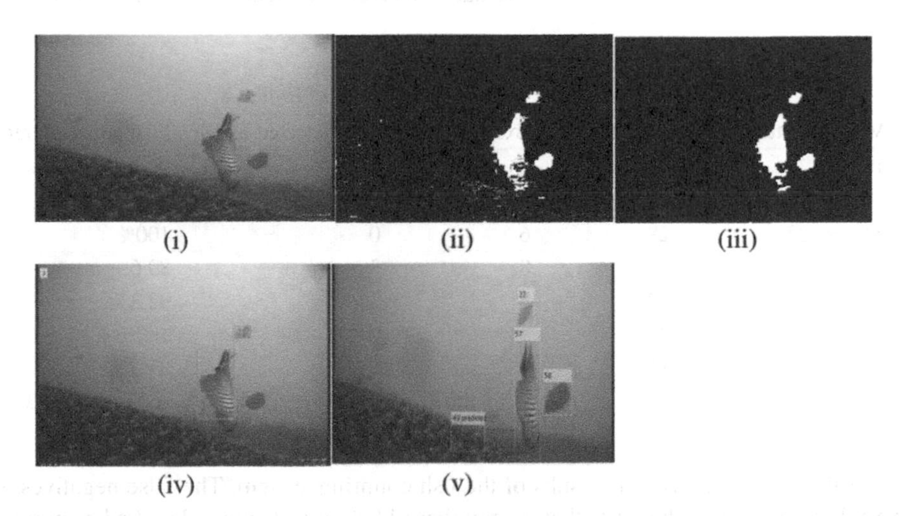


**Fig. 3.** (i) Video Frame, (ii) Foreground, (iii) Foreground after post-processing, (iv) Object Detection, (v) Object Tracking

$$\text{Fish Detection Rate} = \frac{\text{Correctly Detected Fish Count}}{\text{Actual Fish Count}} = \frac{TP}{TP + FN}$$

Where,

True Positive(TP): Number of fishes are accurately classified.
False Positive(FP): Number of non-fish objects missclassified as fish object.
True Negative(TN): Number of non-fish objects accurately classified.
False Negative(FN): Number of fish objects missclassified as non-fish object

**Table 1.** Fish detection results

| Video | Actual count | Detected fishes | FN | FP | Fish detection rate |
|---|---|---|---|---|---|
| 1 | 334 | 348 | 20 | 34 | 94% |
| 2 | 329 | 381 | 26 | 78 | 92% |
| 3 | 190 | 308 | 6 | 124 | 96.8% |
| 4 | 154 | 180 | 9 | 35 | 94% |
| 5 | 220 | 250 | 12 | 52 | 90% |
| | | | | | Average = 93.36% |

Table 1 summarizes the results of the fish detection system. There are false positives in that table as the quality of videos is low and some of them have moving background. Average fish detection rate is 93%.

$$\text{Fish Counting Rate} = \frac{\text{Correctly Counted Fish Count in Video}}{\text{Actual Fish Count in ideo}} = \frac{TP}{TP + FN}$$

**Table 2.** Fish counting system results

| Video | Actual count | Detected fish | Extra counted | Not detected as fish | Fish counting rate |
|---|---|---|---|---|---|
| 1 | 28 | 33 | 6 | 1 | 96.4% |
| 2 | 14 | 10 | 1 | 5 | 64.2% |
| 3 | 22 | 28 | 6 | 0 | 100% |
| 4 | 29 | 34 | 8 | 3 | 89.6% |
| 5 | 26 | 32 | 8 | 2 | 92.3% |
| 6 | 6 | 7 | 1 | 0 | 100% |
| 7 | 8 | 11 | 4 | 1 | 87.5% |
| | | | | | Average = 90% |

Table 2 summarizes the results of the fish counting system. The false negatives are very low in that which means that the number of fish objects miss classified as non-fish objects is low. The Average fish counting rate is 90%.

### 4.2 Fish Classification

$$\text{Fish Classification Rate} = \frac{\text{Correctly Classified Fish species objects}}{\text{Actual Fish species Count}} = \frac{TP}{TP + FN}$$

Here TP, FN values are calculated from confusion matrix, which is the output of the multi-class support vector machine classifier.

**Table 3.** Fish classification system results

| No | Species name | Total count | Correctly classified | Incorrectly classified | Fish classification rate |
|---|---|---|---|---|---|
| 1 | Abudefduf vaigiensis | 306 | 276 | 30 | 90% |
| 2 | Acanthurus nigrofuscus | 2511 | 2310 | 201 | 92% |
| 3 | Amphiprion clarkii | 2985 | 2687 | 298 | 90% |
| 4 | Chaetodon lunulatus | 2494 | 2394 | 100 | 96% |
| 5 | Chaetodon trifascialis | 24 | 22 | 2 | 92% |
| 6 | Chromis chrysura | 3196 | 2940 | 256 | 92% |
| 7 | Dascyllus aruanus | 904 | 877 | 27 | 97% |
| 8 | Dascyllus reticulatus | 3196 | 2876 | 320 | 90% |
| 9 | Myripristis kuntee | 3004 | 2673 | 331 | 89% |
| 10 | Plectrogly-Phidodon dickii | 2456 | 2013 | 443 | 82% |
| 11 | Zebrasoma scopas | 271 | 184 | 87 | 68% |
|  | Average | 21,346 | 19,252 | 2094 | 88.9% |

Table 3 summarizes the results of the fish classification system. The classification rate of Zebrasoma scopas fish species is less as the texture of that species is similar to texture. The average fish classification rate is 88.9%.

## 5 Conclusion and Future Work

This paper proposes an approach that gives an accuracy of 93% for detection and 90% for classification. The non-linear motion of the fishes coupled with a complex background results in high false positives. The future work could focus on the reduction of false positives through video enhancement techniques and texture features. Most recent

AI methods, for example, deep learning could be used to increase the efficiency of the fish classification system.

**Compliance with Ethical Standards.**

✓ All authors declare that there is no conflict of interest

✓ No humans/animals involved in this research work.

✓ We have used our own data.

## References

1. Boudhane, M., Nsiri, B.: Underwater image processing method for fish localization and detection in submarine environment. J. Vis. Commun. Image Represent. **39**, 226–238 (2016)
2. Spampinato, C., et al.: Detecting, tracking and counting fish in low quality unconstrained underwater videos. VISAPP **2**, 514–519 (2008)
3. Prabowo, M.R., Hudayani, N., et al.: A moving objects detection in underwater video using subtraction of the background model. In: 2017 4th International Conference on Electrical Engineering, Computer Science and Informatics (EECSI), Yogyakarta, pp. 1–4 (2017). https://doi.org/10.1109/eecsi.2017.8239148
4. Palazzo, S., Kavasidis, I., Spampinato, C.: Covariance based modeling of underwater scenes for fish detection. In: 2013 IEEE International Conference on Image Processing. IEEE (2013)
5. Rathi, D., Jain, S., Indu, S.: Underwater fish species classification using convolutional neural network and deep learning. In: 2017 Ninth International Conference on Advances in Pattern Recognition (ICAPR). IEEE (2017)
6. Shortis, M., Ravanbakhsh, M., et al.: An application of shape-based level sets to fish detection in underwater ımages. In: GSR (2014)
7. Fabic, J.N., et al.: Fish population estimation and species classification from underwater video sequences using blob counting and shape analysis. In: 2013 IEEE International on Underwater Technology Symposium (UT). IEEE (2013). In Proceedings International Conference on Communication and Signal Processing, ICCSP 2017 (2017)
8. Ghani, A.S.A., Isa, N.A.M.: Enhancement of low quality underwater image through integrated global and local contrast correction. Appl. Soft Comput. **37**, 332–344 (2015)
9. Benson, B., et al.: Field programmable gate array (FPGA) based fish detection using Haar classifiers. American Academy of Underwater Sciences (2009)
10. Prabhakar, P., Kumar, P.: Underwater image denoising using adaptive wavelet subband thresholding. In: Proceedings IEEE ICSIP 2010, Chennai, India, pp. 322–327 (2010)
11. Feifei, S., Xuemeng, Z., Guoyu, W.: An approach for underwater image denoising svia wavelet decomposition and high-pass filter. In: Proceedings IEEE ICICTA 2011, Shenzhen, China, pp. 417–420 (2011)
12. Wren, C.R., Azarbayejani, A., Darrell, T., Pentland, A.P.: Pfinder: real-time tracking of the human body. IEEE Trans. Pattern Anal. Mach. Intell. **19**(7), 780–785 (1997)
13. Evans, F.H.: Detecting fish in underwater video using the EM algorithm. In: Proceedings of the 2003 International Conference on Image Processing (ICIP), vol. 3, p. III–1029. IEEE (2003)
14. Khanfar, H., et al.: Automatic Fish Counting in Underwater Video Cuenta Automática de Peces en Video Submarino Comptage Automatique de Poisson dans la Vidéo Sous-Marine. Preprocessing. Indian Journal of Science and Technology, pp. 1170–1175 (2014)

15. Singhal, N., Singhal, N., Kalaichelvi, V.: Image classification using bag of visual words model with FAST and FREAK. In: 2017 Second International Conference on Electrical, Computer and Communication Technologies (ICECCT), Coimbatore, pp. 1–5 (2017)
16. Aarthi, R., Arunkumar, C., RagheshKrishnan, K.: Automatic isolation and classification of vehicles in a traffic video. In: 2011 World Congress on Information and Communication Technologies, Mumbai, pp. 357–361 (2011)
17. Venkataraman, D., Mangayarkarasi, N.: Computer vision based feature extraction of leaves for identification of medicinal values of plants. In: 2016 IEEE International Conference on Computational Intelligence and Computing Research (ICCIC). IEEE (2016)
18. https://www.imageclef.org/lifeclef/2015/fish

# Plant Leaf Disease Detection Using Modified Segmentation Process and Classification

K. Nirmalakumari[1(✉)], Harikumar Rajaguru[1], and P. Rajkumar[2]

[1] Department of ECE, Bannari Amman Institute of Technology, Sathyamangalam, Tamilnadu, India
{nirmalakumarik, harikumarr}@bitsathy.ac.in

[2] Robert Bosch Engineering and Business Solutions Pvt. Ltd, Coimbatore, Tamilnadu, India
rajkumar.ppt@gmail.com

**Abstract.** Agriculture is essential in our Indian economy and the plant infections greatly affect the production rate in terms of quality and quantity. Humans get exhausted of monitoring the diseased plants and hence there is a necessity to monitor the plant growth everyday by using some software techniques. The proposed method will routinely detect and classify the plant diseases in exact way. The core process involved in this method comprises of image capturing, preprocessing, segmentation, extraction of informative features and finally classification part to conclude healthy or diseased. The proposed method utilizes a modified segmentation algorithm that improves the segmentation results, thus in turn improves the classification accuracy of the Artificial Neural Network classifier.

**Keywords:** Image capturing · Preprocessing · Segmentation · K-means clustering · Feature extraction · Artificial neural network

## 1 Introduction

Agriculture is very essential in India and the farmers have a decision to choose the required crops. One of the important issues in agriculture is plant disease which affects the amount, quality and development of plant growth. Position and categorization of plant infections are simple procedure to improve plant yield and financial improvement. The suitable pesticides for the plant have to be identified to suppress the disease and increase the production. In-order to raise the production with good quality the plant needs to be monitored regularly since the plant disease leads to decrease of the product. For effective cultivation, one should observe the health as well as the infection of the plant. Diseases in plant form the basis for substantial loss of the product. India is the world's highest producer of pomegranate. India exports 54000 tons of Pomegranates which makes 1.55% of total export in the whole world. Pomegranate is a vital fruit crop, because of its health benefits. 10–31% of production reduces because of the disease found on the leaf of the pomegranate plant. Henceforth the initial stage identification of the disease is essential, endorsing farmers to prevent the damage in crop production for yield growth. Plants suffer from leaf diseases due to bacteria, fungal and viral infections. The various spots and patterns on the different parts like branch, leaf and fruit are noted and that indications are considered for disease detection.


S. Smys et al. (Eds.): ICCVBIC 2019, AISC 1108, pp. 442–452, 2020.
https://doi.org/10.1007/978-3-030-37218-7_50

In ancient days, the expertise person would manually monitor and analyze the plant disease that involved huge work and also was time consuming. Image processing is one such procedure that is employed to detect the disease in plant. There are few countries where farmers are not having enough facilities to contact the expertise since it is time consuming and at the high cost. In such situation, image processing technique comes into picture and is beneficial in monitoring the field. The symptoms on the leaves will help in automatically detecting the disease, which is easy and also cheap. The technique involved in processing of images measures the affected area of infection, determines the texture, color and shape of the affected area, and also classifies the infection. The method proposed in this paper will solve the above issues by means of image processing techniques. The second section describes the literature works that had been carried out in this field and third section describes the proposed procedure with its block diagram. Section four discusses the results obtained by the proposed method and finally section five provides the conclusion of this research.

## 2 Literature Review

Ghaiwat et al. [1] developed a review on diverse techniques for classification that can be employed for disease sorting of plant leaves. For particular test example, k-nearest-neighbor method works good and it is simple one for class prediction. If non linearly separable data during training makes hard to define optimal parameters, which is one of the drawback in SVM. Authors in paper [2] define four main steps for detecting the leaf diseases. In first step, the input color image with three planes namely red, green and blue layers are converted to HSI, a color transformation structure and can be used as color descriptor. In step two, pixels containing green values are screened and eliminated by employing suitable threshold level. In step three, the useful segments from the above step are obtained using pre-computed threshold value and segmentation is performed in final step.

Mrunalini et al. [3] offers the system to categorize and recognize diverse infections that affect plants. The recognition can be performed with machine learning that reduce the currency, computation and time. The method involved for extraction of features is the color co-occurrence method. For programmed leave disease detection, neural networks are utilized. The method projected can considerably provide a provision for accurate leaf disease detection, and appears to be vital approach, in event of root and stem diseases, placing less effort in calculation. In paper [4] explained the identification step that includes four key steps that are described: Initially, in a color image, consider the color alteration arrangement and applyig a exact threshold, the green pixels are screened and detached, and again additionally tracked by segmentation task, and for receiving beneficial segments, the texture information are calculated. The final step involves the classifier to classify the extracted features for disease diagnosis. The strength of the proposed algorithm is shown by means of results for totally 500 leaves of different plants.

Kulkarni et al. developed a procedure for initial and correct plant disease detection with ANN (Artificial Neural Network) and different image processing methods. As the projected technique uses the ANN classifier and Gabor filter for categorization and extracting features respectively and it produces up to 91% recognition rate. Thus the

ANN utilizes the combination of features to detect the disease [5]. Authors discussed disease recognition in Malus domestica over an active procedure by means of texture, clustering with k-means and analysis of colors [6]. For classifying and distinguishing various agricultural plants, it customs the features containing texture and colors that vary in normal and abnormal plant leaves. In future, the task of classification can be achieved by employing clustering, Bayes and principal component classifiers.

Rastogi [7], developed an approach for rating the infections in plant leaves in an automatic way. It employs Machine Vision Techniques and Artificial Neural Network (ANN) for automatic detection of leaf disease with grading. This system are very supportive as it is effective than normal procedure. The measure of eulidean and clustering are used for segmentation of leaf image for identifying foreground and background areas. Finally the plant disease affected area is estimated and the leaves are graded into different classes. Islam [8], presented the leaf disease detection using two classifiers namely GA and PNN which provides accurate and less computational work. The experiment results indicate that the accuracy of classifier with GA performs good than the PNN.

Tripathi [9], proposed a system for plant leaf disease detection by processing images and the application of NN techniques. The color and texture features are extracted and they are provided as inputs to train the NN classifier for successive detection of leaf diseases. Muthukannan [10], employed three classifiers namely Feed Forward Neural Network (FFNN), Learning Vector Quantization (LVQ), and Radial Basis Function (RBF) for classification of leaf disease of two types namely bitter gourd and bean based on their evaluated performance measures. Papers [11] describes the different review that has been carried out for feature extraction of plant leaf diseases and paper [12] explains the crop disease identification using image segmentation technique.

**Shortcoming of present works:**

- The application further require improvement in accuracy for certain cases. Additional effort to optimize the solution is required.
- Data should be known in advance for necessary segmentation.
- Database addition is desired for achieving higher accuracy.
- Only rare diseases are being addressed. Many more diseases has to be found.
- The imaginable causes that promote misclassifications includes various infections in plants, optimization is required and extra training trials are desirable for further cases and to envisage the disease in an accurate way.

## 3 Proposed Methodology

Leaves and its parts are acquired using high quality digital cameras and those images are processed in an efficient way to obtain essential features that can be used in the analysis.

**Algorithm:**

- The foremost process is to acquire the image of a plant with digital camera.
- Input image requires preprocessing for quality improvement and noises present in images. Image of particular part of plant leaf is cropped to obtain the attentive image area and then smoothening is finished by means of the smoothing filter. Enhancement of image is done for improving the contrast. The input RGB image is converted to HSV format.
- The green colored pixels are screened in the third step.
- A threshold value is initialized and if the green component intensity value is lower than initial threshold value, then those pixels are removed.
- Acquire the worthy segments using morphological and K-means clustering opeartions.
- Estimating the features using GLCM.
- Classify the leaf disease and its affected area computation using neural network.

The steps involved for plant leaf disease detection is shown in Fig. 1. The image of various types of leaves are acquired using digital camera and preprocessing is performed to remove noise. The preprocessing process is done for obtaining the valuable features. Finally a classifier is employed to identify and categorize the type of disease.

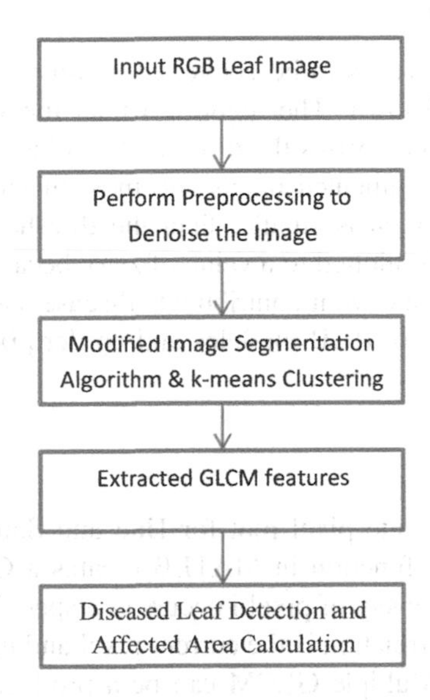


**Fig. 1.** Block diagram for disease identification in plants

### 3.1 Image Acquisition

The blocks involved for detecting the infections in plants are shown in Fig. 1. The Leaf images are acquired using webcam camera with maximum image resolution of 5500 pixels wide and 3640 pixels height. The images are preprocessed and it is resized to 640 * 480 pixels. The input image is a color and is changed to gray image for further process.

### 3.2 Image Pre-processing

This step helps in the elimination of lower frequency levels related noise. The disturbances and noise levels prevailing in the image are removed by means of image clipping, smoothing and enhancement steps. A color space transformation structure is formed that explains the RGB to Hue Saturation Value (HSV) color space conversion. Hue defines the perceived pure color by a viewer. Saturation represents the quantity of white light supplemented to Hue and light's amplitude is described by Value. The Hue component is considered for further process as it provides more valuable information than the Saturation and Value components.

### 3.3 Modified Segmentation

The infected area in a leaf is obtained with segmentation. Permitting the ROI, the image is divided into some significant region. The proposed method includes morphological operations prior to k-means clustering which will enhance the result of segmentation. The main morphological operations include Erosion, Dilation, Opening and Closing. The morphological processes such as opening is performed to fill the holes and closing is done to remove the unwanted areas.

### 3.4 K-Means Clustering

The separated parts of leaves is to be converted as four dissimilar clusters. It is beneficial if the k value is known. The algorithm puts the information of pixel in the clusters. One among four comprises the diseased leaf cluster. Then a definite threshold rate is calculated for the segmented pixels and most of the green pixels are screened when pixel intensity of green is smaller than the threshold value. Further the RGB modules of the this pixel is allotted to a value of zero, because it would characterize the healthiest parts of the leaf and wont contribute to disease that leads to computation time reduction. The zero values of RGB modules and border pixels of diseased cluster gets totally eliminated.

### 3.5 GLCM Features

GLCM is formed for separate pixel plot for Hue and Saturation images of diseased cluster. The graycomatrix function in MATLB creates a GLCM matrix by operating how frequently a exact intensity of pixel i occurs in a specified spatial association to a j pixel value. The spatial correlation is the current pixel and its straight right pixel, which is default in MATLAB. Multiple GLCM can be a produced using offset that rapidly shows the pixel associations of varying distance and direction. The three directions in horizontal, vertical and diagonal directions can be taken. The following GLCM features are considered.

Contrast: Produces the contrast intensity amount between a pixel and its end-to-end pixels in entire image.

Range = [0 (size (GLCM, 1)-1) ^2]. The contrast is computed by the formula given below:

$$\text{Contrast} = \sum_{i,j=0}^{N-1} (i,j)^2 C(i,j) \tag{1}$$

Where, the pixel information at location (i,j) is denoted by C(i,j).

Energy: Produces the sum of squared pixel values in the GLCM matrix. The Range is within 0 and 1 and the value is equivalent to one for a persistent image. The energy can be computed using the below formula:

$$\text{Energy} = \sum_{i,j=0}^{N-1} C(i,j)^2 \tag{2}$$

Homogeneity: Computes the element distribution closeness value to the GLCM diagonal. Range = [0 1]. The value is one for a diagonal GLCM matrix and its formula is given below:

$$\text{Homogeneity} = \sum_{i,j=0}^{N-1} C(i,j)/(1 + \left(i - j)^2\right) \tag{3}$$

Correlation = Amount of correlated pixel to its neighboring pixel over the complete image. Range = [− 1 1].

$$\text{Correlation} = \sum_{i=0}^{G-1} \sum_{j=0}^{G-1} \frac{(i\ X\ j) X\ P(i,j) - \left(\mu_x - \mu_y\right)}{\sigma_x\ X\ \sigma_y} \tag{4}$$

The four mentioned statistics brings data about image textures.

### 3.6 Neural Network Based Classification

The features extracted are provides data to pretrained NN for automatic classi disease categorization. Artificial Neural network (ANN) is an eminent classification tool and also active classifier for numerous real applications. The process of training and validation are the vital steps in evolving an exact process model by means of NNs. The dataset is partitioned into training and validation sets with their features to demonstrate the proficient ANN model. The type of training is definite for getting perfect classification of plant infections.

## 4 Experimental Results and Observations

The four different banana leaves are considered to show the results of our proposed method. Out of four, three are diseased leaves namely Banana Anthracnose, Banana Leaf streak, Banana Spot leaf and a Normal banana leaf. The input RGB image to gray image is done and then denoising is performed to remove the noise followed by segmentation using morphological operations and k-means clustering. Finally the type of disease detection and its affected area are computed using the ANN Classifier. The Fig. 2, the banana anthracnose input color image and it is converted to grayscale. Figure 3 shows the modified segmentation algorithm results and its corresponding output on the color image.

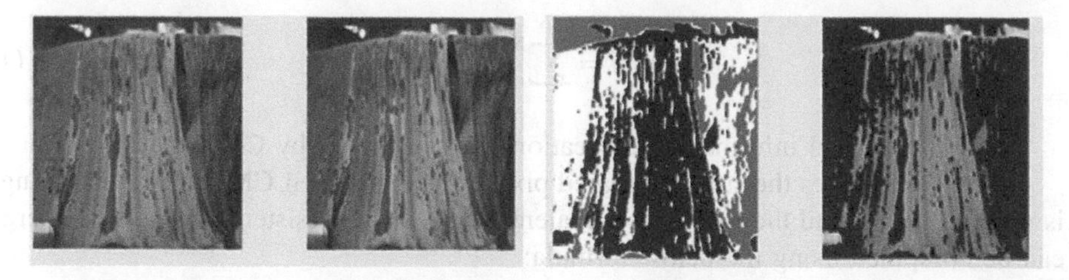

**Fig. 2.** Banana Anthracnose & its grayscale [3] **Fig. 3.** Modified segmentation & its color

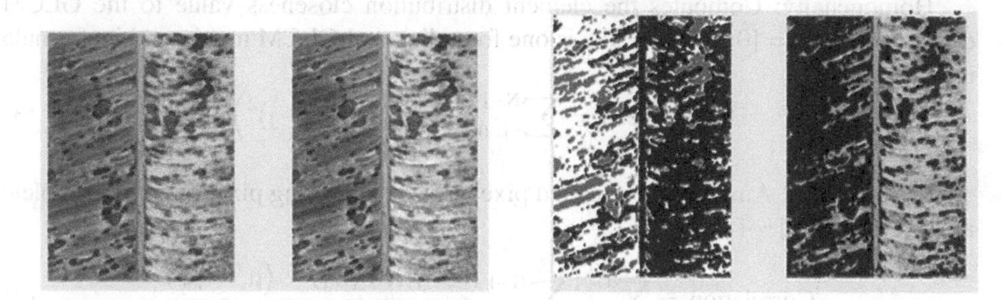

**Fig. 4.** Banana leaf streak and its grayscale [7] **Fig. 5.** Modified segmentation & its color

The image of banana leaf streak and its corresponding grayscale image are depicted in Fig. 4. The Fig. 5 shows the segmented output and its results on the color image. Another infected banana leaf with spots is shown in Fig. 6 and its corresponding segmentation output results for gray and color images are shown in Fig. 7.

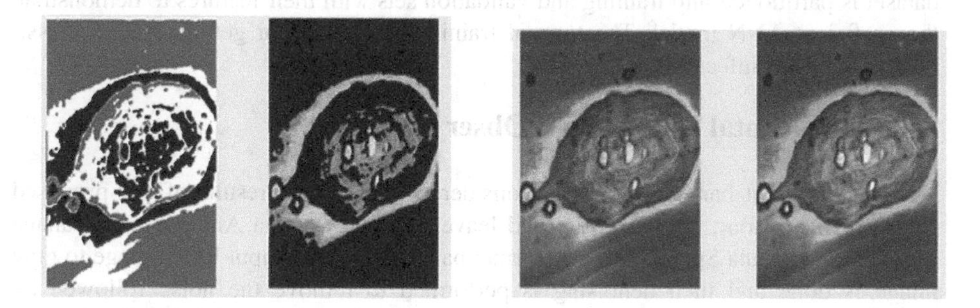

**Fig. 6.** Banana spot leaf and its grayscale [9] **Fig. 7.** Modified segmentation & its color

The Fig. 8 shows the normal banana leaf without infections and the segmentation results are shown in Fig. 9. From the resultant images, it is clear that the modified approach of segmentation aids in finding the infected area in an exact way, which will improve the detection accuracy of the ANN classifier.

The Table 1 shows the computation of affected areas of banana leaves with four GLCM features and its percentage calculation shows the spread of infections in the plant leaf. The normal banana leaf is classified with zero affected area. Hence this proposed algorithm is very robust in the detection of plant leaf diseases.

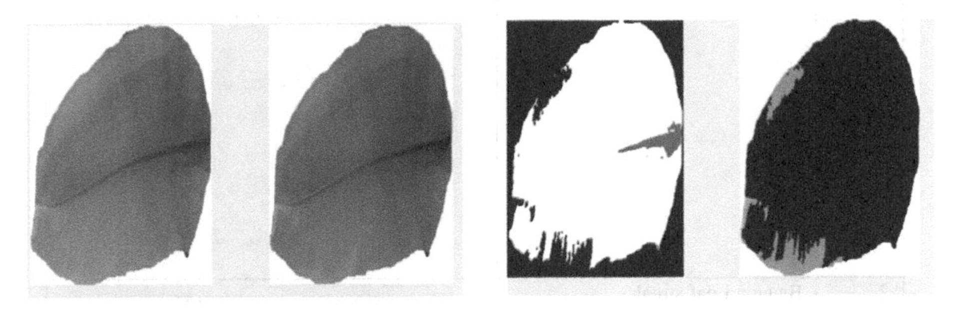

**Fig. 8.** Normal banana leaf [11]

**Fig. 9.** Modified segmentation & its color

**Table 1.** Percentage of affected area calculation for four types of banana leaves

| S.No. | Type of leaf | GLCM features | Affected Area in sq mm | Total area in sq mm | Percentage of affected Area |
|---|---|---|---|---|---|
| 1. | Banana Anthracnose | Contrast = 3.5842<br>Correlation = 0.8504<br>Energy = 0.4433<br>Homogeneity = 0.9360 | 74529 | 196282 | 37.9703 |
| 2. | Banana Leaf streak | Contrast = 2.8351<br>Correlation = 0.8842<br>Energy = 0.4458<br>Homogeneity = 0.9494 | 83120 | 191851 | 43.3251 |
| 3. | Banana Spot leaf | Contrast = 1.3098<br>Correlation = 0.9257<br>Energy = 0.6142<br>Homogeneity = 0.9766 | 16608 | 183560 | 9.04772 |
| 4. | Normal banana leaf | Contrast = 0.7446<br>Correlation = 0.9645<br>Energy = 0.5573<br>Homogeneity = 0.9867 | NIL | 196585 | 0 |

From Table 2, it is identified that the correlation feature with respect to horizontal offset produces a non linear curve for the infected leaves and for the normal banana leaf the graph approximately shows a linear line. Thus using GLCM features and ANN, the plant leaf disease can be detected in an accurate way.

**Table 2.** Texture correlation graph vs offset for four banana leaves

| Sl.No. | Leaf Type | Texture Correlation Graph |
|---|---|---|
| **1.** | Banana Anthracnose | Figure 4<br>File Edit View Insert Tools Desktop Window Help<br>Texture Correlation as a function of offset<br>Result<br>TYPE :Banana Anthracnose<br>AFFECTED AREA = 37.970369 in Sq mm<br>OK<br>Correlation: 0.1 0.2 0.3 0.4 0.5 0.6 0.7 0.8 0.9<br>0 5 10 15 20 25 30 35 40<br>Horizontal Offset |
| **2.** | Banana Leaf streak | File Edit View Insert Tools Desktop Window Help<br>Texture Correlation as a function of offset<br>Result<br>TYPE :Banana Leaf Streak<br>AFFECTED AREA = 43.325269 in Sq mm<br>OK<br>Correlation: 0.2 0.3 0.4 0.5 0.6 0.7 0.8 0.9<br>0 5 10 15 20 25 30 35 40<br>Horizontal Offset |
| **3.** | Banana Spot leaf | Texture Correlation as a function of offset<br>Result<br>TYPE :Banana Spot Leaf<br>AFFECTED AREA = 9.047723 in Sq mm<br>OK<br>Correlation: 0.1 0.2 0.3 0.4 0.5 0.6 0.7 0.8 0.9 1<br>0 5 10 15 20 25 30 35 40<br>Horizontal Offset |
| **4.** | Normal banana leaf | File Edit View Insert Tools Desktop Window Help<br>Texture Correlation as a function of offset<br>Result<br>TYPE :Banana Normal Leaf<br>AFFECTED AREA : NILL<br>OK<br>Correlation: 0.5 0.55 0.6 0.65 0.7 0.75 0.8 0.85 0.9 0.95 1<br>0 5 10 15 20 25 30 35 40<br>Horizontal Offset |

## 5 Conclusion

The proposed method employs the usage of image processing that performs preprocessing, segmentation and extraction of useful features. The morphological operators are included in the segmentation part that improves the detection of affected area to a greater level and the classification of different plant diseases is done using ANN classifier. This proposed method is very significant in detection of leaf disease in a precise and simple way. k-means clustering along with morphological operators namely opening and closing aids in segmentation. The GLCM features extracted after this step are the important features and they are applied to neural network classifier to identify the affected area and the type of plant disease. The proposed method can be utilized for all kinds of diseased plant and also the percentage of affected area of the disease is computed, hence the user can select the appropriate pesticides to control and prevent the disease infection in plants. The future scope of this project will be the application of different algorithms in segmentation, feature extraction and classification parts.

**Compliance with ethical standards**

✓ All authors declare that there is no conflict of interest

✓ No humans/animals involved in this research work.

✓ We have used our own data.

## References

1. Ghaiwat, S.N., Arora, P.: Detection and classification of plant leaf diseases using image processing techniques: a review. Int. J. Recent. Adv. Eng. Technol. **2**(3), 2347–2812 (2014). ISSN
2. Dhaygude, S.B., Kumbhar, N.P.: Agricultural plant leaf disease detection using image processing. Int. J. Adv. Res. Electr. Electron. Instrum. Eng. **2**(1), 599–602 (2013)
3. Badnakhe, M.R., Deshmukh P.R.: An application of K-means clustering and artificial intelligence in pattern recognition for crop diseases. In International Conference on Advancements in Information Technology, vol. 20. IPCSIT (2011)
4. Arivazhagan, S., Newlin Shebiah, R., Ananthi, S., Vishnu, V.S.: Detection of unhealthy region of plant leaves and classification of plant leaf diseases using texture features. Agric. Eng. Int. CIGR. **15**(1), 211–217 (2013)
5. Kulkarni Anand, H., Ashwin Patil, R.K.: Applying image processing technique to detect plant diseases. Int. J. Mod. Eng. Res. **2**(5), 3661–3664 (2012)
6. Bashir, S., Sharma, N.: Remote area plant disease detection using image processing. IOSR J. Electron. Commun. Eng. **2**(6), 31–34 (2012). ISSN: 2278-2834
7. Rastogi, A., Arora, R., Sharma, S.: Leaf disease detection and grading using computer vision technology & fuzzy logic. In: 2nd International Conference on Signal Processing and Integrated Networks (SPIN), IEEE, pp. 500–505 (2015)
8. Islam, M.N., Kashem, M.A., Akter, M., Rahman, M.J.: An approach to evaluate classifiers for automatic disease detection and classification of plant leaf. In: International Conference on Electrical, Computer and Telecommunication Engineering, RUET, pp. 626–629 (2012)

9. Tripathi, G., Save, J.: An ımage processıng and neural network based approach for detection and classification of plant leaf diseases. Int. J. Comput. Eng. Technol. IJCET **6**(4), 14–20 (2015)
10. Muthukannan, K., Latha, P., Pon Selvi, R., Nisha, P.: Classıficatıon of diseased plant leaves using neural network algorithms. ARPN J. Eng. Appl. Sci. **10**(4), 1913–1918 (2015)
11. Pandey, S., Singh, S.K.: A review on leaf disease detection using feature extraction. Int. Res. J. Eng. Technol. (IRJET),06(01) (2019)
12. Kumar, K.V., Jayasankar, T.: An identification of crop disease using image segmentation. Int. J. Pharm. Sci. Res. **10**(3), 1054–1064 (2019)

# Neuro-Fuzzy Inference Approach for Detecting Stages of Lung Cancer

C. Rangaswamy[1,2](✉), G. T. Raju[3], and G. Seshikala[4]

[1] REVA University, Bangalore 560064, India
crsecesait@gmail.com
[2] ECE Department Sambrama Institute of Technology, Bangalore 560097, India
[3] CSE Department RNSIT, Bangalore 560098, India
gtraju1990@yahoo.com
[4] ECE School, Reva University, Bangalore 560064, India
seshikala.g@reva.edu.in

**Abstract.** A hybridized diagnosis system called Neuro-Fuzzy Inference System (NFIS) uses neural network for the classification of lung nodule into benign or malignant and then a fuzzy logic for detecting various stages of lung cancer is proposed in this paper. Using fuzzy logic based algorithms such as Enhanced Fuzzy C-Means (EFCM) and Enhanced Fuzzy Possibilistic C-Means (EFPCM), the required features from the lung CT scan image are segmented and extracted using GLCM and GLDM matrix. Then the features are selected using DRASS algorithm. These features are fed as input for Radial Basis Function Neural Network (RBFNN) classifier with k-means learning algorithm for detecting the lung cancer. Once the lung cancerous nodules are detected, the result of RBFNN is combined with fuzzy inference system that determines appropriate stage of the lung cancer. Experiments have been conducted on ILD lung image datasets with 104 cases. Results reveal that our proposed NFIS effectively classify lung nodule into benign or malignant along with the appropriate stage with considerable improvement in respect of Recall/Sensitivity 94.44%, Specificity 92%, Precision 96.22%, Accuracy 93.67%, and False Positive Rate 0.08.

**Keywords:** Fuzzification · Inference · Defuzzification · K-means · Neuro-Fuzzy

## 1 Introduction

Lung cancer caused more deaths than any other disease in men and women [2]. The identification of early lung cancer is essential for effective action and can rise the chance of survival of the patient [1, 3]. A few logical intercessions have utilized to enhance early detection. Lately, frameworks fueled by machine learning apparatuses have been conveyed to help medicinal work force in breaking down CT image and distinguishing the event of lung disease with significant achievement. Artificial Neural Networks(ANNs) have been used by researchers for classification of lung nodule into benign or malignant. However, ANNs would not state the appropriate stage at which the cells are pretentious [11]. Or, Fuzzy logic is capable of determining the amount of the illness, though it is not adequately practiced in erudition and generalization [4].


S. Smys et al. (Eds.): ICCVBIC 2019, AISC 1108, pp. 453–462, 2020.
https://doi.org/10.1007/978-3-030-37218-7_51

In this paper, we propose a hybridized NFIS to achieve better performance in the classification and determination of the stage of cancer to avoid the limits of these methods. It uses a gray level matrix for the extraction of features.

## 2 Related Work

ANN was used to classify according to the extracted characteristics. Fatma, et al. projected a technique for early detection of lung growth [1]. The Bayesian classification method for the pre-processing of image data was used using histogram analysis. The diagnosis was carried out using an ANN and a SVM to classify cells [6]. Ekta and others also suggested a pulmonary cancer detection and classification system using curve let transformation and NN [2]. Hamad proposed a pulmonary cancer diagnostic system using fuzzy logic and neural networks [3]. Al-Amin [12] presented work using GLCM and NN's to categorize cells as each malignant or benign, similar to the previous work presented by John [10]. Table 1 provides a summary of the contributions of various authors.

**Table 1.** Contributions Summary

| Authors/Reference | Classifier/ Methodology | Datasets | Features | Performance |
|---|---|---|---|---|
| Taher, et al. [1] | Bayesian with histogram analysis | NCI | geometric, and chromatic | accuracy-97% |
| Solanki [2] | Curvelet transform and neural networks | VIA/IELCAP | – | accuracy-90% |
| Hamad [3] | Fuzzy logic and neural networks | LIDC | GLCM | accuracy-89.6% |
| Al-Amin [12] | Neural networks | NCI and ACS | contrast, correlation, energy, homogeneity | accuracy-96.67% |

## 3 Neuro-Fuzzy Inference System

The Fuzzy framework is exceptional in light of the fact that it can deal with numerical information and semantic learning all the while, much the same as the procedure of human derivation [5]. It gives a surmising morphology that enables evaluated human thinking aptitudes to be connected to information based frameworks and furthermore offers a numerical capacity to comprehend the vulnerabilities related with human scholarly advancements, for example, considering and thinking [8]. The Fuzzy rationale worldview can basically be abridged as a lot of phonetic principles connected to the idea of Fuzzy inclusion and compositional tenets of induction. The point is to have a model fit for mapping semantic control rules from the contribution of master learning to a programmed control yield [9]. Fluffy rationale can be demonstrated to decipher the properties of neural systems and give increasingly precise portrayal of their execution [7].

Figure 1 shows the architecture of NFIS system. It consists of RBFNN and FIS. FIS is categorized by fuzzy sets and fluid *if-then* form rules: *"If x is A then y is B"* where A and B are *fuzzy sets and x and y are sets members*. The FIS contains 4 key portions: fozzifier, rules, defuzzifier and *inference engine*. FIS performs input fuszification and determines the extent to which the input is part of the fuszy set. Using the Fuzzy operator, the result of ***If part*** of the rule is combined to find the degree of support for the rule and applies to the output function. It uses the support degree for the whole rule to determine the fuse output set. The part of the fuzzy rule ***Then part*** assigns the output to a whole fuzzy set. Aggregation is used to combine the output of all the rules into a single fuzzy set and the appropriate decisions are made.

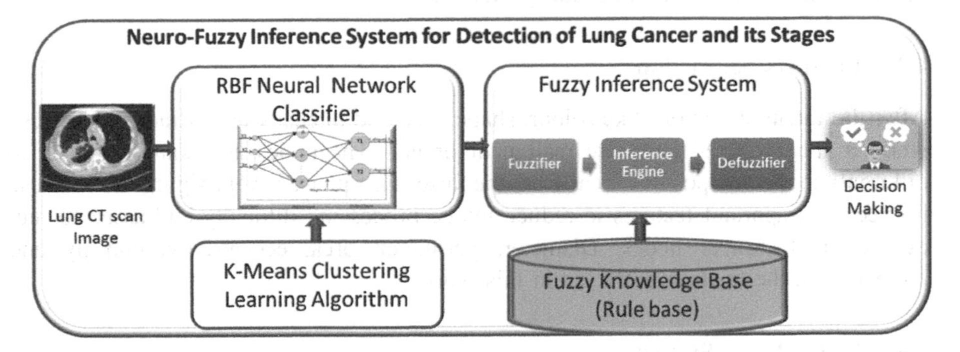


**Fig. 1.** Architecture of NFIS [4]

The FIS has four algorithmic steps:

***Step 1: Initialization***
**Initialize the linguistic variables and terms**
**Define the membership functions**
**Build the rule base (initialization)**

***Step 2: Fuzzification***
**Input data are converted to fuzzy values using the defined membership functions**

***Step 3: Inference***
**Rules are evaluated by constructing a decision structure involving the combination of membership functions with control rules resulting in fuzzy output**

***Step 4: Defuzzification***
**Resulting fuzzified output is converted into non-fuzzy values**

# 4 Methodology

We have modeled an smart diagnostic scheme by means of neural networks and fuzzy logic. Image data sets are handled and transmitted to the RBFNN to train and classify lung cells using the learning method k-means. The cancer cells were then transferred to a FIS in order to decide the stage of lung cancer.

## 4.1 Data Collection

ILD CT image data set with 104 cases, out of which 79 cases have been used for training and 25 cases used for testing.

### 4.2 Features of the System

A CT image from the data set undergoes several procedures in the training model, such as preprocessing, improvement, segmentation and extraction of features. The outputs of these procedures are transmitted to the RBFNN as input training data. In the diagnostic model, the extracted features of the training model were transferred to the neural network for classification. In order to determine the stage of cancer, the classified output is then transferred to the fuzzy system.

### 4.3 Features Extraction

After the intention regions like colour, shape, size and intensity are extracted, it checks whether the extracted areas are malignant or not. The nodes are examined to once differentiate the true positive areas from the false. This phase is very significant because it excerpts important features to reduce image processing difficulty and assist to recognize small cancer nodes. Diameter, perimeter, area, eccentricity, entropy and intensity are the features extracted in this work.

### 4.4 Fuzzy Logic System

The Fuzzy scheme consists of 3 theoretical mechanisms: the membership function used as a rule base in the Fuzzy rule, which includes the selection of the Fuzzy rule, the rule base and the cognitive device, which calculates the inference procedure on the rules in order to reach a reasonable conclusion. We have used the Fuzzy system in our work to understand the productivity of neural network and to describe values precisely.

The language term with the subsequent membership function is defined:

$$MF = \begin{cases} T1, & if\ 2.5\,cm < T \le 3\,cm \\ T2, & if\ \ 3\,cm < T \le 5\,cm \\ T3, & if\ \ 5\,cm < T \le 7\,cm \\ T4, & if\ T > 7\,cm \end{cases}$$

Organizing includes assessing the measure of a malignant growth and its invasion into neighboring tissues and the presence or non-presence of metastasis in lymph hubs or different organs. Stage One implies that the development of disease stays restricted to the lung. On the off chance that the malignant growth is as yet restricted to the lung yet is near the lymph, organize 2 is included. It's stage 3 if it's in the lung and the lymph hub. In the event that the malignancy has penetrated different parts of the body, we state it is as of now in stage 4.

### 4.5 Fuzzy Rules

The FIS uses the Fuzzy rule base to make decisions about the diagnosis of disease, and the system's efficiency depends on the Fuzzy rules. Systems based on rules are mainly used for medical diagnostics. The knowledge base stores all information on the characteristics and the disease in the form of rules. The rules are generated according to data collected by a domain expert.

$$Rule\,1 : if\ T1\ Then\ Stage\ I$$
$$Rule\,2 : if\ T2\ Then\ Stage\ II$$
$$Rule\,3 : if\ T3\ Then\ Stage\ III$$
$$Rule\,4 : if\ T4\ Then\ Stage\ IV$$

### 4.6 Decision Making in NFIS

Neural network's output was combined with the fuzzy system to make the final decision on the stage of a lung cancer. If the output value neural network does not exceed the threshold value, it displays 'no sign of lung cancer'. Otherwise it display 'lung cancer is predicted' and adjusts the fuzzy system to produce the crisp value based on the rule match. The NFIS makes final decision based on the crisp value as follows:

*If crisp value* $\leq 2.75$
{
*Stage-I is predicted*
}
*Else if crisp vale* $>2.75$ *&& crisp value* $\leq 5.0$
{
*Stage-II is predicted*
}
*Else if crisp vale* $>5.0$ *&& crisp value* $\leq 7.0$
{
*Stage-III is predicted*
}
*Else Stage IV is predicted*

## 5 Experimental Results

The outcomes of proposed NFIS are shown in Table 3.

**Table 2.** Extracted Features

| Image | Diameter | Perimeter | Entropy | Intensity | Eccentricity |
|---|---|---|---|---|---|
| Img-1 | 2.7 | 598 | 4.0427 | 250 | 0 |
| Img-2 | 3.1 | 84 | 4.7536 | 250 | 0 |
| Img-3 | 3.7 | 69 | 4.1936 | 250 | 0 |
| Img-4 | 5.6 | 64 | 4.7971 | 250 | 0 |
| Img-5 | 3.5 | 252 | 4.7755 | 244.22 | 0 |
| Img-6 | 6.7 | 79 | 4.7747 | 117.65 | 0 |
| Img-7 | 7.4 | 175 | 4.5695 | 250 | 0 |
| Img-8 | 2.2 | 180 | 4.7904 | 224.14 | 0 |
| Img-9 | 8.2 | 349 | 4.7896 | 248.95 | 0 |

**Table 3.** Results of NFIS

| Image | RBFNN-Classification | Fuzzy-Stage |
|---|---|---|
| Img-1 | Abnormal | Stage I |
| Img-2 | Abnormal | Stage II |
| Img-3 | Abnormal | Stage II |
| Img-4 | Abnormal | Stage III |
| Img-5 | Abnormal | Stage II |
| Img-6 | Abnormal | Stage III |
| Img-7 | Abnormal | Stage IV |
| Img-8 | Normal | None |
| Img-9 | Abnormal | Stage IV |

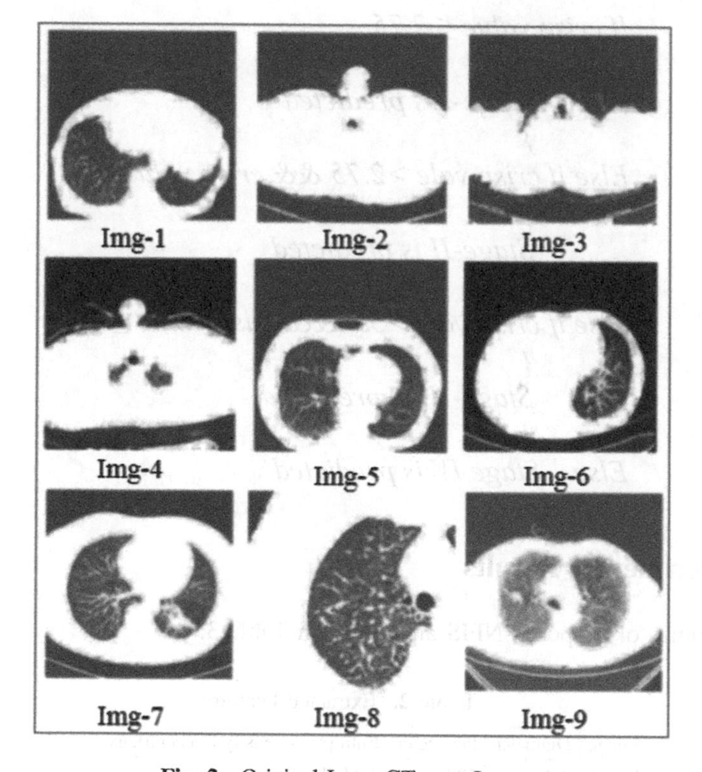


**Fig. 2.** Original Lung CT scan Images

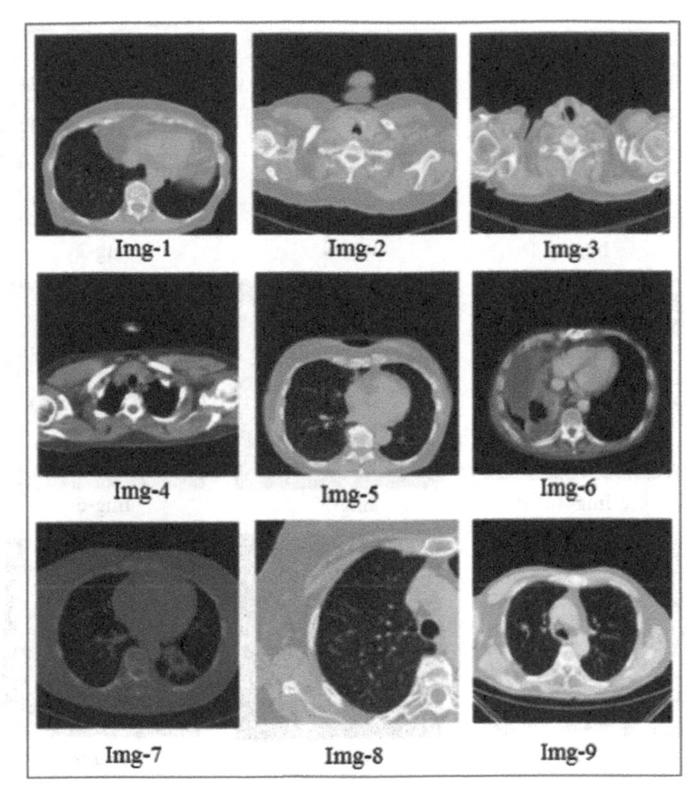


**Fig. 3.** Preprocessed lung CT scan Images

**Table 4.** Performance of RBFNN Classifier

| Data | Actual | Predicted | | Total |
|---|---|---|---|---|
| | | Positive(present) | Negative(absent) | |
| Training | Positive | TP - 51 | FP - 2 | 79 (76%) |
| | Negative | FN - 3 | TN - 23 | |
| Testing | Positive | TP - 14 | FP - 0 | 25 (24%) |
| | Negative | FN - 3 | TN - 8 | |
| Total | | | | 104 |

After the lung CT scan regions have been separated into positive and negative regions, the negative regions are rejected while the positive regions are maintained, feature extraction has been performed. The features like area, perimeter, eccentricity and diameter are take out. Table 2 presents the results of some of the extracted features. The results of the NFIS suggested are shown in Table 3 (Figs. 2, 3, 4, 5, 6) (Tables 4 and 5).

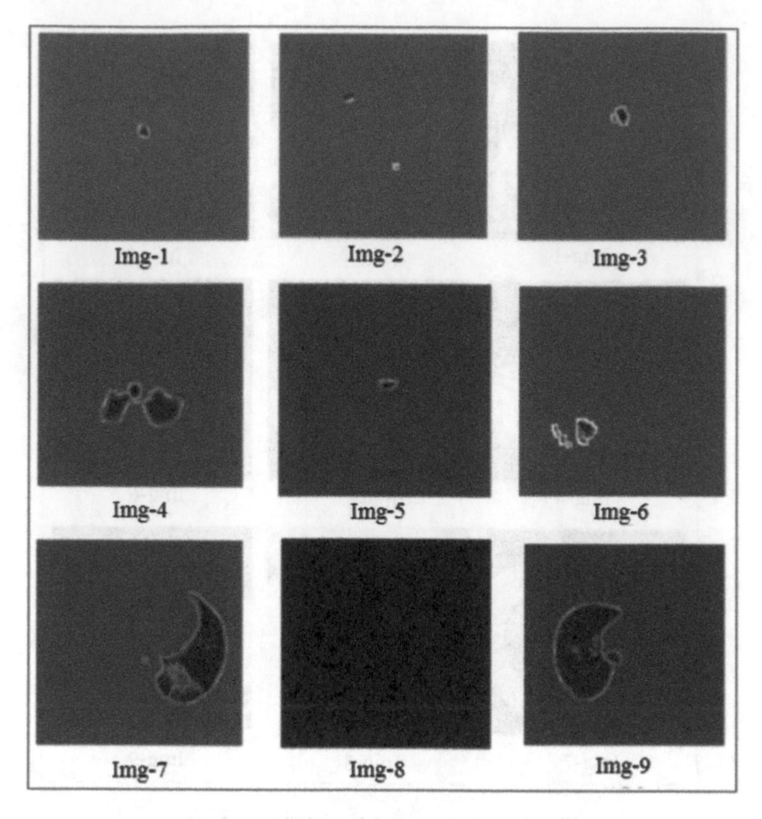


**Fig. 4.** Segmented lung CT scan Images

***TP rate*** or ***Recall*** or ***Sensitivity*** = TPs/(TPs + FNs) = **94.44%**
***TN rate*** or ***Specificity*** = TNs/(TNs + FPs) = **92%**
***Positive Predictive Value*** or ***Precision*** = TPs/(TPs + FPs) = **96.22%**
***Accuracy*** = (TPs + TNs)/(TPs + TNs + FNs + FPs) = **93.67%**
***False Positive Rate (FPR)*** = FPs/(TNs + FPs) = **0.08**

**Table 5.** Results of RBFNN Classifier

| Actual Diagnosis | RBFNN Classification | | | Total |
|---|---|---|---|---|
| | Normal | Benign | Malignant | |
| Normal | 23 | 2 | – | 25 |
| Benign | 2 | 24 | – | 26 |
| Malignant | 2 | 3 | 23 | 28 |
| Total | 27 | 29 | 23 | 79 |

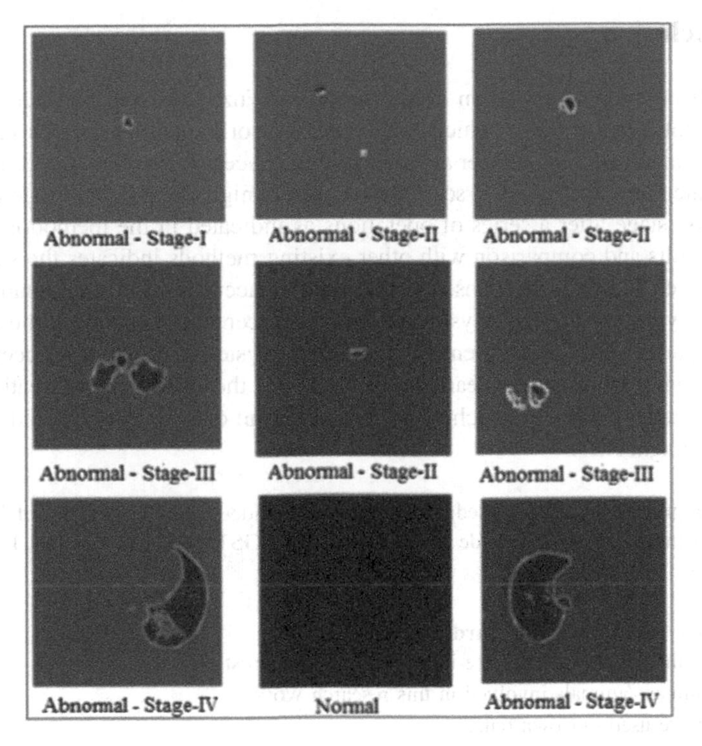


**Fig. 5.** Classified lung CT scan Images – Results1

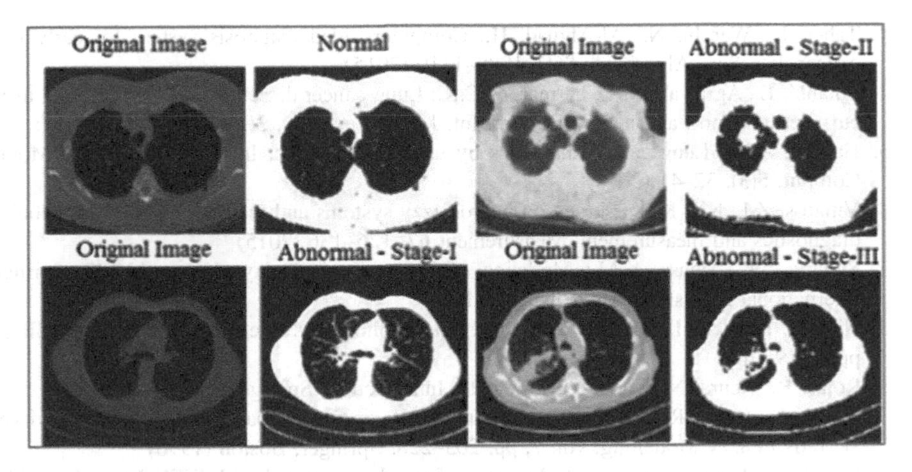


**Fig. 6.** Classified lung CT scan Images– Results2

## 6 Conclusion

A hybridized diagnostic system called the Neuro-Fuzzy Inference System (NFIS) has been presented in this paper, which uses neural networks and fuzzy logic to classify lung cells into cancer and non-cancer and detect lung cancer stages. The proposed NFIS was able to categorize the lung CT scan images into benign or malignant together with the appropriate stage after a series of operations as indicated in the methodology. Experimental results and comparison with other existing methods indicates the superiority of our proposed method, in terms of classification accuracy and computational speed. Proposed system can assist physicians in medical centers in assessing the risk of lung cancer in patients. But the challenge in designing a system is in providing accurate results for detection of lung cancer in early stages to reduce the complexity of treating cancer in advanced stages. Future research could be carried out on a mobile Android platform for real-time diagnostics.

**Acknowledgment.** We acknowledge the support extended by VGST, Govt. of Karnataka, in sponsoring this research work vide ref. No. KSTePS/VGST/GRD-684/KFIST(L1)/2018. Dated 27/08/2018.

**Compliance with ethical standards**

✓ All authors declare that there is no conflict of interest

✓ No humans/animals involved in this research work.

✓ We have used our own data.

## References

1. Taher, F., Werghi, N., Al-Ahmad, H.: Computer aided diagnosis system for early lung cancer detection. Algorithms **8**(4), 1088–1110 (2015)
2. Solanki, E., Agrawal, M.A., Parmar, M.C.K.: Lung cancer detection and classification using curvelet transform and neural network. Int. J. Sci. Res. Dev. **3**(3), 2668–2672 (2015)
3. Hamad, A.M.: Lung cancer diagnosis by using fuzzy logic. Int. J. Comput. Sci. Mobile Comput. **5**(3), 32–41 (2016)
4. Viharos, Z.J., Kis, K.B.: Survey on neuro-fuzzy systems and their applications in technical diagnostics and measurement. Measurement **67**, 126–136 (2015)
5. Suparta, W., Alhasa, K.M.: Modeling of Tropospheric Delays Using ANFIS. Springer International Publishing, Heidelberg (2016)
6. Nielsen, F.: Neural Networks algorithms and applications. Niels Brock Business College, pp. 1–19 (2001)
7. Rojas, R.: Neural Networks: A Systematic Introduction. Springer, Heidelberg (2013)
8. Nauck, D., Kruse, R.: Designing neuro-fuzzy systems through backpropagation. In: Pedrycz, W. (eds.) Fuzzy Modelling, vol. 7, pp. 203–228. Springer, Boston (1996)
9. Lee, C.-C.: Fuzzy logic in control systems: fuzzy logic controller. I. IEEE Trans. Syst. Man Cybern. **20**(2), 404–418 (1990)
10. Jang, J.-S.R.: ANFIS: adaptive-network-based fuzzy inference system. IEEE Trans. Syst. Man Cybern. **23**(3), 665–685 (1993)
11. Ansari, D., et al.: Artificial neural networks predict survival from pancreatic cancer after radical surgery. Am. J. Surg. **205**(1), 1–7 (2013)
12. Al-Amin, M., Alam, M.B., Mia, M.R.: Detection of cancerous and non-cancerous skin by using GLCM matrix and neural network classifier. Int. J. Comput. Appl. **132**(8), 44 (2015)

# Detection of Lung Nodules Using Unsupervised Machine Learning Method

Raj Kishore[1(✉)], Manoranjan Satpathy[2], D. K. Parida[3], Zohar Nussinov[4], and Kisor K. Sahu[5]

[1] Virtual and Augmented Reality Centre of Excellence, Indian Institute of Technology Bhubaneswar, Bhubaneswar 752050, India
rkl6@iitbbs.ac.in

[2] School of Electrical Sciences, Indian Institute of Technology Bhubaneswar, 752050 Bhubaneswar, India
manoranjan@iitbbs.ac.in

[3] Department of Radiotherapy, All India Institute of Medical Sciences, Bhubaneswar 751019, India
drdkparida@gmail.com

[4] Department of Physics, Washington University in Saint Louis, St. Louis, MO 63130-4899, USA
zohar@wuphys.wustl.edu

[5] School of Minerals, Metallurgical and Materials Engineering, Indian Institute of Technology Bhubaneswar, Bhubaneswar 752050, India
kisorsahu@iitbbs.ac.in

**Abstract.** Machine learning methods are now becoming a popular choice in many computer-aided bio-medical image analysis systems. It reduces the efforts of a medical expert and helps in making correct decisions. One of the main applications of such systems is the early detection of lung cancerous nodules using Computed Tomography (CT) scan images. Here, we have used a new method for automated detection of lung cancerous/non-cancerous nodules. It is a modularity maximization based graph clustering method. The clustering is done based on the different region's grayscale values of the CT scan images. The clustering algorithm is capable of detecting nodules of size as small as 4 pixels in two dimension (2D) or 9 voxels in three dimensional (3D) data. The advantage of nodule detection is that it can be used as an extra feature for many supervised learning algorithms especially for those Convolutional Neural Networks (CNN) based architectures where pixel-wise segmentation of data might be required.

**Keywords:** Machine learning · Lung cancer · Graph clustering · Modularity · CNN · Segmentation · CT scan

## 1 Introduction

In 2018 there were 17 million peoples affected by cancer and more than 50% (9.6 million) were died [1, 2]. The most deaths (around 33%) were caused by smoking and use of tobacco, which directly affects the lung. These statistics reveal that lung cancer


S. Smys et al. (Eds.): ICCVBIC 2019, AISC 1108, pp. 463–471, 2020.
https://doi.org/10.1007/978-3-030-37218-7_52

continues to be the major cause of death. The survival probability of a lung cancer patient is inversely proportional to the size of the cancerous nodule at the time of detection. So the early detection of a cancerous nodule in the lung can help medical experts to take timely therapeutic intervention and reduce mortality. There are various algorithms have already been developed for image segmentation. For example, a fast-RCNN algorithm [3] is often used for a region of interest (ROI) detection in CT scan images. These ROIs can be used for cancer detection. Mourchid *et al.* [4] have proposed a new perspective of image segmentation using community detection algorithms for better segmentation. Zhang *et al.* [5] has given hisFCM method which was an improvised form of Fuzzy c-mean (FCM) for image segmentation based on the histogram of the given image. There are several other algorithms [6, 7] developed for segmenting the image for object detection. In the present article, a new state-of-the-art technique is introduced for detecting cancerous/non-cancerous cells/nodules inside the lung. We have used an unsupervised machine learning method for nodule detection which was previously developed for the clustering of spatially embedded networks [8]. It clustered the given network based on the similarity between nodes. This similarity parameter can code different node properties. For example, this parameter can measure similarities between the degrees or intensities of individual nodes/pixels or, in a physical context, depending on the similarities of the stresses or force acting on disparate nodes. In this article, we have detected nodules by clustering the given CT scan images based on the different pixels grayscale values of the CT scan images. These detected nodules can further be classified as cancerous or non-cancerous by passing it through a trained CNN based architecture. The morphological information of nodules detected by the clustering method can also be used as an extra feature during the training of these algorithms. We can pass it as a second channel in CNN based architectures for training. We are expecting that this type of superimposition of information from unsupervised machine learning tool on the top of a supervised learning algorithm will increase the accuracy of these algorithms because it will help in identifying the boundaries of nodules. The mapped information of nodules location in the image data will aid especially those CNN based algorithms which are based on pixel-wise segmentation data [9, 10]. In this article, we have shown that our graph clustering method is detecting the cancerous nodules (labelled by medical experts) along with other non-cancerous nodules.

## 2 Method

In this section, we will discuss the unsupervised machine learning method which is used for detection of the cancerous/non-cancerous nodules in the CT scan lung data. This cancer database contains a fairly large number of training data for lung cancer, which is truly valuable for supervised machine learning for cancer research. However, these machine learning schemes are not truly based on sound scientific physiological considerations, but rather purely rely on the mechanistic understanding of computer vision. Therefore, before really applying these schemes on the Indian ethnic population, one needs to ascertain the applicability of these training data. In the absence of such high quality medically verifiable large training data, such kind of extrapolation will be always questionable. Therefore, using an unsupervised scheme parallel to supervised learning might be useful to remove the ill-effects of such training bias.

We have used the NSCLC dataset [11] because a lot of people are using this data for the task of cancer segmentation. The flow chart of the proposed method is shown in Fig. 1. The first step is converting the '.dcm' files from raw data into a readable '*.tiff*' format because '*.dcm*' the file is not an ideal file type for generating networks. The second step (explained in details in Sect. 2.1) is the separation of the lung portion from the CT scan image data because it contains many irrelevant body portions which unnecessarily increases the number of pixels to be studied. So this lung segmentation process reduces the actual nodule detection time. The third step (explained in details in Sect. 2.2) is network generation from the separated lung portion by considering each pixel/voxel as node and edge between first neighbouring pixels. The fourth step (explained in details in Sect. 2.3) is a clustering of the network generated in the previous step using unsupervised clustering method. The clustering method assembles the closely related pixels (in terms of their grayscale values) in one community. In the fifth step, we assign each community as separate nodule except for the single-pixel communities which belong to the black portion of the image. The sixth and seventh step is for detecting 3D lung nodules. For this purpose, we first convert the grayscale CT image into a binary image by assigning 0 and 1 for nodule pixels and other than nodule pixels respectively. After binary conversion, we stacked these 2D binary images sequentially (based on their Z location) for detecting 3D lung nodules.

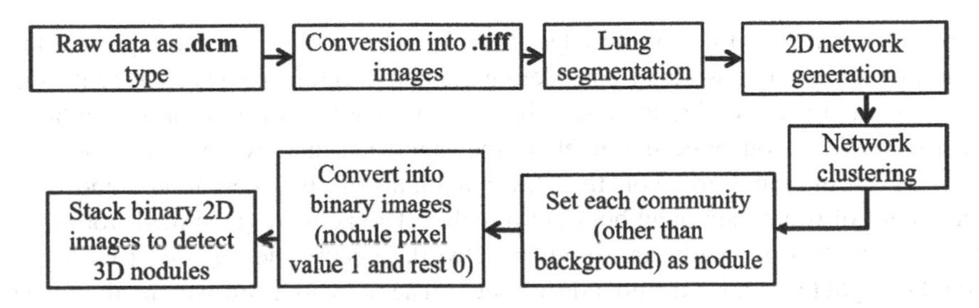


**Fig. 1.** Flow chart of lung nodule detection using unsupervised machine learning method

## 2.1 Lung Segmentation

A single slice in a CT scan consists of 262144 pixels out of which, most of the pixels are not lung portion. It consists of a ribcage, stomach and other organs which are not important for lung cancer detection. Thus segmentation of lung portion is an important step in cancer detection. The pixel value corresponds to the electron density inside the body and varies from −1024 to +1000. All the nodules, organs and blood vessels have a high value of electron density compared to soft tissues. We have segmented the lung region from scans by using thresholding and morphological operations such as dilation, erosion, etc. [12] because there is a considerable amount of intensity change across the boundary of the lungs. Pure thresholding will not segment the boundary regions properly. Watershed algorithm [13] with the asymmetric kernel is used in creating a mask where the regions in the mask with value +255 corresponds to lung region and the remaining region is non-lung portion. The lung portion varies as we move from top to bottom thus segmentation of lung portion from original Dicom images, reduces the average number of pixel/node per image from 262,144 to approximately 16000 pixels/node which is nearly 16 times smaller. Due to this reduction in node size, the overall clustering time reduces without compromising with the accuracy (Fig. 2).

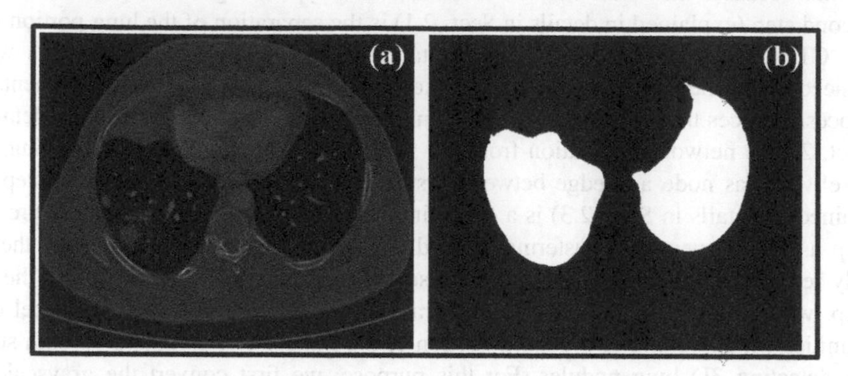


**Fig. 2.** Segmentation of lung region from CT scan image. (a) The CT scan image and (b) The segmented lung region (white region).

### 2.2 Network Generation

A network consists of nodes and edges between these nodes. We can represent the CT scan images into a network by considering the pixel (2D) or voxels (3D) centres as the node position and an edge is present between a node to its first nearest neighbouring pixels. We have converted the pixels of the segmented lung region into a network by considering the same protocol. In 2D, a maximum of eight neighbours and in 3D, a maximum of twenty-six neighbour are possible. The network generated from the lung pixels (Fig. 3(b)) is nearly an 8-regular network because the degree of each node is equal to eight (except for the boundary nodes). The network is un-weighted, as we have used the weight of each edge as unity. So the node degree cannot be used as a clustering parameter (clustering parameter helps decide the community membership of each pixel). Since the pixels greyscale values are different and have a large range from −1024 to +1000, it is used as the clustering parameter for such networks.

### 2.3 Network Clustering

The network generated from the lung portion is clustered into "communities" by optimizing (i.e. maximizing) a modularity function developed by Raj et al. [8] for spatially embedded networks. It was successfully applied on the granular network and was able to determine the "best" resolution scale at which the community structure has the highest associability with the naturally identifiable structures. The modularity function is given below:

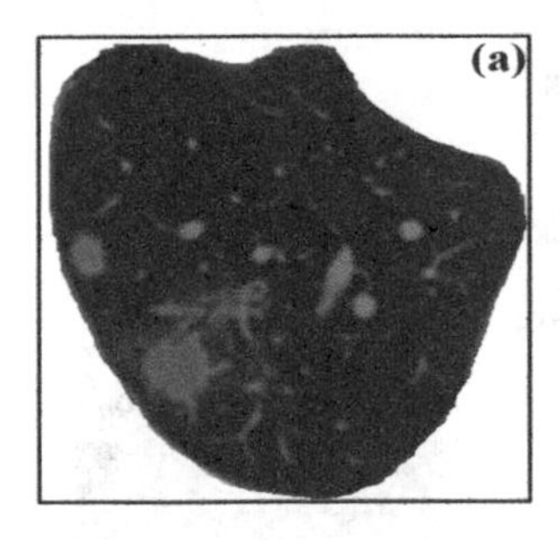

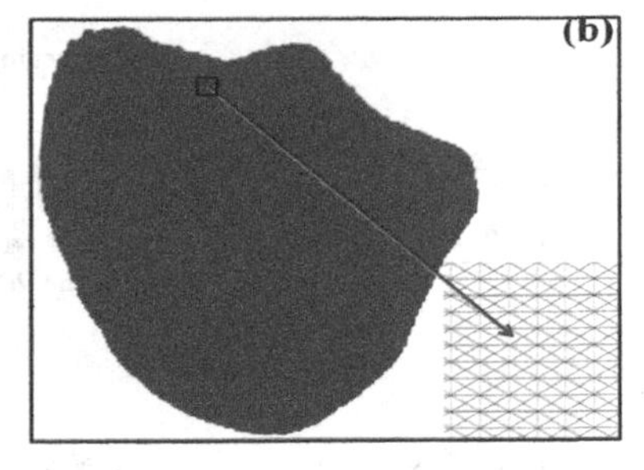


**Fig. 3.** Network generation from (a) the segmented lung portion. (b) The network formed by considering each pixel center as node and edge between first nearest neighbor pixels.

$$Q(\sigma) = \frac{1}{2m} \sum_{i,j} \left( a_{ij} A_{ij} - \theta(\Delta x_{ij}) |b_{ij}| J_{ij} \right) \left( 2\delta(\sigma_i, \sigma_j) - 1 \right) \tag{1}$$

Here $m$ is the total number of edge, $A_{ij}$ is the adjacency matrix, $J_{ij} = 1 - A_{ij}$, δ is the Kronecker delta function, σ is the community membership. The $a_{ij}$ and $b_{ij}$ are the strength of connected and missing edges between nodes $i$ and $j$ respectively. It is the difference between the local average greyscale value (a clustering parameter) of node $i$ and $j$ and the global average greyscale value $<k>$ of the given network,

$$a_{ij} = b_{ij} = \left( \frac{k_i + k_j}{2} - <k> \right) \tag{2}$$

Where, $<k> = \frac{1}{N} \sum_{r=1}^{N} k_r$ and $N$ is the total nodes and $k$ is the node degree.

Our method has mainly two distinctions over other existing methods of modularity calculation. First, it does not use any null model. Second, it considers the effect of inter-community edges while detecting the most modular structure. It also restricts the over penalization for missing links between two distant nodes within the same community by using Heaviside unit step function $\theta(\Delta x_{ij})$. It defines the neighborhood and penalizes only for the missing links inside the neighborhood. Here $\Delta x_{ij} = x_c - |\vec{r}_i - \vec{r}_j|$ is the difference in Euclidian distance between nodes $i, j$ and a predefined cutoff distance for neighborhood $x_c$, which is chosen as $x_c = 1.05(R_i + R_j)$ for the present study (where $R$ is the distance between two touching pixels).

$$\theta(\Delta x_{ij}) = \left\{ \begin{array}{ll} 1 & \Delta x_{ij} > 0 \text{ (within cut off)} \\ 0 & \text{otherwise (outside cut off)} \end{array} \right\} \tag{4}$$

The clustering algorithm used here tries to merge a node with any of its connected neighbors to increases the overall modularity of the network. This process continues until no more merging is possible i.e. none of the merging increases the modularity of the network. The pictorial representation of the algorithm is given in Fig. 4.

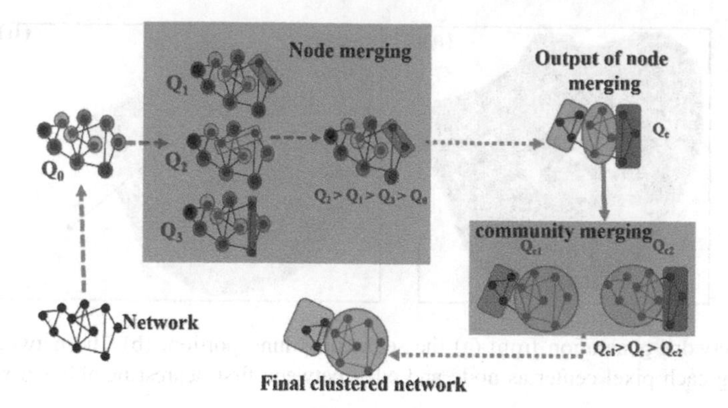


**Fig. 4.** Pictorial representation of clustering algorithm. Q represents the modularity of given partition

The algorithm consist of mainly three steps:

(i) *Initialization*: The connection matrix which is also called as adjacency matrix is calculated. Initially, the number of community is same as the number of nodes in the network (*i.e.* single node community). This type of initial community distribution is called symmetric distribution. Initial partition has modularity $Q_O$.

(ii) *Node merging:* In this step, each node is selected iteratively and checked for possible merging sequentially with all its connected neighbor communities. Modularity is calculated after each merging (*e.g.* $Q_1$, $Q_2$, $Q_3$ *etc.*: refer "node merging" in Fig. 4). Finally, the merge which has the highest positive change in the modularity of network after and before merging in the community is selected. This node merging continues until no further merging increases the modularity of the system.

(iii) *Community merging:* The node merging method used in step (ii) often converged at local maxima in the modularity curve. To remove any local maxima trap in the modularity curve, we merge any two connected community. If this merging increases the modularity then we keep these merging else reject it. The output of community merging is the final most modular clustered network.

## 3 Results

We have detected nodules in various 2D slices of CT scan data of different patients. The pixels of the given image are segregated in different communities/nodules based on their grayscale values using the network clustering algorithm discussed in Sect. 2.3. Three different 2D slices of a patient CT scan image at different Z locations are shown in Fig. 5(a–c). The communities/nodules detected by the given modularity function in these images are shown in Fig. 5(d–f). The red circled regions in all the three CT scan images are the cancerous cells suggested by medical experts after performing different diagnosis processes including biopsies. It is clear that in all the three images, clustering through modularity function can detect the cancerous nodule as a single community along with other nodules. It has captured not only the big size cancerous nodules but the small ones as well. These nodules are the potential regions need to be investigated.

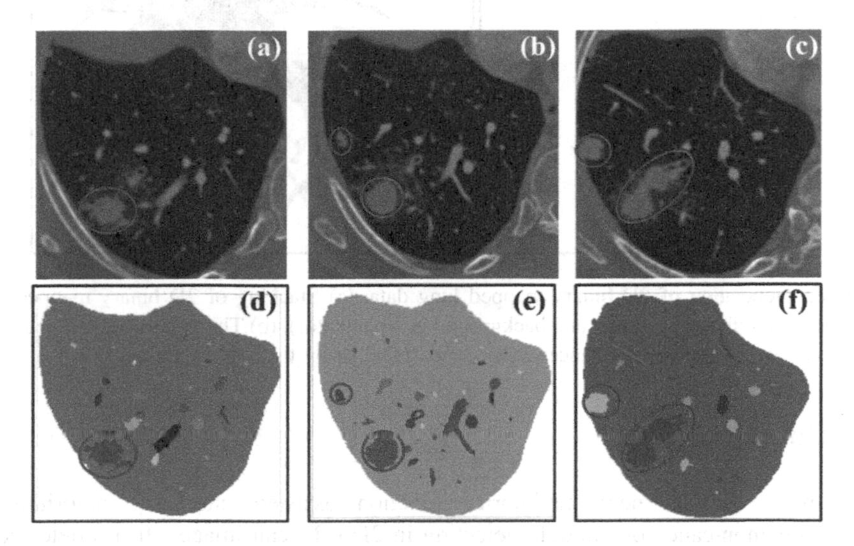


**Fig. 5.** Nodule detection in 2D slices of CT scan data. (a–c) The 2D slices at different Z position of the lung data. The red circled regions are cancerous cells (labelled by the medical expert) and (d–f) are the corresponding colored images which contained different size of nodules. The color assigned to each detected nodule and the background is randomly selected. The same cancerous region as in panel (a–c) are circled in red color (Data source from Luna data: https://luna16.grand-challenge.org/Data/)

We have converted these clustered 2D images into binary mapped images by assigning nodule pixels as black and other than nodule (background) pixels as white. Then, all these 2D binary mapped images of a patient are stacked sequentially to generate the 3D binary mapped image (Fig. 6). The 3D binary mapped image represents mainly the opaque objects inside the lung and can be used for various 3D CNN architectures as the second channel along with the original unclustered 3D CT scan images for training purpose. It may increase the learning efficiency of the architectures because of providing extra morphological information of nodules.

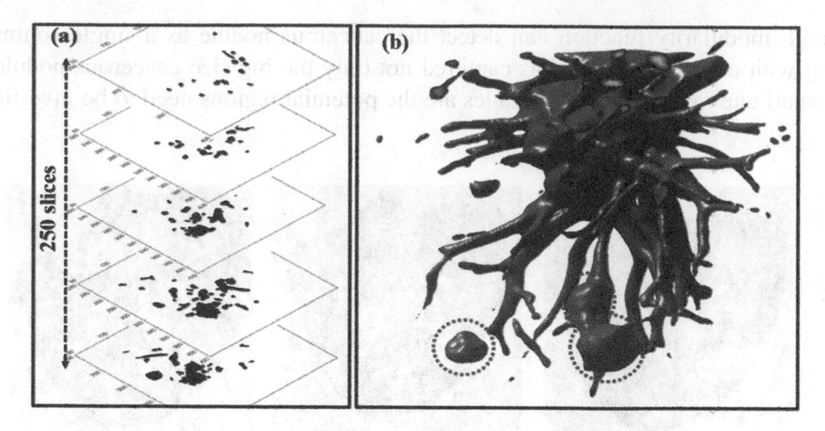


**Fig. 6.** Generation of 3D binary mapped lung data. (a) Stacking of 2D binary mapped slices (detected nodules are in black and background in white color). (b) The 3D binary mapped image. The red circled nodules are cancerous nodules (labelled by the medical expert)

## 4 Conclusions

We have used a new modularity function assisted clustering algorithm for cancerous/non-cancerous nodule detection in 2D CT scan images. It has detected the cancerous pixels as separate nodules. It has detected the nodules of bigger as well as small sizes. The clustered image is binary mapped (black for nodule region and white for the background) and can be used as an extra feature during the training of a CNN based architecture which depends on pixel-wise image segmentation data. This mapping contains the morphological information of nodules and thus might help reduce training as well as validation loss. However, more samples need to be studied for establishing the model. In future, the detected nodules can be used as a region of interest (ROI) which is mostly computed using RCNN. The advantage of the present method over RCNN is that it does not require any labelled data which is sometimes difficult to acquire.

**Acknowledgement.** This project is funded by Virtual and Augmented Reality Centre of Excellence (VARCoE), IIT Bhubaneswar, India.

## References

1. Cancer Research UK. https://www.cancerresearchuk.org/health-professional/cancer-statistics/worldwide-cancer. Accessed August 2019
2. Ardila, D., Kiraly, A.P., Bharadwaj, S., Choi, B., Reicher, J.J., Peng, L., Tse, D., Etemadi, M., Ye, W., Corrado, G., Naidich, D.P.: End-to-end lung cancer screening with three-dimensional deep learning on low-dose chest computed tomography. Nat. Med. **25**(6), 954 (2019)

3. Ren, S., He, K., Girshick, R., Sun, J.: Faster r-CNN: towards real-time object detection with region proposal networks. In: Advances in Neural İnformation Processing Systems, pp. 91–99 (2015)
4. Mourchid, Y., El Hassouni, M., Cherifi, H.: An image segmentation algorithm based on community detection. In: International Workshop on Complex Networks and their Applications, pp. 821–830. Springer, Cham, November 2016
5. Zhang, X., Zhang, C., Tang, W., Wei, Z.: Medical image segmentation using improved FCM. Sci. China Inf. Sci. **55**(5), 1052–1061 (2012)
6. Hu, D., Ronhovde, P., Nussinov, Z.: Replica inference approach to unsupervised multiscale image segmentation. Phys. Rev. E **85**(1), 016101 (2012)
7. Browet, A., Absil, P.A., Van Dooren, P.: Community detection for hierarchical image segmentation. In: International Workshop on Combinatorial Image Analysis, pp. 358–371. Springer, Heidelberg, May 2011
8. Kishore, R., Gogineni, A.K., Nussinov, Z., Sahu, K.K.: A nature ınspired modularity function for unsupervised learning involving spatially embedded networks. Sci. Rep. **9**(1), 2631 (2019)
9. Kamal, U., Rafi, A.M., Hoque, R., Hasan, M.: Lung cancer tumor region segmentation using recurrent 3D-denseunet. arXiv preprint arXiv:1812.01951 (2018)
10. Abraham, N., Khan, N.M.: A Novel Focal Tversky Loss Function with Improved Attention U-Net for Lesion Segmentation. In: 2019 IEEE 16th International Symposium on Biomedical Imaging (ISBI 2019), pp. 683–687 (2019)
11. https://wiki.cancerimagingarchive.net/display/Public/NSCLC +Radiogenomics#39fc8e84140 54aaaa88e56b88bb061f6
12. Sharma, D., Jindal, G.: Identifying lung cancer using image processing techniques. In: International Conference on Computational Techniques and Artificial Intelligence (ICCTAI), vol. 17, pp. 872–880 (2011)
13. Sankar, K., Prabhakaran, M.: An improved architecture for lung cancer cell identification using Gabor filter and intelligence system. Int. J. Eng. Sci. **2**(4), 38–43 (2013)

# Comparison of Dynamic Programming and Genetic Algorithm Approaches for Solving Subset Sum Problems

Konjeti Harsha Saketh and G. Jeyakumar(✉)

Department of Computer Science and Engineering, Amrita School of Engineering, Coimbatore, Amrita Vishwa Vidyapeetham, Coimbatore, India
cb.en.u4cse17036@cb.amrita.students.edu,
g_jeyakumar@cb.amrita.edu

**Abstract.** Albeit Evolutionary Algorithms (*EAs*) are prominent, proven tools for resolution of optimization problems in the real world, appraisal of their appropriateness in solving wide variety of mathematical problems, from simple to complex, continues to be an active research area in the domain of Computer Science. This paper portrays an evaluation of the relevance of Genetic Algorithm (*GA*) in addressing the Subset Sum Problem (*SSP*) of Mathematics and providing empirical results with discussions. A *GA* with pertinent mutation and crossover operators is designed and implemented to solve *SSP*. Design of the proposed algorithm are clarified in detail. The results obtained by the proposed *GA* are assessed among different instances with different initial population by the intermediary solutions obtained and the execution time. This study also adapted the traditional Dynamic Programming (*DP*) approach, pursuing a bottom-up strategy, to solve the *SSP*. The findings revealed that the *GA* approach would be unpreferred on account of its longer execution time.

**Keywords:** Genetic Algorithm · Dynamic Programming · Subset Sum Problem · Algorithm design · Mutation and crossover

## 1 Introduction

Evolutionary Computation (*EC*) has a repository of algorithms, termed as Evolutionary Algorithms (*EAs*), commonly applied in pursuit of answers to global optimization problems. *EAs* are inspired by the biological evolutionary principle of Darwin's theory. They follow a generate-and-test methodology (trial and error) of searching for optimal solutions. In the exercise of *EAs*, an initial set of solutions (initial population) is generated, to start the search; this set undergoes an evolutionary search process that generates transient solutions, in their quest for the required optimal solution. Evolution of the population of potential solutions takes place at each consecutive step of the generation. At each step, the less desired candidates in the populations are stochastically removed, and promising candidates are retained. This emulates the natural selection theory of 'Survival of the fittest'. Each generation of *EA* follows a series of steps, termed as: fitness evaluation,


S. Smys et al. (Eds.): ICCVBIC 2019, AISC 1108, pp. 472–479, 2020.
https://doi.org/10.1007/978-3-030-37218-7_53

parent selection, reproduction (mutation and crossover) and survivor selection. The survivors (surviving candidates) are identified based on their fitness, which is calculated using a fitness function relevant to the problem. Two main processes that create diversity in the population are mutation and crossover.

The Subset Sum Problem (*SSP*) is defined as "Given a set of non-negative integers, and a value, *sum*, determine if there is a subset of the given set, whose aggregate is equal to the given *sum*". In the domain of Computer Science, *SSP* is an important decision-making problem, which is *NP*-Complete. Using *SSP* as a benchmarking problem, to test the working of new algorithms proposed by researchers, is a common practice among the computer science research community.

*GA*, an instance of *EA*, follows the general algorithmic framework of *EAs*. *GA* is popular for its wide applicability [1, 2], as it supports encoding with all the type of depictions such as binary, integer, real, and permutation-based representations. This paper elucidates details of the design of an effective *GA*, for solutions to popular mathematical problems. The problem chosen for this study is Subset Sum Problem (*SSP*).

## 2 The Subset Sum Problem

The *SSP* issue can be mathematically stated as follows: Given a set of *n* positive integers, $S = \{x_1, x_2, \ldots, x_n\}$ and a non-negative integer, *sum*, find a subset, *SS* of *S*, with *m* integers, whose aggregate is equal to *sum*.

For the given, $S = \{x_1, x_2, \ldots, x_n\}$, $SS = \{y_1, y_2, \ldots, y_m\}$, $y_i \in S$ for $i = 1 \ldots m$, is a valid sub set iff $\sum_{i=1}^{m} y_i = sum$ [iff = if and only if].

To formulate an *SSP* as an optimization problem, the objective function and the constraints need to be defined. They are given below.

The objective function is Minimize $f(SS) = sum - \sum_{i=1}^{m} y_i$

$$\text{subject to } y_i \in S \text{ for } i = 1 \ldots m$$

The optimal solution is a $SS^*$ with $f(SS^*) = 0$.

## 3 Related Works

This section presents few recent investigations reported in literature, explaining different approaches for solving *SSP*. Time entailed in solving *SSP* problems can be reduced by restricting the number of iterations executed by the algorithm, and altering the process of solving *SSP* with suitable changes in the algorithm. [3] suggests that number of iterations can be restricted by the user, if required, and can find all subsets with the sum equivalent to the required sum. The approach used in this work for solving *SSP* is as follows

i. Arrange all the elements of the array in ascending order.
ii. Remove the elements greater than required sum.
iii. Fill the bit map with zero and find the first partial solution.
iv. If the partial solution's sum is equal to required sum, stop the process, else continue to find the first 1-bin in the bit map until finding the partial solutions with required sum.

Also, if there is a restriction on number of iterations then process might be terminated without finding a solution.

*SSPs* can be solved in pseudo-polynomial time, by Dynamic Programming [4]. In this method, a Boolean 2-dimensional array, *A,* with size of $x \times y$ is created, where $x =$ *required sum* and $y =$ *size of main array.* Dynamic programming method can be used in two ways: either bottom-up or top-down approach. The value of any element, *A[i] [j]* is set to *True* if the sum of *A[i] [0 to j − 1]* is equal to *I*, where *I* is less than or equal to the given *sum,* otherwise set to *False*. Whether there exists a subset can be tested, by checking the element *A [required sum] [size of main array]* i.e. if it is set to *True*, it can be inferred that there indeed exists a subset whose elements aggregate to required sum.

A work combining the *L* shortest path algorithm, and the finite-time convergent recurrent neural network, to form a new algorithm, for the *L* smallest *k*-subsets sum problem is developed and presented in [5]. This work formed the solution to the subset sum problem, by combining the solutions from a set of sub-problems. [6] presented priority algorithm approximation ratios for solving *SSP*; this probe is focused on the power of revocable decisions, where accepted data items can be rejected later, to maintain the feasibility of the solution.

Besides the various approaches proposed to solve *SSP*, the *SSP* is also used to study and analyze different aspects of optimization problem. For example, [7] analyzed the properties of *SSP*; the authors highlighted the correlation between the dynamic parameters of the problem and the resulting movement in the global optimum. It is also shown the degree of movement of global optimum is relative the problem dynamics. An article details the algorithms in *FLSSS*, an *R* package, for solving various subset sum problems is there in [8]. [9] works on a series of assumptions. First, the given array is sorted in ascending order using Quick Sort. Next, a variable called *limit point* is set such that elements after that point are greater than the required sum, hence need not be used for the algorithm. Initial population is generated on the basis of cumulative frequency. Limit point in the initial population is found and candidates from the population are taken according to the genes of the assumed solutions. Then evolutionary process is performed until a subset with required sum is obtained.

An improved *GA* for solving *SSP* is presented in [10]. A work analyses the effect of changing the parameters of *GA* on the results of solving *SSP* is presented in [11]. A work tried to find the approximate solution of *SSP* using *GA* along with rejection of infeasible offspring is presented in [12]. This work introduces a penalty function to discard the candidates not fit for next population.

In spite of the previously reported existence of many *GA* and non-*GA* approaches to solve *SSP*, this paper intends to describe solving the *SSP* issue with a classical *GA* and a Dynamic Programming approach. Algorithms with both these approaches are implemented to solve the *SSP,* and a comparative study on their performance is presented in this paper.

## 4 Design of Experiments

The design of experiment includes describing the parameters, and set their values in the chosen problem domain and the algorithm domain. The parameters to be set in the problem domain of *SSP* are: size of the main array, elements of the main array and required sum. Under the assumption that all the elements in the main array are distinct, a sample setup of *SSP* parameters is presented in Table 1.

**Table 1.** A sample parameter set up of *SSP*.

| Parameter | Meaning | Values |
|---|---|---|
| *Size_Ma* | Size of the main array | 5 |
| *Elements* | Elements in the main array | 12, 17, 3, 24, 6 |
| *R_Sum* | Required sum | 21 |

In the algorithmic domain, the experimental set up of this paper includes solving *SSP* by classical dynamic programming methodology and *GA*. In dynamic programming method the process of finding the subset is done, as follows, with a Boolean array.

- A 2-dimensional Boolean array, *dp* [*R_Sum*+1] [*Size_Ma*+1] is created. Every row of $0^{th}$ column is initialized to *True,* and every column of $0^{th}$ row is initialized to *False*.
- Method of bottom up approach is used to fill the remaining elements of the *dp* array.
- A variable *val* is assigned with *array* [i – 1]. The *dp[i] [j]* is assigned to *dp [i – 1] [j]* until *j* is less than *val*. Remaining values of the particular row are assigned as follows: *dp[i] [j]* = *dp [i – 1] [j – value],* until i = *R_Sum*+1 to *Size_Ma*+1.
- If the value of *dp [R_Sum] [Size_Ma]* is *True,* then it can be inferred that there is a subset that sums to the *R_Sum,* else there is no subset.
- If there is a subset, the values in the subset are printed using the method of backtracking, in which we check from *dp [R_Sum] [Size_Ma]* to *dp [0] [0]*. The compound condition, which is to be true for an element is present in the subset is (*j – val* > 0 && dp *[i – 1] [j – val]*).

As *GA* follows the common template of *EA*, the design parameters, commonly used in any *EA,* are to be set. The design of an *EA*, to solve an optimization problem includes determination of the following parameters: Population representation (with population size and size of each candidate), Fitness evaluation, Logic of selecting parents for reproduction, Mutation operator, Crossover operator, Logic of selecting survivor and Termination criteria. The values set for design of the *GA,* in this study's experiment are explained below.

- The population size is set equivalent to *Size_Ma*. An initial population with the *Size_Ma* candidates is initialized. In this population, the size of each candidate is also fixed random. The components of each candidate are taken at random, from the main array such that no two elements, among the candidates, are equal.
- The fitness of the candidates is calculated as the absolute difference between the sum of the subset and the required sum. If the fitness of a candidate is equal to 0 then it a solution for *SSP,* whose sum of elements is equal to *R_Sum*.
- Top two fittest candidates in the population are selected as parents for the reproduction. Thus the parents are the candidates whose fitness values is closer to the *Rossum,* compare to other candidates in the population.

- A *K*-point crossover is used during the reproduction process. The crossover point is randomly generated between the maximum of the smallest size of the fittest and second fittest chromosomes in the population. The crossover is carried out in such a way that no two components in the candidates are similar, before and after the crossover.
- Mutation is done on a random basis. If a particular candidate is selected for mutation then the mutation point is generated random in the range 0 to size of the fittest chromosome. The component at that random point is changed with a random value.
- During the stage of survivor selection, the least fittest candidate is swapped with the newly generated candidate.
- The above steps are repeated until a candidate has been found with fitness value equivalent to zero.
- The number of generations and execution time, consumed by *GA* to arrive at a solution of the *SSP*, are recorded for the comparative analysis.

# 5 Results and Discussions

The results obtained, in our intent to solve the *SSP* by the dynamic programming methodology and *GA*, are presented in this chapter.

## 5.1 Dynamic Programming (*DP*) Approach

For the given input array and required *sum*, by using bottom-up strategy of the *DP* approach, a 2-dimensional Boolean array *dp* is populated with *True* or *False*, and the value of *dp[R_Sum][Size_Ma]* resulted in *True*. Next, using the method of backtracking, elements of the subset are printed. The results, obtained by the Dynamic programming approach, are presented in Table 2, and the content of the Boolean array used is shown in Fig. 1.

## 5.2 Classical *GA* Approach

The results obtained, for the same parametric set up of *SSP* by *GA*, are presented in Table 3. The performance metrics used for the analysis are

(1) Success Rate (*SR*)
(2) Average Number of Generation (*AvgNoG*)
(3) Average Execution Time (*AvgET*)

- The *SR* is calculated using the Eq. (1)

$$SR = (NoSR\#/ToR\#) * 100\% \tag{1}$$

where, *NoSR#* is the number successful runs and *ToR#* is the total number of runs. *Successful run* is an instance of runs, where *GA* able to obtain Solution subset with the sum of elements = *R_Sum.*

- The *AvGNoG* is calculated using the Eq. (2)

$$AvgNoG = \left(\sum_{R=1}^{MaxR} G_R\right)/MaxR \tag{2}$$

where, *R* is the run number
*MaxR* is the maximum number of runs (*MaxR* is set 10).
$G_i$ is the number of generations taken by *GA* at run *i*.

- The *AvgET* is calculate using the Eq. (3)

$$AvgET = \left(\sum_{R=1}^{MaxR} ET_R\right)/MaxR \tag{3}$$

where $ET_R$ is the execution time taken by *GA* at run *i*.

**Table 2.** Results for *SSP* by *DP* approach

| Serial no. | *Size_Ma* | Elements | *R_Sum* | Solution subset | Time (s) |
|---|---|---|---|---|---|
| 1 | 5 | 12, 17, 3, 24, 6 | 21 | 3, 6,12 | $1.07 \times 10^{-3}$ |

```
Row 1 -> false false false false false false false false false
false false true false false false false false false false
false false

Row 2 -> false false false false false false false false false
false false true false false true false false false false
false false

Row 3 -> false false false false false false false false false
false false true false false true false true false false false
false

Row 4 ->false false true false false false false false false
false false false false false true false false true false true
false

Row 5 ->false false true false false false false false false
false false false false false true false false true false true
false

Row 5 ->false false true false false true false false true
false false false false false false false false false false
false true
```

**Fig. 1.** Output of the 2-dimensional Boolean array after computing using *DP*

The measured performance metrics, for *GA* on its performance at solving the *SSP* designed in the experiment, is presented in Table 3. The values of *AvgNoG* and *AvgET* are measured only for the successful runs, and presented in Table 4. The results show that the *SR* in solving the *SSP* is 70%. Its *AvgNoG* is 7.14 and *AvgET* is 74.28 ms. The findings validate the applicability of *GA* in solving *SSP* with a simple experimental set up.

On comparing the performance of the *GA* approach versus the classical Dynamic Programming methodology (Tables 2 and 3), for solutions to the *SSP*, the downside of *GA* is found to be the increased execution time, ensued by the stochastic processes at all stages. While Dynamic programming approach consumed approximately 2 ms to find the solution, *GA* took around 75 ms. However, the performance of *GA* can still be augmented to achieve higher success rate with lesser *AvgNoG* and *AvgET,* by careful selection of its parametric values.

**Table 3.** Results for *SSP* by *GA* approach

| Run# | *SSP* solved (yes/no) | Number of generations | Time (s) | Solution subset |
|---|---|---|---|---|
| 1 | Yes | 19 | $82 \times 10^{-3}$ | 12, 6, 3 |
| 2 | No | – | – | – |
| 3 | No | – | – | – |
| 4 | Yes | 11 | $69 \times 10^{-3}$ | 6, 12, 3 |
| 5 | Yes | 0 | $58 \times 10^{-3}$ | 12, 6, 3 |
| 6 | Yes | 4 | $60 \times 10^{-3}$ | 3,12, 6 |
| 7 | Yes | 1 | $75 \times 10^{-3}$ | 3, 12, 6 |
| 8 | No | – | – | – |
| 9 | Yes | 7 | $83 \times 10^{-3}$ | 6, 12, 3 |
| 10 | No | 8 | $93 \times 10^{-3}$ | 6, 3, 12 |

**Table 4.** The performance of *GA* at solving *SSP*

| Performance metric | Value measured |
|---|---|
| $SR$ | 70% |
| $AvgNoG$ | 7.14 |
| $AvgET$ | $74.28 \times 10^{-3}$ s |

# 6 Conclusions

The paper presented the details of the design and implementation of a *GA* framework to solve the *SSP*. The performance of *GA* is appraised on three metrics, for different sample runs. As a comparative study, the *SSP* is also solved using the classical Dynamic Programming approach. *GA* is able to solve the *SSP* with 70% success rate. The experimental results, comparing both the approaches, reveal that the execution time of *GA* based approach is much higher than the classical *DP* approach.

This work can be extended further analyzing the impact on performance, with modifications to the *GA* approach, with varying values of its inherent parameters.

**Compliance with Ethical Standards**

✓ All authors declare that there is no conflict of interest
✓ No humans/animals involved in this research work.
✓ We have used our own data.

## References

1. Janani, N., Shiva Jegan, R.D., Prakash, P.: Optimization of virtual machine placement in cloud environment using genetic algorithm. Res. J. Appl. Sci. Eng. Technol. **10**(3), 274–287 (2015)
2. Raju, D.K.A., Shunmuga Velayutham, C.: A study on genetic algorithm based video abstraction system. In: Proceedings of World Congress on Nature Biologically Inspired Computing (2009)
3. Isa, H.A., Oqeili, S., Bani-Ahmad, S.: The subset-sum problem: revisited with an improved approximated solution. Int. J. Comput. Appl. **114**, 1–4 (2015)
4. Jain, E., Jain, A., Mankad, S.H.: A new approach to address subset sum problem. IEEE Explorer (2014)
5. Gu, S., Cui, R.: An efficient algorithm for the subset sum problem based on finite-time convergent recurrent neural network. Neuro-Computing **149**, 13–21 (2015)
6. Ye, Y., Borodin, A.: Priority algorithms for the subset-sum problem. J. Comb. Optim. **16**(3), 198–228 (2008)
7. Rohlfshagen, P., Yao, X.: Dynamic combinatorial optimization problems: an analysis of the subset sum problem. Soft Comput. **15**, 1723–1734 (2011)
8. Liu, C.W.: FLSSS: a novel algorithmic framework for combinatorial optimization problems in the subset sum family. J. Stat. Softw. **VV**(II), 1–22 (2018)
9. Bhasin, H., Singla, N.: Modified genetic algorithms based solution to subset sum problem. Int. J. Adv. Res. Artif. Intell. **1**(1), 38–41 (2012)
10. Wang, R.L.: A genetic algorithm for subset sum problem. Neuro-Computing **57**, 463–468 (2004)
11. Oberoi, A., Gupta, J.: On the applicability of genetic algorithms in subset sum problem. Int. J. Comput. Appl. **145**(9), 37–40 (2016)
12. Thada, V., Shrivastava, U.: Solution of subset sum problem using genetic algorithm with rejection of infeasible offspring method. Int. J. Emerg. Technol. Comput. Appl. Sci. **10**(3), 259–262 (2014)

# Design of Binary Neurons with Supervised Learning for Linearly Separable Boolean Operations

Kiran S. Raj(✉), M. Nishanth, and Gurusamy Jeyakumar

Department of Computer Science and Engineering, Amrita School of Engineering, Amrita Vishwa Vidyapeetham, Coimbatore, India
{cb.en.u4cse17430, cb.en.u4cse17441}@cb.amrita.students.edu, g_jeyakumar@cb.amrita.edu

**Abstract.** Though the *Artificial Neural Network* is used as a potential tool to solve many of the real-world learning and adaptation problems, the research articles revealing the simple facts of how to simulate an artificial neuron for most popular tasks are very scarce. This paper has in its objective presenting the details of design and implementation of artificial neurons for linearly separable Boolean functions. The simple Boolean functions viz *AND* and *OR* are taken for the study. This paper initially presents the simulation details of artificial neurons for *AND* and *OR* operations, where the required weight values are manually calculated. Next, the neurons are added with learning capability with perceptron learning algorithm and the iterative adaptation of weight values are also presented in the paper.

**Keywords:** Neural networks · Supervised learning · Weight adaptation · Boolean operations · Binary neurons

## 1 Introduction

In human bodies, the Biological Neural Network (*BNN*) integrates the processing elements called neurons to form the nervious system. These processing elements are to process the information sent to them as external inputs with dynamic nature. Therese neurons are interconnected through many synapses. The neural network in computer science tries to replicate the *BNN* and tries to analyze data inputs to predict the accurate output by Artificial Neural Network (*ANN*). The *ANNs* are designed to learn and to perform a task by analyzing the given examples with learning. The *ANNs* are the frameworks consisting of suitable learning algorithms to analyze the complex data inputs to get the desired outputs. The applications of *ANNs* includes character recognition, image processing, stock market prediction, video analytics etc.

In nature, neurons have several dendrites (inputs), a cell nucleus (processor) and an axon (output). The biological neurons can be modeled as mathematical model to design the artificial neurons. These Artificial neurons are elementary units in an artificial neural network. Artificial neurons are designed to mimic aspects of their biological


S. Smys et al. (Eds.): ICCVBIC 2019, AISC 1108, pp. 480–487, 2020.
https://doi.org/10.1007/978-3-030-37218-7_54

counterparts. Neurons work for producing a desired output with an activation function for a given set of inputs and a set of weights.

This papers presents design and implementation details of artificial neuron for performing common Boolean operations. Initially, the neurons are implemented without the learning capability. Their performances are verified with the expected outcomes of the Boolean operations. Then, the neurons are added with learning capability with perceptron learning algorithm. The working these learning neurons also explained in the paper.

The remaining part of paper is organized in to 6 sections. The Sect. 2 explains the working of artificial neurons, the Sect. 3 presents the works related the work presented in this paper, the Sect. 4 discusses the design and implementation of the neurons without learning, the Sect. 5 explaining the working of neurons with learning, finally the Sect. 6 concludes the paper.

## 2 Artificial Neurons

The artificial neuron receives one or more inputs and sums them to produce an output. Usually each input is separately weighted, and the sum is passed through a nonlinear function known as an activation function or transfer function. Specifically, in an *ANN* (Artificial Neural Network) we do the sum of products of inputs and their corresponding weights and apply an activation function to it get the output of that layer and feed it as input to the next layer (as it is shown in Fig. 1).

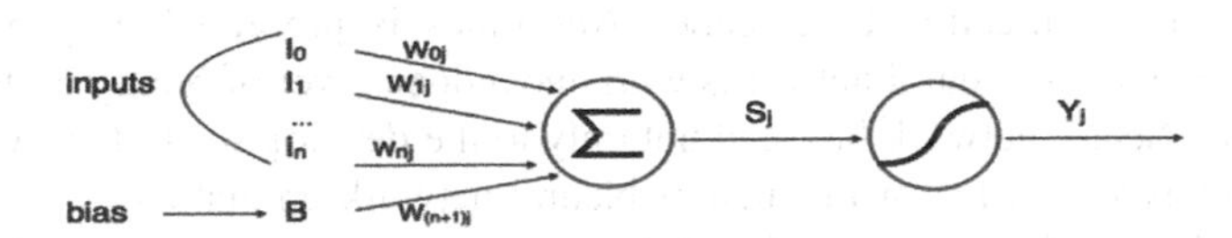


**Fig. 1.** The schematic diagram of an artificial neuron

The main purpose of an activation function in an *ANN* is to convert an input signal of a node in an *ANN* to an output signal. The activation function makes the neural network more powerful and adds ability to it to learn complex and complicated data set and represent non-linear complex arbitrary functional mappings between inputs and outputs.

## 3 Related Works

This section presents the recent studies available in the literature related to our system. There has been great effort and significant progress in the field of neural network recently due to their efficiency and accuracy. For example, a simple neural network design composed of neurons of two types to realize the $k^{th}$ basic Boolean (logic) operation is proposed in [1]. The above neural network performs the Boolean operation based on recursive definition of "basic Boolean operations". A work presenting the modelling of neural Boolean networks by non-linear digital equations is shown in [2], where new memory neural systems based on hyper incursive neurons (neurons with multiple output states for the same input) instead of synaptic weights are designed and implemented. Similar to that, the work presented in [3] depicts the reduction of memory space and computation time using Boolean neural networks in comparison to representation of Boolean functions by usual neural networks.

Threshold functions and Artificial Neural Networks (*ANNs*) are known for many years and have been thoroughly analyzed. An implementation of the basic logic gates of *AND* and *XOR* by Artificial Neuron Network using Perceptron and Threshold elements as Neuron output functions is described in [4]. One of the various applications of Neural network includes the implementation of logic gates with the help of Neurons. An approach for the realization of such logic gates by applying McCulloch-Pitts model has been presented in [5]. Deep neural networks (*DNNs*) have shown unprecedented performance on various intelligent tasks such as image recognition, speech recognition, natural language processing, etc.

An analysis a fully parallel *RRAM* synaptic array architecture that implements the fully connected layers in a convolutional neural network with (+1, −1) weights and (+1, 0) neurons has been made in [6] and fully parallel *BNN* architecture (*P-BNN*) has been proposed in the same work. A research work aimed at implementing a model to generate binarized *MNIST* digits and experimentally compare the performance for different types of binary neurons, *GAN* objectives and network architectures is described in [7]. Neural networks are widely used in classification problems in order to segregate two classes of points with a straight line. In [8], a well-adapted learning algorithm for classification tasks called Net Lines is presented. It generates small compact feed forward neural networks with one hidden layer of binary units and binary output units. Neural network has lead not only to the development of software but also to the development of hardware such as Neural network emulators.

A detailed implementation of a fully digital neural network emulator, mainly composed of 128 serial adders that can emulate neural layers of any size, only depending on the amount of *RAM* is shown in [9]. It can be viewed as a 128-processor specialized *SIMD* machine. The embedded computer vision systems which emulate the human vision are used in wide range of applications. These applications require high accuracy, low power, and high performance where the neural networks come into use. A case study in [10], proposes a neuron pruning technique which eliminates almost part of the weight memory. In that case, since the weight memory is realized by an on-chip memory on the *FPGA*, it achieves a high-speed memory access. Similar examples for research works based on *FPGAs* and neuron pruning technique are [11] and [12].

Considering the wide applicability of Artificial Neural Network as a tool to solve most common learning problems around us [13–18], this paper presents the learning principle of artificial neurons in using them for solving Boolean *AND* and *OR* operations. The details of the design and implementation of the *AND* and *OR* neurons with and without learning are explained forth coming sections.

# 4 Binary Neurons Without Learning

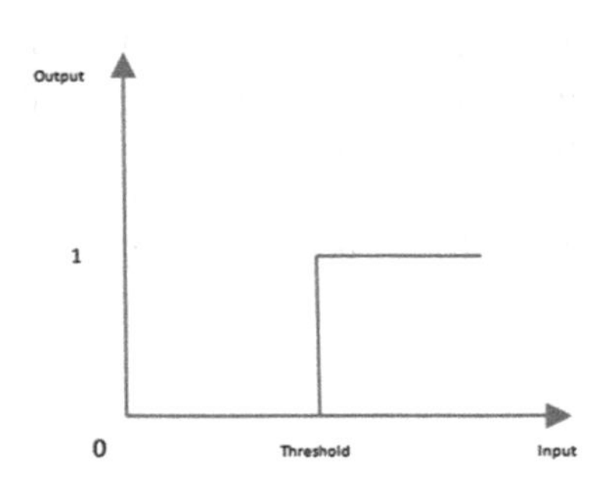


**Fig. 2.** Binary threshold function

A binary neuron which is trained to execute the basic Boolean functions such as Bitwise *AND*, *OR*, *NOR* and *NAND* is implemented with the help of Binary Threshold Functions (Shown in Fig. 2). Following the linear way of manual classification by providing the appropriate weight and the corresponding linear constants (depending upon the input), an equation is constructed to execute the task of performing the bitwise operations. The equation to determine the activation value of the neuron is

$$y = ((W_0 * X_0) + (W_1 * X_1) + (W_2 * X_2)) \quad (1)$$

Assuming weights as (through an Ad-hoc method) $W_0 = -1.5$, $W_1 = 1$ and $W_2 = 1$, the code snippet for realizing the binary *AND* operation with the *AND*-neuron in shown in Fig. 3. The performance of the *AND* neuron is verified with the truth table for all possible inputs of $X_1$ and $X_2$ and found it is correct. The verified results are presented in Table 1. Assuming weights as (through an Ad-hoc method) $W_0 = -0.5$, $W_1 = 1$ and $W_2 = 1$, the code snippet for realizing the binary *OR* operation with the *OR*-neuron in shown in Fig. 4. The performance of the *OR*-neuron is verified with the truth table for all possible inputs of $X_1$ and $X_2$ and found it is correct. The verified results are presented in Table 2.

```
def evaland(x0, x1, x2, w0, w1, w2):
  y=((w0*x0)+(w1*x1)+(w2*x2))
  if y>=0:
    z=1
  else:
    z=0
  return z
x1=int(input("Enter value of x1 \n"))
x2=int(input("Enter value of x2 \n"))
w0=-1.5; x0=1;w1=1;w2=1
ans=evaland(x0, x1, x2, w0, w1, w2)
print("Output : ", ans)
```

**Fig. 3.** Code Snippet for Binary AND neuron.

```
def evalor(x0, x1, x2, w0, w1, w2):
  y=((w0*x0)+(w1*x1)+(w2*x2))
  if y>=0:
    z=1
  else:
    z=0
  return z
x1=int(input("Enter value of x1 \n"))
x2=int(input("Enter value of x2 \n"))
w0=-0.5;x0=1;w1=1;w2=1
ans=evalor(x0, x1, x2, w0, w1, w2)
print("Output : ",ans)
```

**Fig. 4.** Code Snippet for Binary OR neuron.

**Table 1.** Verification of the results for *AND*-neuron

| $X_O$ | $X_1$ | $X_2$ | $W_O$ | $W_1$ | $W_2$ | $f$ |
|---|---|---|---|---|---|---|
| 1 | 0 | 0 | −1.5 | 1 | 1 | 0 |
| 1 | 0 | 1 | −1.5 | 1 | 1 | 0 |
| 1 | 1 | 0 | −1.5 | 1 | 1 | 0 |
| 1 | 1 | 1 | −1.5 | 1 | 1 | 1 |

**Table 2.** Verification of the results for *OR*-neuron

| $X_O$ | $X_1$ | $X_2$ | $W_O$ | $W_1$ | $W_2$ | $f$ |
|---|---|---|---|---|---|---|
| 1 | 0 | 0 | −0.5 | 1 | 1 | 0 |
| 1 | 0 | 1 | −0.5 | 1 | 1 | 1 |
| 1 | 1 | 0 | −0.5 | 1 | 1 | 1 |
| 1 | 1 | 1 | −0.5 | 1 | 1 | 1 |

## 5 Binary Neurons with Learning

A distinguishing is made to segregate the expected inputs to the expected outputs. Falling into two categories which correspond to the values where the outputs would be ‘0’ and ‘1’ respectively. On segregation, a learning algorithm is applied which calculates the cost of error at each iteration and correspondingly minimizes the error from the original requirement. The learning rate coefficient is a factor by which the error of the function is minimized and hence the neuron is trained to perform the particular task. The code snippet for *AND*-neuron with learning is show in Fig. 5.

```
import random
def eval(x0, x1, x2, w0, w1, w2):
    y=((w0*x0)+(w1*x1)+(w2*x2))
    if y>=0:
        z=1
    else:
        z=0
    return z
pset=[(1,1)]
nset=[(0,0),(0,1),(1,0)]
actual=-1
x0=1
x1=int(input("Enter value of x1 \n"))
x2=int(input("Enter value of x2 \n"))
y1=(x1, x2)
if y1 in pset:
    actual=1
else:
    actual=0
w0=11
w1=-8
w2=7
print("Before learning weights: ", w0, w1, w2)
guess = eval(x0, x1, x2, w0, w1, w2)
error = actual - guess
while True:
  if error == 0:
    break
  else:
    while True:
      if y1 in pset and guess <= 0:
        w0=w0+x0
        w1=w1+x1
        w2=w2+x2
      elif y1 in nset and guess > 0:
        w0=w0-x0
        w1=w1-x1
        w2=w2-x2
      guess = eval(x0, x1, x2, w0, w1, w2)
      error = actual - guess
      if error == 0:
        break
print("After learning W0, W1, W2 are: ", w0,
w1, w2)
ans= eval(x0, x1, x2, w0, w1, w2)
print("Output after learning : ", ans)
```

**Fig. 5.** Code Snippet for Binary *AND* neuron with learning.

The adaptation of weights using the learning algorithm is done using the Eq. (2).

$$W_{i+1} = W_i \pm \eta X_i \tag{2}$$

where $i$ is the iteration index, and the $\eta$ is the learning rate.

An illustrative execution of the *AND* neuron with perceptron learning algorithm is shown in Table 3, and the code snippet is shown in Fig. 5. The initial values assigned for the weights are $W_0 = 11$, $W_1 = -8$, $W_2 = 7$. The table shows the weight adaptation for the input values of $X_1 = 1$ and $X_2 = 0$, as a sample. It is seen from the results that the *AND*-neuron is able to learn the required weight values in three iterations itself. The adapted weight value for the correct operation of Boolean *AND* operation, in this case, are $W_0 = 9$, $W_1 = -10$ and $W_2 = 7$.

**Table 3.** Verification of the results for *AND*-neuron with learning

| Iteration # | $X_0$ | $X_1$ | $X_2$ | $W_0$ | $W_1$ | $W_2$ | Actual output | Desired output | Error | Error % |
|---|---|---|---|---|---|---|---|---|---|---|
| 0 | 1 | 1 | 0 | 11 | −8 | 7 | 1 | 0 | −1 | 50 |
| 1 | 1 | 1 | 0 | 10 | −9 | 7 | 1 | 0 | −1 | 50 |
| 2 | 1 | 1 | 0 | 9 | −10 | 7 | 0 | 0 | 0 | 0 |

An illustrative execution of the *OR* neuron with perceptron learning algorithm is shown in Table 4, the code snippet is shown in Fig. 6. The code explains that the learning happens iteratively, untill the error becomes zero. The initial values assigned for the weights are $W_0 = 1$, $W_1 = -8$, $W_2 = 9$. The table shows the weight adaptation for the input values of $X_1 = 1$ and $X_2 = 0$, as a sample. It is seen from the results that the *OR*-neuron is able to learn the required weight values in five iterations. The adapted weight value for the correct operation of Boolean *OR* operation, in this case, are $W_0 = 5$, $W_1 = -4$ and $W_2 = 9$.

```
import random
def eval(x0, x1, x2, w0, w1, w2):
  y= ((w0*x0)+(w1*x1)+(w2*x2))
  if y>=0:
    z=1
  else:
    z=0
  return z
pset=[(1, 1),(0,1),(1,0)]
nset=[(0, 0)]
actual =-1
x0=1
x1=int (input("Enter value of x1 \n"))
x2=int (input("Enter value of x2 \n"))
y1=(x1, x2)
if y1 in pset:
  actual=1
else:
  actual=0
w0=random.randint(-1, 1)
w1=random.randint(-1, 1)
w2=random.randint(-1, 1)
print("Before learning weights: ", w0, w1, w2)
while True:
 if error == 0:
   break
 else:
   while True:
     if y1 in pset and guess <= 0:
       w0=w0+x0
       w1=w1+x1
       w2=w2+x2
     elif y1 in nset and guess > 0:
       w0=w0-x0
       w1=w1-x1
       w2=w2-x2
     guess = eval(x0, x1, x2, w0, w1, w2)
     error = actual - guess
     if error == 0:
       break
print("After learning W0, W1, W2 are: ", w0,
w1, w2)
ans= eval(x0, x1, x2, w0, w1, w2)
print("Output after learning : ",ans)
```

**Fig. 6.** Code Snippet for Binary *OR* neuron with learning.

**Table 4.** Verification of the results for *OR*-neuron with learning

| Iteration # | $X_0$ | $X_1$ | $X_2$ | $W_0$ | $W_1$ | $W_2$ | Actual output | Desired output | Error | Error % |
|---|---|---|---|---|---|---|---|---|---|---|
| 0 | 1 | 1 | 0 | 1 | −8 | 9 | 0 | 1 | 1 | 50 |
| 1 | 1 | 1 | 0 | 2 | −7 | 9 | 0 | 1 | 1 | 50 |
| 2 | 1 | 1 | 0 | 3 | −6 | 9 | 0 | 1 | 1 | 50 |
| 3 | 1 | 1 | 0 | 4 | −5 | 9 | 0 | 1 | 1 | 50 |
| 4 | 1 | 1 | 0 | 5 | −4 | 9 | 1 | 1 | 0 | 0 |

## 6 Conclusions

This paper presented the details of simulating artificial neurons to implementing Boolean functions. The implementation details for *OR* and *AND* functions are shown in this paper. This paper demonstrated the use of perceptron learning algorithm for artificial neuron to adapt their weight vector to perform the required task. Thus the paper showed the results obtained with different simulation setups: *OR* without learning, *AND* without learning, *OR* with learning and *AND* with learning. The details presented in this paper reveals that the artificial neuron can be simulated for all type of Boolean operations.

The work presented in this paper can be extended further to include simulation of all the Boolean functions vix *NAND*, *XOR* and *XNOR* with and without learning.

## References

1. Mandziuk, J., Macukow, B.: A neural network performing boolean logic operations. Opt. Mem. Neural Netw. **2**(1), 17–35 (1993)
2. Dubois, D.M.: Hyperincursive McCulloch and Pitts neurons for designing a computing flip-flop memory. In: Computing Anticipatory Systems: CASYS 1998 - Second International Conference, pp. 3–21 (1999)
3. Kohut, R., Steinbach, B.: Boolean neural networks. In: Proceedings Kohut 2004 (2004)
4. Ele, S.I., Adesola, W.A.: Artificial neuron network implementation of boolean logic gates by perceptron and threshold element as neuron output function. Int. J. Sci. Res. **4**(9), 637–641 (2013)
5. Srinivas Raju, J.S., Kumar, S., Sai Sneha, L.V.S.S.: Realization of logic gates using MccullochPitts neuron model. Int. J. Eng. Trends Technol. **45**(2), 52–56 (2017)
6. Sun, X., Peng, X., Chen, P.-Y., Liu, R., Seo, J., Yu, S.: Fully parallel RRAM synaptic array for implementing binary neural network with (+1, −1) Weights and (+1, 0) neurons. In: 23rd Asia and South Pacific Design Automation Conference (ASP-DAC), pp. 574–579 (2018)
7. Dong, H.-W., Yang, Y.-H.: Training generative adversarial networks with binary neurons by end-to-end back propagation. Comput. Res. Repository (2018). arXiv:1810.04714
8. Torres-Moreno, J.-M., Gordon, M.B.: Adaptive learning with binary neurons. arXiv:0904.4587 (2009)
9. Skubiszewski, M.: A hardware emulator for binary neural networks. In: International Neural Network Conference, Springer, Dordrecht (1990)

10. Fujii, T., Sato, S., Nakahara, H.: A threshold neuron pruning for a binarized deep neural network on an FPGA. IEICE Trans. Inf. Syst. **101**(2), 376–386 (2018)
11. Kohut, R., Steinbach, B.: The structure of boolean neuron for the optimal mapping to FPGAs. In: Proceedings of the VIII-th International Conference CADSM 2005, pp. 469–473 (2005)
12. Training with states of matter search algorithm enables neuron model pruning, Technical report - Kanazawa University, November 2018
13. Bindu, K.R., Aakash, C., Orlando, B., Parameswaran, L.: An algorithm for text prediction using neural networks. In: Lecture Notes in Computational Vision and Biomechanics, vol, 28, pp. 186–192 (2018)
14. Kumar, P.N., Seshadri, G.R., Hariharan, A., Mohandas, V.P., Balasubramanian, P.: Financial market prediction using feed forward neural network. In: Communications in Computer and Information Science, pp. 77–84 (2011)
15. Suresh, A., Harish, K.V., Radhika, N.: Particle Swarm Optimization over back propagation neural network for length of stay prediction. Procedia Comput. Sci. **46**, 268–275 (2015)
16. Senthil Kumar T., Sivanandam, S.N.: An improved approach for detecting car in video using neural network model. J. Comput. Sci. **10**, 1759–1768 (2012)
17. Padmavathi, S., Saipreethy, M.S., Valliammai, V.: Indian sign language character recognition using neural networks. Int. J. Comput. Appl. **1**, 40–45 (2013)
18. Anil, R., Manjusha, K., Sachin Kumar, S., Soman, K.P.: Convolutional neural networks for the recognition of Malayalam characters. In: Advances in Intelligent Systems and Computing, vol. 328, pp. 493–500 (2015)

# Image De-noising Method Using Median Type Filter, Fuzzy Logic and Genetic Algorithm

S. R. Sannasi Chakravarthy(✉) and Harikumar Rajaguru

Department of ECE, Bannari Amman Institute of Technology, Sathyamangalam 638401, Tamil Nadu, India
elektroniqz@gmail.com, harikumarrajaguru@gmail.com

**Abstract.** The goal of de-noising the medical images is to get rid of the distortions occurred in the noisy medical images. A new methodology is proposed to overcome the impulse noise affected mammogram images by using hybrid filter (HF), fuzzy logic (FL) and genetic algorithm (GA). The above said method is implemented in three steps: The primary step includes denoising of noisy mammogram images using median filter and adaptive fuzzy median filter respectively. The intermediate step intends to compute the difference vector using the above two filters and it is then given to a fuzzy logic-based system. The system utilizes triangular membership function to generate the fuzzy rules from the computed difference vector value. The last step makes use of genetic algorithm to select the optimal rule. Peak signal to noise ratio (PSNR) value is needed to be found for each population. For obtaining the best fitness value, the new population is formed repeatedly with the help of genetic operator. The performance of the method is measured by calculating the PSNR value. The proposed implementation is tested over mammogram medical images taken from Mammogram Image Analysis Society (MIAS) database. The experimental results are compared with different exiting methods.

**Keywords:** Mammogram · Impulse noise · Median · Fuzzy · Genetic algorithm

## 1 Introduction

For any type of imaging modality, the manual or automatic inspection is required for its final classification or decision making. But if any medical images are affected by noise, then the inspection is of no use which leads to increased risk of human lives. Generally, the noise is uninvited information included in the image and the noise existent in the modality is either additive or multiplicative [1]. The process of restoring medical images assumes that the degradation model is a known one or it can be calculated. During the process, we can consider that the noise is not dependent on spatial coordinates and not correlated relating to the information on an image. The noises are categorized based on the distribution of pixel values or through histogram analysis [1].

Filters are commonly employed to get rid of undesirable noise information from the original image. The paper investigates the mammogram images which are contaminated with salt and pepper (impulse) noise. The work simply involves the pre-processing of


S. Smys et al. (Eds.): ICCVBIC 2019, AISC 1108, pp. 488–495, 2020.
https://doi.org/10.1007/978-3-030-37218-7_55

medical imaginings (mammograms) which are needed for fruitful classification or appropriate diagnosis of the disease. A mammogram image is obtained as an x-ray photograph of the breast and it is used to diagnose the breast cancer among females without any signs or else symptoms [2]. The mammograms considered in the proposed work is taken from the MIAS database.

If every rules created from fuzzy logic system be used, then it put away more time for processing; thus simply the optimal rules are selected using genetic algorithm. This will lessen the execution period, upsurge the visual quality and also upturns the PSNR value. The flowchart of the system is given in Fig. 1.

## 2 Materials and Methods

### 2.1 Impulse Noise

Impulse noise is a kind of acoustic distortion that has near-instantaneous (impulse-like) unwanted sharp and sudden distortions. The common cause for this category of noise is electromagnetic interference, unwanted scratches on the recording films, explosions like gunfire and bad synchronization in digital means of recording and transportation [3]. If the transmission channel is not good, then the noise variance will distort the image signal with very high negative (minimum) or with very high positive (maximum) value. Thus the mammogram image will distorted with some black (pepper) and white (salt) pixel spots [4]. This type of noise is referred as impulse noise. The noise model for any image is mathematically defined as

$$S_{i,j} = \begin{cases} X_{i,j} & \textit{with probability} \quad p \\ Y_{i,j} & \textit{with probability} \quad 1-p \end{cases} \tag{1}$$

where $X_{i,j}$ specifies the pixel values in original image (noiseless) and $Y_{i,j}$ denotes the distorted pixel values changed on behalf of original noiseless pixel values in a mammogram image. The degree at which the mammogram image is distorted is specified by the parameter $p$. The fixed-valued model is embraced of ruined pixels in which the pixel values are ruined with either extreme (salt) or minimum (pepper) of the tolerable pixel range.

### 2.2 Median and Adaptive Fuzzy Median Filter

The median filter is a kind of non-linear technique, commonly used for digital filtering to remove noise from any type of images. Such noise removal is a popular pre-processing stage to increase the results of further processing (like edge detection and classification on an image) [5]. Since the median filter preserves the edges during noise removal, it is widely employed in digital and medical image processing.

The filtering technique assumes that all pixels in the image and looks at its adjacent pixels to check whether it is representative of its environment or not. The technique calculates the middle value among the pixels and simply replaces it using the computed median of those values [6]. The median value is determined by first arranging every pixel based on their numerical values from the surrounding neighborhood. The particular pixel being considered is then replaced with the middle (median) pixel value. When the neighborhood being taken have an even quantity of pixels in the taken image, the average between the two center values of the pixel is considered [7].

The adaptive fuzzy median filtering method is intended to overcome the limitation of the above said median filtering technique. Since the neighborhood size of a standard median filter is fixed, its performance gets lessen whenever there is an increase in the density of spatial noise [8]. However, in the adaptive fuzzy median type of filter, there is an adaptation (change) in the neighborhood size during the removal process. This change or variation depends on the median value of pixels present in the current window. The size of this window is more whenever the median value is an impulse [9].

If any image is getting affected highly by noise, then the entire pixels present in the filtering window are same as the extreme values. This will increase the size of the window with a maximum predefined size ($W_{max}$) for noise removal processing, after that the replacement of $I(x, y)$ by the calculated median value (determined using the filtering window). While determining the median value, if the pixel values are almost same as the extreme findings in filtering window, then the calculation of pixel values present in filtering window is done irrespective of the extreme values used to determine the median one [8]. At this stage, the fuzzy corrupted value is calculated and then used to check whether or not the filtering window is highly affected by impulse-like noise. The median filters are also used for smoothening (blurring) the input images. Consequently the efficiency in suppression of noise is done at the cost of blurred or distorted image characteristics [10]. Thus an appropriate way to evade this problem is to integrate necessary decision-making control in the filtering process that leads to adaptive fuzzy median filter.

### 2.3 Fuzzy Logic and Genetic Algorithm

The word *fuzzy* denotes the one which is not clear or vague. Generally in the practical world, we came across many situations that are not predicted as either true or false [11]. The fuzzy logic gives a very valuable and flexible reasoning in these situations. The system generates rules for the available inputs based on the type of membership function used. Fuzzification and Defuzzification denote the conversion of crisp into fuzzy values and vice-versa [12].

The genetic algorithm is a one used for optimizing either constrained or unconstrained problems. This optimization depends on the selection of natural ones, this implies the driving of biological evolution [13]. The algorithm changes the population of respective solutions periodically. At each stage, the technique randomly chooses the individuals from the present population as parents and thereby utilizes these parents to generate the children for subsequent generation. This process of generation "evolves" for finding an optimal solution. The algorithm is used widely for the applications having discontinuous objective function, non-linear, stochastic or non differentiable [14]. It can also solve the problems of mixed-integer programming with fewer things are not allowed to be integer.

## 3 Proposed Method

### 3.1 Calculation of Impulse Noise Difference Using Hybrid Filter

The paper includes the usage of two separate filters: median and adaptive fuzzy median type in the first stage for the better filtering of input medical images (mammograms). The output of median and adaptive fuzzy median type is denoted by $Md(x,y)$ and $FMd(x,y)$. The difference between the outputs of this hybrid filter is

$$d_1 = I(x,y) - Md(x,y) \tag{1}$$

$$d_2 = I(x,y) - FMd(x,y) \tag{2}$$

where $I(x,y)$ is the input mammogram to the proposed method. These $d_1$ and $d_2$ are given as input to the fuzzy logic system for the generation of fuzzy rules.

### 3.2 Fuzzy Logic for the Generation of Rules

The FL system utilizes the mamdani method along with max-min composition to get rid of unwanted impulse noise in the acquired mammograms. The process of fuzzification yields a $F_1$ score and using this score value, the decision of concluding that the detected pixels are either affected or not by impulse noise [15]. The decision is made using

$$\begin{array}{l} \textit{if } F_1 \textit{ Score} \geq T, \textit{then it is impulse noise distorted} \\ \textit{if } F_1 \textit{ Score} \leq T, \textit{then it is not distorted} \end{array} \tag{3}$$

Using the above condition, the obtained score value is assessed for the impulse noise detection in mammograms. The process of fuzzification is done by using the triangular membership functions and the membership of a component $X$ in crisp set $C$ is termed using a characteristic function as [16]

$$\mu_C(X) = \begin{cases} 1\,(True)\ \textit{if}\ X \in C \\ 0\,(False)\ \textit{otherwise} \end{cases} \tag{4}$$

### 3.3 Genetic Algorithm for Optimization

After defuzzification, the fuzzy output is taken to the genetic algorithm for the purpose of optimization of result. The generation of fuzzy rules are more for the problem and so genetic algorithm is employed for the optimization of fuzzy rules. GA gives good removal of noise even with the different noise content (large state and multimodal state space) [17]. The process of optimization depends on its structure and behaviour of chromosomes for different images. The steps of GA is given as [18]

1. The initial population $(x)$ of GA is taken as randomly selected obtained fuzzy rules.
2. Compute the fitness score of above population $(x)$

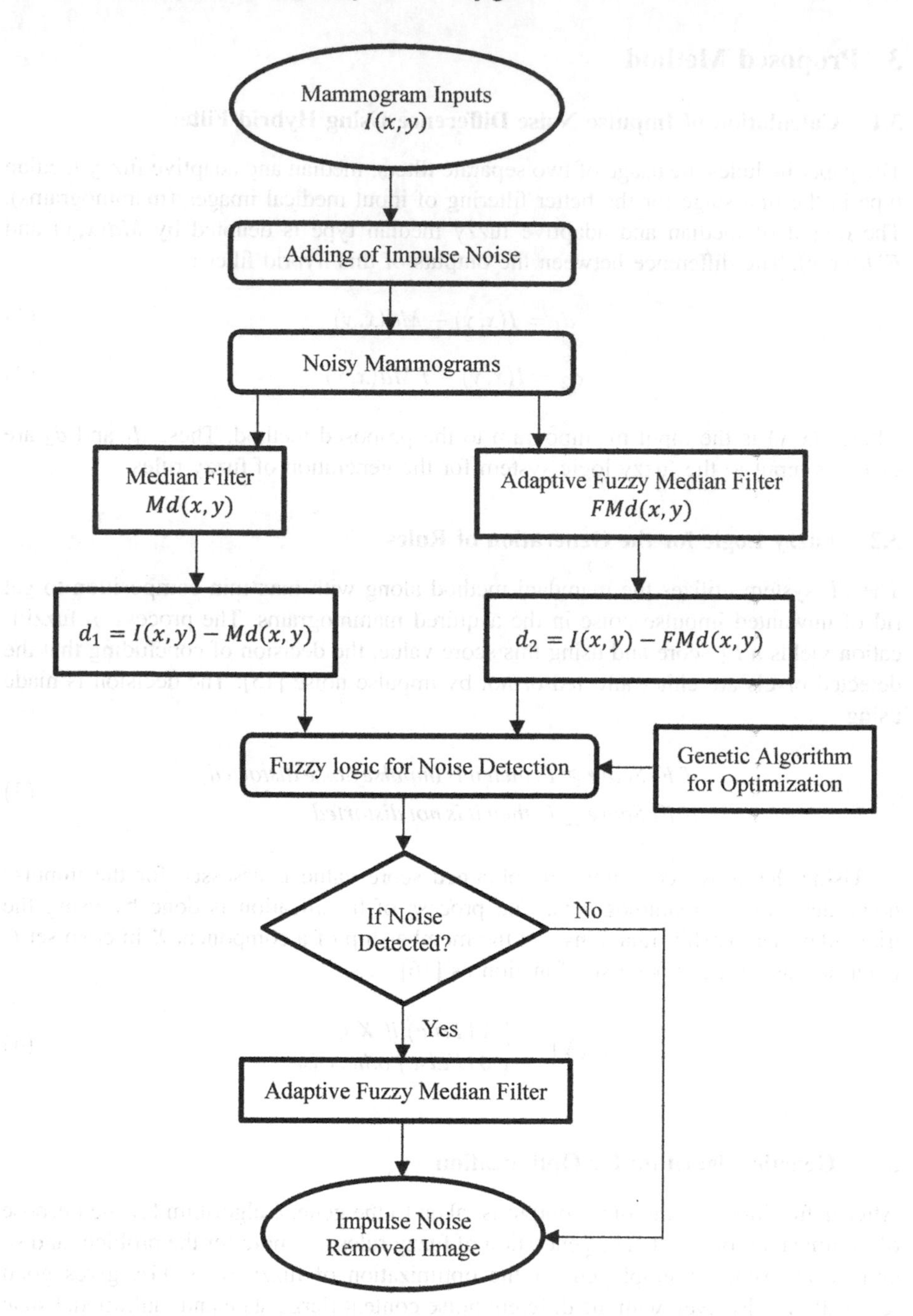


**Fig. 1.** Flowchart of proposed system

3. Repeat until
   a. Select the parent rules from $(x)$
   b. Form the population $(x+1)$ during the cross-over process on parent rules.
   c. Perform the mutation on $(x+1)$.
   d. Testing of fitness value is done for the problem of $(x+1)$.
4. The best optimized fitness value is obtained.

After the generation of initial population using above steps, determining the fitness function is performed in order to test the proposed method. The problem considers the parameter, PSNR as a fitness function [19]. The PSNR as a fitness function is given as [20]

$$PSNR = 10\,log_{10} \frac{255^2}{\frac{1}{xy}\sum_{x=1}^{M}\sum_{y=1}^{N}\left(u_{i,j} - I_{i,j}\right)} dB \tag{5}$$

where $u(i,j)$ and $I(i,j)$ is the pixel elements of the output and original input images and $M, N$ is the input size $(M \times N)$ respectively. The denominator term of above equation gives the Mean Square Error (MSE).

## 4 Experimental Results

The results are taken from the GA and it is evaluated using standard PSNR metric where its calculation is based on the Eq. (5). The results are calculated for various noise ratio as in the Table 1.

**Table 1.** PSNR values for mdb063 image at different noise density

| Methods | PSNR (dB) | | | |
|---|---|---|---|---|
| | 10% | 30% | 50% | 70% |
| Mean filter | 20.36 | 17.55 | 11.84 | 7.87 |
| Decision based filter | 29.52 | 25.51 | 22.63 | 19.94 |
| Wavelet filter | 31.53 | 29.01 | 27.52 | 23.48 |
| Proposed method | 58.76 | 56.94 | 56.31 | 55.46 |

Form the above table, the proposed method works well than other noise removal systems. And the results are plotted graphically as in Fig. 2.

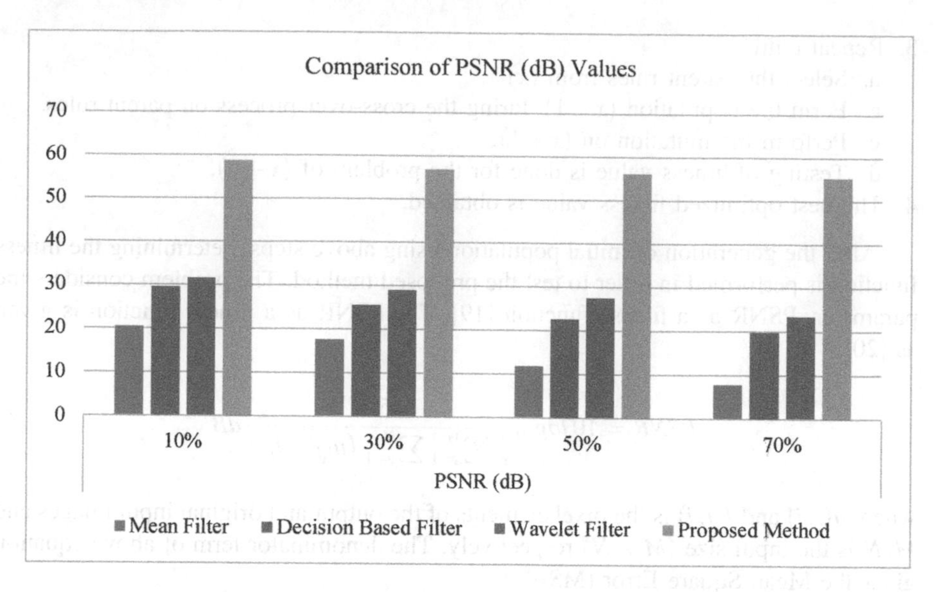


**Fig. 2.** PSNR value comparison of proposed method

## 5 Conclusion and Future Work

A methodology for the removal of impulse noise using hybrid filters is proposed. The filters used are median and adaptive fuzzy median filters. After taking the differences, fuzzy rules are generated using fuzzy logic system. This step is for the detection of impulse noise in the mammogram images. But there is a need for optimization of these generated fuzzy rules, which is done by genetic algorithm. The comparative simulations shows that efficiency of the proposed work is higher than the existing filters. The future work of this work is implementation of the proposed system for clinical images with different optimization algorithms.

**Compliance with Ethical Standards**

✓ All authors declare that there is no conflict of interest.
✓ No humans/animals involved in this research work.
✓ We have used our own data.

## References

1. Leipsic, J., LaBounty, T.M., Heilbron, B., Min, J.K., Mancini, G.J., Lin, F.Y., Earls, J.P.: Adaptive statistical iterative reconstruction: assessment of image noise and image quality in coronary CT angiography. Am. J. Roentgenol. **195**(3), 649–654 (2010)
2. Abirami, C., Harikumar, R., Chakravarthy, S.S.: Performance analysis and detection of micro calcification in digital mammograms using wavelet features. In: 2016 International Conference on Wireless Communications, Signal Processing and Networking (WiSPNET), pp. 2327–2331. IEEE (2016)

3. Liu, L., Chen, C.P., Zhou, Y., You, X.: A new weighted mean filter with a two-phase detector for removing impulse noise. Inf. Sci. **315**, 1–16 (2015)
4. Gupta, V., Chaurasia, V., Shandilya, M.: Random-valued impulse noise removal using adaptive dual threshold median filter. J. Vis. Commun. Image Represent. **26**, 296–304 (2015)
5. Chakravarthy, S.S., Subhasakthe, S.A.: Adaptive median filtering with modified BDND algorithm for the removal of high-density impulse and random noise. Int. J. Comput. Sci. Mob. Comput. **IV**(2), 202–207 (2015)
6. Gan, S., Wang, S., Chen, Y., Chen, X., Xiang, K.: Separation of simultaneous sources using a structural-oriented median filter in the flattened dimension. Comput. Geosci. **86**, 46–54 (2016)
7. Chen, Y.: Deblending using a space-varying median filter. Explor. Geophys. **46**(4), 332–341 (2015)
8. Wang, Y., Wang, J., Song, X., Han, L.: An efficient adaptive fuzzy switching weighted mean filter for salt-and-pepper noise removal. IEEE Signal Process. Lett. **23**(11), 1582–1586 (2016)
9. Chen, Y., Zhang, Y., Shu, H., Yang, J., Luo, L., Coatrieux, J.L., Feng, Q.: Structure-adaptive fuzzy estimation for random-valued impulse noise suppression. IEEE Trans. Circ. Syst. Video Technol. **28**(2), 414–427 (2016)
10. Erkan, U., Gökrem, L., Enginoğlu, S.: Different applied median filter in salt and pepper noise. Comput. Electr. Eng. **70**, 789–798 (2018)
11. Nguyen, H.T., Walker, C.L., Walker, E.A.: A First Course in Fuzzy Logic. CRC Press (2018)
12. Suganthi, L., Iniyan, S., Samuel, A.A.: Applications of fuzzy logic in renewable energy systems–a review. Renew. Sustain. Energy Rev. **48**, 585–607 (2015)
13. Metawa, N., Hassan, M.K., Elhoseny, M.: Genetic algorithm based model for optimizing bank lending decisions. Expert Syst. Appl. **80**, 75–82 (2017)
14. Gai, K., Qiu, M., Zhao, H.: Cost-aware multimedia data allocation for heterogeneous memory using genetic algorithm in cloud computing. IEEE Trans. Cloud Comput. **1**, 1–1 (2016)
15. Zadeh, L.A.: Fuzzy logic—a personal perspective. Fuzzy Sets Syst. **281**, 4–20 (2015)
16. Nayak, P., Devulapalli, A.: A fuzzy logic-based clustering algorithm for WSN to extend the network lifetime. IEEE Sens. J. **16**(1), 137–144 (2015)
17. Li, X., Gao, L.: An effective hybrid genetic algorithm and tabu search for flexible job shop scheduling problem. Int. J. Prod. Econ. **174**, 93–110 (2016)
18. Ding, Y., Fu, X.: Kernel-based fuzzy c-means clustering algorithm based on genetic algorithm. Neurocomputing **188**, 233–238 (2016)
19. Cheng, J.H., Sun, D.W., Pu, H.: Combining the genetic algorithm and successive projection algorithm for the selection of feature wavelengths to evaluate exudative characteristics in frozen–thawed fish muscle. Food Chem. **197**, 855–863 (2016)
20. Fardo, F.A., Conforto, V.H., de Oliveira, F.C., Rodrigues, P.S.: A formal evaluation of PSNR as quality measurement parameter for image segmentation algorithms. arXiv preprint arXiv:1605.07116 (2016)

# Autism Spectrum Disorder Prediction Using Machine Learning Algorithms

Shanthi Selvaraj(✉), Poonkodi Palanisamy, Summia Parveen, and Monisha

Department of Computer Science and Engineering, SNS College of Technology, Saravanampatti, Coimbatore 641035, Tamilnadu, India
psshanthiselvaraj@gmail.com, poonkodi.cse@gmail.com, monish.ragu@gmail.com, summiaparveen@yahoo.in

**Abstract.** The objective of the research is to foresee Autism Spectrum Disorder (ASD) in toddlers with the help of machine learning algorithms. Of late, machine learning algorithms play vital role to improve diagnostic timing and accuracy. This research work precisely compares and highlights the effectiveness of the feature selection algorithms viz. Chi Square, Recursive Feature Elimination (RFE), Correlation Feature Selection (CFS) Subset Evaluation, Information Gain, Bagged Tree Feature Selector and k Nearest Neighbor (kNN), and to improve the efficiency of Random Tree classification algorithm while modelling ASD prediction in toddlers. Analysis results uncover that the Random Tree dependent on highlights chosen by Extra Tree calculation beat the individual methodologies. The outcomes have been assessed utilizing the execution estimates, for example, Accuracy, Recall and Precision. We present the results and identify the attributes that contributed most in differentiating ASD in toddlers as per machine learning model used in this study.

**Keywords:** Machine learning · Random Tree · Accuracy · Autism Spectrum Disorder · Feature selection · Classification

## 1 Introduction

Communication and behaviour is affected by a neuro developmental disorder termed as Autism Spectrum Disorder (ASD) [1]. Primary identification can significantly decrease ASD but, waiting time for an ASD diagnosis is long [2]. The rise in the amount of ASD affected patients across the globe reveals a serious requirement for implementation of easily executable and efficient ASD prediction models [3–6]. Currently, very inadequate autism datasets are accessible and several of them are genetic in nature. Hence, a novel dataset associated with autism screening of toddlers that contained behavioural features (Q-Chat-10) along with the characteristics of other individual's that is effective in detection of ASD cases from behaviour control is collected [3–6].

Clinical information mining [7] is the utilization of information mining methods to clinical information. The classification in healthcare is a dynamic area that provides a deeper insight and clearer understanding of causal factors of diseases.


S. Smys et al. (Eds.): ICCVBIC 2019, AISC 1108, pp. 496–503, 2020.
https://doi.org/10.1007/978-3-030-37218-7_56

In machine learning classification [8, 9] refers to a factual and numerical strategy for predicting a given information into one of a given number of class values. Feature Selection [8] is a mathematical function that picks a subset of significant features from the existing features set in order to improve the classifier performance and time and space complexity.

In this paper, we have used the ASD Toddlers Dataset to evaluate the performance of six feature relevance algorithms viz. Chi Square, Recursive Feature Elimination (RFE), Correlation Feature Selection (CFS) Subset Evaluation, Information Gain, Bagged Tree Feature Selector and k Nearest Neighbor (kNN), and Random Tree classification algorithm. Existing research in the field of machine learning and ASD research is succinctly presented in the following paragraphs.

Recently machine learning methods are widely used to improve the diagnosis process of ASD [10–12]. [13] used different machine learning algorithms like Tree models (random forest, decision tree) and Neural Network and Tree models in order to reduce the ASD diagnostic process time duration [14]. Piloted an experimental review for comparison of six major machine learning algorithms viz. Categorical Lasso, Support Vector Machine, Random Forest, Decision Tree, Logistic Regression, and Linear Discriminant Analysis, in order to choose the best fitting model for ASD and ADHD.

[15] studied the common issues associated to psychiatric disorders inclusive of external validity, small sample sizes, and machine learning algorithmic trials without a strong emphasis on autism. The authors suggested that one of these classifiers, used individually or in combination with the previously generated behavioural classifiers clearly focused on discriminating autism from non-autism category that is encompassing, could behave as a valuable triage tool and pre-clinical screening to evaluate the risk of autism and ADHD [14].

[16] used FURIA (Fuzzy Unordered Rule Induction Algorithm) on ASD dataset for forecasting autistic symptoms of children aged between 4-11 Years. The results exposed that FURIA fuzzy rules were able to identify ASD behaviours with 91.35% accuracy of classification and 91.40% sensitivity rate; these results were better than other Rule Induction and Greedy techniques.

[17] and [18] applied Naïve Bayes, SOM, K-Means, etc. on a set of 100 instances ASD dataset collected using CARS diagnostic tool [19]. Used Random Forest algorithm to predict ASD among 8-year old children in multiple US sites. The results revealed that the machine learning approach predicted ASD with 86.5% accuracy. The above appraisals have concentrated on the comprehensive domain of mining clinical data, the trials encountered, and the results inferred from analysis of clinical data.

## 2 Methodology

The proposed methodology for ASD prediction in toddlers is portrayed in Fig. 1. In this research, ASD patient data collected using Quantitative Checklist for Autism in Toddlers (Q-CHAT) data provided by ASD Tests app have been used for investigation [3–6]. The dataset used to train the classifier and evaluate its performance were downloaded from Kaggle [20]. The dataset contains 1054 samples with 17 features inclusive of class variables. According to WHO report, around 1% of the population has ASD, but this study samples get around 69% Toddlers of the data with positive ASD. It is because the test parameters have only qualitative properties of ASDs.

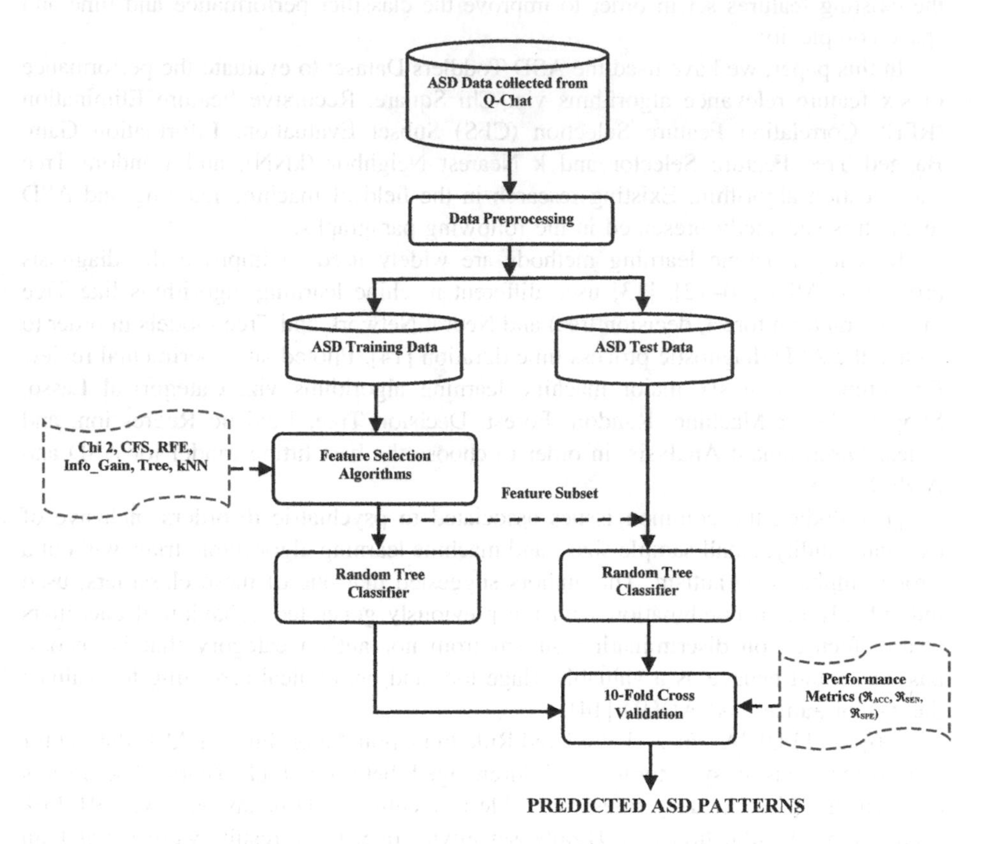


**Fig. 1.** Proposed prediction model for ASD data

In order to identify the most significant features for classification, six feature selection algorithms viz, Chi Square Co-efficient [21, 22], Information Gain [23–26] based on ranked features, Recursive Feature Elimination (RFE) [27, 28], CFS [23–26, 29] using Best First search method, Tree Classifier [30] and kNN [31] were used. We utilized the feature subsets returned by the feature selection algorithms to estimate the performance of the Random Tree Classifier [23–26], in ASD prediction.

An in-depth evaluation revealed the relevance of the features and the predictive power of the classifier in ASD prediction. The confusion matrix [8] for ASD Prediction is given in Table 1.

For a binary classification problem (ASD_Positive, ASD_Negative), the following performance parameters [4, 8] were utilized for the classifier performance evaluation [8, 23–25]: Accuracy ($\Re_{ACC}$), Sensitivity ($\Re_{SEN}$) and Specificity ($\Re_{SPE}$). The measures are represented as follows:

**Table 1.** Confusion matrix of ASD class variables

| Actual class | Predicted class | |
|---|---|---|
| | ASD_ Positive | ASD_ Negative |
| ASD_Positive | True Positive (TP) | False Negative (FN) |
| ASD_Negative | False Positive (FP) | True Negative (TN) |

$$\Re_{\mathrm{ACC}} = Accuracy(\%) = \frac{|TP + TN|}{|TP + TN + FP + FN|} \tag{1}$$

$$\Re_{\mathrm{SEN}} = Sensitivity\,(\%) = \frac{|TP|}{|TP + FN|} \tag{2}$$

$$\Re_{\mathrm{SPE}} = Specificity\,(\%) = \frac{|TN|}{|TN + FP|} \tag{3}$$

The detailed description of obtained results is discussed in the subsequent sections.

## 3 Results and Discussion

The proposed experimentations were carried out on Python Scikit machine learning tool [32] and executed in a PC with Intel Core i5 processor with 1.8 GHz speed and 6 GB of RAM. The objective of this research work is to improve the performance of the classification algorithm using effective feature selection algorithms for ASD prediction in toddlers. The results are discussed in the subsequent sub sections.

### 3.1 ASD Toddlers Dataset

The analysis was carried out using ASD Toddlers dataset obtained from Kaggle [Kaggle]. The details of the datasets used for the study are described in Sect. 3.

### 3.2 Classification Accuracy Without Feature Selection

Random was used to predict ASD positive in toddler's dataset. It used the entire feature set of the dataset. The accuracy of the Random Tree algorithms without feature selection are listed in Table 2.

With all the predictive features, Random Tree classifier predicts ASD in Toddlers with 95.1% & 94.3% accuracies using training and testing datasets respectively.

### 3.3 Significant Features Selected by the Feature Selection Algorithms

The efficiency of six feature selection algorithms were investigated to identify the most relevant features for ASD prediction in toddlers. All the algorithms used default parameters. The Chi-Square algorithm selected the features using the Chi Square

**Table 2.** Performance of classification algorithm with all features

| Classifier | Predictive performance measures in % | | | | | |
|---|---|---|---|---|---|---|
| | Training set (790 samples) | | | Test set (264 samples) | | |
| Random tree | Accuracy ($\Re_{ACC}$) | Sensitivity ($\Re_{SEN}$) | Specificity ($\Re_{SPE}$) | Accuracy ($\Re_{ACC}$) | Sensitivity ($\Re_{SEN}$) | Specificity ($\Re_{SPE}$) |
| | 95.1 | 89.3 | 97.6 | 94.3 | 93.9 | 94.5 |

Significance score and the bagged tree algorithm fits a number of randomized decision trees to rank the features based on their significance [30]. Table 3 depicts the features selected by the feature selection algorithms on the toddler ASD dataset. The symbol '*' in the table indicates that the feature is selected as a significant feature by the corresponding feature selection algorithm. The number specified in the () denotes the number of features selected by the respective feature selection algorithm.

**Table 3.** Significant features selected by feature selection algorithms

| Feature # | Feature name | Chi Square (14) | RFE (15) | CFS (14) | Info_Gain (12) | Bagged Tree (10) | kNN (14) |
|---|---|---|---|---|---|---|---|
| $f_1$ | A1 | * | * | * | * | * | * |
| $f_2$ | A2 | * | * | * | * | * | * |
| $f_3$ | A3 | * | * | * | * | | * |
| $f_4$ | A4 | * | * | * | * | * | * |
| $f_5$ | A5 | * | * | * | * | * | * |
| $f_6$ | A6 | * | * | * | * | * | * |
| $f_7$ | A7 | * | * | * | * | * | * |
| $f_8$ | A8 | * | * | * | * | * | * |
| $f_9$ | A9 | * | * | * | * | * | * |
| $f_{10}$ | A10 | * | * | * | * | * | * |
| $f_{11}$ | Age_Mons | | | * | * | | |
| $f_{12}$ | Sex | * | * | * | | | * |
| $f_{13}$ | Ethnicity | | * | * | * | | |
| $f_{14}$ | Jaundice | * | * | * | | * | * |
| $f_{15}$ | Family_Member_With_ASD | * | * | | | | * |
| $f_{16}$ | By_Whom | * | * | | | | * |

From Table 3 it is observed that features $f_1$ to $f_{10}$ and $f_{14}$ are the most Significant features compared to all other features and plays vital role in predicting ASD in toddlers. RFE selected the maximum of 15 features and bagged tree selected the minimum of 10 features as significant features among 16 features. The optimal selection of these feature selection algorithms is substantiated by the Random Tree classification algorithm with 10 folds cross validation. The performance metrics of the

Random Tree classifier with the features selected by the feature selection algorithms are listed in Table 4.

**Table 4.** Performance of classification algorithm with significant features selected by feature selection algorithms

| Feature selection algorithms | Predictive performance measures in % | | | | | |
|---|---|---|---|---|---|---|
| | Training set (790 samples) | | | Test set (264 samples) | | |
| | Accuracy (ℜACC) | Sensitivity (ℜSEN) | Specificity (ℜSPE) | Accuracy (ℜACC) | Sensitivity (ℜSEN) | Specificity (ℜSPE) |
| Chi Square | 94.9 | 90.2 | 97.1 | 94.3 | 93.9 | 94.5 |
| RFE | 95.2 | 90.2 | 97.4 | 94.3 | 93.9 | 94.5 |
| CFS | 93.5 | 88.1 | 96 | 94.3 | 95.1 | 94 |
| Info_Gain | 95.1 | 89.8 | 97.4 | 93.9 | 92.7 | 94.5 |
| Bagged Tree | 95.7 | 90.2 | 98.2 | 97 | 93.9 | 98.4 |
| kNN | 98 | 98 | 98 | 96.2 | 96.3 | 96.2 |

The results in Table 4 reveals that Random Tree classifier gave better accuracy of 98% with training set and 97% with the testing set using kNN and Bagged Tree feature selection algorithms respectively. kNN algorithm selected the k-value based on the error rates over the iterations. Based on the error rates 15 was selected as the k-value with low error rates.

The classification accuracy of the Random Tree classifier with the significant features selected by the feature selection algorithms improved (2.9% for training set and 2.7% for test set) compared to the model without feature selection. Careful evaluation of the questionnaire and Jaundice history are the significant features which decide ASD positive in Toddlers.

## 4 Conclusion

In this paper, we predicted ASD in toddlers using Random Tree classification algorithm and feature ranking algorithms (Chi Square, RFE, CFS, Information Gain, Bagged Tree and kNN). The results revealed that, Random Tree classifier using kNN feature selection algorithm gives high accuracy of 98% for training data set and Random Tree classifier through the features obtained by bagged tree algorithm gives high accuracy of 97% for test data set. The model improved the accuracy from 95.1% to 98% for the training set and from 94.3% to 97% for the test set. We conclude that Random Tree classifier using the relevant features selected by kNN and bagged tree algorithms outperforms other feature selection algorithms while modelling ASD in toddlers. From the rules generated by the Random Tree classifier, we observe that careful observation of kids and properly answering the ASD questionnaire will help to diagnose ASD and treat the toddlers in advance. Thus, these issues need to be concentrated to decrease the ASD positive rate in toddlers.

**Compliance with Ethical Standards**

✓ All authors declare that there is no conflict of interest.

✓ No humans/animals involved in this research work.

✓ We have used our own data.

## References


1. National Institute of Mental Health. https://www.nimh.nih.gov/health/topics/autism-spectrum-disorders-asd/index.shtml
2. Penner, M., Anagnostou, E., Ungar, W.J.: Practice patterns and determinants of wait time for autism spectrum disorder diagnosis in Canada. Mol. Autism **9**, 16 (2018)
3. Thabtah, F., Kamalov, F., Rajab, K.: A new computational intelligence approach to detect autistic features for autism screening. Int. J. Med. Inform. **117**, 112–124 (2018)
4. Tabtah, F.: Autism spectrum disorder screening: machine learning adaptation and DSM-5 fulfillment. In: Proceedings of the 1st International Conference on Medical and Health Informatics, pp. 1–6, Taichung City. ACM (2017)
5. Thabtah, F.: ASDTests a mobile app for ASD, screening (2017). www.asdtests.com
6. Thabtah, F.: Machine learning in autistic spectrum disorder behavioural research: a review. Inform. Health Soc. Care J. **44**(3), 278–297 (2017)
7. Miotto, R., Wang, F., Wang, S., Jiang, X., Dudley, J.T.: Deep learning for healthcare: review, opportunities and challenges. Brief. Bioinf. **19**(6), 1236–1246 (2018)
8. Han, J., Kamber, M.: Data Mining: Concepts and Techniques. Academic Press, Cambridge (2012)
9. Breiman, L.: Classification and Regression Trees. Routledge, New York (1984)
10. Wall, D.P., Dally, R., Luyster, R., Jung, J.Y., Deluca, T.F.: Use of artificial intelligence to shorten the behavioural diagnosis of autism. PLoS One **7**(8), e43855 (2012)
11. Wall, D.P., Kosmiscki, J., Deluca, T.F., Harstad, L., Fusaro, V.A.: Use of machine learning to shorten observation-based screening and diagnosis of Autism. Transl. Psychiatry. **2**(2), e100 (2012)
12. Lopez Marcano, J.L.: Classification of ADHD and non-ADHD Using AR Models and Machine Learning Algorithms. (Doctoral dissertation), Virginia Tech (2016)
13. Duda, M., Ma, R., Haber, N., Wall, D.P.: Use of machine learning for behavioral distinction of autism and ADHD. Transl. Psychiatry. **9**(6), 732 (2016)
14. Wolfers, T., Buitelaar, J.K., Beckmann, C.F., Franke, B., Marquand, A.F.: From estimating activation locality to predicting disorder: a review of pattern recognition for neuroimaging-based psychiatric diagnostics. Neurosci. Biobehav. Rev. **57**, 328–349 (2015)
15. Al-diabat, M.: Fuzzy data mining for autism classification of children. Int. J. Adv. Comput. Sci. Appl. **9**(7), 11–17 (2018)
16. Pratap, A., Kanimozhiselvi, C.: Soft computing models for the predictive grading of childhood Autism—a comparative study. Int. J. Soft Comput. Eng. (IJSCE) **4**(3), 64–67 (2014). ISSN 2231-2307
17. Pratap, A., Kanimozhiselvi, C.S., Vijayakumar, R., Pramod, K.V.: Predictive assessment of autism using unsupervised machine learning models. Int. J. Adv. Intell. Para. **6**(2), 113–121 (2014). https://doi.org/10.1504/IJAIP.2014.062174
18. Maenner, M.J., Yeargin-Allsopp, M., Van Naarden Braun, K., Christensen, D.L., Schieve, L. A.: Development of a machine learning algorithm for the surveillance of autism spectrum disorder. PLoS One **11**(12), e0168224 (2016)
19. Autism Screening. https://www.kaggle.com/fabdelja/autism-screening-for-toddlers

20. Ramani, G., Selvaraj, S.: A novel approach to analyze a combination of I x J categorical data for estimating road accident risk. Asian J. Inf. Technol. **15**(12), 2005–2015 (2016)
21. Liu, H., Setiono, R.: Chi2: feature selection and discretization of numeric attribute. In: Proceedings of the Seventh IEEE International Conference on Tools with Artificial Intelligence, 5–8 November, p. 388 (1995)
22. Ramani, G., Selvaraj, S.: A pragmatic approach for refined feature selection for the prediction of road accident severity. Stud. Inf. Control **23**(1), 41–52 (2014)
23. Shanthi, S., Geetha Ramani, R.: Classification of vehicle collision patterns in road accidents using data mining algorithms. Int. J. Comput. Appl. **35**(12), 30–37 (2011)
24. Shanthi, S., Geetha Ramani, R.: Classification of seating position specific patterns in road traffic accident data through data mining techniques. In: Proceedings of Second International Conference on Computer Applications, vol. 5, pp. 98–104 (2012)
25. Latkowski, T., Stanislaw, O.: Data mining for feature selection in gene expression autism data. Expert Syst. Appl. **42**, 864–872 (2015)
26. Pedregosa, F., Varoquaux, G., Gramfort, A., Michel, V., Thirion, B., Grisel, O., Blondel, M., Prettenhofer, P., Weiss, R., Dubourg, V., Vanderplas, J., Passos, A., Cournapeau, D., Brucher, M., Perrot, M., Duchesnay, E.: Scikit-learn: machine learning in Python. J. Mach. Learn. Res. **12**, 2825–2830 (2011)
27. Hoeft, F., Walter, E., Lightbody, A.A., Hazlett, H.C., Chang, C., Piven, J., Reiss, A.L.: Neuroanatomical differences in toddler boys with fragile X syndrome and idiopathic autism. Arch. Gen. Psychiatry **68**(3), 295–305 (2011)
28. Price, T., Wee, C.Y., Gao, W., Shen, D.: Multiple-network classification of childhood autism using functional connectivity dynamics. In: Medical Image Computing and Computer-Assisted Intervention – MICCAI 2014. LNCS, vol 8675 (2014)
29. Geurts, P., Ernst, D., Wehenkel, L.: Extremely randomized trees. Mach. Learn. **63**(1), 3–42 (2006)
30. Altay, O. Ulas, M.: Prediction of the autism spectrum disorder diagnosis with linear discriminant analysis classifier and K-nearest neighbor in children. In: 6th International Symposium on Digital Forensic and Security (ISDFS), pp. 1–4, Antalya (2018)
31. Machine Learning in Python. https://scikit-learn.org/stable/
32. Kotsiantis, S.B.: Supervised machine learning: a review of classification techniques. Informatica **31**, 249–268 (2007)

# Challenges in Physiological Signal Extraction from Cognitive Radio Wireless Body Area Networks for Emotion Recognition

A. Roshini(✉) and K. V. D. Kiran

Department of Computer Science Engineering,
Koneru Lakshmaiah Education Foundation, Vaddeswaram,
Guntur 522502, Andhra Pradesh, India
{roshinicse22,kiran_cse}@kluniversity.in

**Abstract.** Advancement in Wireless Technology has taken its track to greater level in communication, where Wireless Sensor Networks has its own identity in gathering Event-driven environmental parameters to rule out natural disasters. Transition has taken these networks from monotonous applications to diverse areas of research in healthcare and medical fields for continuous supervision of ailed patients for early recovery and to provide pertinent treatment. Such networks are called as Wireless Body Area Networks (WBAN) where the sensor nodes either implanted or remain wearable. Proximate technology following WBAN is appending the Cognitive Radio (CR) perception which overcomes the spectral issues to increase the performance and Life time of the Networks. This study gives a knowledge on the difference between the traditional WBAN and CR-WBAN in extracting the physiological signals from human body. Challenges owing from data acquisition, opportunistic spectral access and Multiple channel Utilization for effective signal processing in emotion classification and other medical benefits are focussed. A Centralized approach in a CR environment for body signal extraction, is covered and its challenges are presented. Finally, this paper details on the uncovered research challenges that are yet to be addressed with respect to Physiological signal extraction efficiently for end user benefit.

**Keywords:** Body area networks · Cognitive radio networks · Spectral sensing

## 1 Introduction

Multiple advances in Wireless Technologies has resulted in gathering environmental parameters with respect to event driven activities. Beginning from wireless technologies like Electromagnetic waves to recent 5G communication every system has a challenge of providing efficient data transfer between end to end systems covering all the QoS (Quality of Service) parameters. Essentials of assessing Human Emotions towards diverged Psychophysiological changes is now an emerging field which is still infancy with respect to research, but its acceleration is high with respect to Human-computer interaction.


S. Smys et al. (Eds.): ICCVBIC 2019, AISC 1108, pp. 504–513, 2020.
https://doi.org/10.1007/978-3-030-37218-7_57

Internet of Things (IOT) has paved way through Sensors to capture the societal abnormalities in order to prevent catastrophe due to spatial and temporal dependencies [1]. Alike to event driven changes, human Psychophysiological data can be sensed, gathered and forwarded for early detection of Physical Imbalance leading to mental disorders. Emotional influence takes an important place in today's life, since people are exposed to different stress in their life maybe because of family, workplace or due to any other reasons ranging from infant to adults.

Wireless Sensor Networks (WSN's) has its own identity in gathering physiological data acquisition by using wearable or on wearable sensors for processing physiological monitoring. Context aware sensors are incorporated to explore the vital changes in recording the Electroencephalogram (EEG), electrocardiogram (ECG), skin conductance, electromyogram (EMG), Heart rate variability (HRV), Respiration rate (RR), temperatur (T) and through some of the behavioural modalities like walking, talking and gestures [2].

Extraction of such Physiological signals is the result of the technological advancement in medical field which has coined to the next term Medical wireless body area networks (MWBANS). The application of sensors are broadly classified based on the various health care monitoring issues as Fig. 1.

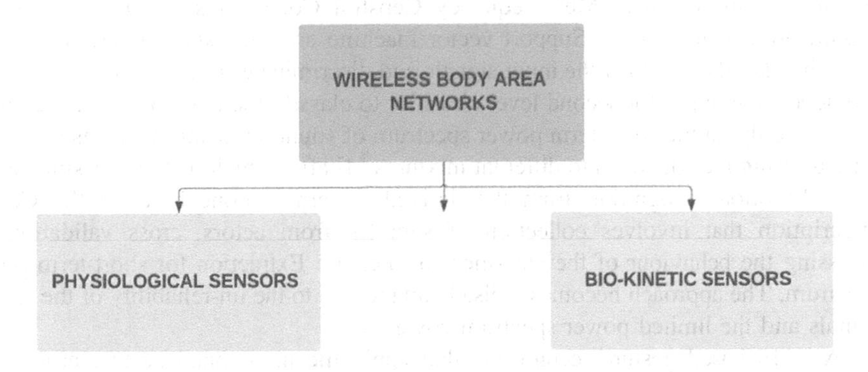


**Fig. 1.** Classification of body area networks for health monitoring

a. *Physiological Sensors*
   These sensors are enabled to capture physiological signals that are used to analyse the human stress levels and conclude the emotional status like sad, anger, anxiety, disgust and trust.
b. *Bio-Kinetic Sensors*
   The acceleration level and angular rotation resulting from the movement of human body are detected from these sensors.

Some of the issues unaddressed in MWBAN are the efficient electromagnetic spectrum usage in terms of primary and secondary users, which could be alleviated, preserving the strength of the Physiological signal irrespective of the person's mobility and scaling down the issues raised due to multipath fading. One single approach which addresses all the issues can be the implementation of Cognitive Radio-Mobile wireless Body Area Networks (CR-MWBAN) [3].

The design of CR-MWBAN mainly focus on moderating the electromagnetic propagations raised due to the patient's abnormalities. The spectrum scarcity issues that hinders the data gathering process from human body remains as an obstacle inspite of the Industrial, scientific and Medical (ISM) bands used for communication. This scarcity creates the Interference issues and it is dependent on the location of placement of sensors whether implanted or wearable.

In this survey, Sect. 2 describes an overview of various techniques used for receiving the Physiological signals will be covered. Section 3 covers the advantages of CR-WBAN over Wireless Body Area Networks (WBAN). Section 4 covers the challenges of CR-WBAN in the medical applications with respect to emotion recognition. Section 5 concludes with the various areas that are uncovered with respect to cognitive radio networks.

## 2 Literature Survey

Emotion Recognition by gathering the speech signals, with a two-tier hierarchical classifier is made where Mel Frequency Cepstral Coefficients are used for speech classification with suitable Support vector machine and Gaussian Mixer. At the first level the classifiers extract the input signals and discriminate function values which are provided as an input for second level classifier to classify the emotions. The technique focusses only on the short-term power spectrum of sound by using K-means algorithm for clustering the sounds in to different mixtures [4] after which pattern classification is done. Emotion recognition from the derived pattern is done base on the Corpus Description that involves collection of samples from actors, cross validation for assessing the behaviour of the classifier and Feature Extraction for short-term power spectrum. The approach becomes a disadvantage due to the un-reliability of the speech signals and the limited power spectrum coverage.

A WiFi based gesture recognition that apply the in-air hand gesture around the user's mobile device is used. The technique strengthens the WiFi signal strength without any training for gesture recognition. System debacle's various networking issues like signal prone to noisy data, type and detection of gesture, reducing the faulty data due to high human interference and accustom to the signal change. WiGest method uses three steps to determine the gestures and their attributes from segmentation, pattern encoding and action mapping [5]. The performance analysis is environment oriented where the system is compared with the location of the access point in a empty room, with a one layer obstacle and two layered obstacle. This distance variation however leads the sustainability of the accuracy to certain extent. The signal extraction is done only in a 2D-motion.

Frequency modulated carrier wave (FMCW) method is used to track the 3D motion tracking and localizing the systems, through WiTrack system from the radio signal reflected from the human body. System does not insist to wear the any device, rather Kinect and imaging systems are used to track the motion with the physical components of WiTrack like the Transmission Antenna and Receiving Antenna [6]. The implementation is done as three applications that involve tracking of human motion through wall, elderly fall detection in case of various activities like walking, standing and sitting. The system also averts the multipath fading of the signal by monitoring the distance between the object and WiTrack device on a periodic basis. The limitation of such system lies on the number of persons tracked at a time who fall behind the directional antennas.

Ultra-wide band and Radio frequency sensor are used to monitor the heart rate and lung motion without the influence of human body contact. The main challenges fall on the hindrances of the signal penetration issues due to the tissues and the body muscles in the body. Second challenge is the detection of impulse response of the radar sensor [7] system since it is not known well before that is used to resolve the reflection points. This approach does not make use of dynamic compressive sensing algorithm that will enable temporal correlation of signal.

Apart from extracting only the physiological signals a direct interaction of emotions and physiological signals without implanting a device to record the signals for processing emotion recognition is made [8]. The process includes EQ-radio that transmits an RF signal after which the signal is reflected off the human body carrying the heart beat signal to recognize the emotions from each individual beat. From the frequency shift of the RF signal (Breathing cycle) and ECG signal which extracts minute reflections due to the variation.

An ambulatory assessment of Nervous system activation in emotion is used, where physiological signals are extracted from the daily routine activities of human beings by using a in-field data acquisition technique. It enables to gather eve the minute change of human behaviour that varies over time. It focusses on the individual's current and recent state behaviours to collect multiple ratings of a behaviour that can be event-based, [9] interactive or continuous.

All the above methods for extracting Physiological signals to detect emotions has its own advantages and disadvantage, where some of the method can be enhanced with the emerging wireless technology viz Cognitive Radio for medical applications that is used for opportunistic spectrum access, where sensing of human body signals can overcome the trouble while gathering the raw data. Several security issues and wireless communication between other medical devices is also an emerging area where research must be focussed. The Telemedicine field which involves remote patient monitoring for early diagnosis also requires efficient data gathering function which is done by prioritizing the signal based on the requirement.

## 3 Comparison of Wireless Sensor Networks and Cognitive Radio Wireless Sensor Networks

Wireless sensor networks are those that comprise of autonomous tiny sensor nodes used for sensing, data processing and communicating between the network devices on an occurrence of a event. Transition from Adhoc networks to sensor networks is mainly due to the scalability issues for the deployment of the sensing nodes in the network. A typical event driven approach is followed in a wireless sensor network which responds to various environmental abnormalities to prevent natural calamities.

Advancements from environmental monitoring of sensor network has explored and paved way to next level as Wireless Body Area Networks (WBAN) for health care monitoring to lead a quality life. Bio-signals like Arterial Blood pressure, Capnogram, Electrocardiogram, Electrodermal activity, Impedance cardiography and temperature are acquired by Body Area Networks for Physiological and Psychophysiology monitoring. The system utilizes one of the above-mentioned signals to keep track of the physical and mental fitness, whereby the mortality rate and the Psychological effects due to various stress in day today life can be reduced.

On the other side of the coin, WBAN has several detriments in transferring the signal across the spectrum and through the channel. Data transfer issues can be channel based, where a physical medium undergoes issues on bandwidth, noise and attenuation or its can be spectrum based, that refers to the range of frequencies used by a communication system or standard. This paper will focus on the spectrum-based issues while observing the Human body signals as shown in Fig. 2. The WBAN includes sensors that are implanted on the human body.

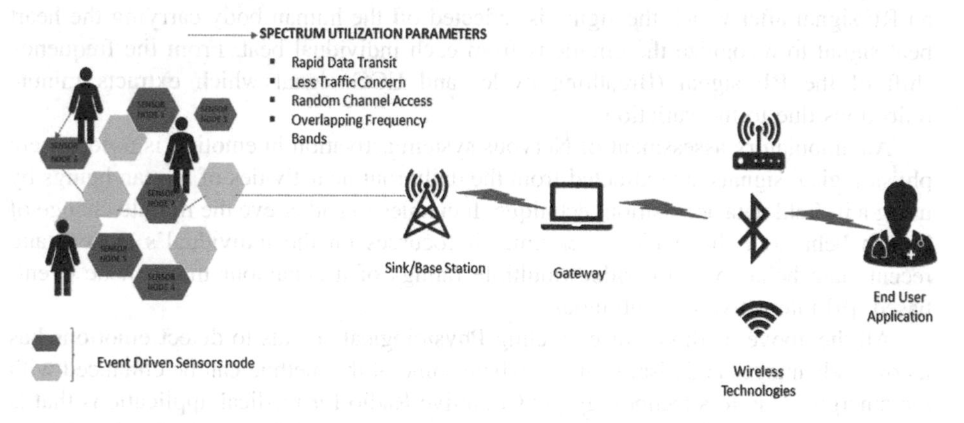


**Fig. 2.** Traditional WBAN for physiological signal extraction

Conventional WBAN operates in Industrial Scientific and Medical (ISM) bands, which allows multiple signals to be transmitted at the same time, whereby the performance of the wireless Networks gets degraded, which also affects the transmission power, extended coverage area and wide deployments due to the overlapping frequency bands. The sensed and gathered signals are then forwarded to the base station in a decentralized manner, where each sensor nodes compete with the other to randomly access the available spectrum.

Cognitive radio is a prototype of next generation communication system that overcomes the disadvantages of the traditional Wireless sensor Networks. It enables the sensing ability of the spectrum used for data transmission, whereby the network traffic can be reduced by following opportunistic spectrum utilization capability. The priority process occurs when an event is triggered, and the signal is generated to access the spectrum. It is not necessary that the licensed spectrums can be accessed only by the primary users (PU), whenever a secondary user (SU) tries to access the spectrum, it first senses the occupancy of the spectrum, if it is free then the SU utilizes the spectrum in an opportunistic mode.

In a similar way the WBAN can be implemented with the Cognitive Radio technique added to it. Alike Wireless sensor Networks, medical applications towards health care. It is useful in the emerging medical field called the Telemedicine where medical specialist in those areas are relatively low. As shown in Fig. 3 a centralized WBAN architecture is used where the scope of sensing is limited in terms of distance (within a room), where the sensors implanted respond to the abnormalities of the Physiological measures.

When such a case is undergone, the spectrum will be utilized in an opportunistic manner and the signals will be gathered by the central controller. The central controller then transmits these signals through the sink and gateway, after which through networking it is passed to the end user or physician who makes a proactive action to the subject. Such an approach in Spectrum sensing will provide an optimal spectral access, spectrum sharing, retaining the signal strength, reduced multipath fading and effective channel utilization, thereby increasing the network performance.

Such type of CR-WBAN can be used in various applications where monitoring of human being is a major issue. Quality issues related to in the implementation of WBAN includes maintaining the reliability of the network, maintaining the security of the data related to patients, maintain an optimized transmission power in order to balance the energy consumption of the sensor nodes to increase the Network Lifetime.

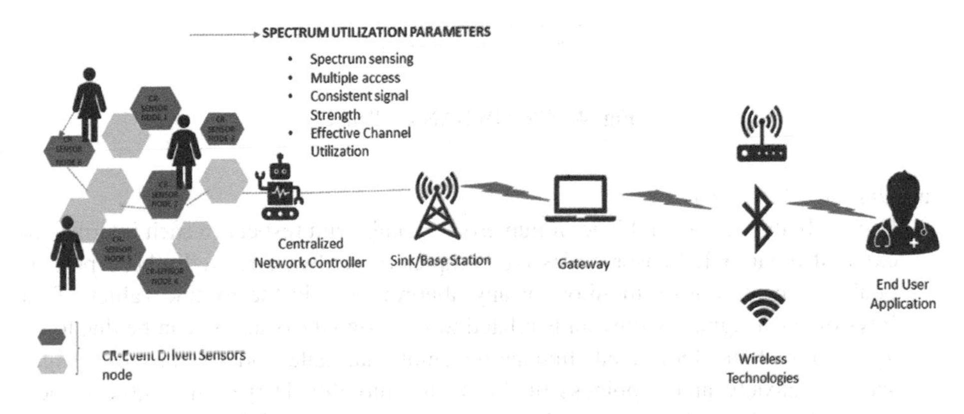


Fig. 3. Cognitive radio-WBAN for physiological signal extraction

# 4 Challenges of Cognitive Radio-Medical Wireless Body Area Networks

Wireless Body Area Networks (WBAN) appended with Cognitive Radio abilities used explicitly for medical applications constitute Medical Wireless Body Area Networks (CR-MWBAN), although there exists similarities between a typical WSN and WBAN, extraction of environmental parameter differs from extracting Physiological signals whereby there derives unique challenges that need to be addressed for efficient and accurate processing of the data. They include:

a. Optimize Data Acquisition
b. Efficient Spectrum Sensing
c. Opportunistic Spectral Access
d. Reduce Multipath Fading

This section describes about the objectives of these challenge to be addressed by the sensor nodes implanted on a human body for an improved system for medical field (Fig. 4).

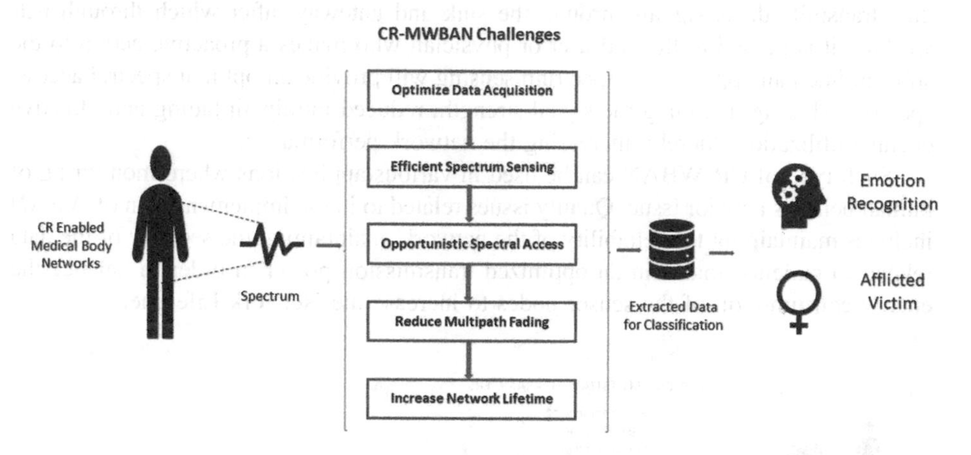


**Fig. 4.** CR-MWBAN challenges

a. *Optimize Data Acquisition*
Human body is an assemblage of numerous signals with respect to each internal and external organ, CR-Sensor nodes are completely hardware related whose primary goal is to continuously monitor for any abnormalities in the routine values of the Physiological signals. Constraints related to such signal extraction can be due to the age factor of the Individual, fluctuating emotional state and feelings like anger, sadness, anxiety and happiness or due to the mobility [12] of the subject under observation. Since the sensor nodes are battery oriented, which has its limitations in the durability, this creates a delay or sometimes lead to unreliability in extracting thee signals. This parameter is to be addressed in a cognitive Network set up, which is required to guarantee an optimized data acquisition.

b. *Efficient Spectrum Sensing*
Spectrum scarcity is a main issue in wireless technology, for reliable communication. In CR-MWBAN the hike in transmitting the signals from the sensor node to the central controller or to the base station, undergoes various interference issues in

medical sectors due to high sensitivity of the Biomedical devices. To reduce such interference the spectrum has to be sensed for free slot out of several spectrum usage. The secondary user (SU) in need of a spectrum to transfer the Physiological signal, utilizes the idle spectrum after sensing for the availability and pre-empt it once the primary user (PU) requires to access the spectrum. This sensing issue needs to be carried out by both the medical devices [3] and the sensor nodes in the MWBAN. Accurate Sensing is to be carried out for the effective utilization of the spectrum in data transmission.

c. *Opportunistic Spectral Access*
Sensing of the spectrum which results in the increased throughput to occupy the available bandwidth for Physiological signal transit. Greater the probability detection of a free spectrum the greater should be the success rate. Along with the multiple signals in under transmission the sensed signal takes its position in an opportunistic way. Since the importance of medical signals is meant for early detection of health issues and mental fitness, this opportunistic access must be accelerated abruptly which is not addressed in CR-MWBAN.

d. *Reduced Multipath Fading*
When physiological signals are carried cross various objects before reaching the base station, it experiences fairly large deviations causing impact in MWBAN [13]. It is highly dependent on the static and mobile characteristics of the subject, whereby the signal strength is degraded thereby reducing the quality of the signal for processing. Hence all the reflector object that fall on the line of flight must be contributed for the overall received signal. The "flash effect" [6] process which incorporates the removal of the unwanted signals needs to be addressed in the usage of Cognitive networks.

Addressing all the above challenges has a direct impact on the Network Lifetime and it results in the increased energy, power consumption and transmission power. Since there is a necessity that sensor networks should always possess robust and reliable quality feature for efficient monitoring. Optimizing the data acquisition process will ed to processing of the signals in timely manner. The reduction of multipath fading contributes to the volatility of wireless links. Followed by the data collection considering various factors the information is stored in the server for pre-processing and efficient Machine Learning Algorithm is used for emotion recognition.

Traditional Wireless Body Area Networks fails to choose the correct spectrum unit for signal transmission, that results in delayed information arrival, unlike Cognitive Radio Body Area Networks makes use of opportunistic spectral access that sends the abnormalities in the parameter value for fast recognition (Fig. 5)

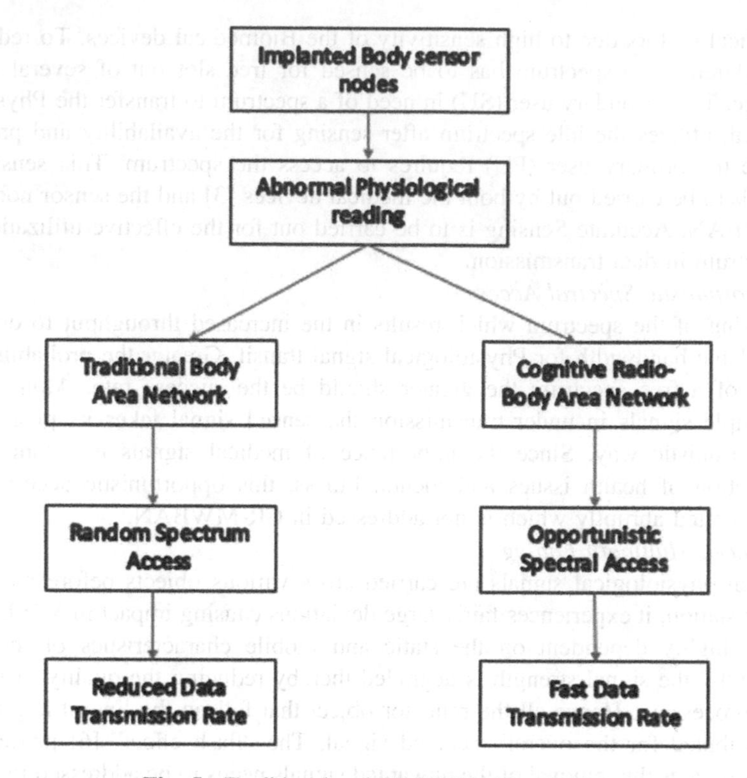


**Fig. 5.** Traditional and CR-MWBAN variations

## 5 Conclusion

Wireless Body Area Networks with the Cognitive Radio ability in medical sector applications through effective utilization of the spectrum for data gathering is presented. Comparing with the traditional Wireless Sensor networks, an explicit technique for physiological signal extraction is done where the issues with respect to mental wellbeing is focussed. Various challenges that needs to be addressed in CR-MWBAN is highlighted, irrespective of a specific medical signal. The study can be still narrowed to specific signal and can be extended.

**Compliance with Ethical Standards**

✓ All authors declare that there is no conflict of interest.

✓ No humans/animals involved in this research work.

✓ We have used our own data.

## References


1. Elnahrawy, E., Nath, B.: Context-Aware Sensors, pp. 77–93. Springer, Heidelberg (2004)
2. Marzencki, M., Hung, B., Lin, P., Huang, Y., Cho, T., Chuo, Y., Kaminska, B.:: Context-aware physiological data acquisition and processing with wireless sensor networks. IEEE (2010). ISBN 978-1-4244-6290-2
3. Sodagari, S., Bozorgchami, B., Aghvami, H.: Technologies and challenges for cognitive radio enabled medical wireless body area networks, pp. 2169–3536. IEEE (2018)
4. Vasuki, P., Aravindam, C.: Improving emotion recognition from speech using sensor fusion techniques (2012)
5. Abdelnasser, H., Youssef, M., Harras, K.: WiGest: a ubiquitous WiFi-based gesture recognition system (2015)
6. Adib, F., Kabelac, Z., Katabi, D., Miller, R.C.: 3D tracking via body radio reflections. In: The Advanced computing Systems, April 2014. ISBN 978-1-931971-09-6
7. Gao, J., Ertin, E., Kumar, S., al Absi, M.: Contactless sensing of physiological signals using wideband RF probes. IEEE (2013). ISBN 978-1-4799-2390-8/13
8. Zhao, M., Adib, F., Katabi, D.: Emotion recognition using wireless signals. ACM (2016). ISBN 978-1-4503-4226-1/16/10
9. Wac, K., Tsiourti, C.: Ambulatory assessment of affect: survey of sensor systems for monitoring of automatic nervous system activation in emotion. IEEE Trans. Affect. Comput. **5**(3), 251–272 (2014)
10. Amir, Z, Hussain, S., Fernando, X., Grami, A.: Cognitive wireless sensor networks: emerging topics and recent challenges. IEEE (2009). ISBN 978-1-4244-3878-5/09
11. Joshi, G.P., Nam, S.Y., Kim, S.W.: Cognitive radio wireless sensor networks: applications, challenges and research trends. Sensors **13**(9), 1424–8220 (2013)
12. Mujdat, S., Cicibas, H.: Real-time data acquisition in wireless sensor networks. Res. Gate (2014). https://doi.org/10.5772/10457
13. Puccinelli, D., Haenggi, M.: Multipath fading in wireless sensor networks: measurements and interpretation. ACM (2006). ISBN 1-59593-306-9/06/0007
14. Swetha, K., Kiran, K.V.D.: Survey on mobile malware analysis and detection. Int. J. Eng. Technol. **7**(2.32), 279–282 (2018)
15. Anguraj, D.K., Smys, S.: Trust-based intrusion detection and clustering approach for wireless body area networks. Wirel. Pers. Commun. **104**(1), 1–20 (2019). https://doi.org/10.1007/s11277-018-6005-x
16. Roshini, A.: Secure quantum key distribution encryption method for efficient data communication in WBAN. Int. J. Eng. Technol. **7**(2.32), 331–335 (2017)

# Pancreatic Tumour Segmentation in Recent Medical Imaging – an Overview

A. Sindhu(✉) and V. Radha

Avinashilingam Institute for Home Science and Higher Education for Women, Coimbatore, India
sindhu@psgrkcw.ac.in, radhasrimail@gmail.com

**Abstract.** Pancreatic tumour is one of the deadliest diseases, which is the fourth leading cause of cancer death worldwide. Detecting pancreatic cancer at an early stage may increase the life of the patients. Pancreatic tumour segmentation is one of the difficult challenges in medical field. Accurate and Efficient segmentation in medical images are emerging as a challenging task during radiotherapy planning. Various medical modalities like MRI, CT and PET are widely used for diagnosing the abnormalities present in the medical images. Image segmentation plays an important part for the exact detection of the tumour in diagnosing, detecting, treatment and planning. In this review paper, various algorithms are used for segmenting the pancreatic tumour in medical images were discussed.

**Keywords:** Pancreatic · PET-CT · MRI · CT · PET · Segmentation

## 1 Introduction

With the increasing use of medical images like MRI, CT, PET, PET-CT, for diagnosing the disease and for treatment planning. The techniques applied for segmenting the pancreatic tumour is specific to application and medical modality. Limited research work has been done in recent past years for segmenting pancreatic tumour in medical modality. It is found from the survey that pancreatic tumour detection is done based on patient history and the symptoms but not applying any techniques using image processing. The procedure for segmenting the pancreatic tumour by using various algorithms plays an important part in medical field for further evaluation of disease and it is most important for detecting the pancreatic cancer at an early stage. In this survey, authors applied various segmentation method in pancreatic tumour and achieved better results. For the purpose of literature survey, the articles related to the work from Pubmed, IEEEXplore, Google Scholar, and ScienceDirect are reviewed from 2010 to 2019. In this review, Sect. 2 describes the Image Segmentation in medical image modalities, Sect. 3 discussed the various methods for pancreatic tumour segmentation in images and in Sect. 4, conclusion is discussed


S. Smys et al. (Eds.): ICCVBIC 2019, AISC 1108, pp. 514–522, 2020.
https://doi.org/10.1007/978-3-030-37218-7_58

## 2 Image Segmentation in Medical Image Modality

Many Imaging techniques are performed on various Medical modalities like MRI, CT, Ultra sound, PET etc. Accurate segmentation is the important step in medical images for finding the abnormalities [20]. Segmentation is the method of splitting an image into areas with comparable characteristics such as gray level, colour, texture, brightness and contrast. The objective of Medical Images segmentation is.

- Study the structure of anatomy and function.
- Identify region of interest (location of tumour, lesion and other abnormalities)
- Measure the size of the tumour
- Treatment planning aid (Tables 1 and 2).

**Table 1.** Common segmentation methods used in Medical Images.

| PET/CT | CT and MRI |
|---|---|
| Manual Segmentation<br>Thresholding Based Segmentation | Threshold Based<br>Region Based |
| Region Based Segmaitation | Fuzzy Based |
| Stochastic and Learning Based Methods | Neural Network Based |
| Boundary Based Methods | |
| Joint Segmentation Methods | |

## 3 Literature Survey

### 3.1 Tumour Segmentation on PET-CT Images

Co-segmented approach of tumour segmentation is applied to attained functional information and anatomical information from both PET and CT. Graph-cut method is used to minimize the problem caused by the MRF in PET-CT. Two-Subgraphs are built and adpative context cost is added between the sub-graphs and this method recognized that quantitative and qualitative results shows efficient improvement. This proposed method was applied individually on each modalities [13].

MRF method is proposed to segment the tumour concurrently from PET and CT and obtain optimal solution time. The technique formulates the job of co-segmentation in an input image as a binary labelling of MRF. The co-segmentation based on Graph-based algorithm obtained global energy function solution. Their findings using co-segmentation show better results when compared with PET and CT images alone [16].

MAP-MRF model was proposed for co-segmentation in PET-CT images of brain and was accomplished by solving MAP issue using EM (Expectation Maximum) algorithm. This technique endured lengthy execution times [17].

Fuzzy Markov random field for lung tumour segmentation is proposed on PET-CT images and it utilizes joint posterior probabilistic model which is better than Gaussian joint distribution and it takes advantage of both PET and CT for the identification of tumour volume and the results shows effective segmentation [18]. A New method for detecting liver tumour using colour segmentation in PET-CT also proposed. Clustering algorithm based on binary tree quantization is used for segmentation [19].

**Table 2.** Advantage and Limitations of Tumour segmentation in Medical Images

| | MRI | CT | PET | PET-CT |
|---|---|---|---|---|
| Advantages | • Excellent capability for soft tissue imaging<br>• High signal to noise ratio<br>• Variable contrast can be achieved by using different pulse sequences for segmentation | • Effective in reducing the variability and shape for location of the tumor<br>• Preserves more boundaries and prevents segmentation leakage | • Ensure segmentation speed<br>• Improves segmentation precision<br>• Improves the efficiency of image processing<br>• Reduce the time of image segmentation<br>• Early detection of the disease<br>• Segmentation accuracy is still limited due to the spatial resolution of the image data | • Can provide complementary functional and anatomical information<br>• Can able to define tumour volume<br>• Avoid inter-and intra- observer variability caused by manual method<br>• Different features can be extracted from PET and CT inside one single N- dimensional |
| Limitations | • Acquisition take longer time<br>• Difficult to obtain uniform image quality<br>• Difficulties in segmenting complex structure with variable shape, size, and properties | • Selection of proper value of threshold is difficult<br>• Performance is affected by the presence of artifacts<br>• Fake edges or weak edges may be present in the detected edges | • Segmentation accuracy is still limited due to the spatial resolution of the image data | • Nearby organs with similar intensities to tumor tissues in PET and CT and it is difficult to segment the tumors<br>• Preforms on a manually selected regions rather than whole volume enclosed by image set |

### 3.2 Tumour Segmentation on CT Images

Efficient pancreas segmentation algorithm has been proposed from CT Images. Pancreas Region can be labelled using Region – growing algorithm and return the result. To evaluate the segmentation performance, Jaccard index between the original image and the extracted image, which was manually defined by the medical experts were computed. The results show better when compared with other recent methods [1].

Three unique techniques have been proposed, spatial standardization and atlas guided segmentation is developed in finding the variability in shape and location. The algorithm is validated with CT volumes distributed in a pancreas region and it was enhanced by the proposed algorithm. Jaccard Index between the extracted region and the original one was found to be 57.9% [5].

A semi automatic method has been implemented to extract the pancreatic region from abdominal CT images. It consists of fast-marching level set method that creates initial region of pancreas and to overcome this problem distance regularized level set method is used to extracts pancreatic region. This method overwhelms the over segmentation process at weak boundary and extract pancreas from CT images exactly [11].

### 3.3 Tumour Segmentation on MRI Images

Pre-mature stage of pancreatic cancer focus on attributed based analysis of Magnetic Resonance Image (MRI). Image texture characters are proposed and statistical evaluation is done for the identification of pancreatic tumour. The system is made automatic using machine vision algorithms to avoid wrong diagnosis in patients. Texture features are used and Bacterial Foraging Algorithm (BFA) is used to classification [3].

The images of the patients mounted through MRI, CT scans of pancreatic cancer can be diagnosed by using CAD system, Haar wavelet transform and clustering methods are used to process the image of pancreatic tumour. The estimated range of each image is based on the threshold value. For the accurate detection of pancreatic cancer, this approach found to be reasonable [10].

### 3.4 Tumour Segmentation on PET Images

The segmentation mostly used in PET imaging is based on techniques of thresholding. SBR (Signal to Back Ground ratio) is frequently used to calculate the limit value based on mean background and lesion signal [15].

Image segmentation plays a significant role in disease identification in order to distinguish the neighbouring tissues in the pancreas. Research shows that the technique of segmentation and quantification of lesions is accelerated by a dramatic rise in PET and hybrid imaging. They evaluated state-of-the-art methods of image segmentation and latest techniques developed on PET images [14].

PET image segmentation can be categorized into region generation and boundary edge detection. Visual saliency model is suggested to identify the region of an image. It is also suggested to presegment the image of PET and optimize the function map For foreground and background region, Gaussian mixture model is initialized [4].

Grabcut algorithm is used for segmenting the salient images in PET and the proposed methods shows better results [6]. A Fast PET tumour segmentation method using super pixel is applied. The distance vector of each super pixel is calculated by PCA, finally k-means clustering is applied to recognize tumour and non-tumour pixels. Rapid execution time is accomplished by the amount of super pixels and the distance vector size [8, 9] (Tables 3 and 4).

**Table 3.** Segmentation Algorithm for pancreatic tumour in Medical Images.

| Images | Region Based Algorithm | SIFA | EM-MAP |
|---|---|---|---|
| CT | • Randomly choosing a point in pancreas region as seed point<br>• Calculate Mean value of Pancreas region<br>• Find new seed point in the nearest intensity to mean value<br>• Segmented region as a logical matrix | • Decomposing the image in to set of binary images<br>• Fractal dimensions computed to describe the segmented texture patterns<br>• SIFA starts detects edges and comers in an image<br>• SIFA finds Region of Interest | • Registration of Three phase CT volume<br>• Extraction of liver and spleen (EM-MAP)<br>• MAP based Segmentation of Pancreas<br>• Fine extraction of pancreas with classifier ensemble [12] |
| PET | Visual Saliency Model | PCA and K-means Clustering | Graph Based Segmentation |
| | • Generate salient Map region for Foreground and background area<br>• Set the value of mask image in different area<br>• Calculate all pixel co-ordinates<br>• Energy function of image and get the segmentation | • Principal Component analysis is applied on super pixels to reduce dimensionality<br>• Apply K-means algorithm to localize the tumour | • Random walk segmentation applied to found the weak boundaries<br>• FCM applied to localize the tumour |
| PET-CT | K-means clustering and subtractive clustering algorithm | Fuzzy Markov Random Field | Binary Tree quantization clustering algorithm |
| | • Load the image to be segmented<br>• Apply partial contrast stretching<br>• Initialize number of cluster, k<br>• Update the pixel based on cluster center<br>• Find the Euclidean distance of each centroid [21]<br>• Reshape the cluster into image | • Set of pixels to be segmented<br>• Membership degree of the voxel at location to tumour class<br>• Image features of the voxel at location extracted from PET-CT images<br>• CT image intensity and SUV derived from PET are calculated to segment the tumour | • Clusters of the pixels are computed<br>• Local optimization strategy in selecting the splitting axis<br>• Splitting axis passes through the centroid of all the colors in the region<br>• RQI is selected using GUI<br>• K-means algorithm is applied to group the objects with similar characteristics to form the cluster |

**Table 4.** Survey on pancreatic tumour segmentation using CT, MRI and PET images.

| S. No | Title of the paper | Year | Medical Modality | Segmentation Approach (Proposed) | Results |
|---|---|---|---|---|---|
| 1 | Efficient Pancreas Segmentation in Computed Tomography Based on Region-Growing | 2015 | CT | Region Growing Algorithm | Efficient and better than the other methods available in literature |
| 2 | Automated pancreas segmentation from three-dimensional contrast-enhanced computed tomography | 2010 | CT | Spatial Standardisation and atlas guided segmentation | Method was effective in reducing the variability in the shape and location of the pancreas |
| 3 | Soft Computing Analysis for Detection of Pancreatic Cancer Using MATLAB | 2018 | CT | STFA (Segmentation Based Fractal Texture Analysis) | Detecting edges and corners in the image and tries to find points that are differentiating from other images [2] |
| 4 | Detection of Pancreatic Tumor using Bacterial Foraging Algorithm | 2019 | MRI | Texture Extraction with BFA | Texture feature Extraction with BFA is used. Results show better by extracting the MRI images of pancreas with benign and Malignant tumour |
| 5 | A Fast Segmentation Algorithm of PET Images Based on Visual Saliency Model | 2016 | PET | Optimized GrabCut algorithm | Based on Visual saliency Model, image is segmented to simplify the operation steps, reduce the processing time and improve the segmentation effect |
| 6 | A Graph-Theoretic Approach for Segmentation of PET Images | 2011 | PET | Graph based (RW) random walk segmentation | Provides globally optimum delineations. Weak object boundaries can be found by RW method. Compared with FLAB and FCM and proposed method shows high reproducibility |
| 7 | A Bottom-up Approach for Pancreas Segmentation using Cascaded Superpixels and (Deep) Image Patch Labeling | 2016 | CT | Cascaded Super pixel segmentation (automated Bottom up approach) | This approach may generate super pixels that adequately pres erve more boundaries, including very weak semantic object boundaries and prevent segmentation leakage [7] |
| 8 | Fast PET Scan Tumor Segmentation Using Superpixels. Principal Component Analysis and K- Means Clustering | 2018 | PET | PCA and K-means Clustering | PCA was applied on super pixels to reduce the dimensionality and speed up the execution time of the segmentation algorithm. K-means clustering was used and is able to localize and delineate the tumour with satisfactory accuracy |

(*continued*)

**Table 4.** (*continued*)

| S. No | Title of the paper | Year | Medical Modality | Segmentation Approach (Proposed) | Results |
|---|---|---|---|---|---|
| 9 | Detection of Pancreatic Cancer using Clustering and Wavelet Transform Techniques | 2015 | MRI and CT | K-Means Clustering and Haar wavelet transform | Provided CAD system by making us e of Haar wavelet and clustering techniques. The analysis is based on threshold value, if it falls with in the range estimated for each image. Results obtained are found to be the early and accurate detection of cancer cells |
| 10 | A Hybrid Method for Pancreas Extraction from CT Image Based on Level Set Methods | 2012 | CT | Marching Level set method and modified distance regularized level set (MDRLS) | MDRLS can overcome the shortage of over segmentation in weak boundary region. Energy decrement algorithm is applied to preserve pancreas region from leaking into nearby tissues. This hybrid method is time effective and able to achieve accurate segmentation |

# 4 Conclusion

Image Segmentation is the important step in medical research for classifying the abnormalities. Many algorithms have been implemented in previous research work for the segmentation of medical images. In this review paper, segmentation algorithms have been proposed for pancreatic tumour with different medical images have been studied and showed the results. From this developed algorithms, it was identified that, not much work has been done for segmenting pancreatic tumour with different medical modalities. Efficient algorithms has to be developed and it needs more research work in PET-CT images to help radiologist for detecting pancreatic tumour at an earlier stage.

**Future Research enhancement**

Current Methods for the segmentation of tumour in the medical field resulting in lengthy execution time and over segmentation process. As per the survey it was identified that not much work has been done on PET-CT pancreatic tumour segmentation. In future, many segmentation methods can be implemented based on salient object detection to find the exact location of tumour in medical images particularly for PET-CT.

**Compliance with Ethical Standards**

✓ All authors declare that there is no conflict of interest.

✓ No humans/animals involved in this research work.

✓ We have used our own data.

## References


1. Tam, T.D., Binh, N.T.: Efficient pancreas segmentation in computed tomography based on region-growing (2015)
2. Balakrishna, R., Anandan, R.: Soft computing analysis for detection of pancreatic cancer using MATLAB. Int. J. Pure Appl. Math. **119**(18), 379–392 (2018). ISSN 1314-3395
3. Sujatha, K., Ponmagal, R.S., Yasoda, K.: Detection of pancreatic tumor using bacterial foraging algorithm. Int. J. Recent Technol. Eng. (IJRTE) **8**(1S4) (2019). ISSN 2277-3878
4. Lu, L., Xiaoting, Y., Bo, D.: A fast segmentation algorithm of PET images based on visual saliency model. In: 2nd International Conference on Intelligent Computing, Communication and Convergence, Procedia Computer Science, vol. 92, pp. 361–370 (2016)
5. Shimizu, A., Kimoto, T.: Automated pancreas segmentation from three-dimensional contrast-enhanced computed tomography. Int. J. CARS. https://doi.org/10.1007/s11548-009-0384-0
6. Bağci, U., Yao, J., Caban, J.: A graph-theoretic approach for segmentation of PET images. In: 33rd Annual International Conference of the IEEE EMBS Boston, Massachusetts, USA, 30 August–3 September 2011
7. Farag, A., Lu, L.: A bottom-up approach for pancreas segmentation using cascaded superpixels and (deep) ımage patch labeling. IEEE Trans. Image Process. https://doi.org/10.1109/tip.2016.2624198
8. shah, J., Surve, S., Turkar, V.: Pancreatic tumour detection using ımage processing. In: ICAC3 2015 Elsevier Procedia Computer Science, vol. 49, pp. 11–16 (2015)
9. Hagos, Y.B., Minh, V.H.: Fast PET scan tumor segmentation using superpixels, principal component analysis and K-means clustering. MDPI **1**(1), 7 (2018)
10. Reddy, C.K.K., Raju, G.V.S., Anisha, P.R.: Detection of pancreatic cancer using clustering and wavelet transform techniques. In: International Conference on Computational Intelligence and Communication Networks (2015)
11. Jayasri, S., Prabha, R.S.: Survey on pancreatic tumour segmentation. Int. J. Eng. Res. Technol. (IJERT), **7**(04) (2018). ISSN 2278-0181
12. Rueckert, d., Schnabel, J.A.: Registration and segmentation in medical maging. In: Registration and Recognition in Images and Videos. Studies in Computational Intelligence, vol. 532, p. 137. Springer, Heidelberg (2014). https://doi.org/10.1007/978-3-642-44907-9_7
13. Song, Q., Bai, J., Han, D.: Optimal Co-segmentation of tumor in PET-CT images with context information. IEEE Trans. Med. Imag. **32**(9), 1685–1697 (2013)
14. Black, Q.C., Grills, I.S., Kestin, L.L., Wong, C.Y., Wong, J.W., Martinez, A.A., Yan, D.: Defining a radiotherapy target with positron emission tomography. Int. J. Radiat. Oncol. Biol. Phys. **60**(4), 1272–1282 (2004)
15. Foster, B., Bagci, U., Mansoor, A.: A review on segmentation of positron emission tomography ımages. Comput. Biol. Med. **50**, 76–96 (2014)
16. Han, D., Bayouth, J., Song, Q.: Globally optimal tumor segmentation in PET-CT ımages: a graph-based co-segmentation method. In: Information Processing in Medical Imaging, August 2011

17. Xia, Y., Wen, L., Eberl, S., Fulham, M., Feng, D.: Segmentation of dual modality brain PET/CT images using the MAP-MRF model. In: 2008 IEEE 10th Workshop on Multimedia Signal Processing, pp. 107–110. IEEE (2008)
18. Guo, Y., Feng, Y., Sun, J.: Automatic lung tumor segmentation on PET/CT images using fuzzy Markov random field model. Comput. Math. Methods Med. (2014). https://doi.org/10.1155/2014/401201
19. Bangar, N., Sharma, A.: A proposal for color segmentation in PET/CT-guided liver images. In: Advances in Intelligent Systems and Computing book series (AISC), vol. 249. Springer (2014)
20. Sharma, N., Aggarwal, L.M.: Automated medical image segmentation techniques. J. Med. Phys. **35**(1), 3–14 (2010). https://doi.org/10.4103/0971-6203.58777
21. Dhanachandra, N., Manglem, K., Chanu, J.Y.: Image segmentation using K-means clustering algorithm and subtractive clustering algorithm. In: Procedia Computer Science, (IMCIP-2015), vol. 54, pp. 764–771 (2015)

# A Normal Approach of Online Signature Using Stroke Point Warping – A Survey

S. Prayla Shyry(✉), P. Srivanth, and Derrick Thomson

Department of Computer Science Engineering,
Sathyabama Institute of Science and Technology, Chennai, India
suja200165@gmail.com, srivanth.stg@gmail.com,
derrickthomson260@gmail.com

**Abstract.** In today's world, the method to identify the online verification technique is appeared with the help of the signature device. Some common human biometrics include iris recognition, retinal scanner, voice detection, fingerprint detector, and thumbprint detection. Fingerprint detector is used in the mobile devices. Traditionally, due to the in-built inconsistency of the collected signature, a normal procedure known as Stroke Point Warping (SPW) was recommended. In the proposed system, a normal methodology called DSUM (Distributed Data Split up Model) for signature verification is used to improve the accuracy level of the extracted features. In this DSUM model, the signature area is being extracted and splitted into N number of blocks. For studying the features accurately each N blocks are being fetched up for multi modal feature extraction. The neural network (decision making) is being done by using the multiple scores and featured vectors. Nowadays, neural networks and deep learning gives good solution for problems in image identification, voice detection and natural language processing.

**Keywords:** Biometrics · Distributed Data Split Up Model (DSUM) · Inherent variability · Neural network · Stroke Point Warping (SPW)

## 1 Introduction

The measurement of the difference between the physiological or the behavioral traits for recognizing a person is known as biometric. The important feature of biometric is handwritten online signature which is gained by a pen-sensitive tablet. Two techniques has been emerged over here that is the methodology of identifying the signature and obtain the electronic signature device. Few common types of human biometrics are fingerprint detector, iris recognition, retinal scanner, voice detection and thumbprint. The dynamic signature of each user can be captured in a framework for online signature verification.


S. Smys et al. (Eds.): ICCVBIC 2019, AISC 1108, pp. 523–528, 2020.
https://doi.org/10.1007/978-3-030-37218-7_59

## 2 Related Works

Urvashi et al. (2018) proposed about the online signature verification which is classified on the fuzz modeling. The modification of the fuzzy function is done by using the structural parameters. These structural parameters make the users signature to be verified easily. The proposed paper verifies the extraction of the sample signatures which are gathered from the users based on the time gap and different situations. Two different methods is classified to obtain the structural parameters. The first type is based on the Shannon entropy function which is being joined with an objective function in terms of the error. Another type is based on the communication being collected from the input fuzzy sets computing in s-norms. Both these signatures that is genuine and forged signatures is being examined for the advanced techniques.

Rahardika et al. (2018) proposed about the online signature verification by collecting 1210 global features of the dynamic signatures. The signatures are being collected from the previous research techniques and new global features are also added from the fresh frameworks. The main aim is to select the same signatures from the different feature sets. Two methods have been suggested to obtain the score of each feature- Information Gain Ratio and Correlation. Among this, each feature is being ranked up based on the score. Once the ranking process is done these 1210 features is being divided in the form of 10, 20, 30 and so on until all the 1210 feature sets are arranged in the ascending order. Then each feature are being testified by using the classifiers like Random Forest, SVM and Naive Bayes with the help of 10 folds cross validation. This theory is being used in SVC2004 dataset task 1. The final result shows that 120 features which was tested first are the most suitable features. These 120 global features shows somewhat better result for the dynamic signature verification.

Alaei et al. (2017) proposed a skilled offline-signature verification which is determined on different figural representation. The extraction step consists of a set of Local Binary Pattern (LBP) which are used on the feature from both the image of the signature and its sample signature. The interval figural data is created for every signature of the user. A set of signature consisting of the interval values is being verified for each handwritten signature. A fuzzy model process was used which proposes about the similarity between the sample signature and the corresponding signature. The proposed verification approach is being evaluated as a benchmark of English signature and a dataset which is composed of Bangla. While comparing the signature verification methods a number of errors has been resulted and hence it is derived that this method always performs when the training samples of the signature is eight or more.

Liu et al. (2015) proposed a technique for verification of online signature which is based on sparse representation and discrete cosine transform (DCT). A new attribute of DCT is being introduced which is used to achieve a firm demonstration of the signatures by using fixed count of coefficients. This leads to the signature corresponding procedure and provides an alternate which deals with the time series of different lengths. After that, a new method that is sparse representation is applied for the verification of the online signature. A specific method is used for constructing the dictionaries and the sparse-representation is being extracted. Finally both the features that is the energy and sparse are connected to form a feature vector. These preparatory are done on the Sabanci University's Signature Database (SUSIG) – SVC2004 and Visual databases. The final result proposes us that the signature verification is being authenticated reliably and is done much better than the other methods.

Pirlo et al. (2015) proposed about the online signature verification which achieves the potential of consistent information in handwritten signatures. According to earlier models, this model specifies a signature which uses a domain strategy. First the user signature is fragmented into various parts depending on the consistent user model. According to consistent model, for every part the representation for verification of signature is detected. After that in the verification process, the dependability of every part of the unknown signature is determined in several commercial domains of representation. Then the dependency of unknown signature is regulated by linking the signature verification. This experiment was then accomplished in SUSIG database and the result, acquired shows the efficiency of advanced theory compared to other experiments.

## 3 Existing System

In the existing system, the variability of the gathered signature is obtained by the stroke point warping (SPW) technique. The SPW technique was introduced and the normalization of the coefficients of correlation of two signatures obtained by the same user was examined. A selection strategy was used by merging the normalization of signature and the SPW method. The defect found was that the image pixel is not as same as the genuine and the forged signature because of point warping. The point warping is broken up as compared to the genuine signature.

**Fig. 1.** Signature pad used for data acquisition {author's ow photo}

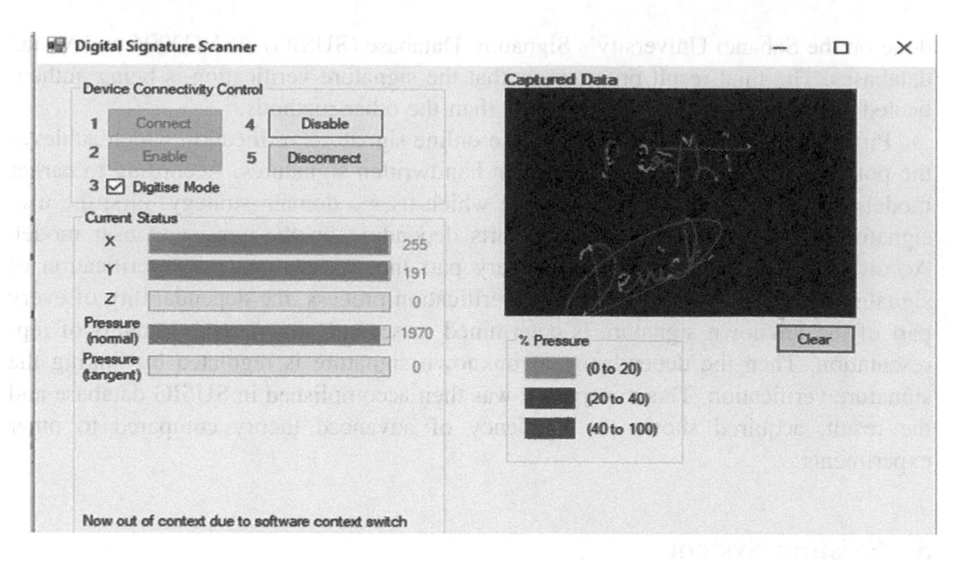


**Fig. 2.** Range of pressure calculation (input data).

## 4 Proposed System

Figure 1 shows the process of signature input. In the above Fig. 2, few signatures are collected for experiments. The shape of signature and its various functions are represented in the Fig. 2. The signatures are gathered using the signature pad and Visual Studio. The X-axis and the Y-axis of the user signature that is the horizontal and the vertical lines are being initiated. When the user signs on the signature pad, the X-coordinate and Y-coordinate are being determined horizontally and vertically. Then the writing pressure the user has given is being checked out. The pressure value checks out how much pressure the user has being given while writing on the signature pad. The pressure value consists of the three levels the user has given. In the above Fig. 2 you can see the percentage of the writing pressure of the user. If the user has given a light pressure on the signature pad, then green colour (0 to 20%) would be displayed. If the user has given an average pressure, then blue colour (20 to 40%) would be displayed. If the user has given an immense pressure, then red colour (40 to 100%) would be displayed.

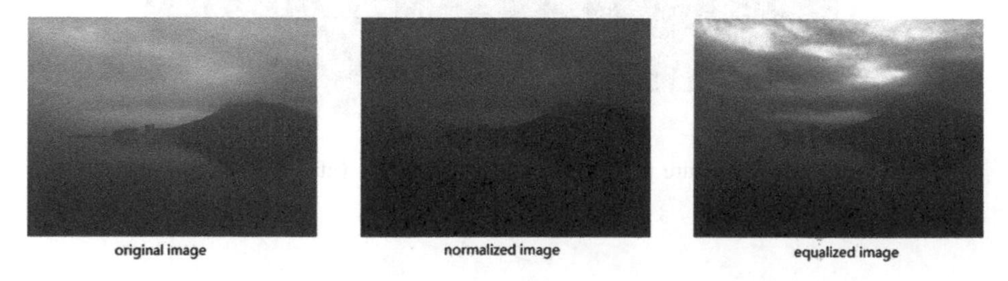


**Fig. 3.** Normalization technique (deep learning) [7].

In the above Fig. 3, the image is extracted and splitted up into N number of blocks. Each N blocks has been fetched up for the multi-modal feature extraction so that the features can be studied accurately. Finally using multiple scores and Feature vectors the Deep Neural network classification is done for decision making. Two useful attributes are being based on signature and instant phases are developed. Modification is our implementation. We are implementing multimodal based user verification system. Neural network and back propagation algorithm is used for signature verification.

In the above Fig. 4, testing process is carried out that is testing of the signature is done. The signature which is given is being tested. After the testing process takes place, then the signature input process takes place. The input signature is being given for testing. Then after the input signature process takes place, normalization of the signature is done. DSUM Normalization is used for the normalization in the online signature verification technique. After the normalization of the signature, extraction of the signature takes place that is stroke extraction process traces the both the signatures and is traced part by part and checks whether the signature is matching or not. If the signature is matched in the stroke extraction process then it is sent for the multi-modal extraction. Multi- modal feature extraction is the format of the image that is jpeg, gif, etc. It describes about the format or the way the image must be stored.

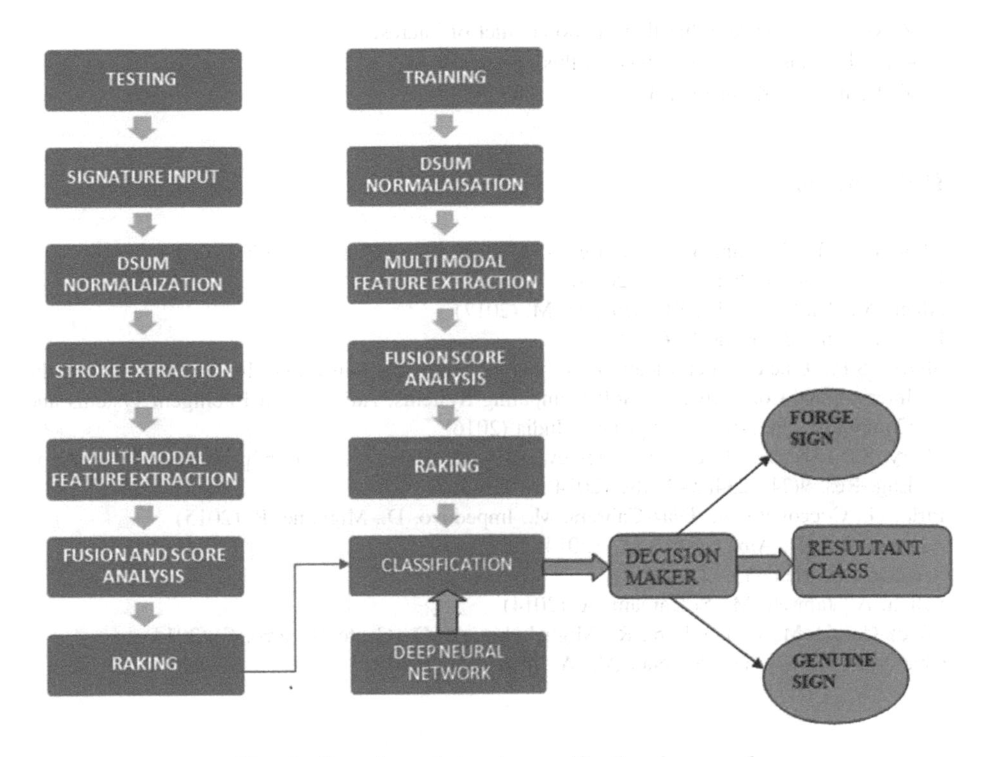


**Fig. 4.** Overview of signature verification framework.

After storing the image in the proper format then fusion and analysis of the image takes place. Fusion and score analysis process is used for joining all the parts again which was splitted up for the extraction. Then collection of all the signatures takes place in raking. Finally classification of the signature is done through deep neural network which results in decision making and checks whether the signature is forged or genuine and the final resultant class is proposed.

## 5 Conclusion

We conclude about the normal approach of online signature verification. Here we use Distributed Data Up Model for the feature extraction and the accuracy level of the signature can be proved. Through the data split model, normalization is done. After normalization of the signature, extraction takes place. The image is being stored in the multi-modal feature and fusion and analysis of the signature takes place which results in raking and the classification of the signature is done. The work can be extended with different classifiers and the same can be analyzed by varying the parameters.

**Compliance with Ethical Standards**

✓ All authors declare that there is no conflict of interest.
✓ No humans/animals involved in this research work.
✓ We have used our own data.

## References

Choudhary, U., Woungang, I., Rodrigues, J.J.P.C., Dhurandher, S.K. (2018)
Rahardika, A.R., Tjahyanto, A. (2018)
Alaei, A., Pal, S., Pal, U., Blumenstein, M. (2017)
Liu, Y., Yang, Z., Yang, L. (2015)
Shyry, S.P.: Efficient identification of bots by K-means clustering. In: Proceedings of the International Conference on Soft Computing Systems, Advances in Intelligent systems and Computing, pp. 307–318. Springer, India (2016)
Shyry, S.P., Sheeba, M.: Literature review on the detection of bots in p2p network. Int. J. Appl. Eng. Res. **9**(24), 23485–23489 (2014)
Pirlo, G., Cuccovillo, V., Diaz-Cabrera, M., Impedovo, D., Mignone, P. (2015)
Argones-Rúa, E., Alba-Castro, J.L. (2014)
Sae-Bae, N., Memon, N.D. (2014)
Fallah, A., Jamaati, M., Soleamani, A. (2014)
López-García, M., Ramos-Lara, R., Miguel-Hurtado, O., Cantó-Navarro, E. (2014)
Chatterjee, A., Fournier, R., Naït-Ali, A., Siarry, P. (2010)

# Spectral Density Analysis with Logarithmic Regression Dependent Gaussian Mixture Model for Epilepsy Classification

Harikumar Rajaguru(✉) and Sunil Kumar Prabhakar

Department of ECE, Bannari Amman Institute of Technology, Coimbatore, India
harikumarrajaguru@gmail.com

**Abstract.** One of the serious disorders causing seizures in the neurology is the Epilepsy. The formation of the attacks happens due to the unusual activities of the neurons. The EEG is utilized in the effective observation of the brain abnormalities. The EEG can effectively analyze the different sorts of the status of the physiological in the brain and can provide valuable data about any neurological disorder. Therefore, EEG is quite a powerful diagnostic tool for analyzing many neurological disorders like epilepsy, dementia, paralysis, sleep disorders etc. As the EEG recordings for epileptic patients are quite long, the most important features based on the Power Spectral Density (PSD) are extracted using the Logarithmic Regression dependent Gaussian Mixture Model to know the risk of epilepsy. Results showed that when PSD is classified with Logarithmic Regression Gaussian Mixture Model, an appropriate accuracy of 95.835% in the classification along with an appropriate Performance Index of 91.58% is acquired.

**Keywords:** Epilepsy · EEG · PSD · Gaussian Mixture

## 1 Introduction

Epilepsy is a long term persisting neurological problem and it causing disorders to a numerous of people in the world [1]. The seizures occur spontaneously and it is the characteristic trait of epilepsy. Many parts of the body are affected with this seizure disorder. For many patients, surgery and continuous medications does not give a remedy and it only produces side effects. The anxiety of having a seizure attack at any time during day to day activities disturbs the mindset and health of the patient surely [2]. With the boon of computers and mathematical algorithms, it has become easy for the acquisition of EEG signals, managing and storing it [3]. EEG helps in the measurement and analysis of brain waves. The diagnostic tools and the state of the art equipment for analysis of epilepsy has to be improved so that a better health provision can be supplied. Generally, the EEG recordings are examined and analyzed by the neurophysiologist visually for the identification of epileptic seizures in the EEG signals [4]. It is quite time consuming and very expensive in order to inspect the epileptic signals in such a standardized procedure. False detection and diagnosis of epilepsy by the humans causes further agony to human life. Therefore the requirement for a system


S. Smys et al. (Eds.): ICCVBIC 2019, AISC 1108, pp. 529–535, 2020.
https://doi.org/10.1007/978-3-030-37218-7_60

that is intelligent is entailed to identify and classify the epilepsy properly. A few important works both in the processing and classification of EEG signals are detailed in the following.

For the EEG detection and prediction, a novel image extraction prototype utilizing different kernel neural networks and variance feature attainment techniques was analyzed by Ge and Zhang [5]. A modification to the Sparse Representation classification technique was implemented for epileptic EEG analysis and reported in [6]. To analyze the medical big data for seizure detection, a cloud-based Brain Computer Interface (BCI) methodology was introduced by Hosseini [7]. The Metric Multidimensional Scaling and Aggregation Operators were utilized in epilepsy classification and reported in [8]. The specialized EEG features was extracted and classified with machine learning techniques for seizure detection by Wang and Lyu [9]. A modification in Linear embedding methodology along with various kinds of modified Principal Component Analysis (PCA) was utilized for epilepsy classification [10]. The seizure prediction in a low-complex manner was done from EEG signals with the help of spectral power analysis in [11]. The probabilistic mixture model with certain feature reduction techniques was analyzed conceptually for epilepsy classification in [12]. By combining the time-domain and inter-channel analysis of EEG features, the pre-ictal and inter-ictal stages was classified by Lin et al. [13]. For epilepsy classification, a distance remote monitoring prototype was developed in [14]. The structure of the work is written as follows; the materials and methods are elaborated in Sect. 2 followed by the usage of Logarithmic Regression dependent GMM post classification in Sect. 3. The results and discussion is detailed in Sect. 4 and concluded in Sect. 5. The block diagram illustration is shown in Fig. 1.

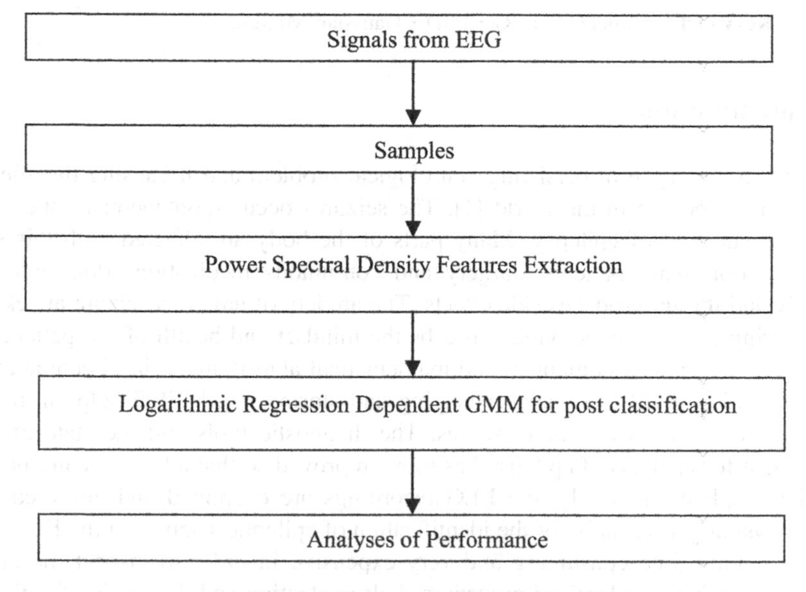


**Fig. 1.** Illustration of the paper

## 2 Experimental Methodology

The EEG database analyzed here is for a single patient is acquired from the University of the Bonn in Germany that is available for the public and research use available online. The data available to public and researchers consists of 5 different sets. In this work, the details shown in Table 1 are measured.

**Table 1.** Nature of the subject analyzed

| Consideration of subject | Specific features |
|---|---|
| Patients nature | Ictal activity based on seizure |
| Placement of electrodes | Inside the zone of the epileptogenetic |
| Utilization of electrodes | Intracranial section |
| No. of epochs | 100 |
| Time of each epoch | 23.6 (Sec) |

There are about 100 single channel EEG epochs which had a total duration of 23.6 s in the utilized EEG database. The epochs of EEG has a duration of 23.6 s. With the help of 128 channel amplifiers module, the EEG recordings was performed. The sampling rate considered here is 173.61 Hz and it had a 12 bit A/D resolution for digitization.

A. **Extraction of Power Spectral Density Features**

Most of the signals utilized in real time applications are in such a manner that its future variation cannot be known easily and exactly. It is only feasible to interpret probabilistic statement about the particular variation. To describe such a signal, a random sequence is utilized which consists of an ensemble of all the possible realizations where each of it has some recurring probability. From the entire group of realizations, a single signal realization can be observed by the experimenter. A random signal generally has average power which is finite in nature and so it is assessed by an average Power Spectral Density [15]. The discrete time signal $\{z(p); p = 0, \pm 1, \pm 2, \ldots\}$ is assumed to be a specific order of random variables which has a zero mean as explained as $E\{Z(p)\} = 0$ for all $p$, where $E\{\bullet\}$ explains the main function of expectation operator. The autocovariance sequence (ACS) or the covariance function of $z(p)$ is written as

$$l(h) = E\{z(p)z^*(p - h)\}$$

The covariance function is simply assumed to be based only on the delayed lag in between the average of 2 consecutive samples. The PSD is expressed as the Discrete Time Fourier Transform (DTFT) of the covariance function expressed as

$$\varphi(w) = \sum_{h=-\infty}^{\infty} l(h)e^{-jwh}$$

## 3 Logarithmic Regression Gaussian Mixture Model for Post Classification

The extracted features using the PSD are fed inside the Logarithmic Regression dependent Gaussian Mixture Model so that the epilepsy can be classified from EEG signals. The density is designed as the classical GMM [16] and is expressed as

$$s_{K,c,\Sigma,d}(w) = \sum_{k=1}^{K} \pi_{d,k} \Phi_{c_k,\Sigma_k}(w)$$

where $K \in N^*$ is the entire no. of mixture variate elements; $\varphi_{c,\Sigma}$ denotes the Gaussian density comprising of mean $a$ and covariance matrix $\Sigma$, and it is expressed as

$$\Phi_{c,\Sigma}(w) = \frac{1}{\sqrt{(2\pi)^p |\Sigma|}} e^{-\frac{1}{2}(w-c)' \Sigma^{-1} (w-c)}$$

and the mixture weights are obtained from a $K$ – tuple $(w_1, \ldots, w_K)$ with a specific logistic idea as follows

$$\pi_{d,k} = \frac{e^{d_k}}{\sum_{k'=1}^{K} e^{d_{k'}}}$$

In this classifier, a system is analyzed where both the weight mixture and the means can rely upon a covariate factor. A typical '$n$' pair of randomly obtained variables $((G_i, W_i))_{1 \leq i < n}$ is observed where covariates $G_i s$ are independent and $W_i s$ are conditionally independent to the $G_i s$. The conditional density $s_0(\bullet|g)$ has to be analyzed considering the Lebesgue measure of $W$ given $G$. This conditional density is modeled by a mixture of Gaussian regression which has varying logistic weights expressed as

$$s_{K,c,\Sigma,d}(w|g) = \sum_{k=1}^{K} \pi_{d(g),k} \Phi_{c_k(g),\Sigma_k}(w)$$

where $(c_1, \ldots, c_k)$ and $(d_1, \ldots, d_k)$ represent the $K$ – tuples of functions chosen in the set $\gamma_K$ and $D_K$. The aim is then to estimate these functions $c_k$ and $d_k$, the number of classes $k$ and the covariance matrices $\Sigma_k$ so that the error is minimized to the core in between the conditional density, for both true and estimated values.

## 4 The Results with Discussion

When the significant features of PSD are obtained and when it is sorted with Logarithmic Regression dependent GMM. The parameters like, Accuracy, Specificity Performance Index and Sensitivity, the appropriate results are determined as shown in Table 2. The mathematical formulae framed in its accord are given below.

$$PI = \left( \frac{PC - MC - FA}{PC} \right) \times 100$$

Where perfect classification is represented by the PC the missed classification is represented by MC and the false alarm denoted as the FA. The following expressions give the sensitivity (Se), accuracy and the specificity (Sp)

$$Se = PC/(PC + FC) \times 100$$

$$Sp = PC/(PC + MC) \times 100$$

$$Accuracy = \frac{Se + Sp}{2}$$

**Table 2.** Performance of PSD with Logarithmic Regression Gaussian Mixture Model

| Name | Average values |
|---|---|
| PC (%) | 91.67 |
| MC (%) | 0 |
| FA (%) | 8.33 |
| PI (%) | 91.58 |
| Specificity (%) | 100 |
| Sensitivity (%) | 91.67 |
| Accuracy (%) | 95.83 |

The Bland Altman Plot for the Logarithmic Regression Gaussian Mixture Model Classifier for Epileptic EEG signal after the PSD feature extraction is shown in Fig. 2. It is a famous technique to plot the data in order to analyze the agreement between two different classifiers. The Bland-Altman plot determines whether one Classifier can replace another. In this study, the question is whether the Logarithmic Regression Gaussian Mixture Model Classifier can replace the Gold standard. The Bland-Altman Plot indicates 97% good agreement (±1.96) of Confidence level.

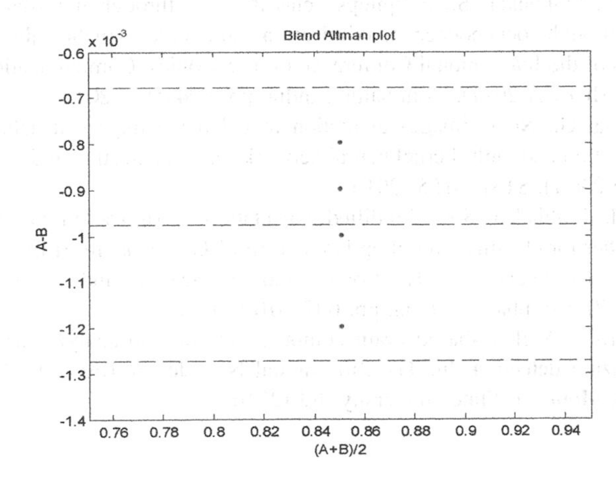


**Fig. 2.** Depiction of Bland Altman Plot

## 5 Conclusion

Due to the rapid, random, excessive and abnormal rhythmic discharges occurring in the brain, lead to the occurrence of the seizure. The fluctuations in electrical events of the brain could be analyzed using Electroencephalogram. So the paper utilizes the, PSD to extract the features and later it is classified using Logarithmic Regression dependent Gaussian Mixture Model. In this paper, an appropriate accuracy of 95.83% in classification and an appropriate performance index of 91.58% are attained. In future, the main aim is to work with epilepsy classification from Electroencephalogram signals using the soft computing methods.

**Compliance with Ethical Standards**

✓ All authors declare that there is no conflict of interest.
✓ No humans/animals involved in this research work.
✓ We have used our own data.

## References

1. Prabhakar, S.K., Rajaguru, H.: Application of linear graph embedding as a dimensionality reduction technique and sparse representation classifier as a post classifier for the classification of epilepsy risk levels from EEG signals. In: Proceedings of the International Conference on Graphic and Image Processing (ICGIP), October 23–25, Singapore (2015)
2. Prabhakar, S.K., Rajaguru, H.: Performance comparison of fuzzy mutual information as dimensionality reduction techniques and SRC, SVD and approximate entropy as post classifiers for the classification of epilepsy risk levels from EEG signals. In: Proceedings of 2015 IEEE Student Symposium in Biomedical Engineering and Sciences (ISSBES), November 4. Universiti Teknologi Mara, Malaysia (2015)
3. Rajaguru, H., Prabhakar, S.K.: Non linear ICA and logistic regression for classification of epilepsy from EEG signals. In: IEEE Proceedings of the International Conference on Electronics, Communication and Aerospace Technology (ICECA 2017), Coimbatore, India, pp. 577–580 (2017)
4. Rajaguru, H., Prabhakar, S.K.: Epilepsy classification through multi-label dimensionality reduction through dependence maximization and elite genetic algorithm. In: IEEE Proceedings of the International Conference on Electronics, Communication and Aerospace Technology (ICECA 2017), Coimbatore, India, pp. 594–597 (2017)
5. Ge, J., Zhang, G.: Novel images extraction model using improved delay vector variance feature extraction and multi-kernel neural network for eeg detection and prediction. Technol. Health Care **23**(s1), S151–S155 (2015)
6. Rajaguru, H., Prabhakar, S.K.: Modified expectation maximization based sparse representation classifier for classification of epilepsy from EEG signals. In: IEEE Proceedings of the International Conference on Electronics, Communication and Aerospace Technology (ICECA 2017), Coimbatore, India, pp. 607–610 (2017)
7. Hosseini, M.-P.: A cloud-based brain computer interface to analyze medical big data for epileptic seizure detection. In: The 3rd Annual New Jersey Big Data Alliance (NJBDA) Symposium. Montclair State University, NJ (2016)

8. Rajaguru, H., Prabhakar, S.K.: Metric multidimensional scaling and aggregation operators for classifying epilepsy from EEG signals. In: IEEE Proceedings of the International Conference on Electronics, Communication and Aerospace Technology (ICECA 2017), Coimbatore, India, pp. 567–570 (2017)
9. Wang, N., Lyu, M.R.: Extracting and selecting distinctive EEG features for efficient epileptic seizure prediction. IEEE J. Biomed. Health Inf. **19**(5), 1648–1659 (2015)
10. Prabhakar, S.K., Rajaguru, H.: Expectation maximization based PCA and Hessian LLE with suitable post classifiers for epilepsy classification from EEG signals. In: Proceedings of the 8th International Conference on Soft Computing and Pattern Recognition (SoCPaR). VIT University, Vellore (December 2016)
11. Zhang, Z., Parhi, K.K.: Low-complexity seizure prediction from iEEG/sEEG using spectral power and ratios of spectral power. IEEE Trans. Biomed. Circuits Syst. **10**(3), 693–706 (2016)
12. Prabhakar, S.K., Rajaguru, H.: Conceptual analysis of epilepsy classification using probabilistic mixture models. In: 5th IEEE Winter International Conference on Brain-Computer Interface, January 9–11, South Korea (2017)
13. Lin, L.-C., Chen, S.C.-J., Chiang, C.-T., Wu, H.-C., Yang, R.-C., Ouyang, C.-S.: Classification preictal and interictal stages via integrating interchannel and time-domain analysis of EEG feature. Clin. EEG Neurosci. **48**(2), 139–145 (2016)
14. Prabhakar, S.K., Rajaguru, H.: Development of patient remote monitoring system for epilepsy classification. In: 16th International Conference on Biomedical Engineering (ICBME), Singapore, December 7–10 (2016)
15. Rajaguru, H., Prabhakar, S.K.: Power spectral density and KNN based adaboost classifier for epilepsy classification. In: IEEE Proceedings of the International Conference on Electronics, Communication and Aerospace Technology (ICECA 2017), Coimbatore, India, pp. 441–445 (2017)
16. Prabhakar, S.K, Rajaguru, H.: ICA, LGE and FMI as dimensionality reduction techniques followed by GMM as post classifier for the classification of epilepsy risk levels from EEG signals. In: 9th IEEE European Modelling Symposium 2015, October 6–8, Madrid, Spain (2015)

# Comprehensive Review of Various Speech Enhancement Techniques

Savy Gulati(✉)

Lovely Professional University, Phagwara, India
savygulati99@gmail.com

**Abstract.** In this modern era, effective speech enhancement is the foremost need as almost every device from tiny to large incorporate speech functions in one or the other way. It is essential part for numerous applications such as speech recognition, speech translation, speech verification and many more, because speech related applications are incomplete without enhancing the speech quality and intelligibility. Computer Aided Diagnosis systems based on speech are also becoming prevalent. Apart from this, communication is the essence area where degradation in speech cannot be afforded. Thus, there is need of efficient algorithms which reduces or eliminates the effect of various noise sources. In this paper, review of various speech enhancement algorithms has been carried out in a comprehensive manner.

**Keywords:** Speech · Speech enhancement · Various noise sources · Clean speech

## 1 Introduction

Speech enhancement is basically a technique to improve the quality and intelligibility of the speech. Speech may get degraded due to various unwanted sources so there is a prime need to clean and refine speech [1]. Speech enhancement is a challenging task as it is difficult to eliminate noise without affecting the desired speech signal, so efficient segmentation of both noisy components from the desired components is demanded. Other than this, there are numerous noise sources each having its different characteristics so to deal with these various types of noise sources is demanding task. In spite of such difficulties speech enhancement techniques has always remained in vogue [1], as these are exploited in many applications such as Telecommunication, speech recognition, speech verification, hearing aids, VoIP and many more [1, 2].

Apart from this, speech recognition is also needed for human-machine interaction systems, where quality of the speech matters a lot for effective communication. Furthermore, single channel systems are more likely to face problems of signal degradation with the addition of the noise from various external as well internal factors. Rest of the paper is structured as follows: In Sect. 2, literature survey has been carried out, Sect. 3 includes comprehensive review on various techniques used for speech enhancement and Sect. 4 provides the conclusion.


S. Smys et al. (Eds.): ICCVBIC 2019, AISC 1108, pp. 536–540, 2020.
https://doi.org/10.1007/978-3-030-37218-7_61

## 2 Literature Survey

Yelwande, Dixit and Kansal [3] enhance speech by performing noise cancellation with use of Adaptive Wiener filter incorporating Normalized Least Mean Square Error algorithm. Proposed work includes the addition of construction, musical drums and jet noise to the speech signal under observation. It is observed from the results that mean square error is least in case of jet noise thus can be used to remove jet noise from the speech signal. Bahrami and Faraji [4] propose modelling of clean speech by using Weibull PDF. Results are compared with that of Rayleigh and Gamma PDF in terms of Segmental SNR and PESQ and it is seen that Weibull outperform other two. In [5], Zhao and Zhu proposed speech enhancement approach. Initially division of frames into many sub-bands is done by Wavelet Packet decomposition. Then voice activity detector differentiate between voice and noise based on two parameters namely energy of frame and spectral flatness. Afterwards, thresholding is performed on basis of VAD which results in noise removal and then to further enhance the speech iterative based on Kalman filter is applied. Proposed method outperformed other three methods (AT, IKF, S-IKF) in SNR and PESQ terms, as the computation complexity has been decreased also pre-processing leads to the refinement of speech signal which makes the task of IKF easy and effective.

Rehr and Gerkmann [6] proposed reduction of the residual noise from the speech with the help of super-Gaussian estimators. This approach yields better results than the Gaussian estimators. Apart from this, use of this estimator inclined the performance of the MLSE method. Also, the results carried out using Listening experiment. Wang, Du, Dai and Lee [7] carried out speech enhancement using 2 stage DNN, first stage involves the acceptance of many features and provides multi-vector of three features. Further these are given as an input to the other stage, which provides three clean speech features, followed by the post processing. Results compared with the DNN and LSTM targeting methods, which reveal that proposed method provide an edge over the others. Hou et al. [8] introduces joint framework for both audio and visual using DNN. This method provides enhanced speech and reconstructed images at the output. Method has also been compared to other speech enhancement algorithms, in which it is clearly observed that the proposed method outperform all. Other than this, method's performance is also checked in response to that of only audio based DNN and it has been seen that former produce better results than the latter. Also, the training has been carried out in end to end manner.

Furthermore, it is observed that lip shapes can be proved as effective features for the voice activity detection Jang and Liu [9] exploited DNN and CASA with dual based microphone. Training and testing is done using DNN. For training of DNN, recording data is used. Apart from this, both matched and unmatched training and testing environments are being considered. Moreover dual based hardware is also proposed which does not affect speech quality much. Stahl and Mowlaee [10] calculated harmonic parameters with the use of pitch-synchronous signal representation. Then, harmonic amplitudes are modeled based on Gamma distribution. In last, different estimators are obtained for different models (voiced, unvoiced and absence of speech). Then based on the Bayesian risk function mutual detection and estimation is found out. This proposed system is based on the stochastic-deterministic approach. Thus, observed that proposed method enhanced speech quality. Upadhyay, Jaiswala [11] carried out speech enhancement using wiener filter. And, the noise is estimated past and present values of power spectral with use of smoothening filter. Results of the proposed method compared with spectral subtraction and hence obtain better results. Yong, Chan and Nordholm [12], proposed method for speech enhancement based on the group of neural networks, where each network is dedicated for a particular band and SNR. Also the method is compared to the other state-of art techniques which are based on MMSE.

## 3 Review of Various Speech Enhancement Techniques

Comprehensive review on various speech enhancement techniques is provided in Table 1 as follows:

**Table 1.** Comprehensive review of various speech enhancement techniques

| Cited in | Publisher | Year | Dataset | Noise considered | Techniques used for speech enhancement |
|---|---|---|---|---|---|
| [3] | IEEE | 2017 | – | From jet, construction and musical drum | Adaptive wiener filter using NMS algorithm, done in time domain |
| [4] | IEEE | 2017 | TMIT | White, non-stationary and street noise | Modified MMSE-based algorithm by using Weibull PDF for fitting, in frequency domain |
| [5] | IEEE | 2017 | TSP | Non-stationary & babble noise from NOISEX-92 | Wavelet packet thresholding followed by usage of IKF |
| [6] | IEEE | 2018 | TMIT | Traffic and babble noise And some other random noise sources | Use of super-gaussian estimators based MLSE approach |
| [7] | IEEE | 2017 | – | 15 unseen noise sources | Two stage DNN framework – first MOL-DNN for learning, second MOE-DNN for ensemble |
| [8] | IEEE | 2018 | Car noise from AVICAR | 91 noise types including, car noises under five driving conditions | Audio and visual based DNN based on learning multi-task |
| [9] | IEEE | 2017 | – | – | DNN and CASA based with smoothing parameter, Dual |

(*continued*)

**Table 1.** (*continued*)

| Cited in | Publisher | Year | Dataset | Noise considered | Techniques used for speech enhancement |
|---|---|---|---|---|---|
| | | | | | based microphone also proposed |
| [10] | IEEE | 2018 | TMIT | Factory and babble noise | Stochastic deterministic based approach, utilizes pitch signal representation |
| [11] | Science Direct | 2016 | Speech of male and female speakers from NOIZEUS | White, musical, pink and additive noise | Wiener filter with recursive noise estimation |
| [12] | IEEE | 2017 | NOIZEUS | Pink noise | Multiple neural networks where each is operated on a particular band and SNR |

## 4 Conclusion

Speech Enhancement is a basic and crucial task which is to be carried out for various speech related applications. Apart from speech related applications, other fields like Communication, Artificial intelligence, Robotics etc. also require speech enhancement techniques. Speech Enhancement is a challenging task as there are numerous noise sources which corrupt the desired speech signal and each of these noise signal show different characteristics which makes this task even more difficult. Random characteristics of different noise makes the work of enhancement demanding. In this paper, comprehensive review of various techniques employed to carry out speech enhancement is provided. And it is has been evident that still there is a scope of improvement of efficiency and effectivity in speech enhancement task.

**Compliance with Ethical Standards**

✓ All authors declare that there is no conflict of interest.
✓ No humans/animals involved in this research work.
✓ We have used our own data.

## References


1. Upadhyay, N., Karmakar, A.: An improved multi-band spectral subtraction algorithm for enhancing speech in various noise environments. In: International Conference on Design and Manufacturing (IConDM). Procedia Engineering. Elsevier, vol. 64, pp. 312–321 (2013). https://doi.org/10.1016/j.proeng.2013.09.103
2. Zhang, Y., Zhao, Y.: Real and imaginary modulation spectral subtraction for speech enhancement. Speech Commun. **55**, 509–522 (2012). https://doi.org/10.1016/j.specom.2012.09.005

3. Yelwande, A., Kansal, S., Dixit, A.: Adaptive wiener filter for speech enhancement. In: International Conference on Information, Communication, Instrumentation and Control (ICICIC), pp. 1–4. IEEE Press, Indore (2017). https://doi.org/10.1109/icomicon.2017.8279110
4. Bahrami, M., Faraji, N.: Speech enhancement by minimum mean-square error spectral amplitude estimation assuming Weibull speech priors. In: Artificial Intelligence and Signal Processing Conference (AISP), pp. 190–194. IEEE Press, Shiraz (2017). https://doi.org/10.1109/aisp.2017.8324079
5. Zhao, M., Zhu, W.P.: Adaptive wavelet packet thresholding with iterative Kalman filter for speech enhancement. In: Global Conference on Signal and Information Processing (GlobalSIP), pp. 71–75. IEEE Press, Montreal (2017). https://doi.org/10.1109/globalsip.2017.8308606
6. Rehr, R., Gerkmann, T.: On the importance of super-Gaussian speech priors for machine-learning based speech enhancement. IEEE/ACM Trans. Audio Speech Lang. Process. **26**, 357–366 (2018). https://doi.org/10.1109/TASLP.2017.2778151
7. Wang, Q., Du, J., Dai, L.R., Lee, C.H.: A multi-objective learning and ensembling approach to high-performance speech enhancement with compact neural network architectures. IEEE/ACM Trans. Audio Speech Lang. Process. **26**, 1185–1197 (2018). https://doi.org/10.1109/TASLP.2018.2817798
8. Hou, J., Wang, S., Lai, Y., Tsao, Y., Chang, H., Wang, H.: Audio-visual speech enhancement using multimodal deep convolutional neural networks. IEEE Trans. Emerg. Top. Comput. Intell. **2**, 117–128 (2018). https://doi.org/10.1109/TETCI.2017.2784878
9. Jiang, Y., Liu, R.: A dual microphone speech enhancement method with a smoothing parameter mask. In: 10th International Congress on Image and Signal Processing, BioMedical Engineering and Informatics (CISP-BMEI), pp. 1–5. IEEE Press, Shanghai (2017). https://doi.org/10.1109/cisp-bmei.2017.8302095
10. Stahl, J., Mowlaee, P.: A pitch-synchronous simultaneous detection-estimation framework for speech enhancement. IEEE/ACM Trans. Audio Speech Lang. Process. **26**, 436–450 (2018). https://doi.org/10.1109/TASLP.2017.2779405
11. Upadhyay, N., Jaiswala, R.K.: Single speech enhancement using wiener filtering with recursive noise estimation. In: 7th International Conference on Intelligent Human Computer Interaction (IHCI). Procedia Computer Science. Elsevier, vol. 84, pp. 22–30 (2015). https://doi.org/10.1016/j.procs.2016.04.061
12. Yong, P.C., Chan, K.Y., Nordholm, S.: Utilizing neural network and critical band processing for speech enhancement. In: Asia-Pacific Signal and Information Processing Association Annual Summit and Conference (APSIPA ASC), pp. 1300–1303. IEEE Press, Kuala Lumpur (2017). https://doi.org/10.1109/apsipa.2017.8282232

# An Approach for Sentiment Analysis Using Gini Index with Random Forest Classification

Manpreet Kaur(✉)

Department of Computer Science and Engineering,
Amritsar College of Engineering and Technology, Amritsar, Punjab, India
manpreetkaur0389@gmail.com

**Abstract.** Person to person communication locales have turned out to be famous and basic spots for sharing wide scope of feelings through short messages. These feelings incorporate joy, pity, tension, dread, and so forth. Dissecting short messages helps in distinguishing the sentiment expressed by the group. Sentiment Analysis on movie reviews recognizes the general estimation or sentiment communicated by a commentator towards a movie. Numerous analysts are dealing with pruning the sentiment analysis model that plainly recognizes and distinguishes between a positive review and a negative review. In the proposed work, demonstrate that the utilization of features optimized by Gini index feature selection are then concatenating with Machine Learning classification gives better results both in terms of accuracy and class details parameters when tested against classifiers like Gini index with SVM, correlation with random forest and information with random forest. The proposed model unmistakably separates between a positive reviews and negative reviews.

**Keywords:** Sentiment analysis · Opinion mining · Random forest · SVM · Correlation · Information gain · Gini index

## 1 Introduction

Sentiments denote the feelings, emotions, views, ideas of individuals for the specific product. Sentiment analysis of opinion mining is very difficult. There are many complications like natural language processing for computerized extracting, categorizing and summarizing opinions which are expressed in online. Sentiment analysis is a study that is accomplished by various organizations for identifying user feedback about the products. This makes the other users for knowing the perfect selection of their favoured product.

Sentiment analysis is commonly employed in opinion mining for knowing sentiments, subjectivities moreover sensitive states in online texts. The process was accomplished on product evaluation by organizing the products attributes. At the present time, sentiment polarity analysis is utilized in an extensive range of domains like finance.

This concentrates on examining the direction-based text that involves text containing statements or opinions. The process of sentiment classification investigates whether the specific text is subjective or objective or if the text constitutes both the


S. Smys et al. (Eds.): ICCVBIC 2019, AISC 1108, pp. 541–554, 2020.
https://doi.org/10.1007/978-3-030-37218-7_62

feelings of positive or negative. This classification method has many numbers of essential qualities that may include various process, jobs, techniques, attributes and also application domains.

There exist many numbers of jobs in the classification of sentiment polarity. There are three major characteristics of classification are a class, level besides assumption with respect to sentiment sources as well as targets. The distinctive two class problem incorporates the categorization of sentiments as positive or negative. Furthermore, changes include organizing messages as subjective/objective.

Sentiment analysis concentrates on the specification of user's point of view with respect to specific area. The point of view involves assessment, perception or even emotional stages. The most prime job in sentiment or opinion analysis is the categorization of the polarity of specific text at the levels of features, document, sentences etc. After polarities are classified, emotional stages like "happy", "angry" and "sad" are also identified.

The classifications of polarities are the main task in opinion mining and it happens at the time of a portion of text stating an opinion on a single issue is categorized as one among the two conflicting sentiments. Few examples for polarity classification are "like" vs. "dislike" or "thumbs up" vs. "thumbs down". This classification also finds the advantages and disadvantages of statements in online reviews and assists in making the assessment of products more reliable. Another form of binary sentiment classification is agreement detection.

Agreement detection decides whether two text documents should obtain the similar or dissimilar sentiment associated labels. By identifying the classification of polarity this may allocate positivity degree to the polarity. It may place the opinion on a sequence in between positive and negative. This can also categorize the multimedia resources as stated by emotional or mood contents for intention like troll filtering, emotional human-machine communication and cyber-issue detection.

In order to construct an outline of these features, initially, features are found out and the opinions of positive and negative on those features are accumulated. The features of a product may include components, attributes and some other characteristics of a product. The summarization of opinion does not outline the reviews by choosing the subgroup or revises few of the unique statements from the reviews to catch the most important points as the classic text characterization.

Sentiment analysis aims at determining opinions, attitude and emotions in various sources such as news, documents, blogs, customer's reviews, social networking sites such as Twitter, Facebook and so on. Sentiment analysis is basically a classification problem that aims at classifying text on the basis of polarity that is positive, negative or neutral. Due to the emerging Internet technology, many people put forward their opinions, views and attitude towards government policies, products, movie reviews, sports and much more (Pang, Lee et al. 2002; Blitzer et al. 2007). Moreover, social media acts as a platform for users to give opinions and share their views. Nevertheless, Social Media Online Sites can provide valuable data and hidden inferences in real time that can be processed for decision making. Henceforth, this data can be utilized to perform sentiment analysis. The field of sentiment computing has grabbed the attention of various business-oriented organizations to track the sentiments of a user towards products so that they can make further enhancements according to the view of the user.

Moreover, it can be useful for election prediction and detecting liking towards a particular election candidate. Therefore, it acts as a very effective decision-making technique.

It is a kind of mode of communication that can be utilized by any individual lives in any area of the world. With such all-inclusiveness and high-speed information, sentimental analysis on such systems has been most focused on research subjects in NLP in the previous decade. The fundamental point of sentimental analysis is to identify the extremity of the content. The web has drastically changed the manner in which individuals express their perspectives and sentiments [1]. Big information is trending analysis space in engineering science and sentiment analysis is one amongst the foremost necessary a part of this analysis space. Big information is taken into account as terribly great amount of information which might be found simply on internet. Social media communication, remote detecting data and medical records etc. in form of structured, semi-structured or unstructured data and we can utilize these data for sentiment analysis. Sentimental Analysis is all on the brink of get the $64000 voice of individuals towards specific product, services, organization, movies, news, events, issues and their attributes. Sentiment Analysis includes branches of engineering science like linguistic communication process, Machine Learning, Text Mining and Information Theory and Coding. By using approaches, methods, techniques and models of defined branches, we can be order (classify) our information might be in type of news stories, online journals, tweets, film surveys, item audits and so on into positive, negative or neutral sentiment according to the sentiment is expressed in them.

## 2 Background

Utilization of Hybrid Feature Extraction Method (HFEM) makes the model increasingly effective in terms of accurate classification by adding the advantages of individual feature extraction method. HFEM improves the space unpredictability by decreasing the input space to negligible number of features that are adequate to represent the review content. Subsequently, the results got are highly promising both in terms of space complexity and classification accuracy [2].

Anand et al. attempted to perform ABSA survey on motion picture review data. Dissimilar to different spaces, for example, camera, PCs cafés and so forth, a noteworthy piece of film surveys is committed to portraying the plot and contains no data about client interests. The overhead of growing physically labeled data to assemble the classifier is stayed away from and the subsequent classifier is demonstrated to be powerful utilizing a little physically constructed test set. Furthermore, they proposed a scheme to identify aspects and the corresponding opinions utilizing a lot of hand created guidelines and viewpoint i.e. rules and aspect clue words. Three schemes for choice of perspective clue information words are investigated - manual labeling (M), clustering (C) and review guided clustering (RC). The aspect and sentiment detection utilizing all the three schemes is empirically evaluated or observationally assessed against a manually built test set. The experiments build up the effectiveness or viability of manual labeling over cluster based methodologies however among the cluster based approaches, the ones utilizing the review guided clue words performed better [3].

Catal et al. exploit a sentiment classification model on basis of Vote troupe classifier utilizes from three individual classifiers: Bagging, Naïve Bayes & Support Vector Machines. Moreover, in bagging they utilized SVM as base classifier. The main focus of this research is to enhance the execution of machine learning classifiers for feeling grouping of Turkish audits and documents. Their experimental results show that multiple classifier system based approaches are much better for sentiment classification of Turkish documents. They performed experiments on three different domains such as book review, movie reviews and shopping reviews. The authors concluded that this approach is not restricted to just one domain and can be extended to several other domains as well [4].

Pannala et al. explained aspect based sentiment analysis concept in machine learning era. This paper mainly explored sentiment analysis in light of the prepared informational collection to give the positive, negative and unbiased surveys for various items in the advertising scene. In angle based estimation examination (ABSA) the point is to recognize the parts of substances and the notion communicated for every perspective. A definitive objective is to have the capacity to produce rundowns posting every one of the angles and their general extremity. To prepare the application for the given informational collections SVM (bolster vector machine) and ME (Maximum Entropy) arrangement calculations have been utilized. Performance of algorithms is analyzed based on precision, recall and F Measure [5].

Guha et al. presented a detailed description of our system, that stood 4th in Aspect Category subtask. Experimented with Ensemble Learning technique for slot 3, which we want to explore and improve further. They submitted unconstrained systems for the Restaurants dataset [6].

Yang et al. exploit a novel approach for extracting emotions. They used graphical emoticons, punctuation expressions along with a compact lexicon to label data. They provided a multi-label emotion classification algorithm (MEC) for analyzing emotions in short text of Weibo, which is a very famous online social networking site in china (just like Twitter). The approach they used is phycology independent for it worked well on different phycology theories for emotion classification. In the proposed approach they exploit K-nearest neighbors (KNN) for tweet level analysis and Naïve Bayes for Word level analysis of emotion. Moreover, their approach outperformed various state-of-art methods as discussed in experiment and results. The dataset contained tweets about Malaysia 370 missing flight and they concluded from their approach that the episode of Anger has a postponement subsequent to limit of Sadness [7].

Nithya, this paper represents Sentiment analysis that mainly on subjective and polarity detection. A proposed work include: (i) Feature Extract- Commonly, Sentiment analysis uses AI calculation machine learning algorithm and a technique to extract features from written texts and after that train the classifier. (ii) Preprocessing-stemming refers reducing words to their roots. Porter's stemming algorithm used for removing stop words. Mostly, adjective words have sentiment. (iii) Product aspects - Textstat is a freely available that can be used for extracting pattern. (iv) Find polarity of opinionated sentence- here SentiStrength lexicon-based classifier used to detect sentiment strength. Here, 575 reviews have been taken from shopping sites. Tanagra1.4 tool used for data mining. Naïve bayes classification done through this tool based on each individual features such as display, accessories, battery life, weight and cost. Results shows that 'battery life' have most positive value so it improves branding and 'cost' have very low positive value that indicate seller to concentrate more on reputation and product quality [8].

Javier et al., paper introduces a COSMOS stage for sentiment and tension analysis (investigation) on twitter dataset. Tool utilized for sentiment analysis is SentiStrength. To flow application dependent on cloud environment, it utilizes virtualized Hadoop Clusters in Open Nebula. This system configuration utilized for execution or performance aspects should be dispersed as to lessen fluctuation in the analysis performance. It likewise shows the design or architecture for information preparing of COSMOS using OpenNebula and Hadoop [9].

Rehman, Text classification is thought to be one of the tremendous zones of research in data mining and machine learning. It can be viewed as normal machine learning issue in which reports must be recognized into different known classes. Specialists have paid very concentration especially to some particular issues in content order. We utilized Machine learning calculations like Nearest Neighbors, k-Means, Support Vector Machines and Naïve Bayes to recognize content records. The Performance of a content grouping undertaking is straightforwardly influenced by portrayal of information. When highlights are chosen fittingly straightforward classifiers may deliver great arrangement comes about even. The most ordinarily utilized elements to speak to words, Term Frequency (TF) and Inverse Document Frequency (IDF), may not be constantly suitable. In this paper our concentration is to enhance portrayal of content archives by characterizing new term weights for terms in content reports with the goal that both precision and execution can be made strides [10].

## 3 Existing Approach

The existing approach utilizes random forest as an ensemble classification technique. It is an unpruned classification arrangement or regression trees, start from bootstrap tests of the training data, using irregular element choice (random feature selection) in the tree introduction (presentation) process. Expectation (Prediction) is made by accumulate (majority vote in favor of classification or averaging for regression) the forecasts of the troupe. Random forest generally illustrate the most part represent significant execution improvement over the single tree classifier, for example, CART and C4.5. It requests generalization blunder rate i.e. error rate that differentiates positively to Ada boost, yet is increasingly vigorous to commotion. Notwithstanding, similarly to most classifiers, RF can also likewise experience the ill effects of the scourge of gaining from an incredibly imbalanced training data set. As it is formed to limit the overall error rate, it will in general spotlight more on the forecast accuracy of the larger part class, which regularly brings about poor accuracy for the minority class.

To beat the issue, balanced random forest is proposed.

## 4 Proposed Work

The proposed work consists of hybrid model comprising Gini index-based feature selection with balanced random forest as a classification technique for predicting the sentiments of movie reviews. The proposed Gini index feature selection include the issues of movement of prior class probability and worldwide decency of a component in two phases. In the first place, it changes the examples space into a component explicit standardized examples space without bargaining the intra-class include circulation. In the second phase of the system, it distinguishes the highlights that segregates the classes most by applying gini coefficient of disparity. Also, Balanced **Random Forest Algorithm** is used for classification which handle missing values using median for numerical values or mode for categorical values. The proposed work leads to the selection of relevant attributes for prediction and less error rate and accuracy in the result.

Firstly, the gathering of raw data from Imdb movies review site and afterward separate filtering techniques are applied to create that raw data into organized structured format. Filtration of textual movie reviews consist of following phases:

- Word parsing and tokenization: In this stage, every client survey parts into words of any natural processing language. As motion picture audit contains block of character which are alluded to as token.
- Removal of stop words: Stop words are the words that contain little dataset should have been evacuated. Stemming: It is described as a technique to decrease the induced words to their one of a kind word stem. For example, "talked", "talking", "talks" as dependent on the root word "talk". We have utilized Snowball stemmer to inferred word to their starting point.

After pre-processing and filtration feature selection using Gini Index is done to rank the words according to the gini index functionality (Fig. 1).

Selected features data is then classified by using Balanced Random Forest

- For every cycle in random forest, draw a bootstrap test from the minority class. Arbitrarily draw a similar number of cases, with substitution, from the dominant part class.
- Induce a classification tree from the information to most extreme size, without pruning. The tree is prompted with the Random Tree algorithm, with the accompanying adjustment: At every hub, rather than scanning through all factors for the ideal split, just hunt through a lot of m-try randomly chosen variables.
- Repeat the two stages above for the occasions wanted. Total the expectations of the gathering and make the last forecast.

Evaluate and analyze the performance using k cross validation model on the basis of Recall and Precision of existing algorithms and new proposed algorithm.

The proposed methodology follows the below steps:

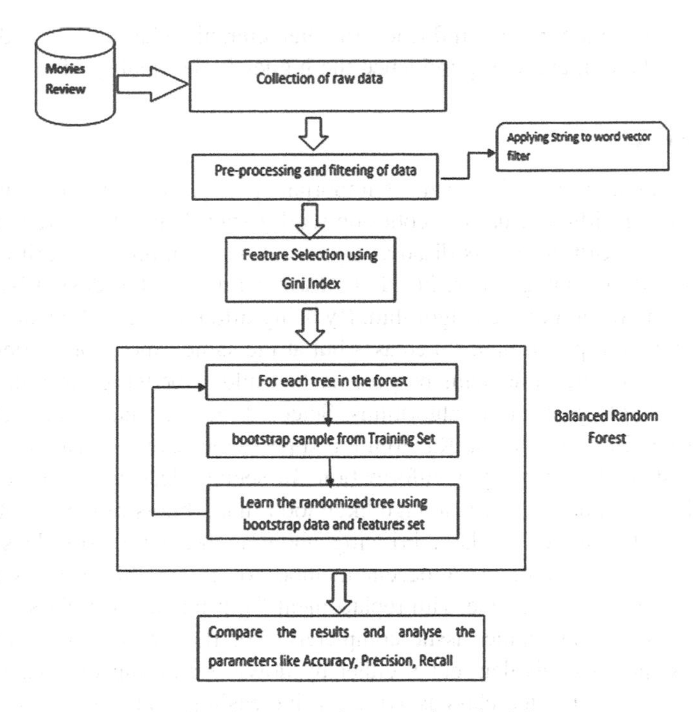


**Fig. 1.** Flowchart of proposed methodology

### 4.1 Preprocessing and Filtration

Preprocessing and Filtration is performed to normalize the data. Filtering techniques like StringToWordVector filter, which converts the string qualities into a lot of characteristics speaking to word event data from the content contained in the strings, has been utilized.

### 4.2 Feature Selection

It is utilized to as a splitting method. In this algorithm we accumulate informational data sets for testing. Let A be the sampling, m be the isolated no of subsets, P be the probability and Ci be the various classes then $GiniIndex(A) = 1 - \sum_{i=1}^{m} P_i^2$.

Right when the base of GiniIndex(A) is zero then it implies that every one of the records have a spot with a alike classification at this gathering; it shows that the best accommodating information can be procured. At that point exactly when all of the instances of accumulation have a standard flow to a particular specific classification, GiniIndex(A) accomplishes most prominent, exhibiting the base profitable data got. For gini index the little contaminating impact is the higher the quality. Then again,

$$GiniIndex(A) = \sum_{i=1}^{m} P_i^2$$

measuring the contaminating influence of characteristic classify method, the more prominent is the contaminating influence the higher is the quality of attribute.

### 4.3 Classification

The classification process is the most important process in sentiment analysis. The classification algorithms if used in cohesion work better than if they are used as single because every algorithm has its disadvantages and when different algorithms are used together their disadvantages is reduced. To further improve the design bagging technique can be used for a chosen algorithm. By using different algorithms in an ensemble learner method, the performance increases but at the same time a lot of time is utilized for training. But since the main purpose is to build a better performing design as compared to single performing algorithms. Selected features data is then classified by using Balanced Random Forest. Random forest persuades every constituent tree from a bootstrap test of the preparation information. In seeing (learning) exceedingly overburden unbalanced data, there is a huge likelihood that a bootstrap test contains few or even none of the minority class, bringing about a tree with horrible showing for foreseeing the minority class. A decent method for fixing this issue is to utilize a stratified bootstrap; i.e., sample with replacement from inside each class. This still yet does not tackle the imbalance issue completely. For the tree classifier, misleadingly making class priors equivalent equal either by down-examining the majority class or oversampling the minority class is typically increasingly successful as for a given exhibition estimation, and that down inspecting appears to have an edge over overtesting. In any case, down-examining the majority class may bring about loss of data, as a huge piece of the dominant part class isn't utilized. Random forest inspired us gathering trees initiated from balanced down-tested information. Balanced Random Forest consolidates the down inspecting greater part class strategy and the gathering learning thought, misleadingly modifying the class dissemination so classes are spoken to similarly in each tree.

### 4.4 Algorithm: Each Tree is Grown as Following Algorithmic Steps

**Step 1:** Consider M and N which represents the number of training cases and number of variables in classifier.
**Step 2:** To find a decision at node of tree, n of input variables are to be used and n < N.
**Step 3:** Training set for tree is to be picked m times with substitution from M training cases that are accessible. By estimating their classes, we can use utilize whatever is left of the cases to approximate the error of the tree.
**Step 4:** *n* factors are to be picked arbitrarily for every node of tree on which we make choice at that node. On the basis of *n* variables present in training data, calculate the superlative split.
**Step 5:** Every tree is developed to the greatest degree and there is no pruning.

### 4.5 Building and Evaluating Model

Model is built and tested using cross validation 10 folds techniques. In this method, the full informational data set is taken as an input, at that point separate it into k no's of proportionate sets (k1, k2, ..., k10 for instance for 10-overlap CV) without spreads. By then at the essential run, take k1 to k9 as preparing set and develop a model. Utilize that model on k10 to get the presentation. Next comes k1 to k8 and k10 as preparing set. Build up a model from them and apply it to k9 to get the presentation. Along these lines, utilize every one of the folds where each crease all things considered 1 time is utilized as test set. At that point midpoints the performances of the considerable number of models in type of exactness (accuracy). This will be the final performance of an algorithm using k cross validation model and the parameter Recall, Precision, accuracy etc. are calculated and compared with the existing algorithm.

## 5 Results and Discussion

The proposed technique is simulated in java netbeans using weka as an external library. The evaluation parameters of the proposed procedure technique are compared with the existing techniques.

**Table 1.** Accuracy comparison of proposed technique with the existing technique

| Algorithms | Accuracy |
|---|---|
| IG with Random Forest | 81.63 |
| Correlation with Random Forest | 79.37 |
| Gini Index with SVM | 76.22 |
| Gini Index with BRF | 90.69 |

The Table 1 above demonstrates the comparison of accuracy of the proposed procedure with the current method or existing technique. The technique with more accuracy is superior than the other technique. Accuracy of the proposed technique is 90.69 whereas accuracy of the Gini index with SVM is 76.22, Correlation with Random Forest is 79.37 and that of Information gain with Random Forest is 81.63.

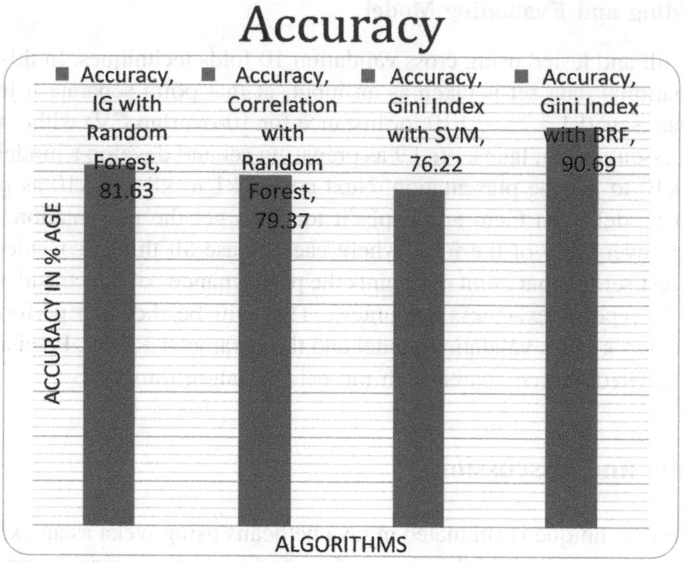

**Fig. 2.** Showing the accuracy comparison of existing with the proposed algorithm

The Fig. 2 above shows the accuracy comparison of the existing technique and proposed technique. The chart obviously demonstrates that the proposed calculation performs better as its accuracy is 90.69.

### 5.1 Precision Comparison

**Table 2.** Precision comparison of proposed technique with the existing technique

| Algorithms | Precision |
|---|---|
| IG with Random Forest | 0.777 |
| Correlation with Random Forest | 0.779 |
| Gini Index with SVM | 0.724 |
| Gini Index with BRF | 0.909 |

The Table 2 above shows the comparison of precision of the proposed technique with the existing technique. The technique with more precision is superior to the other technique. Precision of the proposed technique is 0.909 whereas precision of the Gini index with SVM is 0.724, Correlation with Random Forest is 0.779 and that of Information gain with Random Forest is 0.777.

The Fig. 3 below shows the precision comparison of the existing technique and proposed technique. The graph clearly shows that the proposed algorithm performs better as its precision is 0.909.

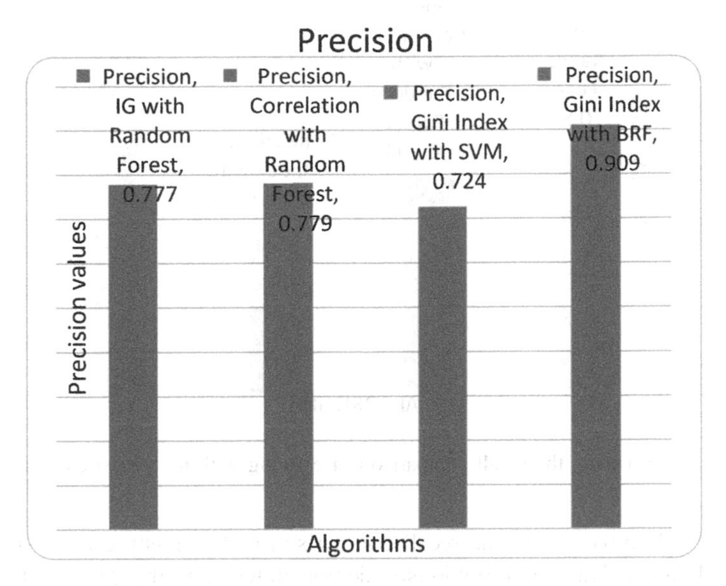


**Fig. 3.** Showing the precision comparison of existing with the proposed algorithm

## 5.2 Recall Comparison

**Table 3.** Recall comparison of proposed technique with the existing technique

| Algorithms | Recall |
|---|---|
| IG with Random Forest | 0.816 |
| Correlation with Random Forest | 0.794 |
| Gini Index with SVM | 0.762 |
| Gini Index with BRF | 0.907 |

The Table 3 above shows the examination of recall of the proposed technique strategy with the existing technique. The technique with more recall is better than the other technique. Recall of the proposed technique is 0.907 whereas recall of the Gini index with SVM is 0.762, Correlation with Random Forest is 0.794 and that of Information gain with Random Forest is 0.816.

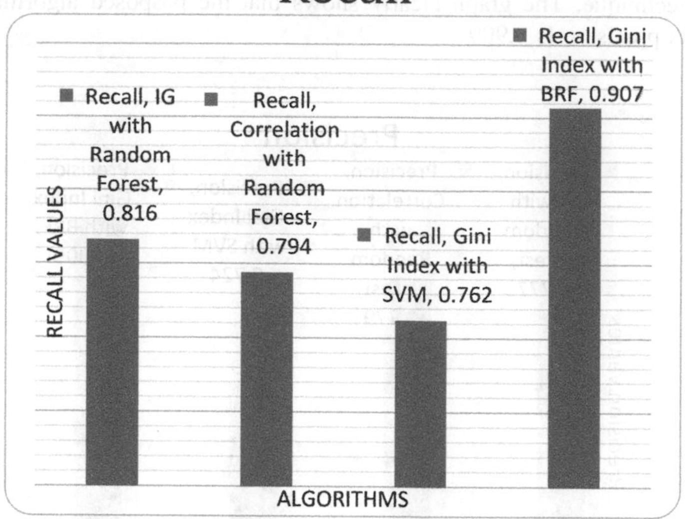


**Fig. 4.** Showing the recall comparison of existing with the proposed algorithm

The Fig. 4 above shows the recall comparison of the existing technique and proposed technique. The chart obviously demonstrates that the proposed calculation performs better as its recall is 0.907.

### 5.3 F-Measure Comparison

**Table 4.** F-measure comparison of proposed technique with the existing technique

| Algorithms | F-measure |
|---|---|
| IG with Random Forest | 0.785 |
| Correlation with Random Forest | 0.785 |
| Gini Index with SVM | 0.715 |
| Gini Index with BRF | 0.902 |

The Table 4 above shows the comparison of f-measure of the proposed technique with the existing technique. The technique with more f-measure is better than the other technique. F-measure of the proposed technique is 0.902 whereas f-measure of the Gini index with SVM is 0.715, Correlation with Random Forest is 0.785 and that of Information gain with Random Forest is 0.785.

The Fig. 5 below shows the f-measure comparison of the existing technique and proposed technique. The chart obviously demonstrates that the proposed calculation performs better as its f-measure is 0.902.

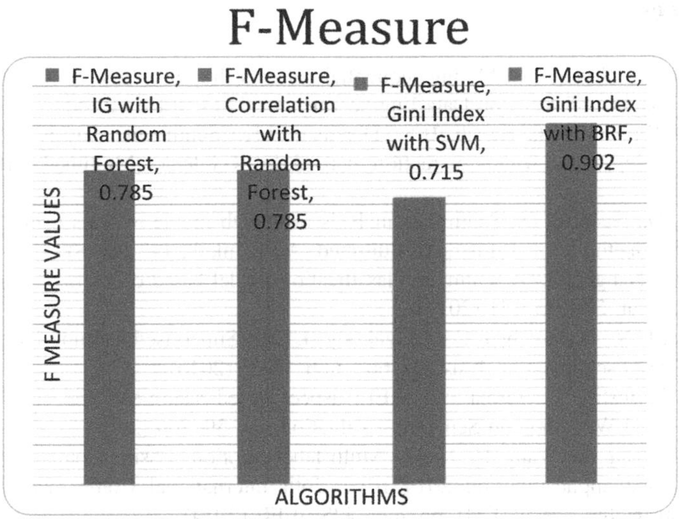

F-Measure

■ F-Measure, IG with Random Forest, 0.785
■ F-Measure, Correlation with Random Forest, 0.785
■ F-Measure, Gini Index with SVM, 0.715
■ F-Measure, Gini Index with BRF, 0.902

F MEASURE VALUES

ALGORITHMS

**Fig. 5.** Showing the f-measure comparison of existing with the proposed algorithm

## 6 Conclusion

Sentiment analysis aims at conclusive opinions, attitude and emotions in various sources such as news, documents, blogs, customer's reviews, social networking sites such as Twitter, Face book and so on. This proposed work has implemented Gini Index feature selection based Balanced Random Forest Classification of movies reviews collected from ImDb site. Further the proposed classification model is compared with three existing techniques including Information gain with random forest, correlation with random forest and Gini Index with random forest, The results shown in the above results and discussion section clearly represents the proposed sentiments prediction model is better than the all the three existing techniques with respect to accuracy, precision, recall and f-measure. The proposed work is more efficient than previous algorithms.

In future, any other more efficient artificial intelligence technique can be used for more optimized results like as (ACO) ant colony optimization, particle swarm optimization etc. to improve the features set.

**Compliance with Ethical Standards**

✓ All authors declare that there is no conflict of interest.
✓ No humans/animals involved in this research work.
✓ We have used our own data.

## References

1. Agrawal, R., Srikant, R.: Mining sequential patterns. In: 1995 Proceedings of the Eleventh International Conference on Data Engineering, pp. 3–14. IEEE (1995)
2. Keerthi Kumar, H.M., Harish, B.S., Darshan, H.K.: Sentiment analysis on IMDb movie reviews using hybrid feature extraction method. Int. J. Interact. Multimed. Artif. Intell., 1–7 (2018)
3. Anand, D., Naorem, D.: Sentiment analysis on IMDb movie reviews using hybrid feature extraction method. Int. J. Interact. Multimed. Artif. Intell., 1–7 (2018)
4. Catal, C., Nangir, M.: A sentiment classification model based on multiple classifiers. Appl. Soft Comput. **50**, 135–141 (2017)
5. Pannala, U.N., Nawarathna, C.P., Jayakody, J.T.K.: Supervised learning based approach to aspect based sentiment analysis, pp. 662–666. IEEE (2016)
6. Guha, S., Joshi, A., Varma, V.: SIEL: aspect based sentiment analysis in reviews. In: International Workshop on Semantic Evaluation, pp. 59–766 (2015)
7. Yang, J., Jiang, L., Wang, C., Xie, J.: Multi-label emotion classification for tweets in weibo: method and application. In: 2014 IEEE 26th International Conference on Tools with Artificial Intelligence (ICTAI), pp. 424–428. IEEE (2014)
8. Nithya, R., Maheshwari, D.: Sentiment analysis on unstructured review. In: International Conference on Intelligent Computing Application, pp. 367–371. IEEE, March 2014
9. Farhadloo, M., Rolland, E.: Multi-class sentiment analysis with clustering and score representation. In: 13th International Conference on Data mining Workshops, pp. 904–912. IEEE, December 2013
10. Rehman, A., Babri, H.A., Saeed, M.: Feature extraction algorithms for classification of text documents. In: International Conference on Communications and Information Technology, Aqabajordan, pp. 1148–1154 (2012)

# Mining High Utility Itemset for Online Ad Placement Using Particle Swarm Optimization Algorithm

M. Keerthi(✉), J. Anitha, Shakti Agrawal, and Tanya Varghese

Department of Computer Science and Engineering,
DSATM, Bangalore 560082, Karnataka, India
keerthiimohan@gmail.com, anitha.jayapalan@gmail.com,
shaktigrwl@gmail.com, tanyavarghese4@gmail.com

**Abstract.** The process of obtaining relevant content from the raw datasets, present in huge databases is called data mining. It consists of four stages: Data Sources, Data Gathering, Modeling, and Deployment. These databases are searched thoroughly using various data mining techniques. Association rule mining, also known as ARM is the significant method of mining the data that tells us about the relationships of the items that are available in the proceedings. Traditionally, the mining of data focuses on the association between the Frequent Itemset that are present in large numbers in the databases. Mining High Utility Itemset (HUI) is an advancement of the problem of Frequent Itemset Mining (FIM). In the mining of HUI, the aim is to find itemsets that have more utility use than that of the threshold utility value. In this paper, Particle Swarm Optimization (PSO) algorithm is used to mine high utility itemsets.

**Keywords:** Data mining · High utility itemset · Particle Swarm Optimization · Ad placement · Ad clicks and impressions

## 1 Introduction

Data mining is a significant way of obtaining appropriate and coherent data from various forms of data repositories. It is useful in exploring bulk amounts of data to find patterns and analyse them. The first step in data mining is the collection of the dataset from different databases which can be used for defining and solving the problems. This gathering of all the relevant information i.e. the data sources forms the initial phase in this process. In this particular phase, the data is combined or collected from various data sources, as data is present in different forms in distinct areas. The data can be stocked up in text files, database and so on. Metadata is used in order to scale down the errors in the process of integration of data. In addition the other problem encountered is redundancy of data. In a circumstance like this, we may find identical data in distinct tables which is a part of the same database. The integration of data aims to scale down the repetition to the utmost and also ensures that it does not affect the efficiency and dependability of the data. The second stage is the discovery, examination, and study of data to extract useful data out of all data gathered. The data which is needed for the


S. Smys et al. (Eds.): ICCVBIC 2019, AISC 1108, pp. 555–562, 2020.
https://doi.org/10.1007/978-3-030-37218-7_63

analysis, in this process, is mined from the database. It needs bulk amount of data which had been collected previously for analysis. The repository of data with the information that had been integrated, consists of a lot of data than what was needed. The required data needs to be chosen and accumulated from the total available data. The third stage involves constructing a model and selecting suitable algorithms to evaluate the expected outcome in all the test cases. In this stage, we have administered methods to mine the patterns from the collected datasets. Also, here the mining consists of various works, such as clustering, prediction, classification, time series analysis, etc. The fourth stage is the evaluation of the results in the context of the business goal. In this phase, new business requirements can arise due to the discovery of new patterns during the evaluation of data. The acquiring of business insights is a repetitive process in data mining. The decision to continue with the process must be made at this particular time before moving the project on to the deployment phase. In short, it involves the analysis of the products or outcomes and taking adequate actions based on that. Frequent itemset mining (FIM) has acquired an ample amount of consideration and also being administered in various realms as it is one of the starting stage of ARM which stands for association rule mining [11, 12]. The main task of FIM is to mine all those item sets which have their occurrence frequency above the threshold value specified by the users. The patterns which appear intermittently within a given data set are the frequent patterns. The itemset consisting of such kind of patterns are called frequent itemset and thus the mining or exploration of such item sets are called frequent itemset mining. To solve such problems, an upcoming topic in the field of data mining is Utility Mining. The major purpose of utility mining is to find out the item sets with highest efficiency or utility based on the choices specified by the user. High Utility Itemsets Mining is the discovery of all those item sets that have utility above the threshold value defined by the user. Itemset Utility Mining is an enhancement of Frequent Itemset Mining, which explores the itemsets that occur frequently. There have been many algorithms related to mining that have been put forward in order to analyse the HUIs. Some of them are the techniques based on pattern growth and methods like level-wise candidate Generation-and-Test [7, 13].

## 2 Related Work

There is a lot of experimentation which has been carried out previously based on the ads and ad slots or ad placements [1, 2, 4–6]. A form of online advertising within which the publishers are paid by the advertisers for putting up the graphical ads on their websites is called display advertising. The normal methodology of marketing of the display advertising has been agreed upon previously as contracts between publishers and advertisers [2]. Thus there have also been various studies based on conversion prediction in display advertising [2, 3, 9, 10].

The above studies led to the making of a classification model which categorized based on whether a conversion will take place on an ad call or not. Such models need ample amount of data regarding the present user (like their location, age, gender), the ads and ad slots (like category, type, textual information etc.) in order to have more accuracy in the studies. It is difficult to slide through the ad slots in a mobile application as compared to the web pages where we can slide to obtain the text information. Also the information about the user may not be available due to privacy concerns.

Thus in order to overcome these, an ad slot mining algorithm was proposed which helped to mine new and relevant ad slots for each of the ads. [8] And additionally this algorithm doesn't need any information like user data or textual information. This proposed algorithm is similar to the method of click prediction which was proposed by Menon et al. [4]. An algorithm of Click Through Rate (CTR) prediction was proposed by them which combines the matrix factorization and collaborative filtering techniques. Also it makes use of the hierarchical data of the websites along with the ads with the algorithm using matrix factorization in order to enhance the efficiency of the click through rate prediction method.

## 3 Proposed System

In the proposed system, a dataset of different advertisements are considered that are displayed on one website to extract the High Utility Itemset. The high utility itemset (HUI) for the proposed system are the ads that have the conversion rate higher than the specified threshold value of conversion rate, which is used to predict the types of ads that results in the generation of higher revenue.

These HUIs are extracted from the dataset by calculating the conversion rate of the number of impressions on an ad to the number of clicks on that ad for the optimal ad placement on a website. In Sect. 1, we introduce the Particle Swarm Optimization algorithm to find out the HUI from different types of ads in such a way that the time and space with respect to the number of iterations will be optimized and results will be obtained more efficiently. In Sect. 2, the strategy used for extracting the HUI with the extended version of the PSO algorithm is explained. The extended PSO compares the local best conversion rate of an ad to the global best conversion rate of that ad to compute the HUI. In Sect. 3, we discuss the attributes used in the selected dataset where each ad has its own type, number of impressions, clicks, and the revenue generated that is tested with the algorithm and the results obtained are cross-checked with the possible outcomes.

### 3.1 Particle Swarm Optimization Algorithm

In this section, the structure of the PSO algorithm which is a Bio-Inspired algorithm is used to explain the mining of high utility itemset. PSO is similar to the Genetic Algorithm (GA) with respect to the advanced calculation techniques. The PSO system is loaded with a finite number of random solutions and the optimal solution is found by the updation of the generations. But contrary to GA, the PSO algorithm does not use any operators for evolution like mutation or crossover. The possible solutions in PSO which are referred to as particles are computed throughout the problem space by successively using the current optimum particles. A framework of this algorithm is illustrated in Fig. 1. PSO may possibly sound difficult but it is very easy to execute. A collection of variables have their values aligned to the member which has its value closest to the target at any given moment, over a series of iterations. The algorithm keeps on calculating and storing three global variables which are: (i) the condition which specifies the number of iterations (ii) gBest or global best that indicates each particle's utility in the present iteration and which is closest to the target (iii) the counter or stopping value which stops the execution if the target is not found.

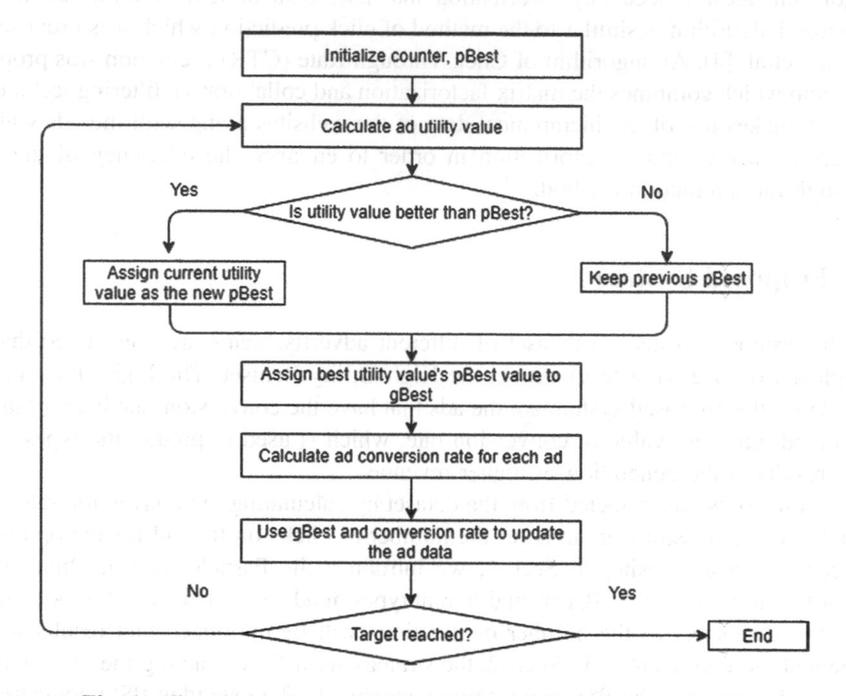


**Fig. 1.** Flowchart of extended Particle Swarm Optimization algorithm

Here, each itemset consists of the particle's position, its velocity and the number of iterations to be done to calculate the gBest value for each iteration and compare it with the pBest. Then we check, if the value of the gBest is greater or better than that of the value of pBest then we change the current value of pBest to gBest and follow the algorithm until we get the target result.

The data which we are using here is the data in which the particles are nothing but the different types of ads, the utility for the system is the conversion rate of the number of impressions to the number of clicks on an ad and the particle's velocity is the clicks on an ad.

### 3.2 Algorithm Description

In the proposed system, we use an extended version of the Particle Swarm Optimization algorithm to obtain the optimized results by performing number of iterations on the dataset. We find the best ads displayed on a website by a publisher that generates higher revenue using this system. In our algorithm, different types of ads shown on online websites are used instead of particles in PSO to gain high utility ads and we are using utility or fitness as the conversion rate from impressions to clicks on any ad to compute optimal results. The velocity of the particles in the proposed system is shown as the number of clicks on the ads which increases the revenue. The formula to calculate the utility or the conversion rate is given below:

$$\frac{1D436}{1D43C} \times \boldsymbol{1D7\,CF1D7\,CE1\,D7CE} \tag{1}$$

In (1), C is the clicks on an individual ad and I is the impressions on that same ad. In the extended version of PSO, for all the itemsets the counter and pBest value is assigned. The pBest value is the specified threshold value of the conversion rate. Then the iterations are performed until the specific condition is fulfilled for all the ads. In all the iterations each time, the utility or the conversion rate is calculated and then compared with the pBest value. If the current conversion rate is greater or better than that of the value of pBest, the value of pBest is set to the current conversion rate. Else if the value of pBest is better than the conversion rate then the gBest value is set to the pBest value.

```
For each ad
{
   Initialize counter, pBest
}
Do until maximum number of iterations
{
   For each ad
   {
      Compute ad Utility (conversion rate)
      If the value of utility is better than pBest
      {
         Set pBest to the current value of utility
      }
      If pBest is better than gBest
      {
         Set gBest = pBest
      }
   }
   For each ad
   {
      Compute ad conversion using the above formula
      Use gBest and conversion rate for updation of ad Data
   }
```

### 3.3 Experimental Environment and Datasets

The data which is used to find out the results and effectiveness of the method is real time data of ads published on a website is shown in Table 1.

**Table 1.** Characteristics of the dataset

| Type of ad | Impressions | Clicks |
|---|---|---|
| 1 | 136693 | 812 |
| 2 | 120529 | 693 |
| 3 | 107492 | 676 |

The attributes which are used to conduct the experiments are type of an ad, impressions and clicks which are extracted from the database, which was carried out with the algorithm proposed and run for different threshold values.

## 4 Experimental Results

Firstly, we ran the experiments on a MacBook Pro mid 2012 with 2.5 GHz Intel Core i5 processor and 8 GB DDR3 RAM. The dataset and results were verified with a python program developed on the basis of the proposed algorithm. The ad platform that we used to run the experiments consists of different types of ads and the impressions on each ads with the number of clicks, which is used to calculate the efficiency of ad to increase the revenue model for publishers. The number of iterations performed in our experiments divides the dataset and then the itemset in each division are pushed into the program for the experiments and the minimum utility threshold value is set as 2 which is average value of conversion from impression to clicks and pBest is initially set to 2 only. Out of 3,000 itemsets we extracted the values in which the conversion rate of the impression to click is high and thus we gain the high utility itemset which can tell us which types of ads are better to display on any website online and which types of ad can generate more revenue. Also we calculated the time complexity and the number of HUI's obtained by changing the values of the maximum number of iterations.

The number of high utility itemsets which are extracted from running the python code was plotted on a graph against the different values of maximum iterations is represented in Fig. 2. Here, when the number of iterations were changed the number of HUIs obtained were different.

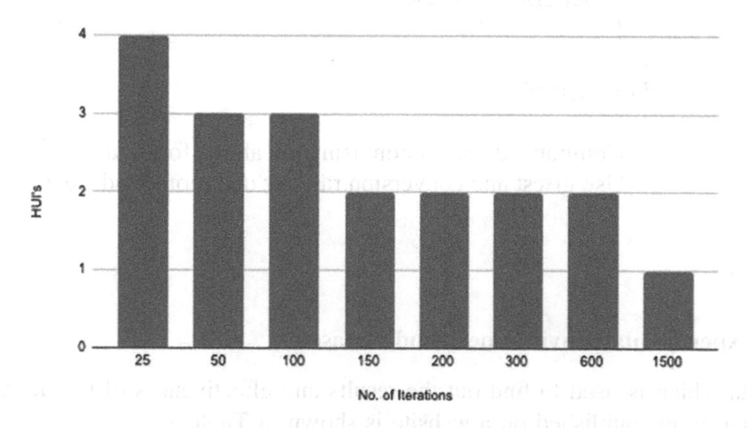


**Fig. 2.** Representation of HUI's obtained for different values of maximum iterations

So, with the help of the above graph, we can conclude that the number of high utility itemsets were decreasing when there is an increase in the number of iterations. Also in order to derive the time taken by the values of iterations, considered above, from the high utility itemsets obtained as a result of the above algorithm, a graph is plotted for these iterations against time as shown in Fig. 3.

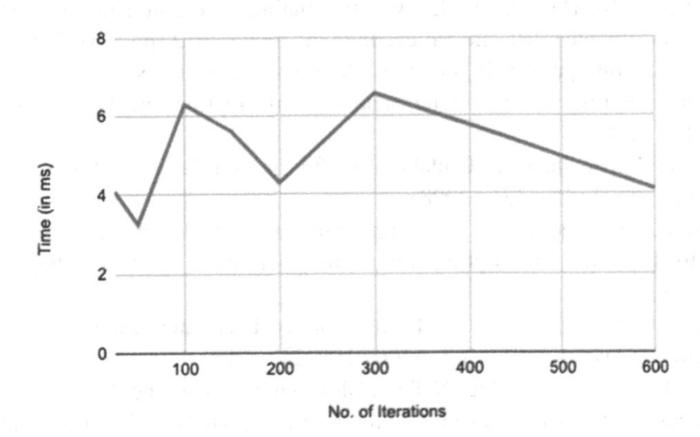


**Fig. 3.** Time taken for different values of maximum iterations

So, with the help of the above graph, we can conclude that the time taken to perform various iterations has no specific rule or order to complete the iteration.

## 5 Conclusion

In the proposed system, the dataset is obtained from a website and the impressions to clicks conversion rate was calculated. This conversion rate helped to determine the high utility ads for the ad placement online on a website using the most efficient Bio-Inspired algorithm that is Particle Swarm Optimization algorithm.

For future enhancements, if we are able to collect the position attribute for the ads for a particular website, then we can find the optimal position where high utility ad can be placed for generation of high revenue.

**Compliance with Ethical Standards**

✓ All authors declare that there is no conflict of interest.

✓ No humans/animals involved in this research work.

✓ We have used our own data.

## References

1. Vargiu, E., Giuliani, A., Armano, G.: Improving contextual advertising by adopting collaborative filtering. ACM Trans. Web (ACM TWEB) **7**(3), 13 (2013)
2. Chapelle, O., Manavoglu, E., Rosales, R.: Simple and scalable response prediction for display advertising. ACM Trans. Intell. Sys. Tech. (ACM TIST) **5**(4), 6:11–6:134 (2014)
3. Lee, K., Orten, B., Dasdan, A., Li, W.: Estimating conversion rate in display advertising from past performance data. In: Proceedings of KDD 2012, pp. 768–776 (2012)
4. Menon, A.K., Chitrapura, K.P., Garg, S., Agarwal, D., Kota, N.: Response prediction using collaborative filtering with hierarchies and side-information. In: Proceedings of KDD 2011, pp. 141–149 (2011)
5. Banerjee, S., Ramanathan, K.: Collaborative filtering on skewed datasets. In: Proceedings of WWW 2008, pp. 1135–1136 (2008)
6. Theocharous, G., Thomas, P.S., Ghavamzadeh, M.: Personalized ad recommendation systems for life-time value optimization with guarantees. In: Proceedings of IJCAI 2015, pp. 1806–1812 (2015)
7. Li, Y.C., Yeh, J.S., Chang, C.C.: Isolated items discarding strategy for discovering high utility itemsets. Data Knowl. Eng. **64**(1), 198–217 (2008)
8. Taniguchi, K., Harada, Y., Duc, N.T.: Adslot mining for online display ads. In: 2015 IEEE International Conference on Data Mining Workshop (ICDMW) (2015). https://doi.org/10.1109/icdmw.2015.82
9. Ahmed, A., Das, A., Smola, A.J.: Scalable hierarchical multitask learning algorithms for conversion optimization in display advertising. In: Proceedings of WSDM 2014, pp. 153–162 (2014)
10. Rosales, R., Cheng, H., Manavoglu, E.: Post-click conversion modeling and analysis for non-guaranteed delivery display advertising. In: Proceedings of WSDM 2012, pp. 293–302 (2012)
11. Song, W., Huangi, C.: Mining high utility itemsets using bio-inspired algorithms: a diverse optimal value framework. IEEE Access (2018). https://doi.org/10.1109/access.2018.2819162
12. Agrawal, R., Srikant, R.: Fast algorithms for mining association rules. In: Proceedings International Conference on Very Large Data Bases, pp. 487–499 (1994)
13. Liu, Y., Liao, W.K., Choudhary, A.N.: A two-phase algorithm for fast discovery of high utility itemsets. In: Proceedings 9th Pacific-Asia Conference on Advances in Knowledge Discovery and Data Mining, pp. 689–695 (2005)

# Classification of Lung Nodules into Benign or Malignant and Development of a CBIR System for Lung CT Scans

K. Bhavanishankar[1] and M. V. Sudhamani[2(✉)]

[1] Department of CSE, RNSIT, Bengaluru, India
bsharsh@gmail.com
[2] Department of ISE, RNSIT, Bengaluru, India
mvsudha_raj@hotmail.com

**Abstract.** Computer Aided Diagnosis (CAD) systems provide a great assistance to the radiologists in screening the CT scans of lung cancer patients in identifying and classifying the lung nodules. The decision of the CAD systems can be further improved with the help of Content Based Image Retrieval (CBIR) systems which provide better visualization assistance by displaying most similar images for a given query image thus helping the radiologists with better visualization of the instances. This paper has two proposals: (i) Deep learning approach-autoencoder, to classify lung nodules in to benign or malignant (ii) CBIR system that accepts an input query image of a nodule and retrieves most similar images. The proposed classification approach is evaluated with sensitivity, specificity and accuracy and obtained values (%) are 89.4, 91.4, and 87.4 respectively. Similarly, the CBIR system's performance is examined with respect to precision, recall and F-score whose obtained values (%) are 92%, 81% and 85% respectively. These results are compared with existing works and found to be performing better.

**Keywords:** Autoencoder · Classification · Feature extraction · CAD · CBIR

## 1 Introduction

Lung cancer (small cell and non-small cell) remains the leading cause of death among all other cancers. According to the estimates of American Cancer Society, around 2,34,030 new cases and about 1,54,050 deaths have been reported due to lung cancer in 2018 in USA [1]. Numerous research contributions have time and again proved that early detection and diagnosis of lung cancer enhances both the survival rate and quality of patient's life. Because of its accuracy and ability to render both detailed and high-quality images Computed Tomography (CT) scans have taken the leading place among other imaging modalities. To handle the quantum of images generated by CT scanners, Commuter Aided Diagnosis (CAD) systems have been used by the radiologists. Though the radiologists do lot many clinical trials, the CAD systems do assist them in diagnosis. Talking about the limitations, conventional CAD systems follow black box approach, wherein the focus is only on the input and the output. The radiologists may focus on some structures and miss other important (may) artifacts. The conventional


S. Smys et al. (Eds.): ICCVBIC 2019, AISC 1108, pp. 563–575, 2020.
https://doi.org/10.1007/978-3-030-37218-7_64

CAD systems doesn't provide any scope for comparison with other patients details to study the similar cases. To overcome this limitation, the CAD and Content Based Image Retrieval (CBIR) systems can be designed to work in sync. CBIR approach has been well accepted in detection and diagnosis of lung lesions by fetching the similar cases form the huge repositories of medical images [2, 3].

It is worth noting that, though the CAD-CBIR systems are very popular in detection and diagnosing breast cancers in mammography, they are also extensively used in lung cancer detection and classification. This paper proposes a two-stage methodology: in the first stage a classification based on deep neural network-autoencoder is proposed which accepts images of nodules as input and categorizes them as benign or malignant. In the second stage, number of features for the nodules are extracted and stored in the feature repository. When a query image is given as input, a distance measure is used to compare the features of the query image and those of the images in the repository to retrieve the most similar images. These resultant images would serve the radiologist to make a better decision in detecting and diagnosing the lung cancer.

This paper is organized as follows,

Section 2 reviews the literatures in the field of CAD and CBIR systems, Sect. 3 illustrates the proposed methodology for classification of nodules, Sect. 4 illustrates development of CBIR system, Sect. 5 highlights the experimental results and discussion and Sect. 6 concludes the paper with future scope of the work.

## 2 Literature Review

This section highlights the significant and recent contributions made by researchers in the area of development of CAD and CBIR systems for medical applications in general and lung cancer in particular.

Autoencoder and binary decision tree-based classifier was designed to build a CAD system by [4] for lung cancer classification and their experimentation was on LIDC repository. [5] Proposed a Deep Convolutional Neural Network (DCNN) for detecting and classifying lung nodules. A CAD system was developed based on contextual clustering, GLCM/LBP features with SVM and k-NN classifiers by [6]. Their experimentation was on LIDC-IDRI data repository. An image based CADx systems was developed by [7] using CNN and R-CN algorithms for detecting and classifying lung nodules and diffuse lung diseases. Both the algorithms were tested on data set with and without data augmentation. Rule based classifier was implemented by [8] in designing a CAD system to detect and classify lung nodules. The resultant system was tested on 400 CT scans of LIDC data set. Convolutional Neural Network based classifier was implemented by [9] for detection and classification of lung lesions. Extensive experiments were conducted on both LIDC-IDRI and LUNA-16 data set. A multi-section CNN architecture was proposed by [10], their approach did not make use of the spatial annotation and the experiments were conducted on LIDC data sets. [11] Proposed a multi-level CNN paradigm for malignancy classification with three CNNs for extracting multiscale features of lung nodules in CT scans. A multi view collaborative deem models was presented by [12] to separate the benign nodules from malignant ones from the chest CT data collected from LIDC-IDRI repository. Further, extensive research articles can be found in the research database on classification on lung nodules.

CBIR has been very fascinating area of research since last few decades for active researchers. Abundant research articles published in this duration has proposed diverse methodologies to solve research problems in different domain of applications [13, 14]. Though CBIR systems have successfully proved their worth in non-medical domains, they have been gaining a sheer momentum in medical domains and especially in investigative radiology [15]. A CBIR system for identifying the lung nodules according to the malignancy levels was proposed by [16]. Two representative features namely nodule density level and nodule lesion density heterogeneity are used in their approach and the experiments were on LIDC data sets [17]. [18] Proposed an automatic learning method that learns the similarity between textual distance from the radiology reports. Their work was on the images captured by EMOTION DUO of Siemens healthcare. A CBIR paradigm based on texture features and Mahalanobis distance metric was implemented by [19] on LIDC dataset containing both benign and malignant nodules. Deep neural network-based feature learning method to automatically learn the features was proposed by [20]. The CNN model was used to learn the malignancy level of the nodules obtained from LIDC repository. Amazon Web Services were used for storing, computing and retrieval of data over internet. [21] proposed a two-stage configuration to retrieve CT images of lung based on texture features (GLCM) and shape features (Zernike). Euclidian distance was used to compute the similarity Their experiments were on CT slices collected from cancer hospital.

## 3 Proposed Methodology - Nodule Classification

### 3.1 Classification of Nodules

The nodule classification phase of a CAD system is shown in Fig. 1. Each of the processing phases are elaborated in the following paragraph.

### 3.2 Segmentation

The extraction of lung parenchyma from the input CT scans of thorax is done through series of steps including thresholding, connected component analysis, morphological operations and a rule based algorithm [22].

### 3.3 Candidate Nodule Detection

From the segmented lung several hundreds of unsolicited components are filtered to identify, detect and retain the candidate lung nodules. Image noise and small components of irregularities are filtered using minimum area filters. Blood vessels which appears as cylindrical structures in 3-D view are eliminated using layer-based filters. The spherical and spiculated artifacts are retained by using a rule-based approach based on the diameters of the components under observation [23].

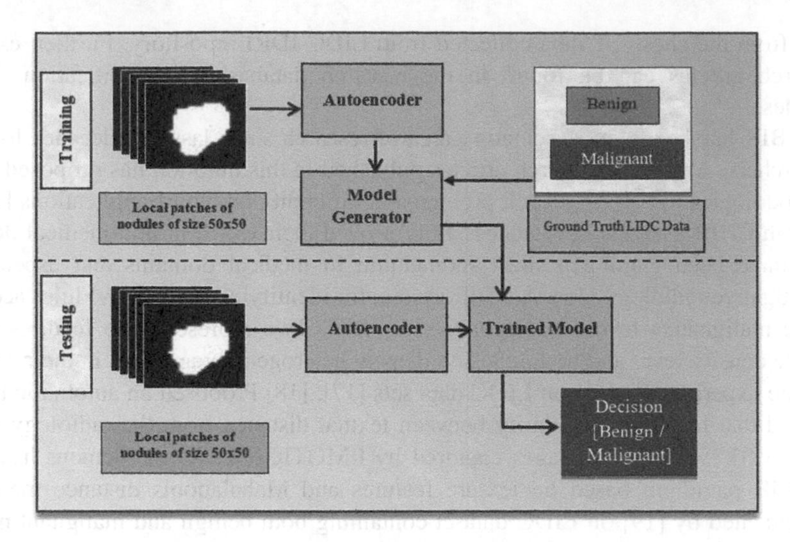


**Fig. 1.** Nodule classification phase of a CAD

### 3.4 Candidate Nodule Classification

The candidate lung nodules are further classified as nodules or non-nodules using a deep neural network- autoencoder model. A four stacked autoencoder was modeled which accepts the cropped images of candidate lung nodules of size 50 * 50 and the ground truth data processed from LIDC database and yields a binary classified result. Hence the input candidate nodules are categorized into nodules or non-nodules.

### 3.5 Classification of Nodules

This section illustrates the classification of the nodules into benign or malignant. Autoencoder model – a multilayer deep neural network-based approach has been employed for the classification. An autoencoder network comprises of an input layer, an encoding layer and a decoding layer. The encoding layer (hidden layer) has the capability to discover the most relevant and descriptive features of the input data. In this approach the autoencoder network is stacked to create a deep network of stacked autoencoder with three stacks. The working model of the entire model is shown in Fig. 2.

The input to the autoencoder model is $X$, which is a feature vector obtained by cropping the nodule images of size $50 \times 50$. Along with $X$, the ground truth data processed from LIDC repository is also fed to the network for the purpose of training. The feature vector $X$ can be represented as (1)

$$X = \{x_1, x_2, x_3 \ldots, x_n\} \quad (1)$$

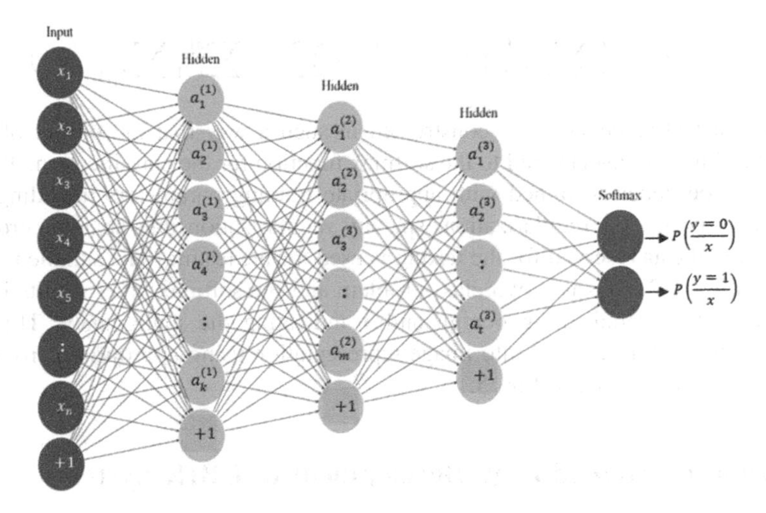


**Fig. 2.** Proposed three stack autoencoder network

Let $x_i$ be the input data in the neuron $i$, $a_i^j$ be the activation of neuron $i$ in layer $j$, $w_{ik}^j$ be the weight mapping between neuron $i$ in $j^{th}$ layer and neuron $k$ in $(j+1)^{th}$ layer and $b_i^j$ be the bias related to neuron $i$ in layer $j$. The sample articulation of activation for neuron 3 in layer 3 and for input feature vectors $x_0$ through $x_4$ can be expressed as follows (2).

$$a_3^3 = f\left(w_{30}^2 x_0 + w_{31}^2 x_1 + w_{32}^2 x_2 + w_{33}^2 x_3 + w_{34}^2 x_4 + b_3^2\right) \tag{2}$$

It should be noted that the autoencoder model differs from the conventional neural network in a way that, the output of the autoencoder $\widehat{X}$ is always similar to the input $X$. The output $\widehat{X}$ is computed using (3) and (4).

$$A = f(W^e X + b) \tag{3}$$

$$\widehat{X} = f\left(W^d A + b'\right) \tag{4}$$

Where $e$ represents encoder and $d$ represents decoder of the autoencoder network.

Input layer and hidden layers are considered to be part of encoder, where the hidden layer transforms the input feature vector $X$ in to feature vector $A$ which is a vector of more learned features. The decoder includes both hidden layer and output layer and transforms the descriptive feature vector $A$ into $\widehat{X}$ which is the output feature vector. $W^e$ and $W^d$ are the weight matrices of encoder and decoder respectively. The function $f(\cdot)$ is a sigmoid function used to activate the neuron in each layer of the autoencoder network. Moment the condition $X \approx \widehat{X}$ is achieved, the input feature vector $X$ is believed to be regenerated from $A$, which is compressed feature vector. The standard cost function to estimate the gap between $X$ and $\widehat{X}$ is given in (5).

$$J(W,b) = \frac{1}{N}\sum_{i=1}^{N}\frac{1}{2} \parallel \widehat{x}_i - x_i \parallel^2 + \lambda\sum_{l=1}^{N_l-1}\sum_{i=1}^{M_i}\sum_{J=1}^{M_{i+1}}(W_{ij}^l)^2 \quad (5)$$

The stacked autoencoder is constructed in such a way that the output of the last stack (third hidden layer) is fed to the softmax function for the classification. The entire stacked autoencoder is trained with appropriate training parameters including hidden size, max epochs, L2 regularization, sparsity regularization and sparsity proportion. A set of input data reserved for the testing is used for the testing. The sample results are shown in Table 2. This is a two-class problem and the softmax layer results in binary classification as 0 indicating benign and 1 indicating the malignancy. This output vector is further transformed into image representation using appropriate program and those images are shown in Fig. 5.

## 4 Proposed Methodology Development of CBIR System

The following section illustrates the process of building a CBIR system. The architecture of the proposed CBIR systems is shown in Fig. 3 and its working is explained as follows.

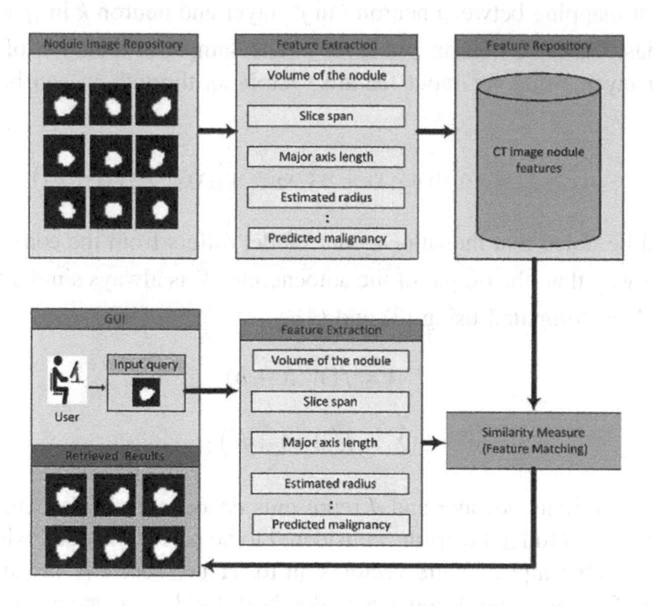


**Fig. 3.** Architecture of the proposed CBIR system

### 4.1 Feature Extraction and Feature Repository

The image repository contains the images of benign and malignant nodules obtained from the classification approach proposed in the previous section. The features of these images are extracted and stored in the feature repository. The feature extraction process involves extracting following features and is highlighted as follows.

- Volume of the nodule
  Since a nodule would span multiple slices, it is reasonable to consider the area of the nodules in all the slices of its presence and is referred as volume which is given in (6)

$$volume = V(c)\sum\nolimits_{z=0}^{l}\sum\nolimits_{y=0}^{m}\sum\nolimits_{x=0}^{n}c(x,y,z) \tag{6}$$

  where c is component in binary image
- Slice span
  This 3-D feature provides the details of the number of slices in which the nodule is present as given in (7) and (8).

$$\sum\nolimits_{z=1}^{L}f(c,z) \tag{7}$$

$$f(c,z) = \begin{cases} 0, \; if \; c \; does \; not \; exist \; in \; slice \; z \\ 1, other \; wise \end{cases} \tag{8}$$

  where c is the component under consideration
- Major axis length
  It is a geometric feature, which gives the longest diameter, i.e. the line segment passing through the center connecting two widest points on a perimeter
- Estimated radius
  It is essentially the radius of the component, if the pixels were arranged as a sphere. It gives an approximate average distance from the center of the component to the edge and is expressed in (9)

$$R = \sqrt[3]{3V/4\pi} \tag{9}$$

- Predicted malignancy
  The image repository contains the images of benign and malignant nodules. Since the input to the CBIR system will be either benign or malignant query, the predicated malignancy feature is also extracted and is represented in (10)

$$M(c) = \begin{cases} 0, if \; c \; is \; classified \; as \; benign \\ 1, if \; c \; is \; classified \; as \; malignant \end{cases} \tag{10}$$

  where $M(c)$ is degree of malignancy of a component $c$.

### 4.2 Similarity Measure

The main aim of the CBIR system is to retrieve the similar images based on query. As discussed above, a binary image of a component is given as the query image by the user to the CBIR system. The query image is from the results of classification approach proposed. The features mentioned above are extracted for this query image and given as input to the similarity measure module. Along with this, the features of individual components from the feature repository are also given as input to similarity measure module. Similarity measure compares the weighted distance between these two feature sets against a threshold and decides if the two components are similar or not and finds the degree of similarity.

The distance measure used to compute the similarity measure is Minkowski and is given in (11).

$$D(c1, c2) = \sum\nolimits_{i=1}^{n} \sqrt{f^2(c1, i) - f^2(c2, i)} * w(i) \quad (11)$$

Where,

- $D(c1, c2)$ = Weighted distance between 2 components $c1, c2$
- $n$ = No. of features,
- $f(c, i)$ = normalized $i^{th}$ feature of component $c$,
- $w(i)$ = weight of $i^{th}$ feature

The similarity measure is computed using (12)

$$\text{Similarity} = \begin{cases} 0, & \textit{if } D(c1, c2) > T \\ 1, & \textit{otherwise} \end{cases} \quad (12)$$

For the input query image, all the components from the feature repository that are found to be similar are displayed in the output window.

# 5 Experimental Results and Discussion

The following section illustrates the experimental details of both classification of nodules and CBIR system.

## 5.1 Classification of Nodules

**Data Set**

The proposed method was validated on the CT scans obtained from Lung Image Database Consortium (LIDC) and Image Database Resources Initiative (IDRI) repository available in public domain. The repository contains 1018 cases each having an average of 250 scans. Each of these cases include set of CT scans and corresponding XML files which contains the annotations carried out by four experienced radiologists. Each radiologist examined each of the scans as part of two stage blinded and unblended review process and marked the nodule into different categories. Since four radiologists have independently marked the nodules, the data set considered for the experiments are based on consensus method. Well circumscribed nodules are included whereas Ground Glass Opacity (GGO) nodules Juxta-pleural Nodules are excluded for the experiments.

From this repository 600 CT studies each having an average of 250–300 scans are considered for the experimentation purpose and the details are summarized in Table 1.

**Table 1.** Summary of data set

| | |
|---|---|
| Number of cases considered for experiments | 600 |
| Number of images in each scan (approx.) | 250 |
| Image format | DICOM |
| Image dimension | 512 × 512 |
| Voxel size | 16 bits per voxel |

As discussed in the Sect. 3, the input given to the stacked autoencoder model is the cropped images of nodule and the ground truth data processed from LIDC repository. The network is trained rigorously considering different values for the hyper parameters including hidden size, L2 regularization, sparsity regularization, sparsity proportion and maximum epochs. By keeping one hyper parameter constant and varying others, several experiments are conducted. for the data set with randomly selected images for training and testing. While training it was decided to use three stacks as part of stacked autoencoder.

The final outcome of these experiments is tabulated as in Table 2 with sensitivity, specificity and accuracy. The table also highlights the optimum result obtained during testing of the model. The optimum result (%) obtained with respect to sensitivity, specificity and accuracy is 89.4, 91.3 and 87.4 respectively. The plot of these values is shown in the Fig. 4. The result of the classification process is depicted in Fig. 5 as well.

**Table 2.** Summary of the results obtained from the proposed autoencoder model

| Exp. no | Sensitivity TP/(TP +FN) | Specificity TN/(TN +FP) | Accuracy (TP+TN)/(TP +TN+FP +FN) |
|---|---|---|---|
| 1 | 82.6% | 79.6% | 78.1% |
| 2 | 87.0% | 81.6% | 81.3% |
| 3 | 86.6% | 84.9% | 83.4% |
| 4 | 87% | 88.3% | 85.6% |
| 5 | 89.4% | 91.3% | 87.4% |

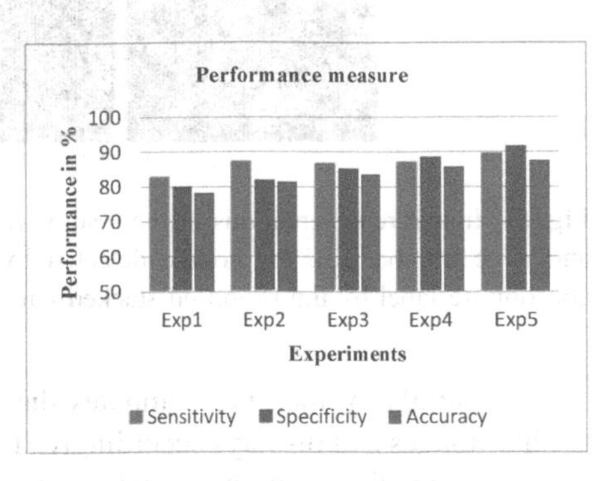


**Fig. 4.** Result analysis of proposed autoencoder model

## 5.2 Results of Proposed CBIR System

Results of classification process of proposed CAD system is used as the input to CBIR system. The features of both benign and malignant nodule images are extracted and stored in feature repository. When an input query image (benign or malignant) is given, the features of the query image are extracted and are compared with those of the images in the feature repository.

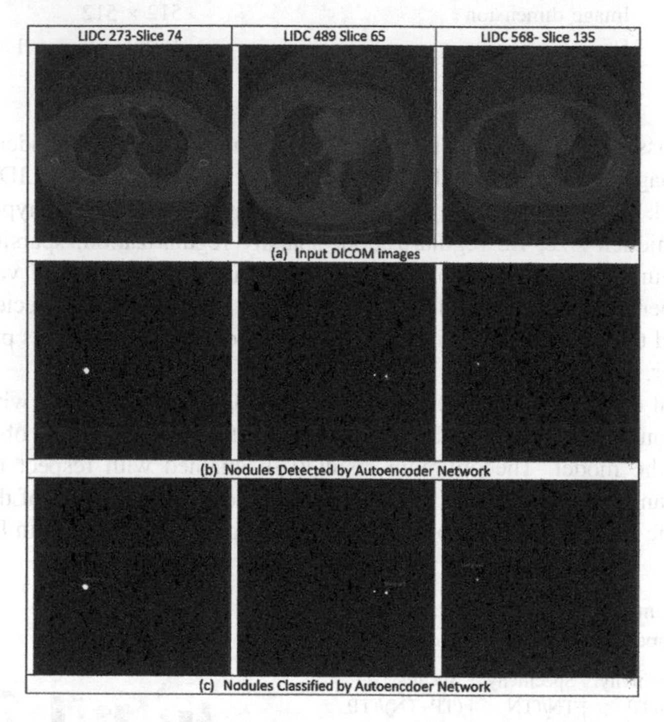


**Fig. 5.** Image representations of the results. (a) Input DICOM images with LIDC case number and slice number [6]. (b) nodule detected by autoencoder model (c) Classified nodule with appropriate label by the proposed stacked autoencoder model

The similarity measure computes the degree of similarity and retrieves the most similar images and displays accordingly. The sample display of the retrieved images for the corresponding malignant query image is shown in the Fig. 6. The figure also displays the precision and recall metrics.

The CBIR system is evaluated against precision and recall. The precision is the percentage of retrieved images that are actually similar subjective to the features extracted. So precision is expressed as (13). Whereas recall is the percentage of actually similar images that have been retrieved and is expressed as (14). F1 measure is also calculated using these results and is expressed as (15).

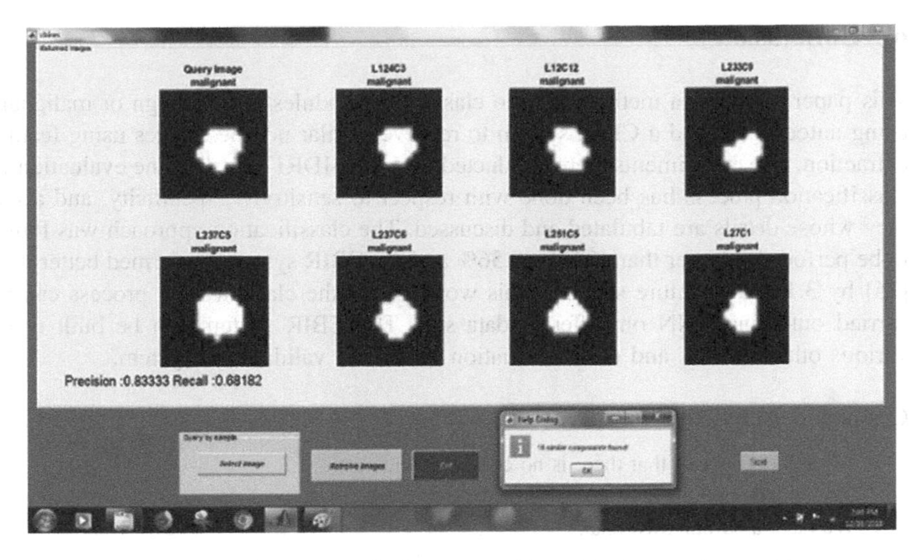


**Fig. 6.** Sample result of proposed CBIR system with similar images for a malignant query

$$precision = \frac{number\ of\ similar\ images\ retrieved}{total\ number\ of\ retreiced\ images} \tag{13}$$

$$recall = \frac{number\ of\ similar\ images\ retreived}{number\ similar\ images\ in\ the\ repostory} \tag{14}$$

$$F1score = 2 \times \frac{precision \times recall}{precesion + recall} \tag{15}$$

Several experiments are carried out with different input query images and results are tabulated in Table 3 and shown in Fig. 7.

**Table 3.** Evaluation of proposed CBIR system

| Exp. no. | LIDC cases | Precision | Recall | F1-score |
|---|---|---|---|---|
| 1 | LIDC 3 | 0.95 | 0.83 | 0.87 |
| 2 | LIDC 31 | 0.92 | 0.9 | 0.90 |
| 3 | LIDC 40 | 0.92 | 0.81 | 0.85 |
| 4 | LIDC 126 | 0.92 | 0.83 | 0.86 |
| 5 | LIDC 171 | 0.91 | 0.88 | 0.89 |
| 6 | LIDC 298 | 0.96 | 0.90 | 0.92 |
| Average | | **0.93** | **0.85** | **0.88** |

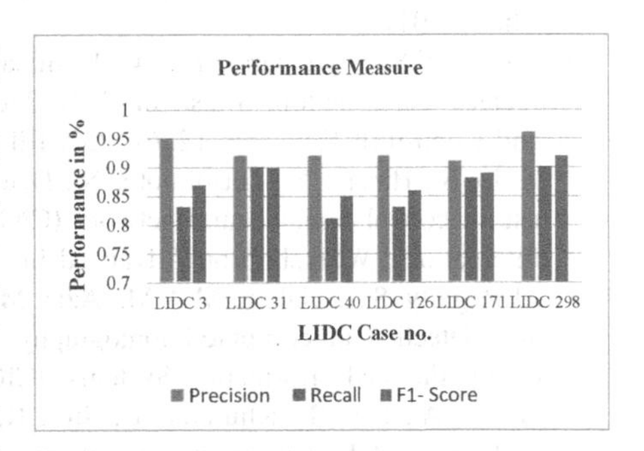


**Fig. 7.** Results of proposed CBIR system.

## 6 Conclusion

This paper proposed a methodology to classify the nodules in to benign or malignant using autoencoder and a CBIR system to retrieve similar nodule images using feature extraction. The experiments were conducted on LIDC-IDRI data set. The evaluation of classification process has been done with respect to sensitivity, specificity, and accuracy whose details are tabulated and discussed. The classification approach was found to be performing better than [24] by 0.56% and the CBIR system performed better than [25] by 3.1%. The future scope of this work is that the classification process can be carried out using CNN on different data sets. The CBIR system can be built using various other features and other evaluation metrics to validate the system.

**Compliance with Ethical Standards**

✓ All authors declare that there is no conflict of interest.
✓ No humans/animals involved in this research work.
✓ We have used our own data.

## References


1. https://www.cancer.org/cancer/non-small-cell-lung-cancer/about/key-statistics.html
2. Wei, G., Ma, H., Qian, W., Qiu, M.: Similarity measurement of lung masses for medical image retrieval using kernel based semi supervised distance metric. Med. Phys. **43**(12), 6259–6269 (2016)
3. Dhara, A.K., Mukhopadhyay, S., Dutta, A., Garg, M., Khandelwal, N.: Content-based image retrieval system for pulmonary nodules: assisting radiologists in self-learning and diagnosis of lung cancer. J. Digit. Imaging **30**, 63–77 (2016)
4. Kumar, D., Wong, A., Clausi, D.A.: Lung nodule classification using deep features in CT images. In: 12th Conference on Computer and Robot Vision (CRV), pp. 133–138. IEEE (2015)
5. Ding, J., Li, A., Hu, Z., Wang, L.: Accurate pulmonary nodule detection in computed tomography images using deep convolutional neural networks. In: International Conference on Medical Image Computing and Computer-Assisted Intervention, pp. 559–567. Springer, Cham (2017)
6. Baboo, S.S., Iyyapparaj, E.: A classification and analysis of pulmonary nodules in CT images using random forest. In: 2018 2nd International Conference on Inventive Systems and Control (ICISC), pp. 1226–1232. IEEE (2018)
7. Kido, S., Hirano, Y., Hashimoto, N.: Detection and classification of lung abnormalities by use of convolutional neural network (CNN) and regions with CNN features (R-CNN). In: International Workshop on Advanced Image Technology (IWAIT), pp. 1–4. IEEE (2018)
8. El-Regaily, S.A., Salem, M.A.M., Aziz, M.H.A., Roushdy, M.I.: Lung nodule segmentation and detection in computed tomography. In: 8th International Conference on Intelligent Computing and Information Systems (ICICIS), pp. 72–78. IEEE (2017)
9. Gupta, A., Das, S., Khurana, T., Suri, K.: Prediction of lung cancer from low-resolution nodules in CT-scan images by using deep features. In: International Conference on Advances in Computing, Communications and Informatics (ICACCI), Bangalore, India, pp. 531–537 (2018)

10. Sahu, P., Yu, D., Dasari, M., Hou, F., Qin, H.: A lightweight multi-section CNN for lung nodule classification and malignancy estimation. IEEE J. Biomed. Health Inform. **23**, 960–968 (2018)
11. Lyu, J., Ling, S.H.: Using multi-level convolutional neural network for classification of lung nodules on CT images. In: 40th Annual International Conference of the IEEE Engineering in Medicine and Biology Society (EMBC), pp. 686–689. IEEE (2018)
12. Xie, Y., Xia, Y., Zhang, J., Song, Y., Feng, D., Fulham, M., Cai, W.: Knowledge-based collaborative deep learning for benign-malignant lung nodule classification on chest CT. IEEE Trans. Med. Imaging **38**, 991–1004 (2018)
13. Muller, H., et al.: A review of content-based image retrieval systems in medical applications —clinical benefits and future directions. Int. J. Med. Inform. **73**(1), 1–23 (2004)
14. Smeulders, A.W.M., et al.: Content-based image retrieval at the end of the early years. IEEE Trans. Pattern Anal. Mach. Intell. **12**, 1349–1380 (2000)
15. Akgül, C.B., Rubin, D.L., Napel, S., Beaulieu, C.F., Greenspan, H., Acar, B.: Content-based image retrieval in radiology: current status and future directions. J. Digit. Imaging **24**(2), 208–222 (2011)
16. Wei, G., Ma, H., Qian, W., Zhao, X.: A content-based image retrieval scheme for identifying lung nodule malignancy levels. In: 29th Chinese Control and Decision Conference (CCDC), pp. 3127–3130. IEEE (2017)
17. Armato, S.G., McLennan, G., Bidaut, L., et al.: The lung image database consortium (LIDC) and image database resource initiative (IDRI): a completed reference database of lung nodules on CT scans. Med. Phy. **38**(2), 915–931 (2011)
18. Ramos, J., Kockelkorn, T.T.J.P., Ramos, I., Ramos, R., Grutters, J., Viergever, M.A., van Ginneken, B., Campilho, A.: Content-based image retrieval by metric learning from radiology reports: application to interstitial lung diseases. IEEE J. Biomed. Health Inform. **20**(1), 281–292 (2016)
19. Wei, G., Ma, H., Qian, W., Jiang, H., Zhao, X.: Content-based retrieval for lung nodule diagnosis using learned distance metric. In: 39th Annual International Conference of the Engineering in Medicine and Biology Society (EMBC), pp. 3910–3913. IEEE (2017)
20. Ma, L., Liu, X., Zhou, C., Zhao, X., Zhao, Y.: A two-stage sliding window method for region-based lung CT image retrieval. In: 5th International Workshop on Pulmonary Image Analysis, pp. 153–160 (2013)
21. Ibanez, D.P., Li, J., Shen, Y., Dayanghirang, J., Wang, S., Zheng, Z.: Deep learning for pulmonary nodule CT image retrieval—an online assistance system for novice radiologists. In: Data Mining Workshops (ICDMW), pp. 1112–1121. IEEE (2017)
22. Bhavanishankar, K., Sudhamani, M.V.: 3-D segmentation of lung parenchyma in computed tomography scans. Int. J. Appl. Eng. Res. (IJAER) **86**, 477–481 (2015)
23. Bhavanishankar, K., Sudhamani, M.V.: Filter based approach for automated detection of candidate lung nodules in 3D computed tomography images. In: International Conference on Cognitive Computing and Information Processing, pp. 63–70. Springer, Singapore (2017)
24. Dey, R., Lu, Z., Hong, Y.: Diagnostic classification of lung nodules using 3D neural networks. In: IEEE 15th International Symposium on Biomedical Imaging (ISBI 2018), Washington, DC, pp. 774–778 (2018)
25. Wei, G., Ma, H., Qian, W., Jiang, H., Zhao, X.: Content-based retrieval for lung nodule diagnosis using learned distance metric. In: 39th Annual International Conference of the IEEE Engineering in Medicine and Biology Society (EMBC), Seogwipo, pp. 3910–3913 (2017)

# Correlation Dimension and Bayesian Linear Discriminant Analysis for Alcohol Risk Level Detection

Harikumar Rajaguru(✉) and Sunil Kumar Prabhakar

Department of ECE, Bannari Amman Institute of Technology, Coimbatore, India
harikumarrajaguru@gmail.com

**Abstract.** Alcohol consumption always leads to a lot of physical and mental health problems. The constant use of alcohol affects almost every part of the human body especially the brain, liver, pancreas, heart and the immune system. Also, there are numerous social and economic challenges like weakness in memory and concentration, lack of proper decision-making skills, emotional and cognitive impairments etc. Therefore, the alcohol detection level of a particular patient has to be known in order to assess the capability of the alcoholic patient to perform a specific task. In this research work, the concept of Correlation Dimension (CD) and the Bayesian Linear Discriminant Analysis (BLDA) is employed to assess the risk rate of alcohol for a single alcoholic patient with the aid of Electroencephalography (EEG) signals. The enumerated Results report a total classification accuracy of 82.016% when CD is used and an average classification accuracy of 84.97% is obtained when BLDA is utilized.

**Keywords:** Alcohol · EEG · CD · BLDA

## 1 Introduction

Alcoholism is a prevalent neurological disease and many people across the globe suffer from this disease [1]. People who constantly drink alcohol suffer from problems like impaired memory and sleep, blurred vision, difficulty in walking and talking, appetite disorders etc. Therefore alcoholism not only affects the mobility and cognitive impairments but also severely damages the entire brain and other absolutely essential organs in the human body [2]. To discover the risk level of alcohol for a particular patient, EEG signals are used and when incorporated with advanced signal processing techniques it serves as a boon to the clinicians for an in-depth analysis [3]. Moreover it is quite a convenient and inexpensive technique and therefore it plays a very significant role in the analysis of patients concerned with neurological disorders. A few important works where EEG signal processing has dealt with alcoholic risk level detection is discussed here.

The human brain was analyzed after the consumption of alcohol based on EEG signal by Wu et al. [4]. The analysis of EEG signals based on feature extraction was done by Sun et al. for both alcoholics and non-alcoholic patients in [5]. Based on the concept of both decision trees and graph entropy, Principal Component Analysis was


S. Smys et al. (Eds.): ICCVBIC 2019, AISC 1108, pp. 576–582, 2020.
https://doi.org/10.1007/978-3-030-37218-7_65

implemented to it to determine whether a person is alcoholic of not by Wang et al. [6]. (SVM) and Neural Networks were used to classify the alcoholics and non-alcoholics separately by Kousarrizi et al. [7]. Spectral entropy was employed with Neural Network classifiers for EEG based detection of alcoholics by Shri and Sriraam [8]. A graph theoretic analysis showing the disturbed pattern of EEG networks in chronic alcoholism was explained by Cao et al. [9]. Spectral Density (PSD) Analysis was implemented to investigate the EEG signals by Malar et al. in [10]. Based on the graph entropy, the alcoholic EEG signals were analyzed by Zhu et al. [11]. In this paper, the alcoholic EEG data was analyzed thoroughly with the help of CD and BLDA classifiers. The illustration of the work is shown in Fig. 1.

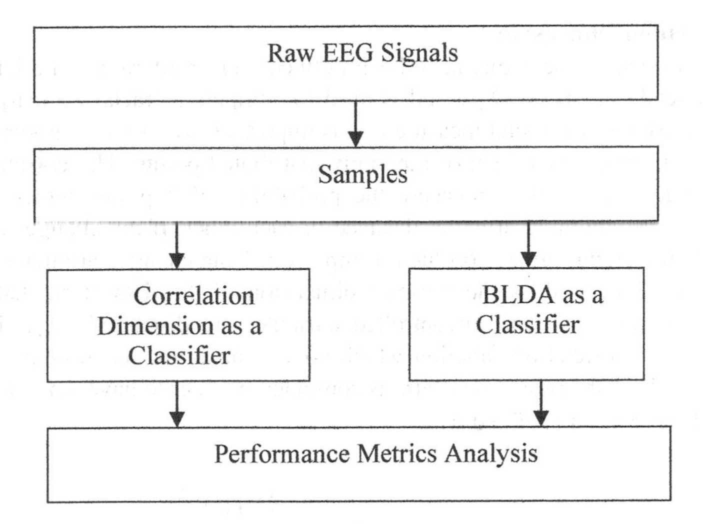


**Fig. 1.** Block diagram of the work

The remaining paper is arranged with the. Section 2 the works related to the proposed. Correlation Dimension and BLDA as post classifiers for alcoholic detection from EEG signals is elaborated in Sect. 3. Section 4 details the evaluation results and discussion. Section 5 gives the conclusion.

## 2 Materials and Methods

One alcoholic patient is subjected through single trial 64 channels and with the help of three electrodes, the EEG signals are obtained. A 12 bit ADC samples the signals into 256 samples/sec implemented with signed representation. For 10 s duration, each channel is acquired and so ($256 \times 10 = 2560$) samples per particular channel are obtained. Therefore 2560 samples are assumed to be a bin and so as there are 64 channels, 1, 63,840 samples are present which can be totally grouped in 64 bins.

The human brain is surely affected by the consumption or influence of alcohol. According to the [12], Acharya had evaluated the correlation dimension of an alcoholic patient to be 5.6. However, in this paper, correlation dimension itself is used as a post classifier and the gold standard or the target value for the single chronic alcoholic patient is set as 0.45 in order to assess the risk level of alcohol from the patient by analyzing the EEG signals.

## 3 Post Classification with Correlation Dimension and Bayesian LDA

The post classification algorithms for alcohol risk level detection from EEG signals used here are Correlation Dimension and Bayesian LDA

**A. Correlation Dimension**

A series of fractal dimensions have been defined well in literature and have the correlation fractal dimension $D_{corr}$ and is used to compute a fractal dimension or measurement [13]. From a fractal measure $\mu$, it is important to compute a fractal dimension measurement from a given set of randomly distributed points. The correlation fractal dimension $D_{corr}$ is used to measure the probability of 2 points which are chosen randomly to be within a particular distance of each other. If the changes are made in the correlation fractal dimension then it implies that the changes are made in the data set. Here $D_{corr}$ is used as the intrinsic dimension of the dataset so that correlated attributes are identified and uncontrolled attributes are discarded. $D_{corr}$ is reckoned with the aid of correlation function which covers the entire set of cells which has a given size $\varepsilon$. Then the probability $p_i(\varepsilon)$ is computed in order to have a point of the set in the $i^{th}$ cell. The CD is defined as

$$D_{corr} = \lim_{\varepsilon \to 0} \sum_i \log \frac{\left( \sum_i p_i(\varepsilon)^2 \right)}{\log(\varepsilon)}$$

In a cell of size $\varepsilon$, the probability to find a pair of points is given by the quantity $\sum_i p_i(\varepsilon)^2$. For small value of $\varepsilon$, the Euclidean distance in between every pair of points is applied. For large sets which has $N$ number of points, the approximation is done by the correlation sum mentioned as follows

$$C(\varepsilon) = \lim_{N \to \infty} \frac{1}{N^2} \times \{no.of.pairs(x_i, x_j)\}$$

$$C(\varepsilon) = \lim_{N \to \infty} \frac{1}{N^2} \sum_{i,j}^{N} H\left(\varepsilon - \|x_i - x_j\|_2\right) = P\left(\|x_i - x_j\|_2\right)$$

where $H$ denotes the Heaviside step function which has a value of 0 or 1.

There correlation dimension can be expressed as

$$D_{corr} = \lim_{\varepsilon \to \infty} \frac{\log(C(\varepsilon))}{\log(\varepsilon)}$$

B. **Bayesian Linear Discriminant Analysis (BLDA)**

Minimization of the risk which is related to the classification decision is very important for BLDA. The main advantage of using BLDA is that it can easily deal with noisy and high dimensional datasets [14]. The targets are assumed as $t$ and the feature vectors are denoted as $z$ and both are related to each other in a linear manner to the additive white Gaussian noise $n$ in the Bayesian regression, i.e.,

$$t = v^T z + n$$

where the representation of a weight vector is denoted as $v$. The likelihood representation of the weights $v$ practically used in regression is written as follows $p(F/\beta, v) = \left(\frac{\beta}{2\pi}\right)^{N/2} \exp\left(-\frac{\beta}{2}\|Z^T v - t\|^2\right)$

where a vector having the regression targets is mentioned by $t$ and $Z$ indicates the matrix which is obtained by stacking the training feature vectors horizontally [14]. $F$ specifies the pair $\{Z, t\}$. For the noise, the inverse variance is expressed as $\beta$ and the total number of samples present in the training set is denoted as $N$. Performing inference in a Bayesian environment is important and so for the specific latent variable, a prior distribution is preferred [14]. The expression for prior distribution is given as

$$p(v/\alpha) = \left(\frac{\alpha}{2\pi}\right)^{\frac{F}{2}} \left(\frac{\varepsilon}{2\pi}\right)^{\frac{1}{2}} \exp\left(-\frac{1}{2} v^T J'(\alpha) v\right)$$

where the number of features is denoted as $F$ and $J'(\alpha)$ represents the $(F+1)$ dimensional diagonal matrix expressed as

$$J'(\alpha) = \begin{bmatrix} \alpha & 0 & .. & 0 \\ 0 & \alpha & .. & 0 \\ : & : & : & : \\ 0 & 0 & .. & \varepsilon \end{bmatrix}$$

where $\varepsilon$ is set to a very minor value. If both the prior and the likelihood are given, then using the Bayes rule the computation of the posterior distribution is done as

$$p(v/\beta, \alpha, F) = \frac{p(F/\beta, v) p(v/\alpha)}{\int p(F/\beta, v) p(v/\alpha) dv}$$

The posterior is Gaussian as both likelihood and prior are Gaussian in nature and from likelihood, its parameters can be easily obtained [14]. Using the following equation, the mean '$m$' and Covariance '$C$' of the posterior is calculated as

$$m = \beta(\beta ZZ^T + J'(\alpha))^{-1}Zt$$
$$C = (\beta ZZ^T + J'(\alpha))^{-1}$$

The predictive sharing of resources for a novel input vector $\hat{z}$ is obtained as follows:

$$p(\hat{t}/\beta, \alpha, \hat{z}, F) = \int p(\hat{t}/\beta, \hat{z}, v)p(v/\beta, \alpha, F)dv$$

Therefore the linear discriminant function is obtained as $q = m^T\hat{z}$, where the EEG vector which is to be categorized is denoted as $\hat{z}$ and the input of the BLDA classifier is mentioned as $q$.

## 4 Results and Discussion

If the correlation dimension and the BLDA are utilized to assess the risk level detection of alcohol from EEG signals, then consideration of factors like False Alarm (FA), Missed Classification (MC) and Perfect Classification (PC) are essential. Analyzing these factors, Classification performance index, specificity, Sensitivity, and accuracy are calculated and the average results are depicted in Table 1.

**Table 1.** Performance analysis of CD and BLDA Classifiers for alcohol risk level detection

| Name | CD as a Classifier | BLDA as a Classifier |
|---|---|---|
| PC (%) | 67.2 | 69.94 |
| MC (%) | 28.2 | 14.32 |
| FA (%) | 4.6 | 15.73 |
| PI (%) | 51.19 | 53.61 |
| Specificity (%) | 70.44 | 85.67 |
| Sensitivity (%) | 93.59 | 84.26 |
| Accuracy (%) | 82.016 | 84.97 |

## 5 Conclusion

As the traditional techniques of visually inspecting the EEG signals is time consuming and requires highly qualified experts and medical professionals, automated EEG classification systems are a boon in arenas of clinical and research fields. Thus, in this paper, the alcohol risk level detection from EEG signals for a single alcoholic patient was investigated with the help of two classifiers such as Correlation Dimension and Bayesian Linear Discriminant Analysis. Results showed that when CD is used as a post classifier, an approximate classification accuracy of 82.016% is acquired with an approximate performance index of 51.19. When BLDA is used as a post classifier, an average classification accuracy of 84.97% is enumerated with an average performance index of 53.61. Future works is to analyze the utilization of various post classifiers for the alcohol risk rate detection from signals acquired from the EEG.

**Compliance with Ethical Standards**

✓ All authors declare that there is no conflict of interest.
✓ No humans/animals involved in this research work.
✓ We have used our own data.

## References


1. Ziya, E., Akif, A., Mehmet, R.B.: The classification of EEG signals recorded in drunk and non-drunk people. Int. J. Comput. Appl. **68**(10), 40–44 (2013)
2. Rajaguru, H., Prabhakar, S.K.: Softmax discriminant classifier for detection of risk levels in alcoholic EEG signals. In: IEEE Proceedings of the International Conference on Computing Methodologies and Communication (ICCMC 2017), Erode, India (2017)
3. Prabhakar, S.K., Rajaguru, H.: Development of patient remote monitoring system for Epilepsy classification. In: 16th International Conference on Biomedical Engineering (ICBME), Singapore, 7–10 December 2016
4. Wu, D., Chen, Z.H., Feng, R.F., Y., Li, G., Luan, T.: Study on human brain after consuming alcohol based on EEG signal. In: Proceedings of 2010 3rd IEEE International Conference on Computer Science and Information Technology, vol. 5 (2010)
5. Sun, Y.G., Ye, N., Xu, X.H.: EEG analysis of alcoholics and controls based on feature extraction. In: Proceedings of 8th International Conference on Signal Processing, vol. 1 (2006)
6. Wang, S., Liu, Y., Wen, P., Zhu, G.: Ananlyzing EEG signals using graph entropy based principal component analysis and J48 decision tree. Int. J. Signal Process. Syst. **4**(1), 67–72 (2016)
7. Kousarrizi, M.N., Ghanbari, A.A., Gharaviri, A., Teshnehlab, M., Aliyari, M.: Classification of alcoholics and non-alcoholics via EEG using SVM and neural networks. In: 3rd International Conference on Bioinformatics and Biomedical Engineering, 2009. ICBBE 2009, pp. 1–4. IEEE, June 2009
8. Shri, P.T.K., Sriraam, N.: EEG based detection of alcoholics using spectral entropy with neural network classifiers. In: 2012 International Conference on Biomedical Engineering (ICoBE), pp. 89–93. IEEE, February 2012
9. Cao, R., Wu, Z., Li, H., Xiang, J., Chen, J.: Disturbed connectivity of EEG functional networks in alcoholism: a graph-theoretic analysis. Biomed. Mater. Eng. **24**(6), 2927–2936 (2014)
10. Malar, E., Gauthaam, M., Chakravarthy, D.: A novel approach for the detection of drunken driving using the power spectral density analysis of EEG. Int. J. Comput. Appl. **21**(7), 10–14 (2011). (0975 – 8887)
11. Zhu, G.H., Li, Y., Wen, P.P.: An efficient visibility graph similarity algorithm and its application for sleep stages classification. In: Proceedings of 2012 International Conference on Brain Informatics, Macao, 4–7 December 2012

12. Acharya, U.R., Subburam, V.S., Chattopadhya, S., Suri, J.: Automated diagnosis of normal and alcoholic EEG signals. Int. J. Neural Syst. 1–9 (2012)
13. Theiler, J.: Efficient algorithm for estimating the correlation dimension from a set of discrete point. Phys. Rev. A **36**(9), 4456–4462 (1987)
14. Zhou, W., Liu, Y., Yuan, Q., Li, X.: Epileptic seizure detection using Lacunarity and Bayesian linear discriminant analysis in intracranial EEG. IEEE Trans. Biomed. Eng. **60**(12), 3375–3381 (2013)

# A Smart and Secure Framework for IoT Device Based Multimedia Medical Data

Shrujana Murthy[(✉)] and C. R. Kavitha

Department of Computer Science and Engineering, Amrita School of Engineering, Bengaluru, Amrita Vishwa Vidyapeetham, Bengaluru, India
murthy.shrujana@gmail.com, cr_kavitha@blr.amrita.edu

**Abstract.** Internet of Things (IoT) is the act of associating or registering gadgets/devices/things like advanced cells, clothes washers, and wearable gadget with the internet. IoT organizes and interfaces 'things' and 'individuals' together by making a connection between two individuals, humans-to-things or things-to-things. As the quantity of gadget association expands, it builds a serious security hazard. Security is an all-time concern for IoT at any organizations over the globe. IoT applications and administrations may be actuated subsequent to confirming that all security keys are available. In this way, a conventional model for actualizing security involves a mix of security norms and relating security necessities heading on the useful design of IoT [8]. There are numerous hitches with respect to security matters for IoT, which is still to be addressed, including RFID label security, digital security, remote security, transmission security, protection insurance and so forth [8]. The proposed model aims to provide a secure way to share multimedia medical data over the internet in IoT devices that takes various security concerns in a medical organization into consideration.

**Keywords:** Gmail SaaS · Private cloud · Medical data security · Data security · Advanced Encryption Standard (AES) algorithm · Multimedia medical data · IoT security · Secure data sharing · IoT

## 1 Introduction

Web of Things give benefits by connecting the diverse type of gadgets. They have the restriction in giving good administration. Web of things gadgets is heterogeneous and incorporates remote sensors to less value compelled gadgets. These gadgets are inclined to equipment/programming and system assaults. If not appropriately anchored, it might prompt safety problems like protection and secrecy. To determine the above issue, a Reliable Security Framework for therapeutic information in IoT Devices was proposed. The Internet of Things (IoT) [13] is phasing from a brought along structure to a puzzling system of decentralized shrewd gadgets. IoT gadgets and administrations stand' defenseless to Denial of Service assaults (DoS) [12]. Spy intrigue may be a large danger to remote correspondence security. Spy intrigue may be a large danger to remote correspondence security. Web of Things is a system of heterogeneous gadgets, which convey over remote systems and play out the assignment of detecting and


S. Smys et al. (Eds.): ICCVBIC 2019, AISC 1108, pp. 583–588, 2020.
https://doi.org/10.1007/978-3-030-37218-7_66

activation. IoT goes for associating every single machine and gadget to give an omnipresent availability. In this situation, security and protection of data assume a crucial job. The examination done in the field of IoT security [14] is classified according to the security issue concerned. The proposed and existing strategies are checked on. To securely send and receive any form of multimedia clinical data and prove confidentiality and integrity via the internet by means of IoT devices – mobile to mobile or PC- mobile or vice-versa [10].

Without a far-reaching wellbeing IT protection and security structure, patients will participate in "protection defensive" practices, which may incorporate retention significant wellbeing data from suppliers or keeping away from treatment [7]. The results are noteworthy – for individual and in addition populace wellbeing. There is a need to embrace a thorough security and security system for insurance of wellbeing information as data innovation is progressively used to help trade of therapeutic records and other wellbeing data. Protection and security insurances will construct open trust, which is essential if the advantages of wellbeing data innovation (wellbeing IT) are to be figured it out [11]. Execution of an extensive protection and security structure will require a blend of authoritative activity, control and industry responsibility and must consider the multifaceted nature of the advancing wellbeing trade condition. The system configuration ought to encourage trade not through centralization of information, yet rather through a "system of systems." [9] This dispersed engineering is bound to secure data. The system should likewise accommodate interoperability and adaptability, which bolster development and make open doors for new participants. To assemble shopper trust in e-wellbeing frameworks, it is important that all substances be considered responsible for consenting to the protection and security structure. Patient's personal record ought to be unbroken non-public and the data's confidentiality and integrity ought to be protected against patient's Wireless Personal Area Network (WPAN) from access of any unauthorized personnel. Hence, it's required to develop secure sensor/WPAN/Wireless Local Area Network (WLAN) security or such spec which might dependably and firmly monitor the health application on a person mobile of the patient's while not harming their health or life habits.

## 2 Related Work

A Healthcare and Environment Safety (HES) framework for medical services is structured for gathering information from a remote body WBAN system [1]. This restorative information is then transmitted with the help of a remote sensor and distributed through entryway it into a WPAN system that is a remote individual territory. Enhanced execution, protection and security are attained by the test assessments that are fast and hypothetical for the framework. The hunt criteria is considered as a vital element vector rather than a phrase. Based on this phenomena, sound and sight data are made as a disorganized component by handling intensive scale closeness. The authors proposed a smaller and faster seek file support system that offers better efficiency and record refreshes. Homo-morphic encryption and multiparty calculation are some of the applications.

## 3 Proposed Work

Multimedia file encryption and decryption happens here by using JAVASCRIPT as the scripting language and JAVA as the front end. At the server any file (it could be of any format - .doc, .pdf, .jpg, .mp3, etc.) to be encrypted by the doctor is chosen and encrypted in the JPG format. The encrypted file is sent Gmail to the client who is another doctor or a patient. For encryption we have used DES Algorithm (Data Encryption Standard). At the client, the received JPG file, via mail is downloaded and decrypted. The decrypted file further cannot be viewed even after decryption. This is because the client, i.e., another doctor/patient is supposed to know the correct format of the encrypted file sent by the doctor at the server end. Only then can the client view the decrypted file.

Here, at the server, encryption of the file and sending that encrypted file via Gmail is happening and at the client, downloading the received file and viewing of the decrypted file is happening. From the code point of view, there are two files which has to be run. One is image.java and the other is imagecrypto.java. Most importantly, before we start the execution antivirus in the system has to be disabled. This is because Gmail doesn't allow unknown sources to send any mail. But here we are allowing JAVA to send mail, which is against the policies of Gmail. Since this encryption and decryption is end to end, we are using two mail ids, one for the server and the other for the client. First imagecrypto.java is run at the server. Then image.java is run to send the encrypted file via mail and then imagecrypto.java is run again to decrypt the file received by the client. This is the working of this project.

## 4 Result Analysis

Multimedia decryption and encryption for medical data occurs here. So, decryption and encryption of medical data saved in any file formats was tested. For example, a prescription saved as a word document in '.doc' file format or a scanned copy of a report in the JPG file format saved as '.jpg' file format. It could even be an audio recording of a patient/doctor at the server and doctor/doctor at the client regarding any medical issue the patient is facing or some second opinion needed by a doctor with another doctor for further dealings with a patient's case, which is saved as '.mp3' file format. It could also be in the PDF file format or the data could also be a video which the doctor/patient would like to share secretly.

Few of the test cases which were tried are as follows:

1. Word Document (.doc file format)
2. Document (.pdf file format)
3. Image (.jpg file format)
4. Audio (.dct file format)
5. Video (.mp4 file format)

The various types of file formats were first chosen to encrypt, encrypted and then sent in JPG file format via Gmail to the receiver. The files were then downloaded, decrypted and viewed in the right format to get access to the actual data sent by the sender. To read a word document, Microsoft Office Word needs to be installed in the

PC/mobile wherever it needs to be viewed. Also, to read a document, an Adobe Acrobat Reader or a PDF Reader installed in the system. To see the image, an Image Viewer needs to be installed in the system. Likewise, to hear the audio file, Express Scribe Translator software needs to be installed in the system because the audio file is in the '. dct' format. Using this software, whatever the doctor has to say is recorder. Similarly, to be able to see a video file, a VLC Media Player application needs to be installed in the system. This has been tested with authentic medical audio data recorded through the Express Scribe Translator software, original medical text data converted to it from the recording through the Express Scribe Translator software, actual medical document data converted to it from the word document from the recording through the Express Scribe Translator software, real scanned medical image and an actual medical video.

Below are figures showing how an encrypted file looks after downloading it, after decrypting it and after viewing the decrypted file in the right format (format that the sender has sent in) (Figs. 1, 2, 3 and 4).

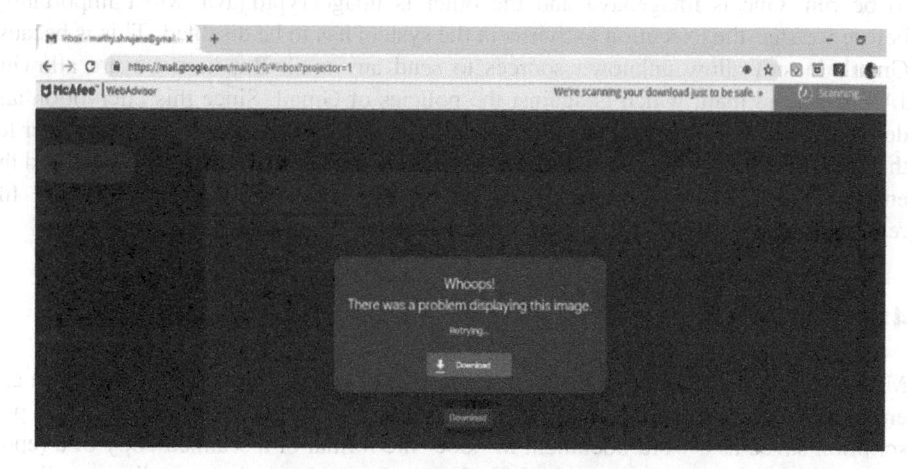


**Fig. 1.** Encrypted file is downloaded

**Fig. 2.** Encrypted file after decrypting

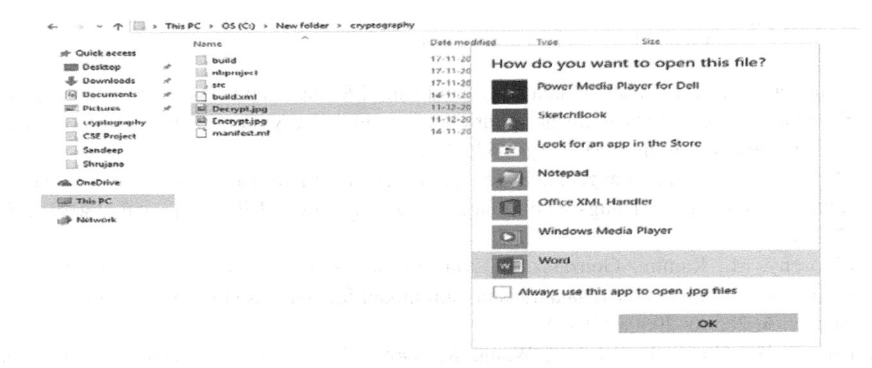


**Fig. 3.** Decrypted file being chosen to view in correct format

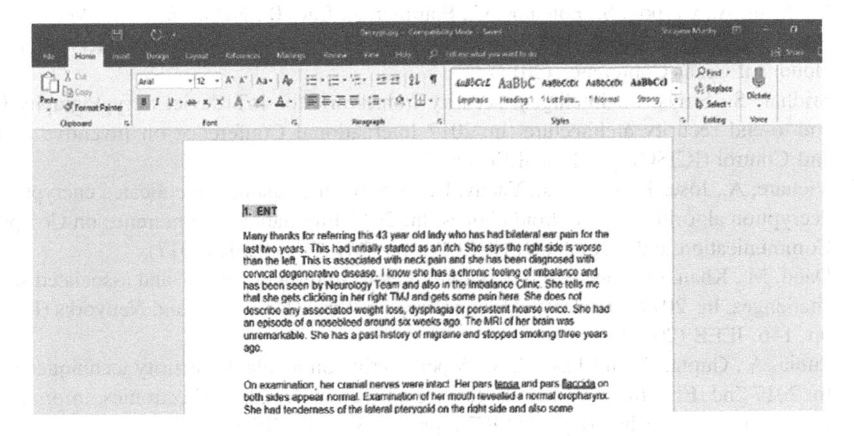


**Fig. 4.** Decrypted file when viewed in original sender format

## 5 Conclusion and Future Work

The available data security techniques are analyzed and the various file format types that are required for safety of personal data linking the details of the patient to the doctor along with the associated hospital or pharmacy. The existing techniques that offer security just support either one of the data file formats. None of the techniques available till now support all the data file formats. Neither has anyone included mail and tried securing the data. Future work is focused on further improvement of security. Also, various other encryptions can be done on the already encrypted file which is like double encryption to improve the decryption complexity and to ensure more safety of the medical data shared.

### Compliance with Ethical Standards

- ✓ All authors declare that there is no conflict of interest.
- ✓ No humans/animals involved in this research work.
- ✓ We have used our own data.

## References

1. Avudaiappan, T., Balasubramanian, R., Pandiyan, S.S., Saravanan, M., Lakshmanaprabu, S. K., Shankar, K.: Medical image security using dual encryption with oppositional based optimization algorithm. J. Med. Syst. **42**(11), 208 (2018)
2. Zheng, D., Wu, A., Zhang, Y., Zhao, Q.: Efficient and privacy-preserving medical data sharing in Internet of Things with limited computing power. IEEE Access **6**, 28019–28027 (2018)
3. Elhoseny, M., Ramírez-González, G., Abu-Elnasr, O.M., Shawkat, S.A., Arunkumar, N., Farouk, A.: Secure medical data transmission model for IoT-based healthcare systems. IEEE Access. **6**, 20596–20608 (2018)
4. Huang, H., Gong, T., Ye, N., Wang, R., Dou, Y.: Private and secured medical data transmission and analysis for wireless sensing healthcare system. IEEE Trans. Ind. Inform. **13**(3), 1227–1237 (2017)
5. Canteaut, A., Carpov, S., Fontaine, C., Fournier, J., Lac, B., Naya-Plasencia, M., Sirdey, R. et al.: End-to-end data security for IoT: from a cloud of encryptions to encryption in the cloud. In: Cesar Conference (2017)
6. Sridhar, S., Smys, S.: Intelligent security framework for IoT devices cryptography based end-to-end security architecture. In: 2017 International Conference on Inventive Systems and Control (ICISC), pp. 1–5. IEEE (2017)
7. Vichare, A., Jose, T., Tiwari, J., Yadav, U.: Data security using authenticated encryption and decryption algorithm for Android phones. In: 2017 International Conference on Computing, Communication and Automation (ICCCA), pp. 789–794. IEEE (2017)
8. Daud, M., Khan, Q. Saleem, Y.: A study of key technologies for IoT and associated security challenges. In: 2017 International Symposium on Wireless Systems and Networks (ISWSN), pp. 1–6. IEEE (2017)
9. Punia, A., Gupta, D. and Jaiswal, S.: A perspective on available security techniques in IoT. In: 2017 2nd IEEE International Conference on Recent Trends in Electronics, Information & Communication Technology (RTEICT), pp. 1553–1559. IEEE (2017)
10. Wang, Q., He, M., Du, M., Chow, S.S., Lai, R.W., Zou, Q.: Searchable encryption over feature-rich data. IEEE Trans. Dependable Secure Comput. **5**(3), 496–510 (2016)
11. Talwana, J.C., Hua, H.J.: Smart world of Internet of Things (IoT) and its security concerns. In: 2016 IEEE International Conference on Internet of Things (iThings) and IEEE Green Computing and Communications (GreenCom) and IEEE Cyber, Physical and Social Computing (CPSCom) and IEEE Smart Data (SmartData), pp. 240–245. IEEE (2016)
12. Al Farhan S., Kavitha C.R.: End-to-end encryption scheme for iot devices using two cryptographic symmetric keys. Int. J. Control Theory Appl. (2016)
13. Krishnan, S., Anjana, M.S., Rao, S.N.: Security considerations for IoT in smart buildings. In: 2017 IEEE International Conference on Computational Intelligence and Computing Research (ICCIC), Coimbatore, pp. 1–4 (2017)
14. Rakesh, N.: Performance analysis of anomaly detection of different IoT datasets using cloud micro services. In: 2016 International Conference on Inventive Computation Technologies (ICICT), Coimbatore, pp. 1–5 (2016)

# Exploration of an Image Processing Model for the Detection of Borer Pest Attack

Yogini Prabhu[1(✉)], Jivan S. Parab[1], and Gaurish M. Naik[2]

[1] Department of Electronics, Goa University, Taleigao Plateau, Goa, India
elect.yogini@unigoa.ac.in, jsparab@unigooa.ac.in

[2] Faculty of Natural Sciences and Department of Electronics, Goa University, Taleigao, Goa, India
gmnaik@unigoa.ac.in

**Abstract.** This paper presents an attempt to utilise an image processing technique for automatic detection of the infestation stage of borer pest attack in cashew tree. Earlier numerous studies have employed various image processing techniques and achieved different RGB models for determining plant stress due to invasion by pathogens and also for monitoring the general plant health (Estimation of N and lipids). However, the techniques were focused on the detection of diseases with symptoms of deformations in plant parts (consumed or inhabited by pest or microbes) like leaf or stem, which are even visible. On the contrary, in case of cashew stem and root borer, its *modus-operandi* is such that there are no visible (external) indications of its attacks, even though it has been tunnelling inside the trunk. Also, the recovery measures *viz.* applications of chemicals become unsuccessful when the boring in the trunk circumference is more than 50%; this calls for immediate control action, discernibly through detection at early stages of the borer attack. After tackling the problem statement by resorting to spectroscopic methods, we have hereby applied an RGB-based model to the images of leaves (for two cases of illumination scenarios: day-light and incandescent lamp, and two image capturing devices: a digital camera i.e. Panasonic LUMIX DMC-SZ10 and a cell phone camera) to analyse to possibility of identification of the pest attack.

**Keywords:** Image processing · RGB model · Cashew stem and root Borer (CSRB) · Pest · Plant stress · Infestation stages · Chlorophyll

## 1 Introduction

One of the major causes of infection of the trees in forests and cash crops in plantations, leading to their death in period of six months is known to be borers [1–5]. A borer attack in farms is tackled by clearing of root collar region and applying some chemicals to regions in and out the spotted borrowings by the beetles.

In tropical regions like Goa, the Cashew Stem and Root Borer (CSRB)- scientific name: *Plocaederus ferrugineus* pest, is one prevalent pest which attacks cashew (Fig. 1) and mango trees. It begins the attack on collar region of the tree and advances into the trunk by tunnelling within it [6]. Also the pest being nocturnal is not visible at day-time. Thus, the incidence of damage is not perceived unless observed at individual


S. Smys et al. (Eds.): ICCVBIC 2019, AISC 1108, pp. 589–597, 2020.
https://doi.org/10.1007/978-3-030-37218-7_67

trees, which is a tedious task for a vast cultivation area). So, since the cashew (*Anacardium occidentale L.*) is a crucial commercial crop with a large potential for foreign currencies, the severe borer pest attacks leads to significant depletion in cashew nut yield, eventually causing a calamity to the farmers.

### 1.1 Earlier Relevant Works

For detection of plant stresses due to invasion by pathogens and also for monitoring the general plant health (Estimation of N and lipids), earlier works have employed various image processing techniques and achieved different RGB models. However, the techniques were focused on the detection of diseases with symptoms of deformations in plant parts (consumed or inhabited by pest or microbes) like leaf or stem, which are even visible.

In the case of detection and monitoring of adult-stage whitefly (*Bemisia tabaci*) and thrip (*Frankliniella occidentalis*) in green-houses, there was combination of an image processing algorithm and artificial neural networks [7]. Detection of the objects in the images, segmentation, and morphological and color property estimation was performed by an image- processing algorithm for each of the detected objects. Finally, classification was achieved by means of a feed-forward multi-layer artificial neural network. The proposed whitefly identification algorithm achieved high precision (0.96), recall (0.95) and F-measure (0.95) values, whereas the thrip identification algorithm obtained similar precision (0.92), recall (0.96) and F-measure (0.94) values.

An RGB (Red-Green-Blue) based Image analysis method was developed for rapid and non-invasive determination of chlorophyll content of leaves of micro-propagated potato plants [8]. The chlorophyll content predicted by their model showed significant correlation with the one measured by chlorophyll content meter. To identify stress level from leaf colour [9], the CIE chromaticity diagram was used to transform leaf colour information in RGB into wavelength (in nm). Here also they also performed Arnon method to formulate empirical relation between chlorophyll a $C_{c\text{-}a}$ and its wavelength $\lambda_{avg}$. Yuzhu et al. [10] has shown that G/(R + G + B) gives the beast correlation with chlorophyll content. Kawashima and Nakatani [11] had given (R-B)/(R + B) using colour image processing.

Ali has given the formula (Eq. 1) to determine chlorophyll [12] changes from plant to plant and also with the environmental condition. The algorithm here non-linearly maps normalised value of G, with respect to R and B, using a logarithmic sigmoid transfer functions as follows:

$$\text{Chl} = logsig\frac{G - \left(\frac{R}{3}\right) - \left(\frac{B}{3}\right)}{255} \tag{1}$$

where Chl is Chlorophyll estimation, G is value of Green pixel, R is value of Red pixel and B is value of Blue pixel.

In another Image processing technique not only chlorophyll content was detected but also size of leaf [13] was determined by using Haar transform, for detection of distortion of the leaf boundaries by pest consumption.

A three colour analysis [14] for determining chlorophyll a and lipids as well as growth dynamics, when compared with experimental results of standard methods, demonstrated a squared correlation coefficient ($R^2$) of 0.99 for chlorophyll a and lipids. The RGB model corresponded very well with previous studies. Also, it was verified in real cultivations of the microalgae in a photo bioreactor for its reliability.

### 1.2 Our Contribution

In order to design a system which will be capable to detect the presence of borer infestation in early stage, we had performed initial analysis on the samples of leaves and bark, in the past. Absorbance [6] as well as reflectance spectro-photometry was utilised to figure out the changes in spectra peaks for different health conditions: Healthy, Infested Start, Infested Middle and Infested End. There was a clear distinction in the level of peaks, which is sufficient enough to be utilised as critical to decide the category of the health status of a given tree.

However during the literature studies based on plant health, we found that there is scope for attempting Image processing techniques for the CSRB pest. Hence, we selected a promising RGB-model out of several others, applied it to the obtained database of leaves images for four study conditions and, derived conclusions observing the data of output values.

The rest of the paper is organized as follows: Sect. 2 presents the description of the materials and methods, including flowchart of the procedure steps. Section 3 is focussed on the details of results obtained. Section 4 presents the conclusion of the four study cases. Section 5 mentions the scope of further works.

## 2 Materials and Methods

### 2.1 Study Area

The study area is a private commercial plantation in place known as Mollem in South Goa. It spans to an area of 28000 $m^2$ between 74.226294°E (longitude) and 15.379047°N (latitude) with elevation 16 m (52 ft). The climate here is tropical. Mollem has significant rainfall most months, with a short dry season. The climate here is classified as Am (i.e. tropical monsoon) by the Köppen-Geiger system [15]. The average temperature in Mollem is 27.3 °C. The average annual rainfall is 2574 mm. The travelling duration by car to the laboratory from the site location is 70 ± 10 min. The plantation has got 85% cashew trees, while remaining 15% constitute mangoes, Indian Blackberry, areca nuts and coconuts.

### 2.2 Data Acquisition

Field data was collected in early summer of 2017. A total of 40 trees were selected, 10 each in category of healthy, and infestations - initial, middle and end. A thorough site survey was conducted to identify individual category, by spotting of dry, boring dust, which had been burrowed out of the bark surfaces during the tunnelling by borer.

**Fig. 1.** Pictures of the leaf samples for each study category – (a) healthy condition: (in clockwise order) NI-DC, LI-DC, LI-CC, NI-CC (b) infested end condition: (in clockwise order) LI-CC, NI-CC, NI-DC, LI-DC; where LI is Lamp Illumination, NI is Normal Illumination, CC is cell phone camera and DC is Digital Camera.

Leaves were plucked from the identified trees and placed in a labelled zip-locked bag. All bags were placed in a box filled with frozen ice-packs [16] so as to curb any effect of temperature rise during the day progress during transportation in the bio-chemical properties of the leaves. At the laboratory, all the leaves were wiped off any water drops and dust, by using clean cotton before proceeding to data processing.

The database was obtained for four combinations comprising of two cases of illumination scenarios: day-light and incandescent lamp, and two image capturing devices: a digital camera i.e. Panasonic LUMIX DMC-SZ10 and a cell phone camera (specifications of each device are stated Tables 1 and 2).

### 2.3 Data Processing

For every leaf, six pixels were selected in leaf area and their RGB values extracted using MATLAB Image Processing functions. Then the selected RGB model i.e. Equation is applied to the each set of RGB values to obtain six RGB model outputs. All the six output values were averaged to obtain a consolidated value for a given leaf.

This was done for ten leaves, in each of the categories: Healthy, Infested Start, Infested Middle and Infested End. The entire procedural steps of the data processing phase are stated in the form of flowchart as shown in Fig. 2.

**Table 1.** Digital camera Panasonic LUMIX DMC-SZ10

| | | |
|---|---|---|
| Pixels | Camera effective pixels | 16 Megapixels |
| Sensor | Sensor size/Total pixels/Filter | 1/2.33-inch CCD sensor/Total pixel number 16.6 Megapixels/Primary color filter |
| Lens | Aperture | F3.1-6.3/2-step (F3.1/7.8 (W), F6.3/16.3 (T)) |
| | Optical zoom | 12x |
| | Focal length | f = 4.3–51.6 mm (24–288 mm in 35 mm equiv.) |

**Table 2.** Specifications of cell-phone camera - Obi Boa 503

| Pixels | Camera effective pixels | 8 Megapixels |
|---|---|---|
| Others | | 3264 × 2448 pixels; 30 fps |

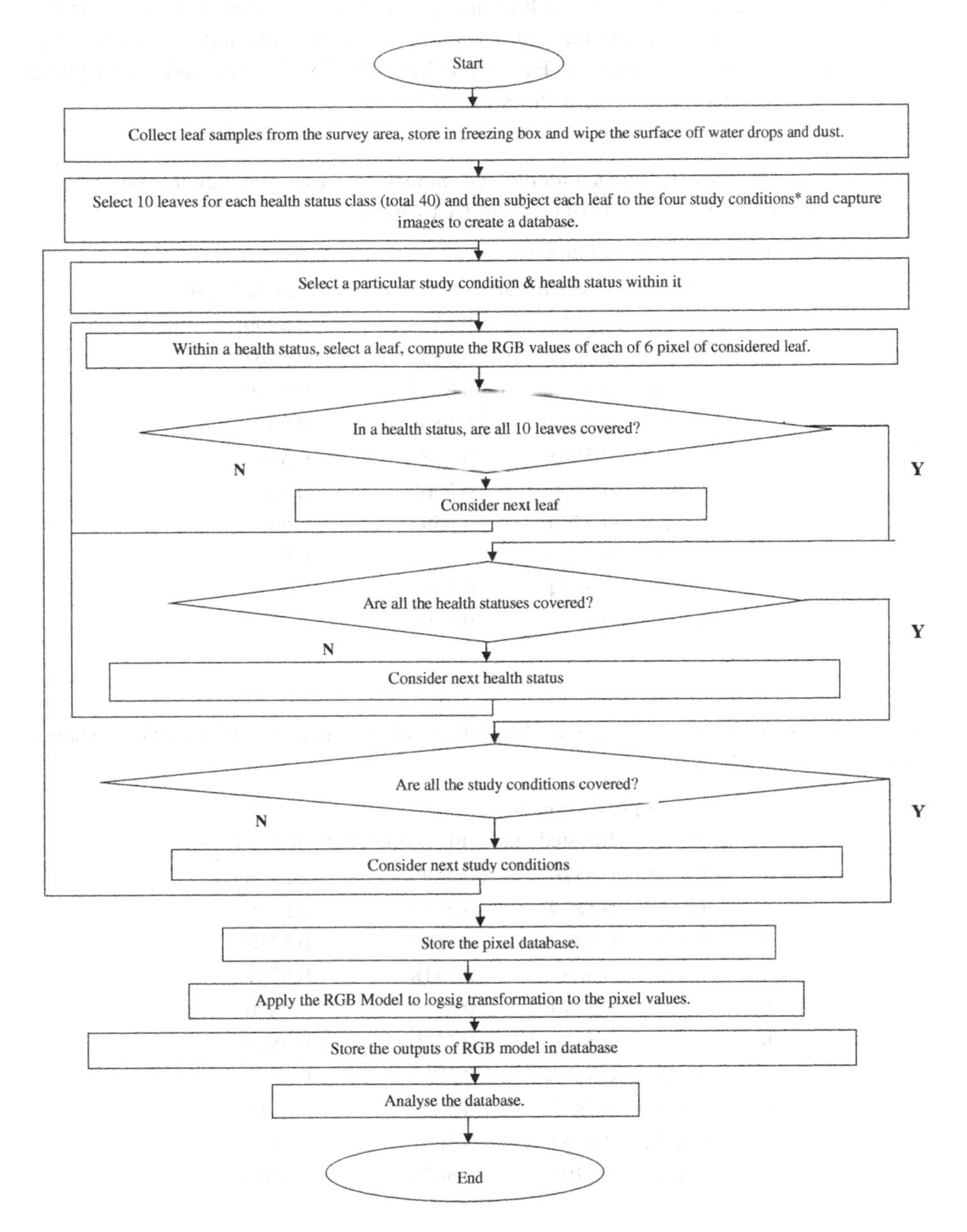


**Fig. 2.** Flowchart of the method and data processing.

## 3 Results

The chlorophyll values were obtained as output of the RGB model, when applied to the six sets of RGB components of leaf pixels for each of the health status of cashew trees: Healthy, Infested Start, Infested Middle and Infested End.

These experiments carried out to find any possibility of detection in case of; two types of illumination scenarios: day-light and incandescent lap, and two image capturing devices: a digital camera i.e. Panasonic LUMIX DMC-SZ10 and a cell phone camera are tabulated in Tables 3, 4, 5 and 6.

**Table 3.** Chlorophyll values obtained for the RGB model; for the case of illumination scenario - day-light and digital camera i.e. Panasonic LUMIX DMC-SZ10.

| Sr. no. | Chlorophyll content | | | |
|---|---|---|---|---|
| | Healthy | Infested start | Infested middle | Infested end |
| 1 | 0.5527 | 0.5514 | 0.5518 | 0.5499 |
| 2 | 0.5531 | 0.5537 | 0.5434 | 0.5504 |
| 3 | 0.5543 | 0.5664 | 0.5432 | 0.5489 |
| 4 | 0.5583 | 0.5574 | 0.5566 | 0.5484 |
| 5 | 0.5558 | 0.5565 | 0.548 | 0.5630 |
| 6 | 0.5602 | 0.5509 | 0.5508 | 0.5564 |
| 7 | 0.5499 | 0.5594 | 0.5595 | 0.5457 |
| 8 | 0.5583 | 0.5629 | 0.5478 | 0.5487 |
| 9 | 0.5683 | 0.5624 | 0.5542 | 0.5476 |
| 10 | 0.5564 | 0.5688 | 0.5493 | 0.5522 |

**Table 4.** Chlorophyll values obtained for the RGB model; for the case of illumination scenario - day-light and a cell phone.

| Sr. no. | Chlorophyll content | | | |
|---|---|---|---|---|
| | Healthy | Infested start | Infested middle | Infested end |
| 1 | 0.5889 | 0.5826 | 0.5599 | 0.5656 |
| 2 | 0.6001 | 0.5879 | 0.5695 | 0.5740 |
| 3 | 0.5763 | 0.5897 | 0.5587 | 0.5718 |
| 4 | 0.5749 | 0.5855 | 0.5718 | 0.5721 |
| 5 | 0.5817 | 0.5774 | 0.5773 | 0.5836 |
| 6 | 0.5799 | 0.5870 | 0.5670 | 0.5818 |
| 7 | 0.5831 | 0.5968 | 0.5858 | 0.5778 |
| 8 | 0.5773 | 0.5897 | 0.5702 | 0.5715 |
| 9 | 0.5851 | 0.5855 | 0.5651 | 0.5799 |
| 10 | 0.5824 | 0.5792 | 0.5670 | 0.5780 |

**Table 5.** Chlorophyll values obtained for the RGB model; for the case of illumination scenario-incandescent lamp and digital camera.

| Sr. no. | Chlorophyll content | | | |
|---|---|---|---|---|
| | Healthy | Infested start | Infested middle | Infested end |
| 1 | 0.5355 | 0.532 | 0.5323 | 0.5438 |
| 2 | 0.5331 | 0.5384 | 0.5292 | 0.5759 |
| 3 | 0.5407 | 0.5389 | 0.5425 | 0.5401 |
| 4 | 0.5354 | 0.5440 | 0.5308 | 0.5600 |
| 5 | 0.5369 | 0.5359 | 0.5372 | 0.5439 |
| 6 | 0.5312 | 0.5435 | 0.5326 | 0.5337 |
| 7 | 0.5245 | 0.5428 | 0.5386 | 0.5471 |
| 8 | 0.5368 | 0.5443 | 0.5354 | 0.5567 |
| 9 | 0.5374 | 0.5278 | 0.5358 | 0.5569 |
| 10 | 0.5412 | 0.5420 | 0.5384 | 0.5452 |

**Table 6.** Chlorophyll values obtained for the RGB model; for the case of illumination scenario-incandescent lamp and cell phone camera.

| Sr. no. | Chlorophyll content | | | |
|---|---|---|---|---|
| | Healthy | Infested start | Infested middle | Infested end |
| 1 | 0.5393 | 0.5425 | 0.5389 | 0.5533 |
| 2 | 0.5364 | 0.5378 | 0.5264 | 0.5488 |
| 3 | 0.5356 | 0.5381 | 0.5401 | 0.5389 |
| 4 | 0.5367 | 0.5370 | 0.5314 | 0.5406 |
| 5 | 0.5379 | 0.5367 | 0.5413 | 0.5518 |
| 6 | 0.5399 | 0.5457 | 0.5309 | 0.5352 |
| 7 | 0.5474 | 0.5471 | 0.5332 | 0.5494 |
| 8 | 0.5418 | 0.5462 | 0.5360 | 0.5339 |
| 9 | 0.5334 | 0.5455 | 0.5413 | 0.5513 |
| 10 | 0.5371 | 0.5449 | 0.5387 | 0.5491 |

As observed from the tables i.e. Tables 3, 4, 5 and 6, the values do not have any particular trend i.e. increase or decrease, either along the stages of progress of infestations or for different combinations of illuminations and the devices, noticeably no distinction even.

## 4 Conclusion

Given a certain chlorophyll content value, it does not assert any corresponding infestation stage. This will create an uncertainty in the decision-making about the status of CSRB attack.

Thus we say that, the results obtained with a well-known computational model of RGB do not achieve recognition and classification of the infestation stages categories or levels of CSRB infestation. Image processing was attempted to explore any possibility of detection of the borer pest and thus found to be not relevant for the purpose of timely of detection of beetle CSRB pest attack in cashew trees.

## 5 Scope of Further Works

The CIElab method of colour analysis can be attempted as a scope for future works. This method can be implemented as an altogether new approach complete in itself, or incorporated into the existing algorithm for an hybrid approach, to possibly derive a solution relevant for the purpose of timely of detection of beetle CSRB pest attack in cashew trees, and also further improve the performance.

**Acknowledgment.** Authors would like the acknowledge the financial support received from Department of electronics and Information Technology (DeitY), New Delhi towards support of fellowship under the Visveswaraya PhD Scheme for the period of 2015–2020.

## References

1. Hofstetter, R.W., Mahfouz, J.B., Klepzig, K.D., Ayres, M.P.: Effects of tree phytochemistry on the interactions among endophloedic fungi associated with the southern pine beetle. J. Chem. Ecol. **31**(3), 539–560 (2005)
2. Moraal, L.G.: Infestations of the cypress bark beetles Phloeosinus rudis, P. bicolor and P. thujae in The Netherlands (Coleoptera: Curculionidae: Scolytinae). Entomologische Berichten **70**(4), 140–145 (2010). ISSN 0013–8827
3. Berg, A.R., Heald, C.L., Hartz, K.E.H., Hallar, A.G., Meddens, A.J.H., Hicke, J.A., Lamarque, J.-F., Tilmes, S.: The impact of bark beetle infestations on monoterpene emissions and secondary organic aerosol formation in western North America. Atmos. Chem. Phys. **13**(6), 3149–3161 (2013)
4. Zhao, T., Axelsson, K., Krokene, P., Borg-Karlson, A.-K.: Fungal symbionts of the spruce bark beetle synthesize the beetle aggregation pheromone 2-methyl-3-buten-2-ol. J. Chem. Ecol. **41**(9), 848–852 (2015). https://www.youtube.com/watch?v=d8TCT5UkjaU. Accessed 20 Sept 2015. Do pine beetle fan the flames in western forest.mp4
5. Do pine beetle fan the flames in western forest.mp4. https://www.youtube.com/watch?v=d8TCT5UkjaU. Accessed 20 Sep 2015
6. Yogini, P., Jivan, P., Rajendra, G., Gourish, N.: Farmer-friendly portable system for diagnosis of pest attack in cashew trees. CSI Trans. ICT (2018). https://doi.org/10.1007/s40012-018-0198-8
7. Espinoza, K., Valera, D.L., Torres, J.A., López, A., Molina-Aiz, F.D.: Combination of image processing and artificial neural networks as a novel approach for the identification of Bemisia tabaci and Frankliniella occidentalis on sticky traps in greenhouse agriculture. Comput. Electron. Agric. **127**, 495–505 (2016). https://doi.org/10.1016/j.compag.2016.07.008

8. Yadav, S.P., Ibaraki, Y., Gupta, S.D.: Estimation of the chlorophyll content of micropropagated potato plants using RGB based image analysis. Plant Cell Tissue Organ Cult. (PCTOC) **100**(2), 183–188 (2010). https://doi.org/10.1007/s11240-009-9635-6. 1573-5044
9. Shibghatallah, M.A.H., Khotimah, S.N., Suhandono, S., Viridi, S., Kesuma, T.: Measuring leaf chlorophyll concentration from its color: a way in monitoring environment change to plantations. In: AIP Conference Proceedings, vol. 1554, pp. 210–213 (2013). https://doi.org/10.1063/1.4820322
10. Yuzhu, H., Xiaomei, W., Shuyao, S.: Nitrogen determination in pepper (Capsicum frutescens L.) Plants by colour image analysis (RGB). Afr. J. Biotechnol. **10**(77), 17737–17741 (2011)
11. Kawashima, S., Nakatani, M.: An algorithm for estimating chlorophyll content in leaves using a video camera. Ann. Bot. **81**(1), 49–54 (1998)
12. Ali, M.M., Al-Ani, A., Eamus, D., Tan, D.K.: A new image processing based technique to determine chlorophyll in plants. Am.-Eurasian J. Agric. Environ. Sci. **12**(10), 1323–1328 (2012)
13. Arora, A., Menaka, R., Gupta, S., Mishra, A.: Haar transform based estimation of chlorophyll and structure of the leaf. ICTACT J. Image Video Process. **03**(04), 612–615 (2013). ISSN: 0976-9102(ONLINE)
14. Su, C.H., Fu, C.C., Chang, Y.C., Nair, G.R., Ye, J.L., Chu, I.M., Wu, W.T.: Simultaneous estimation of chlorophyll a and lipid contents in microalgae by three-color analysis. Biotechnol. Bioeng. **99**, 1034–1039 (2008). https://doi.org/10.1002/bit.21623
15. https://bigladdersoftware.com/epx/docs/8-3/auxiliary-programs/koppen-climate-classification.html
16. Abdullah, H., Darvishzadeh, R., Skidmore, A., Groen, T., Heurich, M.: European spruce bark beetle (Ips typographus, L.) green attack affects foliar reflectance and biochemical properties. Int. J. Appl. Earth Observ. Geoinf. **64**, 199–209 (2018). https://doi.org/10.1016/j.jag.2017.09.009

# Measurement of Acid Content in Rain Water Using Wireless Sensor Networks and Artificial Neural Networks

Sakshi Gangrade(✉) and Srinath R. Naidu

Department of Computer Science and Engineering,
Amrita School of Engineering, Amrita Vishwa Vidyapeetham, Bengaluru, India
sakshigangrade1996@gmail.com,
r.srinath@blr.amrita.edu

**Abstract.** Acid rain is the subject of concern from last many years. Acid rain is rain polluted by acid which has been dissolved into the atmosphere. Knowing the pH of water is very important to reduce the same by reducing the pollution levels. This project performs the predictions of pH of future rainfalls using Artificial Neural Networks. Dataset for temperature, humidity, windspeed and air quality are collected. Tuple of this dataset is provided as an input for training the artificial neural network. On considering the dataset for past one week as input, we have performed the predictions of pH of rainwater and checked the same by implementing a hardware to measure the pH, temperature and turbidity of water. These values are then displayed on an LCD.

**Keywords:** Artificial Neural Networks · Feedforward networks · Wireless Sensor Networks · Back propagation

## 1 Introduction

Rainfall is a most important water resource. As all the water needs are fulfilled by rainwater whether the source of water is sea, river, lake or groundwater. Now a days due to increase in industrialization, the pollution levels are also increasing. This pollution causes the emissions of harmful gases such as Sulphur Dioxide (SO2) and Nitrogen Dioxide (NO2) into the atmosphere, which combines with oxygen and water molecules to form acid rain.

Acid rain is rain polluted by acids which are released into the atmosphere due to emissions from factories and vehicles. Whenever these gases combine with water, they form nitric acid (HNO3), sulphuric acid and carbonic acid (HCO3). As these acids are strong, they dissociate in water to give H +ve ions. Hence, the presence of these acids increases the concentration of H +ve ions and thus increase the acidity of water and reducing the pH.

The pH of water is a measure of the concentration of H +ve ions. The pH value for neutral water is 7, for acidic water is in the range of 0 to 6 and for basic water is in the range of 8 to 14. Acid rain has various harmful effect on environment, such as damage to aquatic wildlife, damage to soil, dying of trees and damage to monuments. Knowing


S. Smys et al. (Eds.): ICCVBIC 2019, AISC 1108, pp. 598–605, 2020.
https://doi.org/10.1007/978-3-030-37218-7_68

the pH of rainwater is the primary step to reduce the effects of acid rain. Knowing the value of pH of rainwater, measures can be taken to reduce its after effects by covering the crops, keeping plants inside the house and covering sensitive equipment such as marble statues which get corroded due to acid rain. The system describes the working of an Artificial Neural Network, which is created by providing the dataset for New Brunswick, New Jersey as the input data.

In this project, Artificial Neural Network is implemented to perform predictions for the value of pH of rainwater. Tuple of data which includes temperature, humidity, windspeed and air quality for New Brunswick, New Jersey is collected. This data is provided as input to train the Artificial Neural Network and to predict future values of ph. On considering this data for past one week as input to train the network, predictions for the values of pH and turbidity are performed. The neural network is created using New Jersey data, this data is provided to validate the model.

As, by providing New Jersey data for the input, values of pH can be predicted, similarly by collecting the dataset for these parameters for Indian context a neural network can be created which will forecast the pH values for India. For this to achieve, a hardware needs to be constructed, which consists of sensors to measure the temperature, pH and turbidity of rainwater. The values obtained from these sensors will then be used as data to train a similar neural network and hence value of pH can be forecasted for Indian data. Therefore, a hardware is implemented which consists of pH, temperature and turbidity sensors to measure these values for rainfall in real time and these values are then displayed on a server. The next section focuses on the literature survey, in Sect. 3, the implementation of the project is described, Sect. 4 shows the obtained results and the work is concluded with Sect. 5.

## 2 Literature Survey

This section describes the literature survey on the area of Wireless Sensor Networks and Artificial Neural Networks. In [1], a system is proposed which describes the method to monitor rainfall. A tipping bucket rain gauge is used to measure the amount of rainfall, temperature and humidity sensors are placed in the rain gauge station to measure the temperature and humidity of rainwater. This collected information is then transferred to the GPRS module. Transfer of information from the rain gauge station to the web server is performed by the GPRS module. On reaching the data collection station the collected data is displayed on the server. Through this website the user can get the information about the amount of rainfall during a particular interval of time.

According to [2], a long-term monitoring is conducted from 2003 to 2007 on the Ijira lake catchment area which was found to be acidic. Dry and wet deposition was measured separately. For wet deposition precipitation samples were collected and values of pH, conductivity, concentration of SO2(2-), NO3(-), Cl (-), Ca (2+), NH4(+), K (+), Mg (2+) and Na (+) were measured. For dry deposition analysis of forests, soil and land water was performed. Graphs are plotted which indicate that as the concentration of NO3(-) and SO2(2-) increases, pH value decreases linearly.

In [3], operations are performed on the surface water to provide early warning in flood situations. Cluster heads as well as sensor nodes are placed near river Damodaran.

The data from sensor nodes are transferred to the cluster heads for monitoring of river parameters. Cluster heads collect the information about the river characteristics which are transferred by the sensor nodes. Neural network algorithm is used at each cluster head to process the information obtained from the sensor nodes. Multilayer perceptron having feed forward network is used. The output indicates the flood warning parameters indicating no flood, alarming, low, high and very high flood.

In [4], a system is proposed an ANN model which predicts the nitrate concentration in groundwater. Samples for 24 wells are collected monthly for 1 year. pH, temperature, conductivity and ground water level are provided as input for training the network. Backpropagation algorithm is used for training. Nitrogen values are predicted which form the output of the network.

In [5], a paper is proposed to determine pH, temperature and dissolved oxygen in water. This data is displayed on the website so that user can login and access the data. Sensors are used to measure these parameters and data from these sensor nodes are transferred to the cluster head which collects data from various sensor and then transmit it to the server. Zigbee communication module is used for data transfer. Output is displayed in the form of graphs representing time on X axis and temperature on y axis.

## 3 Implementation

This section discusses the working of the system. In this project, Artificial Neural Networks are implemented to perform forecasting the values of pH of rainwater. Dataset for maximum temperature, minimum temperature, humidity, wind speed and air quality are provided as input to train the Artificial Neural Network. This data is collected from New Brunswick, New Jersey, for one year on a daily basis. The output of the neural network consists of pH of water. The system trains the Artificial Neural Network with the input data and perform predictions of the values of ph. This system is implemented on Lenovo idea pad 320 laptop.

The collected daily data is stored in an excel sheet. The input data includes the maximum temperature, minimum temperature, humidity, wind speed and air quality for one year on a daily basis and the target includes pH of the rainwater. A neural network is implemented by providing the input parameters, training the network and obtaining the output Fig. 1. An artificial neural network consists of input layer, hidden layer and output layer. Neurons are present in these layers which are interconnected to perform prediction of the pH values. It consists of 5 neurons in the input layer representing the 5 input parameters and 1 neuron in the output layer representing the output parameter. Training of the network is performed with 70% of the dataset and the remaining 30% of the dataset is used for testing of the neural network. Random initial weights are assigned to the network. Non-linear transfer function is applied to the input. The network is then trained by varying the weights, learning rate and number of neurons in the hidden layer, until optimised outputs are obtained. Weights represent strength of connection between units. Once a minimum validation error is obtained, the weights are fixed and testing of the network is performed with the same weights and output values are obtained. Regression plots are formed. The regression value determines the closeness between the target value and the obtained output. The choice for the number of hidden layers and the number of neurons in a hidden layer is an important task while training a Neural Network. In this work, various tests were performed in which different values were provided for the number of hidden layers, number of neurons in those hidden layers, learning rate of the network and the transfer function to obtain the best fit. Network with 1 hidden layer having 6 neurons was fixed as it provided highest accuracy.

In this model supervised learning is performed by providing the pH values on a daily basis for 1 year as the target. Training of the network is performed for 1 year on the input parameters providing the target values and varying the number of neurons in the hidden layer until the outputs obtained from the network are equal to the target values. When the target value of the pH and the obtained output value of pH from the neural network match, the neural network is fully trained. Fit net (Function Fitting Neural Network) with tan sig transfer function is used for training the neural network. Fit net constructs a function fitting neural network for regression with the size of hidden layer which is 6 in this case and training function. Tan sig transfer function is applied to the hidden layer as it is a non-linear function and ranges from −1 to 1. Bayesian Regularization backpropagation is used as the training function, as it generalizes well because it minimizes combination of squared error and weights and determines their correct combination.

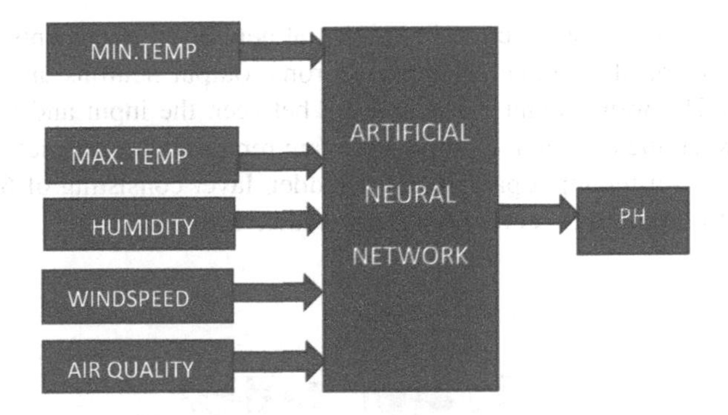


**Fig. 1.** Flow diagram

The learning rate which provided accurate output is 0.01. The entire data is divided into training and testing dataset. Training is done on 70% of the data and testing is done on 30% of the data. Training dataset is the one on which the network performs training and the testing dataset is the unknown dataset provided to the network. During training of the neural network, 70% of the dataset is provided as input to the neural network and the target values corresponding to these 70% inputs is provided. The obtained outputs are compared with the target values and based on the difference between the target values and output values, weights and biases of the network is modified. Values of weights which provide optimised regression for training and testing is selected After training is performed the testing dataset determines the amount of fit on the test dataset.

# 4 Result

The network in this project is a function fitting neural network with tan sig transfer function. The input weights are randomly generated. The performance of the system is measured in mean square error. The obtained output represents a graph between the output and target values.

On varying the values of number of neurons present in the hidden layer and the learning rate, optimized outputs are obtained. Several activation functions were tested to obtain the desired fit. Tan sig activation function with Bayesian Regularization backpropagation was then fixed as it provided the best fit.

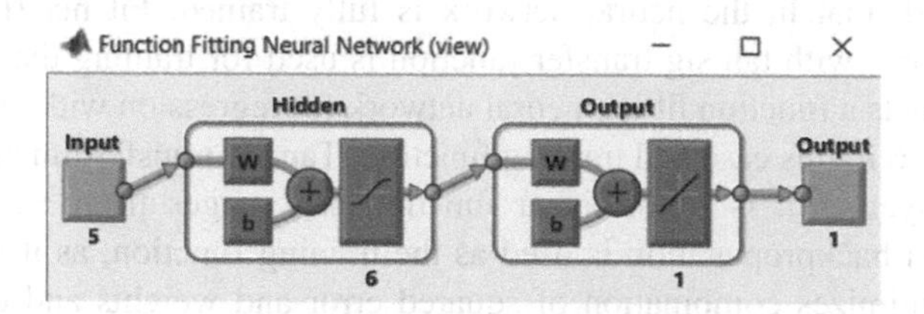


**Fig. 2.** Function fitting neural network

Figure 2 represents the function fitting neural network. It represents the structure of the neural network, the number of input neurons, output neurons and neurons in the hidden layer. The weights and biases present between the input and the hidden layer and also between the hidden and output layer are represented. This network consists of 5 input neurons for the input parameters, 1 hidden layer consisting of 6 neurons and 1 neuron in the output layer representing the predicted pH values.

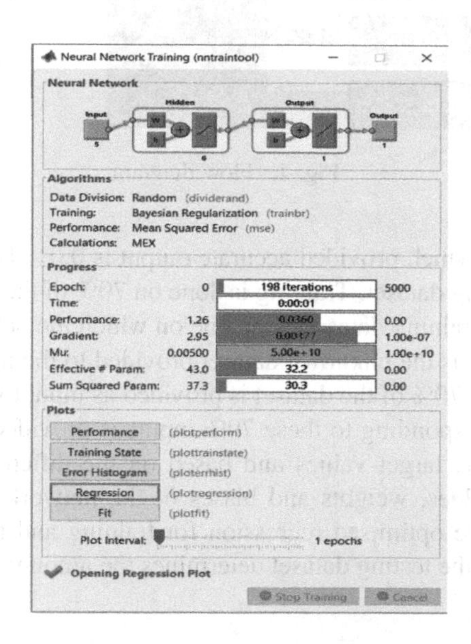


**Fig. 3.** Neural network tool box

Figure 3 represents the neural network toolbox indicates the way in which data is divided, the training algorithm used and the performance parameter. In this project Bayesian Regularization is used as the training algorithm and mean square error are declared as the performance parameter.

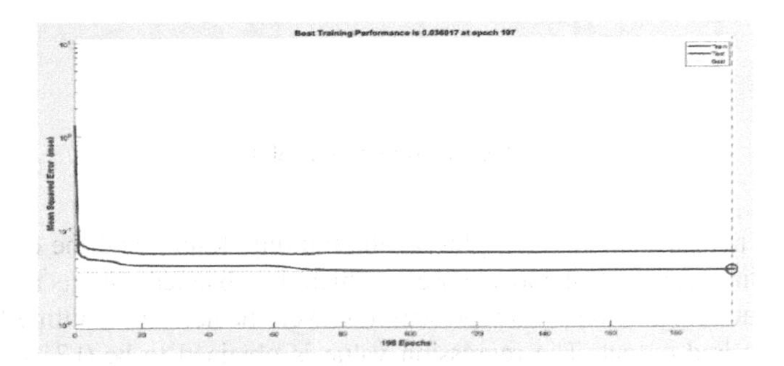

**Fig. 4.** Performance plot

Figure 4 represents the performance plot of the neural network represents a graph between the number of epochs and the mean square error for the training and testing dataset. The red coloured line is for the testing data and blue line is for the testing data.

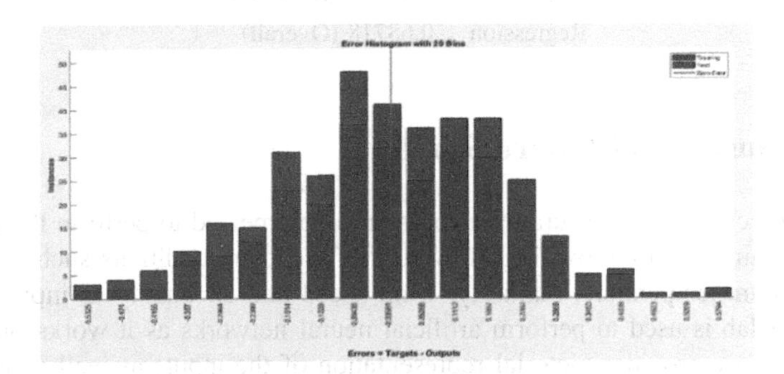

**Fig. 5.** Error histogram

The error histogram represents the error. Error is the difference between the target value and the obtained output. Figure 5 represents an error histogram. It consists of 20 bins which indicate the error at different instances. The orange line represents zero error and blue and red colours represent training error and testing error respectively. The error corresponding to maximum instances ranges between −0.0160625 to 0.1441575.

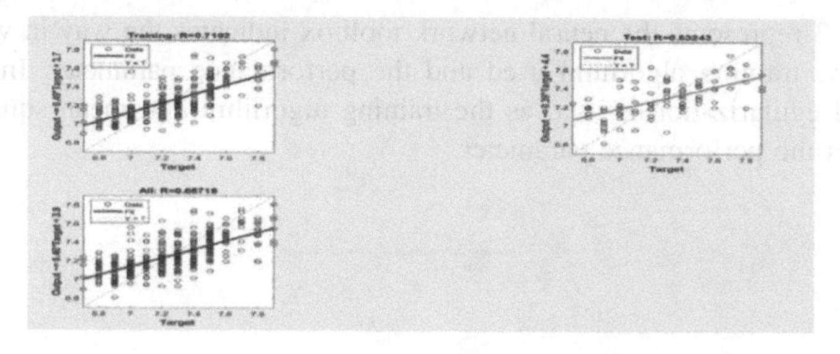

**Fig. 6.** Regression plot

Figure 6 represents regression plot for the training, testing and the overall dataset. The dashed line represents the ideal case in which the obtained outputs are same as the desired outputs. Regression plots represent how well the network is fitting the inputs to obtain the desired output. The regression value is obtained to be 0.7102 for training dataset and 0.63243 for testing dataset (Table 1).

**Table 1.** Overall results

| Parameter | Value |
|---|---|
| Performance | 0.0360175 (mse) |
| Error | −0.0307 to 0.0248775 |
| Regression | 0.68718 (Overall) |

## 5 Conclusion and Future Scope

In this project, Artificial Neural Networks are implemented to perform the predictions on pH of rainwater by providing the values of weather conditions such as maximum and minimum temperature, humidity, wind speed and air quality as input to the network. Mat lab is used to perform artificial neural networks as it works on net-based GUI and hence provides pictorial representation of the inputs as well as the outputs. Training and testing of the dataset are performed and outputs are obtained. The obtained Regression value for training data is 0.7102 and for testing dataset is 0.63243. Hardware is implemented to measure the values of pH, turbidity and temperature of rainwater.

In future work, data for temperature, humidity, windspeed and air quality will be collected for India by constructing a hardware including sensors to measure the temperature and humidity. This data will then be provided to train an artificial neural network to predict the values of pH of rainwater.

**Compliance with Ethical Standards**

✓ All authors declare that there is no conflict of interest.

✓ No humans/animals involved in this research work.

✓ We have used our own data.

## References

1. Mangundu, E.M., Mateus, J.N., Zodi, G.-L., Johson, J.: A wireless sensor network for rainfall monitoring, using cellular network: a case for Namibia. Paper Presented at 2017 Global Wireless Summit, GWS 2017, pp. 240–244, January 2018. https://doi.org/10.1109/GWS.2017.8300469. www.scopus.com
2. Akimoto, H., Oda, T., Dokiya, Y., Ogura, N.: Report of Long Term National acid Deposition Monitoring in Japan (JFY 2003–2007)
3. Roy, J.K., Gupta, D., Goswami, S.: An improved flood warning system using WSN and Artificial Neural Network. Paper Presented at 2012 Annual IEEE India Conference, INDICON 2012, pp. 770–774 (2012). https://doi.org/10.1109/INDICON.2012.6420720. www.scopus.com
4. Yesilnacar, M.I., Sahinkaya, E., Naz, M., et al.: Neural Network prediction of Nitrate in groundwater of Harran plain, Turkey. Environ. Geol. **56**, 19 (2008). https://doi.org/10.1007/s00254-007-1136-5
5. Alkandari, A., Alabduljader, Y., Moein, S.M.: Water monitoring system using Wireless Sensor Network (WSN): case study of Kuwait beaches. In: 2012 Second International Conference on Digital Information Processing and Communications (ICDIPC). IEEE (2012)

# Modified Gingerbreadman Chaotic Substitution and Transformation Based Image Encryption

S. N. Prajwalasimha(✉), Sidramappa, S. R. Kavya, A. S. Hema, and H. C. Anusha

Department of Electronics and Communication, ATME College of Engineering, Mysuru, India
prajwalasimha.snl@gmail.com

**Abstract.** In this chapter, a combined Pseudo Hadamard transformation and modified Gingerbreadman chaotic substitution based image encryption algorithm is proposed. Intrinsic properties of images such as high inter-pixel redundancy and bulk data capacity, encryption is done in two stages: transformation and substitution. Pseudo Hadamard transformation reduces correlation between the adjacent elements in the host image and entropy is increased by subjecting it to modified Gingerbreadman chaotic substitution. The initial conditions for modified Gingerbreadman chaotic generator is considered from 128 bits secrete key. The random sequence generated by modified Gingerbreadman chaotic generator is introduced in the substitution stage of encryption process to diffuse the pixel values of cipher image after transformation. The cipher images are subjected for various security analysis and the results obtained are better compared to many existing techniques.

**Keywords:** Transformation · Encryption · Correlation · Entropy · Security · Substitution · Redundancy

## 1 Introduction

Significant and swift maturation in technology has made advancement in the past decade in the field of coding and communication for digital multimedia. Authorization to end users is more essential in order to facilitate successful commercialization of multimedia information [1]. In context of image encryption, the four classes of attacks in cryptanalysis [2] are: Chosen ciphertext attack, Chosen plaintext attack, Known plain text attack, and Ciphertext only attack. In the ciphertext only attack, the attacker has access over ciphertext image and cryptanalyze the same to retrieve plaintext image. In the identified plaintext attack, the invader has access over certain plaintext images and their equivalent ciphertext images to disclose the plaintext image by cryptanalyzing the algorithm. In selected plaintext attack, the attacker has temporary access over the encryption algorithm and select few known plaintext images to produce corresponding ciphertext images. In selected ciphertext attack, the attacker has provisional access over the decryption algorithm and select few known ciphertext images to generate equivalent plaintext images.


S. Smys et al. (Eds.): ICCVBIC 2019, AISC 1108, pp. 606–614, 2020.
https://doi.org/10.1007/978-3-030-37218-7_69

Traditional and conformist encryption algorithms, such as DES, RSA or AES cannot be applied for images, due to rate adaptation in heterogeneous networks for multimedia transmission and for multimedia content in image extraction [3]. In order develop an efficient encryption algorithm, chaos based encryption schemes are more popular, now a day [4–9].

A combined Pseudo Hadamard transformation (PHT) and modified Gingerbreadman Chaotic (GBC) replacement based image encryption algorithm has been proposed in order to provide two stage security per each round.

## 2 Proposed Scheme

Encryption is performed in two phases: substitution and transformation. In the transformation phase adjacent pixels are shuffled using Pseudo Hadamard transformation (PHT) and both original (host) and replacement images are exposed to PHT. In the substitution phase, the transformed pictures of host and replacement are exposed for bitwise XOR operation. Random sequence generated by modified Gingerbreadman chaotic key sequence generator is used along with secrete key to construct S-box. Final cipher image is obtained by substituting elements of S-box with each element of cipher image obtained from first stage.

### 2.1 Encryption Algorithm

**Step 1:** The original (host) and substitution images are first subjected for Pseudo Hadamard transformation.

$$O'(\varepsilon,\omega) = O((\mu+\delta)\ mod\ 2^n, (\mu+2\delta)\ mod\ 2^n) \tag{1}$$

$$S'(\varepsilon,\omega) = S^t((\mu'+\delta')\ mod\ 2^n, (\mu'+2\delta')\ mod\ 2^n) \tag{2}$$

$$1 \le \mu, \mu', \delta, \delta' \le 2^n$$

Where, $O$ is the original (host) image of size $2^n \times 2^n$
$S$ is the substitution image of size $2^n \times 2^n$
$O'$ is the transformed image (host) of size $2^n \times 2^n$
$S^t$ is the transformed image (substitution) of size $2^n \times 2^n$

**Step 2:** Both transformed images of host and substitution are subjected for bitwise XOR operation

$$C(\varepsilon,\omega) = O'(\varepsilon,\omega) \oplus S'(\varepsilon,\omega) \tag{3}$$

Where, $C(\varepsilon,\omega)$ is the cipher image from first stage of size $2^n \times 2^n$

**Step 3:** Transformed and substituted image from previous stage is subjected for bitwise XOR operation with the predefined elements of S-box 1.

$$C''(\varepsilon, \omega) = C(\varepsilon, \omega) \oplus S\ box\ 1 \tag{4}$$

**Step 4:** The cipher image from first stage is subjected for substitution with S-box 2 created using random sequences generated by modified Gingerbreadman chaotic generator.

$$x'_n = (1 - y_n + |x_n| + k1) \bmod 2^n \; 1 \leq n \leq 8 \tag{5}$$

$$y'_n = (x_n + k2) \bmod 2^n \tag{6}$$

Where $x_n$ = present key value

$$C'(\varepsilon, \omega) = C''(\varepsilon, \omega) \oplus S\ box\ 2 \tag{7}$$

Where, $C'(\varepsilon, \omega)$ is the cipher image after diffusion of size $2^n \times 2^n$ (Figs. 1 and 2).

### 2.2 Decryption Algorithm

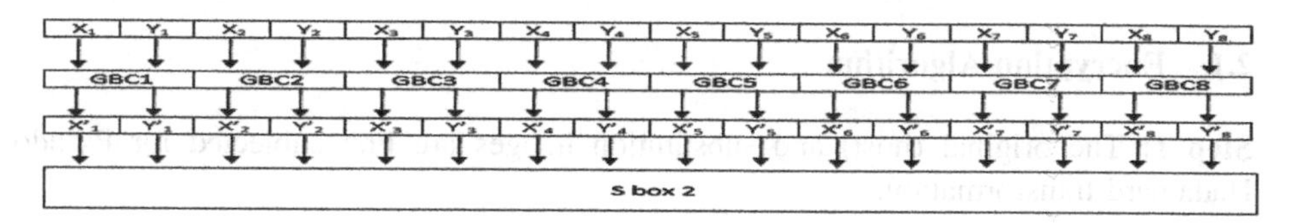


**Fig. 1.** Key sequence generation algorithm

**Step 1:** The obtained cipher image is first XORed with the elements of S-box 2 used for encryption.

$$x'_n = (1 - y_n + |x_n| + k1) \bmod 2^n \; 1 \leq n \leq 8 \tag{8}$$

$$y'_n = (x_n + k2) \bmod 2^n \tag{9}$$

Where $x_n$ = present key value

$$C''(\varepsilon, \omega) = C'(\varepsilon, \omega) \oplus S\ box\ 2 \tag{10}$$

**Step 2:** The substituted image from previous stage is subjected for bitwise XOR operation with elements of S-box 1.

$$C(\varepsilon, \omega) = C''(\varepsilon, \omega) \oplus S\ box\ 1 \tag{11}$$

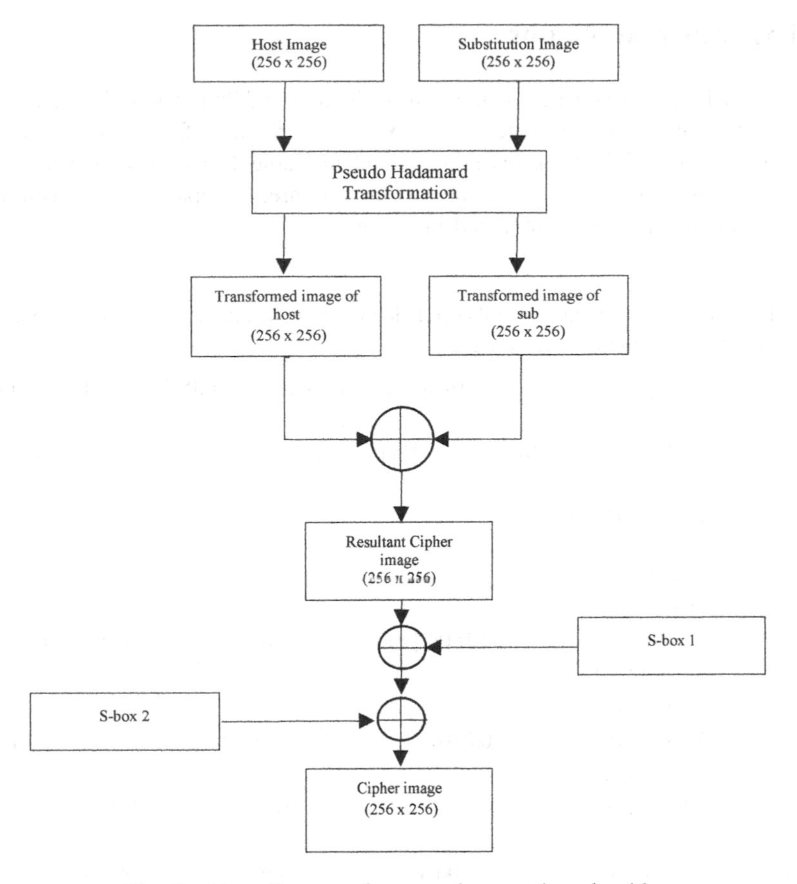


**Fig. 2.** Flow diagram of proposed encryption algorithm

**Step 3:** The substitution image is subjected for PHT and then XORed with cipher image from previous stage.

$$S'(\varepsilon, \omega) = S^t(\mu' + \delta') \, mod \, 2^n, (\mu' + 2\delta') \, mod \, 2^n \tag{12}$$

$$O'(\varepsilon, \omega) = C(\varepsilon, \omega) \oplus S'(\varepsilon, \omega) \tag{13}$$

**Step 4:** the obtained image from substitution stage is subjected for inverse PHT to get original image.

$$O(\rho, \psi) = O'((2\varepsilon - \omega) \, mod \, 2^n, (\omega - \varepsilon) \, mod \, 2^n) \tag{14}$$

## 3 Experimental Results

Security analysis is performed on the basis of Number of Pixel Changing Rate (NPCR), Unified Average Changing Intensity (UACI), Information Correlation and Entropy between host and cipher images as tabulated in Table 1. Experimental analysis and implementation is done using Matlab 2013a software. Comparison of execution time with existing techniques is tabulated in Table 2.

**Table 1.** Comparision of entropy and correlation between standard and encrypted images along with computational time for each image.

| Images | Entropy = 8 [2] | Correlation | UACI = 33.4635% [2] | NPCR = 99.6093% [2] |
|---|---|---|---|---|
| Lena | 5.5407 (Blow Fish) [11] | 0.0021 [11] | 31.00 [10] | 90.21 [10] |
| | 5.5438 (Two Fish) [11] | | | |
| | 5.5439(AES 256) [10] | | | |
| | 5.5439 (RC 4) [9] | 0.1500 [10] | 32.01 [10] | 99.60 [10] |
| | 7.5220 [10] | | | |
| | 7.6427 [12] | | | |
| | 7.9958 [10] | 0.0107 | 33.4201 [10] | 99.5859 [10] |
| | 7.9971 [10] | | | |
| | 7.9972 [9] | | 33.4888 | 99.5697 |
| | 7.9972 | | | |
| Baboon | 7.9947 [10] | −0.0076 | 30.87 [10] | 99.59 [9] |
| | 7.9950 [10] | | | |
| | 7.9971 | | 33.3080 | 99.5804 |
| Peppers | 7.9954 [10] | 5.2745e − 04 | 30.71 [10] | 99.61 [10] |
| | 7.9960 [10] | | 33.4937 | 99.6170 |
| | 7.9976 | | | |
| Plane | 7.9971 | 0.0012 | 33.4261 | 99.6140 |
| Cameraman | 7.9972 | −0.0070 | 33.4416 | 99.6460 |
| Elaine | 7.9973 | 8.0678e − 04 | 33.5144 | 99.5865 |
| Carnev | 7.9977 | −0.0043 | 33.5105 | 99.5895 |
| Donna | 7.9972 | 0.0013 | 33.4451 | 99.6384 |
| Foto | 7.9970 | 0.0064 | 33.6099 | 99.6063 |
| Galaxia | 7.9971 | −0.0011 | 33.5587 | 99.5987 |
| Leopard | 7.9971 | −0.0029 | 33.4287 | 99.6033 |
| Montage | 7.9974 | 0.0044 | 33.5028 | 99.5773 |
| Pallon | 7.9973 | −0.0023 | 33.3346 | 99.6140 |
| Vacas | 7.9971 | −0.0030 | 33.5627 | 99.5850 |
| Fiore | 7.9971 | −0.0017 | 33.2883 | 99.6002 |

(*continued*)

**Table 1.** (*continued*)

| Images | Entropy = 8 [2] | Correlation | UACI = 33.4635% [2] | NPCR = 99.6093% [2] |
|---|---|---|---|---|
| Mapasp | 7.9976 | −0.0027 | 33.4349 | 99.5819 |
| Mare | 7.9973 | −0.0016 | 33.4644 | 99.6078 |
| Mesa | 7.9973 | 0.0056 | 33.6535 | 99.6277 |
| Papav | 7.9973 | −0.0027 | 33.4376 | 99.6307 |
| Tulips | 7.9974 | 0.0020 | 33.6636 | 99.6323 |

The approximation values of UACI and NPCR are very nearby to the ideal values, indicating high efficiency against differential security attacks. After transformation phase, interpixel redundancy in the cipher image is effective reduced and high entropy is observed after substitution phase, very close to the ideal values [2] (Figs. 3 and 4).

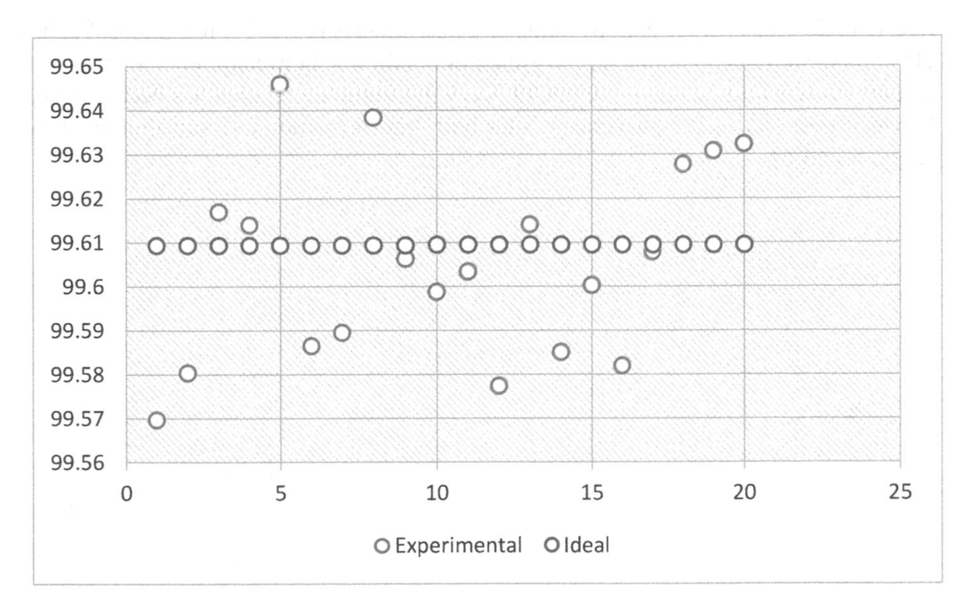


**Fig. 3.** Number of pixel changing rate of cipher images

*Inference 1:* The average value of count of pixel changing rate (NPCR) is 99.6053% for cipher images and is very much close to the ideal value 99.6093% with a hair line difference of 0.004% [2]. The NPCR value floats between 99.5697% to 99.646%. The average value of unified average changing intensity (UACI) is 33.4784% for cipher images and it is greater than ideal value 33.4635% [2]. The UACI value floats between 33.2883% to 33.6636%

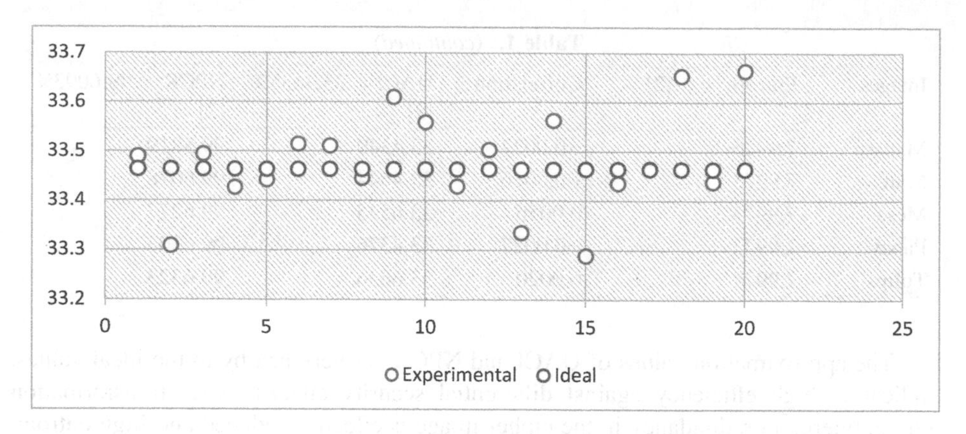


**Fig. 4.** Unified average changing intensity of cipher images

*Inference 2:* The average entropy value of cipher images is 7.9973. It is almost 99.96% close to ideal value. The average correlation coefficient between cipher images with respect to their host images is −0.00019. Very minimum correlation has been observed between host and final cipher image after both transformation and substitution stages (Figs. 5 and 6).

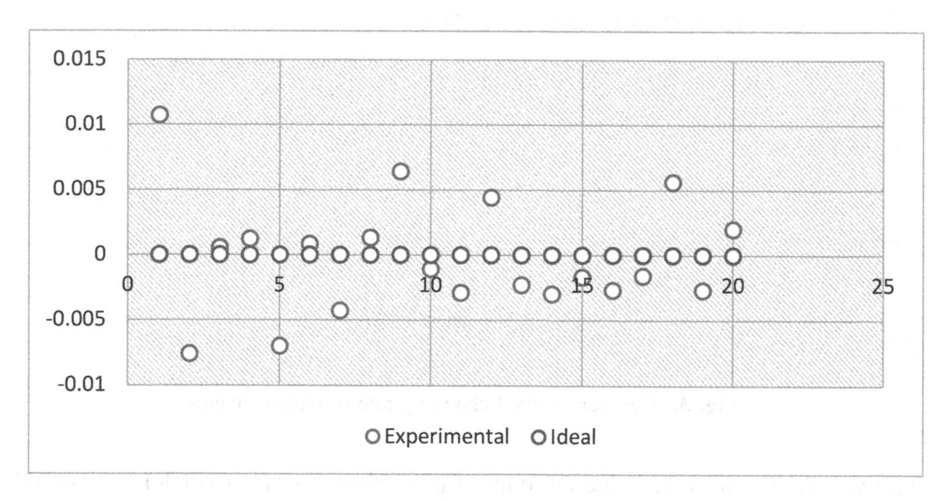


**Fig. 5.** Mean correlation between original and cipher images

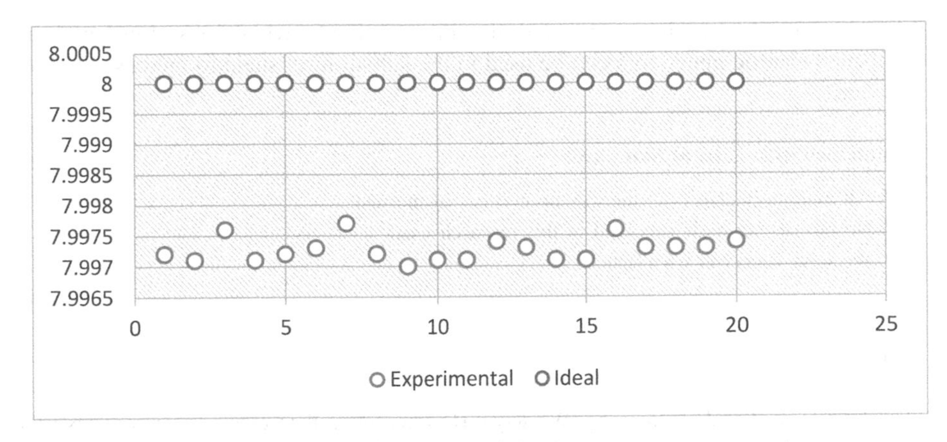


**Fig. 6.** Information entropy of cipher images

**Table 2.** Illustration of Host images, Substitution image, Cipher images after transformation and substitution [4–7]

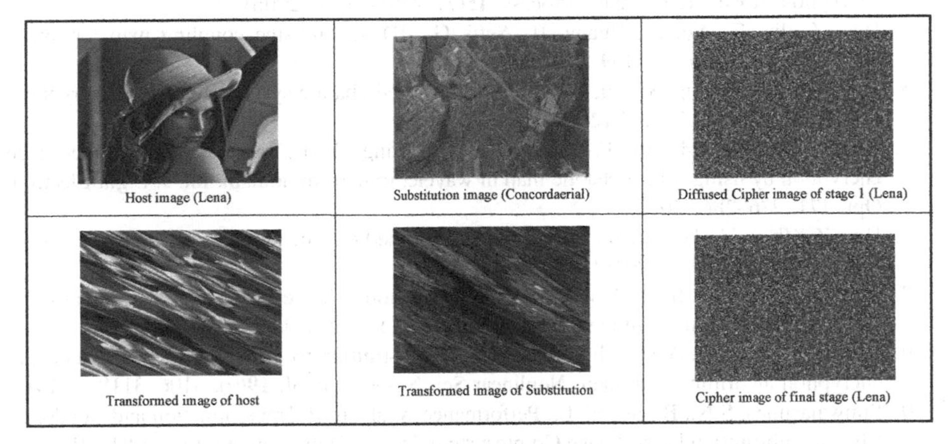


# 4 Conclusion

In the proposed encryption scheme, Pseudo Hadamard transformation (PHT) is used to reduce inter pixel redundancy of original (host) image and modified Gingerbreadman chaotic (GBC) substitution is used to increase entropy in the cipher image. Due to PHT, very minimum correlation between the adjacent pixels is observed and due to modified GBC high entropy in the cipher image is observed. Poor similarity between host and final cipher image is observed with very minimum average correlation co-efficient. Average values of 33.4784% unified average changing intensity (UACI) and 99.6053% number of pixel changing rate (NPCR) are obtained for a set of twenty standard images and they are very close to the ideal values. Secrete key of 128 bits length is used to increase the level of difficulty for brute force attacks. The algorithm makes use of minimized execution time compared to other available techniques. Further, more number of chaotic generators can be used in the substitution stage to increase the level of security.

**Compliance with Ethical Standards**

✓ All authors declare that there is no conflict of interest.
✓ No humans/animals involved in this research work.
✓ We have used our own data.

## References

1. Chuman, T., Sirichotedumrong, W., Kiya, H.: Encryption-then-compression systems using grayscale-based image encryption for JPEG images. IEEE Trans. Inf. Forensics Secur. **14**(6), 1515–1525 (2019)
2. Wang, X., Zhu, X., Zhang, Y.: An image encryption algorithm based on Josephus traversing and mixed chaotic map. IEEE Access Lett. **6**, 23733–23746 (2018)
3. Mao, Y., Wu, M.: A joint signal processing and cryptographic approach to multimedia encryption. IEEE Trans. Image Process. **15**(7), 2061–2075 (2006)
4. Hua, Z., Jin, F., Xu, B., Huang, H., Setti, G.: 2D logistic-sine coupling map for image encryption. Sig. Process. **149**, 148–161 (2018)
5. Lan, R., He, J., Wang, S., Gu, T., Luo, X.: Integrated chaotic systems for image encryption. Sig. Process. **147**, 133–145 (2018)
6. Li, C.-L., Li, H.-M., Li, F.-D., Wei, D.-Q., Yang, X.-B., Zhang, J.: Multiple-image encryption by using robust chaotic map in wavelet transform domain. Int. J. Light Electron. Opt. **171**, 276–286 (2018)
7. Hua, Z., Zhou, Y., Huang, H.: Cosine-transform-based chaotic system for image encryption. Inf. Sci. **480**, 403–419 (2019)
8. Ahmad, J., Hwang, S.O.: A secure image encryption scheme based on chaotic maps and affine transformation. Multimed. Tools Appl. **75**(21), 13951–13976 (2015)
9. Anees, A., Siddiqui, A.M., Ahmed, F.: Chaotic substitution for highly autocorrelated data in encryption algorithm. Commun. Nonlinear Sci. Numer. Simul. **19**(9), 3106–3118 (2014)
10. Prajwalasimha, S.N., Basavaraj, L.: Performance Analysis of Transformation and Bogdonov Chaotic Substitution based Image Cryptosystem. Int. J. Electr. Comput. Eng. **10**(1), 188–194 (2020)
11. Prajwalasimha, S.N., Basavaraj, L.: Design and Implementation of Transformation and non-Chaotic Substitution based Image Cryptosystem. Int. J. Eng. Adv. Technol. **8**(6), 1079–1083 (2019)
12. Prajwalasimha, S.N., Kumar, A.V., Arpitha, C.R., Swathi, S., Spoorthi, B.: On the Sanctuary of a Combined Confusion and Diffusion based scheme for Image Encryption. Int. J. Eng. Adv. Technol. **9**(1), 3258–3263 (2019)

# Thresholding and Clustering with Singular Value Decomposition for Alcoholic EEG Analysis

Harikumar Rajaguru(✉) and Sunil Kumar Prabhakar

Department of ECE, Bannari Amman Institute of Technology, Sathyamangalam, India
harikumarrajaguru@gmail.com

**Abstract.** With the constant use of alcohol, it affects the social and economic lives of the people to a higher extent. To depict the electrical conditions of the brain, Electroencephalography (EEG) signals are used. The signals of EEG show various patterns for different neurological disorders and so it is widely preferred in clinical diagnosis and analysis. In this paper, for a single chronic alcoholic patient, the EEG signals are obtained and then the reckoning of features is done with the help of Singular Value Decomposition (SVD) method. Similarly, when K-means Clustering is utilized a classification accuracy of 61.45% with a Performance Index of 46.15% is attained.

**Keywords:** EEG · SVD · K-means · Clustering · Hard thresholding

## 1 Introduction

A huge interest in the human brain has been fostered since ancient times and a lot of research activities have been done in the past few decades [1]. As the brain is quite complex and uncertain, many unresolved issues are there to be addressed. The electrical activity of the cerebral cortex can be measured with the practical utilizations of EEG signals [2]. EEG signals exhibits the function of the brain and the entire status or condition of the whole body. The EEG signal interpretation and analysis plays a vital role in areas of clinical medicine and psychology [3]. Analyzing the EEG signals is one of the best techniques to monitor the state of the alcoholic patient and so it serves as an effective tool to analyze the problems occurring in the brain due to alcoholism [4]. Alcoholism is a demanding issue leading to a variety of cognitive and other health complications. Because of this problem, the vital parts of the human body are affected severely thereby leading to dire consequences. A lot of works have shown the works related to alcoholism by the EEG signal processing. The analysis of sample entropy of both the alcoholic and normal people was done by Zou et al. [5]. The effects of alcohol on both the neurobehavioural functions and the brain were analyzed by Berman and Marinkovi [6]. A detailed investigation on the human brain after the consumption of alcohol based on the EEG signals was successfully implemented by Wu et al. [7].


S. Smys et al. (Eds.): ICCVBIC 2019, AISC 1108, pp. 615–623, 2020.
https://doi.org/10.1007/978-3-030-37218-7_70

A discriminant methodology by means of using Softmax function for assessing the alcohol status in EEG signals was executed by Rajaguru and Prabhakar [8].

In this paper, SVD is used to extract the features of a single alcoholic patient and then classification is demonstrated by the utility of Thresholding technique (Hard) and Clustering technique (K means) as suitable Classifiers. A schematic representation of the work is expressed in Fig. 1.

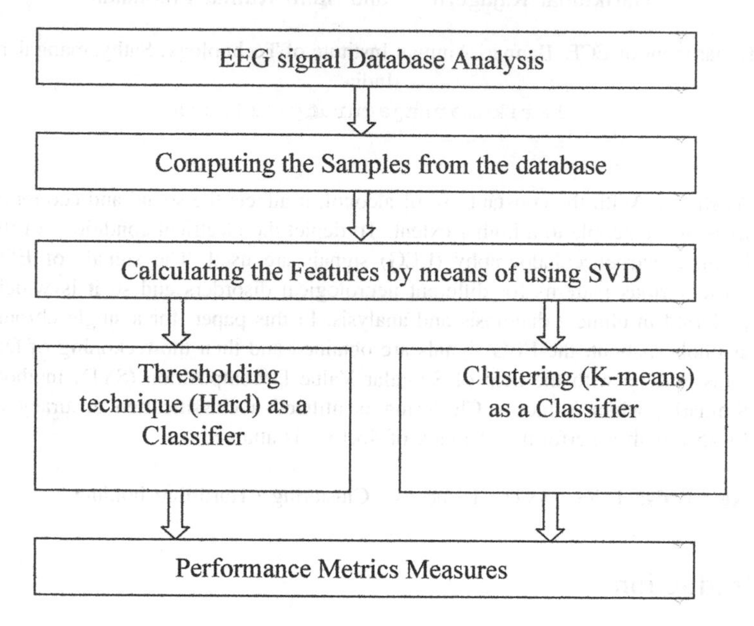


**Fig. 1.** Structure of the work

The work is established as follows. In Sect. 2, the materials and methods are expressed followed by the utility of thresholding and Clustering techniques in Sect. 3. Section 4 explains the results and discussion followed by the conclusion in Sect. 5.

## 2 Materials and Methods

In this section, the data acquisition and the discussion about SVD based feature extraction concept is elaborated. The Table 1 gives us an insight to the alcoholic EEG dataset used in this work.

**Table 1.** Insight of the EEG data (alcoholic)

| | |
|---|---|
| Patients considered for analysis | 1 (one) |
| Channels Utilized here | Single Trial 64 channel |
| Total number of electrodes used | 3 (Three) |
| Sampling of the signal | 256 samples/second |
| Analog to Digital Converter used | 12 bit signed representation |
| Acquirement of channel is at | 10 s duration |
| Every particular channel has samples about | 2560 |
| The obtained number of samples for 64 channels are | 1,63,840 |
| Grouping of samples (in bins) | 64 |

### 2.1 Feature Extraction Using SVD

One of the famous topics in the concept of linear algebra which is highly useful in signal processing theory is SVD [9]. Various values both practical and theoretical are easily framed by SVD. For any real $(a, b)$ matrix, the characteristic feature of SVD can be easily implemented. Initially, a matrix $Z$ with a total number of '$a$' rows and '$b$' columns are assumed, with a rank $r$, and with a suitable condition $r \leq b \leq a$, then $Z$ can be factorized into 3 specific matrices such as

$$Z = PSK^T$$

where $Z$ is an $a \times b$ matrix; $P$ is an orthogonal $a \times a$ matrix $P = [p_1, p_2, \ldots, p_r, p_{r+1}, \ldots, p_a]$ column vector $p_i$, for $i = 1, 2, \ldots, a$ forms an orthonormal set

$$p_i^T p_j = \delta_{ij} = \begin{cases} 1, \ldots, i = j \\ 0, \ldots, i \neq j \end{cases}$$

$K$ is an $b \times b$ orthogonal matrix $K = [k_1, k_2 \ldots k_r, k_{r+1}, \ldots, k_b]$ column vectors $k_i$, for $i = 1, 2, \ldots, b$ forms an orthonormal set

$$k_i^T k_j = \delta_{ij} = \begin{cases} 1, \ldots, i = j \\ 0, \ldots, i \neq j \end{cases}$$

$S$ is an $a \times b$ diagonal matrix which has singular values on the diagonal.

## 3 Thresholding and Clustering Methodologies for Classification

The feature values computed by means of SVD are analyzed as inputs to both the thresholding (hard) and clustering (K means) methodology to assess the alcohol risk level for the specific patient.

### 3.1 Classification Using Hard Thresholding

After the computation of features by means of employing SVD is performed successfully, thresholding (hard) classifier is applied and the below condition detailed in Table 2 is set as gold standard for the obtained SVD values.

**Table 2.** Analysis for thresholding classifier based on gold standard

| 50 < normal |
| --- |
| 50 < alcoholic<147 |
| Epileptic > 147 |

If the SVD values obtained is less than 50, then it is normal. If the SVD values are in the range of 50 to 147, then it is alcoholic and if the SVD values are greater than 147, then it is epileptic.

The step by step methodology is specifically implemented to a particular classifier and therefore the received signal is expressed as $f[n] = h[n] + g[n]$, where $h(n)$ exhibits a signal (unknown) which has to be traced or detected and $g(n)$ indicates a noise factor representing a Gaussian distribution. A plethora of features comprising of high frequency are there in the original signal. The attainment of thresholding methodology comprises of both the thresholding functions respectively (hard and soft) [10].

Soft thresholding is conveyed mathematically as

$$\widehat{H} = \begin{cases} f - \operatorname{sgn}(f)T & \text{if } |f| \geq T \\ 0 & \text{if } |f| < T \end{cases}$$

Hard thresholding is expressed as

$$\widehat{H} = \begin{cases} a & \text{if } |f| \geq T \\ 0 & \text{if } |f| < T \end{cases}$$

Here only Hard thresholding classifier is used and based on Table 2, the classifier correctly classifies the signal into alcoholic for a major part and wrongly classifies as epileptic and normal for a minor part. The classification of the alcoholic EEG data obtained through a single trial 64 channel utilizing Hard Thresholding classifier is depicted in Table 3.

**Table 3.** Results of Hard Thresholding classification

| Allocation of classes | Alcoholic classes | Epileptic classes | Normal classes | Total classes |
| --- | --- | --- | --- | --- |
| Alcoholic | 46 | 9 | 9 | 64 |
| Epileptic | 0 | 0 | 0 | 0 |
| Normal | 0 | 0 | 0 | 0 |

As there are 64 channels, 46 channels comes under alcoholic sector, 9 channels were misclassified under epileptic sector and 9 channels were wrongly classified under normal sector.

### 3.2 Clustering Using K-means

The step wise procedure for K-means Clustering is as follows:

Step 1: The initial centroids are selected as the '$k$' points and the procedure is repeated.
Step 2: All the $k$ clusters are formed by means of specifying all the points to the closest centroid present there.
Step 3: The recomputation of every centroid is done until there is no change in the centroid.

Choosing of the initial centroids is done randomly. The produced clusters always vary from one another. In the entire cluster, the mean of the points is typically the centroid value. Euclidean distance and correlation distance are the two distance metrics used here to measure the closeness. For common similarity measures, the $k$ means algorithm will converge [11]. In the first few iterations itself, majority of the convergence takes place. The Euclidean distance is given as follows:

$$d(i,j) = \sqrt{|x_{i1} - x_{j1}|^2 + |x_{i2} - x_{j2}|^2 + \ldots + |x_{ip} - x_{jp}|^2}$$

The centroid is updated as follows. In order to calculate the '$n$' dimensional centroid points among the '$k$' $n$-dimensional points;

$$CP(x_1, x_2, \ldots, x_k) = \left( \frac{\sum_{i=1}^{k} x^{1st_i}}{k}, \frac{\sum_{i=1}^{k} x^{2nd_i}}{k}, \frac{\sum_{i=1}^{k} x^{nth_i}}{k} \right)$$

Utilizing the Sum of Squared Errors (SSE) is the most commonly preferred technique to evaluate k-mean clusters. For every point, the Sum of Squared Errors is considered as the distance to the closest cluster. To obtain SSE, the errors are squared and then it is summed as follows

$$SSE = \sum_{i=1}^{k} \sum_{x \in C_i} dist^2(m_i, x)$$

where '$x$' denotes a specific data point in cluster $C_i$; $m_i$ is the representative point for all clusters $C_i$. A famous way to drastically reduce SSE is to enhance the number of clusters $k$. A lower SSE is obtained even there is a good clustering but with a smaller $k$ value. A higher SSE is obtained even there is a poor clustering but with a high $k$ values.

Utilizing K-means Clustering, through a lone trial 64 channel, the alcoholic EEG data classification is expressed in Table 4.

**Table 4.** Results of K-means clustering classification

| Allocation of classes | Alcoholic classes | Epileptic classes | Normal classes | Total classes |
|---|---|---|---|---|
| Alcoholic | 26 | 8 | 30 | 64 |
| Epileptic | 0 | 0 | 0 | 0 |
| Normal | 0 | 0 | 0 | 0 |

As 64 channels are considered here, 26 channels were exactly and absolutely classified under the category of alcoholic nature, 8 channels were wrongly interpreted as epileptic and 30 other channels were again wrongly classified as perfectly normal category by the K-means Clustering methodology.

# 4 Discussion of Results

When the significant SVD features are obtained and when it is classified with Thresholding (Hard) and Clustering (K means) ideology, terminology like Performance Index (PI) assessment, Classification Accuracy, Specificity and Sensitivity are computed in Table 5. The expressions for the terms are as:

$$PI = \left( \frac{Perfect.Classify - Missed.Classify - False.Alarm}{Perfect.Classify} \right) \times 100$$

$$Sensitivity = \frac{Perfect.Classify}{Perfect.Classify + False.Alarm} \times 100$$

$$Specificity = \frac{Perfect.Classify}{Perfect.Classify + Missed.Classify} \times 100$$

$$Classification.Accuracy = \frac{Sensitivity \quad factor + Specificity \quad factor}{2}$$

**Table 5.** Thresholding and clustering analysis with SVD features – analysis

| Name | Thresholding (Hard) for classification | Clustering (K-means) for classification |
|---|---|---|
| Perfect Classification (%) | 71.88 | 40.62 |
| Missed Classification (%) | 14.11 | 46.87 |
| False Alarm (%) | 14.11 | 12.5 |
| Performance Index (%) | 60.74 | 46.15 |
| Specificity (%) | 83.59 | 46.43 |
| Sensitivity (%) | 83.59 | 76.47 |
| Accuracy (%) | 83.59 | 61.45 |

The Table 6 shows the Mean and the standard deviation values of the K means clustering for the 64 channels of the single alcoholic patient.

**Table 6.** Mean and standard deviation values utilizing clustering by K-means

| Clustering by K-means | | | | |
|---|---|---|---|---|
| | | Mean | Number | STD |
| 1 | Epileptic | 217.2906 | 8 | 51.18334 |
| 2 | Alcoholic | 105.9849 | 26 | 18.63632 |
| 3 | Normal | 58.443 | 30 | 15.21017 |

The Table 7 shows the Mean and the standard deviation values of the Hard Thresholding for the 64 channels of the single alcoholic patient.

**Table 7.** Means and standard deviation of the Hard Thresholding

| Hard thresholding | | | | |
|---|---|---|---|---|
| | | Mean | Number | STD |
| 1 | Epileptic | 210.741 | 9 | 51.75278 |
| 2 | Alcoholic | 86.75044 | 46 | 22.97233 |
| 3 | Normal | 40.00377 | 9 | 7.982331 |

The comparison of the K means clustering and Hard Thresholding classifiers with respect to the Mean is depicted in Fig. 2. Only the highest values which are obtained for the mean of the K-means clustering are named in the plot.

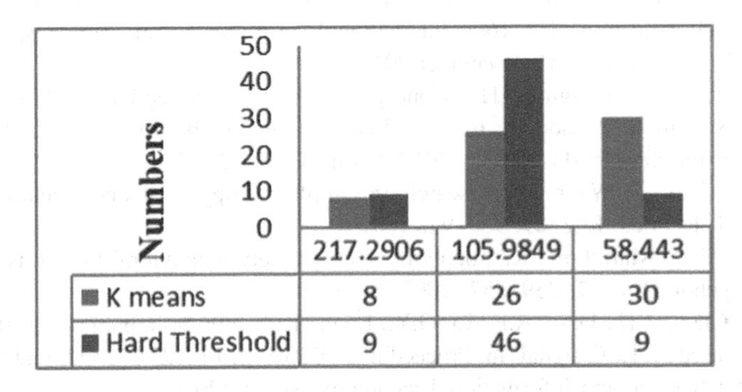


**Fig. 2.** Comparison of K means and hard thresholding classifiers based on mean classifier

# 5 Conclusion

Thus in this paper, the EEG signals which were alcoholic are fully analyzed to trace the risk level of alcohol for a single alcoholic patient. At the initial stage, SVD features were obtained and later it was classified with thresholding and clustering classifiers. A very poor classification accuracy index of 61.45% is attained when clustering through K-means is utilized as a post classifier along with an average performance index assessment of 46.15%. A better classification accuracy of 83.59% along with an average performance index of 60.74% is obtained when Hard Thresholding is used as a post classifier. Future work aims to analyze with various features and other post classification schemes to classify the risk of alcohol in a much effective way.

**Compliance with Ethical Standards**

✓ All authors declare that there is no conflict of interest.
✓ No humans/animals involved in this research work.
✓ We have used our own data.

# References

1. Prabhakar, S.K., Rajaguru, H.: Comparison of fuzzy output optimization with expectation maximization algorithm and its modification for epilepsy classification. In: International Conference on Cognition and Recognition (ICCR 2016), Mysore, India, 30–31 December 2016
2. Prabhakar, S.K., Rajaguru, H.: Expectation maximization based PCA and Hessian LLE with suitable post classifiers for epilepsy classification from EEG signals. In: Proceedings of the 8th International Conference on Soft Computing and Pattern Recognition (SoCPaR), VIT University, Vellore, India, December 2016
3. Prabhakar, S.K., Rajaguru, H.: Development of patient remote monitoring system for epilepsy classification. In: 16th International Conference on Biomedical Engineering (ICBME), Singapore, 7–10 December 2016
4. Prabhakar, S.K., Rajaguru, H.: Conceptual analysis of epilepsy classification using probabilistic mixture models. In: 5th IEEE Winter International Conference on Brain-Computer Interface, 9–11 January 2017, South Korea (2017)
5. Zou, Y., Miao, D., Wang, D.: Research on sample entropy of alcoholic and normal people. Chinese J. Biomed. Eng. **29**, 939–942 (2010)
6. Berman, M.O., Marinkovi, K.: Alcohol: effects on neurobehavioral functions and the brain. Neuropsychol. Rev. **17**, 239–257 (2007)
7. Wu, D., Chen, Z.H., Feng, R.F., Li, G.Y., Luan, T.: Study on human brain after consuming alcohol based on EEG signal. In: Proceedings of 2010 3rd IEEE International Conference on Computer Science and Information Technology, vol. 5 (2010)
8. Rajaguru, H., Prabhakar, S.K.: Softmax discriminant classifier for detection of risk levels in alcoholic EEG signals. In: IEEE Proceedings of the International Conference on Computing Methodologies and Communication (ICCMC 2017), Erode, India (2017)

9. Prabhakar, S.K., Rajaguru, H.: Performance comparison of fuzzy mutual information as dimensionality reduction techniques and SRC, SVD and approximate entropy as post classifiers for the classification of epilepsy risk levels from EEG signals. In: Proceedings of 2015 IEEE Student Symposium in Biomedical Engineering and Sciences (ISSBES), 4 November 2015. Universiti Teknologi Mara, Malaysia (2015)
10. Zhang, Q., Rossel, R.A., Choi, P.: Denoising of gamma-ray signals by interval-dependent thresholds of wavelet analysis. Measur. Sci. Technol. **17**, 731–735 (2006)
11. Prabhakar, S.K., Rajaguru, H.: PCA and K-means clustering for classification of epilepsy risk levels from EEG signals – a comparative study between them. In: Proceedings of the International Conference on Intelligent Informatics and BioMedical Sciences (ICIIBMS), 28–30 November Okinawa, Japan (2015)

# FPGA Implementation of Haze Removal Technique Based on Dark Channel Prior

J. Varalakshmi[1](✉), Deepa Jose[2], and P. Nirmal Kumar[1]

[1] Applied Electronics, Department of ECE, CEG, Anna University, Chennai 600025, Tamilnadu, India
varshajl911@gmail.com

[2] Department of ECE, KCG College of Technology, Chennai 600097, Tamilnadu, India
deepa.ece@kcgcollege.com

**Abstract.** Image dehazing is a much innovative and growing technology in applications of computer vision. Haze removal technique in FPGA using Nexys 4 DDR is implemented in this paper. The dark channel prior (DCP) has been an efficient dehazing technique. However, DCP can induce inaccurate approximation of transmission which results in colour distortion and halo effects in the brighter regions of an image. The proposed algorithm is implemented for haze removal of image to avoid the colour distortion of haze in bright and in non-bright areas with less complexity. The algorithm that is proposed in this paper is compared with the DCP method and the Tarel algorithm using Structural Similarity index (SSIM). The results show that the proposed algorithm removes haze effectively in both bright and non-bright areas of an image and the implementation in FPGA is done with less computational complexity. Implementation of dehazing in FPGA can be used in many applications of computer vision such as surveillance, military and transportation areas.

**Keywords:** Haze removal · FPGA · Structural Similarity index (SSIM)

## 1 Introduction

The images captured during unfavourable weather conditions lose contrast due to the scattering of light while propagation. Many systems for aviation, navigation, surveillance, etc., need clear visibility of the input images. Therefore, the improvisation of image hazes removal method as been subjected to great attention now a days.

Many different methods have been suggested and implemented for haze removal of image. Tarel et al. [4] proposed an algorithm for haze removal of image using a median filter which preserves the edges and corners. The main disadvantage of Tarel algorithm is that it produces colour distortion. He et al. [6] proposed a method of dark channel prior used for removal of haze for images with non-sky. He [6] introduced a model of the hazy image and an algorithm for soft matting interpolation. The disadvantage of the He's algorithm is that it produces halo like artifacts in the case of brighter areas of an image and also it has greater computational complexity. He [2] also proposed a method using the guided joint bilateral filter. This method has a better dehazing effect at scenes


S. Smys et al. (Eds.): ICCVBIC 2019, AISC 1108, pp. 624–630, 2020.
https://doi.org/10.1007/978-3-030-37218-7_71

where the depth changes abruptly. The drawback of this paper is that it fails in the sky or brighter regions of an image. Some improved algorithms [3, 7, 8] were proposed to deal with the disadvantages of the classical He algorithm. In [9], a pipelined hardware architecture with 11-stage was performed for haze removal in real time. An ASIC was designed in [1] for guided image filtering. However, both [9] and [1] will have performance with less efficiency and will demand more memory. The main objective of the proposed algorithm is to remove haze of an image in the case of bright, white cloud or sky areas and also to get a dehazed image without any color distortion. Thus the proposed algorithm gives good results with less computational complexity to be implemented in Field Programmable Gate Array.

## 2 Dark Channel Prior

The dark channel prior is based on the concept that at least one of the colour channels of RGB has very low pixel value i.e.

$$J_{dark}(x) = \min_{c\in\{R,G,B\}}(\min_{x\in\Omega(x)}(J_c(x)) \approx 0 \quad (1)$$

Where $J_c(x)$ is the colour channel of J(x), $\Omega(x)$ denotes a local patch, and $J_{dark}(x)$ is the dark channel of the J(x).

The transmittance t(x) in the atmosphere for the images with haze obtained is relatively inaccurate. The atmospheric transmittance *t*(*x*) is given by:

$$t(x) = 1 - \omega \min_{c\in\{R,G,B\}}\left(\min_{x\in\Omega(x)}\left(\frac{I(x)}{A}\right)\right) \quad (2)$$

where parameter $\omega = 0.95 (0 < \omega \leq 1)$ and I(x) is the input image.

The recovered image J(x) is given by the following:

$$J(x) = \left(\frac{I(x) - A}{\max(t(x), t_0)}\right) + A \quad (3)$$

Where $t_0$ is a threshold for limiting transmittance t(x).

The value of $t_0$ is taken as 0.1, as the transmittance of the hazy images should not be zero. The highest value of the pixel among the three channels of the image is taken as the atmospheric light A and the original haze free image is restored as J(x).

## 3 Proposed Methodology

The proposed methodology's block diagram is given in Fig. 1. Initially the hazy image is taken as the input and then with the help of BRAM in FPGA, the data is accessed and the proposed dehazing algorithm is performed. Finally the dehazed output is displayed on the VGA monitor via the VGA cable.

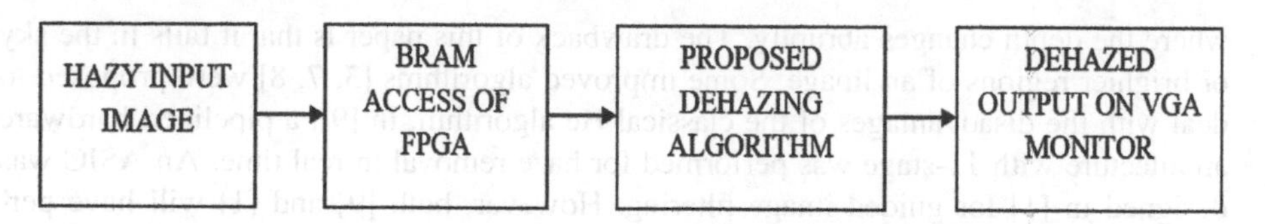


**Fig. 1.** Block diagram of the proposed system

## 3.1 Haze Removal Algorithm

The haze removal methods proposed by He and Tarel remove haze from an image but still have some drawbacks. Dark channel prior has good dehazing results with non-sky areas. However, the He algorithm is not applicable to the bright or sky areas and may lead to colour distortion. Thus, approximation of the transmittance for images with haze having sky regions may be incorrect. To solve these problems, a transmission matrix was devised.

The atmospheric light A is taken as the value with the largest pixel value in the RGB channel of the input image.

### 3.1.1 Atmospheric Transmittance

The dark channel prior method does not work well in the case of brighter regions of an image. The atmospheric transmittance in this proposed algorithm is given by:

$$t = \max\left(\min_c\left(\min_{y\in\Omega(x)}\left(\frac{255 - I(x)}{255}\right), \max\left(\frac{I(x) - 170}{255}\right)\right)\right) \quad (4)$$

The scene recovery operation is done by:

$$J = \left(\frac{I - A}{t}\right) + A \quad (5)$$

Thus the dehazed output is recovered as J using the proposed algorithm based on DCP.

## 3.2 FPGA Implementation

FPGA implementation flow of the system to be proposed is given in Fig. 2. The input image with haze is converted into .coe file and is accessed with the help of BRAM. The Verilog code for the proposed dehazing algorithm based on dark channel prior is written in Xilinx Vivado 2016.4 and the dehazing process is performed. The final output of the dehazed image using the proposed algorithm is displayed in the VGA monitor using the VGA connector. The VGA uses 24 bits to display the image.

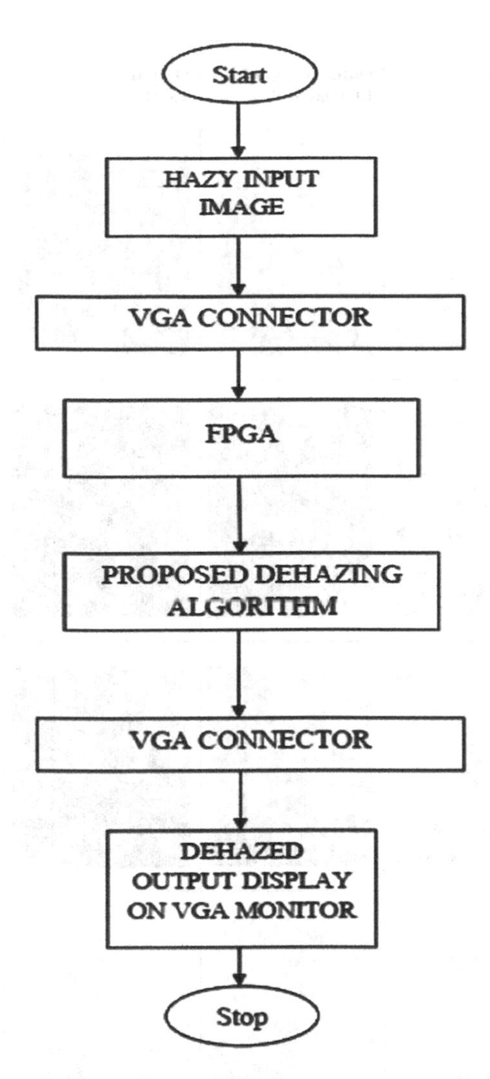


**Fig. 2.** FPGA implementation flow of proposed methodology

## 4 Results and Discussion

In the experimental results, the output images of Dark channel prior method, Tarel method and the proposed technique for haze removal is compared. The performance metric comparison is also done for the different methods.

All experiments are carried out on Intel(R)Core TM i5-3230M CPU @ 2.60 GHz 4 GB memory, Windows 8.1 in MATLAB 2017a and in Xilinx Vivado 2016.4 using Nexys 4 DDR. The source of the dataset is from the website: http://live.ece.utexas.edu/research/fog/fade_defade.html (Fig. 3, Table 1).

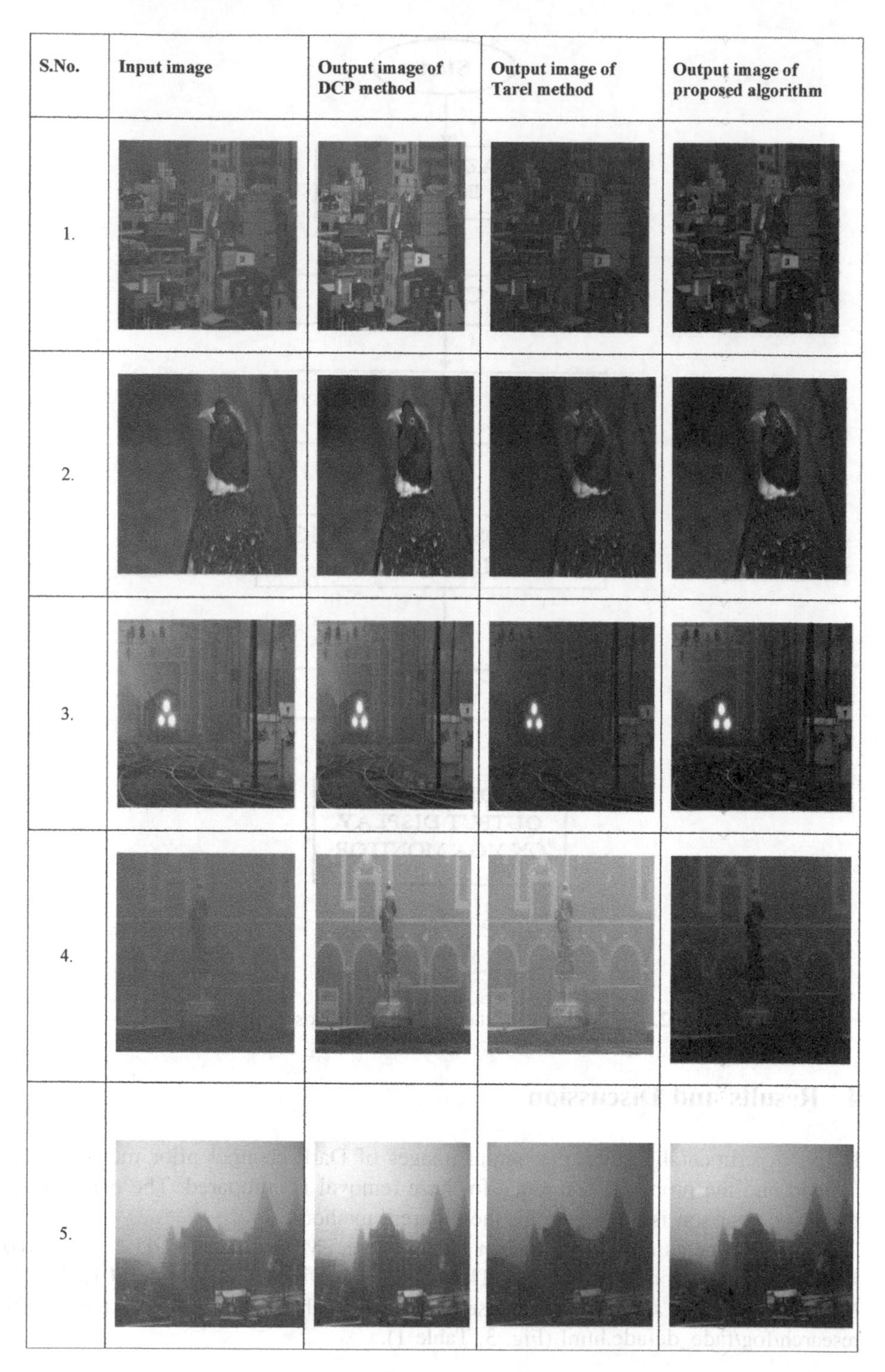


**Fig. 3.** Output comparison of images using different methods [2, 5–7, 9]

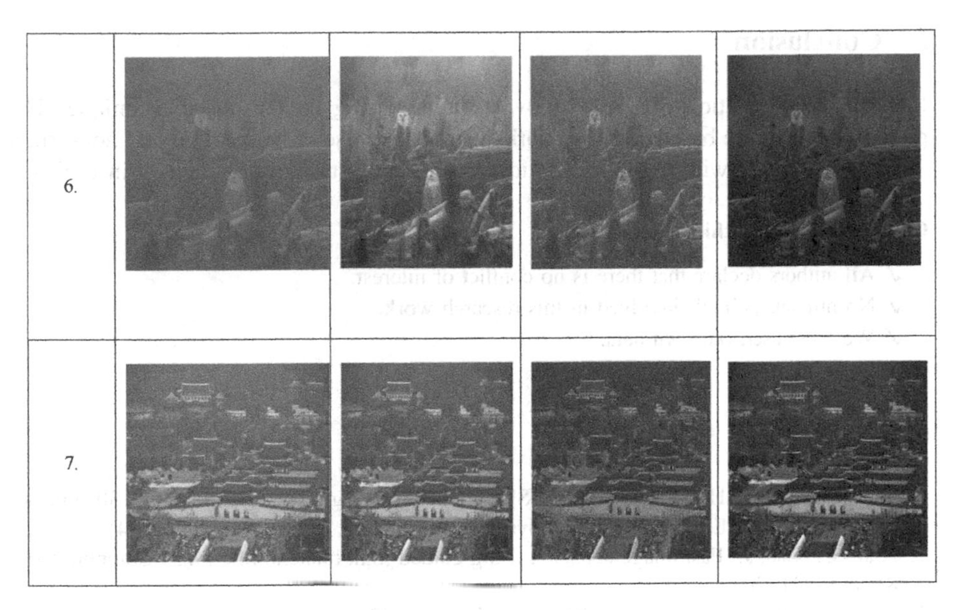


**Fig. 3.** (*continued*)

**Table 1.** SSIM comparison of different methods

| S. No. | DCP method | Tarel algorithm | Proposed algorithm |
|---|---|---|---|
| 1 | 0.8951 | 0.6109 | 0.3144 |
| 2 | 0.6597 | 0.5378 | 0.3616 |
| 3 | 0.9701 | 0.7170 | 0.3356 |
| 4 | 0.8566 | 0.8736 | 0.3973 |
| 5 | 0.9413 | 0.8646 | 0.6260 |
| 6 | 0.6465 | 0.6270 | 0.4017 |
| 7 | 0.9707 | 0.7187 | 0.5203 |

On comparison of the dark channel prior method, Tarel method and the proposed technique, it can be inferred that in the output images of the proposed technique based on DCP, haze is removed well both in bright and non-bright areas of the image when compared with the other methods.

The SSIM performance metric comparison is done for the different methods. From the SSIM values, it can be inferred that the SSIM value is smallest for the proposed method. A smaller value of SSIM signifies that the output image is less similar with the hazy image which gives the inference that it is more similar with the haze-free image. This shows that the algorithm that was proposed can retrieve a dehazed image much better than DCP and Tarel algorithm. On evaluation by the performance metrics, it can be concluded that the proposed technique based on DCP gives a better performance than the He algorithm.

## 5 Conclusion

FPGA implementation of haze removal is done using the proposed technique. The proposed technique based on DCP works well in the case of both bright and non-bright areas of an image with less computational complexity to be implemented in FPGA.

**Compliance with Ethical Standards**

✓ All authors declare that there is no conflict of interest.
✓ No humans/animals involved in this research work.
✓ We have used our own data.

## References

1. Kao, C.C., Lai, J.H., Chien, S.Y.: VLSI architecture design of guided filter for 30 frames/s full-HD video. IEEE Trans. Circuits Syst. Video Technol. **24**(3), 513–524 (2014)
2. Xiao, C., Gan, J.: Fast image dehazing using guided joint bilateral filter. Vis. Comput. **28**(6–8), 713–721 (2012)
3. Meng, G., Wang, Y., Duan, J., Xiang, S., Pan, C.: Efficient image dehazing with boundary constraint and contextual regularization. In: Proceedings of the IEEE International Conference on Computer Vision, pp. 617–624 (2013)
4. Tarel, J.P., Hautiere, N.: Fast visibility restoration from a single color or gray level image. In: Proceedings of the 12th IEEE International Conference on Computer Vision, pp. 2201–2208 (2009)
5. He, K., Sun, J., Tang, X.: Guided image filtering. IEEE Trans. Pattern Anal. Mach. Intell. **35** (6), 1397–1409 (2013)
6. He, K., Sun, J., Tang, X.: Single image haze removal using dark channel prior. IEEE Trans. Pattern Anal. Mach. Intell. **33**(12), 2341–2353 (2011)
7. Tang, K., Yang, J., Wang, J.: Investigating haze-relevant features in a learning framework for image dehazing. In: Proceedings of the IEEE Conference on Computer Vision and Pattern Recognition, pp. 2995–3002 (2014)
8. Kratz, L., Nishino, K.: Factorizing scene albedo and depth from a single foggy image. In: Proceedings of the 12th IEEE International Conference on Computer Vision, p. 1701 (2009)
9. Shiau, Y.H., Yang, H.Y., Chen, P.Y., Chuang, Y.Z.: Hardware implementation of a fast and efficient haze removal method. IEEE Trans. Circuits Syst. Video Technol. **23**(8), 1369–1374 (2013)

# Impulse Noise Classification Using Machine Learning Classifier and Robust Statistical Features

K. Kunaraj(✉), S. Maria Wenisch, S. Balaji, and F. P. Mahimai Don Bosco

Loyola-ICAM College of Engineering and Technology, Chennai, India
k.kunaraj@gmail.com, wenischs@gmail.com, {sbalaji,mahimaidonbosco.fp}@licet.ac.in

**Abstract.** Identifying corrupted pixels in an image helps the denoising algorithm to perform better. In this paper, I propose a machine learning (ML) based classification algorithm which uses random decision forest classifiers and classical pixel-wise statistical parameters. The ML algorithm is trained to identify impulse noise by taking the computed statistical parameters from the corrupted image as input. From the experiment, it is clear that the identification of corrupted pixels depend on both the robustness of the chosen image parameters and the accuracy of the trained classifier. Particularly, random forest classifier which is an ensemble of random decision trees is employed as it suits better for this application. The implemented decision tree which is limited to 10, shows a better classification performance in the images corrupted with impulse noise. The improvement in the classification is better for random impulse noises rather than salt and pepper noises. Also, its significance is visible in low and medium noise intensities rather than high densities and hence I limit the noises by 50% for the discussion.

**Keywords:** Machine learning · Image denoising · Impulse noise · Random forest classifier · Noise detection

## 1 Introduction

Random valued impulse noise estimation and removal is a well-established research area where a handful of algorithms are available with appreciable performance. Identification of corrupted pixels in an image is a necessary step which should be considered before denoising the image. If the corrupted pixels are accurately identified, then the denoising would be efficient as none of the corrupted pixels go unnoticed. Pixel-wise learning algorithms has been found to be efficient for certain classes of image processing problems where the proper learning yielded good results. Both machine learning algorithms and deep neural networks are being used for many image processing applications.

Implementation of image denoising techniques with machine learning and deep learning algorithms is a commonly studied problem in computer vision [1]. With the advancement of deep learning methods, attempt were made to denoise the images by


S. Smys et al. (Eds.): ICCVBIC 2019, AISC 1108, pp. 631–644, 2020.
https://doi.org/10.1007/978-3-030-37218-7_72

learning image features [2, 3]. Already, supervised learning based noise pixel estimators using robust statistical features like median absolute deviation (MAD) and median was implemented in a genetic programming (GP) based impulse noise filter [4]. Singular-value decomposition (SVD) and neural network (NN) are used for noise level estimation of Gaussian white noise and mixed noise [5]. The GP based noise detector sources the maximum possible information from the statistical features extracted from every single window of the image has done better pixel classification [6].

Normally, the denoising algorithms used mean and median filter with different weights which runs all over the image but of less efficient as it changes the uncorrupted pixels [7–9]. By incorporating noise pixel classification, the switching median filters showed better performance when compared to the conventional uniformly applied filters and it is followed by several combined noise detectors and filters like.

- Multistate median (MSM) filter [10]
- Tristate median (TRI) filter [11]
- Adaptive center-weighted median (ACWM) filter [12]
- Pixel-wise median absolute deviation (PWMAD) filter [13]
- Adaptive switching median (ASWM) filter [14]
- Directional weighted median (DWM) filter [15]
- Luo-iterative median filter [16]
- Conditional signal-adaptive median (CSAM) filter [17]
- Rank-ordered logarithmic difference edge-preserving regularization filter (ROLD-EPR) [18]
- Robust Outlyingness ratio detector [19]

Hence, by using robust statistical parameters and machine learning algorithms, the classification of corrupted pixels otherwise, noise detection is made accurate and also it can be improved if the classifier model is properly trained. The existing robust statistical features are utilized for this purpose as they convey important information for the random forest classifiers.

Section 2 highlights about the noise model considered in this research. Section 3 discusses about the robust statistical features called pixel parameters. Machine learning algorithms with emphasis on random forest classifier is discussed in Sect. 4. The corrupted pixel classification methodology is discussed in Sect. 5. A detailed implementation methodology is discussed in Sect. 6. Section 7 shows the simulation results obtained and plots to compare.

## 2 Noise Model

Two additive impulse noise models are considered here. The first of its kind is salt and pepper noise (SPN) which is caused due to the erroneous behavior of the analog-to-digital converter or while transmission. In SPN, the corrupted pixel takes the extreme values pixel like 0 or 255. With probability ‘p’;

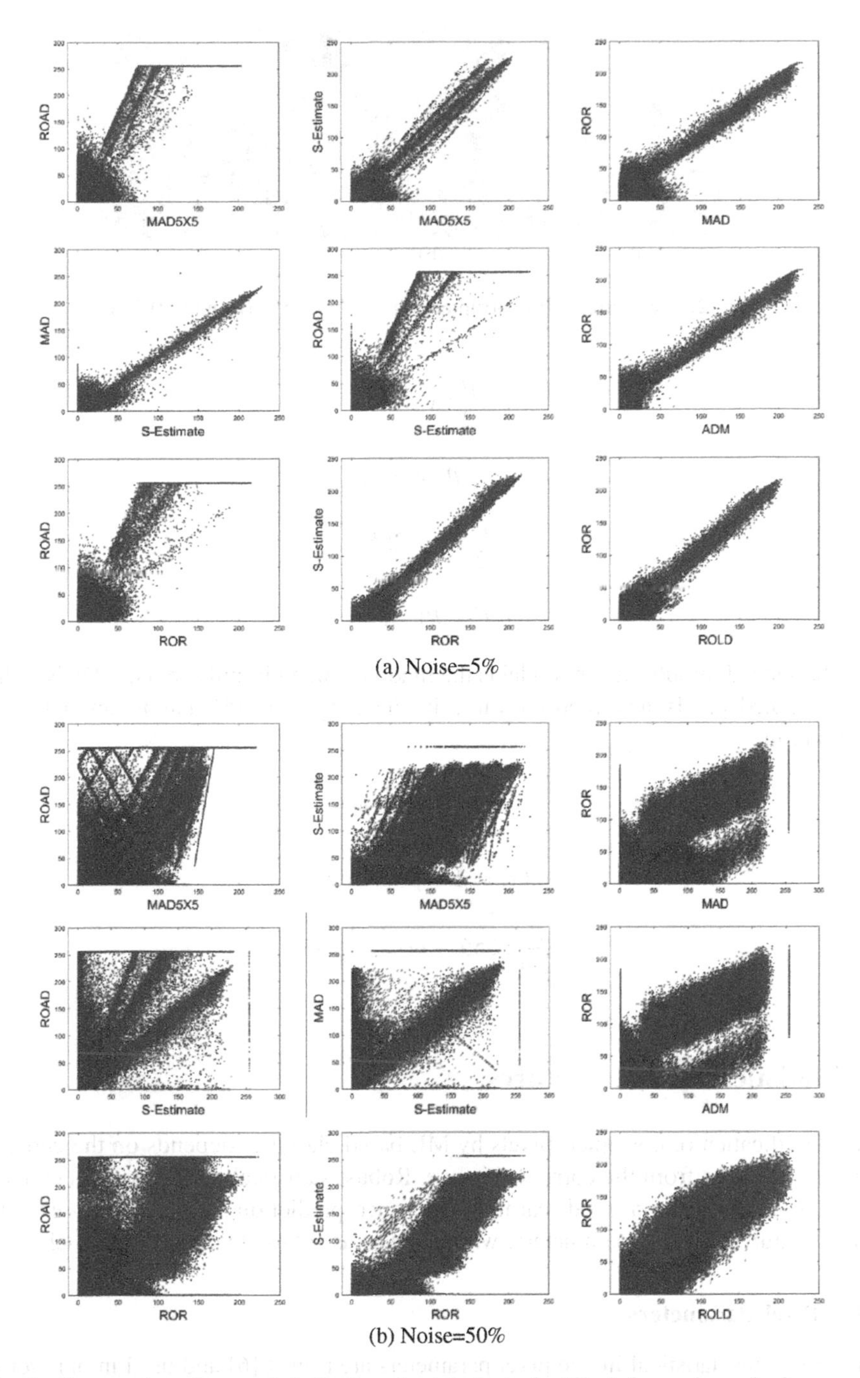


(a) Noise=5%

(b) Noise=50%

**Fig. 1.** Scatter plot of corrupted (Red) and original pixels (Blue) for various pixel parameters (a) Noise = 5% (b) Noise = 50%

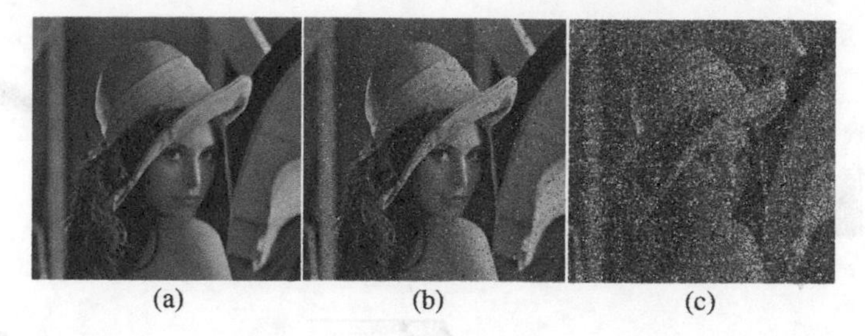

(a) (b) (c)

**Fig. 2.** Lena image (a) corrupted with SPN of 5% (b) and 50% (c)

$$\frac{p}{2} : x = 0$$

$$1 - p : x = y_{i,j}$$

$$\frac{p}{2} : x = 255$$

$$P = p_1 + p_2$$

The second impulse noise model is the random valued impulse noise (RVIN) where the corrupted pixels any random value in the range [0,255] and it has a uniform distribution.

$$\frac{p}{2m} : 0 \leq x < m$$

$$f(x) = 1 - p : x = y_{i,j}$$

$$\frac{p}{2m} : 255 - m < x \leq 255$$

## 3 Statistical Image Features

The classification of corrupted-pixels by ML based classifier depends on the choice of features extracted from the corrupted image. Robust statistical image features extracted from the image make a good learning and hence prediction. I call the chosen set of image features as pixel parameters, which serves as robust features for ML-classifier.

### 3.1 Pixel Parameters

The following statistical image pixel parameters are robust [6] and used in our work to extract the features of each image pixel from the surrounding window.

- Rank-Ordered Absolute Difference (ROAD) [20]
- Rank-ordered logarithmic difference (ROLD) [21]
- Median absolute deviation (MAD) [22]
- S-Estimate [23]
- Absolute Deviation from Median (ADM) [22]
- Robust Outlyingness ratio (ROR) [24]

#### 3.1.1 ROAD

N denoting the pixel coordinate and the (2 N + 1)-by-(2 N + 1) is the current window centered at (0, 0).

i.e., $\Omega_N = \{(s,t)| - N \leq s,t \leq N\}$ and let $\Omega_N^0 = \Omega_N \backslash (0,0)$.

$d_{st}(y_{i,j}) = |y_{i+s,j+t} - y_{i,j}|, \forall (s,t) \in \Omega_N^0$, where $d_{st}$ denotes the absolute difference between the centre pixel and the surrounding pixels in the taken window.

After sorting the elements of $d_{st}$ in the ascending order, the sum of first k elements $(r_k)$ is calculated to have the ROAD value of the current window.

$$ROAD_m(y_{i,j}) = \sum_{k=1}^{m} r_k(y_{i,j})$$

Where $2 \leq m \leq (2N+1)^2 - 2$.

#### 3.1.2 ROLD

To avoid the misclassification of pixel if the corrupted pixel value varies randomly, a logarithm can be considered while finding the absolute differences.

$$\widetilde{D}_{st}(y_{i,j}) = \log_a |y_{i+s,j+t} - y_{i,j}|, \forall (s,t) \in \Omega_N^0$$

Hence for any a > 1, the number $\widetilde{D}_{st}$ is always in $(-\infty, 0]$.

To keep the dynamic range [0,1], a truncation scheme is used with a linear transformation:

$$D_{st}(y_{i,j}) \equiv 1 + \max\{\log_a |y_{i+s,j+t} - y_{i,j}|, -b\}/b; \forall (s,t) \in \Omega_N^0.$$

a and b are positive numbers which controls the shape of the logarithmic function and truncation position respectively. The efficiency of detection depends on the selection on a, b and it can be selected using the function $h_{a,b}(x)$, which is defined as;

$$h_{a,b}(x) = 1 + \max\{\log_a x, -b\}/b, (x \geq 0)$$

#### 3.1.3 MAD

The noise variances can be easily calculated using MAD and at times MAD may not perform well in case of highly corrupted image. The median absolute deviation about the median is given by.

$MAD_n = b \times med_i|x_i - med_j x_j|$, where $med_j x_j$ is the median pixel value in the sub-window.

#### 3.1.4 S-Estimate

This is one of the robust noise estimator which works well even for highly corrupted images with edges.

$$S = med_i\{med_j|t_i - t_j|\}$$

Where, $t_i, i = 1, \ldots, N$ is the pixel window with N pixels.

The inner median of $\{|t_i - t_j|\}$ for each i should be computed first and it is followed by the outer median which is the median of those inner medians.

#### 3.1.5 Absolute Deviation from Median (ADM)

Absolute deviation from median 5 × 5 window:

$$u(x,y) = abs(g(x,y) - med(W_5))$$

Absolute deviation from median 3 × 3 window:

$$u(x,y) = abs(g(x,y) - med(W_3))$$

#### 3.1.6 Robust Outlyingness Ratio (ROR)

$$u(x,y) = \left|\frac{g(x,y) - med(W_5)}{MADN(W_5)}\right|$$

## 4 Machine Learning Algorithms

There are several machine learning algorithms intended for unique applications as they show better performance in their own niche. The focus is on building a robust classifier provided with statistical features extracted from the image. The commonly available classifiers are:

- Classification Trees
- Discriminant Analysis
- Naive Bayes Classifier
- Nearest Neighbors
- Support Vector Machine Classification

The random forest classifier is chosen for our task as it has proven benefits in terms of classification accuracy and computational time. Also the decision forest can be trained with random level of noise intensities and hence the trained model can be expected to work with wide noise ranges.

### 4.1 Random Forest Classifiers

Ensemble or bag of decision trees are used to classify data with higher accuracy [25, 27]. A decision tree is used to make decision based on set of conditions. It can also split a complex problem into simple hierarchical problems which can be solved stage by stage. The tree structure and other parameters can be learnt from the training data set for a complex problem where it is hard to formulate manually.

Initially the decision tree is trained individually using a synthetic dataset formed from the pixel features extracted from the corrupted image in our case. After training, the tree structure and the parameters are fixed and it should be tested for accuracy while it can be used.

Classification forest or decision forest is a group of randomly trained decision tree with each node is split based on the condition and assumes 0 if the condition fails, 1 if it passes $h(v, \theta_j) \in \{0, 1\}$.

The learner parameters are $\theta = (\phi, \psi, \tau)$, where $\psi$ being the geometric primitive for separating data, $\tau$ carries threshold in order to separate classes and $\phi$ selects appropriate features from the vector $v$. The classification forest infers a class label $c \in C$ with $C = \{c_k\}$, after it is trained with the input test data v which is a multi-dimensional vector $v = (x_1, \ldots, x_d) \in R^d$.

The objective is to optimize the energy for a given set of data and their class labels. Each split nodes are optimized by $\theta_j^* = \arg\max_{\theta_j \in \tau_j} I_j$, where $I_j$ is the information gain given by $I_j = H(S_j) - \sum_{i \in \{L,R\}} \frac{|S_j^i|}{|S_j|} H(S_j^i)$ and $H(S)$ is the entropy.

## 5 Proposed Implementation Methodology

The noise detection performance depends on how well the machine learning classifiers are trained. To train the classifiers, the original image is corrupted with various levels of noise intensities. Both the noise models viz., RVIN and SPN are considered during the training phase. Each noise model is classified into several noise intensity levels ranging from 1% to 70% of corrupted pixels. To view the differences of the pixel parameters like ROAD, ROLD, ROR, etc., of the original and corrupted pixels, a scatter plot is shown in Fig. 1. The differences in the pixel parameters between the corrupted and original pixels are clearly visible in Fig. 1.

From the scatter plots, as the regions are distinct for lower noise intensity, it is obvious that the classification of corrupted pixels is less complex and can be carried out by considering few pixel parameters like MAD, ROAD etc. For highly corrupted images, it is hard to define a clear graphical boundary to classify them and the overlap is clearly visible. So, training the classifier and validating the model is complex as the model has to work irrespective of the amount of corrupted pixels. Hence I consider all six pixel parameters discussed in Sect. 3 and I have chosen random forest classifiers with several random classification trees which will be trained to work with wide noise intensities. Also the conducted simulations over different images show that the classification accuracy improves along with number of grown trees. This fact is further discussed and validated in Sect. 5 with a clear graph in Fig. 5. Figure 1a shows the scatter plot of pixel parameters for 5% additive impulse noise to the lena image (Fig. 2) and the dispersion of is minimum. A simple classification tree can be trained to identify these corrupted pixels. As the noise intensity grows, the dispersion of the scatterplot increases as shown in Fig. 1b and hence a set of complex classification trees are needed to classify the corrupted pixels. Random forest classifier with several random complex classification tree will serve this purpose.

The complete experimental setup is divided into two major phases like other machine learning counter parts viz.

1. Training & Cross Validation (CV)
2. Testing

### 5.1 Training and Cross Validation

As shown in Fig. 3, training the model involves the extraction of all the statistical parameters from the corrupted image. For this training process, an image dataset with randomly corrupted pixels are created with different levels of noise intensity.

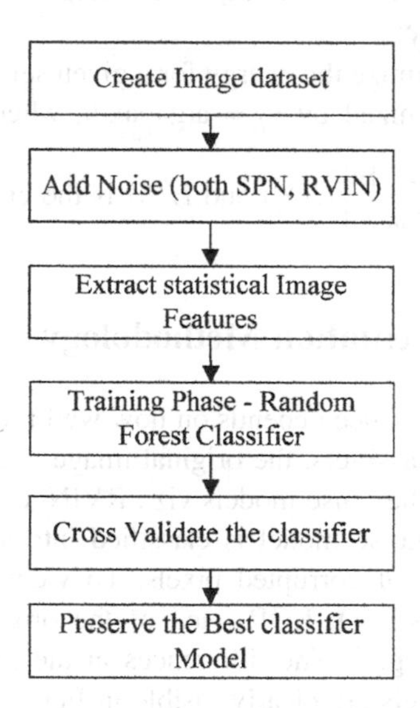


**Fig. 3.** Model creation phase

The random forest classifier is trained with the set of images and cross validated to check if the final model has sufficient classification accuracy. The values in confusion matrix are plotted in Fig. 4 to show the variation of True Positive (TP), True Negative (TN), False Positive (FP) and False Negative (FN) over noise density for a particular trained model. Figure 5 shows the Miss Classification Rate (MCR) of the same model.

### 5.1.1 Training and Cross Validation Results

After the sufficient training of the classification trees with the existing image dataset, it is cross-validated. Table 1 provides CV results for noise density from 10% to 50% while considering only three random trees (N = 3) in the random forest classifier.

**Table 1.** Correlation matrix results from CV (N = 3)

| Noise (%) | TP | TN | FP | FN | Accuracy | MCR |
|---|---|---|---|---|---|---|
| 10 | 9.95% | 89.90% | 0.10% | 0.06% | 99.84% | 0.16% |
| 20 | 19.81% | 79.16% | 0.84% | 0.19% | 98.97% | 1.03% |
| 30 | 29.54% | 67.08% | 2.92% | 0.47% | 96.62% | 3.38% |
| 40 | 38.16% | 54.87% | 5.00% | 1.98% | 93.02% | 6.98% |
| 50 | 47.03% | 41.56% | 8.33% | 3.07% | 88.59% | 11.41% |

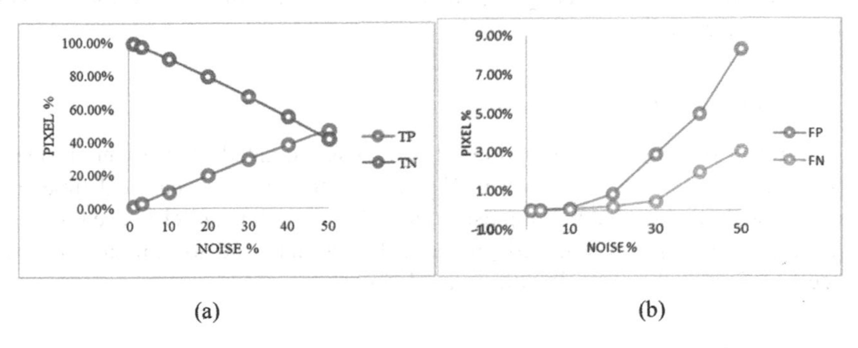


(a) (b)

**Fig. 4.** Confusion matrix plot to show model performance over increasing noise levels (N = 3) (a) TP & TN (b) FP & FN

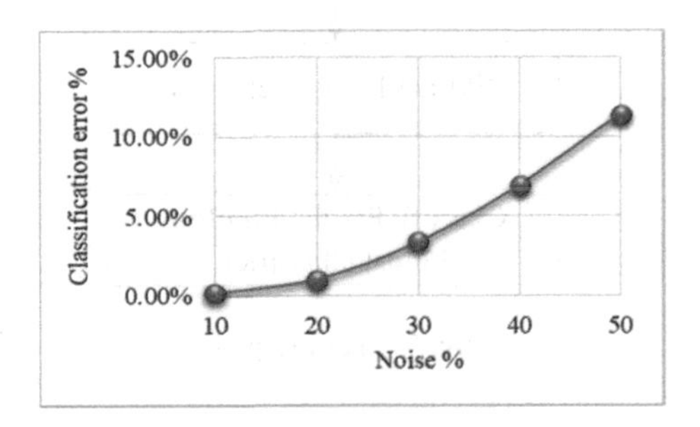


**Fig. 5.** Classification Error (N = 3)

### 5.1.2 Impact of No. of Classification Trees

Apart from the pixel parameters, the classification error also depends on the number of random classification trees used in the random forest classifier. To depict this dependency, two graphs are plotted in the Fig. 6, showing variation in the classification error for various noise intensities.

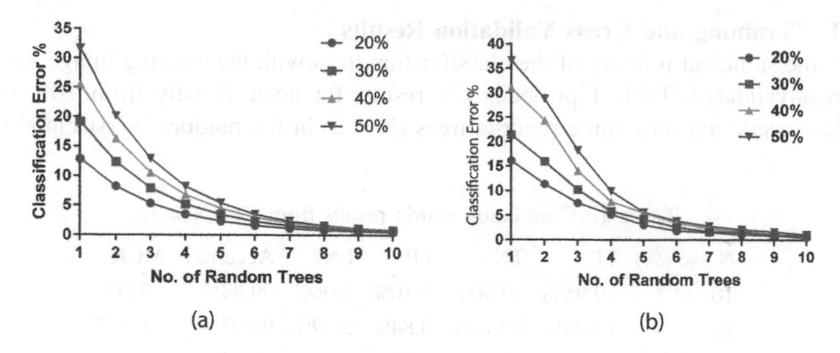


**Fig. 6.** Variation of classification error over no. of trees. (a) SPN (b) RVIN

As the number of trees used in the random forest classifier increases, the classification error decreases exponentially as depicted in Fig. 6 which is true for a maximum of 10 random trees.

### 5.2 Testing Phase

After a model is being created, optimized and cross-validated, it is essential to test the model with test data set as shown in Fig. 7. The test data set has real time images corrupted with random valued impulse noise. The classifier model is used to check if the corrupted pixels are classified properly. To validate and to compare the performance of this model, other classifiers are also used.

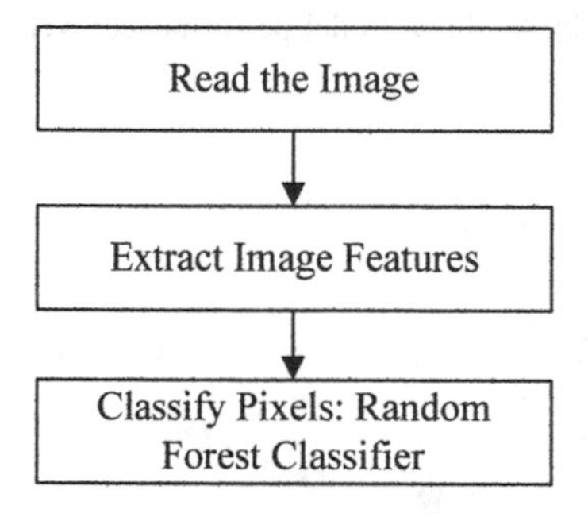


**Fig. 7.** Testing phase

## 6 Experimental Results and Analysis

The simulation results show a steeper reduction in the classification error as the number of grown trees increases. Further, the plot clearly indicates a saturation point, where the classification error is almost same even if more number of classification trees are used. Hence, an optimal number of classification trees should be chosen to train for a given dataset, which is 10 for this problem. Randomly selected 10 classification trees forms the random forest classifier for classifying the corrupted pixels based on the input provided by the pixel parameters. As the pixel parameters are the compilation of the robust statistical parameters extracted from each pixel, the classification accuracy is expected to be high. The noise detection by the proposed method is considerably good for both SPN and RVIN. The robust statistical features along with the random forest classifier make this possible. The following sections discuss the results in detail.

### 6.1 Comparison of Noise Detection Results of SPN

The Table 2 and Fig. 8 compares the performance of the existing noise detection algorithms with the proposed technique taking classification error as the criteria. The proposed ML-based classification of corrupted pixels works reasonably well for identifying SPN even for higher noise densities. This classification can further be improved by improving the robustness of the statistical features and also by improving the performance of the classification tree used.

**Table 2.** Comparison of noise detection results (Lena Image) corrupted with SPN

| Noise detectors | Noise % | | | |
|---|---|---|---|---|
| | 20 | 30 | 40 | 50 |
| SWM [13] | 0.50 | 1.05 | 1.07 | 1.36 |
| ACWM [12] | 0.17 | 0.40 | 0.83 | 1.70 |
| LUO-ITERATIVE [16] | 0.36 | 1.10 | 2.55 | 5.19 |
| DWM [15] | 0.95 | 1.04 | 1.67 | 2.94 |
| ASWM [14] | 2.18 | 2.43 | 2.81 | 4.14 |
| ROR-NLM [19] | 0.37 | 0.38 | 0.45 | 0.98 |
| ML | 0.29 | 0.35 | 0.44 | 0.95 |

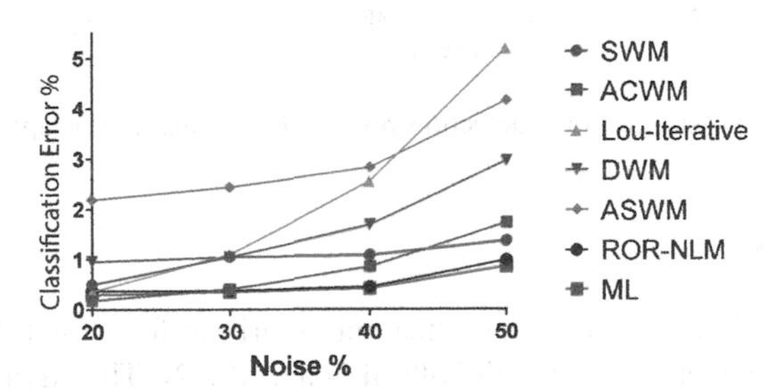


**Fig. 8.** Comparison of noise detection results (Lena Image) corrupted with SPN

### 6.2 Comparison of Noise Detection Results of RVIN

The Table 3 and Fig. 9 compare the performance of the noise detection algorithms. The ML based classifier performs well for low densities of RVIN as it classifies the corrupted pixels more efficiently. Increasing noise densities makes the classification task complex and hence its performance also slows down.

**Table 3.** Comparison of noise detection results (Lena Image) corrupted with RVIN

| Noise detectors | Noise % | | | |
|---|---|---|---|---|
| | 20 | 30 | 40 | 50 |
| SWM [26] | 4.76 | 7.85 | 9.74 | 12.40 |
| TRI [11] | 4.52 | 6.57 | 8.58 | 11.11 |
| MSM [10] | 4.49 | 6.47 | 8.52 | 11.29 |
| ACWM [12] | 3.16 | 4.67 | 6.31 | 8.62 |
| PWMAD [13] | 6.06 | 8.30 | 10.44 | 12.75 |
| Luo-Iterative [16] | 3.70 | 5.88 | 8.27 | 11.22 |
| DWM [15] | 4.97 | 6.75 | 7.95 | 9.30 |
| ASWM [14] | 4.22 | 5.54 | 6.92 | 8.64 |
| ROLD-EPR [18] | 5.14 | 7.15 | 8.21 | 9.27 |
| ROR-NLM [19] | 4.39 | 5.45 | 6.72 | 8.19 |
| ML | 1.03 | 3.38 | 5.79 | 7.98 |

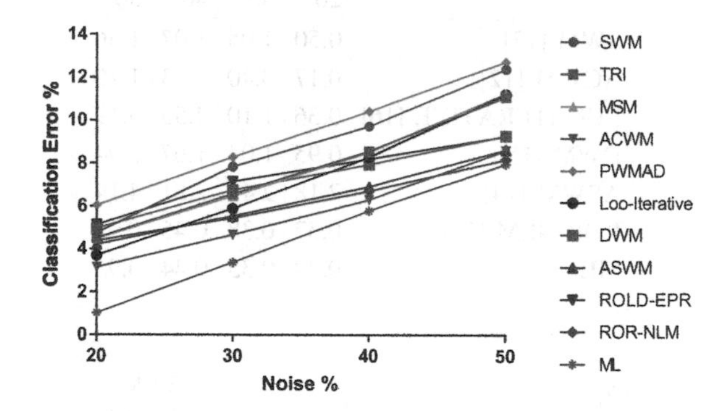


**Fig. 9.** Comparison of noise detection results (Lena Image) corrupted with RVIN

# 7 Conclusion

A random forest classifier based classification algorithm for classifying noise pixels is trained, implemented and tested with both SPN and RVIN. The algorithm uses some of the robust statistical features named pixel parameters along with random forest classifier to classify the corrupted pixels. The identification of corrupted pixels depend on both the statistical features and the chosen classifier which should be carefully trained and cross-validated with vast dataset. The impact of number of random decision trees on the classification accuracy helps to fix the number of trees to limit the complexity in training and CV. Overall, the trained decision forest performs better and the precise improvement in the performance is given in Tables 2 and 3.

## References


1. Zhao, A.: Image denoising with deep convolutional neural networks (2016)
2. Vincent, P., Larochelle, H., Bengio, Y., Manzagol, P.-A.: Extracting and composing robust features with denoising autoencoders. In: Proceedings of the 25th International Conference on Machine Learning (ICML 2008), pp. 1096–1103. ACM, New York (2008)
3. Mao, X.-J., Shen, C., Yang, Y.-B.: Image restoration using very deep convolutional encoder-decoder networks with symmetric skip connections. arXiv preprint arXiv:1603.09056 (2016)
4. Majid, A., Khan, A., Mirza, A.M.: Combination of support vector machines using genetic programming. Int. J. Hybrid Intell. Syst. **3**, 109–125 (2006)
5. Wang, Z., Yuan, G.: Image noise level estimation by neural networks. In: International Conference on Materials Engineering and Information Technology Applications (MEITA 2015) (2015)
6. Javed, S.G., Majid, A., Mirza, A.M.: Multi-denoising based impulse noise removal from images using robust statistical features and genetic programming. Multimed. Tools Appl. **75**, 5887 (2016)
7. Huang, T.S., Yang, G.J., Tang, G.Y.: A fast two-dimensional median filtering algorithm. IEEE Trans. Acoust. Speech Signal Process. **ASSP-27**(1), 13–18 (1979)
8. Brownrigg, D.: The weighted median filter. Commun. ACM **27**(8), 807–818 (1984)
9. Ko, S.J., Lee, Y.H.: Center weighted median filters and their applications to image enhancement. IEEE Trans. Circuits Syst. **38**(9), 984–993 (1991)
10. Chen, T., Wu, H.R.: Space variant median filters for the restoration of impulse noise corrupted images. IEEE Trans. Circuits Syst. II Analog Digit. Signal Process. **48**(8), 784–789 (2001)
11. Chen, T., Ma, K.K., Chen, L.H.: Tri-state median filter for image denoising. IEEE Trans. Image Process. **8**(12), 1834–1838 (1999)
12. Chen, T., Wu, H.R.: Adaptive impulse detection using center-weighted median filters. IEEE Signal Process. Lett. **8**(1), 1–3 (2001)
13. Crnojevic, V., Senk, V., Trpovski, Z.: Advanced impulse detection based on pixel-wise MAD. IEEE Signal Process. Lett. **11**(7), 589–592 (2004)
14. Akkoul, S., Ledee, R., Leconge, R., Harba, R.: A new adaptive switching median filter. IEEE Signal Process. Lett. **17**(6), 587–590 (2010)
15. Dong, Y.Q., Xu, S.F.: A new directional weighted median filter for removal of random-valued impulse noise. IEEE Signal Process. Lett. **14**(3), 193–196 (2007)
16. Luo, W.: A new efficient impulse detection algorithm for the removal of impulse noise. IEICE Trans. Fundam. Electron. Commun. Comput. **E88-A**(10), 2579–2586 (2005)
17. Pok, G., Liu, J.C., Nair, A.S.: Selective removal of impulse noise based on homogeneity level information. IEEE Trans. Image Process. **12**(1), 85–92 (2003)
18. Dong, Y., Chan, R.H., Xu, S.: A detection statistic for random valued impulse noise. IEEE Trans. Image Process. **16**(4), 1112–1120 (2007)
19. Xiong, B., Yin, Z.: A universal denoising framework with a new impulse detector and nonlocal means. IEEE Trans. Image Process. **21**(4), 1663–1675 (2012)
20. Garnett, R., Huegerich, T., Chui, C., Wenjie, H.: A universal noise removal algorithm with an impulse detector. IEEE Trans Image Process **14**(11), 1747–1754 (2005). https://doi.org/10.1109/tip.2005.857261
21. Yiqiu, D., Chan, R.H., Shufang, X.: A detection statistic for random-valued impulse noise. IEEE Trans. Image Process. **16**(4), 1112–1120 (2007)
22. Petrovic, N.I., Crnojevic, X.V.: Universal impulse noise filter based on genetic programming. IEEE Trans. Image Process. **17**(7), 1109–1120 (2008)

23. Petrovic, N.I., Crnojevic, V.: Impulse noise filtering using robust pixel-wise S-estimate of variance. EURASIP J. Adv. Signal Process. **2010**, 830702 (2010)
24. Bo, X., Zhouping, Y.: A universal denoising framework with a new impulse detector and nonlocal means. IEEE Trans. Image Process. **21**(4), 1663–1675 (2012)
25. Criminisi, A., Shotton, J., Konukoglu, E.: Decision forests: a unified framework for classification, regression, density estimation, manifold learning and semi-supervised learning. Found. Trends Comput. Graph. Vis. **7**, 81–227 (2012). ISSN 1572-2740, Now Publishers, ISBN 1601985401, 9781601985408
26. Sun, T., Neuvo, Y.: Detail-preserving median based filters in image processing. Pattern Recognit. Lett. **15**(4), 341–347 (1994)
27. Criminisi, A., Shotton, J.: Decision Forests for Computer Vision and Medical Image Analysis. Springer, London (2013). https://doi.org/10.1007/978-1-4471-4929-3

# An Online Platform for Diagnosis of Children with Reading Disability

P. Sowmyasri, R. Ravalika(✉), C. Jyotsna, and J. Amudha

Department of Computer Science and Engineering, Amrita School of Engineering, Bengaluru, Amrita Vishwa Vidyapeetham, Bengaluru, India
sowmyapaladhi@gmail.com, ravalika.ranga@gmail.com,
{c_jyotsna, j_amudha}@blr.amrita.edu

**Abstract.** Reading disability is a condition in which a person is analyzed with difficulty in reading There are different types of reading disorders, some of them are dyslexia, alexia. Recent research has identified that one of the causes for this disability could be hereditary factors. Detecting on time and proper treatment helps in improving the reading ability of a dyslexic child. Proposed method provides a user friendly approach for analyzing the children dyslexic disorder level using an Eye Tracker. Proposed module uses set of input stimulus to analyze all the parameters that could identify the stage of disability in a child. At the end of experiment a dashboard showing the results that could provide a detailed information for the doctor to decide the treatment that could help in reducing child's reading disability. The proposed system increases the understandability of dyslexic disorder levels of a child as it display all the parameters that are measured to test the level of disorder in the form of a dashboard through a web page.

**Keywords:** Dyslexia · Eye tracking · Saccades · Regressions · Common gateway interface

## 1 Introduction

Scientific research over the years suggests that educators can identify the signs which indicate that a child is at risk of reading disability. Identification of such disability in children at early stages helps not only to improve their academic career but also to overshadow their social life.

Eye tracking is a technique which allows us to analyze eye gaze movement in a person and this analysis help us to predict the reading behavioral pattern of a dyslexic child.

Eye tracking allows us to analyze language comprehension in real time. Usage of this method has been widely spread over the years and hence this method is improvised with better and more accurate machines making it more easy to use this. Information processing is affected by dyslexia. Structuring the information, receiving, holding, retrieving and the speed of the information processing is largely affected. According to equality act 2010 purposes, dyslexia can be considered as a disability. Dyslexia can occur in all the three forms severe, moderate or mild forms. No two persons with


S. Smys et al. (Eds.): ICCVBIC 2019, AISC 1108, pp. 645–655, 2020.
https://doi.org/10.1007/978-3-030-37218-7_73

dyslexia can have exactly the same kind of difficulties and strengths. Dyslexic people face trouble translating the language into thought and also thought into language. It is important for students with dyslexia to understand the difficulties caused by their disabilities. Though children with dyslexia have reading disabilities, their IQ level is higher compared to normal children. Thus proper diagnosis and on time treatment can make a dyslexic child stronger and brighter.

The proposed system is a web based application in which the patient's data collected from the eye tracker is taken as input in the form of a text document. This text document contains the details such as gaze vector, plane dimension, areas of interests, eye positions etc. The data is processed using various algorithms to produce the results onto a web platform which makes the user easily and quickly understand the disorder levels of the patient and helps in making a decision to go for what treatment in order to reduce the dyslexic disorders.

Dyslexia is detected majorly in children. This affects them academically and also socially. 20% of school age children are diagnosed with dyslexia. Around 5–10% of world population has dyslexia, it can get as high as 17%. Thus there is a high necessity of preventing the difficulties raised by this disability in children. Early identification and intervention in children can help them progress.

The content in the paper has been organized into various sections. The current section gives the introduction on dyslexia and overview of the system proposed. This Sect. 2 is on literature survey in which the writings from different papers regarding the mechanisms to control dyslexia have been analyzed, understood and written in the form of a summary. Section 3 describes the high level design of the proposed system. The Sect. 4 gives a brief description on how the steps are taken to successfully implement the proposed system. The results extracted by the implementation of all the steps mentioned in the previous section are presented in Sect. 5. Finally Sect. 6 draws the conclusion of this research.

## 2 Literature Survey

**Diagnosis to Dyslexia Using Computational Analysis**
Diagnosis to dyslexia can be done using computational analysis as described in the paper. Dyslexia is defined as the inability to complete task, having bad memory and feeling different from others. The dataset collected can be analyzed using some machine learning algorithms like kernel density estimation, artificial neural networks. Artificial neural networks have three different layers. The input neurons are fed into the system in the first layer. Processing of input neurons is done in the second layer called hidden layer and the classification is done in third layer. The accuracy achieved using artificial neural networks is higher than the other algorithms. Necessary features can be extracted using the dataset and processed to identify the number of dyslexic persons and number of normal persons [1].

Detecting readers with Dyslexia using Machine learning with eye tracking measures- Tobii eye tracker is used to track the eye movements and find gaze points. The data set for classification consisted of 97 people, 48 with dyslexia ranging from

11–50 years and 49 people without dyslexia with ages from 11–54. They have considered 12 readings with 12 different texts and with 12 different typefaces. This is because of the results of previous experiments that the text presentation and content affect the reading ability of dyslexic people. All the useful features have been extracted and SVM binary classification has been applied for predicting the presence of dyslexia. Also, implementation of 10 cross validation improved accuracy. The results of the experiment is proving the fact that typefaces did not improvise prediction because it would even help normal people. Also, the results included the statement that age when removed, accuracy decreased compared to the one obtained previously. This may be because the data set is not uniformly divided for all set of age groups. So the data set should have been different for each age group. Moreover, the data set should have been larger for better results. Also, the results are valid only for Spanish language [2].

**Diagnosis to Dyslexia Using Eye Tracking**
The eye movement of the normal person is different from the dyslexic person. Based on this key point there was a research paper which used eye tracking as its key technology and conducted experiments on different persons to detect if a person is dyslexic or not. The authors viewed dyslexia as neural developmental reading disability. Google based infrared corneal reflector Ober in collecting the data. The data is collected from 185 subjects. The data displayed to the users has eight lines with ten sentences and an average of five words. This systems records the horizontal, vertical positions of both the eyes of a reader. Then eye movement analysis is done to find the fixations and saccades. These are calculated using average position of the eyes, standard deviation of the average position of the eye and the maximum range between the position of the eyes. The saccades and fixations thus describe the reading capabilities of the readers [3].

**Interactive Screening for Learning Difficulties**
Analyzing visual patterns of reading Arabic scripts with eye tracking- The gaze points of the user has been determined here using the near infrared illuminator to get the reflection patterns and thereby calculate the gaze points. These results will be processed in order to extract essential features. The modules in this system are Filter raw gaze data- This uses the filtering algorithms to get the fixations and saccades, analyze fixations- Calculates essential features like fixation count, fixation duration, regressions, path, format and produce results-Uses visualization techniques to display the results. Explore thresholds-Helps the examiner to compare with the previously recorded results. This module is an interactive system which implies easy operation and management. But this system cannot predict the user as the results have to further processed by the practitioner to give the result. Also, this uses Arabic language [4].

**The Prevention, Diagnosis and Treatment of Dyslexia**
Diagnosis-It includes both reading and spelling disabilities. Reading disabilities involve not being able to associate words with sound, replacing different words with similar ones, deducing the content of sentence along with replacement. Spelling disabilities are of two stages. They are phonetic and orthographic. Through standard tests we will be able to find the presence of disabilities. Treatment-No drug treatment for such disabilities. Regular practice under care of parents and teachers and psychological treatment is necessary and can help in improvising the child. Prevention-Building

schemes in kindergarten for developing linguistic abilities in children. These include activities like guess game, rhyme recognition, clapping syllables etc.

**Developing an Application Using Eye Tracker**
The device called eye tribe is equipped with an eye tracker. This device is used in the experiment conducted [5] to understand the areas of interests of a user. This study can be used for various other purposes such as improving website design or magazine cover etc. The stimulus is provided to the user and the device is used to collect the various parameters such as gaze points, fixations, saccades, areas of interests etc., while the user is looking at the stimulus to perform reading operation. The data collected by the device is analysed using various machine learning algorithms. This study also explained the details on how to setup and use the eye tracker to collect the information from the user's reading operation. This analysed can be used to figure out the areas of interests of the user which could help in website development as this system can figure out which part of the webpage is concentrated more by the users. The outcome of this study is to find out where the user reading the stimulus is concentrating more.

**Eye Tracker Usage and Feature Extraction**
Eye movements are essential parameter for to diagnose the dyslexia presence. Since the gaze co ordinates are the ones which represent the eye movements, we need to use the eye tracker which can give us the required parameters. As the [6] describes the usage of eye tracker and the features extracted out of the results, eye tracker produced, SMI Red n professional eye tracker can be used for to capture the eye movements.

# 3 System Model

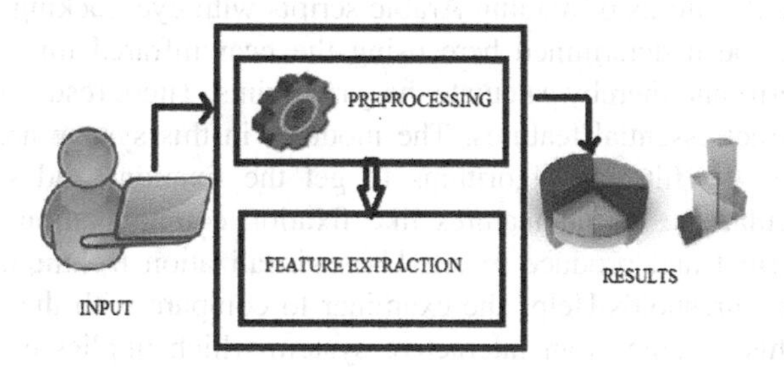


**Fig. 1.** System model

The proposed system model implements the following steps (Fig. 1).

1. Collection of data regarding various measures which on processing can help to categorize a normal reader and a dyslexic reader.
2. Cleaning and Normalization of the raw data to fetch all the required parameters and represent them in the required format.
3. Using the data collected and preprocessed in the above steps as the input to the scripts written in python the features are extracted and the results are produced to

differentiate a good reader and bad reader. Using the common gateway interface module the results produced in the above step are represented through various visualization techniques as a dashboard.

## 4 Implementation

The following section includes brief description about the proposed model and the execution of the proposed interactive architecture (Fig. 2).

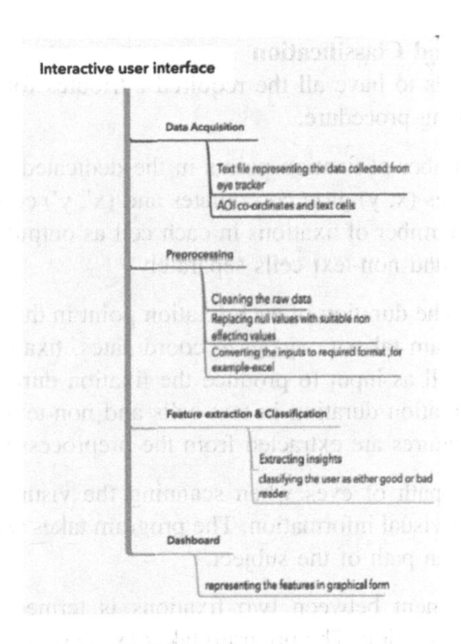


**Fig. 2.** Flow of working model

**GUI-Graphical User Interface**
The model is made interactive by implementing as a web application using wamp, CGI (common gateway interface) and Python. This user interface model provides a login to the user and starts with the step one of taking input data from the user and then to the processing model in which the python scripts are executed to predict the level of dyslexia in the person using the input data and then to the visualization model to display the results to the user in a most understandable manner. The overall implementation can be categorized into 4 important steps.

**Data Acquisition**
The user after authentication will be allowed to input the file. The data from the user is taken in the form of text document which is the most frequent format, an eye tracker gives the result. This data usually includes the details about calibration, tracker and data like diameter, gaze vector, eye position, plane dimensions, frame counter, quality.

Later, the details about the count and dimensions of area of interest(aoi) are acquired. This will be followed by the request to include the details mentioning the count and dimensions of both text and non-text cells of the stimuli used in the diagnosing process.

**Preprocessing**
Preprocessing the data is a very important step to convert the raw data into effective data for further execution. Cleaning of data includes of removal of noisy data and handling the missed values by either removing them or filling them with null values (which do not hold any importance while extracting features). Later the cleaned data is transformed into an excel sheet by formatting the data as per the requirements of further processing.

**Feature Extraction and Classification**
Here the text file needs to have all the required attributes for all the possible events involved in eye tracking procedure.

**Fixation Count:** Number of fixation points in the dedicated region of the text in the grid. The program takes (x, y) gaze coordinates and (x′, y′) coordinates of each cell as input to produce the number of fixations in each cell as output. It gives the number of fixations in text cells and non-text cells separately.

**Fixation Duration:** The duration of each fixation point in the dedicated region of text in the grid. The program takes (x, y) gaze coordinates, fixation duration and (x′, y′) coordinates of each cell as input to produce the fixation duration in each text cell as output. It gives the fixation duration in text cells and non-text cells separately. Using python scripts, the features are extracted from the preprocessed data.

**Scan Path:** It is the path of eyes when scanning the visual field and viewing and analyzing any kind of visual information. The program takes (x, y) gaze coordinates as input and finds the scan path of the subject.

**Saccades:** Eye movement between two fixations is termed as saccade. Backward saccade is called as regression. The program takes (x, y) gaze coordinates as input and finds the number of regressions made by the subject.

**AOI:** It is a tool to select sub regions of the displayed stim- ulus and to extract metrics specifically for these regions. From ground analysis difficult words in the stimuli are identified and are made as AOI's. The dimensions of AOI's taken in the first step are used here for interpretation of fixation counts and durations with respect to AOI. The program takes (x, y) gaze coordinates, fixation duration and (x′, y′) coordinates of each AOI as input to produce the fixation duration in each AOI as output. The program takes (x, y) gaze coordinates and (x′, y′) coordinates of each AOI as input to produce the number of fixations in each AOI as output.

**Classification:** Using KNN classifier the subjects are classified as low risk or high-risk subjects. K nearest neighbors is a simple algorithm that stores all available cases and classifies new cases based on a similarity measure (e.g., distance functions). Out of 20 data samples collected 18 samples are used as training data and 2 samples are used as testing data. The system predicts the class label for the new subject whether he is a low-risk subject or high-risk subject.

The method used for choosing training data set and test data set is random sampling where in each case randomly 18 data sets out of 20 are used for training and 2 data sets are used for testing.

**Dashboard View**
The features extracted are depicted using visualization techniques on the dashboard. The graphical representation helps in easier, quicker and better understanding of the level of dyslexic child. The results are represented in the form of a dashboard using common gateway interface. It is the standard way to produce the results when a request for the view of results of a child's dyslexic report is made after uploading the text files and the details of AOI. As soon as the request is made to the server on clicking the submit button by the client after the completion of uploading the details the scripts for the generation of the results run in the background and project them using different pictorial representations in the form of a dashboard.

## 5 Results and Analysis

Firstly the client is provided with an authentication page through which the details (username and password) are validated using the data stored in the backend and once validated the client is redirected to the following page as shown in Fig. 3.

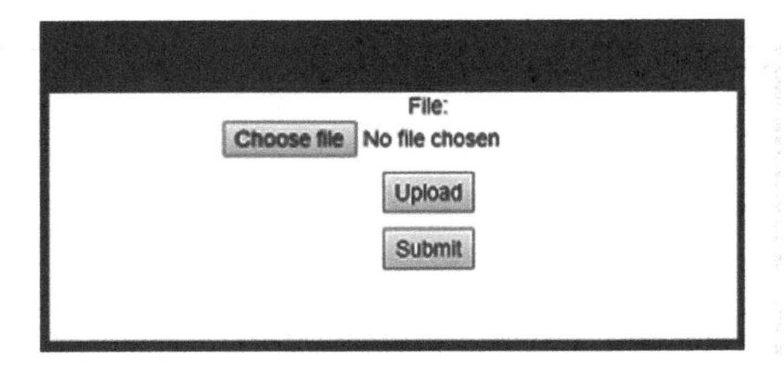


**Fig. 3.** Uploading file

The above diagram represents a sample web page which lets the user to upload the data. The uploaded data can further be processed for insights out of it. The user should upload a text file that contains the observations noted from an eye tracker device. The text file that is taken to process the proposed system is generated using an eye tracker device called SMI Redn Professional. Eye tracking is a technique that follows the movement of the eye mainly the movement of the retina or pupil. The movements are mainly recorded to find out the point of impact of eye gaze. The device used generates many raw data files in the form of text documents. The raw data contains various fields such as fixation points, pupil size etc. These files when uploaded via the web page of the system as shown in Fig. 3 are processed to transform them into an excel document extracting and filtering out the necessary fields from the text documents to make the further steps in the process easier. These conversion is done by importing python libraries like xls, xlrd etc. (Fig. 4).

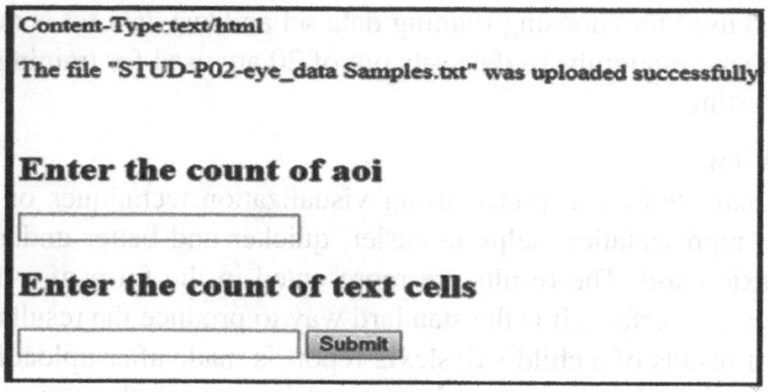

Content-Type:text/html

The file "STUD-P02-eye_data Samples.txt" was uploaded successfully

Enter the count of aoi

Enter the count of text cells

Submit

**Fig. 4.** Inputs-I from user

AOI (Area of interest) are the sub regions in the stimuli shown while capturing the eye tracking movements using the eye tracker. Usually these are the difficult words in the stimuli decided using ground analysis.

The whole frame of stimuli is considered as a grid structure divided into cells of uniform dimensions. Therefore, the cell can be uniquely identified with it's number. Each cell of the grid can either be a text or non text cell.

The above diagram represents a sample page facilitating the user to enter the count of text and AOI (Fig. 5).

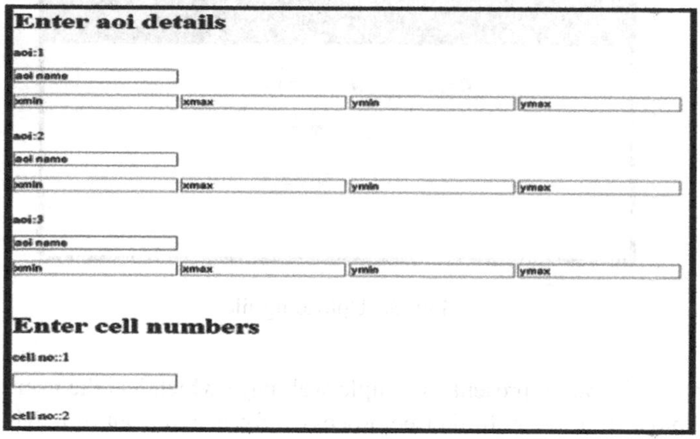

Enter aoi details

aoi:1

aoi name

xmin xmax ymin ymax

aoi:2

aoi name

xmin xmax ymin ymax

aoi:3

aoi name

xmin xmax ymin ymax

Enter cell numbers

cell no::1

cell no::2

**Fig. 5.** Inputs-II from user

The above diagram is sample model representing the page to take in the details like the co-ordinates of AOI, text cell numbers. This facilitates the user to give the co-ordinates of the AOI wherein (xmin, ymin) and (xmax, ymax) representing the range of AOI. Also the cell numbers here represent the number of the cell that the text is present within.

The dashboard depicts the graphs using visualization techniques which are the insights obtained from the data given in the acquisition step. Here, the results are the outcomes of the feature extraction and classification step. The features extracted from the text file are fixation count, fixation durations, scan paths, regression, Fixation count and duration with respect to AOI (Fig. 6).

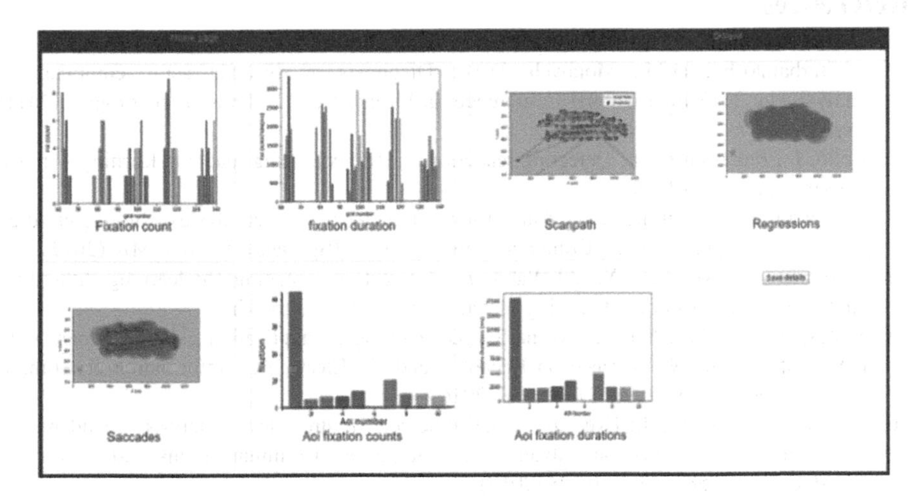


**Fig. 6.** Dashboard

This is the sample dashboard representing the features extracted. These results can be saved for future references using the save button option on the dashboard. All the above results are generated using the algorithms developed in python programming language and these results are displayed to the user of the proposed online platform in the form of a dashboard for better and the quicker analysis that could help in taking further steps in the treatment of a dyslexic child to overshadow his reading disability and help him to progress his/her social life. The results represented in the form of graphs on the dashboard are generated using the static machine learning algorithms developed from the research made to find the third eye assistance for reading disability like dyslexia [19].

## 6 Conclusion

The study proves that reading People with dyslexia have longer fixation duration, more number of fixations and tend to have slower language processing compared to normal readers and their eye movements are very random and are not along the text. The system developed gives the measure of parameters considered in the study and displays the results using various visualization techniques. The readers can go through continuous assessment at regular intervals and assess their improvement. This system can further be used in therapy centers, hospitals for diagnosis purposes.

The system developed can be further improved by providing the features like, the user can store the report of the reader in his account and make comparisons between the reports of various readers. These reports generated can be further sent to specialists for quicker analyzation.

## References

1. Al-Barhamtoshy, H.M., Motaweh, D.M.: Diagnosis of dyslexia using computational analysis. In: 2017 International Conference on Informatics, Health & Technology (ICIHT), Riyadh (2017)
2. Rello, L., Ballesteros, M.: Detecting readers with dyslexia using machine learning with eye tracking measures (2015)
3. Hassanain, E.: A multimedia big data retrieval framework to detect dyslexia among children. In: 2017 IEEE International Conference on Big Data (Big Data), Boston, MA (2017)
4. Al-Edaily, A., Al-Ohali, Y., Al-Wabil, A.: Interactive screening for learning difficulties: analyzing visual patterns of reading arabic scripts with eye tracking
5. Venugopal, D., Amudha, J., Jyotsna, C.: Developing an application using eye tracker. In: IEEE International Conference on Recent Trends in Electronics, Information & Communication Technology (RTEICT). IEEE (2016)
6. Jyotsna, C., Amudha, J.: Eye gaze as an indicator for stress level analysis in students. In: International Conference on Advances in Computing, Communications and Informatics (ICACCI), pp. 1588–1593. IEEE (2018)
7. Costa, M., et al.: A computational approach for screening dyslexia. In: Proceedings of the 26th IEEE International Symposium on Computer-Based Medical Systems, Porto (2013)
8. Shrestha, S., Murano, P.: An algorithm for automatically detecting dyslexia on the fly (2019)
9. Martins, V.F., Lima, T., Sampaio, P.N.M., de Paiva, M.: Mobile application to support dyslexia diagnostic and reading tracking. In: 2008 International Conference on Innovations in Information Technology, Al Ain (2008)
10. Al-Wabil, A., Zaphiris, P., Wilson, S.: Examining visual attention of dyslexics on web navigation structures with eye tracking. In: 2008 International Conference on Innovations in Information Technology, Al Ain (2008)
11. Schulte-Körne, G.: The Prevention, Diagnosis, and Treatment of Dyslexia (2003)
12. da Fonseca, V.: Dyslexia cogniçao e aprendizagem: umaabordagemneuropsicológica das dificuldades de aprendizagem da leitura. Revista Psicopedagogia, **26**(81), 339–356 (2009)
13. Ramus, F., et al.: Theories of developmental dyslexia: insights from a multiple case study of dyslexic adults. Brain **126**(4), 841–865 (2003)
14. Scarborough, H.S.: Very early language deficits in dyslexic children. Child Dev. **61**(6), 1728–1743 (1990)
15. Rello, L., Kanvinde, G., Baeza-Yates, R.: A mobile application for displaying more accessible ebooks for people with dyslexia. Procedia Comput. Sci. **14**, 226–233 (2012)
16. Tibi, S.: Reflections on "Good Practice" Dyslexia in Arabic UNESCO-DITT, February 2010
17. Pavani, M.L., Prakash, A.B., Koushik, M.S., Amudha, J., Jyotsna, C.: Navigation through eye-tracking for human–computer interface. In: Information and Communication Technology for Intelligent Systems, pp. 575–586. Springer, Singapore (2019)

18. Uday, S., Jyotsna, C., Amudha, J.: Detection of stress using wearable sensors in IoT platform. In: Second International Conference on Inventive Communication and Computational Technologies (ICICCT). IEEE (2018)
19. Navya, Y., SriDevi, S., Akhila, P., Amudha, J., Jyotsna, C.: Third eye: assistance for reading disabiltiy. In: International Conference On Soft Computing & Signal Processing, (ICSCSP 2019) (2019)

# Review on Prevailing Difficulties Using IriS Recognition

C. D. Divya(✉) and A. B. Rajendra

Department of CSE, Department of ISE, Vidyavardhaka College of Engineering, Affiliated to Visveswariah Techonological University, Belgaum, Mysuru, India
{divyacd, rab}@vvce.ac.in

**Abstract.** Now a days as we all know that for each and every thing we want biometric verification so keeping that in mind, we are actually studying the one of the biometric that is IriS since when compared to other biometrics its reliability is more. So we just reviewed some of the possibilities that may cause failure to this. A transitory study on certain issues is done here.

## 1 Introduction

Eye which is composed of IriS and pupil plays a very imperative role in recognition process of an individual. So here we are considering it one of the biometrics to be studied. Currently IriS has become one of the vital features in Biometrics. If we want to identify an individual, IriS is one of the biometrics that is engaged into consideration but In this, IriS recognition task also, firm problems that happens, In this paper we are just going to brief about the problems that are faced if we consider IriS as one of main biometrics for recognizing an Individual. Unlike the other biometric modalities involved, assorted feature that are considered from the IriS becomes very important for us to study it in malice of some difficulties.

### 1.1 Obstacles of Drug Possessions on IriS Structure

Drug is one of the ingredients that are actually reported to cause a physiological, psychological change in body. Since pupil is vital in IriS we should ponder that also into account. Pupil contraction, dilation affects IriS recognition in individual. Age, ailment and at hand are plentiful supplementary influences that needs to considered [1]. It is observed that that some agent by name my iatric lessens the recital of recognition. For comparison we can go with a well-organized process aimed at IriS acknowledgment via exemplifying crucial resident disparities [2]. For this if we want to retrieve the database, a database that showed better results when it was being compared with one of the well-known prototypical when expending with some point grounded feature mining technique. This concept proved that drug is actually showing very less mistakes in iriS when allied with pupil. Tropic amide is one of the things that is used to expand pupil, hold on some muscles for treating sickness and doing assessments. Meanwhile some revisions were led to see the consequence of eye dewdrops as well, by which exploring we were able to know the upshot of it in a particular eye disease. Then pupil width was


S. Smys et al. (Eds.): ICCVBIC 2019, AISC 1108, pp. 656–661, 2020.
https://doi.org/10.1007/978-3-030-37218-7_74

restrained by some tools for scrutiny. It gave satisfactory results by applying it in different proportions. By trying out all the possible combinations with this we came to know that lower density of will probably solve our problem [3]. Certain combinations of drops were medications were actually combined to know which gave the best upshot out of them and whether the dewdrops whichever we are using would use in one, both eye for improved outcomes [5]. By taking an example of a actual acquainted disorder in patient captivating antipsychotics for handling schizophrenia. Disorder connected to antipsychotic medications seems fewer, though expending certain medications is conjoint for treating schizophrenia [6]. For surgeries, mixture of numerous agents works better when compared to the usage of a single agent [7]. For this we had an analysis of knowing the outcome of some of the agents on eye surgeries [8]. The risk of certain disease were still observed even after giving a proper treatment by actually working on almost all the possibilities that are actually involved [9]. Application of a gel and its side possessions were keenly observed for the therapy about how much to apply and in what frequency and all for working with singly fixed and widened pupil [10].

### 1.2 Problems Related to Distance of IriS Capturing Device in IriS Capture

Innovative algorithms in the treatment of imperfect images can be solved by applying most known techniques to the problems that are being faced by the people in the industry that are related to the distance of the person from camera [11] Some distance criteria was taken into account which actually worked very well with the population [12] with cooperative environment under visible wavelength illumination. IR system can be constructed using standard equipment, and the enactment of such a system would depend on the nature of the images acquired and about the image quality, the light level turned out to be the best image quality factor followed by focus, reflections, disturbances and level of occlusion and hindrance [13]. CLAHE technique is proven to be the finest image enhancement technique related to two extra techniques by improving the localization accuracy at 7% from the novel image [14]. Incrementing the distance beyond certain range had no impact on performance, components of hardware were used in such a way that even though the entity is far away from the capturing equipment, results should be good if the entity is actually obliging and following the instructions exactly without any instability [15]. For presentation, presentation attack detection which covers different categories for IriS recognition is one of the matters that is being tackled now a days [16], attacks on the lens of different types were observed and came up with the general understanding that it may affect the objective of our system [17]. Brief analysis were conducted on persons who were wearing the lens and they were compared with the people who were not wearing the lens as such [18], further to add on the actual scenario of the lens were tried to analyze with the 3D context, this actually led to the concept of understanding the surface characteristics of the lens in deeper and this became the pioneer in analysis of lens pertaining to shape [19]. Iris was captured at some pre-defined distance with some sources by applying "Direct Least Square Fitting of an Ellipse" algorithm. Evolving dare intended for upcoming devices is that it should be able to provide the IriS pattern in any of the sources without abolishing any contents in the images as such [21].

### 1.3 Complications Associated to Pupil Scope on IriS Recognition

Suppose we want to expand the pupil in an automatic manner requires a deep knowledge of knowing all the characteristics of the eye in depth along with knowing all the processing practices in prior [22]. When we are just discussing about the procedures that may directly affect the iris that is its part that may be pupil it actually depends on plenty of influences and also the upshot on the pupillary area based upon the incremental changes in the environment [23] and pupil control comprises brain regions believed to influence the size of the pupil [24]. Pupil dilation is considered as way of measuring the mental capability of a person [25]. According to the observations made pupil will not only expand, contract only for light upshots but also for the emotional disturbances that occurs in the minds of the individual, so the emotional perspectives also should be hooked on the account [26]. By examining this kind of the pupil expansion using certain biometric device we were able to know in depth about the various features of it that needs to be added as a part of our evaluation study for better outcome [27]. After analyzing all the above possibilities the we had to inculcate the positive and negative upshots of factors on this study and how much they will affect on the patient and what might be their severity in this context [28]. When we tried to analyze pupil activity with respect to the web page the it worked out in some good way but not exactly in the way that we wanted it to that is we wanted to reflect, detect emotions as well by studying or focusing on various actions by the user [29]. Medical field is growing very quick now a days and as all of us know our concept of recognition of iris also plays a role in imaging for thinking as the brain activity. Concurrent actions are observed for a process for a particular and which was further used to process the information. this information was additionally made use of the studies that needs to be carried out [30].

### 1.4 Problems Related to Age Factor on IriS Recognition

It is observed that eyelid drooping increases as age increases [PH]. Increased occlusions in IriS images with the passage of time. Less IriS area proposes fewer bits available for matching, which might yield large comparison score. Consequently, eyelid droop can potentially donate to template ageing [31]. Definite portions of the lens were examined and their outcome was studied with respect to the aged people [32]. We can know the age related problems on pupil expansion by actually considering the subject details at the time of enrolling and verifying [33]. Values need to be set for actually studying the values for accurateness of definition of concept based on the texture [34]. Later on while studying the pupil interaction we dealt with different writings which actually stated that earlier to old age there was not an issue but after like fifty plus of age the effect was showing a negative graph between light and eye part, so we started to take this also into consideration while making the study on light properties [36]. We have many factors to be dealt with while we are in a certain study of concepts so in processing we come up with many of these techniques one of it is a segmentation after doing this actually what happened was the dilation of pupil receptive came down as person crosses a particular time in his life [37]. If we take sensor age into account the it also influences the performance of the system in practical orientation of the application which along with it reflects on the accuracy also. We also observed its nature with respect to the noise and thereby we came to the conclusion that physical condition that the sensor is having also matters a lot [38]. When we studied the above scenarios then we found out that luminance level decreases with time when we compare it with the younger subjects and is not at all concerned with the gender. defects that are found in the pixel density is directly proportional to the images apprehended which can also reveal the capturing date of imager in image forensics [39].

## 2 Conclusion

Herein we have gone through briefing of innumerable factors that might be considered for studying IriS recognition or for considering the study of IriS as one the biometric. Here we have gone through the upshot of pupil, since pupil is one of the most important parts of the eye that needs to be considered while studying IriS. The other consideration is the outcome of drugs on pupil enlargement and along with that distance of IriS capturing device also plays a major role in IriS recognition since we must have clear image of the IriS for recognizing the individual based on his IriS patterns and the last one is age factor. We know that the aged people will be having diseases that secondarily upshots the IriS of the eye so we did a brief study on that. By considering all these factors we can frame the objective that needs to be solved by adapting the proper methods for IriS recognition.

**Compliance with Ethical Standards**

✓ All authors declare that there is no conflict of interest.
✓ No humans/animals involved in this research work.
✓ We have used our own data.

## References


1. Tomeo-Reyes, I., Ross, A., Chandran, V.: Investigating the impact of drug induced pupil dilation on automated IriS recognition. In: 2016 IEEE 8th International Conference on Biometrics Theory, Applications and Systems, BTAS 2016 (2016)
2. Ma, L., Tan, T., Wang, Y., Zhang, D.: Efficient IriS recognition by characterizing key local variations. IEEE Trans. Image Process. **13**(6), 739–750 (2004)
3. Shirzadi, K., Amirdehi, A.R., Makateb, A.L.I., Shahraki, K., Khosravifard, K.: Studying the Upshot of Tropicamide Various, vol. 8, no. 2, p. 88889 (2015)
4. Park, J.-H., Lee, Y.-C., Lee, S.-Y.: The comparison of mydriatic upshot between two drugs of different mechanism. Korean J. Ophthalmol. **23**(1), 40–42 (2009)
5. Novitskaya, E.S., Dean, S.J., Manukau, C., Health, D., Moore, J.E., Mcmullen, T.: Upshots of some ophthalmic medications on pupil size: a literature review, June 2014 (2009)

6. Med, S.J., Rep, C., Makino, S.: Intraoperative floppy-IriS syndrome associated with use of antipsychotic drugs. Sch. J. Med. Case Rep. **5**(2), 127–128 (2017)
7. Ozer, P.A., Altiparmak, U.E., Unlu, N., Hazirolan, D.O.: Intraoperative floppy-IriS syndrome: comparison of tamsulosin and drugs other than alpha antagonists, February 2013
8. Chang, D.F., et al.: ASCRS White Paper: clinical review of intraoperative floppy-IriS syndrome. J. Cart. Refract. Surg. **34**(12), 2153–2162 (2008)
9. Casuccio, A., Cillino, G., Pavone, C., Spitale, E.: Pharmacologic pupil dilation as a predictive test for the risk for intraoperative floppy-IriS syndrome. J. Cart. Refract. Surg. **37** (8), 1447–1454 (2011)
10. Gala, P.K., Henretig, F.M., Alpern, E.R, Sampayo, E.M.: An interesting case of a unilaterally dilated pupil an interesting case of a unilaterally dilated pupil, pp. 7–9, December 2017
11. Nguyen, K., Fookes, C., Jillela, R., Sridharan, S., Ross, A.: Long range IriS recognition: a survey. Pattern Recogn. **72**, 123–143 (2017)
12. Dong, W., Sun, Z., Tan, T.: A design of IriS recognition system at a distance, December 2009 (2017)
13. Ranjan, S., Prabu, S., Swarnalatha, P., Magesh, G., Sundararajan, R.: IriS recognition system, January 2018
14. Hassan, R., Kasim, S., Jafery, W.A.Z.W.C., Shah, Z.A.: Image enhancement technique at different distance for IriS recognition. Int. J. Adv. Sci. Eng. Inf. Technol. **7**(4), 1510–1515 (2017)
15. Fancourt, C., et al.: IriS Recognition at a Distance, pp. 1–13 (2005)
16. Czajka, A., Bowyer, K.W.: Assessment of the state of the art ∗, vol. 0, no. 0 (2018)
17. Gupta, P., Behera, S., Vatsa, M., Singh, R.: On IriS Spoofing using Print Attack
18. Baker, S.E., Hentz, A., Bowyer, K.W., Flynn, P.J.: Contact lenses : handle with Carefor IriS recognition. In: 2009 IEEE 3rd International Conference Biometrics Theory, Applications and Systems, pp. 1–8 (2009)
19. Hughes, K., Bowyer, K.W.: Detection of contact-lens-based IriS biometric spoofs using stereo imaging. In: 2013 46th Hawaii International Conference on System Sciences, Ken Hughes School of Engineering and Computer Science, University of the Pacific Kevin W. Bowyer Department of Computer Science and Engineering, University of Notre Dame, pp. 1763–1772 (2013)
20. Du, Y.: Video based non-cooperative IriS segmentation, vol. 6982, pp. 1–10 (2008)
21. Thavalengal, S., Vranceanu, R., Condorovici, R.G., Corcoran, P.: IriS pattern obfuscation in digital images. In: IEEE International Joint Conference on Biometrics, pp. 1–8 (2014)
22. Jomier, J., Rault, E., Aylward, S.R., Hill, C.: Automatic quantification of pupil dilation under stress. In: International Symposium on Biomedical Imaging (ISBI 2004), The University of North Carolina at Chapel Hill, pp. 249–252 (2004)
23. Ekman, I., Poikola, A., Mäkäräinen, M., Takala, T., Hämäläinen, P.: Voluntary pupil size change as control in eyes only interaction, January 2008
24. Johansson, B., Balkenius, C.: A computational model of pupil dilation ∗, October 2017
25. Kosch, T., Hassib, M., Buschek, D., Schmidt, A.: Look into my eyes: using pupil dilation to estimate mental workload for task complexity adaptation
26. Goldinger, S.D., Papesh, M.H.: Pupil dilation reflects the creation and retrieval of memories (2012)
27. Garzón, N.: Upshot of pharmacological pupil dilation on measurements and iol power calculation made using the new swept-source optical coherence tomography-based optical biometer. J. Gynecol. Obstet. Biol. la Reprod. (2016)

28. Thakkar, K.N., Brascamp, J.W., Ghermezi, L., Fifer, K., Schall, D., Park, S.: Reduced pupil dilation during action preparation in schizophrenia. Int. J. Psychophysiol. **128**, 111–118 (2018)
29. Loyola, P., Martinez, G., Muñoz, K., Velásquez, J.D., Maldonado, P., Couve, A.: Neurocomputing Combining eye tracking and pupillary dilation analysis to identify Website Key Objects. Neurocomputing **168**, 179–189 (2015)
30. Siegle, G.J., Steinhauer, S.R., Stenger, V.A., Konecky, R., Carter, C.S.: Siegle et al, Pupil dilation and fMRI, pp. 1–14 (2003)
31. Browning, K.: Biometric Aging Author (2014)
32. Atchison, D.A., Markwell, E.L., Pope, J.M., Swann, P.G.: Age-related changes in optical and biometric characteristics of emmetropic eyes. J. Vis. **8**, 1–20 (2008)
33. Ortiz, E., Bowyer, K.W., Flynn, P.J., Hall, F., Dame, N.: A linear regression analysis of the upshots of age related pupil dilation change in IriS biometrics. In: 2013 IEEE Sixth International Conference on Biometrics: Theory, Applications, pp. 1–6 (2013)
34. Sgroi, A., Bowyer, K.W., Flynn, P.J.: The prediction of old and young subjects from IriS texture, pp. 1–5
35. Mehrotra, H., Vatsa, M., Singh, R., Majhi, B.: Does IriS Change Over Time? April 2016 (2013)
36. Telek, H.H.: The upshots of age on pupil diameter at different light amplitudes, vol. 3, no. 2, pp. 80–85 (2018)
37. Abidin, Z.Z., Manaf, M., Shibghatullah, A.S, Anawar, S.: Iris recognition failure in biometrics : a review, July 2015 (2013)
38. Bergm, T.: Impact of sensor ageing on IriS recognition
39. Fridrich, J.: Sensor defects in digital image forensic. In: Sencar, H.T., Memon, N. (eds.) Digital Image Forensics, pp. 179–218. Springer, New York (2013)

# Performance Analysis of ICA with GMM and HMM for Epilepsy Classification

Harikumar Rajaguru(✉) and Sunil Kumar Prabhakar

Department of ECE, Bannari Amman Institute of Technology, Coimbatore, India
harikumarrajaguru@gmail.com

**Abstract.** The most frequent neurological disorder that is common among most of the individuals is the Epilepsy. The characteristic feature of epilepsy is the frequent occurrence of seizures due to the undesirable and abnormal neurons firing in the different brain regions. Electroencephalography (EEG) is a powerful tool which helps us to analyze a lot of neurological disorders. These signals contain a vital information about the epileptic seizure detection. For analyzing various physiological states of the brain, EEG signals are highly utilized in both medical literature and academics. The EEG recordings measured are generally very long and establishing it for analysis is quite a tough task to peform. In this work, Independent Component Analysis (ICA) is utilized effectively as a technique for the dimensionality-reduction and classified with two different models namely, Hidden Markov Model (HMM) and Gaussian Mixture Model (GMM) and then the performance is analyzed. The results show that ICA is classification based on the HMM and GMM shows an appropriate classification accuracy of 87.65% and 93.75% respectively.

**Keywords:** Epilepsy · EEG · HMM · GMM

## 1 Introduction

Ever since ancient times, one of the common neurological disorders present with humans is epilepsy [1]. The brain is targeted by epilepsy and because of this, there is a temporal change in the electrical conditions of the brain. Manifestations such as psychic, motor and sensorial are more prevalent and is commonly associated with spasms. To trace the various states of a seizure, a trained specialist is generally required to scan the EEG chart visually [2]. Monitoring such activities in real time without disturbing the patient from doing everyday tasks is quite a technological challenge. For the seizure prediction, detection and classification activities, the analysis of EEG signals is very important [3]. The generated electrical signals by the brain are recorded and analyzed by the EEG. For monitoring and diagnosing the epilepsy related neurological disorders, it is widely used. The epileptic EEG signals which are secured from the scalp are witnessed by a high-amplitude and periodic waveforms which are synchronized in nature [4]. A lot of sharp and spike waves are often observed in between the seizures. By the visual screening and analysis of the EEG recordings, the classification of the activities is quite difficult due to the lack of highly trained doctors and consumption of more time [5]. Because of this, it takes a longer time to diagnose and therefore the risk


S. Smys et al. (Eds.): ICCVBIC 2019, AISC 1108, pp. 662–669, 2020.
https://doi.org/10.1007/978-3-030-37218-7_75

in the patient increases. Medical expenditure too is a big issue in such cases and this promoted the need for the automated seizure detection and classification procedures. A variety of works has been done in the past with respect to the EEG signal processing for epilepsy classification. A application implementing telemedicine scheme for epilepsy classification from EEG was developed by Prabhakar and Rajaguru [6]. The seizure onset was analyzed on the basis of wideband EEG recordings by Bragin et al. [7]. Various types of DA for epilepsy classification was done by Prabhakar and Rajaguru and it was implemented for telemedicine application successfully [8]. For the interictal EEG recordings, the pattern extraction was analyzed towards the detection of electrodes leading to seizures by Cabrerizo et al. [9]. Singular Value Decomposition and Nonlinear regression based on expectation maximization was also performed by Prabhakar and Rajaguru for epilepsy classification in [10]. The detection of seizure onset based on the exact time was done automatically by Chan et al. in intracranial EEG signals [11]. The Expectation Maximization along with its modification was implemented for the fuzzy output optimization for epilepsy analysis in [12]. The chaotic time series analysis of epileptic seizures was performed by Frank et al. [13]. The K nearest neighbor based Adaboost Classifier along with PSD was implemented for epilepsy analysis in [14]. A hybrid classification model with Minimum Relative entropy analysis for epilepsy classification was performed in [15]. In this paper, ICA is used as a method for the DR and it is classified with the help of GMM and HMM models for epilepsy classification. The nature of the paper is as follows. In Sect. 2, the materials and methods are detailed. The Mixture Models in Sect. 3. Section 4 gives the results and discussion followed by conclusion in Sect. 6. The elucidation of the work is described in Fig. 1.

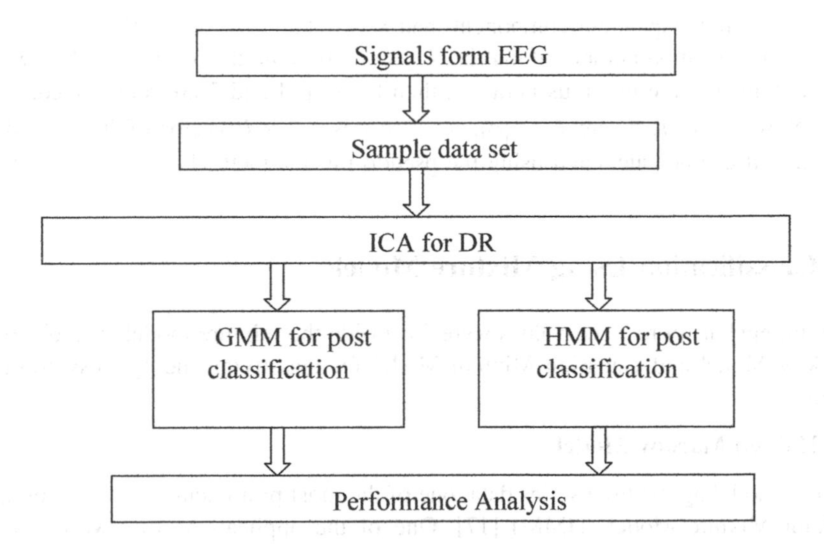


**Fig. 1.** Elucidation of the work

## 2 Materials and Methods

The EEG data from total of twenty epileptic patients was attained in European Data Format (EDF) with the support of many neurologists from Sri Ramakrishna Hospital, Coimbatore. The recordings were attained for different stages like eyes closed and opened, muscle movement, eye movement, hyperventilation state etc. using the ten to twenty international system and the electrodes of sixteen channel. The recordings were done for more than 55 min and so for storage and mathematical analysis purpose, the recordings of the sixteen channels were divided into epochs for convenience. Therefore every channel had three epochs and for every epoch there were four hundred values. For all epochs and channels of the 20 epileptic patients, the samples obtained was too large and it was difficult to process and so ICA is used as a method of dimensionality reduction to eliminate the redundant and unwanted features.

### A. Independent Component Analysis

ICA is utilized efficiently to trace independent components in an intermixed multicarrier noisy signal [16]. The algorithm is stated as follows:

(1) Assume $W$ is a $(A+1)(B+1) \times Q$ observation data matrix. The data matrix $W$ is centered around its mean value $W \leftarrow W - E\{W\}$
(2) The data matrix $W$ is whitened as $V = S\Lambda^{-1/2}S^{H}W$, where $S\Lambda S^{H} = E\{WW^{H}\}$
(3) The random matrix '$k$' is initialized such that $\|k\| = 1$
(4) The '$k$' is updated as
    (a) $k \leftarrow E\{Vf(k^{H}V)\} - E\{f'(k^{H}V)\}k$, where $f(r) = r^{3}$
    (b) $k \leftarrow k/\|k\|$
(5) Now the independent component matrix is obtained as $M = k^{H}V$
(6) If the condition number of matrix $M \leq$ condition number of matrix $M_{prev}$ which is obtained in the previous iteration, then the step 4 and 5 are represented.
(7) Now $\vec{r} = \vec{C}M$, where $\vec{C} = [c_0, c_1, \ldots, c_p]$ is a $1 \times P$ vector of ICA coefficients and it can be calculated using the pseudo-inverse method.

## 3 Classification Using Mixture Models

The dimensionally reduced values were fed inside the mixture models namely Hidden Markov Model and Gaussian Mixture Model for classifying the epilepsy from EEG signals.

### A. Hidden Markov Model

For modeling the time series data, one of the most prominent techniques utilized is Hidden Mixture Model (HMM) [17]. One of the applications of HMM has to be implemented in various domains like speech recognition, computer networks, theoretical physics, computational biology etc. The distribution probability over specific sequence of actions can be represented by this sample dynamic Bayesian network. A particular order of observations $\{q_z\}$ and a particular order of hidden states $\{m_z\}$ is denoted where $z = 1, \ldots, Z$. Two finite sets having conditional independence relations are assumed here. Firstly, if $m_t$ is given, then the observation $q_t$ is independent of all other states and observations. Secondly, the state $m_t$ depends on only $m_{t-1}$, implying that the first order Markov property should be satisfied by the states. These two observations are obtained from these conditional independence relations. The factorization of the joint probability distribution of the observations and states are done as

$$W(m_1..m_T, q_1..q_T) = W(m_1)W(q_1|m_1)\prod_{t=2}^{T} W(m_t|m_{t-1})W(q_t|m_t)$$

The main assumption in HMM is that the hidden state variables are either continuous-valued or discrete-valued, i.e., $m_z \in \{1, .., L\}$. The state vector representation $m_z$ is quite a simple $K$-dimensional vector and here just a single element is having unity and all the rest of the elements are zero. For the elements of the state vector to be unity stage, is should depend on the active levels of the state value. Then $W(m_t|m_{t-1})$ is easily represented by a $L \times L$ state transition matrix indicated by $\Phi$. For the initial state probability $W(m_1)$ is a $K$ dimensional vector and that is denoted by $\pi$. Either a real valued observation or discrete valued observation is allowed by the HMM model. The continuous HMM reflects the real-valued observations and the discrete HMM reflects the discrete-valued observations. Here only a continuous representation of HMM is considered due to the real-valued nature of the EEG. $W(q_t|m_t)$ can be simply framed in various formats such as a mixture of Gaussian, or a simple Gaussian or a neural network for real-valued observation vectors.

The HMM learning can be done in 2 steps, namely inference step and learning step. In the inference step, the calculation of the posterior distribution over the hidden state is performed. In the learning step, the identification of parameters like initial state probability, probability of emission and probability of state transition are traced. To infer the posterior over the hidden state in an efficient manner, the famous forward-backward recursion is utilized.

### B. Gaussian Mixture Model

For GMM, the conditional way of distribution is expressed as follows

$$p(A|Q, \mu, \Sigma) = \prod_{k=1}^{K} N(A|\mu_k, \Sigma_k)^{Q_k}$$

The perceived variable term here is indicated as $A = \{a_1, \ldots, a_N\} \in \Re^{N \times D}$. The assignment variable term of the cluster is indicated as $Q = \{q_1, \ldots, q_N\} \in \{0, 1\}^{N \times K}$. Here $q_n$ is denoted as a 1-of-$K$ binary vector representation. The measurable factor considered here are $\mu = \{\mu_k\}$ and $\Sigma = \{\Sigma_k\}$. A diagonal covariance assumption is considered here as $\Sigma_k = \sigma_k^2 I$. The multivariate Gaussian distribution is expressed as follows

$$N(A|\mu, \Sigma) = \prod_{d=1}^{D} N\left(a_d|\mu_d, \sigma_d^2\right)$$

The conditional distribution can be explained under this assumption as a simple product of $D$ and $K$ highly independent Gaussian univariates [18] as

$$p(A|Q, \mu, \Sigma) = \prod_{d=1}^{D}\prod_{k=1}^{K} N\left(a_d|\mu_{dk}, \sigma_{dk}^2\right)^{Q_k}$$

GMM is further restricted to only a unique parameter representation. It is done with the help of employing a particular situation of sharing the diagonal covariance, $\Sigma_k = I$, i.e., $\sigma_k = 1$.

Now the conditional standard distribution of GMM is expressed as follows

$$p(A|Q, \mu) = \prod_{d=1}^{D}\prod_{k=1}^{K} N(a_d|\mu_{dk})^{Q_k}$$

For every dimension $d$ and $k$, by means of a specific univariate Gaussian probabilities, only one GMM parameter $\mu$ has to be learnt. Here a famous iterative scheme for learning GMM parameter is employed named as the Expectation Maximization (EM) algorithm.

## 4 Results and Discussion

If the ICA is taken as the method of DR and when it is exclusively classified with HMM and GMM models, based on the parameters like Time Delay, Quality Value, Performance Index, Sensitivity, Accuracy and the specificity, the appropriate results are enumerated as shown in Table 1. The mathematical formulae is framed in this regard.

$$PI = \left(\frac{PC - MC - FA}{PC}\right) \times 100$$

Where PC is the perfect classification, MC is the missed classification and the FA is the false alarm.

The Quality Value $Q_V$ is mathematically defined below

$$Q_v = \frac{C}{(R_{fa} + 0.2) * (T_{dly} * P_{dct} + 6 * P_{msd})}$$

**Table 1.** Performance analysis of ICA with mixture models (GMM + HMM)

| Name | ICA + GMM | ICA + HMM |
|---|---|---|
| PC (%) | 87.5 | 75.31 |
| MC (%) | 9.23 | 24.68 |
| FA (%) | 3.26 | 0 |
| PI (%) | 84.32 | 66.61 |
| Specificity (%) | 90.76 | 75.31 |
| Sensitivity (%) | 96.73 | 100 |
| Delay in time (sec) | 2.3 | 2.98 |
| Qv | 20.09 | 16.81 |
| Accuracy (%) | 93.75 | 87.65 |

where C means the constant required in scaling,

$R_{fa}$ denotes the No. of false alarms in a set

$T_{dly}$ means the appropriate delay in classification based on onset (seconds)

$P_{dct}$ specifies the % of PC and

$P_{msd}$ indicates the % of MC

The delay in time is mathematically described below

$$delay\,\mathrm{in}time = 2 * (PC/100) + 6 * (MC/100)$$

## 5 Conclusion

Thus this famous disorder of epilepsy is understood as a common neurological disorder disturbing a lot of human beings and EEG signals show a proven track record for the analysis of this disorder. In this work, ICA is used as a dimensionality reduction method. The classification is based on the HMM and the GMM classifiers. Results show that when ICA is classification based on the GMM and HMM with an average accuracy of 93.75% along with a less time delay of 2.3 s and average quality value of 20.09 and average accuracy of 87.65% along with a less time delay of 2.98 s and HMM with the average quality value of 16.81 respectively Results show that the performance of GMM classifier is far better than HMM classifier when ICA is effectively used as a DR method. Future works tends to analyze the performance of HMM and GMM classifiers with various dimensionality reduction and feature extraction techniques.

**Compliance with Ethical Standards**

✓ All authors declare that there is no conflict of interest.

✓ No humans/animals involved in this research work.

✓ We have used our own data.

## References

1. Prabhakar, S.K., Rajaguru, H.: Code converters with city block distance measures for classifying epilepsy from EEG signals. In: Procedia Comput. Sci. **87**, 5–11 (2016). Fourth International Conference on Recent Trends in Computer Science & Engineering, Chennai, India, 29–30 April 2016
2. Prabhakar, S.K., Rajaguru, H.: Expectation maximization based PCA and Hessian LLE with suitable post classifiers for epilepsy classification from EEG signals. In: Proceedings of the 8th International Conference on Soft Computing and Pattern Recognition (SoCPaR), VIT University, Vellore, India, December 2016 (2016)
3. Prabhakar, S.K., Rajaguru, H.: Performance analysis of ApEn as a feature extraction technique and time delay neural networks, multi layer perceptron as post classifiers for the classification of epilepsy risk levels from EEG signals. In: Proceedings of ICC3 2015, Computational Intelligence, Cyber Security and Computational Models. Advances in Intelligent Systems and Computing, Series, PSG College of Technology, Coimbatore, India, 17–19 December 2015, vol. 412, pp. 89–97. Springer (2015)
4. Prabhakar, S.K., Rajaguru, H.: Particle swarm based sparse representation classifier for classification of epilepsy from EEG signals. In: 14th Electrical Engineering/Electronics, Computer, Telecommunications, and Information Technology Conference (ECTICON 2017) held at Phuket, Thailand, 27–30 June, 2017 (2017)
5. Prabhakar, S.K., Rajaguru, H.: Conceptual analysis of epilepsy classification using probabilistic mixture models. In: 5th IEEE Winter International Conference on Brain-Computer Interface, South Korea, 9–11 January, 2017 (2017)
6. Prabhakar, S.K., Rajaguru, H.: Development of patient remote monitoring system for epilepsy classification. In: 16th International Conference on Biomedical Engineering (ICBME), Singapore, 7–10 December, 2016 (2016)
7. Bragin, A., Wilson, C.L., Fields, T., et al.: Analysis of seizure onset on the basis of wideband EEG recordings. Epilepsia **46**, 59–63 (2005)
8. Prabhakar, S.K., Rajaguru, H.: Epilepsy classification using discriminant analysis and implementation with space time trellis coded MIMO-OFDM system for telemedicine applications. In: IFBME Proceedings, 6th International Conference on the Development of Biomedical Engineering, Ho Chi Minh City, Vietnam, pp. 479–483. Springer (2017)
9. Cabrerizo, M., Adjouadi, M., Ayala, M., Tito, M.: Pattern extraction in interictal EEG recordings towards detection of electrodes leading to seizures. Biomed. Sci. Instrum. **42**, 243–248 (2006)
10. Prabhakar, S.K., Rajaguru, H.: EM based non-linear regression and singular value decomposition for epilepsy classification. In: 6th IEEE ICT International Student Project Conference 2017 (ICT-ISPC), Universiti Teknologi Malaysia, Johor Bahru, Malaysia, 23–24 May 2017 (2017)
11. Chan, A.M., Sun, F.T., Boto, E.H., Wingeier, B.M.: Automated seizure onset detection for accurate onset time determination in intracranial EEG. Clin. Neurophysiol. **119**, 2687–2696 (2008)
12. Prabhakar, S.K., Rajaguru, H.: Comparison of fuzzy output optimization with expectation maximization algorithm and its modification for epilepsy classification. In: International Conference on Cognition and Recognition (ICCR 2016), Mysore, India, 30–31 December 2016 (2016)
13. Frank, G.W., Lookman, T., Nerenberg, M.A.H., et al.: Chaotic time series analyses of epileptic seizures. Phys. D. **46**, 427–438 (1990)

14. Rajaguru, H., Prabhakar, S.K.: Power spectral density and KNN based Adaboost classifier for epilepsy classification. In: IEEE Proceedings of the International Conference on Electronics, Communication and Aerospace Technology (ICECA 2017), Coimbatore, India, pp. 441–445 (2017)
15. Rajaguru, H., Prabhakar, S.K.: A hybrid classification model using artificial bee colony with particle swarm optimization and minimum relative entropy as post classifier for epilepsy classification. In: International Conference on Computational Vision and Bioinspired Computing (ICCVBIC 2017). Springer Lecture Notes on Computational Vision and Biomechanics, Coimbatore, India (2017)
16. Jung, T., Makeig, S., Humphries, C., Lee, T., Mckeown, M., Iragui, V., Sejnowski, T.J.: Removing electroencephalographic artifacts by blind source separation. Psychophysiology **37**, 163–178 (2000)
17. Bruckner, D., Velik, R.: Behavior learning in dwelling environments with hidden markov models. IEEE Trans. Ind. Electron. **57**(11), 3653–3660 (2010)
18. Lim, K.-L., Wang, H.: Learning Gaussian mixture model with a maximization-maximization algorithm for image classification. In: 12th IEEE International Conference on Control and Automation (2016)

# Evaluating the Performance of Various Types of Neural Network by Varying Number of Epochs for Identifying Human Emotion

R. Sofia(✉) and D. Sivakumar

Department of Electronics and Instrumentation Engineering, Annamalai University, Chidambaram, India
sofiame25988@gmail.com

**Abstract.** Epochs represents the number of times a network gets trained with a particular database. In this scenario, when we go for more number of epochs there is a possibility that neural network will memorize those values like a human brain, which should not be the case for an efficient target achievement. Here the emotions identified are based on varying number of epochs by using three types of network they are FFNN (Feedforward neural Network), BRRNN (Bayesian Regularized Recurrent Neural Network) and ANFIS (Adaptive Neuro Fuzzy Inference System) and with the help of these networks six basic emotions has been identified they are happy, sad, angry, fear, disgust and surprise. The network which achieves maximum accuracy with less number of epochs has been analysed based on the performance evaluation of Error Histogram, Error plot, Confusion Matrix, Regression Plot, Mean Absolute Error, Accuracy, Sensitivity, Specificity, Precision, Target Vs Output Plot.

**Keywords:** ANFIS · BRRNN · FFNN · Epochs · Human emotion

## 1 Introduction

As shown in above Fig. 1, emotion of the human can be measured using two methods they are: **Indirect Method** it is based on observer judgments and communication approaches, **Direct Method** used in recent days for identifying human emotion are LBP (Local Binary Pattern), PCA (Principle component Analysis), AU (Action Unit), ASM (Active Shape Model), Gabor filter, ROI (Region of Interest) localization method and their disadvantage as shown in the figure above.

In paper [1] finding out the facial expression based on matching with the Morphable Model, i.e., finding out the pose, shape and expression and making that fit with the Morphable model and finding out the human emotion, and they say that Anger, surprise are less impacted by face registration and system gets difficulty in differentiating between the disgust, fear with anger, surprise [2]. A descriptor is set based on areas and angles of triangles formed by the landmarks from face images. And then these descriptor are used for facial expression classification with conditional Random field and citation KNN classifier [3]. Uses LBP feature extraction technique (uses only


S. Smys et al. (Eds.): ICCVBIC 2019, AISC 1108, pp. 670–678, 2020.
https://doi.org/10.1007/978-3-030-37218-7_76

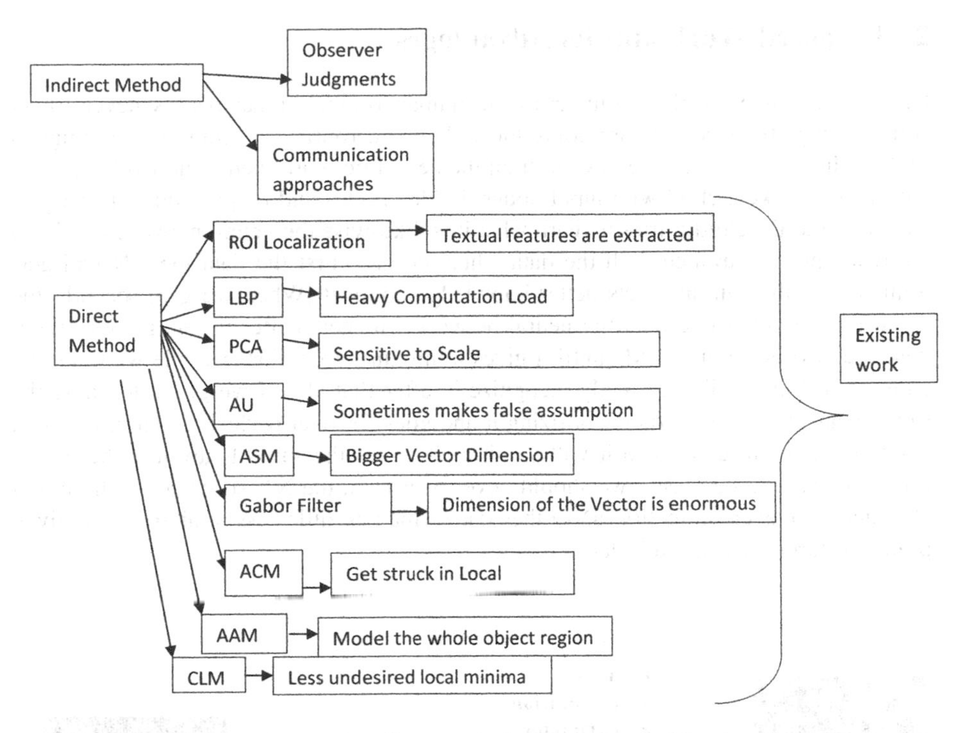


**Fig. 1.** Direct and Indirect method of detecting facial expression

the informative region of face) and classify the human emotion using the SVM Classifier. Here the neutral expression is taken as base and from the neutral facial features how much the facial features of other expression differ is calculated [4]. Gabor filter based feature extraction and classified using Neural Network and the facial attributes are extracted using PCA [5]. LBP is used for Facial Feature extraction and the neural network is used as the classifier [6]. Selected face regions such as mouth, eyebrows and eyes are extracted by using region detection algorithm and classify the expression with Hidden Markov Model (HMM) [7]. Uses various feature extraction techniques such as ACM (Active Contour Model), ASM (Active Shape Model), AAM (Active Appearance Model), CLM (Constrained Local Model) and compares its advantages and disadvantages [8]. Human Emotion has been classified with the help of eyes and mouth using susan edges [9]. Uses Gabor filter for identifying emotion [10]. Appearance based model of eyes and mouth for identifying emotion [11]. Euclidean and Mahattan distance measure for facial expression classification [12]. Full face is considered and based on the Action Unit the emotions are identified. [13–21, 25] are survey paper about various techniques of extracting features and emotion identification using various classifier. [22, 23] are books which tell about the general emotions of human emotions (Fig. 2).

## 2 Proposed Work and Its Advantages

Because neural network is similar to the human brain, human beings never forget certain things because they are doing those things in routine or seeing those things in routine. In the same way we have to train the neural network, neural network what we create first is like a child with input nodes, hidden nodes and output nodes. If we give something to the child it tries to learn what it is, like-wise the neural network will learn the data and gets trained with the data when we give first the data the NN will gets trained with the data and gets better knowledge about it. When we give second time again to view the same data the neural network still gets better in recognition. Like-wise we have to train the NN until it gives minimum error. For example we teach the child something until it correctly recognize it, after that also if we go on teaching the same means the child starts memorizing it and goes for over perfect recognition, when a small change in color also it will say it as it is not the same I saw it is the wrong object. For the same reason we should never over train the NN so what it will do is it also goes for over perfection and other data which is little less similar to the given preferred target will get rejected.

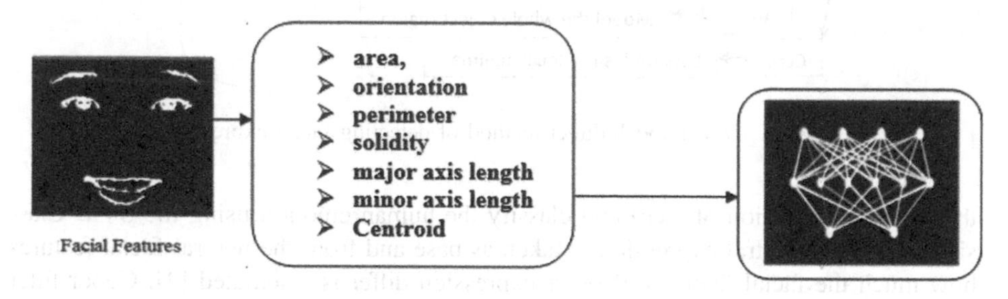


**Fig. 2.** Methodology [2]

We have to give the number of epochs in perfect manner so that we should not do under fitting or over fitting the data. If we under fit the data the neural network will not get trained with the database properly to produce the desired target and if we try to over fit the data the NN will start to memorize the data. So in order to have the perfect NN we have to try giving the number of epochs from minimum in such a way that, where the error is also getting minimum error.

## 3 Performance Evaluation

Here the performance has been evaluated using Error Histogram, Confusion Matrix, Error Plot, Regression Plot, Mean Absolute Error, Sensitivity, Specificity, Accuracy, Precision and Target Vs Output Plot. For confusion matrix it is referred as Happy (A), sad (B), angry (C), fear (D), disgust (E) and surprise (F) (Table 1).

**Table 1.** Performance evaluation of the networks when the epochs = 3

| **Neural Network** | **FFNN** | **BRRNN** | **ANFIS** |
|---|---|---|---|
| Epochs | 3 | 3 | 3 |
| Time consumption for Training and recognition | 10 Seconds | 12 Seconds | 6 Seconds |
| Efficiency of recognition | None of expressions are identified correctly | Only one expression has been identified correctly | Most of the given expressions are identified correctly |
| Error Histogram | Here the maximum error falls between -0.30 to -0.15. Error Span is from -3 to +3. | Here the maximum error falls between -0.375 to -0.275. Error Span is from -2 to 3. | maximum errors fall between -0.1 and-0.03. |
| **Confusion Matrix** | Confusion matrix with the diagonal element of 1 indicates perfect recognition. But here we can see that none of the expressions has been identified correctly. | Here in this matrix only the expression D has been identified properly with 100% efficiency. Expression A,B,E,F has not done been identified and the expression C has been identified upto 23.5% | Here the diagonal element 1 indicate perfect recognition all the 6 expressions. |
| **Error plot** | Here the error span is from -3 to +3 | Here the error span is from -2 to 3 | Here the horizontal line indicates that the system works without error |

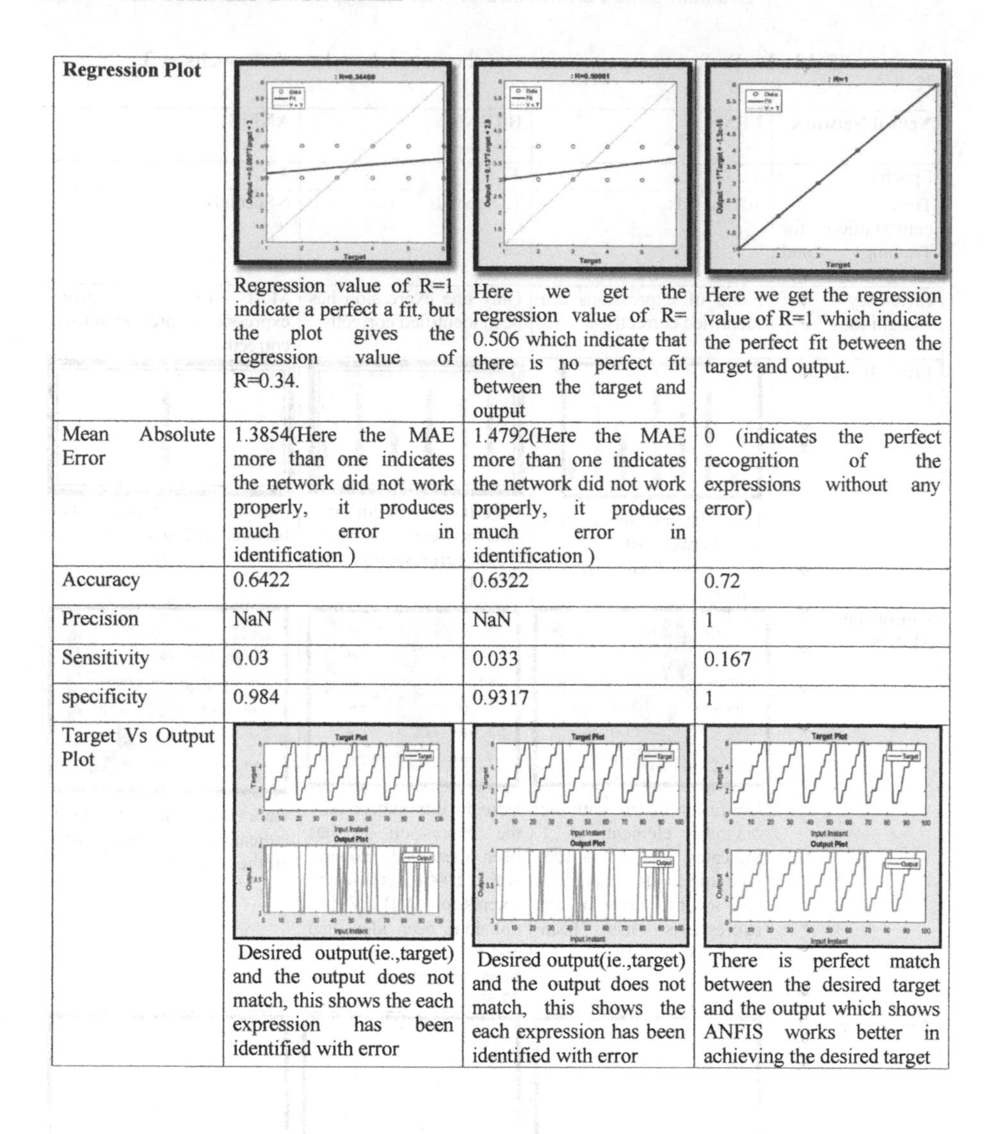

| **Regression Plot** | Regression value of R=1 indicate a perfect a fit, but the plot gives the regression value of R=0.34. | Here we get the regression value of R= 0.506 which indicate that there is no perfect fit between the target and output | Here we get the regression value of R=1 which indicate the perfect fit between the target and output. |
|---|---|---|---|
| Mean Absolute Error | 1.3854(Here the MAE more than one indicates the network did not work properly, it produces much error in identification ) | 1.4792(Here the MAE more than one indicates the network did not work properly, it produces much error in identification ) | 0 (indicates the perfect recognition of the expressions without any error) |
| Accuracy | 0.6422 | 0.6322 | 0.72 |
| Precision | NaN | NaN | 1 |
| Sensitivity | 0.03 | 0.033 | 0.167 |
| specificity | 0.984 | 0.9317 | 1 |
| Target Vs Output Plot | Desired output(ie.,target) and the output does not match, this shows the each expression has been identified with error | Desired output(ie.,target) and the output does not match, this shows the each expression has been identified with error | There is perfect match between the desired target and the output which shows ANFIS works better in achieving the desired target |

Since the ANFIS performance is good with only 3 epochs, so the training for ANFIS has been stopped with only 3, and as the neural network did not perform well for the 3 epochs so its epochs has been increased to 100. And also if we see in the table below the time consumption also gets increases with the number of epochs and also the recognition rate is also not achieved, so for this methodology of identifying the facial expression the ANFIS works better (Table 2).

**Table 2.** Performance evaluation of the networks when the epochs = 100

| **Neural network** | **FFNN** | **BRRNN** | **ANFIS** |
|---|---|---|---|
| Epochs | 100 | 100 | 3 |
| Time consumption for Training and recognition | 43 seconds | 98.15 seconds | 6 seconds |
| Efficiency of recognition | Low recognition rate | Low Recognition rate | Most of the given expressions are identified correctly |
| Error Histogram | 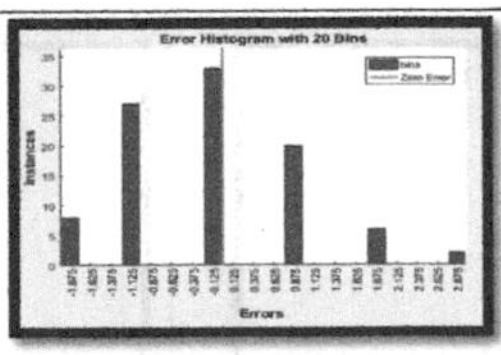 <br>Here the maximum error falls between -0.375 to -0.125. Error Span is from -2 to +? | 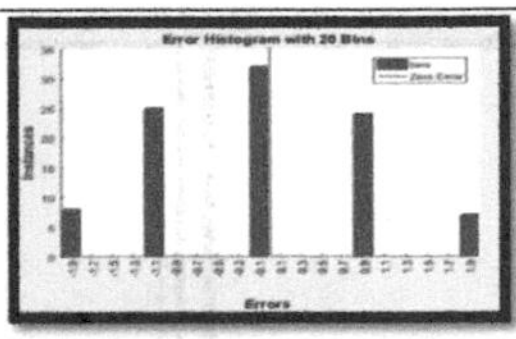 <br>Here the maximum error falls between -0.3 to -0.1. Error Span is from 2 to 13. | 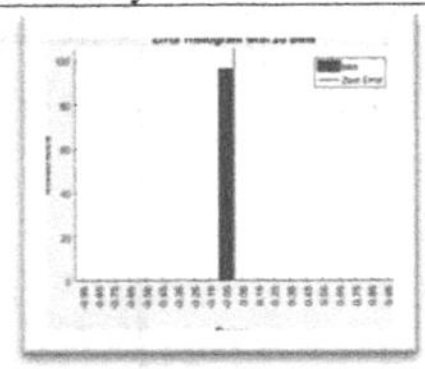 Maximum errors fall botween -0.1 and-0.03. |
| **Confusion Matrix** | 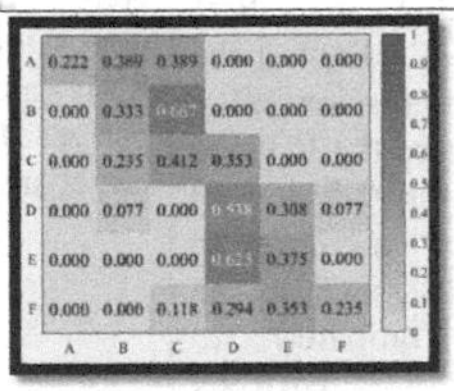 <br>Confusion matrix with the diagonal element of 1 indicates perfect recognition. But here we can see that none of the expressions has been identified correctly Expression A has been identified upto 22%, and similarly Expression B has been identified upto 33%. But when we increase the number of epochs all the expressions has been identified upto certain percentage not like the confusion matrix in above table for the epochs 3. | 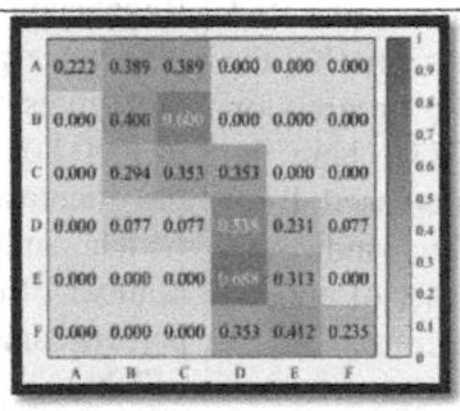 <br>Confusion matrix with the diagonal element of 1 indicates perfect recognition. But here we can see that none of the expressions has been identified correctly Expression A has been identified upto 22%, and similarly Expression B has been identified upto 40%. But when we increase the number of epochs all the expressions has been identified upto certain percentage not like the confusion matrix in above table for the epochs 3. | 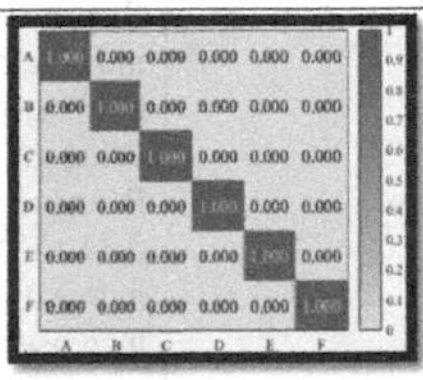 <br>Diagonal element of 1 indicate perfect identification all the expression |

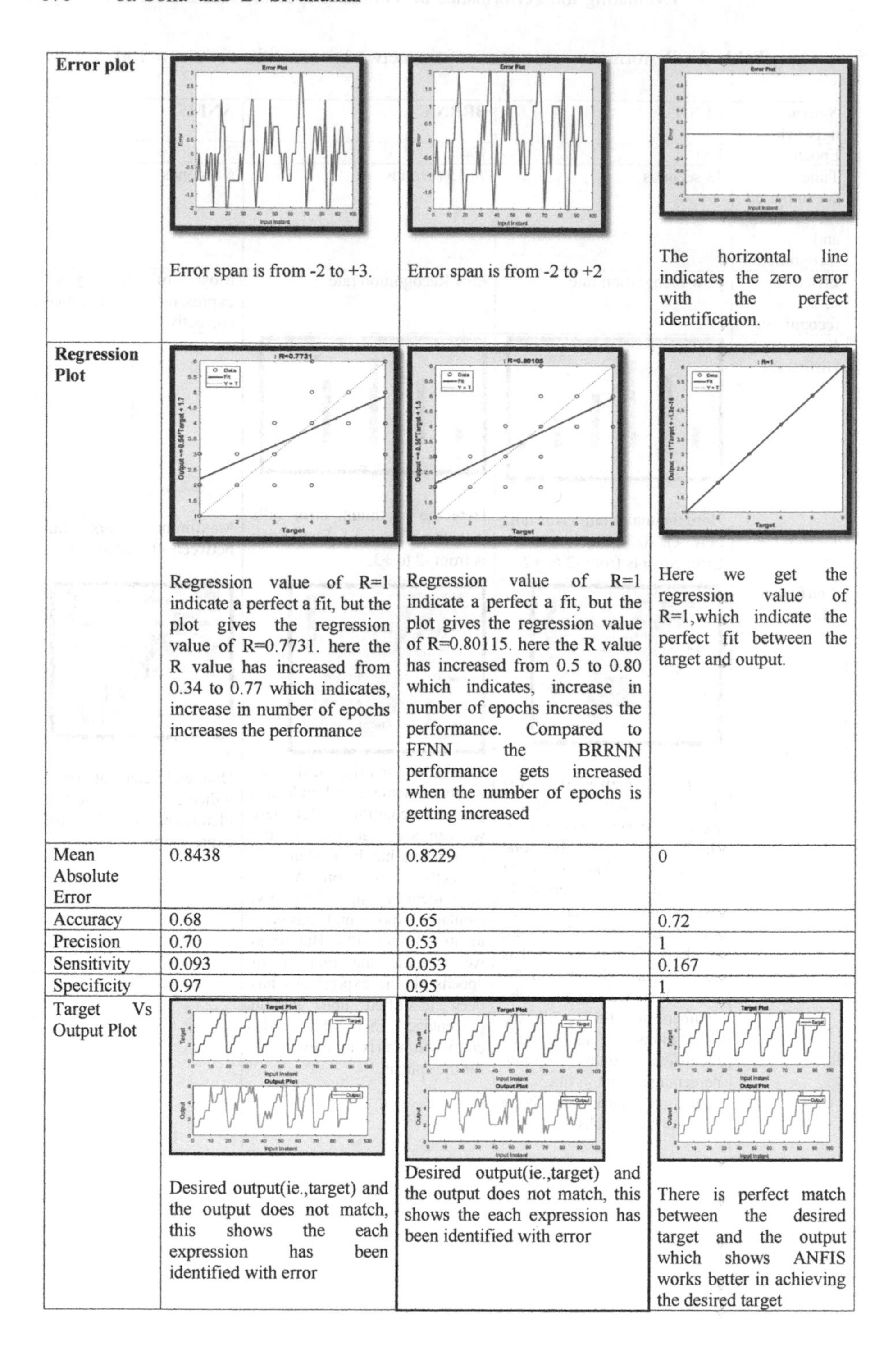

| | | | |
|---|---|---|---|
| **Error plot** | Error span is from -2 to +3. | Error span is from -2 to +2 | The horizontal line indicates the zero error with the perfect identification. |
| **Regression Plot** | Regression value of R=1 indicate a perfect a fit, but the plot gives the regression value of R=0.7731. here the R value has increased from 0.34 to 0.77 which indicates, increase in number of epochs increases the performance | Regression value of R=1 indicate a perfect a fit, but the plot gives the regression value of R=0.80115. here the R value has increased from 0.5 to 0.80 which indicates, increase in number of epochs increases the performance. Compared to FFNN the BRRNN performance gets increased when the number of epochs is getting increased | Here we get the regression value of R=1,which indicate the perfect fit between the target and output. |
| Mean Absolute Error | 0.8438 | 0.8229 | 0 |
| Accuracy | 0.68 | 0.65 | 0.72 |
| Precision | 0.70 | 0.53 | 1 |
| Sensitivity | 0.093 | 0.053 | 0.167 |
| Specificity | 0.97 | 0.95 | 1 |
| Target Vs Output Plot | Desired output(ie.,target) and the output does not match, this shows the each expression has been identified with error | Desired output(ie.,target) and the output does not match, this shows the each expression has been identified with error | There is perfect match between the desired target and the output which shows ANFIS works better in achieving the desired target |

## 4 Conclusion

A Novel Feature extraction technique has been carried out for identifying the facial expression which in turn identifies the human emotion. Here we have used three types of network by varying the number of epochs, and the result observed is when the number of epochs is kept minimum of 3 the performance of FFNN and BRRNN is poor and whereas the performance of ANFIS is good, but as the ANFIS uses both neural network and Fuzzy logic it may work better with minimum number of epochs, so in order to check the performance of other two network, the number of epochs has been increased to 100 which gives a better result compared to number of epochs 3. When we increase the number of epochs the time consumption has increased to 43 s and for BRRNN 98.15 s but for ANFIS we have achieved the best result with reduced number of epochs in reduced time consumption, so we are concluding that for our methodology ANFIS works best, which will be implemented in future for medical applications.

**Compliance with Ethical Standards**

✓ All authors declare that there is no conflict of interest.
✓ No humans/animals involved in this research work.
✓ We have used our own data.

## References

1. Allaerts, B., Mennesson, J., Bilasco, I.M., Djeraba, C.: Impact of face registration techniques on facial expression recognition. Signal Process. Image Commun. **61**, 44–53 (2018)
2. Acevedo, D., Negri, P., Buemi, M.E., Fernandez, F.G., Mejail, M.: A simple geometric-based descriptor for Facial expression recognition. In: 12th International conference on Automatic Face and Gesture Recognition. IEEE (2017)
3. Kumar, S., Bhuyan, M.K., Chakraborty, B.K.: Extraction of informative regions of a face for facial expression recognition. IET Comput. Vis. (2016). https://doi.org/10.1049/iet-cvi.2015.2073
4. Dagar, D., Hudait, A., Tripathy, H.K., Das, M.N.: Automatic emotion detection model from facial expression. In: International Conference on Advanced Communication Control and Computing Technologies (ICACCCT) (2016)
5. Muttu, Y., Virani, H.G.: Effective face detection, feature extraction and neral network based approaches for facial expression recognition. In: International Conference on Information Procession (ICIP), 16–19 December 2015
6. Boruah, D., Sarma, K.K., Talukdar, A.K.: Different face regions detection based facial expression recognition. In: 2nd International Conference on Signal Procession and Integrated Network (SPIN) (2015)
7. Abouyahya, A., El Fkihi, S., Thami, R.O.H., Aboutadine, D.: Feature extraction for facial expression recognition. IEEE (2016)
8. Dileep, M.R., Danti, A.: Human emotion classification based on eyes and mouth using susan edges. Int. J. Sci. Eng. Res. **7**(7) (2016)
9. Reddy, Ch.S., Srinivas, T.: Improving the classification accuracy of emotion recognition using facial expressions. Int. J. Appl. Eng. Res. **11**(1), 650–655 (2016)

10. Kundu, T., Saravanan, C.: Advancements and recent trends in emotion recognition using facial image analysis and machine learning models. In: International Conference on Electrical, Electronics, Communication, Computer and Optimization Techniques (ICEEC-COT) (2017)
11. Greeche, L., Jazouli, M., Es- Sbai, N., Majda, A., Zarghili, A.: Comparison between euclidean and manhattan distance measure for facial expressions classification. IEEE (2017)
12. Tarnowski, P., Kolodziej, M., Majowski, A., Rak, R.J.: Emotion recognition using facial expressions. In: International Conference on Computational Sciences, ICCS, 12–14 June 2017. Procedia Computer Science 108C (2017)
13. Patel, T., Shah, B.: A survey on facial feature extraction techniques for automatic face annotation. In: International Conference on Innovative Mechanisms for Industry Application (ICIMIA 2017). IEEE (2017)
14. Hulliyah, K., Bakar, N.S.A.A., Ismail, A.R.: Emotion recognition and brain mapping for sentiment analysis: a review, pp. 2543–2546 (2016)
15. Kauser, N., Sharma, J.: Automatic facial expression recognition: a survey based on feature extraction and classification techniques. IEEE (2016)
16. Bhardwaj, N., Dixit, M.: A review: facial expression with its techniques and application. Int. J. Signal Process. Image Process. Pattern Recognit. **9**(6), 149–158 (2016)
17. Agrawal, S., Khatri, P., Gupta, S.: Facial expression recognition techniques: a survey. Int. J. Adv. Electron. Comput. Sci. **2**(1) (2015)
18. Deshmukh, S., Patwardhan, M., Mahajan, A.: Survey on real time facial expression recognition techniques. IET Biom. **5**, 155–163 (2016)
19. Brahmbhatt, N.R., Prajapati, H.B., Dabhi, V.K.: Survey and analysis of extraction of human face features. IEEE (2017)
20. Shinde, A.R., Agnihotri, P.P.: Comparitive study of facial feature extraction, expressions and emotion recognition. IBMRD's J. Manag. Res. **3**(2), 66–69 (2014)
21. Rathi, A., Shah, B.N.: Facial expression recognition. Int. Res. J. Eng. Technol. (IRJET) **3**(4), 540–545 (2016)
22. Emotion Recognition, Chapter 5, Introduction to EEG and Speech Recognition, Elsevier (2016). http://dx.doi.org/10.016/B978-0-12-804490-2.00005-1,
23. Measuring Emotions in the Face, Chapter 6 Emotion Measurement, Elsevier (2016). http://dx.doi.org/10.1016/B978-0-08-100508-8.0000-0
24. Schirmer, A., Adolphs, R.: Emotion perception from face, voice and touch: comparisons and convergence. Trends Cogn. Sci. **21**(3), 216–228 (2017)
25. Griffiths, P.E., Walsh, E.: Emotion and expression. In: International Encyclopedia of the Social and Behavioral Sciences, Elsevier, vol. 7, 2nd edn. (2015)

# Bleeding and Z-Line Classification by DWT Based SIFT Using KNN and SVM

R. Ponnusamy[1(✉)] and S. Sathiamoorthy[2]

[1] Department of Computer and Information Science, Chidambaram, India
povi2006@gmail.com
[2] Tamil Virtual Academy, Chennai, India
Ks_sathia@yahoo.com

**Abstract.** Gastrointestinal diseases can be analysed without any surgical procedures by Wireless Capsule Endoscopy (WCE). To interpret the set of large images from WCE by the domain experts will cost a very high time. As an initiative we proposed a method to differentiate the classes like Z-line and Bleeding among the many classes present in GI tract. The challenge in all image analysis is to characterise these images which improves the exact detection of these abnormalities. To solve this issue, a Discrete wavelet transform (DWT) based Scale-Invariant Feature Transform (SIFT) feature with K-nearest neighbour (KNN) and Support Vector Machine (SVM) technique is employed for classifying these two classes. Our model shows an excellent detection accuracy of 98.12 and 94.45 for bleeding and Z-line respectively for the DWT based SIFT with SVM.

**Keywords:** WCE · Bleeding · Z-line · Gastrointestinal tract · SIFT · DWT · KNN · SVM

## 1 Introduction

Wireless capsule endoscopy (WCE) [1] is a diagnosis device by visualising GI tract. In comparison to traditional endoscopes it provides a non-invasive, immediate choice and empowers physicians to explore the general non-open GI tract [2]. Compared to standard endoscopy systems, WCE not only receives complete access to the tiny digestive system [3, 4] but also has a straightforward approach for patients. WCE is a pill-shaped gadget, comprises a brief core length Complementary CMOS (CMOS) cameras consisting of 4 light source, power unit, transmitters and extra tiny sections as shown in the Fig. 1. The container is swallowed by the patient and takes photos for every second at a velocity of two frames (fps). The endoscopic container travels through GI tract and transmits over 55,000 frames to the recorder connected to midriff of patient. Whereas WCE captures GI colour pictures over roughly 8 h and sends it remotely to other gadgets used by the patient around the middle section. Those pictures are then uploaded to a workstation and evaluated to make decision clinicians.

Despite WCE being an expert accomplishment, it takes around 120 min for a performed clinician for investigating the video and clarifying entire inaccurate 50,000 pictures for every patient [2]. The unusual pictures mostly have less than 5% of total


S. Smys et al. (Eds.): ICCVBIC 2019, AISC 1108, pp. 679–688, 2020.
https://doi.org/10.1007/978-3-030-37218-7_77

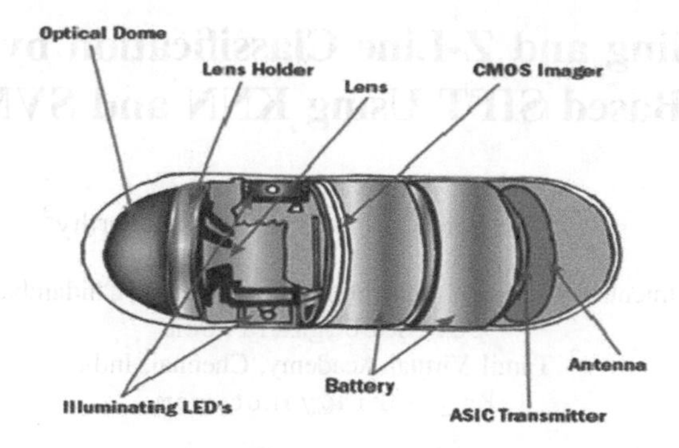


**Fig. 1.** Parts of wireless capsule endoscopy

pictures. A minute variation in shape, surface and spatial details of various GI tract diseases could make it difficult for clinicians to recognize [5]. Hence, an automatic system to analyse images must be need to help clinicians which quickly identifies the abnormal one. More exploration has been done to automatically detect abnormal images in WCE videos. Figure 2 illustrates the sample images of Z-line as well as Bleeding.

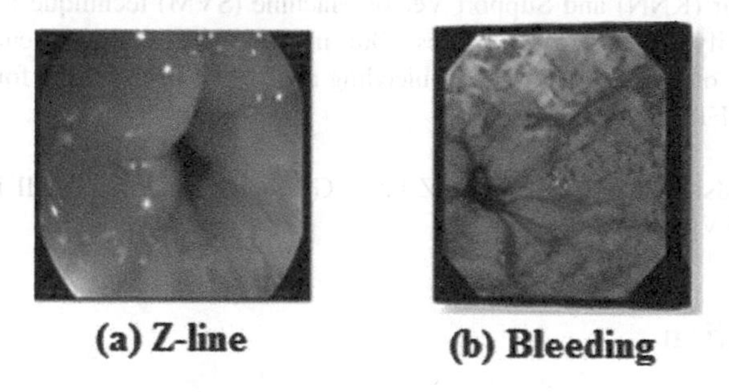


**Fig. 2.** Example images [2, 7]

The upcoming portions are organised here: The relevant works according to these type of problems which are carried out by other authors are discussed in Sect. 2. The new proposed work is discussed in detail in Sect. 3 and subsequently we have discussed the techniques DWT and SIFT, which are efficiently captures the local features. Then, classifier which is used in our model is also discussed in this section. We provided our results and discussion in Sect. 4. In Sect. 5, we depicted the conclusion according to our research.

## 2 Related Works

In many works in WCE videos, automatic multi-abnormality detection has been suggested including Bleeding [6], Polyp [8], Tumour [9], Colon [7], Esophagitis [10] and Ulcer [11] which are most widespread illnesses of GI tract. The earlier works in the detection of abnormality of WCE is only available for one abnormality, such as bleeding or ulcer or tumour, etc. [12, 13]. Since analysis of GI tract from WCE images is more complicated due to the presence of multiple colours, low contrast, blurred areas, complicated backgrounds, lesion shape, texture data, etc, it is suggested in [14] that more effective feature discrimination is required to investigate the class of GI disease. Bleeding is a prevalent indicator for diverse GI illnesses, therefore bleeding recognition is of excellent medical significance in the diagnosis of appropriate illnesses. LiMeng [15] describes chrominance as color characteristic and a local binary pattern (LBP) as textures to identify areas of bleeding in WCE frame. The computed color and texture information is categorized by support vector machine (SVM), Linear discriminant analysis (LDA) and KNN classification models, then the outcome of the three approaches are compared to obtain best performer to categorize the abnormalities. Along the same way, numbers of works hasbeen reported [16, 17]. Nevertheless, sufficient accuracy of categorization is not obtained so far due to the improper feature selection and more apt classification approaches.

A small cluster of proteins on the colon lining is called polyp due to abnormal cell development and is detected in [16] by Karargyris et al. using a pre-processing strategy followed by Log Gabor filter and SUSAN algorithm for edge detection then its geometric information. In [17], a technique called worldwide geometric limitations of polyp and patterns of intensity variability among polyp frontiers is recommended and is studied using convolution neural network. In line with this, several studies are done on [18, 19] and we found that they are limited in accuracy. In [20], ulcer detection in WCE frames is done using Gabor filters with color and texture characteristics and for classification based on neural network is incorporated. The authors [21] suggested AdaBoost method for effective identification of ulcer from the other abnormalities of GI tract diseases and RGB color space is utilized to compute the visual characteristics locally and globally, and probability of a bit plane and wavelet-based characteristics have been computed from desired regions are used to effectively characterize an ulcer disease. However, the accuracy of [22] is limited to some extent. Thus, many numbers of approaches [23, 24] along same direction has been recommended in the literature but they are also limited to ulcer disease only apart from the limitation with the accuracy. Chen et al. [25] addressed hookworm presents in human's stomach by analysing the WCE frames in which color templates from both RGB as well as HSV are incorporated. The RGB is cut as Red, Green and Blue then HSV is cut to Hue, Saturation and Value. In HSV, the Hue and Saturation is utilized for the computation of color template owing to its representation of chrominance information. Although several studies of hookworm from the WCE [26, 27] has been evident in the literature, they are not up to satisfactory level by means of accuracy, computation and storage cost.

## 3 Proposed Image Classification System

We aimed to propose more accurate and unique framework for identification of bleedings and Z-line in which images are transformed to wavelet domain using Discrete Wavelet Transform (DWT) [28] followed by pre-processing strategy [29] then SIFT features are estimated from LL sub-band of discrete wavelet transformed image at decomposition level 3 and KNN and SVM is adopted for classification of proposed feature descriptor. We tested proposed and up-to-date techniques for bleeding and Z-line diseases on standard datasets. The obtained results obviously illustrated that proposed unique framework achieves best results then the others reported in literature. In the proposed approach, we stop the decomposition of images at level 3 based on trial and error approach. Figure 3 shows suggested unique model for classification of bleeding and Z-line diseases in GI tract.

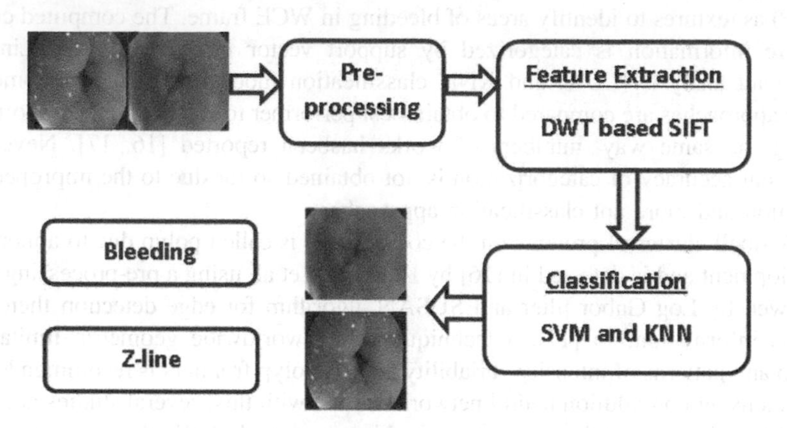


**Fig. 3.** Proposed work for abnormalities detection.

### 3.1 DWT

The proposed system incorporate the advantages of DWT. The images are in RGB space and are resized to 256 × 256 then we have implemented DWT [30] to decompose the images and the levels of decomposition is determined empirically which is 3 for the proposed approach. At decomposition, in every level, we obtain LL, LH, HL, HH sub-bands. In each level, LL sub-bands are further decomposed into 4 sub-bands. LL sub-band at the third level decomposition is considered to compute the proposed SIFT feature descriptor. Figure 4 depicts the decomposition of image into number of sub-bands which use DWT and L and H stands for low and higher frequencies. Figure 5 depicts the process of DWT.

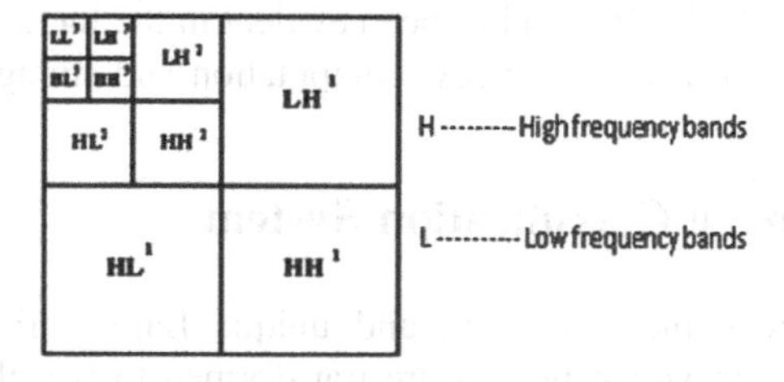


**Fig. 4.** Three level DWT decomposition.

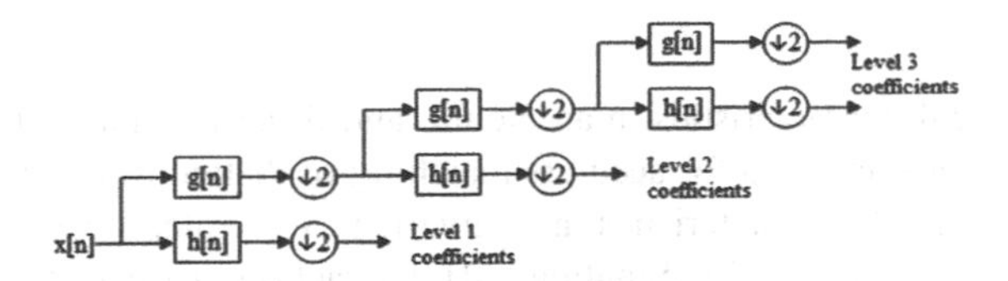


**Fig. 5.** Process of DWT.

DWT is used extensively for the resolution of a multitude of real-life issues in biomedical engineering fields. Wavelet Transform (WT) offers 44 more versatile ways to display signal frequency through use of variable windows. Wavelet transformation uses long windows for a fine, low frequency resolution and short windows for high frequency information. Thus, wavelet transformation provides accurate frequency data at low frequency and accurate time data at elevated frequency. It makes the wavelet suited to evaluation of uneven information patterns, such as pulses that occur at different periods of time [31].

$$y[n] = (x * g)[n] = \sum_{k=-\infty}^{\infty} x[k]g[n-k] \quad (1)$$

Signal representation with lengthy windows to obtain more smooth, low-frequency resolution and short time windows for the purpose of obtaining higher frequency information, and this is done by means of wavelet transformations. Wavelet packet investigation is a widespread form of the DWT. It is a method for extracting non-stationary signal characteristics, which becomes popular rapidly mainly because the generalization of wavelet decomposition offer a greater number of signal analysis opportunities. In wavelet assessment [32], a signal undergo partition to an approximation and detailed one. Multi-resolution analysis breaks down a signal into many data in several resolutions, in which each resolution represents a class of separate physical features within the signal.

The decomposition of a signal into subbands defines the signal and can be treated separated according to its functionalities. Discrete Wavelet Transform (DWT) [33] achieves multi-resolution representation of the filtered active sections. Sample images applied in this process is given in Fig. 6.

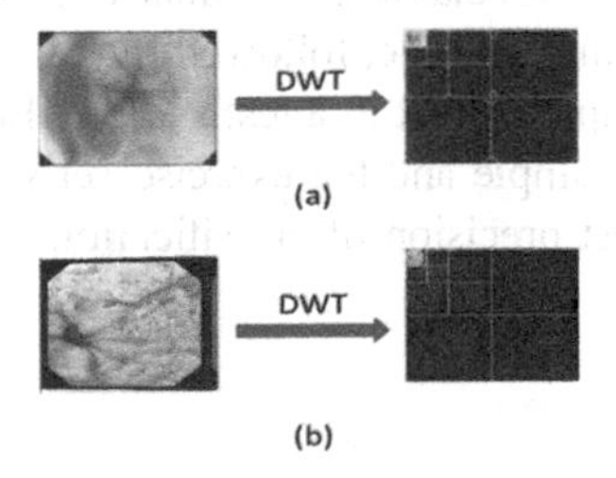


**Fig. 6.** Sample images applied in our DWT process (a) Z-line and (b) Bleeding.

### 3.2 SIFT

For detecting local characteristics in an picture and describe them, SIFT is used. Also these characteristics are used for accurate matching of distinct opinions of the identical objects. The filtered characteristics are invariant to scale, orientation and partly invariant to modifications in illumination [34]. Extraction process of the SIFT function is a involves a set of four steps. Initially, places of prospective interested point is calculated in picture through identifying the extremes in a collection of Gaussian Difference (DOG) filters employed in separate scale-space to the real picture. Then those points of interest situated in lower brightness fields and around the edges will be removed. Afterwards, the remaining points are allocated according to the local picture gradients. At the end, local image characteristics depending upon image gradient is determined at the nearby portions of every key point. Each characteristic is defined in the 4 × 4 nearby of the key points and are vector of 128 components and it is depicted in Fig. 7.

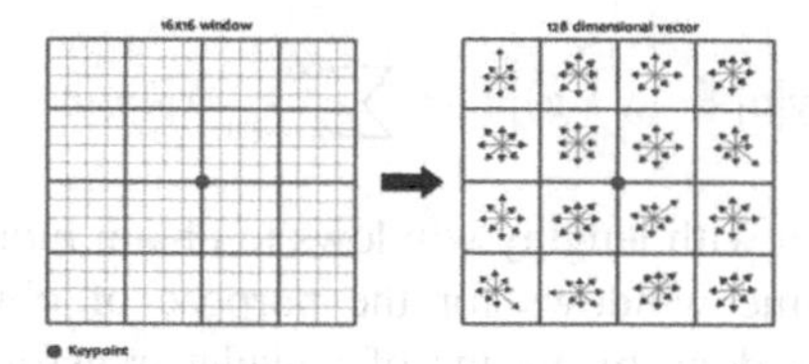


**Fig. 7.** Generation of SIFT descriptor

### 3.3 Classification Using KNN

KNN, a supervised learning method in which feature descriptors are categorized on the basis of nearest samples in feature space [35]. The KNN classification is non-parametric classification technique and no previous understanding of the data structure in training sets is necessary [36].

This classifier calculates the distance to each training information point from unlabeled information and chooses the highest K neighbors on the shortest range. Euclidean distance [37] is used to estimate the distance metric of the KNN, which is used to find out nearest samples of K within a set of training samples (neighbors) of a same type test sample. The KNN classifier calculates the distance between a test sample (feature vector) and all training sample, followed by a majority vote in the class with a K sample of n training sample closest to a test sample. Euclidean distance is the range measurement between the sample and the exercise set samples. Finally, value of K is chosen based on the highest precision of classification among the other value of K.

## 4 Experimental Results and Discussion

The data sets containing pictures from GI tract are gathered from Kvasir [38]. It has 1000 pictures of bleeding and anatomical landmarks such as z-line, each class contains 500 pictures of 40 patients. In each class, the picture set is split into five groups and one is for training and others for testing and thus each group is involved in both training and testing. That is fivefold cross-validation approach was adopted in the proposed strategy.

The SIFT feature is computed from the DWT applied image then the computed SIFT is classified utilizing KNN and SVM and the outcome is illustrated in Table 1. According to our realisation, So far there are no works related to the classification of bleeding and Z-line images. Thus, we have initiated that process for the first time and hence there are no literatures are available for comparative studies. But the works on bleeding have been presented in the past and the recent one results in 97.86% [39] and we utilize the same approach for the Z-line inorder to compare the efficiency of the proposed system. Figure 8 demonstrates that the proposed combination of DWT based SIFT with SVM achieves notably better results than the existing model by Amit kumar kuntu et al. and DWT based SIFT with KNN. Our model produce an accuracy 98.12% and 94.45% respectively for bleeding and Z-Line which is better than existing model by Amit kumar kuntu et al. and DWT based SIFT with KNN. Figure 8 depicts the comparison of our model in terms of accuracy with the existing and proposed approach for bleeding and Z-line disease of GI tract.

**Table 1.** Performance of KNN for bleeding and Z-Line in GI tract

| Method | Bleeding | | | Z-line | | |
|---|---|---|---|---|---|---|
| | Sensitivity | Specificity | Accuracy | Sensitivity | Specificity | Accuracy |
| Amit kumar kuntu et al. | 94.60 | 97.85 | 97.2 | 88.23 | 93.90 | 92.35 |
| DWT with SIFT using KNN | 90.25 | 96.76 | 96.71 | 91.47 | 94.21 | 93.18 |
| DWT with SIFT using SVM | 92.86 | 98.95 | 98.12 | 90.01 | 95.60 | 94.45 |

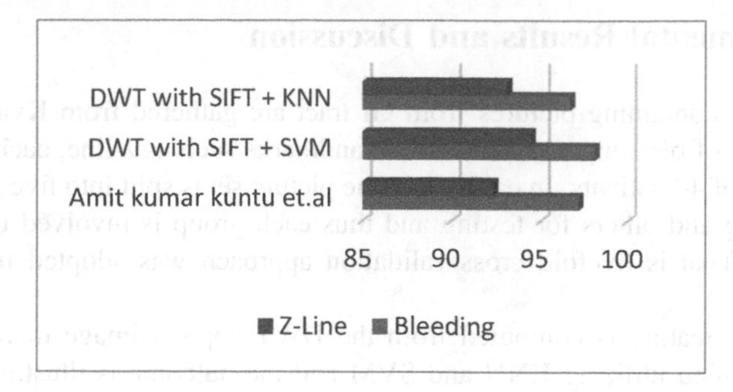


**Fig. 8.** Performance of DWT based SIFT using KNN and SVM

## 5 Conclusion

In this research, we propose an efficient model with the combination of DWT with SIFT and KNN for differentiating the two classes named Z-line and Bleeding, GI tract diseases. The proposed combination of feature vector more effectively captures the texture, color and shape information from the wavelet domain, which reduces the time cost of proposed approach. For checking the efficiency of the suggested method, we takes place on publicly accessible tests on KVasir. The results reveals that proposed combination of DWT with SIFT and SVM outperforms the conventional methods for both bleeding and Z-Line. The accuracy obtained by the combination of DWT with SIFT and SVM are 98.12 and 94.45 for bleeding and Z-line respectively. As a part of future scope, the proposed model is improvised to extra GI tract diseases also.

**Compliance with Ethical Standards**

✓ All authors declare that there is no conflict of interest.
✓ No humans/animals involved in this research work.
✓ We have used our own data.

## References

1. Iddan, G., Meron, G., Glukhovsky, A., Swain, P.: Wireless capsule endoscopy. Nature **405** (6785), 417 (2000)
2. Iakovidis, D.K., Koulaouzidis, A.: Software for enhanced video capsule endoscopy: challenges for essential progress. Nat. Rev. Gastroenterol. Hepatol. **12**(3), 172–186 (2015)
3. Upchurch, B., Vargo, J.: Small bowel endoscopy. Rev. Gastroenterol Disorders **8**(3), 169–177 (2007)
4. Manno, M., Manta, R., Conigliaro, R.: Single-ballon endoscopy. In: Trecca, A. (ed.) Ileoscopy, pp. 79–85. Springer, Milan (2012)
5. Yuan, Y., Li, B., Meng, M.Q.-H.: Improved bag of feature for automatic polyp detection in wireless capsule endoscopy images. IEEE Trans. Autom. Sci. Eng. **13**(2), 529–535 (2016)
6. Charfi, S., El Ansari, M.: Computer-aided diagnosis system for colon abnormalities detection in wireless capsule endoscopy images. Multimed. Tools Appl. 1–18 (2017)

7. Ghosh, T., Fattah, S.A., Wahid, K.A.: Automatic computer aided bleeding detection scheme for wireless capsule endoscopy (WCE) video based on higher and lower order statistical features in a composite color. J. Med. Biol. Eng. **38**(3), 482–496 (2018)
8. Yuan, Y., Meng, M.Q.-H.: Deep learning for polyp recognition in wireless capsule endoscopy images. Am. Assoc. Phys. Med. **44**(4), 1379–1389 (2017)
9. Alizadeh, M., Maghsoudi, O.H., Sharzehi, K., Hemati, H.R., KamaliAsl, A., Talebpour, A.: Detection of small bowel tumor in wireless capsule endoscopy images using an adaptive neuro-fuzzy inference system. J. Biomed. Res. **31**(5), 419–427 (2017)
10. Sivakumar, P., Kumar, B.M.: A novel method to detect bleeding frame and region in wireless capsule endoscopy video. Clust. Comput., 1–7 (2017)
11. Souaidi, M., Abdelouahed, A.A., El Ansari, M.: Multi-scale completed local binary patterns for ulcer detection in wireless capsule endoscopy images. Multimed. Tools Appl., 1–18 (2018)
12. Vasilakakis, M.D., Diamantis, D., Spyrou, E., Koulaouzidis, A., Iakovidis, D.K.: Weakly supervised multi-label classification for semantic interpretation of endoscopy video frames. Evol. Syst., 1–13 (2018)
13. Yanagawa, Y., Echigo, T., Vu, H., Okazaki, H., Fujiwara, Y., Arakawa, T., Yagi, Y.: Abnormality tracking during video capsule endoscopy using an affine triangular constraint based on surrounding features. IPSJ Trans. Comput. Vis. Appl. **9**(3), 1–10 (2017)
14. Iakovidis, D.K., Dimas, G., Karargyris, A., Bianchi, F., Ciuti, G., Koulaouzidis, A.: Deep endoscopic visual measurements. IEEE J. Biomed. Health Inform., 1–9 (2018)
15. Li, B., Meng, M.Q.H.: Computer aided detection of bleeding regions for capsule endoscopy images. IEEE Trans. Biomed. Eng. **56**(4), 1032–1039 (2009)
16. Fu, Y., Zhang, W., Mandal, M., Meng, M.Q.-H.: Computer aided bleeding detection in WCE video. IEEE Trans. Biomed. Eng. **18**, 636–642 (2014)
17. Ghosh, T., Fattah, S.A., Wahid, K.A.: Automatic computer aided bleeding detection scheme for wireless capsule endoscopy (WCE) video based on higher and lower order statistical features in a composite color. J. Med. Biol. Eng. **38**, 482–496 (2018)
18. Karargyris, A., Bourbakis, N.G.: Detection of small bowel polyps and ulcers in wireless capsule endoscopy videos. IEEE Trans. Biomed. Eng. **58**(10), 2777–2786 (2011)
19. Tajbakhsh, N., Gurudu, S.R., Liang, J.: Automatic polyp detection using global geometric constraints and local intensity variation patterns. In: International Conference on Medical image Computing and Computer-Assisted intervention. Springer (2014)
20. Silva, J., Histase, A., Romain, O., Dray, X., Grando, B.: Toward embedded detection of polyps in WCE images for early diagnosis of colorectal cancer. Int. J. Comput. Assist. Radiol. Surg. **9**, 283–293 (2014)
21. Xu, L., Fan, S., Fan, Y., Li, L.: Automatic polyp recognition of small bowel in wireless capsule endoscopy images. In: Proceedings of SPIE, Medical Imaging 2018: Imaging Informatics for Healthcare, Research, and Applications, vol. 10579, p. 1057919, 6 March 2018. https://doi.org/10.1117/12.2303519
22. Htwe, T.M., Shen, W., Li, L., Poh, C.K., Liu, J., Lim, J.H., Ong, E.H., Ho, K.Y.: Adaboost learning for small ulcer detection from wireless capsule endoscopy (WCE) images. In: Asia Pacific Signal and Information Processing Association (APSIPA) Conference (2010)
23. Charisis, V., Tsiligiri A., Hadjileontiadis, L.J., Liatsos, C.N., Mavrogiannis, C.C., Sergiadis, G.D.: Ulcer detection in wireless capsule endoscopy images using bidimensional nonlinear analysis. In: Bamidis, P.D., Pallikarakis, N. (eds.) XII Mediterranean Conference on Medical and Biological Engineering and Computing 2010. IFMBE Proceedings, vol. 29. Springer, Heidelberg (2010)

24. Koshy, N.E., Gopi, V.P.: A new method for ulcer detection in endoscopic images. In: IEEE Sponsored 2nd International Conference on Electronics and Communication System (ICECS 2015) (2015)
25. Chen, H., Chen, J., Peng, Q., Sun, G., Gan, T.: Automatic hookworm image detection for wireless capsule endoscopy using hybrid color gradient and contourlet transform. In: 6th International Conference on Biomedical Engineering and Informatics, Biomedical Engineering and Informatics (BMEI), pp. 116–120 (2013)
26. Vijila Rani, K., Nisha, M.: Hookworm and bleeding detection in WCE images using rusboost classifier. J. Image Process. Artif. Intell. **4**, 13–19 (2018)
27. He, J.Y., Wu, X., Jiang, Y.G., Peng, Q., Jain, R.: Hookworm detection in wireless capsule endoscopy images with deep learning. IEEE Trans. Image Process. **27**, 2379–2392 (2018)
28. Gupta, D., Choubey, S.: Discrete wavelet transform for image processing. Int. J. Emerg. Technol. Adv. Eng. **4**, 598–602 (2015)
29. Ogiela, M.R., Tadeusiewicz, R.: Preprocessing medical images and their overall enhancement. In: Modern Computational Intelligence Methods for the Interpretation of Medical Images. Studies in Computational Intelligence, vol. 84. Springer, Heidelberg (2008)
30. Arunkumar, R., Balasubramanian, M., Palanivel, S.: Indoor object recognition system using combined DCT-DWT under supervised classifier. Int. J. Comput. Appl. **82**, 17–21 (2013)
31. Cvetkovic, D., Ubeyli, E.D., Cosic, I.: Wavelet transform feature extraction from human PPG, ECG, and EEG signal responses to ELF PEMF exposures: a pilot study. Digit. Signal Process. **18**(5), 861–874 (2008)
32. Gokhale, M.Y., Khanduja, D.K.: Time domain signal analysis using wavelet packet decomposition approach. Int. J. Commun. Netw. Syst. Sci. **3**, 321–329 (2010)
33. Mallat, S.: A theory for multiresolution signal decomposition: the wavelet representation. IEEE Trans. Pattern Anal. Mach. Intell. **11**, 674–679 (1989)
34. Banerjee, B., Bhattacharjee, T., Chowdhury, N.: Image object classification using scale invariant feature transform descriptor with support vector machine classifier with histogram intersection kernel. In: International Conference on Advances in Information and Communication Technologies (ICT) 2010, Information and Communication Technologies, pp. 443–448. Springer (2010)
35. Chavan, N.V., Jadhav, B.D., Patil, P.M.: Detection and classification of brain tumors. Int. J. Comput. Appl. **112**(8), 48–53 (2015)
36. Gadpayle, P., Mahajani, P.S.: Detection and classification of brain tumor in MRI images. Int. J. Emerg. Trends Electr. Electron. **5**(1), 45–49 (2013)
37. Murugappan, M., Nagarajan, R., Yaacob, S.: Appraising human emotions using time frequency analysis based EEG alpha band features. In: Conference on Innovative Technologies in Intelligent Systems and Industrial Applications, Monash University, Sunway campus, Malaysia, pp. 70–75, July 2009
38. Pogorelov, K., Randel, K.R., Griwodz, C., Eskeland, S.L., de Lange, T., Johansen, D., Spampinato, C., Dang-Nguyen, D.-T., Lux, M., Schmidt, P.T., Riegler, M., Halvorsen, P.: KVASIR: a multi-class image dataset for computer aided gastrointestinal disease detection. In: Proceedings of the 8th ACM on Multimedia Systems Conference (MMSYS), MMSys 2017, Taipei, Taiwan, 20–23 June 2017, pp. 164–169 (2017)
39. Kuntu, A.K., Fattah, S.A., Rizve, M.N.: An automatic bleeding frame and region detection scheme for wireless capsule endoscopy videos based on interplane intensity variation profile in normalized RGB color space. J. Healthc. Eng. **2018**. Article ID 9423062, 12 pages. https://doi.org/10.1155/2018/9423062

# RNA-Seq DE Genes on Glioblastoma Using Non Linear SVM and Pathway Analysis of NOG and ASCL5

Sandra Binoy[1], Vinai George Biju[1(✉)], Cynthia Basilia[1], Blessy B. Mathew[2], and C. M. Prashanth[3]

[1] CHRIST (Deemed to be University), Bengaluru, India
vinai.george@christuniversity.in
[2] Dayananda Sagar College of Engineering, Bengaluru, India
[3] Acharya Institute of Technology, Bengaluru, India

**Abstract.** Differentially Expressed genes related to Glioblastoma Multiforme as an output of RNASeq studies were further studied to conclude new research insights. Glioma is a type of intracranial tumor (within the skull), which can grow rapidly in its malignant stages. Gene expression in Grade II, III and IV Gliomas is analysed using non linear SVM models. The enriched GO terms were identified GOrilla. Pathways related to NOG and ASCL5 gene were studied using Reactome.

**Keywords:** Differential Gene Expression · RNASeq analysis · Transcriptomic analysis · Glioblastoma Multiforme · SVM

## 1 Introduction

Glioblastoma Multiforme (GBM) is a fast-growing Glioma which occurs in the Glial cells, present in the Central Nervous System (Spinal Cord and/or brain). They are known to be deadly and malignant due to the high mortality rate (5-year relative survival rate of only 5.6%) and results in death typically within 15 months, with medication. Glioma is known to be a very deadly type of brain tumour reporting an estimate contribution of about 80% of all malignant brain [2]. It occurs in the glial cells which can be seen around the neurons. Its main function is to support these nerve cells, with which it can be inferred that it is intracranial [3–5]. It can be very useful to find out the genes being differentially expressed in the cells which could be used in understanding the proteins that are synthesized by the pathways [6–9]. RNA-sequencing is a highly popular technique to find differentially expressed genes and its function in various stages stages of diseases [10–13]. In this study, the focus is particularly on differentially expressed genes through RNA-seq analysis which indicates the disease progression in two grades of Glioma namely Grade II, III and IV [1].


S. Smys et al. (Eds.): ICCVBIC 2019, AISC 1108, pp. 689–696, 2020.
https://doi.org/10.1007/978-3-030-37218-7_78

## 2 Differential Gene Expression

### 2.1 The Dataset and Analysis

The dataset consists of over 17,000 genes which include 325 Glioma samples from various hospitals in China. They have been obtained over a period of 12 years from the patients suffering from the disease. They have been generated as part of the CGGA project to identify oncogenic fusions for a deeper understanding of the progression of Glioma [1]. Differential Gene Expression was performed on the dataset to segregate it into two different files which contain various factors, some of which are p-value, mean, and t-value; for the differentially expressed genes in grades II and IV, and III and IV. Glioblastoma belongs to Grade IV of Glioma, and hence, looking into the genes that are common in these two files would point to the genes belonging to grade IV specifically. The differentially expressed genes had to comply both fold change and p-value measure simultaneously. The unpaired Student's t-test was used to check for p-values and Benjamini and Hochberg algorithm was used to verify the false discovery rate [1]. Two genes NOG and ASCL5 were chosen based on their p-value indicating significant expression difference between grade II and IV. Two genes: NOG (Gene ID: 9241, Location: Chromosome 7) and ASCL5 (Gene ID: 647129, Location: Chromosome 1). Upon further study about them, it was found that they could possibly be associated with the growth of the malignancies, that is, promoting the growth factor of Glioblastoma.

### 2.2 Gene Expression for Grades II and IV Glioma with SVM

Support Vector Machine (SVM) is one among the popular approved machine learning algorithms for Regression and classification tasks [14–18]. We analysed the data for gene expression related to grade II and IV glioma using Radial and Polynomial Kernel. The performance of SVM are indicated in Tables 1 and 2. The radial kernel is found to fit the model better when compared to polynomial kernel considering the ROC, Specificity and Sensitivity values.

**Table 1.** SVM radial kernel

| Sigma | C | ROC | Sensitivity | Specificity |
|---|---|---|---|---|
| 0.01 | 0.25 | 0.888804 | 0.79011 | 0.829796 |
| 0.01 | 0.5 | 0.894826 | 0.794066 | 0.846831 |
| 0.01 | 0.75 | 0.897507 | 0.795385 | 0.848871 |
| 0.01 | 1 | 0.898448 | 0.794725 | 0.852824 |
| 0.01 | 1.25 | 0.898211 | 0.793846 | 0.855561 |
| 0.01 | 1.5 | 0.899409 | 0.790989 | 0.851161 |
| 0.015 | 0.25 | 0.890657 | 0.796703 | 0.833349 |
| 0.015 | 0.5 | 0.896218 | 0.80044 | 0.844463 |
| 0.015 | 0.75 | 0.897405 | 0.799121 | 0.846008 |
| 0.015 | 1 | 0.898867 | 0.803077 | 0.850416 |
| 0.015 | 1.25 | 0.897946 | 0.798022 | 0.843992 |
| 0.015 | 1.5 | 0.897242 | 0.8 | 0.846329 |

**Table 2.** SVM - polynomial kernel

| Degree | Scale | C | ROC | Sensitivity | Specificity |
|---|---|---|---|---|---|
| 2 | 0.001 | 0.25 | 0.862926 | 0.763077 | 0.817796 |
| 2 | 0.001 | 0.5 | 0.866688 | 0.768352 | 0.82418 |
| 2 | 0.001 | 1 | 0.871222 | 0.777143 | 0.830933 |
| 2 | 0.01 | 0.25 | 0.888441 | 0.786813 | 0.845647 |
| 2 | 0.01 | 0.5 | 0.892908 | 0.786154 | 0.857882 |
| 2 | 0.01 | 1 | 0.892672 | 0.789011 | 0.858361 |
| 2 | 0.1 | 0.25 | 0.867404 | 0.769451 | 0.825059 |
| 2 | 0.1 | 0.5 | 0.853755 | 0.765055 | 0.803255 |
| 2 | 0.1 | 1 | 0.847299 | 0.756703 | 0.793294 |
| 3 | 0.001 | 0.25 | 0.866043 | 0.769011 | 0.819804 |
| 3 | 0.001 | 0.5 | 0.870898 | 0.77055 | 0.826933 |
| 3 | 0.001 | 1 | 0.875757 | 0.78 | 0.837302 |
| 3 | 0.01 | 0.25 | 0.892885 | 0.788571 | 0.856769 |
| 3 | 0.01 | 0.5 | 0.890775 | 0.78967 | 0.854345 |
| 3 | 0.01 | 1 | 0.886776 | 0.783077 | 0.855961 |
| 3 | 0.1 | 0.25 | 0.819377 | 0.74022 | 0.775867 |
| 3 | 0.1 | 0.5 | 0.820094 | 0.749231 | 0.770243 |
| 3 | 0.1 | 1 | 0.816739 | 0.741099 | 0.776612 |

## 3 Pathway Analysis

The major pathway analysis tools are Reactome, David and KEGG. KEGG or Kyoto Encyclopedia of Genes and Genomes is the accumulation of assets and manages genomes, medications, infections and compound substances. David Pathway Analysis instrument, is named as David for Annotation, Visualization and Integrated Discovery. Reactome comprises of pathways and reactions of human exercises whereas KEGG has permitting issue. Reactome has crosslinks to different databases and information. An Enrichment Analysis is better when Reactome is utilized. Reactome has better UI and has an elite pathway investigation. Programmatic access for DAVID is by Simple Object Access Protocol (SOAP) or Web Service Description Language (WSDL) and for Reactome is by Representational State Transfer (REST). Reactome analysis is less complex, lightweight, adaptable in light of REST convention and consequently can incorporate to different assets effectively. Reactome apparatus does not force any restrictions on test size. Reactome concentrates great pathways which are refreshed quarterly is steady and gives superior pathway administration and permits intelligent investigation and examination of information [20].

Signalling by BMP: Bone Morphogenetic proteins (BMP) are a member of the TGF-B (Transforming Growth Factor-Beta) family. They bind to kinase receptors that transduce signals signalling pathways. BMP signalling is linked to a wide variety of disorders and even cancer.

TGF-B, the Polypeptide member secreted protein accomplish various functions in the body such as cell differentiation, growth, apoptosis and proliferation. TGF-B is found to be a tumour suppressor in many gastrointestinal cancers. Somatic alterations can trigger TGF-B pathway which, upon activation, stops the process of mutation hence suppressing the growth of tumour. Many non-gastrointestinal malignancies lack these somatic alterations and end up developing reluctance to the impacts of TGF-B. The resistance is considered to be a piece of a signal switch where the TGF-B mislay its growth-impeding factors and is again utilized by the epithelial cells promoting growth [2].

Gorilla is a web-based tool used to identify and visualise the GO enriched terms in hierarchical gene list [19]. Figure 1 displays color-coded trimmed DAG of the GO terms. This helps reduce redundancy in large gene ontology datasets. The color codes show the enrichment p-value of the respective nodes and related genes. Figure 2 displays the pathway analysis of the 58 genes and the pathways are highlighted. Each of those pathways associated are linked to the various functions performed within the cells. When analysing NAG and ASCL5, signal transduction was the function associated, which highlighted proteins such as TGF-beta family members and BMP, that were further studied to learn their role in tumor progression. The major pathways involved are listed in Table 3.

**Table 3.** Pathway analysis of grade IV genes (Pathway names|Application: Reactome)

| Pathway names | Entities | pValue | FDR | Reactions |
|---|---|---|---|---|
| ECM Proteoglycans | 79 | 1.87E-06 | 3.60E-04 | 23 |
| Syndecan Interactions | 29 | 1.07E-05 | 1.03E-03 | 15 |
| Transport glycerol from adipocytes to liver by Aquaporins | 3 | 9.24E-05 | 5.91E-03 | 2 |
| Extracellular matrix organisation | 329 | 1.31E-04 | 6.29E-03 | 316 |
| Non-integrin membrane-ECM interactions | 61 | 1.89E-04 | 7.16E-03 | 22 |
| Histamine receptors | 5 | 2.55E-04 | 7.21E-03 | 3 |
| Assembly of collagen fibrils and multimeric structure | 67 | 2.69E-04 | 7.21E-03 | 26 |
| Collagen degradation | 69 | 3.00E-04 | 7.21E-03 | 34 |
| MET activates PTK2 signalling | 32 | 4.53E-04 | 9.51E-03 | 5 |
| Integrin cell surface Interaction | 86 | 6.84E-04 | 1.30E-02 | 54 |
| NCAM1 interaction | 44 | 1.13E-03 | 1.69E-02 | 10 |
| Collagen chain trimerization | 44 | 1.13E-03 | 1.69E-02 | 28 |
| MET promotes cell motility | 45 | 1.21E-03 | 1.69E-02 | 12 |
| Collagen formation | 104 | 1.38E-03 | 1.74E-02 | 77 |
| Chemokine receptors bind chemokines | 48 | 1.45E-03 | 1.74E-02 | 18 |
| RUNX2 regulates genes in cell migration | 14 | 1.95E-03 | 2.34E-02 | 7 |
| NOTCH2 intracellular domain regulates transcription | 16 | 2.53E-03 | 2.60E-02 | 9 |
| GPCR ligand binding | 643 | 2.60E-03 | 2.60E-02 | 177 |
| NCAM signalling for neurite out-growth | 69 | 4.03E-03 | 3.62E-02 | 23 |
| Signaling by PDGF | 69 | 4.03E-03 | 3.62E-02 | 28 |

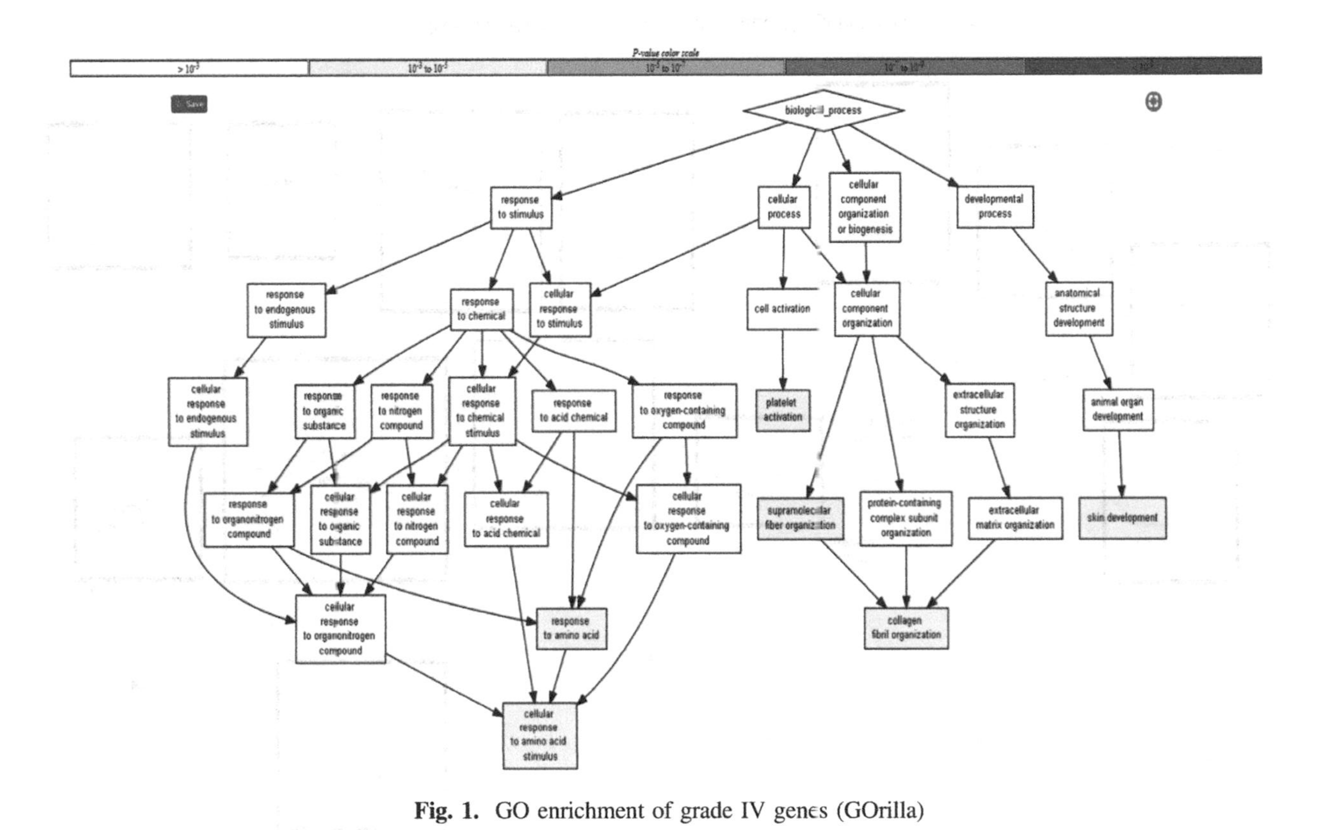


**Fig. 1.** GO enrichment of grade IV genes (GOrilla)

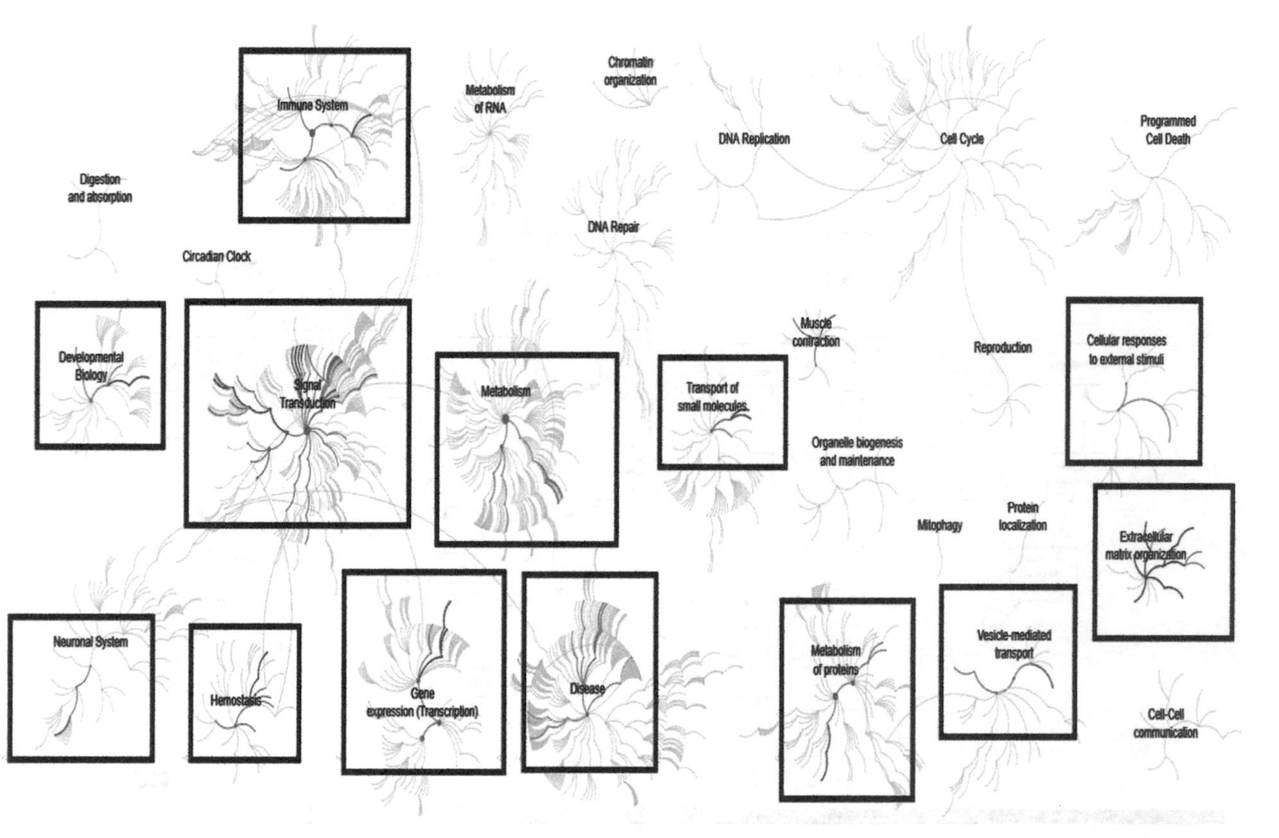


**Fig. 2.** Pathway overview of grade IV genes (Reactome)

The breakdown of the pathways of interest, which have been previously hinged with tumour promotion, is: **ECM proteoglycans**: ECM proteins have been known to play a role in the development and progression of tumour cells [4]. Proteoglycans have also been known to either contain anti-angiogenic properties or directly participate in the promotion of tumor growth by modulating key signaling pathways [5]. **Extracellular Matrix organization**: ECM become disorganised during diseases such as cancer, which validates its effects during the tumor progression as it directly promotes transformation of cells and metastasis [22]. Gene Ontology using GOrilla indicates platelet activation and collagen fibril activation as key characterization under biological process as indicated in Fig. 1. Signal transduction, hemostasis, immune system, metabolism and ECM are the major pathways involved as detected by reactome which is indicated in Fig. 2.

# 4 Conclusion and Future Scope

The study and analysis done on the proteins that are associated with the genes belonging to Grade IV of Glioma cells (Glioblastoma) are further information to what could possibly be the cause of the tumor, in the first place. Having said this, it is also important to understand that proteins could be influenced by each other and the reason for the malfunctionality could be more than just one protein.

NOG gene delivers directions to create a protein called noggin which takes part in the development of body tissues such as nerves, muscles and bones that enable formation of proper joints. The interaction of noggin with bone morphogenetic proteins (BMPs) occurs by receptor attachment that further stimulates various processes in the cell or by inhibiting the receptor binding that reduces the signaling in BMP. Achaete-scute family bHLH transcription factor 5 has been linked to several cancers. The future scope includes protein interaction studies related to these Gene Ontology and pathways detected by both GoRilla and Reactome tool respectively. Modelling of results for protein interaction using machine learning will also be a major future scope of the project.

**Compliance with Ethical Standards**

✓ All authors declare that there is no conflict of interest.
✓ No humans/animals involved in this research work.
✓ We have used our own data.

# References


1. Zhao, Z., Meng, F., Wang, W., Wang, Z., Zhang, C., Jiang, T.: Comprehensive RNA-seq transcriptomic profiling in the malignant progression of gliomas. Nature **4**, 170024 (2017)
2. Yang, L., Moses, H.L.: Transforming growth factor β: tumor suppressor or promoter? Are host immune cells the answer? Cancer Res. **68**(22), 9107–9111 (2008)
3. Park, J., Xu, K., Park, T., Yi, S.V.: What are the determinants of gene expression levels and breadths in the human genome?. Oxford J. 46–56 (2012)

4. Mojares, E., Walker, C., del Río Hernández, A.: Role of extracellular matrix in development and cancer progression. Int. J. Mol. Sci. **19**(10), 3028 (2018)
5. Iozzo, R.V., Sanderson, R.D.: Proteoglycans in cancer biology, tumour microenvironment and angiogenesis. J. Cell Mol. Med. **15**(5), 1013–1031 (2011)
6. Shankar, K., Gupta, D., et al.: Optimal feature-based multi-kernel SVM approach for thyroid disease classification. J. Supercomput. 1–16 (2018)
7. Best, M.G., Sol, N., Kooi, I., et al.: RNA-Seq of tumor-educated platelets enables blood-based pan-cancer, multiclass, and molecular pathway cancer diagnostics. Cancer Cell **28**(5), 666–676 (2015)
8. Cameron, W., Elijah, M., del Río Hernández, A.: Role of extracellular matrix in development and cancer progression. Int. J. Mol. Sci. **19**, 3028 (2018)
9. Iozzo, R.V., Sanderson, R.D.: Proteoglycans in cancer biology, tumour microenvironment and angiogenesis. J. Cell Mol. Med. **15**, 1013–1031 (2011)
10. Darmanis, S., et al.: Single-cell RNA-seq analysis of infiltrating neoplastic cells at the migrating front of human glioblastoma. Cell Rep. **21**(5), 1399–1410 (2017)
11. Bao, Z.-S., et al.: RNA-seq of 272 gliomas revealed a novel, recurrent PTPRZ1-MET fusion transcript in secondary glioblastomas. Genome Res. **24**(11), 1765–1773 (2014)
12. Esteve-Codina, A., et al.: A comparison of RNA-Seq results from paired formalin-fixed paraffin-embedded and fresh-frozen glioblastoma tissue samples. PLoS ONE **12**(1), e0170632 (2017)
13. Patil, V., Pal, J., Somasundaram, K.: Elucidating the cancer-specific genetic alteration spectrum of glioblastoma derived cell lines from whole exome and RNA sequencing. Oncotarget **6**(41), 43452 (2015)
14. Ravale, U., Marathe, N., Padiya, P.: Feature selection based hybrid anomaly intrusion detection system using K means and RBF kernel function. Procedia Comput. Sci. **45**, 428–435 (2015)
15. Chatterjee, R., Yu, T.: Generalized coherent states, reproducing kernels, and quantum support vector machines. arXiv preprint arXiv (2016)
16. Wei, W., Jia, Q.: Weighted feature Gaussian kernel SVM for emotion recognition. Comput. Intell. Neurosci. **11** (2016)
17. Kayzoglu, T., Colkesen, I.: A kernel functions analysis for support vector machines for land cover classification. Int. J. Appl. Earth Obs. Geoinf. **11**(5), 352–359 (2009)
18. Abdollahi, S., et al.: Prioritization of effective factors in the occurrence of land subsidence and its susceptibility mapping using an SVM model and their different kernel functions. Bull. Eng. Geol. Environ. **78**, 4017–4034 (2018)
19. Eden, E., Navon, R., Steinfeld, I., Lipson, D., Yakhini, Z.: GOrilla: a tool for discovery and visualization of enriched GO terms in ranked gene lists. BMC Bioinformatics **10**, 48 (2009)
20. Fabregat, A., et al.: The reactome pathway knowledgebase. Nucleic Acids Res. **44**(D1), D481–D487 (2015)

# Evaluation of Growth and $CO_2$ Biofixation by *Spirulina platensis* in Different Culture Media Using Statistical Models

Carvajal Tatis Claudia Andrea[1(✉)],
Suarez Marenco Marianella María[2], W. B. Morgado Gamero[1],
Sarmiento Rubiano Adriana[2], Parody Muñoz Alexander Elías[3],
and Jesus Silva[4]

[1] Department of Exact and Natural Sciences, Universidad de la Costa, Calle 58#55-66, Barranquilla, Atlántico 080002, Colombia
claucarvajalt@gmail.com, wmorgadol@cuc.edu.co
[2] Department of Nutrition and Diet, Universidad Metropolitana, Calle 76#42-78, Barranquilla, Atlántico 080002, Colombia
nellasuarez@hotmail.com, lusarru@hotmail.com
[3] Engineering Faculty, Universidad Libre, Barranquilla, Colombia
alexandere.parodym@unilibre.edu.co
[4] Universidad Peruana de Ciencias Aplicadas, Lima, Peru
jesussilvaUPC@gmail.com

**Abstract.** This study was proposed for evaluating the $CO_2$ fixation by *Spirulina platensis* in different media, in order to understand the growth dynamics of the photosynthetic microalgae, a useful resource for the mitigation of climate change. The percentage of $CO_2$ fixation by the strain *S. platensis* UTEX LB 2340 was determined during 11 days of sampling, using four (4) culture media. According to the statistical models, spirulina medium represented the best option in terms of cell growth between the tested ones. In this model, the variable day had presented a significant difference, this could be related to the exponential phase of the microorganism used.

**Keywords:** *S. platensis* sp · Carbon dioxide · Fixation · Culture media · Climate change · Statistical models

## 1 Introduction

The increase of atmospheric $CO_2$ concentration (one of the gases responsible for trapping heat in the atmosphere) is changing the energetic balance on Earth, being correlated with the observed increase of the global average temperature [1] generating alterations in the ecosystems of the planet and therefore, affecting the continuous dynamics between climate and biodiversity. It is generally accepted that the increase in the atmospheric levels of greenhouse gases such as $CO_2$ leads to global warming and its direct consequence is the climatological disasters that are increasingly devastating [2]. Point toward to reduce the amount of $CO_2$ emitted to the atmosphere, one of the three main strategies proposal by Conference of the Parties (COP21) that was held in


S. Smys et al. (Eds.): ICCVBIC 2019, AISC 1108, pp. 697–707, 2020.
https://doi.org/10.1007/978-3-030-37218-7_79

Paris is the carbon capture and storage (CCS), the strategy that can be achieved by improving $CO_2$ sinks and implementing fixation systems of this greenhouse gas. In this sense, microalgae can maintain the balance of $CO_2$ in the atmosphere for its ability to perform the process of photosynthesis where the $CO_2$ is used as a carbon source, to be converted into organic compounds by using solar energy [3, 4], their capacity to accumulate inorganic carbon in cytoplasm is one of the reasons to achieve high $CO_2$ uptake efficiencies [1]. Actually, the capture processes conducted by microalgae and cyanobacteria, report an efficiency of $CO_2$ fixation approximately 10–50 times better than plants, and a variety of high-value products, such as dietary supplements for human and animals [5, 6], potentializing their application in the food sector, nutraceuticals, and even in the energy sector using biomass for biofuels [7]. The assimilation of $CO_2$ is highly dependent on some factors like characteristics of the microalgae strain, nutritional needs, $CO_2$ concentration, cultivation system, operating conditions, environmental factors [8], *Spirulina* sp., one of the most studied species, has been identified as the favorable microalgae strains to bio-sequester $CO_2$ [4]. It. is a high source of protein (near 65%). [1, 9], amino acids, carbohydrates, lipids and vitamins A1, B1, B2, B6, B12, C and E [10]. *S. platensis* has shown a high carbon dioxide fixation potential as indicated by its higher kinetic parameters with average and maximum daily fixation rates [3]. Thus, it is essential to understand the growth dynamics of photosynthetic microalgae to determine the $CO_2$ capture ability and overcome the challenges posed by climate change to become a strategy that can be implemented in developing countries because it is an economical and sustainable alternative for $CO_2$ reduction. In biotechnological processes, the constituents of culture media generate high costs and exert a strong influence on microorganism behavior [11]. This study had tested different culture media in order to achieve optimal growth rates and large quantities in outdoor temperature conditions.

## 2 Material and Methods

The experiment to measure the growth and $CO_2$ biofixation *S. platensis* in 4 different culture media was conducted in 4 moments: 1. Preparation of living algal strain; 2. Strain adaptation 3. Inoculation and 4. Cell growth and $CO_2$ biofixation measurement. Figure 1 shows the experiment stages.

### 2.1 Obtaining the Strain and Culture Conditions of the Stock Culture

The microalga used in this study was *Spirulina platensis* UTEX LB 2340, acquired from the UTEX Culture Collection of Algae, the University of Texas at Austin. In the beginning, it was necessary to identify the microscopic features of the strain; as straight and spiral filaments, and gas vesicles [12]. The conditions for stock culture establishment were the same ones used in the UTEX Collection; it means maintenance temperature of 20 °C, pH 8–9 units [3], the light intensity of 3200 lx maximum from cool-white fluorescent lamps, and photoperiod 12/12 h L/D. Constant oxygen air flow was supplied by a fish tank motor with 2 W of potency.

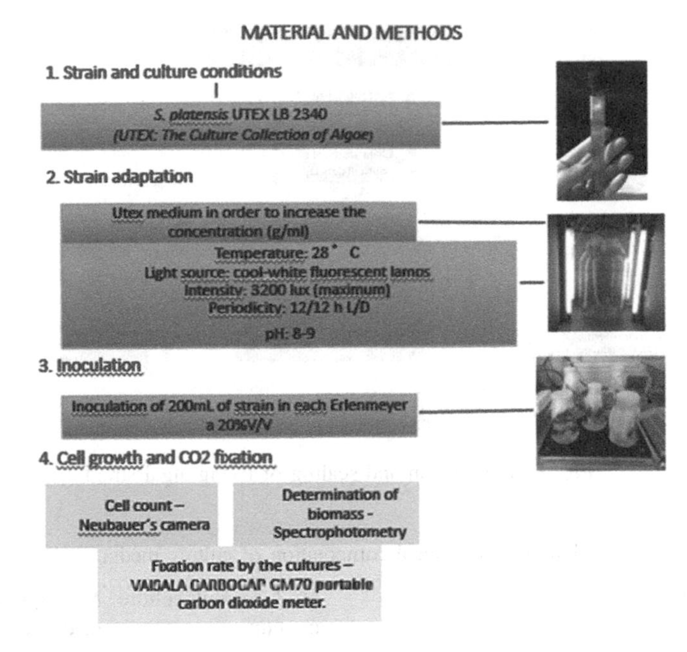


**Fig. 1.** Materials and methods

## 2.2 Strain Adaptation and Scaling

Subcultures (2) from the stock culture were tested at the same conditions with a variation in the maintenance temperature (28 °C) in order to simulate outdoor temperature conditions; in case it was necessary, aliquots of algal strain were maintained as agar slants on Petri dishes, enriched with medium Spirulina. 200 mL from the subculture was transferred to 500 mL Erlenmeyer with the sterilized medium for being used in the experiment. In this moment, the initial biomass concentration of 0.1 $g/L^{-1}$ (25.000 cells/mL) and initial optical density of 0.087 at 610 nm [14, 15]. After 14 days, aliquots (50 mL) were transferred for initiating a fresh inoculum and the bigger portion was used for scaling the cultivate media. Figure 2 shows the procedure to guarantee a large quantity of inoculum for being used after.

## 2.3 Inoculation

In this part, 200 mL of adapted strain in a concentration of 20% V/V, were transferred to 4 jars with a respective culture medium. The chemical composition of Zarrouk culture medium (Medium N° 3) and Spirulina culture medium (Medium N° 4) which need 2 solutions for being prepared (Solution I and II). In culture medium N° 1 preparation, had been used Spirulina culture medium (medium No. 4) removing $NaHCO_3$, $Na_2CO_3$, $K_2HPO_4$, and the addition of 3.5 g of Urea. Medium N° 2 preparation was made with 12,8 g/500 mL of $NaHCO_3$ and 3,5 g/500 mL of urea (Table 1).

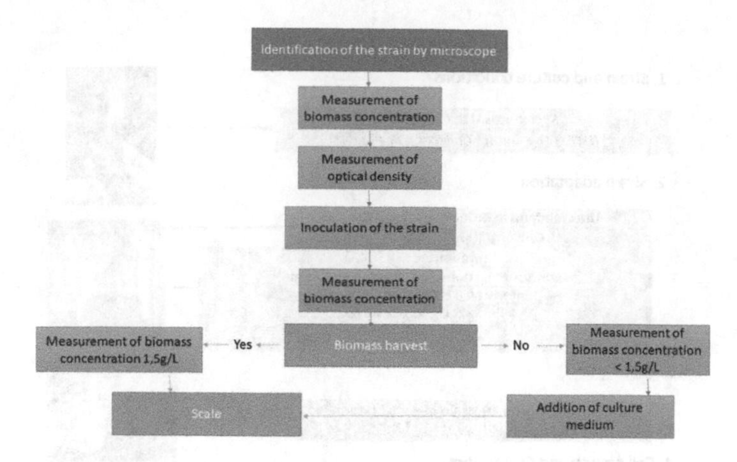


**Fig. 2.** Preparation and scaling of living algal strain

**Table 1.** Chemical composition of culture media.

| Component | N° 3 Zarrouk culture medium | N° 4 Spirulina culture medium |
|---|---|---|
| Solution I | | |
| Sodium bicarbonate - $NaHCO_3$ | 6,805 g/mL | 13,61 g/mL |
| Sodium carbonate - $Na_2CO_3$ | 0,515 g/mL | 4,03 g/mL |
| Dipotasic phosphate - $K_2HPO_4$ | 0,25 g/mL | 0,5 g/mL |
| Solution II | | |
| Sodium nitrate – $NaNO_3$ | 1,25 g/mL | 2,5 g/mL |
| Potassium sulfate – $K_2SO_4$ | 0,5 g/Ml | 1 g/mL |
| Sodium chloride – NaCl | 0,1 g/mL | 1 g/mL |
| Magnesium sulfate heptahydrate $MgSO_4$.7H2O | 0,02 g/mL | 0,2 g/mL |
| Calcium chloride dehydrate – $CaCl_2.2H_2O$ | 0,005 g/mL | 0,04 g/mL |
| Iron sulfate heptahydrate – $FeSO_4.7H_2O$ | 0,025 g/mL | 0,01 g/mL |
| $EDTA.2H_2O$ | | 0,08 g/mL |
| Trace metals solution* | | 1 mL |
| Vitamin solution (Cyanocobalamin, vitaminB12, 05 g/L) | | 1 mL |
| Solution A5* | 1 mL | |
| Solution B6* | 1 mL | |

***Trace metals solution:** Na2EDTA.2H2, 0,8, FeSO4.7H2O, 0,7, ZnSO4.7H2O, 1, MnSO4.7H20, 2, H3BO3, 10, Co(NO3)2.6H2O, 1, Na2MoO4.2H2O, 1, CuSO.5H2O, 0,005 (g/L). ***Solution A5**: H3BO3, 2,86, MnCl2.4H2O, 1,81, ZnSO4.7H2O, 0,22, NaMoO4.2H2O, 0,39, CuSO4.5H2O, 0,079 (g/L). ***Solution B6**: VOSO4.5H2O, 49,6, K2Cr2(SO4)4.2H2O, 96, NiSO4.7H2O, 47,8, Na2WO4.2H2O, 17,9, TiOSO4, 33.3, Co(NO3)2.6H2O, 44.0, Co(NO3)2.6H2O, 0,049 (mg/L)

### 2.4 Analytical Procedures

Three aliquots were taken from each Erlenmeyer at the same time, 24 h interval, for 11 days in order to evaluate the growth, concentration, and $CO_2$% biofixation. An indirect measurement of cellular growth was done by apparent turbidity at a wavelength of 610 nm (optical density – OD) [10] using a UV-vis spectrophotometer (Benchtop); distilled water was used as blank. The cellular concentration was determined by microscopy cell counting by Neubauer's camera without dilution [15]. The cultivation vessel was coupled to sensors for the carbon dioxide measurement in the inlet and outlet of liquid medium. The measure of $CO_2$% was taken at the entrance and then in the container using a measuring device. The percentage of carbon dioxide was measured by an infrared sensor (CARBOCAP VAISALA GM70).

### 2.5 Data Analysis

The statistical treatment of the data focused on the generalized linear model (GLM) for studying the cell concentration, OD and $CO_2$ fixation between the different culture medium used. Simple linear regression models were generated in order to evaluate the $CO_2$ fixation related to the day in every single culture media, (95% confidence). Test LSD (Least significant difference) was applied.

## 3 Results and Discussion

### 3.1 Optical Density Variation

Figure 3 shows a rapid increase in biomass values in the strain that grew up in Medium N° 4, giving the maximum biomass value 0.1736 on the 7th day and the minimum biomass value (0,08) on the 2nd day. A similar growth rate behavior was shown in the strain that grew up in medium N° 1 and N° 3, this was due to the similarity in the media components especially in the salts used; however, the decay first appeared in the medium N° 1 (Fig. 3). In contrast, the growth parameters in Medium N° 2 showed a significant decrease associated with increasing urea concentrations (0.0216 on the 11th day).

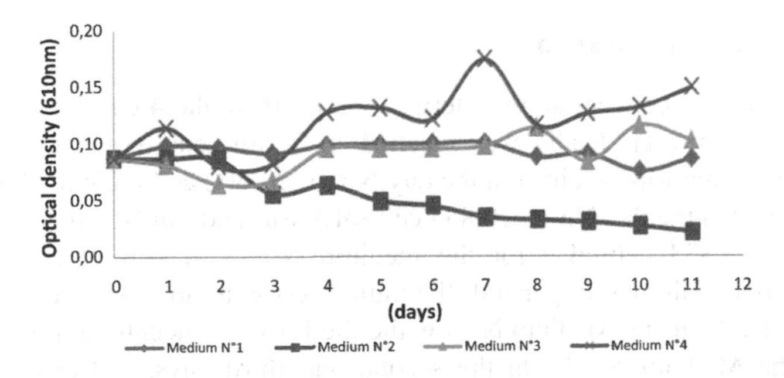


Fig. 3. The optical density of *S. platensis* in the culture media (610 nm)

Table 2 shows statistical Analysis of Variance OD for the generalized. Between the variables evaluated, the variable Medium presents a significant difference (P-value: 0,0000); this is consistent with the optical density data variation in each media. In spite of these results, the variable medium only explained 64% of DO behavior in the experiment.

**Table 2.** Analysis of variance optical density generalized linear model.

| Sum of squares type III | | | | | |
|---|---|---|---|---|---|
| Source | Sum of squares | Gl | Middle-square | F-value | P value |
| Medium | 0,0310417 | 3 | 0,01034723 | 25,82 | *0,0000* |
| Day | 0,000005606315 | 1 | 0,000005606315 | 0,01 | 0,9065 |
| Residual | 0,01563144 | 39 | 0,0004008063 | | |
| Total (corrected) | 0,04667876 | 43 | | | |

R-Square (adjusted by g.l.) = *64%*

According to LSD Fisher test, culture medium No. 4 represents the best option for strain growth due to its high concentration in mineral salts, like sodium bicarbonate, and a discrete increase in pH, both parameter facilitates the selective growth of Spirulina [16, 17]. This could be observed in Fig. 4.

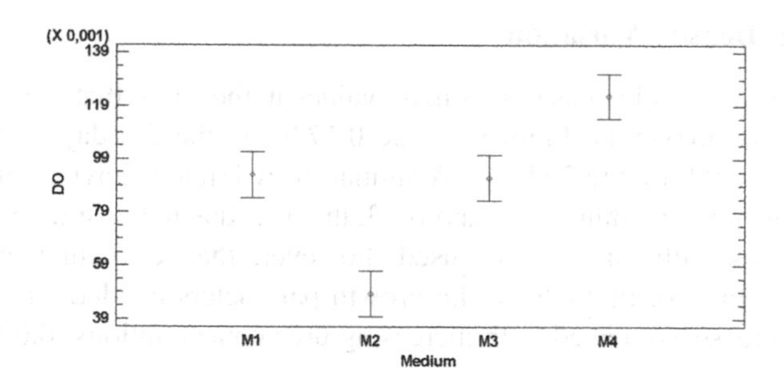


**Fig. 4.** LSD fisher The optical density of *S. platensis* in the culture media

### 3.2 Cellular Concentration

Figure 5 shows the concentration in terms of cell/mL in the 4 culture media. The high concentration per each media was reached in a different moment: in the case of Medium No. 1 this was obtained in the day No. 6 (27.500 cell/mL); for Medium No. 2, it was obtained in the day No. 2 (20.833 cell/mL); for medium No. the high value in the day No. 7 (43.333 cell/mL), for the medium No. 4, it was obtained in the day No. 11 (31.666 cell/mL). In general, the highest concentration was recorded on day 7 (43.000 cells/mL) using Medium N° 4 while the lowest concentration by cells/mL was found in the Medium N° 2. On the second and third days, a decrease in cells per milliliter is observed in all crops, due to the adaptation and acclimation phase of the strain and absorption of nutrients, because the cells in the inoculum require several divisions, to achieve balanced growth due that they were transferred to a new culture medium after have been maintained under constant incubation conditions. The Medium No 2, continued to decrease substantially, owing to the Urea as a nitrogen source, which generated toxicity for the microorganism. Medium N° 3 and Medium N° 1 showed similar concentrations, especially on days 5, 6 and 7, in which a concentration was found between 20.000–28.000 cells/mL (Fig. 5).

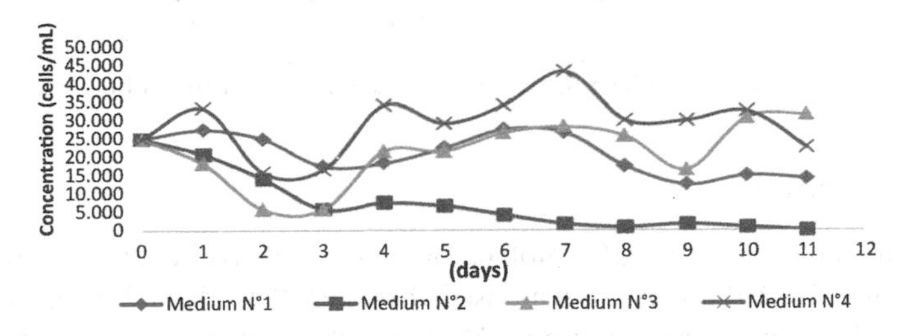


**Fig. 5.** The concentration of *S. platensis* in culture media

According to the sum of squares (Table 3), the Medium had presented a significant difference (P-value: 0,0000). This is consistent with concentration behaviors variation between media showed in Fig. 5. It is important to consider that the media has different components, this is strongly related to the exponential phase initiation because the concentration and dilution of the components vary to a medium to another, this is a critical point for absorption and metabolic rate of the *s. plantensis*.

**Table 3.** Analysis of variance concentration generalized linear model.

| Sum of squares type III | | | | | |
|---|---|---|---|---|---|
| Source | Sum of squares | Gl | Middle-square | F-value | P value |
| Medium | 3,13447E9 | 3 | 1,044823E9 | 18,98 | *0,0000* |
| Residual | 2,201641E9 | 40 | 5,504104E7 | | |
| Total (corrected) | 5,336111E9 | 43 | | | |

R-Square (adjusted by g.l.) = *56%*

According to LSD, in terms of concentration, culture medium No. 4 represents the best option for strain growth. Between the variables evaluated in variance analysis, Day presents a significant difference; this model explains 56% of the experiment behavior (Fig. 6).

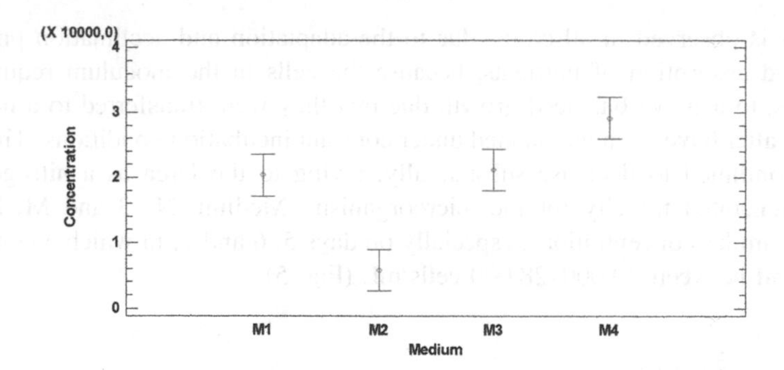


**Fig. 6.** LSD fisher The *S. platensis* Concentration in the culture media

### 3.3 $CO_2$ Biofixation

The maximum percentage of $CO_2$ biofixation was calculated in day 6 (0,0063%) in the culture in which Medium N° 4 was used, between the 7th and 11th day, 0% was reported in the crop where was used a Medium N° 2, this value was obtained by the deficiency of strain in the crop; that is, the low number of cells that dropped from 25,000 cells/mL to 800 cells/mL on day 8 until reaching the total loss of the strain (0 cells/mL) on day 11, without the microorganism capable of fixing the gas, no result was obtained, which indicates that the efficiency is directly proportional to the growth of the microalgae. It should be noted that *S. platensis* is grown at high bicarbonate concentrations and a high pH (15) It is remarkable a greater biofixation in the crop where Medium N° 4 is used because it has nutrients such as metals and vitamins that are necessary for the correct growth of the algae and therefore greater $CO_2$ biofixation, so the importance of these nutrients is observed in the development of the growth of *Spirulina*. Taking into account that selecting the optimal cell concentration is crucial for an efficient sequestration, in this study we found that the concentration of 43,000 cells/mL can generate the best biofixation using a media with a good nitrogen, carbon and vitamins sources, thus, optimum cell concentration, not all light energy is captured by cells, whereas at higher concentrations, a greater proportion of cells are in dark due to self-shading [11, 18–20] (Fig. 7).

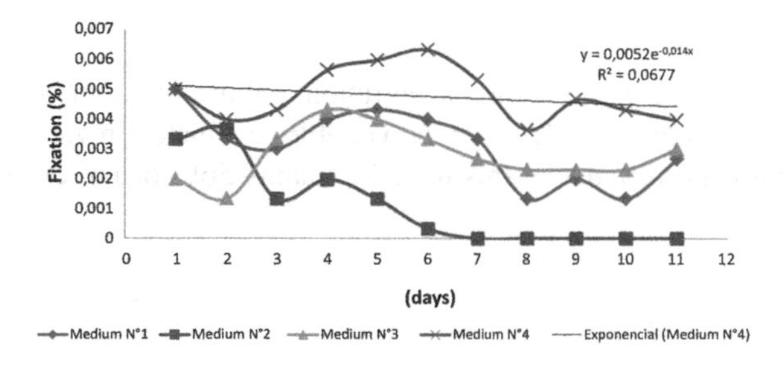


**Fig. 7.** $CO_2$ biofixation of *Spirulina platensis* at 4 culture media

According to the sum of squares (Table 4), the Medium (P-value: 0,0098). and the Day (P-value: 0,0001) had presented a significance difference. It is even more interesting the fact of the interaction Medium-day had presented a significant difference too (P-value: 0,0084). This is coherent evidence of the effect which has the components of the media in the $CO_2$ *s. spirulina* necessity for using it as a carbon source in the increasing phase.

**Table 4.** Analysis of variance $CO_2$ biofixation generalized linear model.

| Suma de Cuadrados Tipo III | | | | | |
|---|---|---|---|---|---|
| Source | Sum of squares | Gl | Middle-square | F-value | P value |
| Medium | 0,00000959831 | 3 | 0,000003199437 | 4,40 | *0,0098* |
| Day | 0,00001324268 | 1 | 0,00001324268 | 18,21 | *0,0001* |
| Medium*Day | 0,00000993005 | 3 | 0,000003310017 | 4,55 | *0,0084* |
| Residual | 0,00002618081 | 36 | 7,272447E-7 | | |
| Total (corrected) | 0,0001275152 | 43 | | | |

R-Square (adjusted by g.l.) = *75%*

According to LSD, in terms of biofixation, culture medium No. 4 represents the best option in terms of $CO_2$ biofixation rate (Fig. 8A) which is consequent with the results obtained in OD and Concentration. Figure 8B shows that the relationship between $CO_2$ biofixation and the day analysis. Results indicated that the relationship is inverted, which means that as the days had passed, the fixation had decreased.

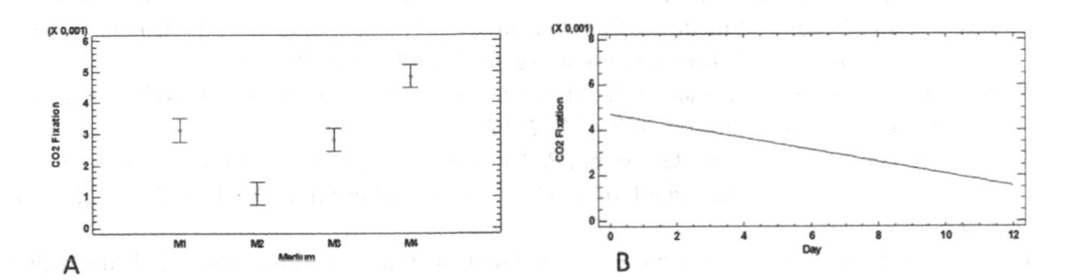


**Fig. 8.** (A) LSD fisher The *S. platensis* $CO_2$ biofixation in the culture media (B) Relationship between CO2 fixation and day

## 4 Conclusion

Statistical analysis has provided objective information [19–21]. However, it is necessary to carry out more in-depth studies in which a prediction of microalgae biomass production can be made in relation to the climatic conditions of different localities. The study has demonstrated the biological capacity of the microalga *S. platensis* for capture the $CO_2$ as the possibility of implementing sinks for the $CO_2$ sequestration of as an economic option and easy to handle and manage for diverse populations of developing countries.

**Acknowledgments.** This study was financially supported by Universidad de la Costa, COLCIENCIAS and Fundación Hospital Universitario Metropolitano.

**Compliance with Ethical Standards**

✓ All authors declare that there is no conflict of interest.
✓ No humans/animals involved in this research work.
✓ We have used our own data.

## References

1. Pires, J.C.: COP21: the algae opportunity? Renew. Sustain. Energy Rev. **79**, 867–877 (2017)
2. Könst, P., Mireles, I.H., van der Stel, R., van Os, P., Goetheer, E.: Integrated system for capturing CO2 as feedstock for algae production. Energy Procedia **114**, 7126–7132 (2017)
3. De Morais, M.G., Costa, J.A.V.: Biofixation of carbon dioxide by Spirulina sp. and Scenedesmus obliquus cultivated in a three-stage serial tubular photobioreactor. J. Biotechnol. **129**(3), 439–445 (2007)
4. Cheah, W.Y., Show, P.L., Chang, J.S., Ling, T.C., Juan, J.C.: Biosequestration of atmospheric CO2 and flue gas-containing CO2 by microalgae. Bioresour. Technol. **4**, 190–201 (2015)
5. Cheng, J., Huang, Y., Feng, J., Sun, J., Zhou, J., Cen, K.: Improving CO2 fixation efficiency by optimizing Chlorella PY-ZU1 culture conditions in sequential bioreactors. Bioresour. Technol. **144**, 321–327 (2013)
6. Chen, C.-Y., Kao, P.-C., Tsai, C.-J., Lee, D.-J., Chang, J.-S.: Engineering strategies for simultaneous enhancement of C-phycocyanin production and CO2 fixation with S. platensis. Bioresour. Technol. **145**, 307–312 (2013)
7. Cardias, B.B., de Morais, M.G., Costa, J.A.V.: CO2 conversion by the integration of biological and chemical methods: Spirulina sp. LEB 18 cultivation with diethanolamine and potassium carbonate addition. Bioresour. Technol. **267**, 77–83 (2018)
8. Soni, R.A., Sudhakar, K., Rana, R.S.: Spirulina-from growth to nutritional product: a review. Trends Food Sci. Technol. **69**, 157–171 (2017)
9. Campanella, L., Crescentini, G., Avino, P.: Chemical composition and nutritional evaluation of some natural and commercial food products based on Spirulina. Analusis **27**(6), 533–540 (1999)
10. Atlas, M.R., Bartha, R.: Microbial Ecology. Fundamentals and Applications, 3rd edn, p. 563. The Benjamin/Cummings Publishing Company, Inc., Redwood City (1993)
11. Castro, G.F.P.D.S., Rizzo, R.F., Passos, T.S., Santos, B.N.C.D., Dias, D.D.S., Domingues, J. R., Araújo, K.G.D.L.: Biomass production by Arthrospira platensis under different culture conditions. Food Sci. Technol. **35**(1), 18–24 (2015)
12. Costa, J.A.V., Colla, L.M., Duarte Filho, P., Kabke, K., Weber, A.: Modelling of S. platensis growth in fresh water using response surface methodology. World J. Microbiol. Biotechnol. **18**(7), 603–607 (2002)
13. Zeng, X., Danquah, M.K., Zhang, S., Zhang, X., Wu, M., Chen, X.D., Ng, I.S., Jing, K., Lu, Y.: Autotrophic cultivation of S. platensis for CO2 fixation and phycocyanin production. Chem. Eng. J. **183**, 192–197 (2012)
14. Zarrouk, C.: Contributionà l étuded une cyanophycée: influence de divers facteurs physiques et chimiques sur la croissance et photosynthese de Spirulina maxima Geitler, Ph.D. Thesis. University of Paris, Paris (1966)

15. Andersen, R.A. (ed.): Algal Culturing Techniques. Elsevier, Amsterdam (2005)
16. Baldia, S.F., Nishijima, T., Hata, Y., Fukami, K.: Growth characteristics of a blue–green alga S. platensis for nitrogen utilization. Nippon Suisan Gakkaishi **57**, 645–654 (1991)
17. De Oliveira, M.A.C.L., Monteiro, M.P.C., Robbs, P.G., Leite, S.G.F.: Growth and chemical composition of Spirulina maxima and S. platensis biomass at different temperatures. Aquacult. Int. **7**(4), 261–275 (1999)
18. Sánchez, M., Bernal-Castillo, J., Rozo, C., Rodríguez, I.: Spirulina (Arthrospira): an edible microorganism: a review. Universitas Scientiarum **8**(1), 7–24 (2003)
19. Madkour, F.F., Kamil, A.E., Nasr, H.S.: Production and nutritive value of S. platensis in reduced cost media. Egypt. J. Aquat. Res. **38**(1), 51–57 (2012)
20. Posso Mendoza, H., et al.: Evaluation of enzymatic extract with lipase activity of Yarrowia lipolytica. An application of data mining for the food industry wastewater treatment. In: Yang, C.N., Peng, S.L., Jain, L. (eds.) Security with Intelligent Computing and Big-data Services, SICBS 2018. Advances in Intelligent Systems and Computing, vol. 895. Springer, Cham (2020)
21. Morgado Gamero, W.B., et al.: Hospital admission and risk assessment associated to exposure of fungal bioaerosols at a municipal landfill using statistical models. In: Yin, H., Camacho, D., Novais, P., Tallón-Ballesteros, A. (eds.) Intelligent Data Engineering and Automated Learning – IDEAL 2018. Lecture Notes in Computer science, vol. 11315. Springer, Cham (2018). https://doi.org/10.1007/978-3-030-03496-2_24
22. Morgado Gamero, W.B., Ramírez, M.C., Parody, A., Viloria, A., López, M.H.A., Kamatkar, S.J.: Concentrations and size distributions of fungal bioaerosols in a municipal landfill. In: Tan, Y., Shi, Y., Tang, Q. (eds.) Data Mining and Big Data, DMBD 2018. Lecture Notes in Computer Science, vol. 10943. Springer, Cham (2018)